# Neuroanatomy
## Selected Papers of Walle J.H. Nauta

CONTEMPORARY NEUROSCIENTISTS
Selected Papers of Leaders in Brain Research

# Neuroanatomy
## Selected Papers of Walle J.H. Nauta

Birkhäuser
Boston • Basel • Berlin

Walle J.H. Nauta
Department of Brain and Cognitive Sciences
Massachusetts Institute of Technology
Cambridge, MA 02139

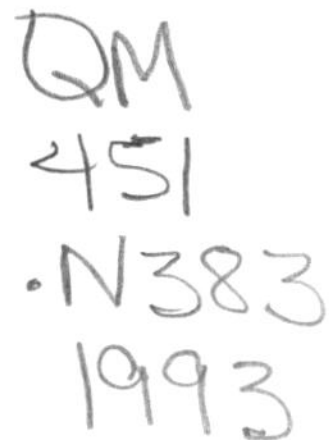


**Library of Congress Cataloging In-Publication Data**
Neuroanatomy : selected papers of Walle J.H. Nauta.
         p.      cm. -- (Contemporary neuroscientists)
      Includes bibliographical references.
      ISBN 0-8176-3539-4 (acid-free )
      1. Neuroanatomy.   2. Title.   I.   Series.
   QM451.N383    1993                          93-10358
   599'.048--dc20                              CIP

# CONTENTS



## I. Degeneration Techniques



## II. Brain and Behavior



## III. General Organization of the Central Nervous System

## IV. Basal Ganglia



## V. Crossroads of Limbic and Striatal Circuitry

## VI. Cerebral Cortex

# Introduction

I received my first introduction to the brain sciences in 1936 and 1937, for me the second and third years of the 7-year medical school curriculum at the University of Leiden. During those years my interest in the subject was aroused in particular by the brilliant lectures of the physiologist G.C. Rademaker – a prominent former member of the Rudolf Magnus school – and the neurohistologist S.T. Bok, noted especially for his histometric studies of the cerebral cortex.

Fascinated as I was by everything I learned about the brain from these outstanding teachers, toward the end of their courses I began to notice conspicuous gaps that separated neurophysiology from neuroanatomy. In fact, I could (or thought I could) detect a reasonable concordance between the two sciences only in case of some sensory and somatic-motor systems. For most other functions anatomical substrates seemed either poorly defined or, as in the case of the central viscero-endocrine system, hardly recognized at all. With all the arrogance of which a 20-year old student is capable I concluded that what the brain sciences needed was a new and more complete anatomy that emphasized in particular the continuity of, and convergences or interconnections between individual conduction systems. And I wistfully mused that perhaps at some time in the future I could make such an endeavour part of my own career.

The confidence I had in the rationality of my reflections on the brain sciences of the 1930s was put to a severe test when I began to try them out on several staff members of the departments of physiology, anatomy, and histology. (At that time in Europe, a student would not think of approaching the Professor himself.) Almost to the man, my respondents reacted in a most discouraging manner. The most common response I evoked amounted to a statement that neuroanatomical tracing of conduction pathways had already been done for more than half a century, and that after all that effort the likelihood of finding anything that had not already been found before seemed very small indeed. In the view of these critics, neuroanatomy, even though it would always remain an important teaching matter, had by and large run its course and was no longer a worthwhile research subject. One of my critics made it a special point to berate me for the arrogance of my apparent belief that I would be able to improve on such a monumental treatise as Santiago Ramón y Cajal's *Histologie du Système Nerveux de l'Homme et des Vertébrés*. That remark really hurt. A fervent admirer of Cajal, I had been awe-struck whenever I consulted his famous textbook. The thought of surpassing the unsurpassable would have seemed preposterous to me. I saw my tentative plans for the future as aimed at a much more modest purpose. It struck me that Cajal, either by choice or because of constraints imposed by the histological techniques available in his time, seemed to have emphasized the analysis of local or short-distance connective intricacies over the tracing of longer lines of conduction. I had occasionally wondered why Cajal had so rarely used the Marchi method, an experimental technique which, we had been told in class, had been designed specifically for the latter purpose. It was not until several years later that

I began to realize the reasons why Cajal might have been reluctant to make more routine use of the Marchi method.

The only faculty member to give me an encouraging response was the associate professor ("prosector") of the anatomy department, Johan Dankmeyer. After hearing me out patiently, he remarked that what I seemed to have in mind sounded like a long-term project, and that I should perhaps first of all give myself a chance to find out whether working in histological laboratories would suit my taste. He suggested that to that end I consider a student-assistantship in the anatomy department. He warned me, however, that my first duty as a student-assistant would be to assist in the Human Dissection course, and furthermore, that in my intended study of the brain I would be on my own, as there was no neuroanatomist on the anatomy staff. Despite these warnings I enthusiastically accepted Dankmeyer's suggestion. And thus, in the fall of 1937, after I had passed the preclinical examination, there began for me what would turn out to be nearly 50 years of involvement with neuroanatomy.

Since an accounting of the first 15 years of that involvement was published recently in the Journal of Neuroscience, and is included in this volume, I can summarize here by characterizing that period as the search for a staining method which would be as selective for degenerating fibers as the Marchi method, but did not share the latter's added selectivity for myelinated fibers. My own discouraging experience with the Marchi method, as well as the consideration that the endings of nerve fibers are invariably unmyelinated had made it seem clear to me that only a myelin-independent technique could offer a reasonable hope of tracing conduction systems into the areas of their terminal distribution. It was not until 1952 that Paul Gygax and I finally succeeded in developing such a method. When it was published in 1954, I was astonished and pleasantly surprised by the interest with which it was received. I had not been aware that so many others shared my view of neuroanatomy as a much neglected subject badly in need of being taken up afresh. Thus, the new axon-degeneration method and its later modifications, in particular the techniques of Fink and Heimer, replaced the Marchi method as the preferred methods of tracing conduction pathways in the brain, a status they maintained until 1972 when they were in turn superseded by the autoradiographic tracing method described by Cowan et al. (1972).

It is a humbling experience to re-read the papers one published some three or four decades ago. Research methods then relied on have meanwhile been replaced by strategies that are not only more convenient to follow but also carry a lesser risk of misinterpretation. All of us who used the axon-degeneration method became stressfully aware of its potential pitfalls. Most of these were directly related to the circumstance that, like the Marchi, the axon-degeneration method required the placement of a brain lesion designed either to interrupt the fiber pathway in question, or to destroy the cell groups from which it originates. In most cases, the lesion had to be placed at a deep-lying brain site by focal, electrolytic tissue destruction around the bare tip of an otherwise insulated, stereotactically guided electrode. Two problems constantly beset us in following this procedure: First, the lesion, beside destroying the cell group of origin of the fiber system at issue also indiscriminately interrupted any unrelated fibers happening to pass through the lesion site. Contamination of one's findings by such fibers of passage was a constant threat that often made elaborate control experiments necessary. Second, and no less vexing, the track left by the electrode in its passage to the intended lesion site was itself a lesion that could result in fiber degeneration unrelated to the fiber system one intended to chart. In the papers assembled in this volume at least one example of such "electrode-track error" can be pointed out: in my 1962 paper on the neural associations of the amygdaloid complex (Chapter 13) I reported evidence of a fiber tract leading to the amygdala from the medial component of the thalamic mediodorsal nucleus. Later studies by Veening (1978), Aggleton et al. (1980), and others by the aid of retrograde cell-labelling with horseradish peroxidase (Lavail et al., 1973) conclusively showed that the thalamo-amygdaloid conection in question does not originate from the mediodorsal nucleus but, instead, from an adjacent midline cell group, the anterior paraventricular nucleus, which in my experiment had been pierced by the electrode.

I have dwelt at some length on these potential sources of error to emphasize that we who tried to trace conduction pathways by the axon-degeneration method did not have an easy time of it, and occasionally even had to conclude that the problem we had set out to explore was unapproachable by this or any other method requiring the placement of a brain lesion. As a case in point, I could mention the question of whether the "non-specific" intralaminar nuclei of the thalamus indeed project directly to the cerebral cortex, as neurophysiological observations had suggested they might. David Whitlock and I could trace degenerating axons to the cat's frontal cortex from small lesions in the internal medullary lamina of the thalamus, but soon afterward discovered that the lamina is a zone of passage for numerous fibers passing to the frontal cortex from the mediodorsal nucleus of the thalamus (Nauta and Whitlock, 1954). At the time, no method was available which could have been used to exclude the distinct possibility that all of the degenerating fibers we had seen in the frontal cortex originated from the mediodorsal nucleus rather than from the intralaminar cell groups, and we therefore had to decide that the answer to our question would have to await a more suitable future strategy. Actually, it was not until twenty years later that such a strategy materialized, in the form of the Lavail et al. (1973) method of retrograde cell-labelling with horseradish peroxidase. By using this new method, Jones and Leavitt (1974) and Kievit and Kuypers (1975) obtained the first conclusive evidence of thalamocortical projections arising from intralaminar thalamic nuclei.

There were further, more singular problems associated with the axon-degeneration method, most notably perhaps its unexplained failure to reveal Wallerian degeneration in the nigrostriatal projection and in the distal, largely synaptic stretch of the climbing fibers that extends into the molecular layer of the cerebellum in contact with Purkinje-cell dendrites. But despite all its shortcomings, I believe it is fair to state that the axon-degeneration method offered a distinct advantage over its 65-year-old precursor, the Marchi method. In the early 1950's, even so modest a technological advance could provide a considerable impetus to the reawakening interest in the brain sciences that marked the post-war years. In neuroanatomy that resurgence gained powerful momentum from the introduction of electron-microscopic techniques painstakingly adapted to the peculiar demands of brain tissue preservation, and from a related renascence of interest in the use of the Golgi method, Cajal's irreplaceable tool that had gradually fallen into disuse during the last two decades preceding the second World War (*cf.* Scheibel and Scheibel, 1971).

Looking back over the years, I consider myself particularly fortunate that the span of my career included the 1950s and 1960s. Throughout the field of the brain sciences those were exciting decades, times of new directions and fresh beginnings, times also of lively dialogue and collaboration among the neuroscience disciplines. It is gratifying that those times now appear as only the beginning of a sequence of developments that has led to the almost bewilderingly rich and diversified scene of today's brain sciences, and still seems to be nowhere near its end.

# References


Aggleton JP, Burton MJ, Passingham RE (1980): Cortical and subcortical afferents to the amygdala in the monkey. (*Macaca mulatta*). *Brain Res.*, 190:347–368.

Cowan WM, Gottlieb DI, Hendrickson AE, Price JL, Woolsey TA (1972): The autoradiographic demonstration of axonal connections in the central nervous system. *Brain Res.*, 37:21–51.

Jones EG, Leavitt R (1974): Retrograde axonal transport and the demonstration of non-specific projections to the cerebral cortex and striatum from thalamic intralaminar nuclei in the rat, cat, and monkey. *J. Comp. Neurol.*, 154:349–378.

Kievit J, Kuypers HGJM (1875): Subcortical afferents to the frontal lobe in the rhesus monkey studied by means of retrograde horseradish peroxidase transport. *Brain Res.*, 85:261–266.

Lavail JH, Winston KR, Tish A (1973): A method based on retrograde intraaxonal transport of protein for identification of cell bodies of origin of axons terminating within the CNS. *Brain Res.*, 58:470–477.

Scheibel ME, Scheibel AB (1971): The rapid Golgi method. Indian summer or renaissance? In: *Contemporary Research Methods in Neuroanatomy*, Nauta WJH, Ebbesson SOE, eds., New York: Springer Verlag, pp. 1–11.

Veening JG (1978): Subcortical afferents of the amygdaloid complex in the rat: An HRP study. *Neurosci. Lett.*, 8:197–202.

BIBLIOGRAPHY I

# Papers Reprinted in This Volume

1. Nauta WJH, Gygax PA (1951): Silver impregnations of degenerating axon terminals in the central nervous system: (1) Technic. (2) Chemical notes. *Stain Technol* 26:5–11

2. Nauta WJH, Ryan LF (1952): Selective silver impregnation of degenerating axons in the central nervous system. *Stain Technol* 27:175–179

3. Nauta WJH, Gygax PA (1954): Silver impregnation of degenerating axons in the central nervous system: A modified technic. *Stain Technol* 29:91–93

4. Glees P, Nauta WJH (1955): A critical review of studies on axonal and terminal degeneration. *Mschr Psychiat neurol* 129:74–91

5. Nauta WJH (1993): Some early travails of tracing axonal pathways in the brain. *J Neurosci* 13:1337–1445

6. Nauta WJH (1946): Hypothalamic regulation of sleep in rats. An experimental study. *J Neurophysiol* 99:285–316

7. Brady JV, Nauta WJH (1953): Subcortical mechanisms in emotional behavior: Affective changes following septal forebrain lesions in the albino rat. *J Comp Physiol Psychol* 46:339–346

8. Nauta WJH (1972): The central visceromotor system: A general survey. In: *Limbic System Mechanisms and Autonomic Functions*, Hockman CH, ed. Springfield, IL: C.C. Thomas

9. Nauta WJH (1958): Hippocampal projections and related neural pathways to the midbrain in the cat. *Brain* 8:319–340

10. Valenstein ES, Nauta WJH (1959): A comparison of the distribution of the fornix system in the rat, guinea pig, cat and monkey. *J Comp Neurol* 113:337–363

11. Mehler WR, Feferman ME, Nauta WJH (1960): Ascending axon degeneration following anterolateral cordotomy: An experimental study in the monkey. *Brain* 83:718–750

12. Nauta WJH (1961): Fibre degeneration following lesions of the amygdaloid complex in the monkey. *J Anat* 95:515–531

13. Nauta WJH (1962): Neural associations of the amygdaloid complex in the monkey. *Brain* 85:505–520

14. Ramon-Moliner E, Nauta WJH (1966): The isodendritic core of the brain stem. *J Comp Neurol* 126:311–335

15. Heimer L, Nauta WJH (1969): The hypothalamic distribution of the stria terminalis in the rat. *Brain Res* 13:285–297

16. Herkenham M, Nauta WJH (1977): Afferent connections of the habenular nuclei in the rat. A horseradish peroxidase study, with a note on the fiber-of-passage problem. *J Comp Neurol* 173:123–145

17. Herkenham M, Nauta WJH (1979): Efferent connections of the habenular nuclei in the rat. *J Comp Neurol* 187:19–47

18. Nauta WJH, Domesick VB (1981): Ramifications of the limbic system. In: *Psychiatry and the Biology of the Human Brain*, Matthysse S, ed. New York: Elsevier

19. Grownewegen HJ, Ahlenius S, Haber SN, Kowall NW, Nauta WJH (1986): Cytoarchitecture, fiber connections, and some histochemical aspects of the interpeduncular nucleus in the rat. *J Comp Neurol* 249:65–102

20. Nauta WJH, Mehler WR (1966): Projections of the lentiform nucleus in the monkey. *Brain Res* 1:3–42

21. Nauta WJH, Smith GP, Faull RLM, Domesick VB (1978): Efferent connections and nigral afferents of the nucleus accumbens septi in the rat. *Neuroscience* 3:385–401

22. Beckstead RM, Domesick VB, Nauta WJH (1979): Efferent connections of the substantia nigra and ventral tegmental area in the rat. *Brain Res* 175:191–217

23. Haber SN, Nauta WJH (1969): Ramifications of the globus pallidus in the rat as indicated by patterns of immunohistochemistry. *Neuroscience* 9:245–260

24. Kelley AE, Domesick VB, Nauta WJH (1982): The amygdalostriatal projection in the rat — an anatomical study by anterograde and retrograde tracing methods. *Neuroscience* 7:615–630

25. Heimer L, Ebner FF, Nauta WJH (1967): A note on the termination of commissural fibers in the neocortex. *Brain Res* 5:171–177

26. Nauta WJH, Karten HJ (1970): A general profile of the vertebrate brain, with sidelights on the ancestry of cerebral cortex. In: *The Neurosciences: Second Study Program*, Schmitt FO, ed. New York: The Rockefeller University Press

27. Nauta WJH (1971): The problem of the frontal lobe: A reinterpretation. *J Psychiat Res* 8:167–187

28. Goldman PS, Nauta WJH (1977a): Columnar distribution of cortical fibers in the frontal association, limbic, and motor cortex of the developing rhesus monkey. *Brain Res* 122:393–413

29. Goldman PS, Nauta WJH (1977b): An intricately patterned prefronto-caudate projection in the rhesus monkey. *J Comp Neurol* 171:369–385

30. Nauta WJH (1989): Reciprocal links of the corpus striatum with the cerebral cortex and limbic system. A common substrate for movement and thought? In: *New Perspectives of Neuropsychiatry*, Mueller J, Yingling C, Zegans L, eds. Basel: S. Karger

# Additional Selected Papers Not Printed in this Volume


Cole M, Nauta WJH (1970): Retrograde atrophy of axons of the medial lemniscus of the cat. *J Neuropath and Exp Neurol* 29:354–369

Cole M, Nauta WJH, Mehler WR (1964): The ascending efferent projections of the substantia nigra. *Trans Am Neurol Assoc* 89:74–78

Faull RLM, Nauta WJH, Domesick VB (1986): The visual cortico-stratio-nigral pathway in the rat. *Neuroscience* 19:1119–1132

Goldman PS, Nauta WJH (1976): Autoradiographic demonstration of a projection from prefrontal association cortex to the superior colliculus in the rhesus monkey. *Brain Res* 116:145–149

Graybiel AM, Nauta HJW, Lasek RJ, Nauta WJH (1973): A cerebello-olivary pathway in the cat: An experimental study using autoradiographic tracing techniques. *Brain Res* 38:205–211

Grove EA, Domesick VB, Nauta WJH (1986): Light microscopic evidence of striatal input to intrapallidal neurons of cholinergic cell group CH4 in the rat: A study employing the anterograde tracer phaseolus vulgaris-leucoagglutinin (PHA-L). *Brain Res* 367:379–384

Haber SN, Groenewegen HJ, Grove EA, Nauta WJH (1985): Efferent connection of the ventral pallidum in the rat. Evidence of a dual striato-pallidofugal pathway. *J Comp Neurol* 235:322–335

Haymaker W, Anderson E, Nauta WJH, eds. (1969): *The Hypothalamus.* Springfield, IL: CC. Thomas

Kaiserman-Abramof IR, Graybiel AM, Nauta WJH (1980): The thalamic projection to cortical area 17 in a cogenitally anophthalmic mouse strain. *Neuroscience* 5:41–52.

Karten HJ, Hodos W, Nauta WJH, Revsin AM (1973): Neural connection of the "visual wulst" of the avian telencephalon. Experimental studies in the pigeon (*Columbia livia*) and owl (*Speotyto cunicularis*). *J Comp Neurol* 150:253–277

Lipp H-P, Collins RL, Nauta WJH (1984): Structural asymmetries in brains of mice selected for strong lateralization. *Brain Res* 310:393–396

Mason JW, Nauta WJH, Brady JV, Robinson JA, Sachar EJ (1961): The role of limbic system structures in the regulation of ACTH secretion. *Acta Neuroveg* 23:4–14

Mehler WR, Nauta WJH (1974): Connections of the basal ganglia and of the cerebellum. *Confin Neur* 36:205–222

Nauta WJH (1951): Ueber die sogenannte terminale Degneration im Zentralvervensystem und ihre Darstellung durch Silberimprägnation. *Schweiz Arch neurol Psychiat* 66:353–376

Nauta WJH (1956): An experimental study of the fornix in the rat. *J Comp Neurol* 104:247–271

Nauta WJH (1957): Silver impregnation of degnerating axons. In: *New Research Techniques of Neuroanatomy*, Windle W, ed., Springfield, IL: C.C. Thomas

Nauta WJH (1964): Some efferent connections of the prefrontal cortex in the monkey. In: *The Frontal Granular Cortex and Behavior*, Warren JM, Akert K., eds. New York: McGraw-Hill

Nauta WJH (1966): Some brain structures and functions related to memory. In: *Neurosciences Research Symposium Summaries. Vol.1*, Schmitt FO, Melnechuk T, eds. Cambridge, MA: MIT Press

Nauta WJH (1972): Neural associations of the frontal cortex. *Acta Neurobiol Exp* 32:125–140

Nauta WJH (1982): The limbic innervation of the striatum. In: *Advances in Neurology, Vol. 35*, Friedhoff AJ, Chase TN, eds. New York: Raven Press

Nauta WJH, Bucher VM (1954): Efferent connections of the striate cortex in the albino rat. *J Comp Neurol* 100:287–296

Nauta WJH, Ebbesson SOE, eds. (1970): *Contemporary Research Methods in Neuroanatomy*. New York: Springer-Verlag

Nauta WJH, Feirtag M (1979): The organization of the brain. *Scientific American* 241:88–111

Nauta WJH, Feirtag M, eds. (1986): *Fundamental Neuroanatomy*. New York: W.H. Freeman and Company

Nauta WJH, Haymaker W (1969): Hypothalamic nuclei and fiber connections. In: *The Hypothalamus*. Haymaker W, Anderson E, Nauta WJH, eds., Springfield, IL: CC. Thomas

Nauta WJH, Kuypers HGJM (1958): Some ascending pathways in the brain stem reticular formation. In: *Reticular Formation of the Brain*, Jasper HH, Proctor LD, Knighton RS, Noshay WC, Costello RT, eds., Boston: Little Brown

Nauta WJH, van Straaten JJ (1947): The primary optic centres of the rat. An experimental study with the "Bouton" method. *J Anat (Lond)* 81:127–134

Nauta WJH, Whitlock DG (1954): An anatomical analysis of the non-specific thalamic projection system. In: *Brain Mechahisms and Consciousness. Symposium of the Council for International Organization of Medical Sciences (UNESCO and WHO)*, Delafresnaye JF, ed., Oxford: Blackwell Scientific Publications

Potter H, Nauta WJH (1979): A note on the problem of olfactory associations of the orbitofrontal cortex in the monkey. *Neuroscience* 4:361–367

Stotijn CPJ, Nauta WJH (1950): Precocious puberty and tumor of the hypothalamus. *J Nerv Ment Dis* 111:207–224

Whitlock DG, Nauta WJH (1956): Subcortical projections from the temporal neocortex in *Macaca mulatta*. *J Comp Neurol* 106:183–212

# I

# DEGENERATION TECHNIQUES

# SILVER IMPREGNATION OF DEGENERATING AXON TERMINALS IN THE CENTRAL NERVOUS SYSTEM: (1) TECHNIC. (2) CHEMICAL NOTES


W. J. H. NAUTA and P. A. GYGAX,[1] *Department of Anatomy, The University of Zürich, and Department of Chemical Technology, The Federal Institute of Technology, Zürich, Switzerland*



Received for publication April 17, 1950



ABSTRACT.—A new silver technic, tested on the brain of the rat, is described, especially suitable for demonstrating terminal degeneration within the central nervous system. It is a modification of the Glees method, designed to avoid use of tap water in preparing solutions. Some of the chemical principles underlying the process of reductive liberation of metallic silver from ammoniacal silver nitrate solutions are discussed.


Since first applied by Hoff (1932), the method of studying central nervous connections in experimental material by silver impregnation of their degenerated synaptic end-portions has been employed on a steadily increasing scale. However, up to the present time a wide application of this research method has been seriously discouraged by technical difficulties.

Recently, Glees (1946) devised a modified Bielschowsky technic which has yielded valuable results in experimental studies on various brain regions (Glees, 1944; Glees, Meyer and Meyer, 1946; Clark and Meyer, 1948; Brodal, 1949). However, we were unsuccessful in an attempt to study hypothalamic connections in the rat by the Glees method, while the cerebral cortex was found to be somewhat resistant also. Since the Glees method involves the use of tap water in the preparation of the reducing fluid, we wish to present a modification which is independent of local variations in the chemical constitution of tap water. As shown by the illustrations, satisfactory pictures of terminal degeneration were obtained by this method in several telencephalic regions such as the neocortex and nucleus caudatus, as well as in the diencephalon. The method has not been tried on experimental material of other regions, but findings in normal sections of the lower brain stem and spinal cord indicate that it will succeed there on experimental material.

## THE STAINING PROCEDURE

1. Fixation in 10% neutral formalin for 2 weeks up to 6 months.
2. Frozen sections of 15–20µ thickness are cut.

---

[1] Now at the research laboratories of J. R. Geigy AG, Basel, Switzerland.


STAIN TECHNOLOGY, VOL. 26, NO. 1, JANUARY 1951

3. Sections are demyelinated by being kept for 6 to 12 hours in 50% alcohol containing 1 ml. of strong ammonia (25% *ammonium hydricum*) per 100 ml. of alcohol. This treatment can be prolonged for some days without ill effect.

4. Sections are passed through at least three dishes of distilled water.

5. Sections are transferred for 12 to 24 hours into a solution composed of:

Silver nitrate ........................ 150 mg.
Distilled water ...................... 10 ml.
Pyridine ........................... 0.5 ml.

6. Without washing, sections are transferred for 2–5 minutes into an ammoniacal silver nitrate solution:

450 mg. silver nitrate are dissolved in
20 ml. of distilled water;
10 ml. of pure ethyl alcohol, 97%, are added; the mixture is allowed to cool to room temperature. From calibrated pipettes
2.0 ml. of strong ammonia (25% *ammonium hydricum*), and
2.2 ml. of a 2.5% aqueous solution of sodium hydroxide are added.

The solution is mixed thoroughly. The container is kept covered to prevent loss of excess ammonia.

7. Without further treatment, sections are transferred quickly into a reducing fluid:

10% alcohol ........................ 45 ml.
10% formalin[2] ...................... 2 ml.
1% citric acid[3] ..................... 1.5 ml.

The sections expand on the surface and rapidly acquire a golden brown color.

8. Sections are transferred for 1 to 2 minutes into a 2.5% solution of sodium thiosulfate.

---

[2]Non-neutralized commercial formalin taken for 100%. The formalin used contained less than 0.03% formic acid.

[3]1 g. of citric acid crystals per 100 ml. aqueous solution.

---

Fig. 1. — Antero-ventral nucleus of thalamus, showing massive degeneration of Vogt's septo-thalamic tract 3 days after lesion of supracommissural part of septal area. Notice contrast with normal capsula interna fibers. ×420.

Fig. 2. — Disintegrating fiber terminals in the inter-anteromedial cell mass of the thalamus. Degeneration time 3 days. ×720.

Figs. 3–5. — Terminal degeneration in medial region of hypothalamus (3), periventricular region of hypothalamus (4) and nucleus caudatus (5). Degeneration time 3 days. ×1200.

Fig. 6. — Normal aspect (6a) of fourth layer of frontal cortex, and (6b) same layer as appearing 4 days after left-sided lesion in rostral part of internal capsule. 6a and 6b represent identical points of right and left side respectively as seen in one and the same section. ×600.

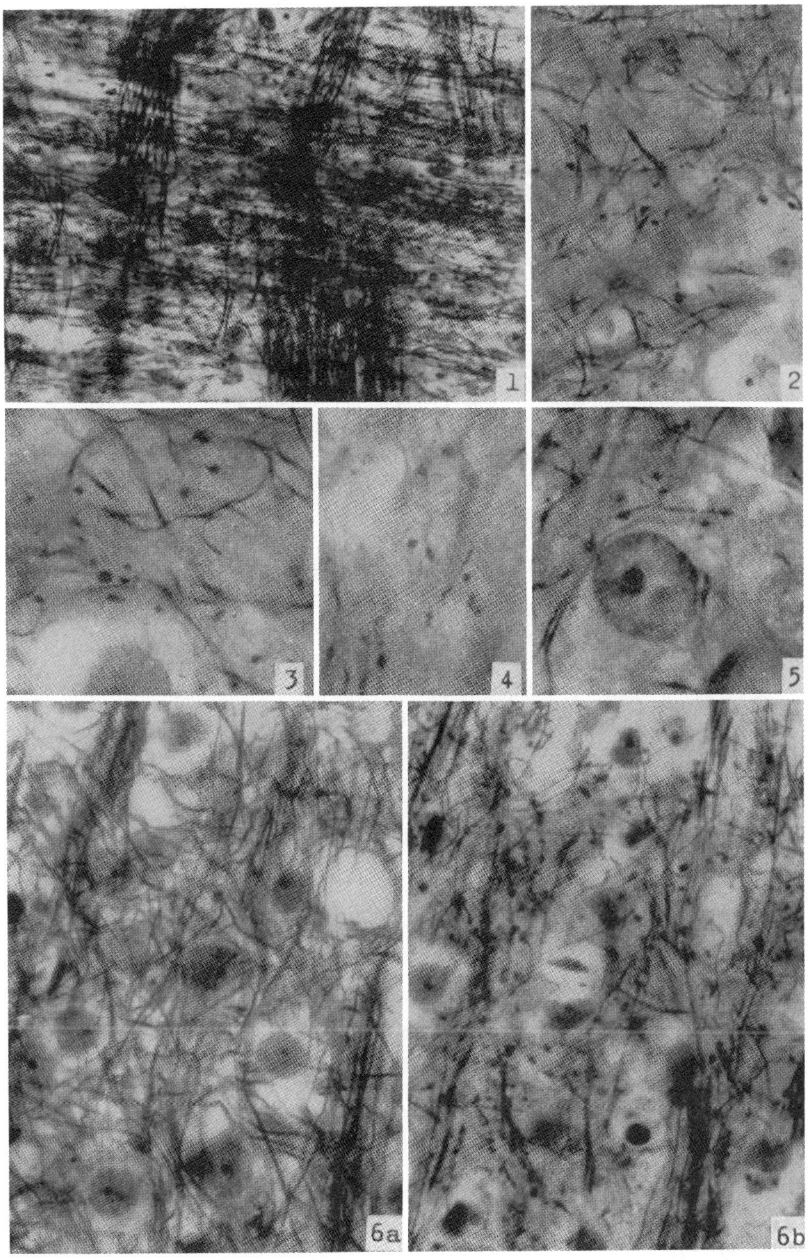

They are washed in at least three dishes of distilled water. Contamination of this wash water by traces of the ammoniacal silver nitrate solution should be avoided, as it is liable to cause silver precipitates on the sections.

9. Finally, sections are passed rapidly through the graded alcohols and xylene. Neutral synthetic media such as clarite or caedax are preferred for mounting.

Throughout the procedure, a glass rod should be used for transferring the sections.

### DISCUSSION

*Fixation and Embedding.* In full agreement with Glees (1946), neutral formalin was found superior to other fixatives. Formalin fixation for more than 6 months, however, diminishes the effectiveness of the staining procedure, and brain tissue fixed for more than a year is usually refractory. Concerning embedding media, by far the best results were obtained in frozen sections. Celloidin embedding is less desirable, and paraffin material is unsuitable.

Since a more detailed knowledge of the physico-chemical aspects of silver deposition in organic material may lead to further improvement of neurological silver technics, some chemical data concerning the solutions employed in the process of impregnation will follow. An account of other important aspects of silver impregnations will be found in the paper of M. Seki (1940).

1. *Ammoniacal silver solutions.* As already found by Prescott (1880), Reychler (1883, 1895) and Draper (1886), addition of ammonia to sufficiently concentrated solutions of silver nitrate leads to precipitation of silver oxide which is progressively and completely redissolved by further addition of ammonia until the ratio is reached of approximately 2 molecules of ammonia to 1 molecule of silver nitrate. Bodlaender and Fittig (1901), continuing upon previous studies of Reychler (1895), Konowaloff (1898) and Berthelot and Delépine (1899), demonstrated the presence of silver in such solutions in the form of a complex cation,

$$[Ag(NH_3)_2]^+ \qquad (I)$$

Britton and Wilson (1933) demonstrated the formation from $NH_3$ and $AgNO_3$ of the complex salt silver-ammonia nitrate

$$[Ag(NH_3)_2]NO_3 \qquad (II)$$

They showed that this salt is not appreciably decomposed by addition of more ammonia, and thus may be regarded as the salt of a base

$$[Ag(NH_3)_2OH] \qquad (III)$$

which is stronger than ammonia. The existence of this base was actually demonstrated by electrometric titration of a solution of silver oxide in ammonia. It is important that the strength of the base (III) was found to be comparable with that of sodium hydroxide.

2. *Instability of the complex* $[Ag(NH_3)_2]^+$. The complex (I) formed in the ammoniacal silver nitrate solution is in equilibrium with its components according to the equations

$$k_1 = \frac{[Ag(NH_3)^+]\,[NH_3]}{[Ag(NH_3)_2]^+} \qquad (IV)$$

$$\text{and} \qquad k_2 = \frac{[Ag^+]\,[NH_3]}{[Ag(NH_3)^+]} \qquad (V)$$

Since the existence of a mono-ammonia-metal complex ("monammine complex ion") as indicated in equation IV was postulated only recently (Bjerrum, 1941; Vosburgh and Stockdale Mc Clure, 1943), only the instability constant or complexity constant K (now called "overall complexity constant") is mentioned in the older literature:

$$K = k_1 \cdot k_2 = \frac{[Ag^+]\,[NH_3]^2}{[Ag(NH_3)_2^+]} \qquad (VI)$$

According to Vosburgh and Stockdale Mc Clure (1943), $k_1 = 4.3 \cdot 10^{-4}$ and $K = 6.2 \cdot 10^{-8}$, while (Bjerrum, 1941) the instability constant $k_2$ of the monammine-ion equals $4.8 \cdot 10^{-4}$.

It is evident from these figures that the concentration of non-combined, "free" silver ($Ag^+$ in equation VI) in ammoniacal silver solutions is extremely low. This is a significant fact, since this concentration determines the electro-chemical potential of the solution, and, in accordance with the equilibrium constants, also the decomposition of the silver complex during reduction. It is important in this connection that addition of ammonia to silver nitrate solutions beyond the point where an initially formed precipitate of silver oxide has been dissolved completely, with formation of the complex salt (II), lowers the concentration of silver ions ("free silver") (Kohlschütter, 1912; Britton and Wilson, 1933). This decrease can be explained thus: the excess of the weak base ammonia does not decompose the salt $[Ag(NH_3)_2]NO_3$ into its free base and ammonium-nitrate; instead, in accordance with the law of mass action, it increases the stability of the complex $[Ag(NH_3)_2]^+$, thereby lowering the concentration of free silver. As shown by Kohlschütter, continued addition of ammonia may even suppress the concentration of free silver to levels where reductive liberation of metallic silver by formaldehyde is no longer possible.

An entirely different effect on ammoniacal silver nitrate solutions is shown by strong bases, such as sodium hydroxide. They tend to liberate the base (III) from the complex salt (II). Base (III) in turn, by releasing the decomposable complex ion $[Ag(NH_3)_2]^+$, increases the concentration of free silver in the solution. Accordingly, NaOH is able to negate the effect on free silver concentration of excess ammonia. (Kohlschütter, 1912).

As the antagonistic effects of excess ammonia and sodium hydroxide respectively were expected to furnish a valuable means of controlling the concentration of free silver in ammoniacal silver solutions, experiments were performed to determine the ratio of silver nitrate, ammonia and sodium hydroxide optimum for the demonstration of terminal degeneration. For the brain of the rat, the formula given above yielded the best results.

Changes in the formula were observed to affect profoundly the subsequent stain. If for instance, appreciably less sodium hydroxide is added, e.g. 1.5 ml. instead of 2.2 ml. as indicated, staining is too weak and many fibers are impregnated irregularly. Too high concentrations of sodium hydroxide, e.g. 4 ml., will suppress the impregnation of degenerating fibers and cause overstaining of the "background", although normal fine fibers may be shown in great detail. Addition of still more alkali may prevent impregnation of degenerating fibers completely, while massive and apparently unselective silver depositions occur.

### Reducing Fluid

The use of a very dilute formalin solution in 10% alcohol proved to be preferable to the conventional much stronger aqueous formalin. Acidification of the solution is essential in avoiding unselective silver deposition. For this purpose citric acid was found to be the most suitable of the various acids tried. Excessive acidification suppresses the reduction process, but in suitable concentration citric acid markedly improves the selectivity of the impregnation, presumably by buffering the liberation of metallic silver from the ammoniacal silver nitrate solution.

### Comments

Although the above described method yielded remarkably constant results in the rat's brain, modifications of the technic may prove necessary for work on other animals. Such modifications will probably be achieved most readily by changing suitably the ratio of ammonia and sodium hydroxide in the ammoniacal silver solution.

N.B. Silver ammonia hydroxide has an erratic tendency to decompose, yielding the violently explosive silver amide:

$$2[Ag(NH_3)_2]OH \rightarrow Ag_2NH + 3NH_3 + 2H_2O.$$

The ammoniacal silver solution must therefore be prepared immediately before use, and remnants should be discarded.

We wish to thank Mr. and Mrs. William G. Lucas for valuable technical assistance. We are also indebted to Mr. Gilbert Michel for helpful criticism in the preparation of the manuscript.

### REFERENCES

Berthelot and Delépine. 1899. Sur l'azotate d'argent ammoniacal. Compt. rend. Acad. Sci., **129,** 326–30.

BIELSCHOWSKY, M. 1904. Die Silberimpregnation der Neurofibrillen. J. Psychol. Neur., **3**, 169–89.

BJERRUM, J. 1941. Metal ammine formation in aqueous solutions. Theory of reversible step reactions. P. Haase and Son, Copenhagen. *Cited from:* Chemical Abstracts, **35**, 6527–34. 1941.

BODLAENDER, G., and FITTIG, R. 1902. Das Verhalten von Molekular-verbindungen bei der Auflösung. Zts. phys. Chem., **39**, 597–612.

BRITTON, H. T. S., and WILSON, B. M. 1933. The constitution of ammoniacal solutions of (a) silver nitrate, (b) silver oxide. J. Chem. Soc. London., **1933**, 1050–3.

BRODAL, A. 1949. Spinal afferents to the lateral reticular nucleus of the medulla oblongata in the cat. J. Comp. Neur., **91**, 259–92.

CLARK, W. E., LEGROS, and MEYER, M. 1947. The terminal connections of the olfactory tract in the rabbit. Brain, **70**, 304–28.

DRAPER, H. N. 1886. *In* Pharm. J. and Transactions, London, **17**, 487. *Cited from:* Jahresbericht über die Fortschritte der Chemie (Braunschweig), 1886, 480–1.

GLEES, P. 1946. Terminal degeneration within the central nervous system as studied by a new silver method. J. Neuropath. Exp. Neur., **5**, 54–9.

GLEES, P. 1944. The anatomical basis of cortico-striate connections. J. Anat. London., **78**, 47–51.

GLEES, P., MEYER, A., and MEYER, M. 1946. Terminal degeneration in the frontal cortex of the rabbit following the interruption of afferent fibers. J. Anat. London., **80**, 101–6.

HOFF, E. C. 1932. Central nerve terminals in the mammalian spinal cord and their examination by experimental degeneration. Proc. Roy. Soc. London., **111**, 175–88.

KOHLSCHÜTTER, V. 1912. Ueber Bildungsformen des Silbers. Annalen der Chemie, **387**, 86–145.

KONOWALOFF, D. 1898. Über die Löslichkeit des Ammoniaks in wässerigen Lösungen von Silbernitrat. J. russ. Phys.-chem. Ges., **30**, 367–71.

PRESCOTT, A. B. 1880. Silver ammonium oxide in solution. Chem. News London., **42**, 31.

REYCHLER, A. 1883. Silbernitrat und Ammoniak. Ber. Deut. Chem. Ges., **16**, 990–4.

REYCHLER, A. 1895. Zur Konstitution der Silberammoniakverbindungen. Ber. Deut. Chem. Ges., **28**, 555–8.

SEKI, MASAJI. 1940. Zur Theorie der histologischen Silberschwärzung. Zts. Zellforsch., **30**, 548–66.

VOSBURGH, W. C., and STOCKDALE MC CLURE, R. 1943. Complex ions. Monammine-silver ion. J. Amer. Chem. Soc., **65**, 1060–3.

# SELECTIVE SILVER IMPREGNATION OF DEGENERATING AXONS IN THE CENTRAL NERVOUS SYSTEM

W. J. H. Nauta[1] and Lloyd F. Ryan, Lt. Col. U.S.A.F.,
*Washington, D. C.*

Received for publication November 24, 1951

Abstract.—Rat and rabbit brains containing surgical lesions of 5–10 days' duration were fixed in 10% formalin (neutralized with calcium carbonate) for 1 week to 6 months. Frozen sections (15–20 μ) were rinsed and then soaked 7 minutes in a 1.7% solution of strong ammonia in distilled water. Subsequent treatment was as follows: rinse; 0.05% aqueous potassium permanganate 5–15 minutes; 0.5% aqueous potassium metabisulfite, 2 changes of 2.5 minutes each; wash thoroughly in 3 changes distilled water; 1.5% aqueous silver nitrate, 0.5–1.0 hr.; 1% citric acid, 5–10 sec.; 2 changes distilled water; 1% sodium thiosulfate, 30 sec.; 3 changes distilled water. Each section is then processed separately. Ammoniacal silver solution (450 mg. silver nitrate in 10 ml. distilled water; add 5 ml. ethanol; let cool to room temperature; add 1 ml. strong ammonia water and 0.9 ml. of 2.5% aqueous sodium hydroxide), 0.5–1.0 min. with gentle agitation. Reduction of about 1 minute is accomplished in: distilled water, 45 ml.; ethanol, 5 ml.; 10% formalin, 1.5 ml.; 1% citric acid, 1.5 ml. Rinsing; 1% sodium thiosulfate, 10 sec.; thorough washing followed by dehydration through graded alcohol and 3 changes of xylene or toluene complete the staining process. Normal nerve fibers are slightly stained to unstained, degenerating fibers, black. The treatment in potassium permanganate is critical since too little favors overstaining of normal fibers and too much abolishes staining of degenerating fibers.

The demonstration of axon degeneration of silver impregnation has proved to be of great value in experimental neuroanatomical research. For certain purposes, however, an important drawback of this research method may lie in the fact that, unlike the much older Marchi technic, the experimental silver technics (Glees, 1946; Nauta and Gygax, 1951) are nonselective, i.e. they impregnate normal and degenerating axons with approximately equal intensity. Although the simultaneous demonstration of both fiber categories is often beneficial, experimental study, particularly that of widely distributing fiber systems, is often rendered extremely laborious by the resultant necessity of examining extensive brain areas under high power magnification. Under such conditions scattered signs of axon degen-

---

[1]Now at the Neuropsychiatry Division, Army Medical Service Graduate School, Walter Reed Army Medical Center, Washington, D. C.

eration are easily overlooked in brain parts where nerve fibers of vastly different caliber are numerous. Furthermore, although disintegrating terminal axon ramifications may be clearly distinguishable, it is often difficult to trace degenerating proximal fibers which run inside massive fiber bundles. For such reasons, it would seem to be of some value to have available a selective silver technic, which would afford a more convenient localization of degenerative axon reactions.

By the procedure described below, it has proved possible to obtain black impregnation of disintegrating axons in the rat's and rabbit's brain with normal axons stained various shades of brown (Fig. 1–3). Since the optimal fixation conditions appear to be the same for this method and the nonselective methods, the two procedures can be applied alternatively to the same material. By virtue of this fact, it has been possible to compare adjacent sections from experimental material stained by either of the two methods respectively. As the areas in which axon degeneration could be detected were the same for both technics, it seems likely that the results of the present method, besides being much more readily interpretable, are equivalent to those of the nonselective silver technics.

PROCEDURE

1. Fix for 1 week to 6 months in 10% formalin, neutralized by calcium carbonate.
2. Cut frozen sections of 15 to 20 µ thickness. Sections can be stored in 10% neutral formalin for weeks.
3. Wash briefly and transfer sections for 7 minutes into dilute ammonia: 0.25 ml. of strong ammonia to 15 ml. of distilled water.
4. Wash briefly and bring sections for 5 to 15 minutes into a 0.05% solution of potassium permanganate, in which they obtain a brown color.
5. Decolorize sections in two consecutive volumes of a 0.5% solution of potassium metabisulfite, 5 minutes.
6. Wash very thoroughly in at least two dishes of distilled water, then:
7. Transfer sections for $\frac{1}{2}$ to 1 hour into a 1.5% solution of silver nitrate.
8. Pass sections through a 1% solution of citric acid, 5 to 10 seconds.
9. Wash in three dishes of distilled water.
10. Transfer sections into a 1% solution of sodium thiosulfate, 30 seconds.
11. Wash in at least three dishes of distilled water.

*Through steps 12 to 14 each section should be processed singly.*

12. Bring section for $\frac{1}{2}$ to 1 minute into an ammoniacal silver solution, to be prepared as follows:

11

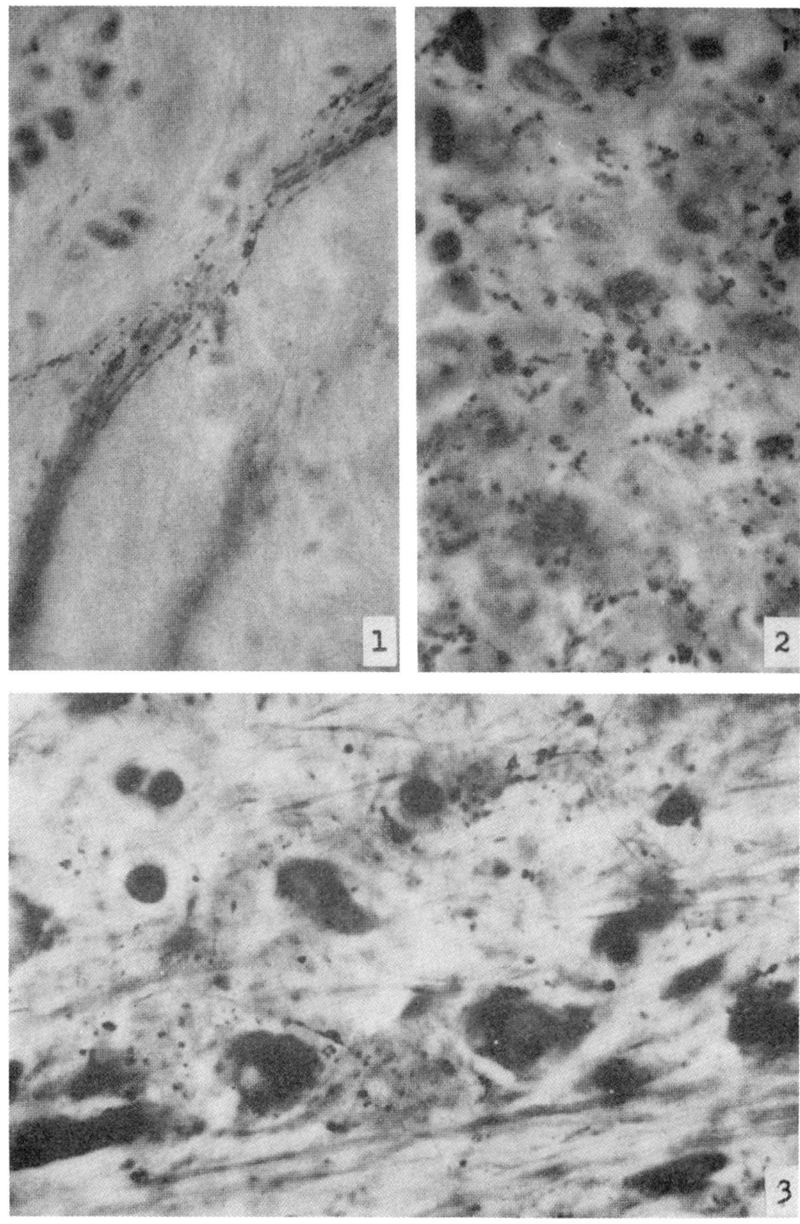

Fig. 1. — Degenerating axons in the capsula interna, 8 days after lesion of the cerebral cortex. × 500. Adult male rat.

Fig. 2. — Profuse terminal degeneration in the medial nucleus of the habenula. Degeneration time 8 days. × 900. Adult male rat.

Fig. 3. — Terminal degeneration in the lateral pontine nucleus, 10 days after lesion of the cerebral cortex. × 900. Adult female rat.

450 mg. of silver nitrate are dissolved in
10 ml. of distilled water.
 5 ml. of pure ethanol are added. The mixture is allowed
        to cool to room temperature, and
 1 ml. of strong ammonia water, and
0.9 ml. of a 2.5% solution of sodium hydroxide are added.
Throughout their stay in this solution, the sections should be
moved about gently with a glass rod.

13. Without washing, transfer section quickly onto a solution composed of:

| | |
|---|---|
| distilled water . . . . . . . . . . . | 45 ml. |
| pure ethanol . . . . . . . . . . . . . | 5 ml. |
| 10% formalin[2] . . . . . . . . . | 1.5 ml. |
| 1% citric acid . . . . . . . . . . . | 1.5 ml. |

For even impregnation, it is essential that the section be allowed to float out rapidly on the surface of this reducing fluid. It acquires a brown color within 1 minute.

14. Wash briefly, pass sections through a 1% sodium thiosulfate solution (approximately 10 seconds) and wash again in three volumes of distilled water.

15. Dehydrate in graded alcohols. Prior to mounting, alcohol must be removed completely by using at least three consecutive volumes of xylene or toluene.

### DISCUSSION

The procedure described above essentially is a modified Bielschowsky method applied to frozen sections previously treated with potassium permanganate and potassium metabisulfite.

Although little is known concerning the actual mechanism of the Bielschowsky impregnation, its analogy with a similar principle known in photochemistry as "physical development" suggests that its two fundamental phases are:

I. Formation in the nerve fibers of silver nuclei, i.e. submicroscopic particles of either metallic silver or a silver compound. These nuclei are formed during the treatment of the section with silver nitrate.

II. Deposition around the nuclei of particles of nascent silver, released from an extraneous source (the ammoniacal silver solution) under the influence of a weak reducer like acidified formalin.

Normally, the silver nuclei formed during phase I are largely insoluble in thiosulfate solutions. Therefore, treatment of the sections with thiosulfate between phases I and II has no profound effect on the result of the procedure. However, in sections which have been treated previously with potassium permanganate and

---

[2]Commercial, non-neutralized formalin taken as 100%.

potassium metabisulfite as indicated in the description of the procedure, normal axons are no longer capable of forming insoluble silver nuclei during phase I, whereas disintegrating axons have retained this capacity. This is an inference from the experience that only disintegrating axons stain black when such sections are subjected to the procedure: phase I — thiosulfate — phase II.

The same difference between normal and degenerating axons was noticed in formalin-fixed sections, demyelinated with ammoniated alcohol, and subjected to phase I and thiosulfate treatment in dark red light. In such sections also, the final phase II fails to blacken the normal axons, while disintegrating axons are stained black. This observation suggests that bright light is one of the factors operating in the production of thiosulfate-insoluble silver nuclei by normal axons during phase I, and that disintegrating axons can form such or similar nuclei in the absence of bright light. Regulation of selective staining by means of the action of light proved to be less practical and less reliable than the technic given, hence the study of the phenomenon has been abandoned for the present.

*Degeneration time.* Significantly, optimal results were obtained only after relatively long survival periods: 5 to 10 days. It was found that for positive results the axon degeneration must have proceeded to the point of drop-like disintegration. Parts of the degenerating axons which have still retained their continuity fail to be impregnated. Mildly positive reactions were observed in the terminal ramifications of cut axons as early as two days after operation, while thicker fibers could not be traced until several days later.

*Permanganate treatment.* This step is one of the crucial parts of the procedure, and it seems subject to various hitherto undefinable factors. For instance, during periods of hot weather results were often entirely negative in this laboratory unless the length of the permanganate treatment was reduced considerably, sometimes even to 30 seconds with a 0.025% solution. Consequently, rather than to rely upon the prescription given, it is advisable to ascertain in a few preliminary experiments, the optimal duration of the treatment for each specimen separately. Generally, too short treatment with permanganate favors co-impregnation of normal fibers, while too intensive treatment prevents impregnation of fibers of either category.

*Demyelination.* Since best results were obtained in material which had not been in contact with alcohol-containing media, it is recommended to omit demyelination of the sections.

LITERATURE

GLEES, P. 1946. Terminal degeneration within the central nervous system as studied by a new silver method. J. Neuropath. Exp. Neurol., **5**, 54–9.

NAUTA, W. J. H., and GYGAX, P. A. 1951. Silver impregnation of degenerating axon terminals in the central nervous system: 1. Technic. 2. Chemical notes. Stain Techn., **26**, 5–11.

14

# SILVER IMPREGNATION OF DEGENERATING AXONS IN THE CENTRAL NERVOUS SYSTEM: A MODIFIED TECHNIC


W. J. H. Nauta and P. A. Gygax, *Army Medical Service Graduate School and Department of Biochemistry, Georgetown University Medical School, Washington, D. C.*



Received for publication Sept. 28, 1953



Abstract.—Frozen sections of formalin-fixed brains containing surgical lesions, were treated with 15% ethanol for 0.5 hr., soaked in 0.5% phosphomolybdic acid for 0.25–1.0 hr., and subsequently treated with 0.05% potassium permanganate for 4–10 min. (The duration of the latter treatment is critical and individually variable). Subsequent procedure is as follows: decolorize in a mixture of equal parts of 1% hydroquinone and 1% oxalic acid; wash thoroughly and soak sections in 1.5% silver nitrate for 20–30 min.; ammoniacal silver nitrate (silver nitrate 0.9 g., distilled water 20 ml., pure ethanol 10 ml., strong ammonia 1.8 ml., 2.5% sodium hydroxide 1.5 ml.) 0.5–1.0 min.; reduce in acidified formalin (distilled water 400 ml., pure ethanol 45 ml., 1% citric acid 13.5 ml., 10% formalin 13.5 ml.) 1 min.; wash, and pass section through 1% sodium thiosulfate (0.5–1.0 min.); wash thoroughly and pass sections through graded alcohols and xylene (3 changes); cover in neutral synthetic resin.


The purpose of this paper is to present a modification of Nauta and Ryan's (1952) silver technic for demonstrating axon degeneration in the central nervous system. Attempts to revise the latter technic were prompted by reports of inconsistent results obtained by its employment in several laboratories.

The procedure here described was developed in this laboratory during the past two years and is now routinely employed with satisfactory results in experiments on the brains of rats, cats, and monkeys. Although the method is not strictly selective for degenerating axons, a considerable proportion of the normal nerve fibers will remain unstained, a circumstance which greatly facilitates the process of tracing axon degeneration. Nevertheless, the lack of complete tinctorial selectivity necessitates a thorough familiarity of the observer with the histological characteristics of axon degeneration, the more so since artificial deformations, mostly in the form of fusiform or rhomboid varicosities, are observed frequently in normal axons. Thus, the only dependable criterion of axon degeneration is represented by the typical drop-like disintegration which is displayed by the axon after a short time following destruction of, or amputation from, its nutrient cell body. In the rat, cat, and monkey most or all axons involved will, as a rule, have reached this stage of degeneration after 5 to 7 postoperative days.

The procedure is as follows:

1. Fix in 10% neutral formalin for 2 weeks to 6 months. Results appear to be optimal after fixation periods of 1–3 months.

2. If embedding is necessary for the preservation of anatomical continuity (for example, in cases of large lesions in the hypothalamus), embed in gelatin as follows: Wash brain slices of 5–10 mm. thickness in running tap water for 24 hours, incubate for 12–18 hours in 25% gelatin at 37° C. (dish covered tightly), wipe off excess gelatin and immerse for 6 hours or longer in cold (40° F.) formalin.

3. Cut frozen sections of 15–25 µ thickness. Assemble sections in 10% formalin, cooled to 70° F. or lower. Sections can be stored in cooled formalin for several weeks, preferably in a refrigerator. Process only a small number of sections at one time; for example, not more than 6 transverse sections of whole cat brain, or 10 of rat brain.

4. Soak sections in 15% ethanol (aq.) for 0.5 hr.

5. Wash briefly in distilled water, and soak sections in 0.5% phosphomolybdic acid for 0.25–1.0 hour.

6. Without washing, transfer sections to 0.05% potassium permanganate for 4–10 minutes. This step is highly critical: too long treatment will inhibit subsequent silver impregnation of all fibers, too short will cause impregnation of too many normal fibers. The optimum duration varies from specimen to specimen and can best be established in a trial run, treating 3 sections in which degenerating axons can be expected for 5, 7, and 9 minutes respectively. During the treatment the sections must be turned over from time to time in order to ensure an even effect.

7. Assemble sections in distilled water and decolorize in a mixture of equal parts of 1% hydroquinone and 1% oxalic acid. Duration 1–2 minutes.

8. Wash thoroughly in at least 3 dishes of distilled water, and soak sections for 20–30 minutes in a 1.5% solution of silver nitrate. *From here on, transfer sections individually.*

9. Wash briefly in distilled water, and transfer sections for 1 minute to an ammoniacal silver nitrate solution, freshly prepared as follows:

| | |
|---|---|
| Silver nitrate | 0.9 g. |
| Distilled water | 20.0 ml. |
| When solution is complete, add: | |
| Pure ethanol | 10.0 ml. |
| Strong ammonia | 1.8 ml. |
| Sodium hydroxide, 2.5% (aq.) | 1.5 ml. |

10. Transfer the section to a reducing fluid composed of the following:

| | |
|---|---|
| Distilled water | 400.0 ml. |
| Pure ethanol | 45.0 ml. |
| Formalin, 10% (non-neutralized) | 13.5 ml. |
| Citric acid, 1% (aq.) | 13.5 ml. |

Allow the section to float evenly out on the surface of the fluid. A brown color will appear within 1 minute. If the stain is too light, add some sodium hydroxide to the ammoniacal silver solution for subsequent sections; if too dark, add some ammonia.

11. Wash briefly; pass sections through 1% sodium thiosulfate, and wash again in 3 changes of distilled water.

12. Mounting is best done prior to dehydration following a gelatin-alcohol method (Albrecht, 1954) described elsewhere in this journal. Subsequently, the sections are passed through graded alcohols and xylene (3 changes). Neutral, synthetic resin is used for applying the cover glass.

### Acknowledgment

The authors wish to acknowledge the valuable technical assistance of Mrs. Mildred H. Albrecht.

### REFERENCES

Nauta, W. J. H., and Ryan, L. F. 1952. Selective silver impregnation of degenerating axons in the central nervous system. Stain Techn., **27**, 175–9.

Albrecht, M. H. 1954. Mounting frozen sections with gelatin. Stain Techn., **29**, 89–90.

# Monatsschrift für Psychiatrie und Neurologie

Revue Mensuelle de Psychiatrie et de Neurologie     Monthly Review of Psychiatry and Neurology

Basle (Switzerland)     S. KARGER     New York

Separatum     Vol. 129, Nr. 1–3 (1955)     Printed in Switzerland

**Festschrift zum 60. Geburtstag von E. Grünthal**

*Glees, P.* and *W. J. H. Nauta :* Mschr. Psychiat. Neurol. *129*, 74–91, 1955

University Laboratory of Physiology, Oxford
and Army Medical Service Graduate School, Washington, DC.

# A Critical Review of Studies on Axonal and Terminal Degeneration

By P. GLEES and W. J. H. NAUTA

*Introduction.*

From a discussion of the application and potentialities of silver methods, which took place at a recent meeting of Dr. *Rioch*'s neurological research unit in Washington, we felt that the time had come to review the work which had been carried out with the *Glees* and *Nauta* silver methods. We also thought an account of the multitudinous sources of error inherent in silver impregnation techniques might be useful to those who lack experience in the interpretation of pictures obtained by such techniques. Careless staining, for instance, may produce effects which might be mistaken for signs of degeneration.

It was not until a short time ago that a simple silver technique for the demonstration of degenerating axons and arborisations became available in the form of a modified *Bielschowsky* technique (*Glees* [1946]). More recently still several further modifications of *Bielschowsky*'s principle of silver impregnation have been employed successfully in the study of degenerating axons and their arborisations (*Nauta* and *Gygax* [1951, 1954], *Nauta* and *Ryan* [1952]).

The *Glees* method, published in 1946, was perfected in the course of various previous studies and has now been used a great deal in *Le Gros Clark*'s laboratory to trace fronto-hypothalamic connections and olfactory connections (*Le Gros Clark* and *Meyer* [1947]). It has also been used by *Meyer* and *Allison* [1949] and by *Beck* [1950] to trace descending frontal connections, and by *Brodal* [1949] and *Brodal* and *Walberg* [1952] to study descending spino-reticular fibres and the terminations in the cerebral cortex of those ascending fibres in the pyramidal system which do not relay in the thalamus.

Whereas the *Glees* method stains normal or degenerated fibres and their terminal arborisations equally well, the *Nauta* method was developed to bring out the pathological changes more or less selectively and to suppress the normal fibre patterns to a varying extent. Because of the suppression of normal fibres, the *Nauta* method impregnates mainly the proximal and pericellular parts of degenerating axons. This method is therefore particularly useful when the degeneration caused by the interruption of one fibre connection e.g. the thalamo-cortical connection) (*Nauta* and *Whitlock* [1954]) can be distinguished only if the enormous number of normal cortical fibres remains largely unstained. For a study of the degeneration on synapses it is advisable, however, to use both methods, for although the areas of terminal distribution can be readily determined by the *Nauta* method, this technique appears to be less suitable than the *Glees* method where cytological details of degenerative changes in the terminal rings or boutons must be identified.

*General remarks on silver impregnation.*

It is as well to keep in mind that silver techniques, used to bring out certain histological structures, are not true staining methods: instead of producing a colour reaction between the dyes and the histological structures, silver techniques are dependent upon the deposit of metallic particles in or on tissue elements. Thus, silver impregnation methods might more accurately be classified as "germination methods" ("Bekeimungsmethoden") than as staining methods. The ultimate distribution of the silver particles will depend on 1. the size of the silver particles formed during the impregnation procedure, and 2. the physical and chemical properties of the tissue elements which are expected to accommodate the particles. It is highly likely that these two factors, extremely complex in themselves, are in large measure responsible for the different effects which can be obtained by the use of different silver techniques. On the other hand, it is a general experience that even the most rigidly standardised silver technique may produce entirely different results in two different brain specimens. Such variations appear, for example, as differences in the intensity of impregnation. The duration and type of fixation, as determinants of physicochemical tissue properties, are more likely to be at the root of such inconsistencies than the too often incriminated impurities of the chemicals employed in the silver impregnation.

Whatever the cause of the characteristic capriciousness of silver impregnation, until proper histochemical methods become available the ever fluctuating histological pictures in the study of axonal and terminal degeneration cannot be avoided, nor can the ill-controllable, non-selective silver deposits, which often convey a deceptive impression of histological structures. Such inconsistencies make it necessary to observe extreme caution and conservatism in interpretation. The present paper is in part intended to clarify some of the principles governing the correct diagnosis of axonal and terminal degeneration and to illustrate in microphotographs both genuine degeneration and those results due to incomplete or faulty impregnation.

### *Terminal or boutonal degeneration (P. Glees).*

A review of the work I have done myself will illustrate the application of a silver method in a study of synapses which derive from one main source. Some new photographs, included in this paper, show the success of a silver technique in revealing the characteristic changes of the terminal portions of axon arborisations.

Realising the difficulty of discriminating degenerated boutons from normal ones, I felt that progress in bouton research would be achieved only if boutonal changes were studied in a system such as the lateral geniculate body, in which probably only one large, conspicuous fibre tract (the optic tract) terminates. The normal neurohistology of the lateral geniculate body of the monkey showed the existence of boutons in the form of very fine rings. These were however by no means easy to find. The terminating rings usually lay

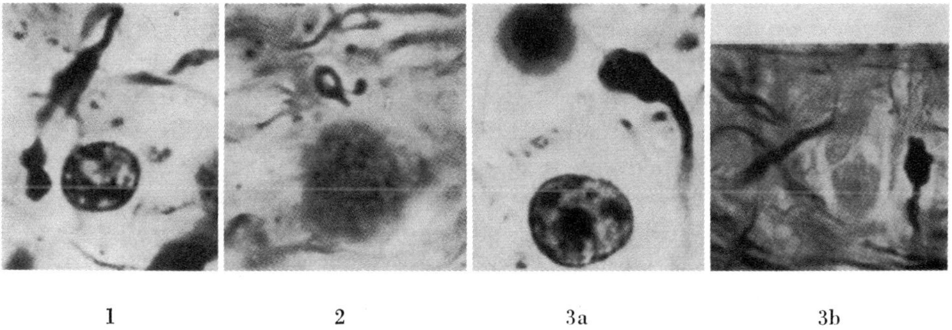

1          2          3a          3b

*Fig. 1.* Lateral geniculate body of monkey. In the centre the nucleus of a nerve cell of the lateral geniculate body; the cell body is not in focus. A degenerated optic preterminal fiber ends in a degenerating bulb (microphotograph, oil immersion). – *Fig. 2.* A thickened degenerating ring-like end formation in contact with the cell body (microphotograph, oil immersion). *Fig. 3a* and *b.* Thickened bulbous end formations in contact with the cell body (microphotograph, oil immersion).

close to the cell body and no end feet could be seen in contact with the dendritic process.

This study (*Glees* and *Le Gros Clark* [1941]) showed very clearly the characteristic degeneration which followed severance of the optic nerve of the monkey. The evidence of terminal degeneration was seen in the thickened rings or bulbous terminal swellings and in the fragmentation of the preterminal fibres. The degeneration of synapses could be shown very clearly, provided the afferents came chiefly from one source and terminated en masse (figs. 1–5). This investigation gave the first direct proof that one optic nerve terminates in alternate layers of the lateral geniculate body. That this might be so had previously been assumed from the indirect evidence of transneuronal *Nissl* changes in the lateral geniculate body following severance of one optic nerve.

The results obtained in a subsequent study of the lateral geniculate body of the cat and rabbit (*Glees* [1941, 1942]) were quite different. Here the evidence suggested that the optic fibre, with its terminal branches, was in contact with at least ten cells, and there was extensive overlap with neighbouring fibres. In the cat I found that terminating optic fibres start as collaterals of the optic tract and end as fine terminal rings. The number of synaptic contacts with one principal cell of the lateral geniculate body was approximately fourteen. I also found synaptic contacts with dendrites as well as with the cell body; the exodendritic contacts were more numerous than the axosomatic (figs. 6 and 7).

An analysis of the quantitative data in monkey, rabbit and cat suggested that the way in which optic fibres terminate in the monkey enables a precise recording of a retinal image to be made at this neural level. On the other hand, the profusion of terminal boutons in the geniculate body of the cat, together with the overlap which appears to exist, would provide an anatomical basis for the high degree of sensitivity at low intensities of illumination. This work has been extended and our results confirmed by *Nauta* and *v. Straaten* [1947].

We concluded from these various studies that terminal degenerations could be a useful mean of identifying fibre tracts and also of establishing whether synapses are essentially axosomatic or axodendritic.

I therefore standardised and stabilised the silver method which I had used for these and other studies. I found that it gave a reliable picture of the normal neurohistology and pathological alterations (such as those caused by experimental lesions) in axons and their terminations. Owing to the completeness of the impregnation it is also possible to estimate the quantitative distribution of terminal arborisations, whether they are normal or pathologically changed. The silver method shows up the actual termination of the fibres and also the place of termination, i.e. either on the cell body or on dendrites. It also shows whether the terminations in one particular fibre system are concentrated on a particular area of the cell surface or are widely dispersed over the cell body or dendrites. The failure to find any typical end formations in a particular nucleus or in parts of the cerebral cortex in the normal state does not mean that further work with silver or other methods may not one day reveal end formations, such as ring-like structures. The degeneration of the fine pericellular fibres (e.g. in the cerebral cortex or corpus striatum) can however easily be identified after sufficient time has elapsed (*Glees* [1944]) and this gives an indication of cerebral connections (*Glees, Meyer* and *Meyer* [1946]). It is of course possible that some boutons can be seen only after they have begun to degenerate and therefore impregnate more strongly in silver nitrate.

To test the scope of the method we continued synaptic studies by cutting the posterior columns of the spinal cord or by severing sensory roots, in order to examine the terminations to an essentially monosynaptic cell station (the nuclei of *Goll* and *Burdach*) (*Glees* and *Soler* [1951]). The preterminal and terminal degeneration was shown very clearly and also the amount of overlap between neighbouring roots (fig. 8).

The observation of numerous ring-like terminal structures on large cells of the reticular substance led me to study the termination of collateral afferents to the reticular substance from the collicular plate. I was joined in this work by Dr. *Pearce*, who communicated the preliminary results at the 1954 meeting of the Anatomische Gesellschaft. For this work we combined the silver method with the *Marchi* method, staining alternate blocks by each technique. The silver method gave us precise information on axosomatic or axodendritic terminations of collicular afferents, while with the *Marchi* method we were able to identify the main stream of fibres. The number of *Marchi* fibres which dropped out gave us a quantitative means of ascertaining the number of fibres which supply various levels of the brain stem.

These, then, are the types of problem we have attempted to solve so far and some of the conclusions we have drawn from our investigations.

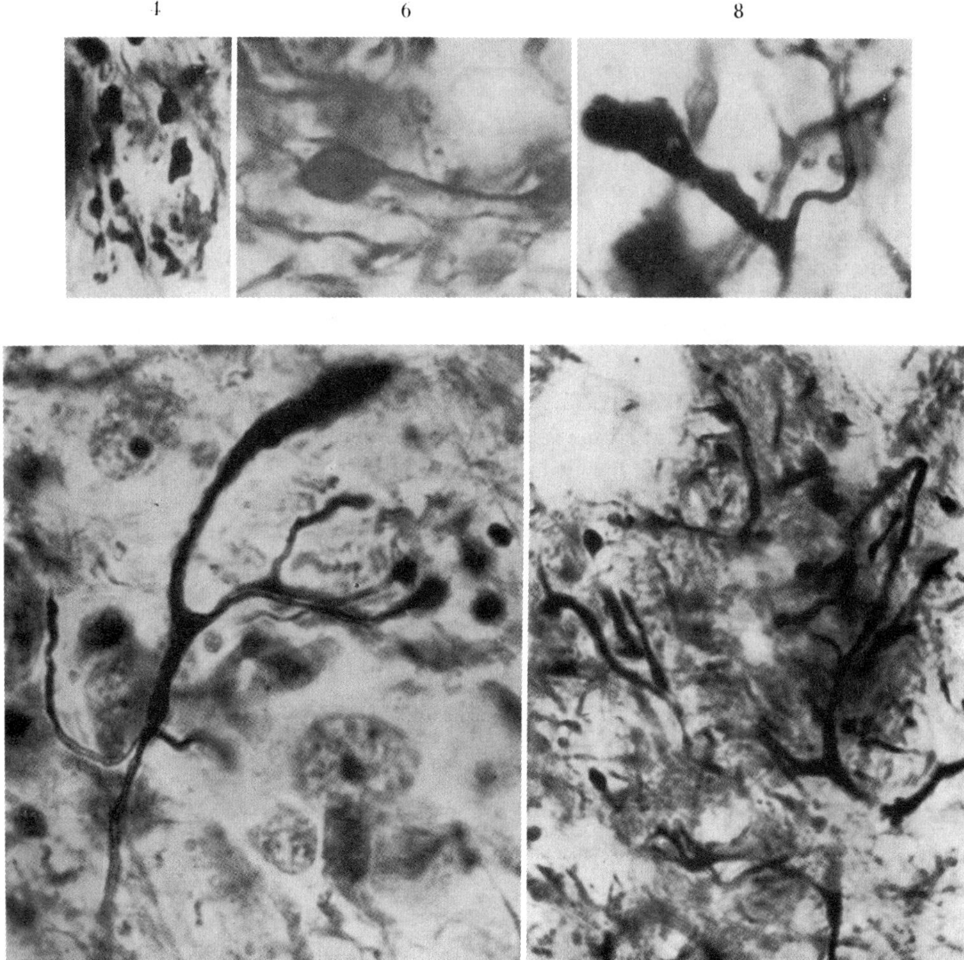



*Fig. 4.* Preterminal degeneration of optic afferents in the lateral geniculate body of the monkey. Note that one preterminal has already reached an advanced stage of degeneration while the other is only thickened. They come from the same stem fibre (microphotograph, oil immersion). – *Fig. 5.* Preterminal and terminal degeneration in the lateral geniculate body of the monkey following severance of the optic tract. Note that the terminal swelling is more advanced than the swelling of the preterminal fibres. The nuclei of the nerve cells are outlined in this picture (microphotograph, oil immersion). – *Fig. 6.* Terminal degeneration of optic fibres in the lateral geniculate body of the cat. Note the gross enlargements of the terminal ring, which is filled with a fibrillary network (microphotograph, oil immersion). – *Fig. 7.* Preterminal and terminal degeneration in the lateral geniculate body of the cat. One stem fibre branches in a great number of preterminal and terminal swellings (microphotograph, oil immersion). – *Fig. 8.* Numerous bulb-like thickenings lying on the surface of a nerve cell of the Nucleus gracilis. These bulbous enlargements are caused either by degenerating boutons or by drop-like disintegration of a pericellular plexus of terminating fibres. This degeneration followed section of posterior roots (microphotograph, oil immersion).

*Axonal degeneration as demonstrated by*
*silver impregnation (W. J. H. Nauta).*

In silver preparations axonal degeneration is characterised by a typical breakdown in the filamentous structure of the axon, resulting in a picture of "droplike disintegration". Apart from the degenerative phenomena observed in boutons terminaux, it is well to keep in mind that *droplike disintegration is the sole dependable criterion of axonal disintegration.* Prior to actual disintegration degenerating axons display local swellings, which cannot however always be distinguished from artificial varicosities in normal axons. Very early degenerative changes of pyramidal axons following cortical ablation in the cat have been briefly described by *Glees* [1948].

Therefore, if it is desired to trace the entire length of a degenerating fibre system, it is necessary to allow of survival periods of sufficient duration to permit the development of drop-like disintegration throughout the length of the fibre system. The time needed for this development appears to vary from species to species and for different fibre systems. For instance, most corticofugal fibres will degenerate to the stage of drop-like disintegration within 5 days in rat and cat, but in the monkey appear to require upward of 7 days to deteriorate to the same stage. With too long survival periods, on the other hand, much of the clarity of the degenerative picture may become lost, especially where fibres of fine calibre are studied. As a rule, the optimum survival time will have to be established for each fibre system separately; however, periods of from five to seven days are usually adequate for the development of complete axon disintegration in the rat and cat.

The characteristics of drop-like disintegration are difficult to describe and can best be appreciated by the aid of representative photographs (figs. 9 and 10). It will be noted that the axons, at the stages illustrated, have broken drown into fragments varying from small single droplets to large lobated complexes. Even if many such axon fragments do not resemble "drops" in the strict sense of the word, the majority displays smoothly rounded contours. Larger or smaller vacuoles are occasionally observed in some fragments.

---

*Fig. 9.* Degenerating dorsal root fibres traversing the dorsal funiculus of the spinal cord, four days after section of L. 7 dors. in the cat. This picture illustrates an early stage of drop-like disintegration of proximal axons (fibres of passage). Compare

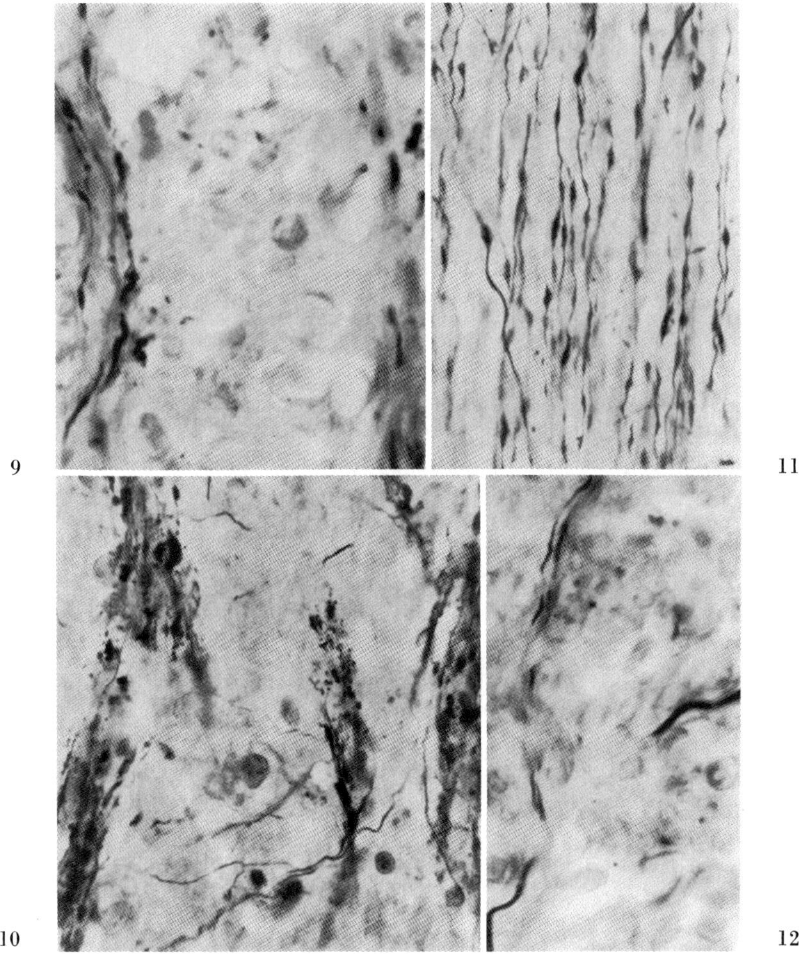

with artefacts shown in figures 11 and 12. ($\times$ 700. Silver impregnation after *Nauta-Gygax*, 1954). – *Fig. 10*. Degenerating fibre bundles of the optic tract, six days after contralateral eye enucleation in a rat. This picture is typical for more advanced degenerative fragmentation of proximal axons. Note the absence of terminal degeneration between the bundles. Compare with figure 14 ($\times$ 700. Silver impregnation after *Nauta-Gygax*, 1954). – *Fig. 11*. Artificial distortion of normal axons in the capsula interna. This common type of artefact is characterised by spindle- and rhomb-shaped swellings in the course of normal axons. In the form illustrated these artefacts can be readily distinguished from fully developed actual axon degeneration. Compare with figures 9 and 10. ($\times$ 700. Silver impregnation after *Nauta-Gygax*, 1954). – *Fig. 12*. Incomplete impregnation of dorsal root fibres in the spinal cord. Note the even appearance of the impregnated parts. ($\times$ 700. Silver impregnation after *Nauta-Gygax*, 1954).

The correct identification of axon degeneration requires great caution on the part of the observer. The most important and persistent source of error is represented by *artificial deformities of normal axons*, a phenomenon which is rather frequently observed following silver impregnation. Such artefacts mostly consist in spindle-shaped or rhomboid varicosities (fig. 11). If present, such irregularities are usually distributed diffusely over all brain parts; occasionally, however, they show a misleading selectivity for one or a few fibre systems. The appearance of artificial varicosities, together with the absence of axon fragmentation, will usually allow a distinction to be made between true degeneration and artefact.

*Incomplete impregnation of axons* may constitute a further source of misinterpretation. This phenomenon is occasionally encountered in axons of large calibre, such as the dorsal root fibres illustrated in fig. 12. Unlike degenerating axons, incompletely impregnated fibres often resemble the broken lines of an ink drawing. In this uncomplicated form these fibres will be easily distinguishable from degenerating axons; in cases, however, where incomplete impregnation occurs in conjunction with artificial varicosity of numerous normal axons, intensely confusing pictures may result which can be interpreted, if at all, only by virtue of considerable experience.

With these limitations in mind, degenerating axons can often be identified with reasonable certainty, provided their orientation coincides with or approximates to the plane of histological sectioning. Much greater difficulty may arise where the fibre system under study appears in transverse section, unless the degeneration is massive or the degenerating axons occupy a relatively isolated position and are not mixed with numerous normal fibres. Cross sections of degenerating axons, scattered diffusely among multitudes of normal black-stained fibres, may not be identifiable at all; under such circumstances it is often necessary to repeat the experiment with the employment of a more suitable plane of sectioning. An example of a more clearly recognisable cross-sectional picture of axon degeneration is given in fig. 13.

*The distinction between proximal and*
*terminal axon degeneration.*

A further important problem is encountered where a differentiation must be made between degeneration of proximal and that of terminal or preterminal parts of the axon. Much of the usefulness of

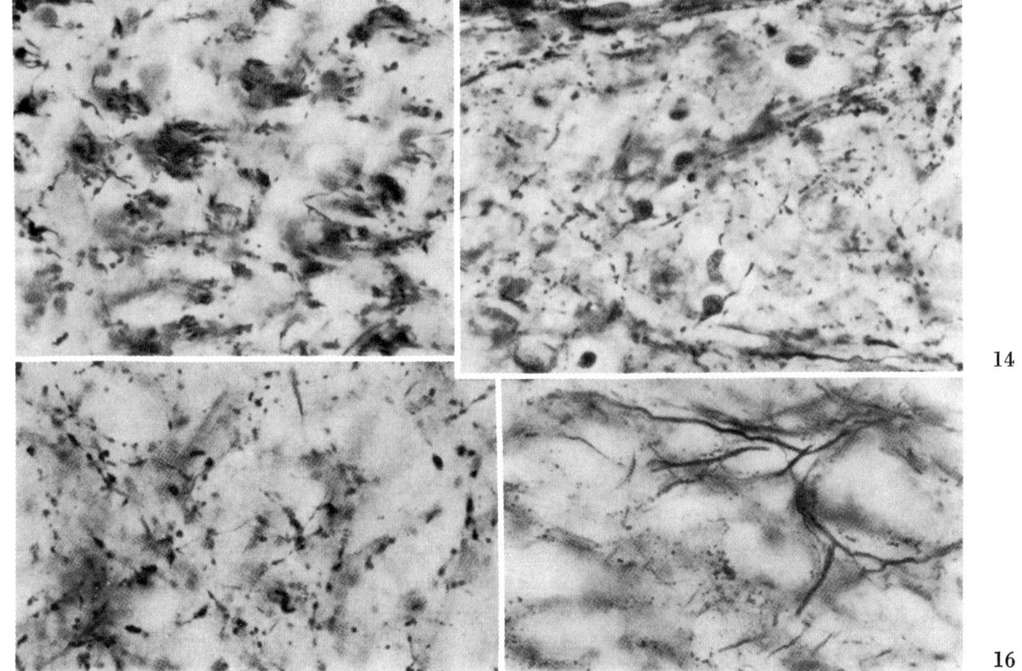

*Fig. 13.* Degenerating axons in transverse section (stratum opticum of superior colliculus, seven days after contralateral eye enucleation in a rat. In this concentration degenerating axons are often readily identifiable even in transverse sections. ($\times 700$. Silver impregnation after *Nauta-Gygax*, 1954.) – *Fig. 14.* Axon degeneration in the lateral geniculate body of the rat six days after contralateral eye enucleation. Note the slender bundles of disintegrating proximal axons, and profuse terminal degeneration characterised by an intricate arrangement of fragmented fine axon ramifications. ($\times 700$. Silver impregnation after *Nauta-Gygax*, 1954.) – *Fig. 15.* Terminal degeneration in the cerebral cortex of the cat five days after thalamic lesion. ($\times 700$. Silver impregnation after *Nauta-Gygax*, 1954.) – *Fig. 16.* Granular silver precipitate in the substantia nigra. Note the disorderliness of this picture as compared with figures 14 and 15. ($\times 700$. Silver impregnation after *Nauta-Gygax*, 1954.)

the techniques used to demonstrate degenerating axons depends on this distinction; it is only by the process of tracing axon degeneration distally that the terminal distribution of many fibre systems can be definitely ascertained. The most important criteria for the identification of the part of the degenerating axons under observation will be clear from a consideration of the normal morphology of the axon.

Of the great majority of axons the proximal part—often referred to as "fibre of passage" or "stem fibre"—follows a comparatively straight, or at most slightly undulating course. Within the area of its terminal distribution the axon usually arborises into a smaller or larger number of tortuous branches of smaller calibre, which in turn may ramify repeatedly and ultimately establish synaptic relationships with one or more cell bodies or dendrites. As demonstrated above all by the *Golgi* technique, such preterminal and terminal axon ramifications may form an intricate plexus enmeshing cell bodies and dendrites.

Degenerating fibres of passage are characterised by a relatively orderly and approximately rectilinear arrangement of comparatively large drop-like fragments, while degenerating terminal portions of the axonal ramifications can be identified by the presence of finer droplets arranged in a seemingly chaotic fashion around and among the cell bodies. Naturally, degenerating terminal ramifications are often observed together with degenerating fibres of passage; where they are abundant the histological picture is markedly different from that presented by degenerating fibres of passage alone (compare fig. 10 with figs. 14 and 15). On the other hand, it is often difficult, if not impossible, to identify a less abundant terminal degeneration scattered among numerous degenerating fibres of passage, especially where the latter appear in transverse section. This condition will be encountered especially in interstitial nuclei of massive fibre systems, such as the lateral hypothalamic region and many parts of the reticular formation of the brain stem.

Even if the picture of terminal degeneration is often at first glance disorderly, closer observation will reveal a well-defined if intricate alignment of the drop-like axon fragments. Unless a well-defined pattern is observed, the structures seen may actually be a random deposit of granular metallic silver (fig. 16). *In general, the diagnosis of terminal degeneration should not be founded on the presence of ill-defined, black-stained structures,* some of which are also frequently encountered in silver impregnated normal brain tissue.

The quantitative aspects of terminal degeneration observed in a given brain region are a further important matter. Although obviously no strict rules can be given, *it is good practice to disregard the occasional observation of only a few solitary degenerating axons,* whether fibres of passage or terminal ramifications. Such sporadic degeneration is often found in normal brains and hence, where observed in experimental material, may not be related to the experimental procedure. Even massive findings, however, are not necessarily attributable to experimental surgery. Especially in rats and cats, focal or diffuse lesions of encephalitic origin are occasionally encountered and should be ruled out before any conclusions are drawn. Thus, as with all other experimental methods, it is always advisable to perform the same experiment in more than one animal.

### Discussion.

With some experience, axon degeneration as it appears in silver impregnated material can usually be identified with reasonable certainty. In spite of much work, however, considerable uncertainty continues to exist in the interpretation of boutons terminaux in normal and pathological material. Some aspects of the existing controversies will be dealt with more fully in this discussion.

Since the discovery by *Held* [1893] that unmyelinated nerve fibres terminating in the trapezoid nucleus of the cat have a special type of ending, comparatively little experimental work has been done to study pathological alterations of end formations (figs. 17 a and b). *Held*'s results, obtained by the use of the *Golgi* method, were confirmed by *Meyer* [1896], who used a methylene blue stain. *Held* [1897] later found end formations also on anterior horn cells, on cells of the reticular formation, on *Purkinje* cells, in the nucleus of *Deiters*, in the ventral cochlear nucleus and in the nucleus of V. For this second study he used other staining methods, such as erythrosin, methylene

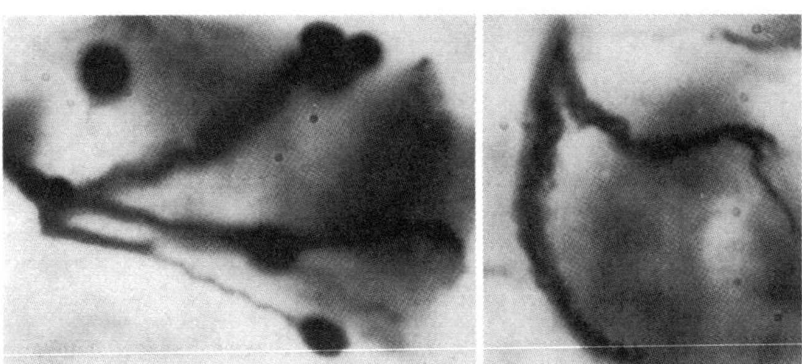

17a      17b

*Fig. 17a and b.* a) The terminations of *Held* around the cells of the trapezoid nucleus in a normal rabbit. Note the bulbous form of termination which might be mistaken for a sign of degeneration were the preterminal fibre not perfectly normal in appearance, thus excluding any degenerative changes (microphotograph, oil immersion, *Glees* impregnation). b) The terminations of *Held* in the normal rabbit. Some may have a more fibrillary structure compared with those shown in figure 17a (microphotograph, oil immersion, *Glees* impregnation).

blue and iron haematoxylin. In 1898 *Auerbach* observed button-shaped end-formations in the spinal cord, and this special type of ending is known as the *Held-Auerbach* end-foot.

The discovery of these end-feet gave added support to the neuron theory, based primarily upon the studies of *Cajal* and formulated by *Forel* [1887] and *Waldeyer* [1891]. The contact between a terminating nerve fibre and a neuron is called a synapse; the origin of this term is discussed by *Fulton* [1949].

*Held*, who had initially supported the neuron theory, altered his opinion when he found that some extracellular fibres around the cell bodies of a trapezoid nucleus penetrated into the cell body and continued as intracellular fibres. The first signs of this change of view are seen in his paper published in 1897. *Cajal* defended the neuron theory throughout his life, and with the *Golgi* method, which he used mainly during the earlier part of his life, he found that the nerve fibres in foetal or newborn animals terminated freely.

When the *Cajal* and *Bielschowsky* silver methods came into use in 1904, it was possible to achieve a reliable impregnation of unmyelinated fibres and these methods showed the fine terminal rings and boutons around the cells of spinal cord and, as *Cajal* found, also around the *Purkinje* cells of the cerebellum.

Apart from the early experiments by *Marinesco* [1904], who occluded the abdominal aorta in rabbits and cats and found degeneration of end bulbs in the spinal cord, no study of synaptic alterations for the purpose of identifying fibre tracts was made until compara-

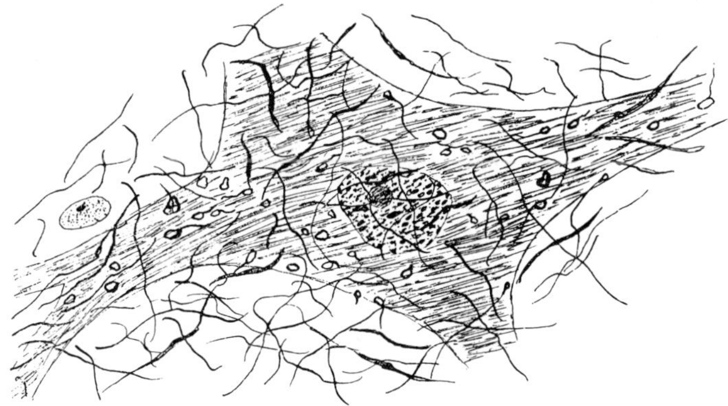

*Fig. 18.* The type of terminals seen on the anterior horn cells of the cat when impregnated with the *Glees* method. Note the thick and reticulated type of terminal. (Pen drawing under oil immersion.)

tively recently, although *Cajal* [1928] observed changes in synapses following cortical lesions and suggested that synaptic changes took place in hibernating animals. *Cajal's* view on synapses in general are set forth in his last work (*Cajal* [1935]).

*Phalen* and *Davenport* [1937] have studied the normal distribution of boutons in the spinal cord of monkeys, dogs, cats, rats and rabbits. Their work showed a great variation in the size and form of these boutons and it was obvious that their appearance and distribution would have to be much more extensively studied before any criteria for the identification of degenerating boutons could be established. *Phalen* and *Davenport* fixed their material in a mixture of pyridin and water, an unsuitable procedure if such very delicate structures as boutons are to be subsequently stained with any degree of reliability. *Block* impregnation is in any case unreliable.

The difference in the appearance of boutons, caused by the block impregnation technique, was not taken sufficiently into consideration by *Hoff* [1932, 1935], whose study of bouton changes in the spinal cord of the monkey following ablation of cortical areas was in many ways pioneer work. The cause of these differences seems to have been disregarded also by *Schimert* [1938] in his otherwise excellent study of boutonal termination in the spinal cord of cats, in which he draws attention to the degeneration of the fine fibres leading to the terminals.

The first systematic study of degenerative changes in the boutons terminaux of the spinal cord was undertaken by *Gibson* [1937], who observed an increased argentophilic and progressive thickening of the normally thin anular boutons in incipient phases of *Wallerian* axon degeneration, followed by a phase of progressive disintegration and ultimate disappearance. It is remarkable that some of the degenerative forms of boutons identified by *Gibson* were observed in normal material by *Phalen* and *Davenport* [1937], a circumstance which tends to complicate seriously the distinction between normal and degenerating boutons in experimental studies, unless, as we have said, fixation or post mortem changes account for this.

The synaptic changes following severance of posterior roots have been examined by *Foerster* et al. [1933], but the changes appeared to be too massive and *Bodian's* [1942] criticism that the experiments had interfered with the vascular supply of the spinal cord seems just. *Bodian* [1937, 1940] devoted two detailed papers to the synapses of vertebrates, studied with his protargol stain. Some of his illustrations, however, give the impression that the club-shaped endings might be the cut surfaces of passing axons rather than endings.

The conclusions drawn from the study of ascending pyramidal fibres by *Brodal* and *Walberg* [1952] should be tentative: only a small number of fibres degenerated in their ascending course and it seems to us extremely difficult to trace their terminations up to the cerebral cortex, quite apart from the fact, which we have already mentioned, that spontaneous cerebral degeneration or degenerative processes may be caused by disease.

In connection with the varied shape of boutons in the spinal cord emphasized *Phalen* and *Davenport* and other workers the hypothesis of *Weber* [1950] is worth mentioning. *Weber* assumes that boutons undergo cyclic changes due to involution or regeneration.

*Hamlyn* [1954] studied the termination of optic fibres in the avian tectum using adult chickens of various breeds and both sexes. One eye was enucleated and survival periods of 3, 5, 7, 9 and 11 days allowed.

At biopsy both optic tecta were removed and frozen sections cut 15 $\mu$. These were stained under identical conditions with a modified *Bielschowsky-Gros* silver method. In view of the complete decussation in the chiasma in *Gallus domesticus* the ipsilateral tectum was used as the control.

In the case of the three day survival period the preterminal fibre pattern of the optic axons showed clearly, but very few terminal formations were seen. At 5 days end formations were found and these had increased in number by 7 days; at 9 days the picture of terminal as well as preterminal degeneration was fully established, numerous swollen endings both of the club and ring types being present.

In 11 day specimens the degenerating fibres and terminals seemed fewer in that part of the degeneration zone deepest to the tectal surface.

The study of the lateral geniculate body of the monkey suggested that only one bouton is in contact with the cell body, for only one large club-shaped ending could be seen when the optic tract was cut. This is then a monosynaptic connection, the ratio of bouton to relay cell being 1:1. Normal preparations, however, show only a very thin anular ending and it seems doubtful whether one such ring would be sufficiently large to fire off the cell body by either electrical or biochemical means. It is possible that the silver method employed does not reveal all the end formations present, since we also found no end formations attached to dendrites, which are usually assumed to be receptive portions of the neuron.

It may well be that the location and scarcity of synaptic endings in the lateral geniculate body of the monkey is unique; in the cat and rabbit the terminal degeneration in the lateral geniculate body, on both cell body and dendrites, was profuse. Another possible explanation for this difference is that the degenerated club-shaped boutons seen in the monkey may have been not true terminals, but degenerated preterminals and that the terminals in the lateral geniculate body of the monkey are too fine to be identified with certainty.

We hope therefore that further studies will be undertaken in the future to clear up this problem.

Apart from this one matter, the *Glees* method will be found useful for a study of terminal changes, since it gives precise information on whether a fibre tract terminates over the whole of one nucleus or only in a portion of it. This method is also suitable for the study of a polysynaptic system of heterogeneous origin, such as the system in contact with anterior horn cells, for which the normal and degenerated end bulbs must be equally stained. It is then possible to locate the spatial relationships of fibre tracts on a given cell body and its dendrites.

Work of this nature, however, requires quantitative studies of the types of bouton, and *Bodian* writes [1942] that any study of precise synaptic relations "must surmount the difficulties presented by a great number of boutons in many cells often from many sources... and the problem of distinguishing with absolute certainty the altered boutons from the normal ones. An attempt to count and classify the boutons in the spinal cord of the cat is at present being made (*Pearce* and *Glees* [1954]). Three main types have so far been identified: thin anular, thick anular and reticulated, and we hope to make a statistical analysis of the distribution of these three types in the normal and pathological state (fig. 18).

Although end feet apparently enlarge during the process of degeneration, whether this is entirely due to the degeneration is questionable; a film of nerve cells and their axons in tissue culture, made by Dr. *Hughes* in Cambridge and shown at the First International Neurochemical Symposium at Oxford in 1954, showed that the swelling is a survival reaction by the terminal portion of the axon. *Hughes* separated the distal portion of the axon from the cell body and found that it continued to grow with its bulbous termination for several hours. The growth cones of the maturing axons, like the boutons terminaux, contain mitochondria (*Bodian* [1942], *Palade* [1954]) and are apparently capable of independent metabolism for a limited period. This might explain the continuous enlargement of the severed portion of the axon in tissue culture or the boutonal swelling when the axon is severed from its cell of origin.

When no special forms of terminations have to be revealed in normal material, apart from the pericellular plexus, both *Glees* and *Nauta* silver methods can be successfully employed. However, when preterminal fibre degeneration and terminal disintegration of the pericellular plexus needs to be brought out clearly against the background of normal fibres, the *Glees* method should be used.

### *Postscript.*

A recent study of axonal degeneration with a silver method originates from *Lance, J. W.*, Brain 77, 314, 1954, and should be consulted as it combines histological and neurophysiological studies of pyramidal fibres.

### *Summary.*

The authors who have developed modifications of the *Bielschowsky* method for impregnating of nerve cells and nerve fibres describe the results achieved with these techniques and discuss the limitations of the methods. Numerous micro-photographs illustrate the various stages of axonal, preterminal, and terminal degeneration. The range of application for each method has been critically evaluated, and attention has been paid to possible artifacts.

### *Zusammenfassung.*

Die Autoren, die beide Modifikationen der Silbermethode von *Bielschowsky* für Nervenzellen und Nervenfasern ausgearbeitet haben, beschreiben die Resultate ihrer Forschungen im Zentralnervensystem. Weiterhin werden die Ergebnisse ihrer Arbeiten und diejenigen anderer Forscher auf diesem Gebiete eingehend diskutiert und Richtlinien für die Erkennung der Degenerationsstadien für axonale, preterminale und terminale Degeneration festgelegt. Die Anwendung der *Glees*- und *Nauta*-Methode für die Analyse bestimmter Faserbahnen sowie eine Beschreibung möglicher Artifakte wird angegeben.

### *Résumé.*

Les auteurs, qui ont élaboré les deux modifications de la méthode argentique de *Bielschowsky* pour les cellules et les fibres nerveuses, décrivent le résultat de leurs recherches dans le système nerveux central. Les résultats de leurs travaux et de ceux d'autres chercheurs dans ce domaine sont discutés en détail et les critères pour reconnaître les stades de dégénérescence axonale, préterminale et terminale sont établis. L'emploi de la méthode de *Glees* et de la méthode de *Nauta* pour l'analyse de certaines voies des fibres est indiqué et une description d'artéfacts possibles est donnée.

## REFERENCES

*Auerbach, L.:* Neurol. Zbl. *17*, 445, 1898. – *Beck, E.:* Brain *73*, 368, 1950. – *Bielschowsky, M.:* J. Psychol. Neurol. *3*, 169, 1904. – *Bodian, D.:* J. comp. Neurol. *68*, 117, 1937; *73*, 323, 1940; Physiol. Rev. *22*, 146, 1942. – *Brodal, A.:* J. comp. Neurol. *91*, 259, 1949. – *Brodal, A.* and *F. Walberg:* Arch. Neurol. Psychiat., Chicago *68*, 755, 1952. – *Cajal, Ramon y:* Z. wiss. Mikr. *20*, 401, 1904; Degeneration and Regeneration of the Nervous System, trans. R. M. May. Oxford Univ. Press, London 1928; Die Neuronenlehre, in Handb. d. Neurol. *1*, 887, 1935. – *Foerster, O., O. Gagel* and *D. Sheehan:* Z. Anat. EntwGesch. *101*, 553, 1933. – *Forel, A.:* Arch. Psychiat. Nervenkr. *18*, 162, 1887. – *Fulton, J. F.:* Physiology of the Nervous System, 3rd ed. Oxford Univ. Press, New York 1949. – *Gibson, W. C.:* Arch. Neurol. Psychiat., Chicago *38*, 1145, 1937. – *Glees, P.:* J. Anat., Lond. *75*, 435, 1941; *76*, 313, 1942; *78*, 47, 1944; J. Neuropath. exp. Neurol. *5*, 54, 1946; Acta anat. *6*, 447, 1948. – *Glees, P.* and *W. E. Le Gros Clark:* J. Anat., Lond. *75*, 295, 1941. – *Glees, P., A. Meyer* and *M. Meyer:* J. Anat., Lond. *80*, 101, 1946. – *Glees, P.* and *J. Soler:* Z. Zellforsch. *36*, 381, 1951. – *Hamlyn, L. H.:* 1954, Personal Communication. – *Held, H.:* Arch. Anat. Physiol., anat. Abt. 201, 1893; 204, 1897. – *Hoff, E. C.:* Proc. roy. Soc. B. *111*, 175, 1932; Arch. Neurol. Psychiat., Chicago *33*, 687, 1935. – *Le Gros Clark, W. E.* and *M. Meyer:* Brain *70*, 304, 1947. – *Marinesco, G.:* Rev. Neurol. *12*, 405, 1904. – *Meyer, M.* and *A. C. Allison:* J. Neurol. Psychiat. *12*, 274, 1949. – *Meyer, S.:* Arch. mikr. Anat. *47*, 734, 1896. – *Nauta, W. J. H.* and *P. A. Gygax:* Stain Tech. *26*, 5, 1951; *29*, 91, 1954. – *Nauta, W. J. H.* and *L. F. Ryan:* Stain Tech. *27*, 175, 1952. – *Nauta, W. J. H.* and *J. J. van Straaten:* J. Anat., Lond. *81*, 127, 1947. – *Nauta, W. J. H.* and *D. G. Whitlock:* in press, 1954. – *Palade, G. E.:* Anat. Rec. *118*, 335, 1954. – *Pearce, G. W.* and *P. Glees:* in press, Verhandlungen d. Anat. Ges. 1954. – *Phalen, G. S.* and *H. A. Davenport:* J. comp. Neurol. *68*, 67, 1937. – *Schimert, J.:* Z. Anat. EntwGesch. *108*, 761, 1938. – *Waldeyer, W.:* Dtsch. med. Wschr. *17*, 1213, 1891. – *Weber, A.:* Bull. Histol. appl. *5*, 73, 1950.

Authors' addresses: Dr. P. Glees,
University Laboratory of Physiology,
*Oxford* (England).
Dr. W. J. H. Nauta, Army Medical Service Graduate School, *Washington*, D.C. (USA).

# SOME EARLY TRAVAILS OF TRACING
# AXONAL PATHWAYS IN THE BRAIN

WALLE J.H. NAUTA

Department of Brain and Cognitive Sciences, MIT, Cambridge, MA 02139

The first great flowering of Neuroanatomy began about 1870, reached its height in the late 1880s and, with some notable exceptions, went into a marked decline during the two decades between the first and the second World War. In this essay I will relate some of my personal experiences during the early stages of a second flowering of Neuroanatomy that began some fifty years ago and ultimately led to the multifaceted Neuroanatomy we now know.

My account begins in 1937, when I had just passed the preclinical examinations at the University of Leiden, and had accepted a two-year student-assistantship in the Anatomy Department. A student-assistantship was a formidable commitment of time and effort, but it also, in addition to free tuition, provided a valuable fringe benefit: student-assistants had access to the histology laboratory, and after receiving basic instruction in histological technique, were allowed to take on a research project of their own. I chose to aquaint myself with the rat brain, as it appeared in serial paraffin sections that I had stained alternately with toluidine blue for cell bodies and with the protargol staining method for nerve fibers published by David Bodian (1936) only the year before. The Bodian method was a great boon to Neuroanatomy: it was the first technique to permit excellent, routine silver staining of nerve fibers in paraffin sections. That was a particular blessing to me, for paraffin sections were the only type of sections I then knew how to cut and handle.

It should be mentioned here that the anatomy of the brain was not a popular research topic in the 1930s. In Europe, at least, the subject was widely regarded as already harvested to exhaustion, and hence hardly worth exploring further. The term "neuroanatomy" did not even exist; the subject was considered to be simply part of general anatomy. It was pursued as an active research subject in only a few Anatomy departments, of which Leiden's was not one.

While slowly becoming better acquainted with the rat brain, I developed a particular interest in the hypothalamus. The literature appearing at that time provided ample reason for such an interest. The first publications of the Scharrers had just come out, reporting their astonishing observation that cells of the magnocellular nuclei of the hypothalamus combine neural with secretory characteristics. The laboratories of S.W. Ranson in Chicago and W.R. Hess in Zurich were reporting a steady stream of intriguing novel findings about the involvement of the hypothalamus in autonomic and endocrine functions, as well as in the sleep-wakefulness cycle. I had also read with admiration Philip Bard's classical paper on "sham rage," a phenomenon clearly implicating the hypothalamus in at least the outward manifestations of anger. And from G.B. Wislocki's laboratories at Harvard had come evidence that the long-sought last link in the efferent pathway connecting the hypothalamus with the anterior pituitary was vascular rather than neural. I thought that these recent findings provided a good reason for an expanded search for the intrinsic and extrinsic fiber connections of the hypothalamus. It seemed to me, moreover, that several of the recent physiological findings were raising anatomical questions that could not be answered in terms of existing anatomical knowledge. And that realization, in turn, led me to think that the anatomy of the brain was a neglected rather than an exhausted subject.

It was not long, however, before it dawned on me that for tracing complex fiber pathways the histological methods with which I had now become familiar would not be enough. Bodian-stained sections might show fibers entering the supraoptic nucleus from the dorsal side, but when one tried to trace those fibers back to their source one would either lose them in the dense hypothalamic fiber plexus or see them exit from the section. An ancient experience even then, and the very reason why, more than fifty years earlier, Vittorio Marchi had introduced his ingenious strategy of tracing nerve fibers by marking them with the anterograde myelin-sheath degeneration induced by severing the fibers from their cells of origin (Marchi and Algeri, 1885). The Marchi method, I thought, ought to be able to provide the answer to questions about the destination of the efferent pathways of any brain structure.

But before I could put that belief to the test, my time in the Anatomy department was up; I passed the much-dreaded clinical examination, and immediately afterward had to begin a 14-month rotating internship. An infinitely greater obstacle to the pursuit of my project was that four months later, in May of 1940, Hitler's armies launched their offensive in the West that in a matter of weeks would secure Germany's full control of the Atlantic coast of the European continent. Holland was overrun after only five days of fighting, and for the next five years had to endure Nazi occupation.

I could not return to my project until two years later, in 1942, when I had at long last completed the rotating internship and had accepted an instructorship in Anatomy at the University of Utrecht. (I could not remain in Leiden because its university had been closed by Nazi authorities a few months earlier, after repeated warnings and accusations that it was "a hornet's nest of ideological subversion.")

In Utrecht, thanks to the generous hospitality of the Pharmacology department, I finally had a chance to place lesions in rat brains, and try my luck with the Marchi method. And to my distress, in the course of that first year in Utrecht the high hopes with which I had started out steadily faded. Although I could easily follow large numbers of degenerating fibers away from lesions placed in the cerebral cortex or thalamus, only a few such fibers showed up around lesions placed in the hypothalamus. Ignorance had been the basis of my disappointed hopes: I had never examined rat brain sections stained for myelin, and so had never realized how sparsely myelinated the hypothalamic fiber plexus is.

It had now become clear to me why the Marchi method had been used so rarely in anatomical studies of the hypothalamus. Obviously, what was needed for tracing hypothalamic fiber connections was a technique that could provide a direct image of the degenerating axon itself, rather than only of its decomposing myelin sheath.

At that point I remembered having come across a reference to a histological method that might reveal degenerating axons directly. That recollection led me to a paper published in 1925 in *Virchow's Archiv* in which the Leningrad neuropathologist J. Rasdolsky described an elaborate staining procedure, the final steps of which stained normal nerve fibers a grayish green with the dyestuff Light Green, and degenerating fibers, including their terminals, a bright red or violet with Fuchsin B. With high expectations, I tried the Rasdolsky method, but to my disappointment had no success with it whatever. In the hundreds of sections I stained with this cumbersome procedure I saw only grayish bundles of normal fibers, and never glimpsed a red-stained degenerating fiber. After several months of persistent failure I decided that another technique for revealing axon degeneration was needed, and that I probably would have to develop it myself.

The more thought about the type of staining method I should concentrate my efforts on — organic-dyestuff adsorption or metallic tissue-impregnation — the likelier it seemed to me that some modification of an already existing silver-impregnation method would be the answer. After all, with the exception of the capricious methylene-blue staining, silver-impregnation thus far had been the only technique capable of providing positive images of axons in tissue sections. And, if Marchi had succeeded in modifying the osmium-staining of normal myelin so that it became a selective indicator of degenerating myelin, might there not perhaps exist some histochemical difference between normal and degenerating axons that could make a second Marchi-type miracle possible? Or so I wishfully argued, as yet without a clue as to where to begin my search.

While I was still pondering how to proceed, our laboratory-supply house sent us a list of items that it could no longer provide, presumably until war's end. Since the list included not only all silver compounds but also such crucial histological necessities as alcohols, xyelene, glass slides and coverslips, all my plans had to be shelved. As events unfolded from here, I did not get back to the "axon Marchi" project until nearly three years later, in early 1946, when the war had ended and I had returned to the Anatomy department at the University of Leiden.

The year 1946 was one of high tension and near-chaos at Leiden, as our teaching program had to accommodate not only the regular student enrollment but also a stream of "returnees:" older students who had lived in hiding for all or part of the war for reasons ranging from involvement in resistance activities to being Jewish, or simply trying to avoid conscription for forced labor by the enemy. Among these returnees was a contentious young fellow who introduced himself to me as Hans Kuypers when I first met him in the dissection laboratory, and who later became one of my cherished colleagues and friends.

Although the year 1946 was one of constant effort to keep the teaching program on track, I did get some staining experiments started, and in one of them I found to my excitement that in rats deprived of one eye and killed 3–5 days after the operation both the Bodian method and Cajal's block-impregnation technique stained a multitude of swollen ring – and club-shaped structures amid a heavily argyrophilic neuropil in the contralateral retinoceptive cell groups (Fig. 1).

Ring-shaped argyrophilic structures had been noted in brain tissue since 1904, when the neurofibrillar silver-staining methods of Cajal and Bielschowsky were introduced (Fig. 2). Cajal (1909) had interpreted the rings as terminal loops of single or bundled neurofibrils. Neither his nor Bielschowsky's method revealed the axoplasmic end-structure in which the neurofibrillar loop was presumably enclosed. Cajal (1909) nonetheless proposed that the neurofibrillar end-ring might be a particular version of the more expansive neurofibrillar network pervading the larger axon terminals that had been discovered some ten years earlier in Golgi preparations and iron-hematoxylin-stained sections of the spinal cord and brainstem, and had been called "end-feet" by Held (1897) and "end-knobs" by Auerbach (1899). (Both names were eventually superseded by Cajal's term *boutons terminaux*.) That the argyrophilic ringlets were – or, at least, marked – axon terminals had become widely accepted after E.C. Hoff (1932) reported finding numerous swollen rings and solid clubs in the spinal dorsal horn after dorsal-root transection. Hoff's finding, suggesting that the boutons terminaux of degenerating axons are marked in silver-stained preparations by swollen end-rings, was confirmed by Gibson (1937), but the reliability of the "bouton method" as a means of determining the ultimate distribution of degenerating fibers was put into question after Phalen and Davenport (1937), Barr (1939), and Minckler (1940) reported considerable variability in the size and shape of end-rings in the normal spinal cord of various vertebrate species. Presumably as a result of these reports, the "bouton method" fell in disuse after 1940. It is noteworthy that the method had been applied exclusively to fiber systems afferent to the spinal cord, and that in doubting its validity, no account was taken of the much stricter criteria for bouton degeneration applied by Szentágothai in a series of three classical studies dealing with dorsal-root and descending afferents to the spinal cord (Schimert, 1938, 1939; Szentágothai–Schimert, 1941). In these studies, using the Gros modification of the Bielschowsky method, Szentágothai disregarded mere differences in size and shape, and counted as degenerating only those axon terminals which had a spongy appearance and were attached to a terminal axon branch exhibiting varicosity and/or fragmentation as evidence of Wallerian degeneration.

The Bielschowsky–Gros method, as used by Szentágothai, appears in retrospect as the light-microscopic forerunner of the electron-microscopic identification of degenerating axon terminals that was introduced some twenty years later. However, since the method seemed unsuitable for my own purpose of long-distance tracing of degenerating axons, I mistakenly paid it scant attention.

To return to my observation of swollen end-rings after unilateral eye enucleation, the fact that they appeared predominantly in the contralateral retinoceptive cell groups

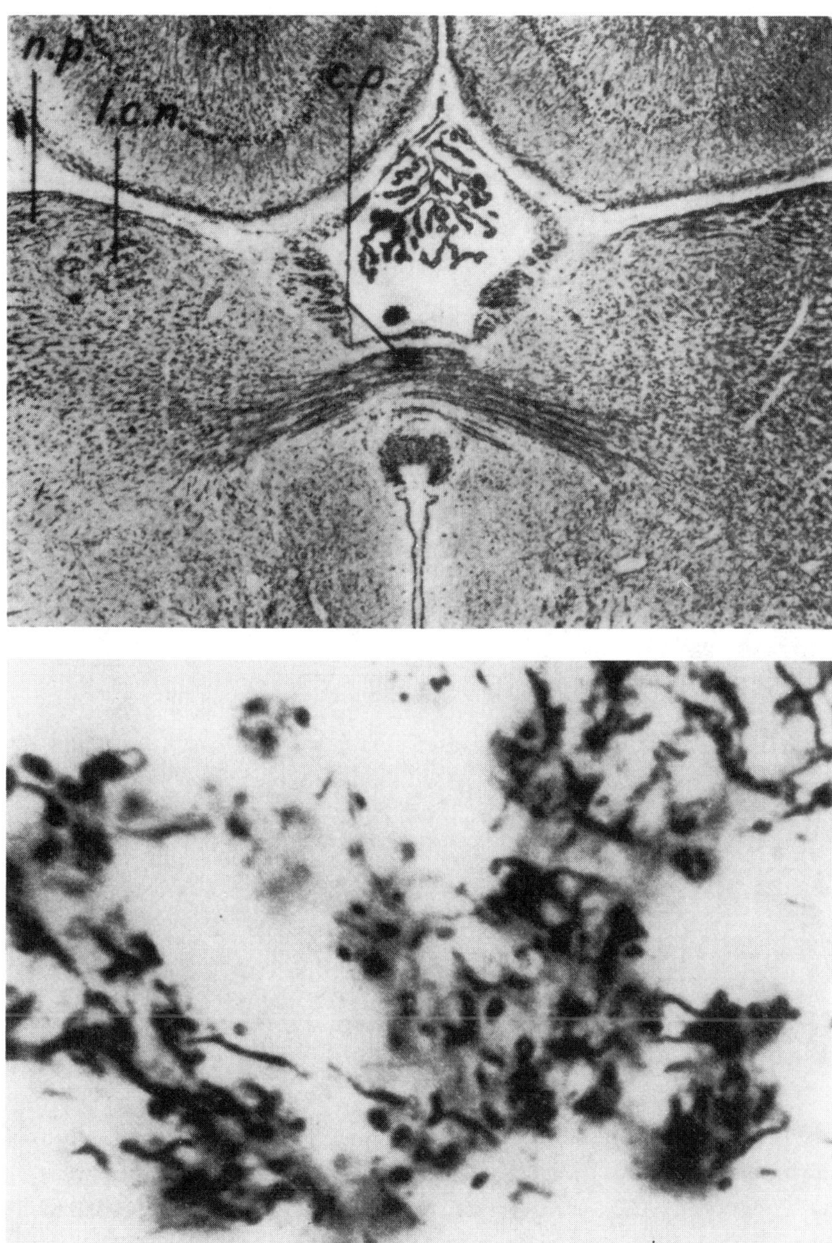


**Fig. 1** Upper panel: low-power view of a frontal section through the pretectal region of a rat killed 3d after enucleation of the right eye. Paraffin section, silver-impregnated by Bodian's (1936) protargol method. On the left side note the oval cluster of heavily stained neuropil in area labelled *l.c.n.* Lower panel: detail of area *l.c.n.* in upper panel, viewed at higher power. In and among nests of hyperargyrophilic neuropil numerous solid club-shaped axon terminals appear. From Nauta and van Straaten (1947), by courtesy of Cambridge University Press.

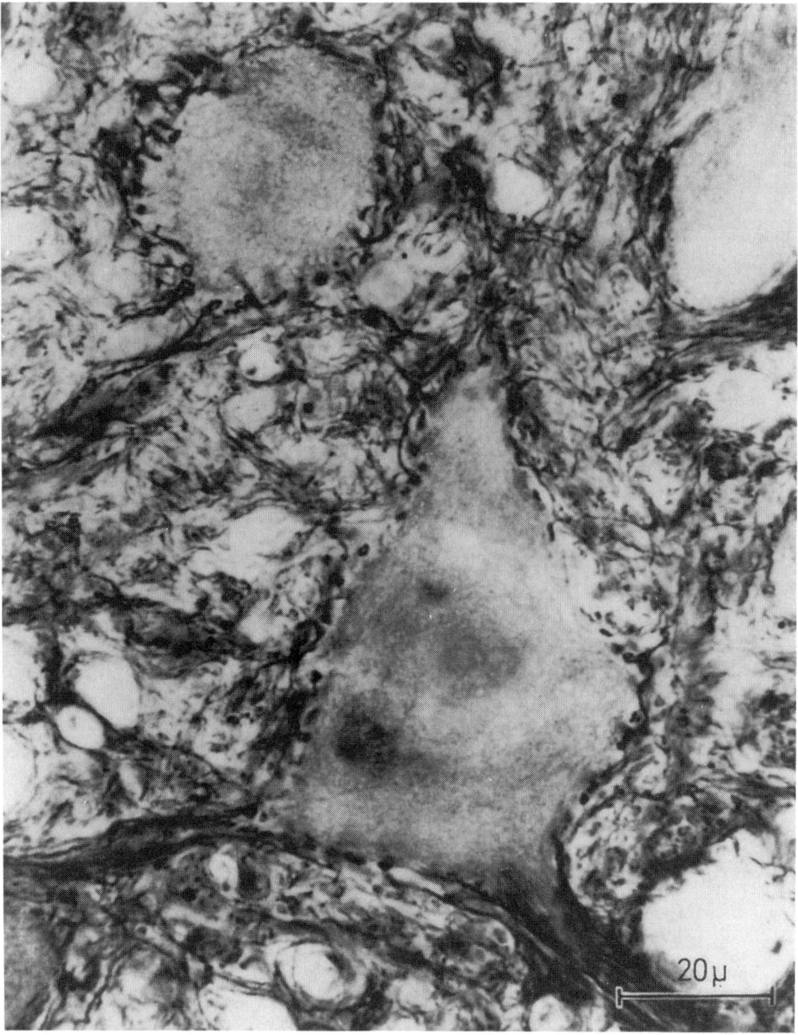


**Fig. 2**  Cells from the base of the dorsal horn of the cat's lumbar spinal cord.  Note the neurofibrillar rings and clubs covering the surface of the neuronal cell body at lower center, and of large dendrite in upper left corner of figure. Tissue stained with Barr's (1939) modification of Cajal's block-impregnation method by A.D. Loewy.  From Guillery (1970), by courtesy of Springer Verlag, New York.

seemed to leave little doubt that they indeed signalled degeneration of retinofugal axons. My initial hope, however, that the Bodian and the Cajal-block method, which had produced these results, could be used for tracing degenerating axons was soon tempered when I could find no such changes in the known termination areas of some other degenerating fiber systems. For example, no darkened neuropil, nor even end-rings of any sort, appeared in the septum or mammillary body following lesions that transected the fornix. Nor could I find such changes in the cerebral cortex after lesions of the thalamus. My optimism dwindled further when I realized that in my eye-enucleation material the axons of the optic tract showed no signs of Wallerian degeneration, even though they had to be the parent fibers of the degenerating boutons in the primary visual cell group. Rather, postoperatively these fibers seemed gradually to lose their argyrophilia; at no time did they exhibit a positive picture of axon disintegration. Puzzled by these discrepancies, I concluded that none of the silver methods thus far used to determine the distribution of degenerating axons actually stained degenerating axoplasm. What they did stain had to be the neurofibrillar element, preferentially its expansion into the axon terminals. And since that neurofibrillar element appeared to be either sparse or absent in some other fiber systems, such as the fornix, I decided that what was needed was a method that could reveal axoplasmic degeneration, not only in the terminals but throughout the course of their parent axons as well.

At the time, the mid-1940s, electron microscopy of the central nervous system still lay ten years in the future. It was not generally recognized then that the crucial details of synaptic morphology, and especially the identity of the postsynaptic element, lie beyond the reach of light-microscopic technology. For instance, Cajal, in a posthumous chapter on the neuron doctrine that appeared as late as 1935 in the *Handbuch der Neurologie*, expressed great uncertainty concerning the existence of boutons terminaux in the cerebral cortex. His doubts were based on the failure of both his own staining technique and the Golgi method to reveal in cortical tissue the ring-or club-shaped end-formations so commonly seen in many subcortical structures (*cf.* also Dempsey's [1957] comments on this subject).

Concepts about the morphology of axon terminals rapidly changed after electron microscopy in the late 1950s began to show the true structure and distribution of synaptic contacts. It is now firmly established that, exactly as proposed by Cajal fifty years earlier, the argyrophilic end-rings are loop-shaped end-stretches of microfilamentous bundles, that such neurofibrillar loops are by no means a standard feature of all fiber systems and are, for example, very sparse in the cerebral cortex, and, most surprising of all, that they are not invariably located within axon terminals and may occur in other parts of the neuron (Guillery, 1970).

The next phase of my benighted search for an "axoplasmic Marchi" method was set at the Anatomy department of the University of Zurich, to which I moved in 1947 for what was meant to be two years but luckily stretched to four.

Through most of my first year in Zurich I peristed in working with the Bodian and the Cajal block-impregnation methods, trying an ever growing variety of fixatives and other chemical tissue pretreatments. All that labor got me nowhere, and I was about to give up the project when I came across a recent article in the *Journal of Neuropathology and Experimental Neurology* in which Paul Glees (1946), then at Oxford, described a modification of the Bielschowsky method that stained degenerating axon terminals, not as rings but as solid blobs. The illustrations of Glees' paper suggested that his method also showed Wallerian degeneration of more proximal parts of the axon, but it was not clear whether the method would permit the tracing of axons over their full length. I tried the Glees method, but to my disappointment found that I could not make it work. The explanation for my failure may have been simple: the Glees method explicitly called for tap water rather than distilled water to make up the 10% formalin that served as the final reducing agent, and it is possible that the tap water in Zurich differed in some crucial respect from that in Oxford and Oslo, the two cities where the Glees method eventually found its widest successful use.

Even though I could not make his method work, Glees' paper had convinced me that the Bielschowsky method offered better prospects than did the Cajal-type methods I had been using up to that time. The principles underlying the Cajal and Bielschowsky procedures differ in important respects. In the Cajal process the final blackening of nerve fibers (in photographic terms: the change from the latent to the visible image by action of the developer) is produced by nascent silver released from the stable salt silver nitrate by a strong reducing agent such as hydroquinone. In the Bielschowsky process, by contrast, the nascent silver is released from a much more labile silver double-salt, silver-ammonia nitrate, by a weak reducing agent, formaldehyde. Comparing the two processes, it now seemed likely to me that the latter one offered the wider opportunity for the sort of fine adjustments that might eventually lead to a method for staining degenerating axons.

In trying to master the original Bielschowsky method, however, I ran into a problem that already had bothered me in my attempts to use the Glees method: as I transferred my sections from the silver-ammonia nitrate solution to the prescribed 10% formalin reducer, they instantly turned an opaque greyish brown, thus obscuring any clear distinction of microscopic detail. Attempts to slow down the release of nascent silver by diluting the reducer and increasing its alcohol content brought little if any improvement.

My results with the Bielschowsky method took a turn for the better after I met Paul André Gygax, a doctoral student of Chemical Engineering at the Federal Institute of Technology in Zurich. Unlike several other chemists whom I had approached earlier, Gygax was immediately challenged when I explained my problem to him. From a survey of the literature on silver double-salts he suggested using sodium hydroxide, in addition to ammonia, in preparing the silver-ammonia nitrate solution. It was an ingenious idea: the weak base, ammonia, suppresses the dissociation of the double-salt, whereas the strong base, sodium hydroxide, promotes it. By adjusting the relative concentrations of the two bases we could now control the rate of release of nascent silver to some extent at least. The best sections we thus produced were sufficiently transparent to provide encouraging glimpses of degenerating fibers among the multitude of normal fibers. Even these sections were not clear enough, however, to allow tracing of degenerating fibers along their entire length.

After months of frustrating attempts to improve the clarity of the sections, the resolution of the problem finally materialized, suddenly and quite by accident. On a Saturday afternoon in the spring of 1949, I discovered that I had used up all of my fresh formalin and had to make up the reducing solution from a half-empty, cork-stoppered bottle of 10% formalin that I found in an out-of-the-way place, where it had been stored for at least six years. In the reducer made up from this stale formalin the sections to my astonishment slowly turned light brown, and when I examined them in the microscope I saw, for the first time in all those years, a picture of axon degeneration that resembled the one I had been hoping for. I excitedly called Paul Gygax, who said he'd be right over. He arrived a short time later, carrying a fresh supply of formalin and a small bottle of formic acid; it had not taken him more than a minute to figure out that the stale formalin I had mentioned must have had a high concentration of formic acid. A few quick trials with fresh formalin acidified with formic acid bore him out.

With this denouement, the project changed from a series of frustrations to a joyride, which, within a few weeks, ended with our conclusion that, of a variety of organic acids we compared, citric acid was the best acidifier for fiber tracing, as it gave the sections a clear, translucent amber background.

The procedure at which we had now arrived (Nauta and Gygax, 1951) stained normal and degenerating fibers with equal intensity (Fig. 3). In a preliminary study of the rat fornix I found it difficult, and sometimes impossible, to trace dispersed degenerating fibers across regions densely filled with normal fibers. Therefore, I decided that it would be necessary to modify the procedure so that it would show degenerating fibers selectively.

In my search for selective visualization I was assisted by Lloyd Ryan, a major in the U.S. Air Force whom I met by chance at a party. Ryan was completing a doctoral dissertation in Aerodynamics at the Federal Institute of Technology, and turned out to have a wide-ranging knowledge of the chemistry of photography. When I told him of my struggles with the Bielschowsky method he exclaimed that its mechanisms must share many features with the photographic process. He asked if I had ever carried it out in the darkroom. No, I said, I hadn't. And so, following his suggestion, we treated sections under a dark-red safety light and found that under these conditions silver stained degenerating fibers more intensely than normal fibers. This modification, however, by no means eliminated all staining of normal fibers, and since it also was very cumbersome to handle free-floating frozen sections in the darkroom, we decided that we needed some chemical alternative to the absence of light as differentiating factor in the silver staining of normal and degenerating fibers.

The search for such a chemical alternative was long and tedious. After months of trying combinations of virtually all mordants ever applied to brain tissue, we finally obtained the hoped-for effect by sequential pretreatment of the sections with potassium permanganate and potassium bisulfite (Nauta and Ryan, 1952). Using this method, and in collaboration with Verena Bucher, a neuroanatomist associated with the distinguished Zurich neurophysiologist W.R. Hess, I completed a charting of the efferent connections of the rat's visual cortex, but when I moved to the Walter Reed Army Institute of Research in Washington, D.C., toward the end of 1951, I was distressed to find that the permanganate-metabisulfate method didn't work there. I never did find out whether that failure was due to differences in the chemicals or in

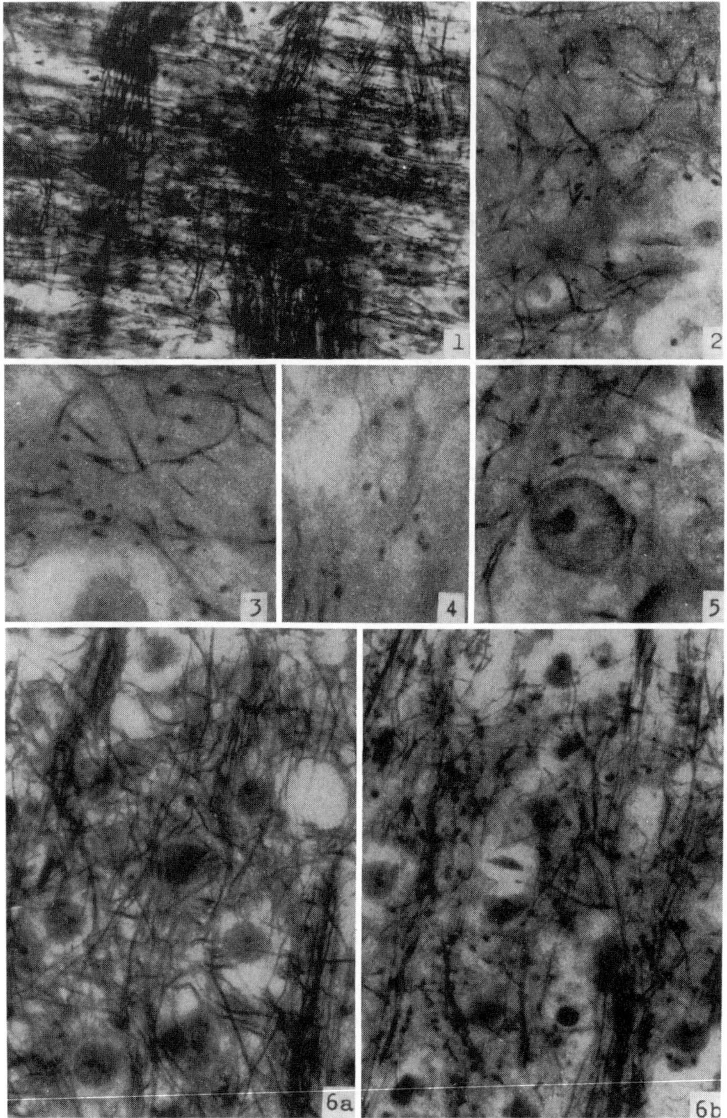

**Fig. 3**  Axon degeneration as appearing in frozen-cut sections stained by the original, non-suppressive Nauta–Gygax method (1951). Panel 1 shows the anteroventral thalamic nucleus exhibiting massive fiber degeneration in the (here horizontally oriented) septothalamic tract of Vogt (1898) in a rat killed 3d following lesion of the supracommisural septum and fornix column. Note the bundles of normal internal-capsule fibers crossing the degenerating tract at nearly right angles. Panels 2–5, from the same specimen as panel 1, show terminal-axon degeneration in the anteromedial thalamic nucleus (2), medial hypothalamus (3), periventricular hypothalamus (4), and nucleus accumbens (5). Panels 6a and 6b show, respectively, left and right frontal cortex (largely layer IV) of a rat killed 4d after receiving a large lesion of the right thalamus and internal capsule. From Nauta and Gygax (1951) by courtesy of Williams and Wilkins, Baltimore, Maryland.

the rats I used in Zurich and Washington. Anyhow, it took the better part of my first year at Walter Reed to get the method to work again, by substituting a treatment with phosphomolybdic acid – later in turn replaced by uranyl nitrate – for the metabisulfite step. In this recovery project once again, I was fortunate to be helped by Paul Gygax, who spent his first year in the U.S.A. in my laboratory at Walter Reed, before moving on to the Dupont Company.

Compared with the original, non-selective method, the selective silver technique we finally published (Nauta and Gygax, 1954) made it easy to trace degenerating axons over long distances, and, in many cases at least, to identify the area of their termination (Fig. 4). As a consequence of that convenience, the "suppressive" method – as it later was often called – came to be used far more widely than its non-suppressive parent method.

Convenient as the suppressive method may have been, it left some important questions of interpretation unanswered. For instance, soon after its introduction, disagreements arose over the interpretation of images of fine-fiber degeneration appearing in the termination areas. Did the method bring out in such areas only the preterminal end-ramifications of degenerating axons, or were at least some of the stained structures actual axon terminals? (*cf.* Fig. 4). In retrospect it seems obvious that this controversy could not be settled definitively by light-microscopic methods alone, even though it should be added that the Bielschowsky-Gros method introduced by Szentágothai in the late 1930s, and abandoned too early, might well have gone some way toward resolving the issue. As neuroanatomical methodology developed, degenerating boutons terminaux were not recognized in the electron microscope until nearly ten years later (see Guillery, 1970, for a brief review). Along with the axon terminals, their postsynaptic correspondents could be identified as well. By combining light- and electron-microscopic techniques it only then, in the early 1960s, became possible to determine not only the distribution but also the mode of termination of degenerating fiber systems.

There were other problems connected with the "suppressive Nauta-Gygax method." The most serious of these was that, as electron microscopy became more widely used, evidence appeared that the selectivity for degenerating fibers had been obtained at a cost. It now seemed that the pretreatment employed to suppress the staining of normal fibers also suppressed, though less drastically, the staining of the finer terminal ramifications of degenerating fibers.

This problem was forcefully brought to my attention in 1965 by Lennart Heimer when he joined us at the Massachusetts Institute of Technology – where I had moved the year before – as visiting associate from Göteborg. At Göteborg, Heimer had been tracing fiber pathways in the rat's olfactory system, and had noted that the terminal degeneration appearing in the piriform cortex following ablation of the olfactory bulb was brought out in very much higher density and greater detail by the non-suppressive than by the suppressive Nauta-Gygax method (Fig. 5). At M.I.T., Heimer set out in quest of a procedure that would selectively stain degenerating axons without suppressing the staining of their terminals. Within a year, in collaboration with Robert Fink, he

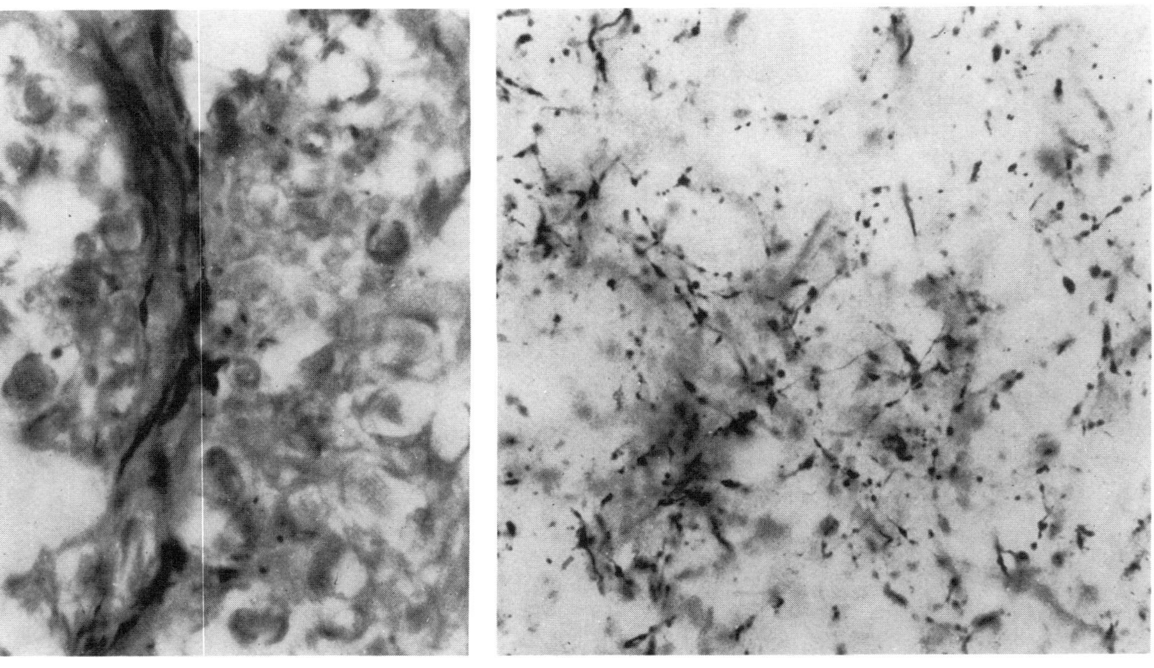

**Fig. 4**   Two examples of selective black staining of degenerating axons by the suppressive Nauta–Gygax method (1954, here slightly modified as described in article referred to below), Left panel: a slender fascicle of degenreating dorsal-root fiers traversing the dorsal funiculus of the spinal cord of a cat killed 5d following transection of a single dorsal root. Right panel: profuse axon degeneration in layer IV of cat parietal cortex, 4d after lesion of lateral thalamus. Note drop-like disintegration of fine preterminal axons encircling cell body toward upper left corner from center. Also note that none of the stained structures can be identified unequivocally as degenerating synaptic terminals. From Windle, WF (ed) New Research Techniques of Neuroanatomy, 1957, by courtesy of Charles C. Thomas, Publisher, Springfield, Illinois.

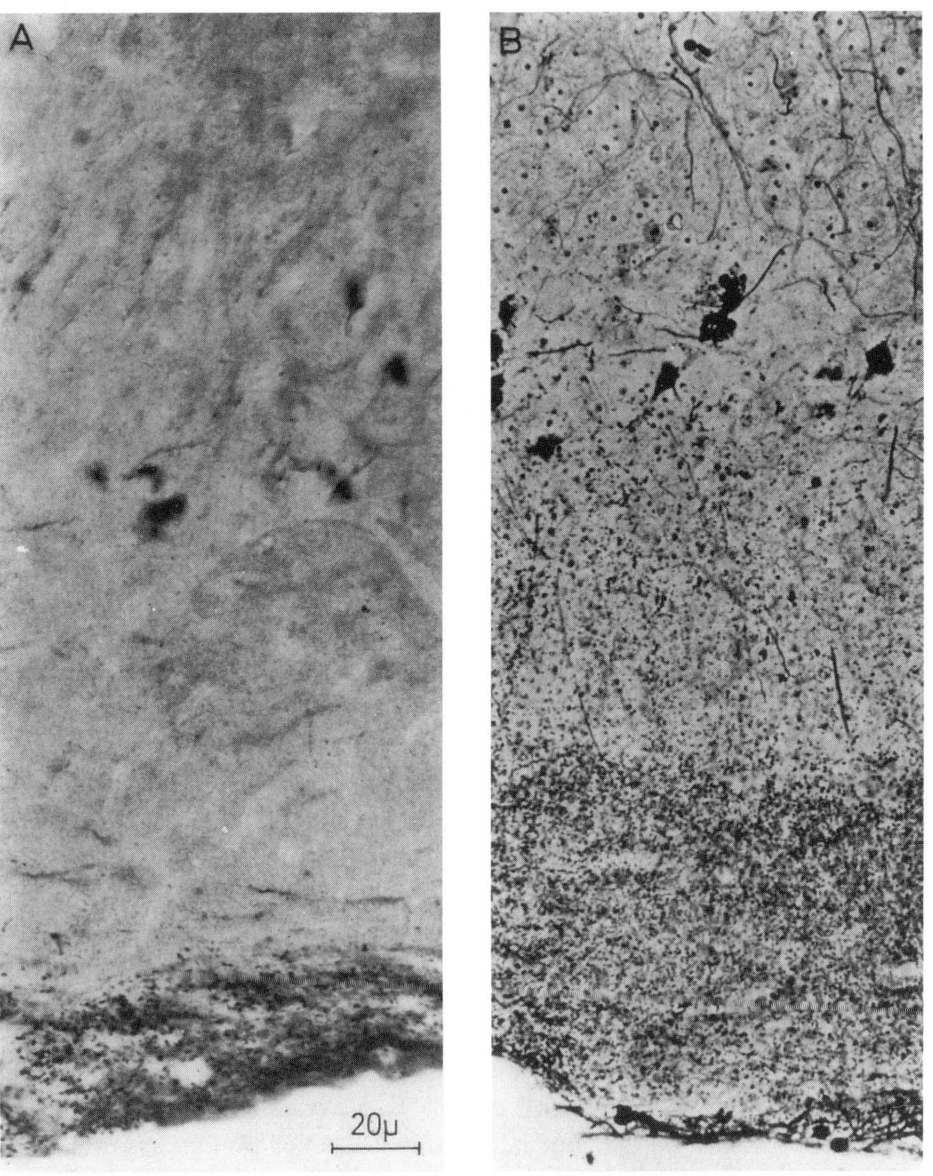


**Fig. 5** Degenerating olfactory-tract fibers in the piriform (olfactory) cortex of a rat, 5d after removal of the ipsilateral olfactory bulb. **A**: a section stained by the suppressive Nauta–Gygax method shows axon degeneration only in the most superficial part of the plexiform layer. **B**: a slightly modified non-suppressive Nauta–Gygax procedure demonstrates massive degeneration of axon terminals throughout the plexiform layer, particularly dense in the layer's superficial (lower) half. Much sparser terminal degeneration appears in the pyramidal and multiform cell layers appearing in the upper third of the photograph. From Heimer (1970a), by courtesy of Springer Verlag, New York.

managed to develop not one, but two such procedures. Remarkably, neither of these required the use of any chemical that had not already been used in the Nauta-Gygax methods. Rather, the main procedural variants were: 1. application of potassium permanganate without previous treatment of the sections with phosphomolybdic acid or uranyl nitrate, and 2. subsequent silver impregnation in a solution of silver nitrate mixed with either uranyl nitrate (Procedure I) or pyridine (Procedure II). (For a full discussion of these modifications see Heimer, 1970a.)

In the piriform cortex of rats deprived of the ipsilateral olfactory bulb 3–5 d. before death the Fink–Heimer methods yielded pictures of abundant terminal degeneration entirely similar to that shown by Fig. 5B. Comparable images appeared in the superior colliculus after contralateral eye enucleation, as well as in the distribution areas of the corpus callosum following hemidecortication, and in the hypothalamic distribution of the transected stria terminalis (*cf.* Heimer, 1970a).

When the two new method were published (Fink and Heimer, 1967), however, they were received with much skepticism. Were those black, spheroidal particles they revealed as densely crowding the plexiform layer of the piriform cortex really silver-stained degenerating boutons? Weren't there far too many of them, and didn't they look much more like the particles of a random silver precipitate? I must admit that I once had harbored the same doubts. Fifteen years earlier, while still working with the original, non-suppressive silver method, I had often encountered such dense accumulations of silver particles, and had decided that they had to be mere random precipitates. I had been wrong, and Heimer (1970b) dispelled all doubts about that by providing electron-microscopic evidence that the black particles in the piriform cortex were indeed silver-stained degenerating axon terminals (Fig. 6). Thereafter, the Fink–Heimer procedures became the methods of choice for tracing degenerating fibers in the central nervous system.

In spite of their promise, the Fink–Heimer methods were destined to compose a brief last chapter in the 90-year long history of tracing fiber pathways by means of experimentally induced Wallerian degeneration. By the time the methods were coming into wide use, Maxwell Cowan, Anita Hendrickson and their associates (1972) were already adapting for routine use the fundamentally different strategy, introduced by Lasek et al. (1968), of tracing nerve fibers by autoradiographic recording of the uptake and intra-axonal transport of radioactively labelled amino acids. This strategy offered enough advantages to replace the Wallerian-degeneration methods almost completely. A few years later, nerve-fiber tracing methods based on the detection of intra-axonal transport of horseradish peroxidase and of plant lectins (Gerfen and Sawchenko, 1984) enriched the neuroanatomical arsenal still further. Today, the student entering Neuroscience encounters a wealth of fiber tracing methods, some purely anatomical, others selectively histochemical in nature, which can aid in the conceptual unravelling of the brain's complexity. As an observer of the present-day scene, I sometimes find it difficult to remember that I once was part of a past scene where there was only the Marchi method to which one could resort for the anterograde tracing of axonal connections in the central nervous system.

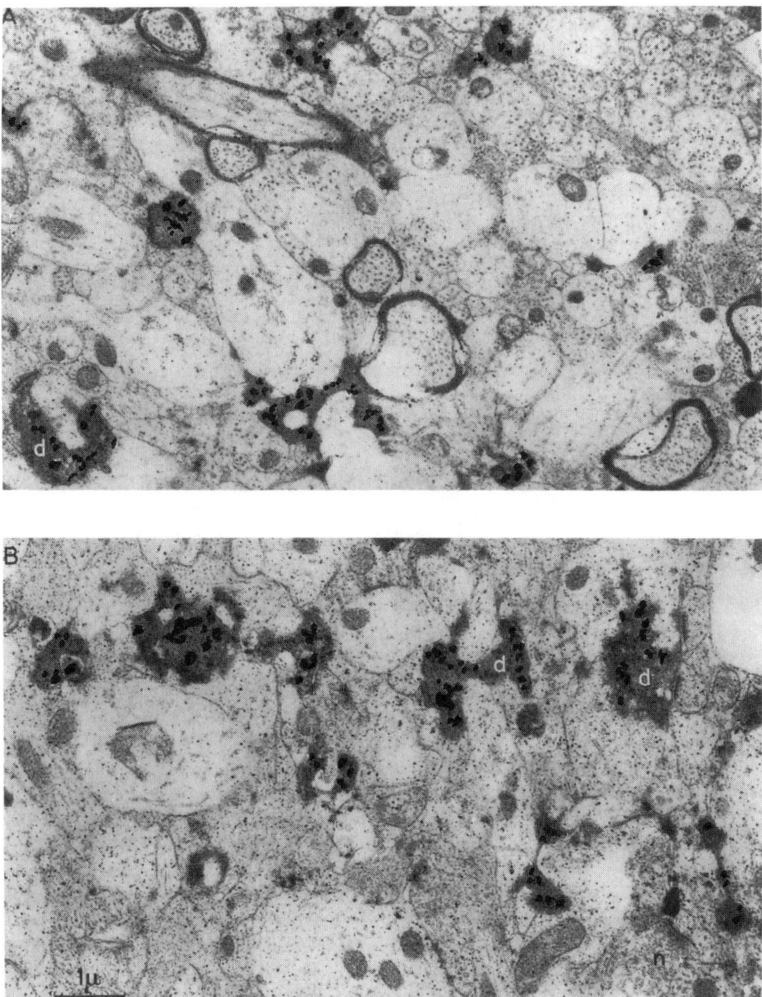

**Fig. 6** Electron micrographs of degenerating axon terminals in the olfactory cortex, silver-impregnated by the Fink–Heimer (1967) method. The procedure applied to obtain these photographs was as follows: (1) frozen-cut sections, 25–30 um thick, were stained for axon degeneration, yielding images such as that shown in Fig. 5B; (2) a small, carefully selected area of about 1 mm square was excised from the stained section, re-fixed, mounted on an empty Epon base, and embedded flat in Epon. (3) The Epon slab so obtained was sectioned on an ultramicrotome, and the resulting ultrathin sections were examined and photographed in the electron microscope. Degenerating axon terminals (*d*) are characterized by high electron-opacity, hence their dark appearance, contrasting markedly with that of normal terminals (*n*, near lower right corner of panel B). While a very fine grain of metallic silver particles appears throughout the sections, note that heavier, coarser-grained silver deposits are limited to the dark profiles of the degenerating boutons terminaux. From Heimer (1970b) by courtesy of Springer Verlag, New York.

51

# References


Auerbach L (1989): Das terminale Nervennetz in seinen Beziehungen zu den Ganglienzellen der Zentralorgane. *Mschr. Psychiat. Neurol.*, 6:192–214.

Barr M (1939): Some observations on the morphology of the synapse in the cat's spinal cord. *J. Anat. (Lond.)*, 74:1–11.

Bielschowsky M (1904): Die Silberimprägnation der Neurofibrillen. *J. f. Psychol. u. Neurol.*, 3:169–189.

Bodian, D (1936): A new method for staining nerve fibers and nerve endings in mounted paraffin sections. *Anat. Rec.*, 65:89–97.

Cajal SR y (1904): Ueber einige Methoden der Silberimprägnierung zur Untersuchung der Neurofibrillen, der Achsenzylinder und derer Endverzweigungen. *Zschr. f. Wissenscftl. Mikrosk.*, 20:401–408.

Cajal SR y (1909): Histologie du Système Nerveux de l'Homme et des Vertébrés, Tome I, p. 74., Paris: Maloine.

Cajal SR y (1935): Die Neuronenlehre. In: *Bumke–Foerster Handbuch der Neurologie,* vol. I, pp. 887–1027.

Cowan WM, Gottlieb DI, Hendrickson AE, Price JL, Woolsey TA (1972): The autoradiographic demonstration of axonal connections in the central nervous system. *Brain Res.*, 37:21–51.

Dempsey EW (1957): New methods versus old. In: *New Research Techniques of Neuroanatomy,* Windle WF, ed., pp. 76–78. Springfield, Illinois, Charles C. Thomas, Publisher.

Fink RP, Heimer L (1967): Two methods for selective silver impregnation of degenerating axons and their synaptic endings in the central nervous system. *Brain Res.*, 4:369–374.

Gerfen CR, Sawchenko PE (1984): An anterograde neuroanatomical tracing method that shows the detailed morphology of neurons, their axons and terminals: Immunohistochemical localization of an axonally transported plant lectin, *Phaseolus vulgaris*-leucoagglutinin (PHA-L). *Brain Res.*, 290:219–238.

Gibson WC (1937): Degeneration of the boutons terminaux in the spinal cord. *Arch. Neurol. Psychiat. (Chicago)*, 38:1145–1157.

Glees P (1946): Terminal degeneration within the central nervous system as studied by a new silver method. *J. Neuropath. Exp. Neurol.*, 5:54–59.

Guillery RW (1970): Light- and electron-microscopical studies of normal and degenerating axons. In: *Contemporary Research Methods in Neuroanatomy,* Nauta WJH, Ebbesson SOE, eds., pp. 77–105. New York: Springer Verlag.

Heimer L (1970a): Selective silver-impregnation of degenerating axoplasm. In: *Contemporary Research Methods in Neuroanatomy,* Nauta WJH, Ebbesson SOE, eds., pp. 106–131. New York: Springer Verlag.

Heimer L (1970b): Bridging the gap between light and electron microscopy in the experimental tracing of fiber connections. In: *Contemporary Research Methods in Neuroanatomy,* Nauta WJH, Ebbesson SOE, eds., pp. 162–172. New York: Springer Verlag.

Held H (1897): Beiträge zur Struktur der Nervenzellen und ihrer Fortsätze. 2[e] und 3[e] Mitteilung. *Arch. f. Anat. u. Physiol. Lpz, Anat. Abt.*, 204–294.

Hoff EC (1932): Central nerve terminals in the mammalian spinal cord and their examination by experimental degeneration. *Proc. Roy. Soc.*, B3:175–188.

Lasek R, Joseph BS, Whitlock DG (1968): Evaluation of a radioautographic neuroanatomical tracing method. *Brain Res.*, 8:319–336.

Marchi V, Algeri G (1885): Sulle degenerazioni descendenti consecutive a lesioni sperimentale in diverse zone della corteccia cerebrale. *Riv. sper. Freniat.*, 11:492–494.

Minckler J (1940): The morphology of the nerve terminals of the human spinal cord as seen in black silver preparations, with estimates of the total number per cell. *Anat. Rec.*, 77:9–25.

Nauta WJH (1957): Silver impregnation of degenerating axons. In: *New Research Techniques of Neuroanatomy*, Windle WF, ed., pp. 17–26. Springfield, Illinios: Charles C. Thomas, Publisher.

Nauta WJH, Bucher VM (1954): Efferent connections of the striate cortex in the albino rat. *J. Comp. Neurol.*, 100:257–286.

Nauta WJH, Gygax PA (1951): Silver impregnation of degenerating axon terminals in the central nervous system: (1) Technic, (2) Chemical notes. *Stain Technol.*, 26:5–11.

Nauta WJH, Gygax PA (1954): Silver impregnation of degenerating axons in the central nervous system: A modified technique. *Stain Technol.*, 29:91–93.

Nauta WJH, Ryan LF (1952): Selective silver impregnation of degenerating axons in the central nervous system. *Stain Technol.*, 27:175–179.

Nauta WJH, van Straaten JJ (1947): The primary optic centres in the rat. An experimental study by the "bouton" method. *J. Anat.* (Lond), 81:127–134.

Phalen GS, Davenport HA (1937): Pericellular end bulbs in the central nervous system of vertebrates. *J. comp. Neurol.*, 68:67–82.

Rasdolsky J (1925): Beiträge zur Architektur der grauen Substanz des Rückenmarks (Unter Benutzung einer neuen Methode der Färbung der Nervenfaskerkollateralen). Virchow's Arch, (*Pathol. Anat.*) 257:356–363.

Schimert J (1938): Die Endigungsweise des Traktus vestibulospinalis. *Zschr. f. Anat. Entwicklungsgesch.*, 108: 761–767.

Schimert J (1939): Das Verhalten der Hinterwurzelkollateralen im Rückenmark. *Zschr. f. Anat. Entwicklungsgesch.*, 109:665–687.

Szentágothai-Schimert J (1941): Die Endigungsweise der absteigenden Rückenmarksbahnen. *Zschr. f. Anat. Entwicklungsgesch.*, 111:322–330.

Vogt O (1898): Sur un faisceau septo-thalamique. *C.R. Soc. Biol.* 50:207–208.

# II

# BRAIN AND BEHAVIOR

Reprinted from
*J. Neurophysiol.*, 1946, *9*: 285-316

# HYPOTHALAMIC REGULATION OF SLEEP IN RATS. AN EXPERIMENTAL STUDY

W. J. H. NAUTA

*Department of Anatomy, University of Utrecht, Holland*

(Received for publication March 4, 1946)

## INTRODUCTION

THE ESSENTIAL role of the nervous system in the regulation of the sleep-waking rhythm of higher animals is now widely recognized. No less a person than von Economo, however, drew attention to the fact that the phenomenon of sleep cannot be accounted for by a mere functional change of the central nervous system from its condition during the waking state. It is indeed an important fact that the function of sleep, instead of being characteristic of higher animals, is also observed in organisms which do not possess a central nervous system, and even in several vegetable species. It is therefore impossible to attribute this mysterious function to any special organ. Since all experimental work on sleep has hitherto been confined to mammals, chiefly to cats and monkeys, practically no data concerning the comparative physiology of this phenomenon are available. Nevertheless, it seems probable that during the phylogenetic development the function of sleep, together with many other mechanisms, was progressively centralized into the nervous system from which organ all changes, characteristic of sleep, were ultimately effected. This centralization proceeded so far that the alternation of wake and sleep seems to be governed in mammals by a circumscribed area of the central nervous system, capable of determining physical and psychical activities. The existence of this "centre" for the regulation of sleep is generally accepted by students of this subject. It has, however, given rise to a number of problems, concerned in the first place with the make-up of the centre, and secondly with its mode of action. Is there one single centre for the regulation of the sleep-and-waking rhythm, or must it be thought of as composed of two antagonistic parts, *viz.*, a sleep and a waking centre? And next, on what structures and along what paths does it primarily exert its influence? These questions have been so divergently answered by various investigators that a brief review of the current opinions seems essential.

The first to recognize a central representation of sleep was the Viennese ophthalmologist Mauthner (32), who from his observations of many cases of Wernicke's disease and of "nona" (which was probably identical with von Economo's lethargic encephalitis) concluded that the area surrounding the oculomotor nucleus was of special importance for the regulation of sleep. In later years evidence in favour of a central regulation of sleep was put forward by many clinicians, of whom von Economo in particular disguished himself by his classical study of the Vienna epidemic of encephalitic lethargica (11). The localisation given by Mauthner has been little changed.

Thus von Economo considered a rather extensive area in the posterior and lateral walls of the third ventricle as the site of the regulating mechanism. A number of observations of cases in which this area was destroyed by tumour seemed to confirm the importance for sleep and waking, ascribed by von Economo to the walls of the third ventricle.

This view could be gained only by careful comparison of many clinical cases in which often very extensive lesions of the central nervous system existed. It is, therefore, not surprising that many investigators were, by less critical observation, led to other conclusions. Trömner (46) concluded from a case of narcolepsy, in which autopsy revealed an extensive abscess of the left thalamus, that the regulation of sleep was a function of the thalamus. In later years Spiegel and Inaba (43) adopted the same view on the basis of their experimental work on rabbits and dogs. The experiments of Ranson on monkeys (36), however, clearly indicate that the thalamus is of no special importance for the alternation of wake and sleep, as even extensive destruction of both thalami did not result in any abnormality of this function. On the other hand, bilateral lesions in the area of the mammillary bodies caused the same marked somnolence which is such an outstanding feature of epidemic encephalitis. These results are in accordance with the clinical view that the vicinity of the third ventricle plays a specific role in the regulation of sleep.

How does the central area (whatever its localisation) so strongly affect the state of our physical and psychical activities? This question has been very differently answered. According to Mauthner (32), the first to accept a central regulation of sleep, an inflammatory edema around the oculomotor nucleus would exert a pressure on the important sensory pathways passing through the midbrain, thereby interrupting the corticopetal flow of impulses and thus causing sleep. From this conception it is evident that the special importance ascribed by Mauthner to the environment of the oculomotor nucleus, was attributed by him only to the topographical relations of this area to the main sensory systems. Moreover, sleep, according to Mauthner, would result from an isolation of the cortex from the outer world, and this opinion has received support from various authorities including Spiegel and Inaba (43), who based their concept on cases of somnolence which they obtained by inflicting lesions to the thalamus. The same stand is taken by Kleitman and Camille (30), who, for instance, claim that imperfect relaxation of the skeletal musculature may cause insomnia by keeping up a continuous stream of proprioceptive impulses to the cortex. An essentially similar opinion was expressed by Trömner in 1912. Whereas Mauthner and others apparently considered various unspecific factors (edema, exhaustion, etc.) the cause of the sensory interruption, Trömner accepted the concept of a nervous centre capable of blocking the sensory relaying centres of the thalamus.

The idea of a nervous centre exerting an active influence on the sleep-and-waking rhythm, thereby formulated for the first time, has since received considerable support, although the details of Tromner's concept of this centre, viz., its localisation in the thalamus and its action via sensibility, have been effectively criticized by all the most competent investigators.

In 1918 von Economo published his report on encephalitis lethargica. In those cases of this disease in which somnolence and ophthalmoplegy were the main symptoms, inflammatory lesions were regularly found in the posterior wall of the third ventricle, extending backward to the level of the oculomotor nucleus. In other cases insomnia was observed, together with chorea. These symptoms von Economo ascribed to inflammation of a more rostrally situated part of the hypothalamus, the tuberal region, and of the adjacent portion of the striate body. The objection that these disorders of sleep might be the result of some toxic influence of the inflammation was rejected by von Economo as the encephalitic sleep was promptly reversible and its interruption did not leave any signs of defective mental lucidity, as would have been the case with intoxication. Therefore von Economo is convinced that the affected areas constitute a specific centre as postulated by Trömner. From the contrasts between the somnolent and the sleepless form of epidemic encephalitis, he concluded that the "Schlafsteuerungszentrum" consists of at least two parts (Fig. 1). The conception lay near at hand that the caudal part (inflammation of which caused somnolence) is essentially a waking centre, while the rostral part must for an analogous reason be supposed to act as a sleep centre.

Sleep, according to von Economo, would result from inhibition of thalamus and cortex by the "Schlafsteuerungszentrum." However, von Economo does not agree with previous workers that sleep is brought about by neutralization of sensory stimuli, because his patients suffering from encephalitis lethargica did not show any decrease of sensibility. The most important element of von Economo's opinion, shared with Trömner, is his belief that sleep is caused by active nervous inhibition of different parts of the central nervous system. This view, however, is contradicted by Ranson and his collaborators.

In the course of Ranson, Barris and Ingram's experiments on cats (27, 37) and Ranson's on monkeys (36), in which they inflicted lesions of great diversity to the hypothalamus, they observed many cases of somnolence following lesions of a well-defined hypothalamic area (which has been referred to on p. 304), but never were able to produce sleeplessness. In their 1939 review of the hypothalamus Ranson and Magoun make the following statement, which illustrates their point of view: "There is no good reason to believe that there is a subcortical centre, which, when active, inhibits the cerebral cortex and causes sleep; but there is abundant evidence that some structure or structures in the region of the third ventricle or aqueduct play an important part in maintaining the waking state, because lesions in this region cause somnolence." By theoretical deductions Salmon (42) arrived at the same opinion.

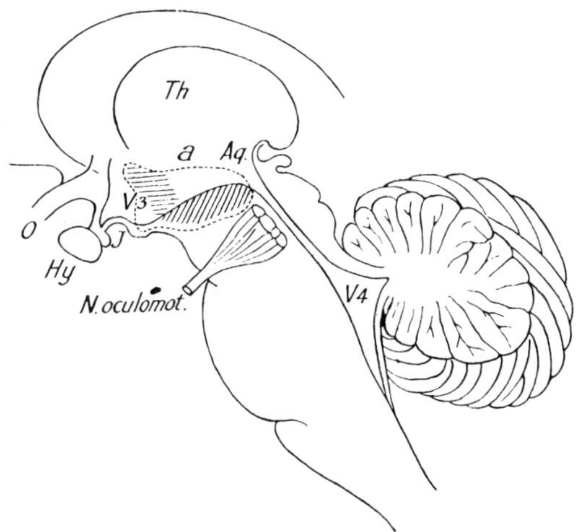


Fig. 1. Von Economo's conception (11) of the localisation of the "Schlafsteuerungszentrum" and its composition in two parts, viz., one (vertically striped) in which inflammatory lesions cause somnolence, and one (horizontally striped) in which similar lesions cause sleeplessness.

The experiments of Hess (23) have often been brought forward in favour of the existence of a sleep centre. By means of a special technique this observer was able to induce sleep in cats by stimulation of points in fore-, tween- and midbrain. These points lay widely scattered over so large an area that it is scarcely believable that Hess invariably stimulated one single centre. Nevertheless, the fact that sleep was obtained by stimulation, i.e., by activation of central nervous tissue, is deserving of our interest. Hess' method of stimulation, however, met with serious criticism from Harrison (21), who claims that it gives rise to electrolytic lesions of the stimulated points. When he altered it so as to avoid destruction of nervous tissue, he never was able to induce sleep. He therefore concludes that sleep in Hess' experiments produced not by stimulation but by lesion. This opinion, however, does not harmonize with the relatively short duration of sleep in Hess' experiments. Whereas the lethargy exhibited by the experimental animals of Ranson and his collaborators persisted for days and weeks, Hess' cats only slept for a few hours. Probably this discrepancy should not be attributed to differences in the amount of nervous tissue destroyed, as even enormous lesions of fore- and tweenbrain fail to interfere with the regulation of sleep if only a circumscribed area of the hypothalamus is left intact. Nevertheless, Harrison's criticism warns us to be reserved in our appreciation of Hess' early findings. Another reason is that in later years Hess seems to have changed his initial opinion. In 1936, for instance, he stated that sleep could never be brought about by stimulation of the hypothalamus, but constantly appeared when the zone of transition between thalamus and subthalamus was stimulated. In 1944 he published some cases of "adynamia" in cats after hypothalamic stimulation, a condition which he distinguished from the sleep referred to in his previous papers by a concomitant plasticity of the animals. Hess' views on the diencephalic regulation of activity have been expressed in a recent review of his results (26), but his conception that sleep may result from activity from some central area still needs confirmation.

We shall not go into the experiments in which disorders of sleep were caused by the introduction of various chemical substances into the brainstem, as their interpretation is too ambiguous to bring us any nearer to a satisfactory concept of sleep.

Looking back on the various opinions about the central regulation of sleep, a considerable amount of agreement now seems to have been reached as to the localisation of the centre of regulation. About its mode of action, however, some fundamental differences of opinion exist. Whereas von Economo agrees with Trömner's original view that sleep results from activity of the regulating centre, which can therefore be thought of as a sleeping centre, Ranson and his school are by their experiments led to the view that the regulating apparatus essentially serves for the maintenance of the waking state, and that sleep appears when it is reduced to inactivity. In an experimental study described below, we hope to reach some conclusions which may contribute to the solution of the problems mentioned in the previous account.

MATERIAL AND METHODS

Only adult albino rats were used in this study. In these animals the effects of experimental lesions of the hypothalamus and adjacent parts of the brain on the regulation of sleep were observed. The choice of the rat as an experimental animal was forced upon us by war conditions and more especially by the impossibility of feeding larger animals during a sufficiently long period. Because of its small size, it was certainly not the most suitable animal for our purpose.

The various functions of the hypothalamus have hitherto chiefly been investigated either by the method of electrical stimulation or by causing extensive destruction of this part of the brain. Much of our knowledge concerning the lower parts of the brainstem has been obtained by a different method, *viz., the infliction of sharply incised wounds*, and it is a striking fact that this procedure has scarcely ever been used in the study of hypothalamic functions. It has been followed only in the research of descending hypothalamic connections by Beattie, Brow and Long (3) and, more recently, by Magoun, Ranson and Hetherington (31). The method was considered a desirable addition to those procedures by which gross lesions of nervous tissues are brought about. It is probably the most suitable method for the tracing of pathways involved in the function under consideration, the damage to the brain being chiefly confined to interruption of fibre connections.

Unfortunately, however, it is still an open question whether the results of any destruction of brain tissue should be attributed to the exclusion of the destroyed or isolated area or to irritation of adjacent parts. Both conceptions can be and have been defended. The fact that many results of lesions of the central nervous system tend to subside in the course of time is often advanced in favour of irritation. It should be stressed, however, that the remarkable recuperative power which is exhibited by autonomic functions after lesion of the brainstem or spinal cord should in all probability be ascribed either to compensatory activity of those centres which subserve the same functions on a lower level ("automatisme étagé") or to the formation of new centres in adjacent regions. As far as the hypothalamus is concerned, there is another reason to doubt the irritative nature of incised wounds. Whereas unilateral electrical stimulation of this part of the brain is sufficient to cause widespread effects, which are eventually markedly bilateral (*e.g.*, pupillary dilatation), incisions in the hypothalmus proved to be effective in our experiments only when bilateral. The sole stimulatory effect observed was a bilateral pupillary dilatation which occurred regularly during the unilateral introduction of the cutting instrument in the posterior part of the hypothalamus, and persisted for a few minutes after it had been withdrawn. Since the pupillary dilatation would, from various observations (38), seem to be a very constant reaction to hypothalamic stimulation, we could not escape the impression that the irritative action of an incision in the hypothalamus, if it exists, is of short duration. All postoperative effects were therefore interpreted as the results of exclusion of certain parts of the brain and not to irritation of the area surrounding the lesion.

For operative purposes the hypothalamus can be reached in several ways, the most usual of which are the subtemporal and the parapharyngeal approaches. Because of the bilaterality of the lesions which are required to obtain disturbances in the regulation of sleep, the subtemporal approach was unsuitable for our purpose. The parapharyngeal method, by which the hypothalamus is reached through the base of the skull, was greatly interfered with by the flat, expanded hypophysis and its encircling blood vessels. In the experiments described in the following paragraphs, transverse lesions of the hypothalamus were inflicted via perforation of the convexity of the brain. Naturally this primitive method, apart from the important advantage of a better general postoperative condition, has a number of disadvantages, the most important of which is the fact that it gives rise to considerable damage to structures which are situated dorsal to the hypothalamus. In fact, this damage is such that the incisions, instead of being restricted to the hypothalamus, extend throughout the dorsoventral diameter of the brain, so that many sagittal fibre connections in neocortex, archicortex and thalamus are interrupted as well as the longitudinal systems of fibres within the hypothalamus. The objection that many postoperative symptoms might be the result of this additional damage is therefore not unreasonable. Consequently, a number of controls was needed to decide whether or not the structures overlying the hypothalamus are involved in the regulation of sleep. These experiments will now be described briefly.

### RESULTS

In a number of animals an incision was made in one of various frontal planes between the anterior and posterior commissures, measuring from $2\frac{1}{2}$ mm. on the right to $2\frac{1}{2}$ mm. on the left side of the median plane, and not extending beyond the ventral border of the thalamus. These wounds corresponded with the most serious accidental lesions encountered (Fig. 2). Animals treated in this way did not develop any disturbances of the sleep-waking rhythm; they rapidly recovered from the operation. It is proved by this observation *that a normal alteration of wake and sleep is kept up so long as the transverse incisions leave the hypothalamus intact.* It does not, however, exclude the possibility that the relevant dorsal structures play some part in the regulation of sleep.

In another group of animals identical lesions were made, but on one side the incision was prolonged to the base of the brain. Consequently only the opposite hypothalamus was left intact (Fig. 3). Although the postoperative mortality among these animals exceeded that of the first group, most of the rats made a rapid recovery. None of them displayed any disorder of sleep. Obviously one intact hypothalamus is sufficient for maintenance of normal sleep rhythm. The reverse experiment was carried out on a third group of rats. In these animals a combination of bilateral section of the hypothalamus and unilateral section of the dorsal structures was obtained by the introduc-

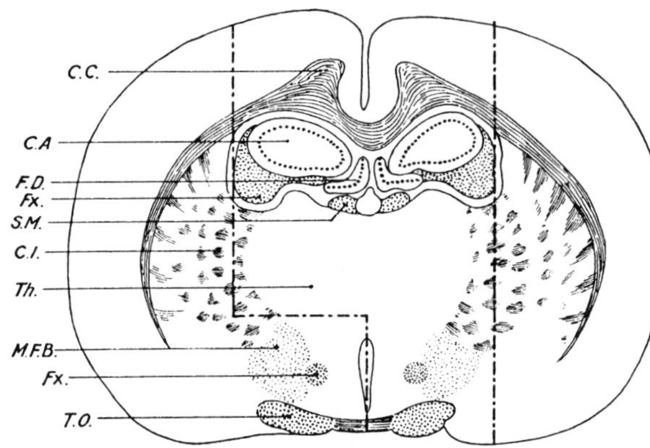


Fig. 2. Bilateral transection (indicated by interrupted lines) of the structures overlying the hypothalamus. C.A., cornu ammonis; C.C., corpus callosum; C.I., internal capsule; F.D., fascia dentata; Fx., fornix; M.F.B., medial forebrain bundle; Th., thalamus; T.O., optic tract.

tion with a "tour de maître" of a small hook-shaped knife, into the hypothalamus depicted in Figure 4. All animals treated in this way developed marked disturbances of the sleep-waking rhythm, which were identical

Fig. 3. Bilateral transection of dorsal structures and unilateral transection of hypothalamus. The lesion is indicated by interrupted lines. Abbreviations: see Fig. 2.

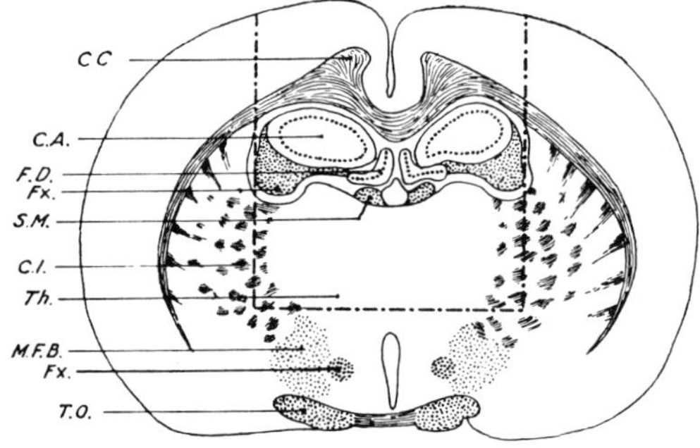


with those observed in cases in which bilateral lesions of the structures overlying the hypothalamus were inflicted.

From these observations it is evident *that the dorsal structures (neocortex, archicortex and thalamus) of one side are unable to keep up the normal regulation of sleep.* The disorders of sleep observed after transverse hypothalamic incisions from above may therefore safely be attributed to the lesions of the hypothalamus and not to the damage to more dorsally situated structures.

As to the operative technique, the operations were all carried out under ether anaesthesia and with aseptic precautions. After median incisions of the skin and periosteum, the latter was pushed aside, and a uni- or bilateral trephine hole of about 4 mm. diameter was made just lateral to the median line, in order to avoid the superior longitudinal sinus. The dura mater was left in situ. In those cases in which the incision was intended to extend to one single lesion through both hypothalami, the small knife, depicted in Figure 4, was used so as to restrict the additional damage to one side. In cases in which it was essenital to leave a medial hypothalamic zone intact—in other words, to restrict the lesions to both lateral hypothalamic areas—two separate incisions were required, and consequently the additional damage to the dorsal structures was inflicted on both sides. In the foregoing

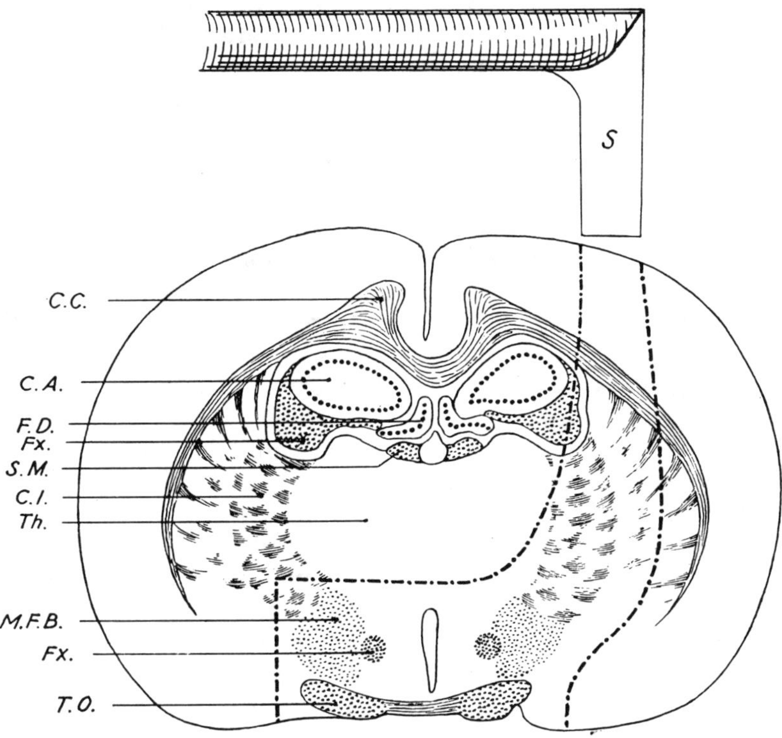


FIG. 4. Unilateral lesion to dorsal structures and bilateral transection of the hypothalamus (interrupted lines). The instrument with which the lesion was inflicted has been indicated with S. For abbreviations see Fig. 2.

account it has already been pointed out that these uni- or bilateral lesions to the structures overlying the hypothalamus had no apparent influence on the regulation of sleep.

After the lesions had been placed and the usual slight hemorrhage had stopped, the periosteum was stitched over the trephine holes and the skin was carefully closed. No dressing was applied. Throughout the operation 3 per cent hydrogen peroxide was used as an antiseptic. The anatomical investigation of the lesions was usually carried out after the animals had died from the results of the operation. When recovery took place the animal was sacrificed some time after the condition had become stationary. After their removal the brains were, as a rule, fixed in 96 per cent alcohol for three days and prepared according to one of Cajal's methods. Sagittal sections were found to be most suitable for the tracing of the transverse incisions. In the following account the disorders of sleep are grouped under two headings, viz., disturbances of the waking capacity and those of the function of sleeping.

*Disorders of waking*

Certain bilateral lesions of the hypothalamus were found to cause a decrease in the waking capacity. The extent of this decrease varied in different cases.

(i) *Eight animals slept uninterruptedly so long as they were not roused by external stimuli.* They lay curled up on one side, breathing regularly, and could promptly be awakened by sufficiently strong stimuli, *i.e.*, by pinching the tail or handling. When they were left alone afterwards, they would *yawn and stretch* and settle down in a comfortable position to go to sleep again. To strong stimuli the animals reacted vigorously, exhibiting all signs of intense emotion. There were no motor disturbances. Apart from the apparent inability to maintain the waking state, the animals displayed other autonomic disorders, which is not surprising in view of the many regulatory functions of the hypothalamus. All sleeping animals developed a marked hypothermia, which often was such that a rectal temperature of 25°C. was found after a stay of 24 hours in an environmental temperature of 18°C. The condition of sleep evidently did not depend on this hypothermia, as it was not less conspicuous in animals which were kept in an incubator of about 30°C. and consequently had rectal temperatures ranging between 35° and 40°C. Details of the temperature regulation cannot be given as a continuous registration of the temperature was not carried out.

The general condition of all hypothermic animals rapidly deteriorated; they constantly contracted a purulent conjunctivitis and rhinitis. To avoid these undesirable complications it was necessary to nurse the animals in a hot box. Moreover, extra attention had to be paid to their feeding. As many of the operated rats, especially the sleeping animals, did not take any food or drink of their own accord, it was essential to feed them artificially. For this purpose about 5 cc. of lukewarm, diluted skimmed milk was administered by tube (a soft catheter with a diameter of two to three millimetres) three times a day. In spite of all these precautions the animals never survived the operation for a long period. Only once was it possible to keep a rat (no. 55) alive during eleven postoperative days, the other animals dying after four to eight days. In rat 55 after the eighth day there were short periods during which the animal no longer lay curled up in its characteristic sleeping attitude, but sat huddled up with half-opened eyes, without, however, showing any spontaneous activity. It is an open question whether or not the capacity of waking is capable of a complete recovery in these sleeping rats. The animals could not be kept alive long enough and even the period of eleven days was too short to allow of an answer. From the results of the experiments of Ingram, Barris and Ranson (27) on cats and those of Ranson (36) on monkeys, it would seem that the capacity of maintaining the waking state does not completely return in these animals after optimal lesions of the hypothalamus. It is, however, improbable that the lesions in these animals were equivalent to those in ours. Future work on animals, better suited to this purpose than the rat, will have to solve the problem whether the total loss of the hypothalamic regulation of sleep can be as satisfactorily compensated as, for instance, the thermoregulation after exclusion of the hypothalamus.

(ii) Another group of animals, instead of developing a complete condition of sleep, exhibited various degrees of drowsiness. They were inactive, and during the first days sat huddled up all the time, with eyes shut to slits. Like the other animals these rats could be roused promptly. After a period which lasted from one to three days, the somnolence tended to decline gradually, but in these animals also the general condition rapidly deteriorated and death resulted within a week so that the waking capacity was never observed to restore itself completely. Only occasional periods were observed in which the animals were in a somewhat more active condition, the eyes, for instance, being opened wider. It is of interest that these drowsy rats were never found to be in a typical sleeping state. Their condition was intermediate between waking and sleeping whenever they were observed. Like the sleeping rats they often developed a hypothermia and purulent infections of the mucous membranes and did not show any tendency to take food or drink of their own accord so that they too had to be nursed with tube feeding and hot box.

(iii) In a third group of rats the operation failed to produce any disorder of the waking capacity; neither did most of these animals

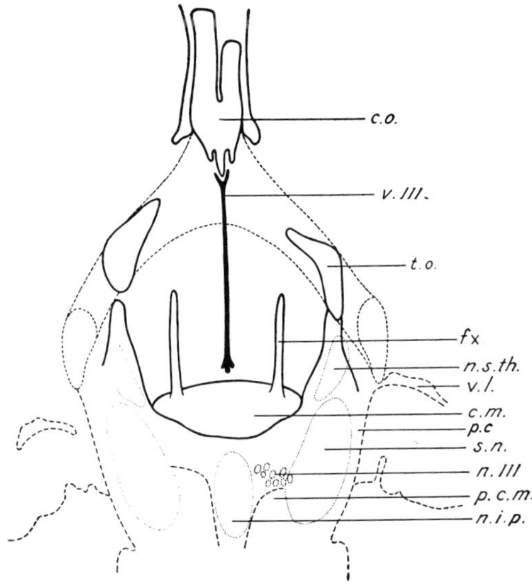


FIG. 5. Key figure to Tables 1–5. The diagram has been composed of two horizontal sections, of which one (solid lines) passes through the dorsal hump of the optic chiasma and through the mammillary bodies. The course of the optic tracts has been indicated by broken lines. The second section (outlines given in broken lines) lies on a more dorsal level and passes through the cerebral and mammillary peduncles (broken lines) and through the substantia nigra, the interpeduncular nucleus and the subthalamic nucleus, outlines of which are stippled. For Tables 1–5 only the basal section has been used. C.M., mammillary body; C.O., chiasma opticum; fx., fornix column; N.I.P., interpeduncular nucleus; N.S.Th., subthalamic nucleus (Luys); N.III, oculomotor nucleus; P.E., cerebral peduncle; P.C.M., mammillary peduncle; S.N., substantia nigra; T.O., optic tract; V.L., lateral ventricle; V.III, third ventricle.

develop any other serious disturbances of the general condition. In order to facilitate a survey of our material we abstained from verbal description of each case. Instead the lesions which were found in every single case were recorded into a diagrammatic horizontal section of the hypothalamus (Fig. 5).

The diagrams obtained in this way have been collected in three tables, *viz.*, Table 1, on which the findings in the completely sleeping rats are depicted; Table 2, showing the cases of drowsiness in as much of a descending order of intensity as was practically possible; and Table 3, which contains those cases in which the operation did not affect the waking capacity. The following points concerning the figures are of special interest.

(a) In all cases of characteristic sleep (Table 1) bilateral lesions were found which ex-

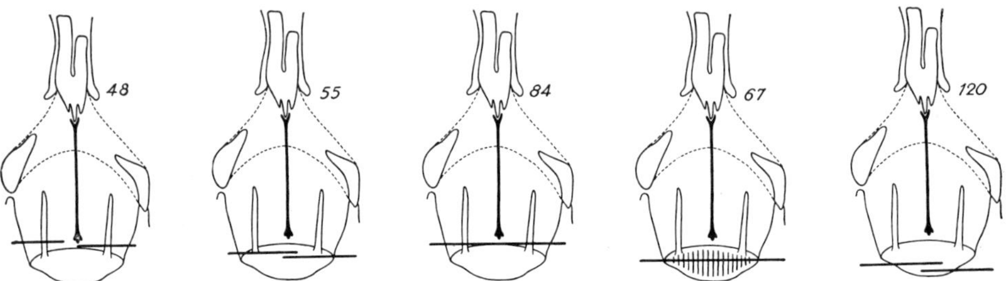

*Table 1. Anatomical findings in five rats exhibiting a typical condition of sleep.*
*For an explanation of the diagram, see Fig. 5.*

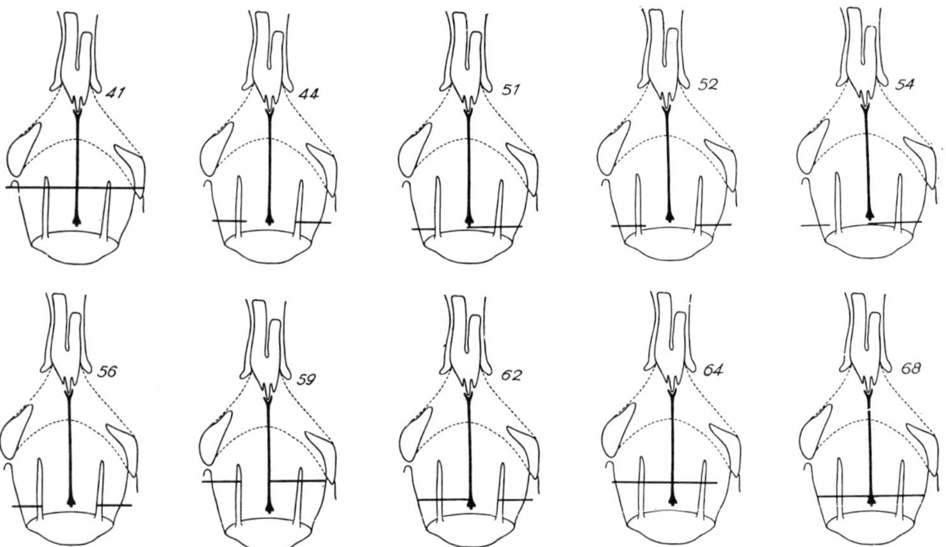

*Table 2. Anatomical findings in ten cases of somnolence. See Fig. 5.*

tended through the entire—or almost the entire (rat 48)—transverse diameter of both hypothalami. All these lesions were situated in the immediate vicinity of the mammillary bodies. In two of the animals the mammillary bodies themselves were involved in the lesions, in rat 67 these nuclear groups were largely destroyed by hemorrhage. In rats 48 and 84 the lesions were situated on the rostral border of the mammillary bodies; in three other cases, of which only rat 120 is shown, they were found slightly caudal to these cell-groups.

(b) Lesions of the mammillary region were also met with in some of the drowsy animals, collected in Table 2. These transections, however, differed from those found in the first group in being *incomplete*. The region medial to the fornix column—comprising the periventricular and the medial hypothalamic areas according to Crosby and Woodburne (cited from 6)—had been left intact on both sides in rats 52 and 56, and on the left side in rat 51. In rat 54 the hypothalamic lesion on the left was confined to the lateral half of the lateral hypothalamic area, leaving the remainder of this area and both inner areas undamaged.

Yet this second group also contains a few cases in which a complete transection of both hypothalami was found. In these cases (rats 68 and 41), however, the lesion was situated farther rostrally than in the animals belonging to the first group. In rat 68, for instance, which was very drowsy, a complete transection of both hypothalami was found

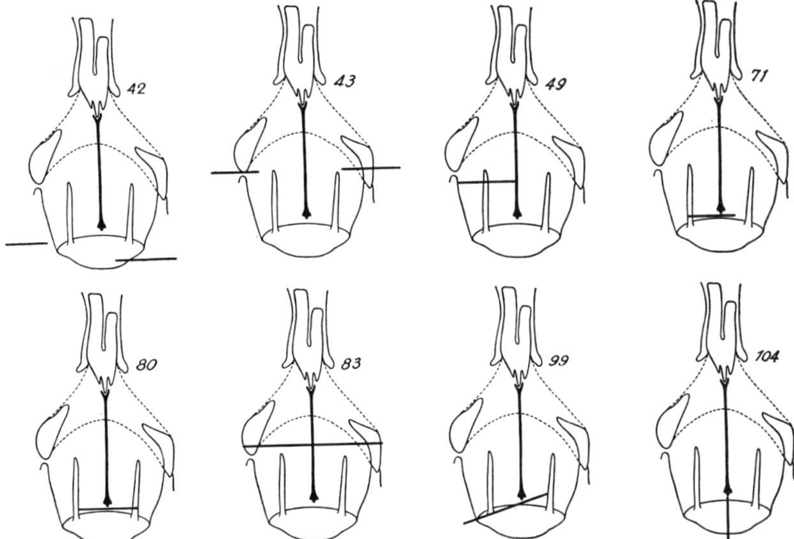

*Table 3. Eight cases in which the operation failed to produce a decrease of the waking capacity. See Fig. 5.*

about 1 mm. anterior to the mammillary bodies. Only a slight somnolence was displayed by rat 41; in this animal a complete bilateral lesion existed about 2 mm. in front of the mammillary bodies. As an identical lesion was found only 0.2 mm. farther ahead in rat 83 (Table 3) which showed no signs of somnolence, it would seem that the area in which hypothalamic lesions must be situated to cause a decrease in the capacity of waking has its rostral limit in a transverse plane about 2 mm. anterior to the mammillary bodies.

It has not been possible to ascertain the caudal limit of the area under consideration. The most posterior lesion which was found to disturb the waking capacity was situated slightly caudal to the mammillary bodies, *i.e.*, in the tegmental region. More caudal lesions resulted in a loss of consciousness which was too little reversible to enable us to make any statements concerning the regulation of sleep in these animals.

(c) The third group (Table 3), containing a number of animals without any apparent disturbance of the waking capacity, is distinguished from the previous groups in various respects. In some of the animals (rats 42 and 49) unilateral lesions of the hypothalamus were found to exist. It has already been pointed out that unilateral hypothalamic lesions, irrespective of their location, are unable to affect the regulation of sleep. Apparently the same holds good for bilateral lesions which are confined to the medial part of the hypothalamus, *viz.*, the zone medial to the fornic column. This is shown by rats 71 and 80, in which only the inner hypothalamic areas (the medial and the periventricular area) were damaged. It would seem to indicate that the medial part of the hypothalamus is of no special importance for the maintenance of the waking state. In the discussion of the second group it was mentioned, however, that the decrease of the waking capacity was less in animals in which the lesions were limited to the lateral area of both hypothalami than in those in which, on the same level, the whole hypothalamus was bilaterally sectioned. This indicates that the inner hypothalamic areas do play a certain role in the maintenance of the waking state. We shall revert to this fact in one of the following sections.

The preponderance of the lateral area in the function of waking, already observed by Ranson (36) in monkeys, is clearly demonstrated by rat 99. No drowsiness was observed in this animal in spite of a transverse lesion in the mammillary region which was complete on the left and had been confined to the inner hypothalamic areas on the right. The lateral area of the right side was the only part of the hypothalamus intact, and it proved capable of maintaining the waking capacity. Evidently one lateral area of the hypothalamus is sufficient for maintaining this function. A lesion of exceptional location was encountered in rat 104. In this animal a median incision was found to exist, which reached the base of the brain between left and right mammillary body. Naturally it had severed the supra-mammillary commissure, which contains the crossing fibres of the hypothalamo-tegmental division of the medial forebrain bundle, in addition to crossing tegmental connections of the fornix column. The animal exhibited a slight drowsiness which entirely disappeared in the course of the first postoperative day. Incidentally, rat 83 was mentioned in the discussion of the second group. The complete bilateral transection of the hypothalamus, found in this case, was apparently situated too far rostrally to interfere with the waking capacity. It will again be reverted to in one of the next sections. A number of supplementary experiments have not been inserted in Table 3. In four rats extensive destructions of both thalami were brought about without causing the animals to show any decrease of the waking capacity. This does not harmonize with the findings of Spiegel and Inaba (43), mentioned in the introduction, and it offers confirmation of Ranson's observations (36).

In addition it should be stressed that bilateral transverse incisions on a level with the mammillary bodies only caused disorders of the function of waking if they involved the basal part of the brainstem, as, for instance, was the case in rat 120 (Table 1). If not extending ventrally beyond the central grey substance around the Sylvian aqueduct, these lesions, intermediate between diencephalon and mesencephalon, fail to interfere with the waking capacity. On this point our findings are in line with the results of Ingram, Barris and Ranson (27). From their experiments, in which somnolence was produced in cats by small bilateral lesions in the basal part of the tegmental area adjacent to the mammillary bodies, we may conclude that the function of waking, as far as it is performed by the midbrain, is localised in the basal part of this structure, *i.e.*, the tegmentum. The conclusions arrived at in the previous account indicate a specific importance of a certain area of the brainstem for the maintenance of the waking state. Certain lesions of this area cause a total loss of this function. In view of the arguments, advanced in the discussion of the operative method pursued in this study, we are inclined to consider this disorder to be a result of the exclusion of a certain centre, which therefore may be termed a waking centre.

### *Connections of the region of the waking centre*

It was pointed out in the preceding section that the region formed by the posterior part of the hypothalamus and a hitherto undefined portion of the adjoining tegmentum mesencephali is likely to contain a waking centre. It is an important fact that the same region of the brainstem, according to the results of Beattie (2), Ranson, Kabat and Magoun (38) and others, is the site of the highest orthosympathetic centre. Its stimulation is followed by a rise of blood pressure, an increase in rate and depth of respiration, pupillary dilatation, pilo-erection, etc., which give an appearance of intense emotional activity.

It has long been recognized that the autonomic balance lies relatively on the orthosympathetic side during the waking state and shifts to the parasympathetic side during sleep. If one combines this fact with the afore-mentioned results of stimulation, it does not seem impossible that there exists at least a partial identity between the waking centre and the ortho-sympathetic centre in the hypothalamus, and that the waking state is merely one of the manifestations of orthosympathetic activity. All ortho-sympathetic phenomena which result from stimulation of the hypothalamus can be brought about only by a descent of hypothalamic impulses to lower levels of the central orthosympathetic system, from where they are con-ducted along the peripheral orthosympathetic pathways to the various end-organs concerned.

Those somatic phenomena which indicate the shift to the orthosym-pathetic side during the waking state are certainly effected along this way. The most striking difference between wake and sleep, however, lies in the degree of consciousness. The question lies near at hand—if perhaps the in-crease of consciousness caused by the activity of the waking centre is brought about by a conduction of impulses via the same lower centres and the same peripheral pathways to the cerebral cortex. Moreover, there is still another possibility for the hypothalamus to stimulate the cerebral cortex, as stimulation of the hypothalamus brings about a production of adrenalin by the suprarenal gland, which tends to raise the level of consciousness.

Experimental evidence, however, seems to indicate that the peripheral autonomic system is not involved in the maintenance of the waking state. The observations of Cannon and his associates on cats in which practically the entire peripheral sympathetic system, including the medulla of both suprarenals, had been removed (8), have proved that under these circum-stances a fairly normal condition can be kept up, so long as the exigencies of the outer world are held within certain limits. They apparently did not observe any change of the sleep-waking rhythm in sympathectomized animals. These facts render it highly probable that the waking centre does not affect the cerebral cortex along peripheral pathways but influences cerebral functions along central corticopetal connections.

At this point the question arises as to what ascending connections may account for the action of the walking centre on the cerebral cortex. As no data concerning this problem could be found in literature, we decided to study the fibre degeneration following lesion of the area in which the waking centre is probably located. For this purpose the mammillary area of the right side was sharply cut across in rat 66. The operation was carried out in the usual way, the wound extending from the dorsal surface to the base of the brain, and con-sequently the structures overlying the hypothalamus were also damaged.

As could be expected in view of the unilaterality of the lesion, the animal did not de-velop any disorder of the sleeping rhythm. It made a rapid recovery, and was killed ten days after the operation. Its brain was treated according to the prescription given for the Marchi technique by Romeis (40), and was embedded in paraffin via the graded alcohols and cedar oil, after which it was serially sectioned in the sagittal plane.

On miscroscopic investigation the location of the lesion was found to be as follows. The dorsal surface of the brainstem was reached at the habenula; after traversing this ganglion the instrument had apparently followed the rostral side of the habenulo-peduncular tract for some distance, but on a more ventral level the wound diverged from this bundle in a rostral direction so that it reached the base of the brain through the caudal one-third of the mammillary body (Fig. 6). In a transverse direction it extended from about 0.2 mm. to so far outside the median plane that it had severed the most medial fibres of the cerebral peduncle as its lateral end. Consequently, the whole transverse diameter of the hypothalamus had been cut, with the exception of a narrow medial zone of about $200\mu$ in width. As a result of this lesion a large part of the fibre connections between the hypothalamus and lower parts of the nervous system had been interrupted. Only those fibres running in the undamaged paramedian zone and including the medial part of the periventricular fibre system of Schütz had been left intact.

As could be anticipated in view of the damage to the mammillary body, degeneration was found in some of the fibre systems connected with this cell group. Vicq d'Azyr's

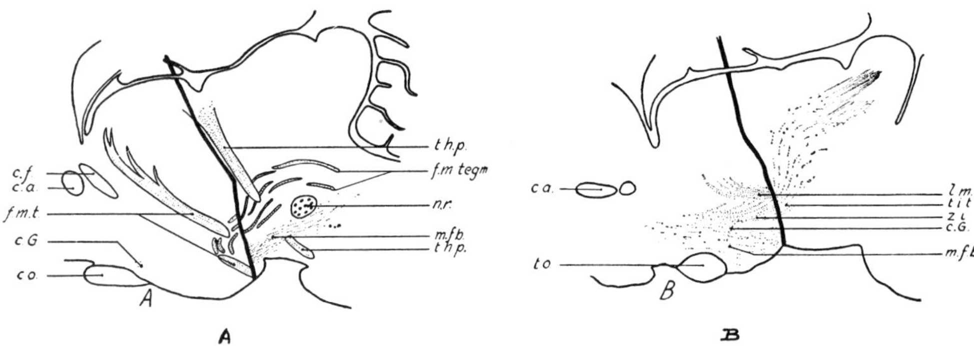


Fig. 6. Sagittal sections through the brain stem of rat 66, Marchi method, ten days after operation. Section A is situated 0.6 mm., section B 1.5 mm., lateral to the median plane. The lesion is indicated with a thick line. Descending (A) and ascending (B) degeneration of the medial forebrain bundle. c.a., anterior commissure; c.f., fornix column; c.G., Ganser's commissure; c.o., optic chiasma; f.m.t., bundle of Vicq d'Azyr; f.m.tegm., mammillo-tegmental fascicle; l.m., lemniscus medialis; m.f.b., medial forebrain bundle; n.r., red nucleus; t.h.p., habenulo-peduncular tract; t.i.t., incerto-tegmental fibres; z.i., zona incerta.

mammillo-tegmental tract was heavily degenerated. Some degeneration was also found in the mammillary peduncle, indicating the occurrence of mammillo-fugal fibres in this bundle, which conflicts with Papez' conception that the system is entirely mammillopetal in nature (34).

Dorsolateral to the mammillary body, in the space between this cell group and the substantia nigra, the lateral hypothalamic area continues into the midbrain tegmentum. Many fibres belonging to this hypothalamic area, chiefly constituents of the so-called medial forebrain bundle, run through this space and connect with tegmental centres. Presumably these fibres, originating in di- and telencephalon, form the first link of a much interrupted pathway to bulbar and spinal autonomic centres. Magoun, Ranson and Hetherington (31) have offered proof that these connections are not arranged into circumscribed bundles in the midbrain but are distributed over almost the entire cross section of its tegmentum. Previously Beattie, Brow and Long (3), working with the Marchi technique, had claimed that the chief descending connections of the hypothalamus course in or near the central grey substance surrounding the Sylvian aqueduct, a position which conforms to that of the periventricular system of Schütz. Probably Beattie *et al.* observed degeneration of this system which connects with the inner part of the hypothalamus, *viz.*, with the periventricular and medial areas. This is in line with the fact that the transverse cut which they inflicted to the hypothalamus was confined to these areas and did not involve the lateral area. They consequently did not observe degeneration of the fibres descending from the lateral hypothalamic area, which are more numerous than those of Schütz's system and, from the results of Magoun, Ranson and Hetherington, would seem to be the chief transmitters of orthosympathetic impulses from the hypothalamus. In our experiment these fibres had been interrupted. By virtue of their degeneration a large number of them could be traced in a caudal direction. Figure 6A shows a rather condensed group of these degenerated fibres, running underneath the red nucleus and radiating in a dorsocaudal direction into the tegmentum. A considerable number, not shown in Figure 6, spreads over more ventral levels of the tegmentum. None of these fibres could be traced into parts lower than the midbrain. A fair number of degenerated fibres ran from the hypothalamus through the supramammillary decussation into the opposite side of the midbrain.

In accordance with the observation of Beattie, Brow and Long (3), a number of osmophilic granules was found in the dorsal longitudinal bundle, *i.e.*, the caudal continuation of the periventricular hypothalamic system of Schütz. As only the lateral part of this system had been damaged, it also contained many normal fibres. Apart from this descending degeneration, a number of degenerated periventricular fibres could be traced in a rostral direction. This proves the occurrence of hypothalamopetal fibres in the system of Schütz, which therefore should not be regarded as completely efferent with regard to the hypothalamus. Identical findings were reported in the opossum by Bodian (7).

A much more extensive ascending degeneration was found in the lateral hypothalamic area. From the lesion a great number of degenerated fibres could be traced rostralward, chiefly occupying the lateral part of the lateral area and obviously belonging to the medial forebrain bundle. On passing forward through the lateral hypothalamic area their number continuously decreased, which suggests a distribution of this ascending system in the hypothalamus. Only a few of the fibres were found to extend farther rostralward, into the septal region, and none could be traced to still higher levels. This ascending system in the medial forebrain bundle conceivably originates in the mammillary region, and farther caudally, in the midbrain tegmentum.[1]

By way of summary we are able to state that our Marchi experiment seems to prove the existence of ascending fibres in several bundles connected with the mammillary body and its adjoining structures, *viz.*, in the mammillo-thalamic bundle of Vicq d'Azyr and in the medial forebrain bundle. The degeneration found in the fornix column cannot give any information concerning the direction in which the fibres of this bundle lead, because both of its terminations—the hippocampal formation and the mammillary body— had been damaged.

In his careful study of the diencephalon of the opossum Bodian (7) did not observe any mammillopetal degeneration in Vicq d'Azyr's fascicle after lesion of this bundle. In contrast to Le Gros Clark and Boggon (9), he concluded that this system is entirely efferent with regard to the mammillary body, a conclusion which seems to be substantiated by Droogleever Fortuyn (15), who in a developmental study of the thalamus observed an outgrowth of fibres only from the mammillary body towards the thalamus and not in the opposite direction. The occurrence of hypothalamo- and septopetal fibres in the medial forebrain bundle has so far not been described.

---

[1] Although it has no apparent bearing on our problem, it is interesting to note that degeneration was found in Ganser's commissure. Fibres of this heavily myelinated system were found to pass rostralwards through Forel's field, whence they curved downwards to cross underneath the third ventricle to the left lentiform nucleus. This finding is in accordance with the recent observation of Glees (19) that Ganser's commissure contains fibres connecting the medial fillet with the opposite globus pallidus.

In conclusion it seems probable that the region in which lesions must be situated to cause a maximal loss of the waking capacity has efferent connections with the anterior nuclei of the thalamus, the lateral hypothalamic area and the septal region.

The next step to take is to decide which of these connections may transmit the impulses from the waking centre to the cortex. None of the aforenamed ascending connections can be traced to the cortex itself. If a corticopetal system originates in the waking centre, it must be supposed to relay in the thalamus, hypothalamus or septum—perhaps in two or all of these structures. It has been pointed out that destruction of the thalamus, despite the conceptions of Trömner and of Spiegel and Inaba and others, is not followed by any decrease of the waking capacity. Bilateral lesions of the septum also fail to bring about alterations of the sleep-waking rhythm, as is shown by rat 114 (Table 5). Moreover, in one rat, which did not show any tendency to somnolence, autopsy revealed a large abscess, which had destroyed the septal region and the anterior half of the thalamus on both sides. Naturally these observations do not exclude the possibility that the septum and the thalamus do play a certain role in the corticopetal transmission of stimuli from the waking centre. At most it proves that these structures are not the only relaying centres involved in this function.

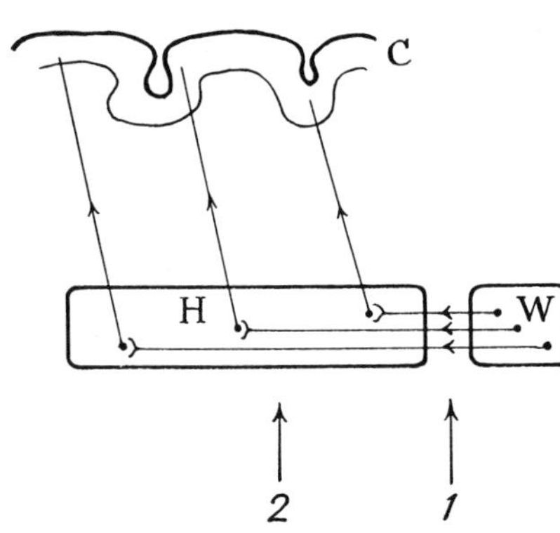


FIG. 7. Diagrammatic representation of the influence of the waking centre on the cortex, as suggested in the text. C., cerebral cortex; H., lateral hypothalamic area; W., waking centre.

It seems reasonable to accept an intensive relay of impulses from the waking centre in the lateral hypothalamic area. The majority of the ascending fibres in the medial forebrain bundle ends here, and it is striking that the degree of somnolence is more or less proportional to the extent to which these ascending fibres have been interrupted, the most effective lesions being situated in the caudalmost part of the hypothalamus and involving its lateral area in which the fibres are contained. If the lateral area of one or both sides partly or completely escapes injury, somnolence is either slight or lacking (cases 64, Table 2; and 43, 80, 99, Table 3). Moreover, the number of ascending fibres severed naturally becomes smaller with more forward localisation of the lesions, and as was pointed out in one of the foregoing sections, the effect of the lesions decreases accordingly. (Cf. Table 1 with cases 68 and 41; Table 2.)

From the fact that a maximal loss of the waking capacity is as well brought about by lesions just in front of the mammillary body as by those closely behind it, it seems probable that the waking centre itself lies in the region caudal to the mammillary body, viz., in the midbrain tegmentum.

We are inclined to suppose that it gives origin to a number of fibres which ascend through the narrow space between mammillary body and substantia nigra which should in this part of their course be considered to be the initial common path of the "waking" stimuli. Rostral to the mammillary body the termination of this system in the hypothalamus—chiefly in the lateral area of this structure—begins, and some fibres even extend farther forward into the septal region, as is demonstrated by our Marchi experiment. Our results seem to indicate that the fibres under consideration synapse in the hypothalamus, whence their stimuli are led off by secondary neurons sideways towards the cerebral cortex. It should, however, be stressed that hypothalamocortical connections, as suggested here, have never been morphologically demonstrated, and that consequently no statements concerning the course of this hypothetic system can at present be made.

Our concept is illustrated by Figure 7, which offers an explanation of the fact that a transection on a level with the mammillary bodies (arrow 1), interrupting the "initial common path," results in greater loss of waking capacity than a similar lesion in a more rostral plane (arrow 2). With more forward location of the lesion a larger number of fibres from the waking centre escapes injury. Finally, at about 2 mm. rostral to the mammillary bodies, their number already ended is sufficient

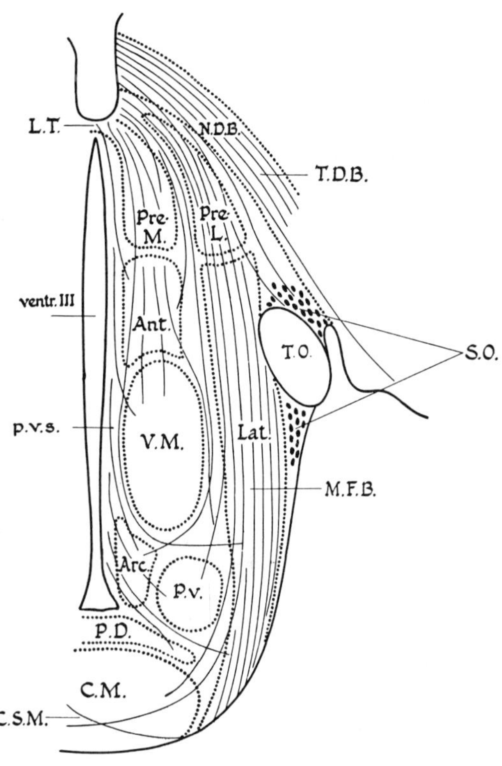



FIG. 8. A diagram of the longitudinal fibre systems in the hypothalamus, as seen in a horizontal section. Note the exchange of fibres between the lateral and the periventricular areas in the premammillary region. Ant., nucleus anterior; Arc., nucleus arcuatus; C.M., corpus mammillare; C.S.M., commissura supramammillaris; Lat., nucleus lateralis; L.T., lamina terminalis; M.F.B., medial forebrain bundle; N.D.B., nucleus of diagonal band; P.D., nucleus praemammillaris dorsalis; Pre.M., nucleus praeopticus medialis; Pre.L., nucleus praeopticus lateralis; P.v., nucleus praemammillaris ventralis; p.v.s., periventricular fibres; S.O., nucleus supraopticus; T.O., tractus opticus; T.D.B., diagonal band of Broca; ventr.III, third ventricle; V.M., nucleus ventromedialis.

to maintain a normal waking capacity, as is suggested by the absence of somnolence after transections on more rostral levels.

It has already been mentioned in the preceding section that the inner

part of the hypothalamus, although certainly of much less importance for
the waking capacity than the lateral area, does seem to play a certain role
in this function. Presumably a small number of fibres from the waking
centre course in this part, which is formed by the periventricular and medial
areas. It is true that no degeneration was found in these areas in rat 66, but
it is possible that the fibres under consideration are unmyelinated. There
certainly exists an extensive exchange of fibres between the lateral and the
two inner areas, especially (Fig. 8) in the region just in front of the mammillary body, and it is possible that a number of ascending fibres deviate
from the lateral area to continue their course in more medial parts of the
hypothalamus. Possibly these fibres account for what remains of the waking
capacity after lesions restricted to the lateral area.

### Sleeplessness

In comparison to the vast clinical and experimental literature concerning
somnolence, astonishingly little is known about the opposite phenomenon:
sleeplessness. It seems that only von Economo (11) has dealt with this subject in extenso. According to his observations inflammation of a rostral part
of the hypothalamus, adjacent to the striate body, may result in insomnia.
It would thus seem that the exclusion of the relevant hypothalamic area
interferes with the function of sleep.

As far as we know, the only experimentally founded conception of insomnia is that of Ranson and Magoun (39) who in the course of many experiments on cats and monkeys never observed any influence of hypothalamic lesions on the capacity of sleeping. Their opinion about the regulating
mechanism is sufficiently illustrated by the statement quoted in our introduction, from which it is evident that they deny the existence of a centre
subserving the function of sleep in a restricted sense.

In the course of our own experiments we arrived at a different opinion.

(a) In a number of rats the lesion of the hypothalamus was followed by a condition
of sleeplessness. After regaining consciousness some of these animals were restless and
irritable, reacting vigorously to minor stimuli. Their condition closely resembled the sham
rage observed by Fulton and Ingraham (16) in cats after prechiasmatic lesions, and described by Bard (1) as a result of decerebrations through the rostral part of the diencephalon. In a number of operated rats no such change of character was observed, the
animals remaining as quiet as before the operation.

In both groups the normal alternation of wake and sleep had completely vanished.
Naturally this fact could only be ascertained by means of a continuous observation of the
animals. The normal difference in activity between day and night—established for the
rat by Szymanski (44) who in a space of 24 hours registered an average of 14 hours of sleep,
distributed over 10 periods, which were longer and more frequent during the day than during the night—was in this way found to have disappeared completely, the rats being awake
whenever they were observed. The animals showed a normal interest in their environment.
Their general condition was excellent at first and they spontaneously took food and drink.
Soon, however, their state deteriorated, which is not surprising considering the large
amount of sleep to which the rat is accustomed. After a period of 24 hours the sleepless
rats usually began to show symptoms of fatigue. They did not eat or drink of their own
accord and their interest in the surroundings decreased. Symptoms of sham rage, if present,
persisted. In spite of the fatigue and even of the succeeding exhaustion, during which the
gait became unsteady, sleep was not forthcoming, the opened eyes and the spontaneous

activity proving that the animals were awake. After a period averaging three days the exhausted animals fell into a state of coma which soon ended in death. A return of the sleeping capacity was never observed in any of the animals.

We did not observe hypothermia in sleepless rats, nor did these animals develop purulent infections of mucous membranes. In a previous section we incidentally mentioned the disappearance of characteristic periods of sleep in drowsy animals. In these cases both the waking and the sleeping capacity seemed to have been disturbed by lesions, situated in the hypothalamic region between the mammillary bodies and a transverse plane about two

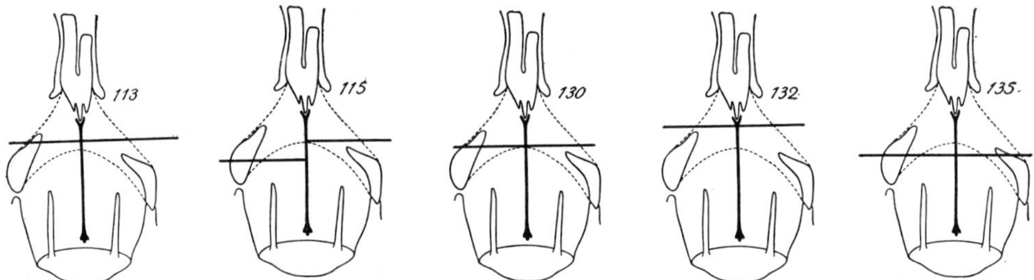

*Table 4. Anatomical findings in five rats exhibiting
sleeplessness. See Fig. 5.*

millimetres in front of these cell groups. The animals referred to in this section, however, showed a loss of the sleeping capacity which was not complicated by any apparent disturbance of the function of waking. In all these cases of uncomplicated asomnia the lesions were found to be situated in the rostral half of the hypothalamus, in contrast to the cases

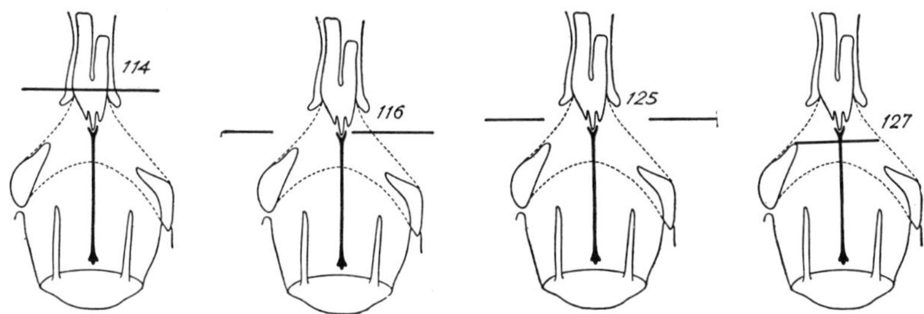

*Table 5. Anatomical findings in four rats which did not
develop sleeplessness. See Fig. 5.*

of sleep or somnolence in which the lesions regularly occupied the caudal half of this part of the brainstem. The border between the two regions lies approximately 2 mm. in front of the mammillary bodies and roughly separates the infundibular and mammillary regions from the more rostrally situated suprachiasmatic and preoptic regions.

Here, also, a long series of verbal descriptions has been omitted. Instead a table is given (Table 4), in which the lesions found in each case are marked in a simple diagram. It will be noticed that in all cases complete bilateral transections were found on various levels in the rostral half of the hypothalamus.

(b) In a second group of animals lesions were found in approximately the same region of the brain. These rats, however, did not develop any apparent disturbance of the sleep-waking rhythm. Four cases of this group are given in Table 5. In one of the animals (rat 114) a transverse section was found in the septal region, slightly rostral to the lamina terminalis. In another rat (no. 125) the paramedian region of the brain, including both hypothalami, had been left intact, while in rat 116 only one of both hypothalami had been cut across. The latter case indicates that the function of sleeping, like that of waking, can be sustained by one hypothalamus.

    (c) We observed only one rat in which the operation resulted in a partial loss of the sleeping capacity. This animal (rat 149, Fig. 9) was sleepless during the first postoperative day, but after that time short periods of sleep occurred with highly irregular intervals. The animal's condition failed only slowly; it could be kept alive for nine days.

The lesions found in this rat are in a sense the reverse of those encountered in rat 127 (Table 5), in which the sleep-waking rhythm was not affected by the operation. Whereas in the latter animal only the medial part of the

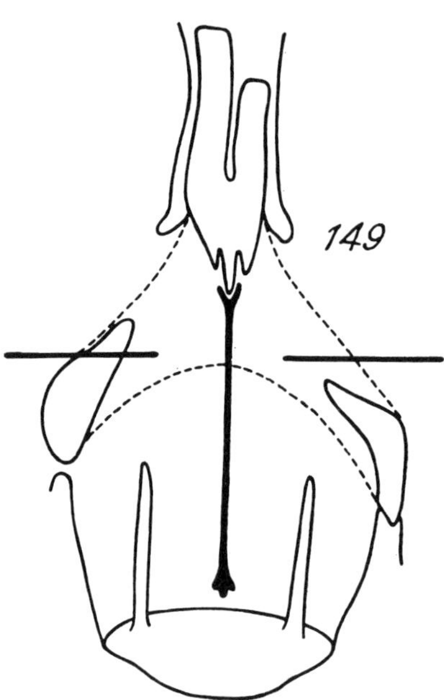


FIG. 9. Lesions in rat 149. Only the lateral area of the suprachiasmatic part of the hypothalamus has been interrupted on both sides.

hypothalamus has been transected, the lesions had been confined to the lateral hypothalamic area in rat 149. Comparison of both cases indicates that the importance of the lateral area of the hypothalamus for the sleeping capacity exceeds that of the inner areas, the latter not being without any significance for this function. It will be noticed that a similar conclusion was reached with relation to the waking capacity. Evidently the rostral half of the hypothalamus, roughly conforming to the suprachiasmatic and preoptic areas, is the site of a nervous structure which is of specific importance for the capacity of sleeping. In the following sections this structure will be referred to as "sleep centre."

This conclusion is in agreement with the conception of von Economo (11) who likewise postulated a rostral hypothalamic centre, to be regarded as the effector of sleep, as proved by the appearance of sleeplessness after its exclusion. In the contradictory opinion of Ranson and Magoun (39), this sleeplessness would not result from destruction of any part of the hypothalamus but from an irritating influence exerted by the inflammation on the adjacent waking centre. Our results, however, do not produce much evidence in support of the latter view. If the sharp transections which we inflicted to the hypothalamus really caused insomnia only by way of irritation of the waking centre, a unilateral lesion of the rostral part of the hypothalamus would already be sufficient to evoke this phenomenon because one hypothalamus is able to produce diffuse and eventually bilateral reactions on stimulation. As, on the contrary, unilateral lesions fail to cause any appreciable change in the alternation of wake and sleep, we are inclined to ascribe the insomnia to the exclusion of a certain function rather than to irritation of the waking centre.

### Action of the sleep centre

The question which was discussed in a previous section with regard to the action of the waking centre also applies to the sleep centre: in what way does it act? For the time being we shall confine this problem to the cerebral manifestations of sleep.

The occurrence of somnolence after exclusion of the waking centre is open to two different interpretations. In the first place, it would seem possible that it results from an unopposed direct action of the sleep centre—the only part of the regulating apparatus left intact—on the cortex. As pointed out in the introduction, this theory was advanced by von Economo. Another possibility is an inhibitory action of the sleep centre on the waking centre, and in this case sleep would result from inactivity of the latter, a concept which fits into the theory of Ranson and his co-workers. If sleep is really brought about by an inhibitory action of the sleep centre on the cortex, the exclusion of this centre would tend to abolish or decrease the somnolence which follows on exclusion of the waking centre. If, on the other hand, this somnolence is essentially the result of the abolition of a certain stimulatory action exerted on the cortex by the waking centre, it would not be affected by exclusion of the sleep centre.

In the preceding sections evidence was presented that exclusion of the waking centre can be obtained by transection of the mammillary part of the hypothalamus, whereas the sleep centre seems to be put out of action by a similar lesion in the suprachiasmatic region. The above-mentioned problem can thus be put in a more practical way as follows: is the somnolence which follows on transection of the mammillary region affected by similar lesion in the suprachiasmatic region?

The experiment which can answer this question would seem to be a simple one. Unfortunately, however, all operations in which a suprachiasmatic lesion was inflicted to a rat previously made somnolent by operation ended in the animal's death during or shortly after the operation. It will be remembered that transections of the mammillary region are not, as a rule, survived for more than 4–8 days, and consequently the lapse of time between the two operations is of necessity too short to allow of sufficient recovery, which explains the unfavourable outcome of this experiment. We were therefore compelled to simplify our method, which could only be attained by performing both operations in one session. With this procedure transverse lesions were inflicted to the mammillary and to the suprachiasmatic region simultaneously. It is clear that the outcome of this experiment depended on the degree in which the somnolence resulted from active inhibition of the cortex by the sleep centre. If the additional suprachiasmatic lesion would fail to produce any differences between these rats and those in which only mammillary lesions were present, it would be plausible to exclude any direct action of the sleep centre on the cortex and other major parts of the nervous system. In this case the sleep centre might be supposed to exert its depressing action on the cortex only via inhibition of the waking centre.

The results of some of our experiments will now be described. In rat 138 the operation resulted in a typical condition of sleep. After waking the animal, which was easily done by pinching its tail, it yawned and made stretching movements, and was soon asleep again. In all respects its behaviour resembled that of the animals shown in Table 1. This condition remained unaltered for seven days, after which the animal unexpectedly died. Autopsy revealed a complete transection of the hypothalamus on a level with the middle of the optic chiasma, and a similar lesion at the rostral border of the mammillary body (Fig. 10). It would appear from this experiment that a transection of the suprachiasmatic region, which in itself is able to cause sleeplessness, fails to interfere with the somnolence resulting from lesion of the mammillary region. Similar observations were made in rat 151, in which identical lesions were found.

*Rat 136.* This animal did not develop a complete condition of sleep. Instead it exhibited a constant drowsiness, from which it could easily be roused and which persisted up to its death four days later. In this rat also transverse lesions were found, one in the suprachiasmatic and one in the mammillary region (Fig. 11). The latter lesion, however, was incomplete, leaving a small part of the right lateral hypothalamic area intact, and, as was mentioned in the discussion of the somnolent animals, incomplete transections of the mammillary region may cause drowsiness. The suprachiasmatic lesion had apparently failed to prevent this decrease of the waking capacity.

*Rat 140.* During the first two postoperative (days this animal displayed a heavy somnolence which in the following period of ten days gradually declined. On the thirteenth day after the operation the animal was completely awake. There were, however, no normal periods of sleep, so that a condition of sleeplessness developed which was comparable to that of the animals of Table 4, and which resulted in death two days later. Two complete transverse lesions of the hypothalamus were found on autopsy. One was situated in the suprachiasmatic region, the other lay about 1 mm. in front of the mammillary bodies (Fig. 12). This case is instructive in various respects. It has already been mentioned that a partial return of the capacity of waking was observed in most of the lethargic animals, but that this return was invariably interrupted by the animal's death. In rat 140, however, the period of survival was longer than usual and, for the first time in this experimental series, permitted us to observe a complete return of the capacity of maintaining the waking state. This observation seems to indicate that the function of waking may completely recover from the disorders which follow on lesions of the infundibular region. We do not see any essential difference between this case and those of the previously described group of somnolent rats, and we are disinclined to ascribe the complete return of its capacity of waking to factors other than its longer survival. In our opinion this case offers a confirmation of the conclusion suggested by our findings in rats 138, 151 and 136, *viz.*, that suprachiasmatic lesions are unable to prevent the results of lesions in the mammillary region. It seems that the results of exclusion of the sleep centre depend on the

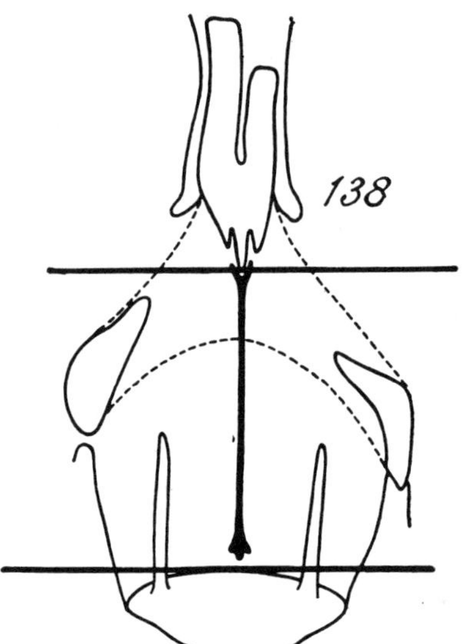


FIG. 10. Hypothalamic lesions in rat 138.

condition of the waking centre, and that it causes sleeplessness only if the latter is intact. If the waking centre is simultaneously put out of action, no sleeplessness results; instead a condition of sleep develops. These facts seem to exclude any direct action of the sleep centre on the cortex and other major nervous structures. It is therefore highly probable that the sleep centre may cause sleep only by inhibition of the waking centre.

It is striking that the survival period of animals in which both transections had been made was longer than that of rats in which only a suprachiasmatic lesion was present. In the first case sleeplessness either does not develop at all, or (rat 140) does not immediately develop; in the second group it does. Apparently the vital importance of sleeping is such that even

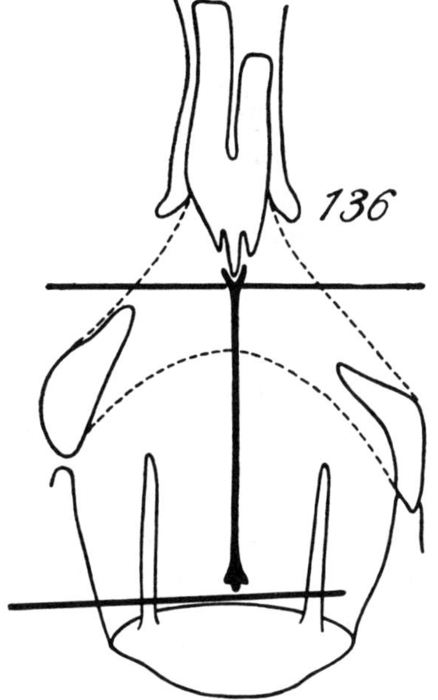

FIG. 11. Hypothalamic lesions in rat 136.

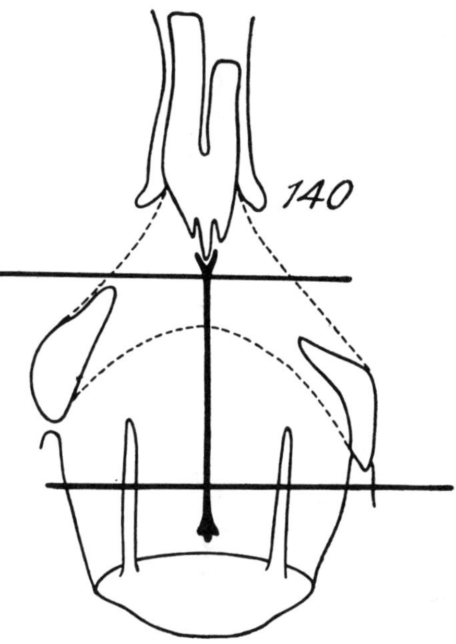

FIG. 12. Hypothalamic lesions in rat 140.

the gross suppression of sleeplessness by means of a second transection of the hypothalamus is able to prolong life.

A large number of fibres which are known to connect the suprachiasmatic with the mammillary region might account for the transmission of impulses from the sleep centre to the waking centre. It would seem from our experiments that these impulses are chiefly conducted through the lateral hypothalamic area, *i.e.*, through the medial forebrain bundle. In one of the preceding sections the possibility was discussed that the impulses from the waking centre are carried in the opposite direction by the same bundle. It thus seems possible that the medial forebrain bundle is serving a double purpose in the regulation of sleep. Our concept is illustrated by Figure 13 which takes into consideration the facts that (i) a transverse lesion on the level of arrow 3 (the suprachiasmatic region) causes sleeplessness; (ii) that a similar lesion in the infundibular region (arrow 2) results in drowsiness combined with fewer periods of normal sleep; while (iii) a transection of the mammillary region (arrow 1) is followed by a complete loss of the capacity of maintaining the waking state, which masks the simultaneous exclusion of the sleep centre.

If Hess was right when claiming that a stimulation of the diencephalon with weak galvanic currents may cause sleep, it is possible that he observed the result of direct or indirect activation of the sleep centre. It should, however, again be stressed that it is not yet clear how Hess' findings should be interpreted.

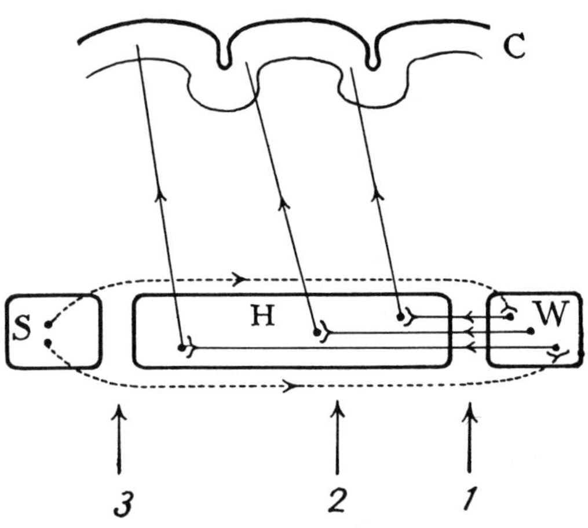



Fig. 13. Diagrammatic representation of the functional relationship between sleep centre and waking centre. Compare with Fig. 7. Fibres carrying "sleep stimuli" have been indicated by interrupted lines. C., cortex cerebri; H., lateral hypothalamic area; S., sleep centre; W., waking centre.

Lastly, von Economo's observation that an initial sleeplessness of patients suffering from epidemic encephalitis usually tends to decline and is gradually replaced by somnolence can be explained on the lines of our concept. Probably the sleeplessness results from inflammation of the sleep centre, and the transition to somnolence seems to indicate a spreading of the process to more caudal hypothalamic parts, finally involving the area of the waking centre. Here, as well as in our experiments, simultaneous exclusion of both centres would result in somnolence. As far as we know, a reverse development of the disease has never been recorded, and this seems to offer a support for our conception that it is the condition of the waking centre which, as far as the regulation of sleep is concerned, determines the manifestations of inflammatory processes of the hypothalamus.

Our point of view with regard to the difference in opinion between the schools of von Economo and Ranson may be summarized as follows. We agree with von Economo that a certain structure in the rostral part of the hypothalamus can be supposed to be of specific importance for the capacity of sleeping. We do not, however, as von Economo did, believe that this structure may cause sleep by active inhibition of the cortex and other parts of the brain. In this respect we fully agree with the American school that sleep results from functional exclusion of the waking centre. Whereas Ranson and his collaborators hold that periods of sleep are caused by more or less intrinsic periodic decreases in activity of the waking centre, we are inclined to attribute these decreases to the inhibitory influence of a sleep centre.

### Hypothalamic localisation of ortho- and parasympathetic centres and their relation to sleep regulation

The localisation of the centres involved in the regulation of sleep suggests a certain antagonism between rostral and caudal part of the hypothalamus. It is interesting to note that certain analogies with the hypothalamic representation of other functions seem to exist. For this purpose we shall turn our attention to the part which is played by the hypothalamus in the regulation of the autonomic balance.

Nearly all investigations of this problem date from the last forty years. After Winkler's observation (47) that stimulation of the hypothalamus of cats gives rise to a production of sweat on the animal's footpads, Karplus and Kreidl (29) started an extensive investigation of the relation between "Gehirn und Sympathicus," which led them to the assumption of an ortho-sympathetic centre in the lateral hypothalamic area, close to the medial side of the subthalamic nucleus of Luys. In later years this conception received support from a number of investigators of whom Beattie *et al.* (2), Ectors (13), Crouch and Elliott (10), Morison and Rioch (33) and Ranson, Kabat and Magoun (39) should be mentioned. From the studies of Ranson and his co-workers it would appear that the orthosympathetic area is chiefly confined to the infundibular and mammillary regions. Stimulation of these areas generally results in a rise of blood pressure, an increase in rate and depth of respiration, and pupillary dilatation. The same symptoms appear on stimulation of points in the midbrain tegmentum, which suggests an extension of the sympathetic area into the mesencephalon.

Concerning the existence of a parasympathetic hypothalamic centre divergent opinions exist. Positive findings have been recorded by Beattie (2) and Beattie and Sheehan (4), who observed a number of parasympathetic phenomena on stimulation of the rostral part of the hypothalamus, *viz.*, augmentation of intestinal peristalsis, gastric secretion, decrease of heart rate and of blood pressure, and pupillary constriction. Like Ranson *et al.*, they were able to produce sympathetic symptoms by stimulating the caudal part of the hypothalamus. On the basis of these findings they divide the hypothalamus into a rostral parasympathetic and a caudal sympathetic part.

Later studies offered little confirmation of their concept. Ectors, Brookens and Gerard (14), and also Crouch and Elliott (10), were unable to evoke other than sympathetic phenomena from both parts of the hypothalamus. Morison and Rioch (33) observed a few parasympathetic phenomena only on stimulation of the rostral pole of the amygdaloid nucleus, the basal part of the septum and the lamina terminalis, *viz.*, a relaxation of the nictitating membrane and a decrease of blood pressure. They were led to the conclusion that these parts of the brain contain a mechanism for inhibition of sympathetic activity. In their admirable investigation of the problem, Ranson, Kabat and Magoun (38) observed some parasympathetic phenomena on stimulation of an area which, although it belongs to the surroundings of the third ventricle, is not regarded as a part of the hypothalamus by Ranson, *i.e.*, the preoptic region. Stimulation of this area resulted in contraction of the bladder and in decrease of blood pressure and of rate and depth of respiration, but was never followed by miosis and intestinal peristalsis, as claimed by Beattie. The depressor reactions could also be elicited by stimulation of several parts of the medial hemispheric wall such as the septum and the genual gyrus. By their results Ranson *et al.* are led to the conclusion that the caudal hypothalamic region contains a complete sympathetic centre, but although some parasympathetic functions seem to be localised in the preoptic region and in adjacent parts of the medial wall of the hemisphere, they are disinclined to regard this area as a complete parasympathetic centre. They further accept the existence of a separate centre in the septal region for contraction of the bladder.

The appearance of parasympathetic phenomena on stimulation of the relevant hypothalamic area is open to two different interpretations. In the first place, one could imagine a conduction of its stimuli to lower parasympathetic centres and from there to the end-organs concerned. Although this mechanism may be involved, it is doubtful whether it offers sufficient explanation for the results of anterior hypothalamic stimulation. There would be little doubt if the parasympathetic picture were more or less complete, but, as pointed out above, intestinal peristalsis, for instance, is lacking. Apart from the contraction of the bladder which seems to have a separate representation in the endbrain, the main results of stimulation of the parasympathetic hypothalamic area are depressor in nature, and this suggests an inhibitory action on the sympathetic centre in the caudal hypothalamic region. It does not seem improbable that the parasympathetic area of the forebrain is essentially a regulator of sympathetic activity. It will be remembered that a similar view is held by Morison and Rioch.

When comparing this concept with that displayed in previous sections, a conspicuous analogy between the regulations of sleep and of autonomic balance is met with. In both functions the rostral part of the hypothalamus seems to exert a depressing action on the caudal part. By this striking congruity in hypothalamic representation one is led to ask what relation exists between the cyclic changes of activity and the autonomic balance. It has already been mentioned that the waking centre and the sympathetic hypothalamic centre are topographically related and perhaps even identical. Moreover, it has long been recognized that the transition from the waking to the sleeping state corresponds to a shift of the autonomic balance from the sympathetic to the parasympathetic side, while the process of awakening is accompanied by a counterchange of autonomic activity. Hess (22) even claims that sleep results from a preponderance of the parasympathetic system, and that the waking state is established by an increase of sympathetic activity. The enhancing influence of adrenalin and related drugs (benzedrine, etc.) on consciousness is in tune with his conception that the waking state is dominated by the sympathetic system, but although it is certain that the autonomic balance lies on the parasympathetic side during sleep, there is little reason to regard sleep as a result of parasympathetic preponderance, because those drugs which are able to imitate parasympathetic activity (acetylcholine and its relatives) fail to produce sleep. Ranson and Magoun (39) are therefore inclined to regard the preponderance of the parasympathetic system during sleep as a result of a decrease of sympathetic activity. This conception is in accordance with our observation that sleep results from inhibition of the caudal hypothalamic area, which contains the highest sympathetic centre.

By way of summary it seems possible to accept a certain topographical and functional congruity between the centres involved in the regulation of sleep and those subserving the regulation of the autonomic balance. Probably both functions are manifestations of one diencephalic mechanism. So far we have dealt only with cerebral symptoms of the cyclic changes of activity. All phenomena belonging to the somatic sphere have been omitted in order to avoid a premature complication of our display, all efforts to explain the somatic manifestations of sleep meeting with considerable difficulties.

It is highly probable that many functions of lower parts of the central nervous system are actively inhibited during sleep. According to Magnus

(17, 18), for instance, the maintenance of the recumbent position during sleep can be explained only by accepting an inhibition of the red nucleus. An important indication of active inhibition of lower centres was recorded by Tarchanow (cited from 35), who demonstrated that the normal inhibition of spinal reflexes during sleep was missing in the distal part of the spinal cord after complete transection in the thoracic region. Apparently the normal suppression of the spinal cord during sleep had been abolished for the segments below the level of the lesion. One is led to ask in which part of the central nervous system this inhibiting mechanism is localised.

According to Gamper (17, 18), no intrinsic changes in intensity are discernible in the decerebrate rigidity which appears after transection below the red nucleus. This would mean that the cyclic changes in muscle tone, which are characteristic of the sleep-waking rhythm, are abolished by decerebration. Gamper arrives at the following conclusion: "Wir können also in den Abschnitten des Zentralorgans, die distal vom Gebiete des roten Kerns liegen, keine den phasischen Ablauf der Lebensvorgänge regulierende Zentralstelle vermuten."

In a child in which the endbrain had not been developed, Gamper observed irregular periods of sleep. On autopsy the hypothalamus turned out to be poorly developed,

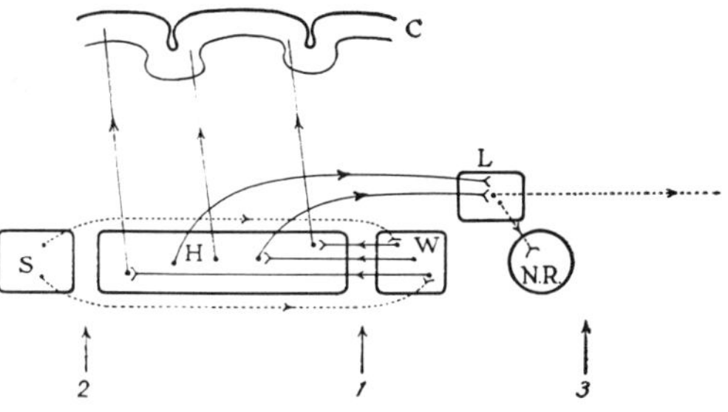

FIG. 14. Diagrammatic illustration of the mechanism of sleep, as proposed in text. C., cortex cerebri; H., lateral hypothalamic area; L., "centre for body sleep"; N.R., nucleus ruber; S., sleep centre; W., waking centre. Fibres conducting "waking" stimuli are indicated by solid lines; broken lines represent fibres involved in the production of sleep.

but the transitional zone between fore- and mid-brain and the whole neural tube caudal to this region were fairly normal. From these facts Gamper concludes that the nervous apparatus for the regulation of sleep must be localised in the rostral part of the midbrain or in the hypothalamus.

It would seem reasonable to suppose that the inhibition of the lower parts of the brain during sleep is a function of the sleep centre in the rostral part of the hypothalamus. Many difficulties, however, arise on trying to answer the question as to along what ways its impulses descend. There is reason to believe that the influence of the sleep centre is not mediated to lower centres by one uninterrupted pathway, because certain lesions in a lower part of the brainstem, *viz.*, the mammillary region, instead of preventing all somatic manifestations of sleep, result in a typical condition of sleep. Clinical observations indicate that the regulating mechanisms of somatic and cerebral sleep are to a certain extent separate. It is, for instance, a well known fact that soldiers on a long march may finally show unmistakable symptoms of cerebral sleep while walking perfectly, which suggests a certain mutual independence of the centres for somatic and cerebral sleep. Cataplexy is perhaps an opposite condition. Normally, however, both manifestations of sleep make a simultaneous appearance, which indicates the existence of a superior correlative mechanism. So long as exact data are lacking, no statements can be made concerning the pattern on which the complete phenomenon of sleep is effected. A mechanism which at present seems conceivable is diagrammatically illustrated by Figure 14. We have accepted the existence of a separate centre for somatic sleep (L), situated caudal to the waking centre and inhibited by it via the lateral hypothalamic area during the waking condition. Under normal conditions it would be able to exert its depressing influence only on lower centres (exemplified in our diagram by the red nucleus) after being freed from the action of the waking centre. As discussed in a previous section, this liberation may result from activity of the sleep centre S.

Our diagram offers an explanation of the facts: (i) that a transection slightly caudal to the sleep centre (arrow 2) causes sleeplessness; (ii) that a similar lesion on the level of arrow 1 (the mammillary region) gives rise to a complete condition of sleep; (iii) that transections below the level of the red nucleus (decerebration arrow 3) are followed by a condition in which all somatic symptoms of sleep are lacking.

No experimental data on dissociation of cerebral and somatic sleep can be found in literature. If our conception of a spatial separation of the sleep centre L from other parts of the regulating apparatus is correct, it would not seem impossible that certain lesions are able to produce isolated disturbances in the regulation of either cerebral or somatic sleep.

## DISCUSSION

In the foregoing account the expressions "waking centre" and "sleep centre" have been repeatedly used. In order to avoid misunderstanding it should be emphasized that the existence is uncertain of smaller or larger groups of specific cells subserving the regulation of sleep only. The function of each of the separate hypothalamic cell groups—with the exception of the supraoptic nucleus, of which the essential role in the regulation of the salt and water balance has been demonstrated by Fisher *et al.* and others—is still unknown. So long as more exact data are lacking, we are able to state only that some structure in the caudal half of the hypothalamus and in the adjacent part of the tegmentum mesencephali is of specific importance for the maintenance of the waking state, while the preoptic region is likely to contain a structure subserving the function of sleeping. For the sake of brevity these structures may be referred to as "waking centre" and "sleep centre" respectively, but these terms should not give the impression that anything more than what has been stated above can be said concerning them. For instance, the possibility that the regulation of sleep is only one of multiple functions of one single nervous apparatus cannot be excluded since there seems to exist a topographical identity between the hypothalamic regions involved in the regulation of sleep and those regulating the autonomic balance. The current conception of "centres" has been criticised by Bethe (5), who, by the extreme plasticity of the nervous system of invertebrates and of the more centralized nervous system of higher animals, was led to reject the thought that nervous functions are carried out by specific centres. Nevertheless, one is forced to the conclusion that the nervous structures normally involved in the regulation of sleep occupy a small area, only lesions of a circumscript location being able to interfere with this regulation. Therefore, if the above restrictions are taken into consideration, the term "centre" would not seem to be inappropriate for indicating these structures.

In the experiments of Ranson *et al.*, as well as in ours, "plasticity" was observed in the function of maintaining the waking state. It is still uncertain whether such recoveries from injuries of nervous tissue must be ascribed to compensatory activity of secondary centres already involved in the function under consideration—a conception which is well illustrated by Ch. Foix's term "automatisme étagé" (45)— or to the formation of a new regulating apparatus in a part of the brain which has afferent and efferent connections similar to those of the destroyed area but is situated outside the shortest route in normal animals, or to some other compensating mechanism. It is a striking fact that no return of the capacity of sleeping could be observed in any of our sleepless rats, not even in rat 140 who survived for 13 days.

Cyclic changes in activity were observed in decorticated dogs by Goltz (20) and Rothmann (41), and by Gamper (17, 18) in an acerebral child. Although it is well known that cortical processes have an important influence on the sleep-waking rhythm, we are justified in denying the existence of a dominating autonomic cortical mechanism for the regulation of sleep. It should be stressed that in animals in which a maximal decrease of the waking capacity had been obtained a normal waking condition could still be evoked by sufficiently strong external stimuli. Only the capacity of *maintaining* this condition had apparently been lost. The concepts of sleep being caused by abolition of sensory stimuli seem to apply to these cases only. It is probable that the waking centre endows the cerebral cortex with a power of maintaining a certain functional "tone" in the absence of external stimuli, and that only sensory stimulation of sufficient strength is able to prevent the onset of sleep when the waking centre is reduced to inactivity.

It scarcely needs to be emphasized that our knowledge concerning the course of the impulses from the waking centre to the cerebral cortex is incomplete. If our concept of an ascending flow of these impulses through the medial forebrain bundle is correct, what connections may account for their further transmission is still obscure. It seems improbable that the thalamus is involved in this transmission to any important extent, which, in conjunction with von Economo's observations on the sensibility of his encephalitic patients, renders it highly probable that the waking centre does not essentially act via sensibility. Of what its influence consists is a question open to further investigation.

Our conception concerning the existence of a sleep centre is based solely on the observation that no periods of sleep could be observed after complete transections in the suprachiasmatic region of the hypothalamus. Kabat, Anson, Magoun and Ranson (28), however, do not mention sleep among the results of stimulation of this region in waking cats, and Ranson *et al.* did not observe sleeplessness after its partial destruction (39). The disagreement between our findings and those of Ranson *et al.* may lie in the difference in techniques. It seems necessary to repeat our experiments in other animals in order to ascertain whether sleeplessness is a general result of anterior hypothalamic transection.


## SUMMARY

1. In the rat complete bilateral transection of the hypothalamus, irrespective of location, interferes with the normal regulation of sleep.

2. The location of lesions which cause disturbances of the function of waking indicates the existence of a structure in the caudal hypothalamic region and in the adjacent part of the midbrain tegmentum, which is of specific importance for the capacity of maintaining the waking state during the absence of external stimuli ("waking centre").

3. There is reason to accept a structure in the preoptic region, which is of specific importance for the capacity of sleeping ("sleep centre").

4. Evidence is offered that sleep is caused by an inhibitory action of the sleep centre on the waking centre.

5. The lateral hypothalamic area seems to be of more importance for the regulation of sleep and waking than the inner areas. It seems probable that the medial forebrain bundle, which occupies this area, is implicated in the transmission of impulses determining the sleep-waking rhythm.

6. The hypothalamic centres involved in the regulation of sleep are topographically identical with those determining the autonomic balance.



## REFERENCES

1. BARD, P. A diencephalic mechanism for the expression of rage with special reference to the sympathetic nervous system. *Amer. J. Physiol.*, 1928, *84*: 490–515.
2. BEATTIE, J. Hypothalamic mechanisms. *Canad. med. Ass. J.*, 1932, *26*: 400–405.
3. BEATTIE, J., BROW, G. R., and LONG, C. H. N. Physiological and anatomical evidence for the existence of nerve tracts connecting the hypothalamus with spinal sympathetic centres. *Proc. roy. Scc.*, 1930, *B106*: 253–275.
4. BEATTIE, J. and SHEEHAN, D. The effects of hypothalamic stimulation on gastric motility. *J. Physiol.*, 1934, *81*: 218–227.
5. BETHE, A. Plastizität und Zentrenlehre. In: *Handbuch der normalen und pathologischen Physiologie*, A. Bethe *et al.* ed., 1931, *15* (2): 1175–1220.
6. BODIAN, D. Studies on the diencephalon of the Virginia opossum. Part I. The nuclear pattern in the adult. *J. comp. Neurol.*, 1939, *71*: 259–323.
7. BODIAN, D. Studies on the diencephalon of the Virginia opossum. Part II. The fiber connections in normal and experimental material. *J. comp. Neurol.*, 1940, *72*: 207–297.
8. CANNON, W. B., NEWTON, H. F., BRIGHT, E. M., MENKIN, V., and MOORE, R. M. Some aspects of the physiology of animals surviving complete exclusion of sympathetic nervous impulses. *Amer. J. Physiol.*, 1929, *89*: 84–107.
9. CLARK, W. E. LE GROS, and BOGGON, R. H. On connections of the anterior nucleus of the thalamus. *J. Anat., Lond.*, 1933, *67*: 215–226.

10. CROUCH, R. L. and ELLIOTT, W. H., JR.   The hypothalamus as a sympathetic centre. *Amer. J. Physiol.*, 1936, *115*: 245–248.
11. ECONOMO, C. VON.   *Die Encephalitis lethargica.* Wien, Deuticke, 1918.
12. ECONOMO, C. VON.   Schlaftheorie. *Ergebn. Physiol.*, 1929, *28*: 312–339.
13. ECTORS, L.   Stimulation of the hypothalamus in chronic hemidecorticated monkeys. *Amer. J. Physiol.*, 1937, *119*: 301–302.
14. ECTORS, L., BROOKENS, N. L. and GERARD, R. W.   Autonomic and motor localization in the hypothalamus. *Arch. Neurol. Psychiat.*, Chicago, 1938, *39*: 789–798.
15. FORTUYN, Ae. B. DROOGLEEVER.   Die Ontogenie der Kerne des Zwischenhirns beim Kaninchen. *Arch. Anat. Physiol.*, Lpz., 1912, 303–352.
16. FULTON, J. F., and INGRAHAM, F. D.   Emotional disturbances following experimental lesions of the base of the brain. (prechiasmal). *J. Physiol.*, 1929, *67*: xxvii–xxviii.
17. GAMPER, E.   Bau und Leistungen eines menschlichen Mittelhirnwesens (Arhinencephalie mit Encephalocele). Zugleich ein Beitrag zur Teratologie und Faserssytematik. *Z. ges. Neurol. Psychiat.*, 1926, *102*: 154–235.
18. GAMPER, E.   *Idem.* II. Klinischer Teil. *Z. ges. Neurol. Psychiat.*, 1926, *104*: 49–120.
19. GLEES, P.   The contribution of the medial fillet and strio-hypothalamic fibres to the dorsal supra-optic decussation. *J. Anat.*, Lond., 1944, *78*: 113–117.
20. GOLTZ, F.   Der Hund ohne Groszhirn. *Pflüg. Arch. ges. Physiol.*, 1892, *51*: 570–614.
21. HARRISON, F.   An attempt to produce sleep by diencephalic stimulation. *J. Neurophysiol.*, 1940, *3*: 156–165.
22. HESS, W. R.   Ueber die Wechselbeziehungen zwischen psychischen und vegetativen Funktionen. *Schweiz. Arch. Neurol. Psychiat.*, 1925, *16*: 285–306.
23. HESS, W. R.   Le sommeil. *C. R. Soc. Biol.*, Paris, 1931, *107*: 1333–1360.
24. HESS, W. R.   Hypothalamus und die Zentren des autonomen Nervensystems: Physiologie. *Arch. Psychiat. Nervenkr.*, 1936, *104*: 548–557.
25. HESS, W. R.   Hypothalamische Adynamie. *Acta Helv. Physiol. Pharmacol.*, 1944, *2*: 137–147.
26. HESS, W. R.   Von dem höheren Zentren des vegetativen Funktionsystems. *Bull. schweiz. Akad. med. Wissensch.*, 1945, *1*: 138–166.
27. INGRAM, W. R., BARRIS, R. W., and RANSON, S. W.   Catalepsy: an experimental study. *Arch. Neurol. Psychiat.*, Chicago, 1936, *35*: 1175–1197.
28. KABAT, H., ANSON, B. J., MAGOUN, H. W., and RANSON, S. W.   Stimulation of the hypothalamus with special reference to its effect on gastrointestinal motility. *Amer. J. Physiol.*, 1935, *112*: 214–226.
29. KARPLUS, J. P. and KREIDL, A.   Gehirn und Sympathicus. II. Ein Sympathicuszentrum im Zwischenhirn. *Pflüg. Arch. ges. Physiol.*, 1910, *135*: 401–416.
30. KLEITMAN, N. and CAMILLE, N.   Studies on the physiology of sleep. VI. The behaviour of decorticated dogs. *Amer. J. Physiol.*, 1932, *100*: 474–480.
31. MAGOUN, H. W., RANSON, S. W., and HETHERINGTON, A.   Descending connections from the hypothalamus. *Arch. Neurol. Psychiat.*, Chicago, 1938, *39*: 1127–1149.
32. MAUTHNER, L.   Zur Pathologie und Physiologie des Schlafes nebst Bemerkungen über die "Nona." *Wien. med. Wschr.*, 1890, *40*: 961, 1001, 1049, 1092, 1144, 1185.
33. MORISON, R. S. and RIOCH, D. McK.   The influence of the forebrain on an autonomic reflex. *Amer. J. Physiol.*, 1937, *120*: 257–276.
34. PAPEZ, J. W.   The mamillary peduncle Marchi method. *Anat. Rec.*, 1923, *25*: 146.
35. RADEMAKER, G. G. J.   Physiologie van den hersenstam. In: *Nederlandsch Leerboek der Physiologie*, 1944, *5*: 234–275.
36. RANSON, S. W.   Somnolence caused by hypothalamic lesions in the monkey. *Arch. Neurol. Psychiat.*, Chicago, 1939, *41*: 1–23.
37. RANSON, S. W. and INGRAM, W. R.   Catalepsy caused by lesions between the mammillary bodies and third nerve in the cat. *Amer. J. Physiol.*, 1932, *101*: 690–696.
38. RANSON, S. W., KABAT, H., and MAGOUN, H. W.   Autonomic responses to electrical stimulation of hypothalamus, preoptic region and septum. *Arch. Neurol. Psychiat.*, Chicago, 1935, *33*: 467–477.
39. RANSON, S. W. and MAGOUN, H. W.   The hypothalamus. *Ergebn. Physiol.*, 1939, *41*: 56–163.
40. ROMEIS, B.   *Taschenbuch der mikroskopischen Technik.* Munich and Berlin, R. Oldenbourg, 1928.

41. ROTHMANN, H.   Zusammenfassender Bericht über den Rothmannschen grosshirn-losen Hund nach klinischer und anatomischer Untersuchung. *Z. ges. Neurol. Psychiat.*, 1923, *87*: 247–313.
42. SALMON, A.   Le sommeil est-il déterminé par l'excitation d'un centre hypnique ou par la dépression fonctionelle d'un centre de la veille? *Rev. Neurol.*, 1932, *1*: 714–720.
43. SPIEGEL, E. A. and INABA, C.   Zur zentralen Lokalisation von Störungen des Wach-zustandes. *Klin. Wschr.*, 1926, *2*: 2408.
44. SZYMANSKI, J. S.   Die Verteilung der Ruhe- und Aktivitätsperioden bei weissen Ratten und Tanzmäusen. *Pflüg. Arch. ges. Physiol.*, 1918, *171*: 324–347.
45. TINEL, J.   *Le système nerveux végétatif.* Paris, Masson et Cie, 1937.
46. TRÖMNER, E.   *Das Problem des Schlafes.* Wiesbaden, Bergmann, 1912.
47. WINKLER, F.   Die zerebrale Beeinflussung der Schweisssekretion. *Pflüg. Arch. ges. Physiol.*, 1908, *125*: 584–594.

Reprinted from The Journal of Comparative and Physiological Psychology
Vol. 46, No. 5, October, 1953
*Printed in U.S.A.*

# SUBCORTICAL MECHANISMS IN EMOTIONAL BEHAVIOR: AFFECTIVE CHANGES FOLLOWING SEPTAL FOREBRAIN LESIONS IN THE ALBINO RAT

JOSEPH V. BRADY AND WALLE J. H. NAUTA[1]

*Army Medical Service Graduate School*

During the past two decades numerous investigators have been concerned with the role of subcortical mechanisms in emotional behavior. Durable changes in affective behavior have been described for animal preparations with transection of the brain stem at the intercollicular level (Sherringtonian decerebration), neocortical ablation, lesions of the hypothalamus, lesions of the amygdaloid complex, and lesions in various parts of the thalamus (4, 11). Surprisingly little attention has been paid, however, to the behavioral effects of lesions in the septum.

The septal region of the forebrain presents a complicated anatomical picture. As shown by Figure 1, its caudal border is formed by the massive fornix column, which contains a large proportion of the fibers originating in the hippocampus, and continues behind the anterior commissure to the mammillary body. Other fibers of hippocampal origin cut through the substance of the septum, subsequently curve down in front of the anterior commissure to distribute, among others, to more anterior parts of the hypothalamus ("precommissural fornix"). In addition to such fibers of passage, the septal region contains numerous nerve cells, grouped into several nuclei. Most of the septal cell groups receive massive efferent connections from the hippocampus, while other fibers appear to reach the septal region from the hypothalamus. The region has extensive efferent connections with the thalamus and hypothalamus. A large number of such fibers leave the region with the fornix column and distribute, among others, to the anterior nuclei of the thalamus and to the habenula. Another contingent of septal efferents joins the precommissural fornix and ends largely in the preoptic area and in the hypothalamus.

By virtue of its paramedian position, the septal region can be destroyed experimentally with only minor involvement of other brain structures. Such experimental lesions must result in interference with most of the hippocampal discharge, and hence with an important contingent of paleocortical activity. Results of lesions in the septal region have been reported by Spiegel, Miller, and Oppenheimer (6) for the cat, and by Krieg (3) for the rat. In both animal species, behavioral changes in the direction of an increased tendency to rage reactions were observed. The present report describes a series of experiments aimed at a more systematic analysis of the behavioral changes exhibited by albino rats with surgical lesions in the septal forebrain region. Specifically, the effects of septal lesions upon the acquisition and retention of a conditioned emotional response, upon the magnitude of the startle response, and upon general emotional reactivity in albino rats have been investigated.

## METHOD

### General Procedure

Thirty male albino rats, approximately 100 days of age at the start of the experiments, served as Ss. The animals were divided into an unoperated control group (6 Ss) and two operated groups, one with lesions in the septal forebrain region ("septal animals," 18 Ss) and one with lesions in the cingulate cortex ("cingulate animals," 6 Ss). In 27 animals, a conditioned emotional response of the "fear" or "anxiety" type was established prior to operation, and retention tests were conducted postoperatively. The remaining 3 animals were operated prior to conditioning in order to evaluate the effects of the lesions upon acquisition of the emotional response. In addition, ratings of emotional reactivity and measurements of startle response magnitude were made in all 30 animals both pre- and postoperatively. Upon completion of the experiments, the animals were sacrificed, their brains fixed, sectioned, and stained in order to verify histologically the location of the lesions.

### Emotional Conditioning

The conditioned emotional response was established in all animals according to a procedure previously described by Brady and Hunt (8). All conditioning trials

---

[1] The authors wish to thank Mr. Irving Geller and Mr. William C. Stebbins for their help in part of this experiment.

took place in the "grill box," a cubical chamber with a grill floor, one transparent wall to permit observation, and an earphone attached to one wall for transmission of a clicking noise. All animals received a series of six conditioning trials distributed throughout a ten-day period so that no more than one trial was given on any one day. Each trial consisted of a 3-min. presentation of a clicking noise, the conditioned stimulus, terminated contiguously by two momentary painful electric shocks to the feet, the unconditioned stimulus, delivered through the grill floor of the apparatus. In this situation, the emotional response develops after a few trials as complete immobility and crouching ("freezing"), usually accompanied by defecation, upon presentation of the conditioned clicker stimulus.

### Rating Scale

In an attempt to evaluate gross changes in emotional reactivity associated with septal injury, a rating-scale technique was employed. Ratings were made of all animals for the following seven behavior components: (a) resistance to capture in the home cage, (b) resistance to handling, (c) muscular tension reaction to capture and handling, (d) squealing and vocalization reaction to capture and handling, (e) urination and/or defecation reaction to capture and handling, (f) aggressive reaction to presentation of forceps in close proximity to the snout, (g) aggressive reaction to prodding with forceps. Previous investigations by Stone (8) and Tryon et al. (9) have demonstrated the efficacy of similar rating-scale techniques in the study of emotional behavior in the rat, and development of the present scale followed a procedure outlined by Tryon et al. (9).

A four-point rating scale was used for each of the seven behavior components. The zero point was fixed by the behavior of tame albino rats such as those conventionally employed in laboratory experiments. The ratings were performed by three independent judges in the relatively standard situation of removing the animal from its home cage and placing it in the emotional conditioning apparatus. Measures of correlation between the judgments of the three raters yielded coefficients of better than .90 for this scale. Ratings of all animals were made on three occasions preoperatively (30, 15, and 1 day before operation) and for 12 consecutive days postoperatively, starting on the first postoperative day.

### Startle Response Measurement

An additional aspect of behavior selected for measurement in the present study was the startle reaction of rats to an explosive auditory stimulus. The method involved the use of a stabilimeter-like apparatus, previously described by Brown, Kalish, and Farber (9), capable of recording the magnitude of a rat's jump to a loud sound. The rat was confined in a small box (2.37 in. wide, 7.37 in. long, 5.5 in. high) surmounted upon the platform of a 5-lb. postage scale. Vertical movements of the platform were transmitted by means of a thread to an ink-writing muscle lever whose deflections were recorded on a 5-in. tape moving at a rate of .53 in. per second. The startle stimulus was produced by a 22-cal. blank pistol fired about 2 ft.

from the confinement box. The exact time at which the startle stimulus occurred was registered on the tape by an electromagnetic signal marker.

Twenty-four hours prior to operation all animals were given a series of five presentations of the startle stimulus, and the heights of all jumps to these stimuli were recorded. Successive trials were administered at intervals ranging from 1 to 2 min., the average interval being about 90 sec. Again, on the day following operation, all animals received a series of five presentations of the startle stimulus according to precisely the same procedure employed preoperatively.

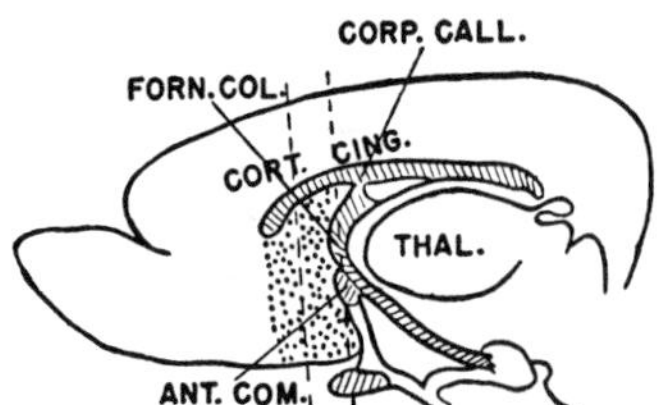


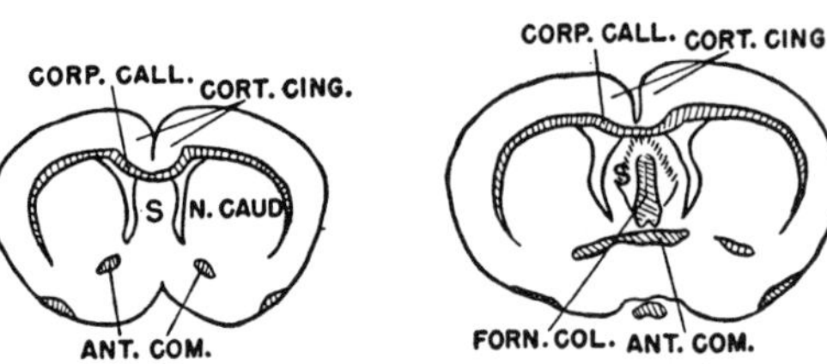


Fig. 1. Semidiagrammatic representation of the rat forebrain. Upper figure: medial aspect with septal region stippled. Lower figures: two frontal sections at levels indicated by broken lines through stippled area of upper figure. ANT. COM., anterior commissure; CORP. CALL., corpus callosum; CORT. CING., cingulate cortex; FORN. COL., fornix columns; N. CAUD., caudate nucleus; THAL., thalamus.

### Operative Procedure

All operations were performed with the animals under ether anesthesia. After incision of the skin, the periosteum of the cranial vault was pushed aside, and a small burr hole was produced in the midline, 4 to 5 mm. behind the frontal pole of the cerebral hemisphere. Through this opening, a slender cutting instrument, essentially a needle with the tip bent at a right angle and sharpened sideways, was sunk between both hemispheres to a measured depth, whereupon it was rotated several times in order to ensure the production of a bilateral paramedian lesion. Upon completion of this procedure, the periosteum was stitched over the trephine hole, and the skin was closed. No bandage was applied. In most cases, the effect of the anesthesia completely disappeared within the first two postoperative hours.

Figure 1 shows a midsagittal and two frontal sections through the areas involved in the operations. On the midsagittal plane, the septum is represented by

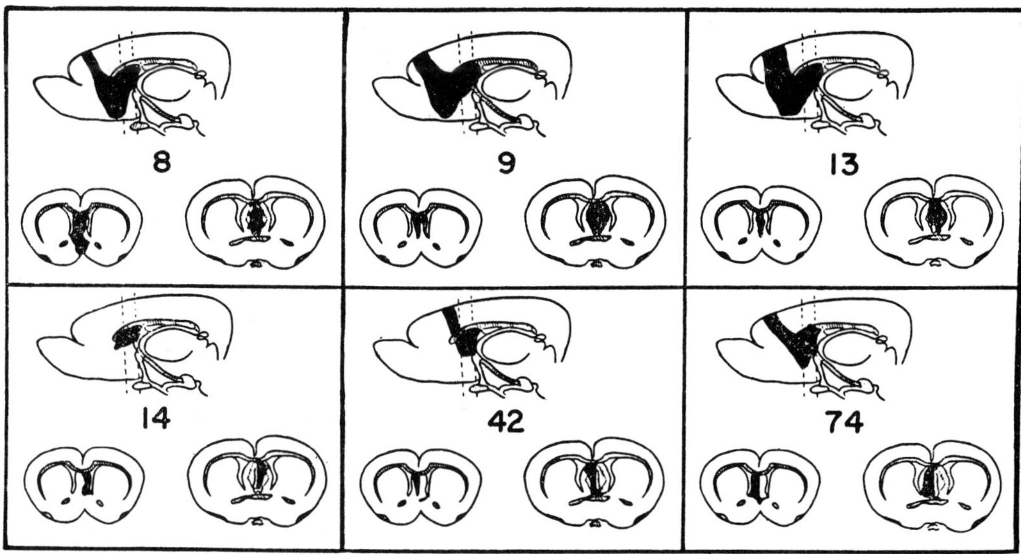

Fig. 2. Medial and frontal reconstructions of septal lesions in experimental Group A. Black indicates location and extent of the lesions in the reconstructions. See Figure 1 for orientation.

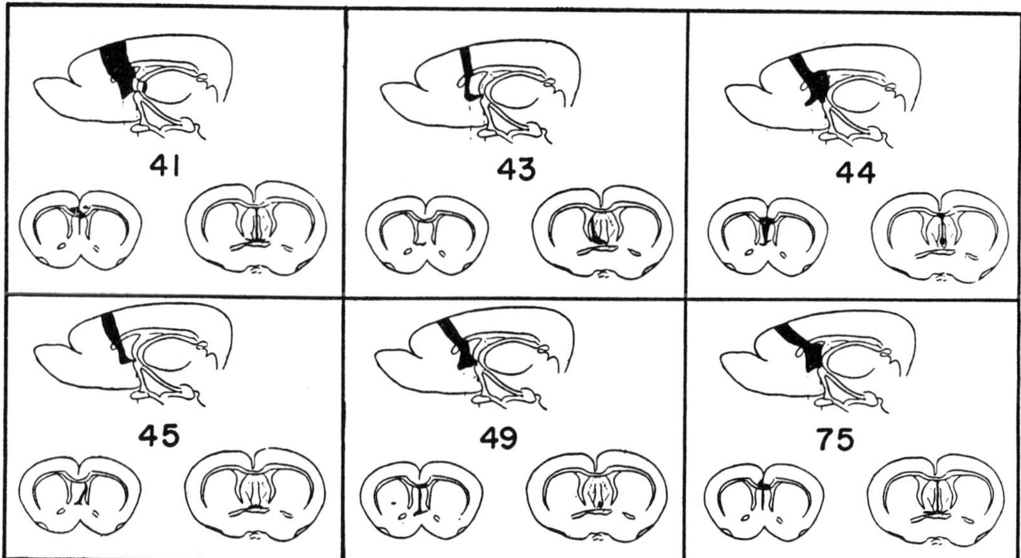

Fig. 3. Medial and frontal reconstructions of septal lesions in experimental Group B. Black indicates location and extent of the lesions in the reconstructions. See Figure 1 for orientation.

the stippled area (the anterior and basal boundaries being somewhat arbitrary), and the two broken lines through this area indicate the level of the two frontal sections depicted below. It is evident from this figure that the instrument used in the surgical procedure, on its way toward the septal region, passed between the opposing surfaces of left and right cerebral hemisphere. In many cases, this area of the cortex (cingulate cortex) sustained more or less extensive damage, while in most cases the corpus callosum was penetrated by the instrument immediately prior to its entrance into the septal region. The six operated control animals in which lesions were produced only in the cingulate cortex and corpus callosum provided a necessary control for such damage as well as for the effects which might be attributable to any operation per se.

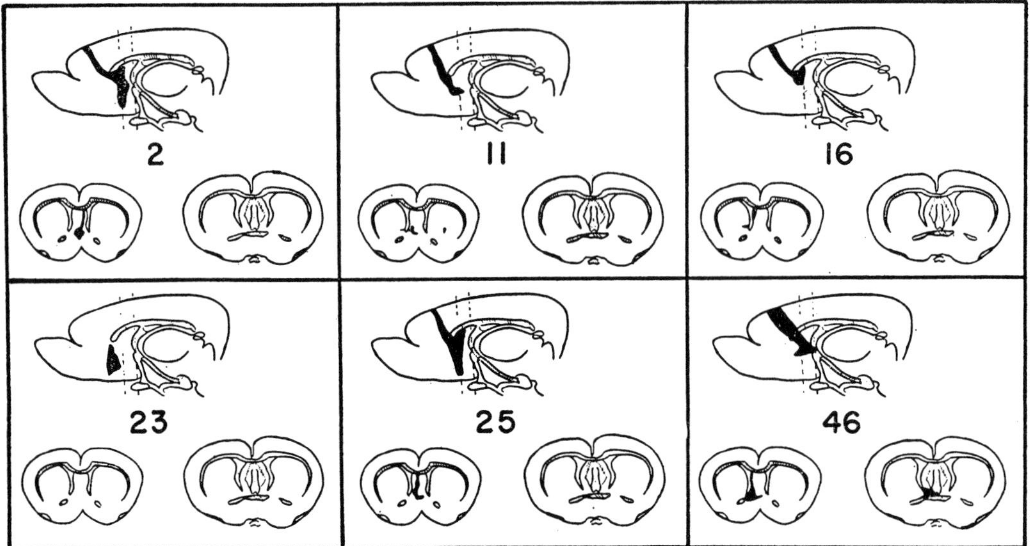

FIG. 4. Medial and frontal reconstructions of septal lesions in experimental Group C. Black indicates location and extent of the lesions in the reconstructions. See Figure 1 for orientation.

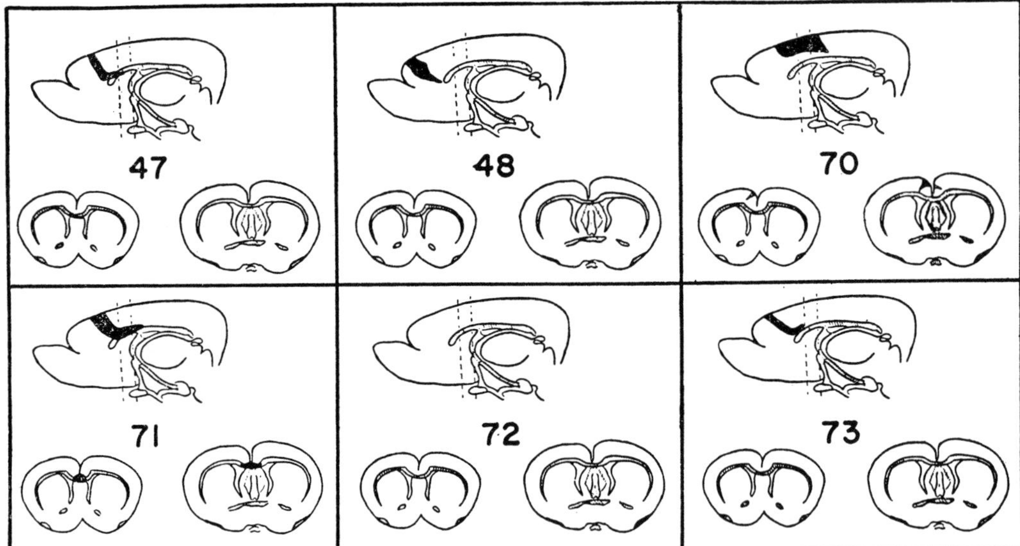

FIG. 5. Medial and frontal reconstructions of cortical lesions in operated control group. Black indicates location and extent of the lesions in the reconstructions. See Figure 1 for orientation.

## Histological Procedure

Histological verification of the lesions was obtained for all 24 operated animals. The brains were fixed in 10 per cent formalin for one to four weeks following removal and were subsequently embedded in paraffin. Serial sections were cut at $20\,\mu$, and, of each brain, two alternating series were prepared, one of which was stained for cells with thionin, and the other with a modified Weil stain for myelinated fibers. Projection drawings were then made of every fifth section of the Weil-stained series. The lesion was drawn in after observations in both the Weil and thionin series. Figures 2, 3, 4, and 5 show two of these drawings at the indicated levels together with a reconstruction of the lesion on the midsagittal plane for each of the 24 operated animals. Figure 1 is intended to serve as a key to the interpretation of these plates. Histological analysis of the brains of two of the six nonoperated control animals was also accomplished and, as would be expected, no lesions were found.

### RESULTS

*General Observations*

Gross and dramatic changes in behavior appeared in the septal animals within 2 hr. after operation. Such animals presented a picture of striking alertness with limbs rigidly extended and eyes intently following the movements of the observer approaching the cage. The sudden presentation of almost any auditory stimulus produced an explosive startle reaction. The typical exploratory response of the normal rat to presentation of such an innocuous object as a pencil was replaced by "freezing" behavior with the vibrissae turned stiffly forward. Rapidly approaching objects were attacked immediately with vigorous biting, while a magnet-like reaction was exhibited when the object was moved slowly before the snout. Attempts to capture or handle the animal were responded to by fierce attacks on $E$ alternating with vigorous flight reactions. In certain instances, merely opening the cage would provoke the animal sufficiently to have it leap from the cage and hop about the room in an odd, rabbit-like fashion. When placed in a group cage, five or six such septal animals would fight continuously and vigorously for extended periods of time with loud and frequent vocalizations accompanying much of the above behavior. None of these phenomena was observed in either the operated or nonoperated control animals at any time during the course of the experiments. A motion-picture record was prepared in conjunction with these experiments to illustrate the character of these changes following septal lesions in the rat.[2]

*Changes in the Conditioned Emotional Response*

By the sixth conditioning trial all animals had acquired the conditioned emotional response. All 30 animals crouched and remained immobile throughout the 3-min. presentation of the clicker, and 28 of the 30 animals defecated in response to the conditioned stimulus. No significant differences in rate of acquisition appeared, however, between the group of 3 animals (rats 2, 14, 25) operated prior to conditioning and the remaining 27 animals.

[2] A review of this film is presently in press in the *Psychological Abstracts*.

On the day following the sixth conditioning trial, septal lesions were produced in 15 more animals (rats 8, 9, 13, 42, 74, 11, 16, 23, 46, 41, 43, 44, 45, 49, 75), lesions of the cingulate cortex were produced in another 6 animals (rats 47, 48, 70, 71, 72, 73), and the remaining 6 animals served as nonoperated controls. Histological reconstruction of the lesions in the two operated groups is shown in Figures 2, 3, 4, and 5. Three days after operation, tests for retention of the conditioned emotional response were conducted with all animals by a 3-min. presentation of the clicker alone. Complete retention of the conditioned emotional response was evident in the 6 operated control animals with cortical lesions only and in the 6 non-operated animals, as well as in the 3 animals operated prior to conditioning. All 15 animals crouched, defecated, and remained completely immobile during the clicker. In contrast, however, the conditioned emotional response appeared significantly diminished in the group of 15 animals in which septal lesions had been produced following conditioning. All 15 animals showed some movement during the 3-min. clicker presentation, and only 3 of the 15 animals defecated in response to the conditioned stimulus. This difference in frequency of defecation between the experimental septal group, on the one hand, and the operated and nonoperated control groups, on the other, is highly significant ($\chi^2 = 17.2$, $p < .01$). The septal lesions appear to have appreciably diminished the strength of the conditioned emotional response on this immediate postoperative test.

*Changes in Emotional Reactivity*

The results of the pre- and postoperative ratings for emotional reactivity in the 18 experimental, 6 operated control, and 6 non-operated control animals are shown in Figure 6. It is apparent that no significant preoperative differences exist between the three groups although the operated control animals show a slightly but consistently higher degree of emotional reactivity than the other two groups. Postoperatively, however, there appears a rather marked change in the emotional reactivity of the experimental septal animals as compared to both their own preoperative level and the level of the two control groups. The difference between each of the three preoperative ratings and each of the first three postoperative ratings for the experimental septal group is significant beyond the .01 level of confidence. Furthermore, the difference between the experimental septal group and the two control groups on each of the first six postoperative ratings is also significant beyond the .01 level of confidence. The figure also shows that there is a gradual decline in the magnitude of the ratings for the experimental

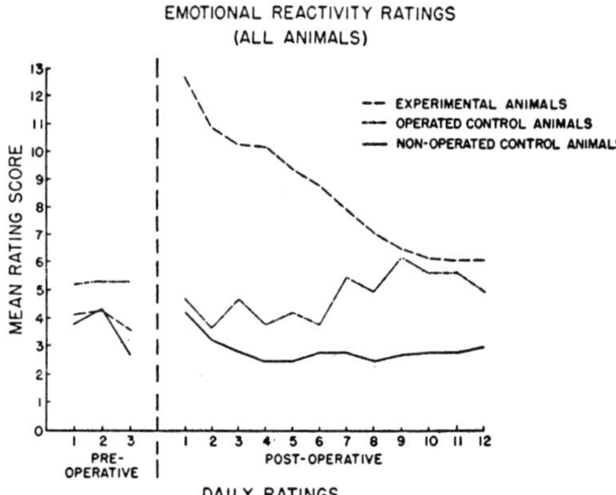


FIG. 6. Changes in emotional reactivity following experimental septal lesions

septal group over the 12-day postoperative rating period, so that by the ninth or tenth day, the curve for this group closely approximates that of the operated controls.

*Changes in Startle Response*

Table 1 shows the results obtained from measurements of the startle response magnitude both pre- and postoperatively in the experimental, operated control, and nonoperated animals. The animals' scores for both the pre- and postoperative tests represent the means of the startle responses elicited on the five trials conducted on each of the two days. Inspection of the tabulated group mean values indicates that there was no significant difference between the three groups on the preoperative test. Postoperatively, however, the experimental septal animals show a marked increase in the magnitude of the startle response, while neither the operated nor the nonoperated control animals show any such gross change in the amplitude of the response. The difference between the pre- and postoperative mean values for the experimental septal group is highly

significant ($t = 3.82$, $p < .01$), as is the difference between the postoperative mean value for the experimental septal group, on the one hand, and the postoperative mean value for operated control group ($t = 3.29$, $p < .01$) and nonoperated control group ($t = 3.76$, $p < .01$), on the other. As may well have been predicted on the basis of the rating scale changes, septal lesions appear to have significantly increased the magnitude of the startle response in the experimental animals.

TABLE 1

Magnitudes of Startle Responses for Experimenta Operated Control, and Nonoperated Control Groups on Pre- and Postoperative Tests

| GROUP | PREOPERATIVE RESPONSE IN MILLIMETERS | | POSTOPERATIVE RESPONSE IN MILLIMETERS | |
| --- | --- | --- | --- | --- |
| | Mean | SD | Mean | SD |
| Experimental | 9.80 | 7.69 | 22.58 | 6.52 |
| Operated control | 10.63 | 8.08 | 9.98 | 7.84 |
| Nonoperated control | 8.25 | 5.98 | 9.32 | 6.87 |

## DISCUSSION

The results of these experiments reflect quite clearly the critical role of certain forebrain mechanisms in the mediation of some forms of emotional behavior. The striking changes in emotional reactivity which appear immediately following lesions of the septal region in the experimental animals do not occur in either the animals with lesions of the cingulate cortex or in the nonoperated animals. The duration of these changes, however, appears to be limited. Although the present data do not permit any precise definition of this time parameter, it can be seen from Figure 6, that considerable diminution in rated emotional reactivity occurs during the first 12 postoperative days. Whether this change can be accounted for on the basis of physiological recovery with time alone, continued daily handling, or both, is presently under investigation.

Within the experimental group, however, differences did appear in both the degree and duration of the changes in emotional reactivity following septal lesions. In an attempt to provide some systematic explanation of this behavioral variation, the 18 septal animals were divided into the three anatomical subgroups of 6 animals each, represented in

Figures 2, 3, and 4. Figure 2 shows that in the animals of Group A, large lesions were found in the supracommissural part of the septal area, involving, among other structures, all or most of the components of the fornix column. Figure 3 shows that in the animals of Group B, the supracommissural part of the septal area was involved to a lesser degree, and considerable parts of the fornix column escaped injury. Figure 4 shows that in the animals of Group C, only minimal damage to parts of the septal area other than those involving the supracommissural septum and the fornix column was sustained.

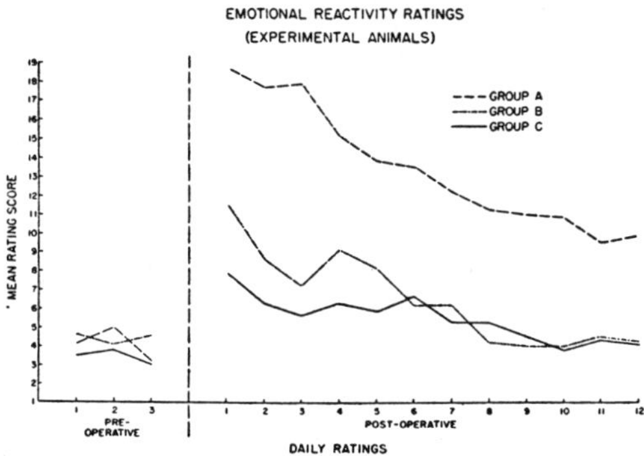


FIG. 7. Changes in emotional reactivity in experimental subgroups A, B, and C

Figure 7 shows the breakdown of the pre- and postoperative ratings for emotional reactivity in experimental Groups A, B, and C. It is apparent that no significant preoperative differences exist between the three groups. Postoperatively, however, the difference between Group A, on the one hand, and Group B and C on the other, is highly significant (.01 level of confidence for the first two postoperative daily ratings). In addition, the difference between Groups B and C on the first four daily postoperative ratings is significant well beyond the .05 level of confidence.

The fact that the intensity of the postoperative effects appears to be roughly commensurate with the extent of injury to the fornix column naturally raises some doubt as to the "septal" nature of the behavioral phenomena observed in this study. The classical concept of the fornix column as consisting of fibers originating in the hippocampus would indeed tend to justify an interpretation of the findings described above as the result of interference with a hippocampal mechanism. As pointed out above, however, the fornix column, at the level where it was severed in the present experiments, receives a large number of nerve fibers arising from the septal gray matter and distributing, among others, to the anterior nuclei of the thalamus (10), to the habenula (7), and to the hypothalamus. The present experiments leave it an open question whether the behavioral effects of septal lesions are due to interruption of such septodiencephalic connections, or to interference with the massive hippocampo-hypothalamic discharge pathway contained in the fornix column, or to both. A possible role of direct hippocampal projections to the brain stem in affective behavior could obviously be demonstrated only by bilateral interruption of the fornix system proximal to the level where it receives its contribution from the septal nuclei. Such experiments, however, will be complicated by substantial and hardly avoidable injury to adjacent brain structures.

The effect of the experimental septal lesions in decreasing the strength of the conditioned emotional response, however, presents an apparent paradox when the marked increase in general emotional reactivity found in the same experimental animals is considered as an effect of such lesions upon the emotional repertoire of the organism considered as a unitary instinctual pattern. At least two possible explanations for this apparent inconsistency present themselves. First, the change in the conditioned emotional response may represent only a nonspecific effect of septal ablation upon previously learned behavior in general, so that the decrease in the strength of the previously conditioned "fear" response may not necessarily be related to the emotional character of the reaction per se. That is, septal lesions might increase general emotional reactivity on the one hand while impairing the retention of any previously learned behavior on the other.

Secondly, the apparent inconsistency between the two "emotional" changes in behavior following septal ablation may be related to a confounding of the two effects in the postoperative test for retention of the conditioned emotional response. That is, postoperative presentation of the conditioned clicker stimulus to the experimental septal animals may have elicited an unconditioned movement reaction specific to their hypersensitive physiological state which tended to occlude somewhat the previously conditioned "freezing" response. Experiments are presently in progress to test these two possibilities, as well as to evaluate the duration of the changes in emotional behavior following septal lesions.

### SUMMARY AND CONCLUSIONS

The present report describes a series of experiments investigating the behavioral changes exhibited by albino rats with surgical lesions in the septal forebrain region.

Thirty male albino rats were divided into an unoperated control group (6 animals), and two operated groups, an "experimental septal" group (18 animals) and an operated control group with lesions of the cingulate cortex (6 animals). Both pre- and postoperatively, data were collected on all animals with respect to (a) the acquisition and retention of a conditioned emotional response (CER) of the "fear" or "anxiety" type, (b) the magnitude of the startle response, and (c) general emotional reactivity as reflected in a seven-item rating scale.

Upon completion of the behavioral observations, the animals were sacrificed and the lesions reconstructed histologically.

1. Significant increases in both emotional reactivity and startle response magnitude were found in the "experimental septal" animals postoperatively. No significant change could be observed in either the operated or nonoperated control animals. In addition, the septal lesions were found to have appreciably diminished the strength of the previously conditioned "fear" response although such lesions appeared to have no effect upon the acquisition of this response.

2. The magnitude of the changes in the behavior of the experimental septal animals appeared to be roughly commensurate with the extent of injury to the fornix column, suggesting the possible importance of this complex paleocortical fiber bundle in the transmission of neural impulses involved in determining affective behavior.

### REFERENCES


1. BRADY, J. V., & HUNT, H. F. A further demonstration of the effects of electro-convulsive shock on a conditioned emotional response. *J. comp. physiol. Psychol.*, 1951, **44**, 204–209.
2. BROWN, J. S., KALISH, H. I., & FARBER, I. E. Conditioned fear as revealed by magnitude of startle response to an auditory stimulus. *J. exp. Psychol.*, 1951, **41**, 317–328.
3. KRIEG, W. J. S. Personal communication to W. E. Le Gros Clark, in *The hypothalamus*. London: Oliver & Boyd, 1938.
4. LINDSLEY, D. B. Emotion. In S. S. Stevens (Ed.), *Handbook of experimental psychology*. New York: Wiley, 1951. Pp. 473–516.
5. SCHREINER, L. H., RIOCH, D. McK., MASSERMAN, J., & PECHTEL, C. Behavioral changes following thalamic injury in the cat. *J. Neurophysiol*, in press.
6. SPIEGEL, E. A., MILLER, H. R., & OPPENHEIMER, M. J. Forebrain and rage reactions. *J. Neurophysiol.*, 1930, **3**, 538–548.
7. SPRAGUE, J. M., & MEYER, M. An experimental study of the fornix in the rabbit. *J. Anat., Lond.*, 1950, **84**, 354–368.
8. STONE, C. P. Wildness and savageness in rats of different strains. In K. Lashley (Ed.), *Studies in the dynamics of behavior*. Chicago: Univer. of Chicago Press, 1932. Pp. 3–55.
9. TRYON, R. C., TRYON, C. M., & KUZNETS, G. Studies in individual differences in maze ability. X. Ratings and other measures of initial emotional responses of rats to novel inanimate objects. *J. comp. Psychol.*, 1941, **32**, 417–473.
10. VOGT, O. Sur un faisceau septo-thalamique. *C. R. Soc. Biol.*, 1898, Tome L, 206–207.
11. WHEATLEY, M. D. The hypothalamus and affective behavior in cats: A study of the effects of experimental lesions, with anatomic correlations. *Arch. Neurol. Psychiat., Chicago*, 1944, **52**, 296–316.


*Received December 19, 1952.*

In: Limbic System Mechanisms and Autonomic Function
Charles H. Hockman (Ed)., Springfield, Thomas, 1972

*Chapter 2*

# The Central Visceromotor System: A General Survey

WALLE J. H. NAUTA

UP TO the present time, the existence of a visceromotor system central to the preganglionic motor neurons has been very largely a physiological concept that appears to have originated shortly after the middle of the nineteenth century with Flourens' notion of a respiratory "noeud vital" in the medulla oblongata, Schiff's identification of a medullary "vasoconstrictor center" and Claude Bernard's medullary "pîqure glycosurique." Only half a century later was the hypothalamus identified as a major visceromotor territory, largely by the fundamental studies of Karplus and Kreidl.[9] The central visceromotor system thus became visualized, as it still is today, as consisting essentially of the hypothalamus and the neural links connecting the latter with the preganglionic motor neurons. In the intervening sixty years, a great store of anatomical and physiological information with respect to the hypothalamus has accumulated, but the anatomical picture of the mesencephalic, bulbar and spinal intermediaries of the central visceromotor system has remained so rudimentary that little more than a few general statements can be made about these levels between the hypothalamus and the preganglionic motor neurons.

The account to follow is intended to serve as a general overview of the anatomical organization of the central visceromotor system. It must be emphasized that it is based largely upon observations made in the course of studies conducted by conventional neuroanatomical methods, in particular Golgi and fiber-degeneration techniques. Explorations by fundamentally different methods, histochemical techniques in particular, may eventually lead to modifications of the present concepts regarding the organization of the central visceromotor system.

## DESCENDING NEURAL PATHWAYS OF THE CENTRAL VISCEROMOTOR SYSTEM

Clinical experience as well as physiological observations [13] indicate that the functional connections linking the hypothalamus with the preganglionic motor neurons are spread widely over the cross section of the midbrain and rhombencephalon. The results of experimental studies by axon-degeneration methods suggest that these conduction lines are interrupted by synaptic junctions at several points along their descent. Contrary to earlier descriptions,[3] such experimental studies have failed to demonstrate direct hypothalamic efferents extending caudalward beyond the mesencephalic tegmentum and circumaqueductal grey substance.[8,17]

### Descending Connections of the Hypothalamus

The neural conduction routes descending from the hypothalamus have been found to follow two major trajectories. Foremost among these connections is a descending component of the medial forebrain bundle which originates chiefly in the lateral hypothalamic zone and can be subdivided into a medial division distributed largely to paramedian mesencephalic cell groups, and a lateral component that traces a wide path through the adjoining, more lateral region of the mesencephalic tegmentum approximately corresponding to the zone indicated by Taber[28] as the subcuneiform area.[17] A second conduction route leading caudalward from the hypothalamus is provided by descending constituents of the dorsal longitudinal fasciculus of Schütz which appear to originate in more medial hypothalamic cell groups and are distributed over the ventral half of the central grey substance of the midbrain. Both the dorsal longitudinal fasciculus of Schütz and the medial subdivision of the medial forebrain bundle connect the hypothalamus with a paramedian region of the midbrain composed of the central grey substance and several tegmental cell groups including the large and complex raphe nucleus (nucleus centralis tegmenti superior) of Bechterew and two circumscript round cell groups, the nuclei tegmenti dorsalis and ventralis of Gudden near the pontomesencephalic border. The efferent connection of the hypothalamus with this paramedian mesencephalic territory (the "limbic midbrain area")[17] is reciprocated by at least two circumscript fiber systems: the ascending component of Schütz's bundle and the mammillary peduncle. The "limbic midbrain area" thus can be considered to represent, in part at least, the caudal pole of a hypothalamomesencephalic circuit. It is very likely, however, that the descending components of this circuit are also involved in the transmission of hypothalamic discharge to the preganglionic motor neurons. Morest[15] has traced fiber degeneration elicited by lesions placed caudally in the central grey substance caudalward through the periventricular grey matter of the fourth ventricle as far as the medulla oblongata; the ultimate synaptic distribution of this periventricular fiber system remains to be determined. A further possible pathway for the transmission of hypothalamic impulses to the motoneuronal level would seem to be provided by the central tegmental radiation of Weisschedel, a diffuse and radially disposed fiber system linking the central grey substance with a large cross-sectional area of the mesencephalic tegmentum.

The foregoing account can be summarized by the statement that the currently available data suggest the existence of two major pathways descending from the hypothalamus, one of which directly involves a wide region of the midbrain tegmentum, whereas the other follows a trajectory confined to the central grey substance. There almost certainly exist numerous cross-links between the two conduction channels, both in the form of long dendrites extending in both directions across the border between the tegmentum and central grey substance and in the form of axonal connections between these two major subdivisions of the mesencephalon. Neither of the two descending hypothalamic pathways appears to extend beyond the caudal border of the midbrain, and it therefore seems likely from the current anatomical data that the impulse transmission from the hypothalamus to the motoneuronal level largely, if not indeed entirely, depends on conduction systems originating in the mesencephalon.

### Descending Connections from the Mesencephalon and Rhombencephalon

It is fair to say that beyond the descending hypothalamic efferents the trail of the central visceromotor pathways becomes very largely lost. With the exception of the periventricular fibers traced to the medulla oblongata by Morest—a conduction pathway which likely represents only a relatively small component of the visceromotor system—no single anatomical structure in the midbrain or rhombencephalon can be identified with any certainty as a link in the chain of conduction toward the preganglionic motor neurons. It seems certain that the pathways in question follow a wide course through the brain stem reticular formation, and in view of the latter's known characteristics,[25] it appears likely that the downward conduction involves numerous synaptic junctions. This notion is borne out by observations made in experimental studies of fiber systems descending to the spinal cord from the brain stem. According to our observations, and in confirmation of previous reports by Kuypers *et al.*[12] and Nyberg-Hansen,[20] the massive fiber degeneration elicited in the reticular fiber systems by lesions in the mesencephalic tegmentum rapidly diminishes in volume as it descends; in the cat and monkey even the longest of the mesencephalic reticular fibers appear not to extend caudally beyond the medulla oblongata. Lesions in the pontine and medullary reticular formation, by contrast, lead to reticulospinal fiber degeneration that extends throughout the length of the spinal cord and is the more massive the farther caudal the lesion is located. Considered together, these observations sug-

gest that the preganglionic visceral motor neurons of the spinal cord may be accessible to the neural efflux of the hypothalamus only (or nearly so) by way of *successive* synaptic articulations at mesencephalic and rhombencephalic levels.

Findings such as those mentioned above, although suggesting a highly complex neural organization, are still quite general and leave some elementary questions unanswered. It is, for example, uncertain whether and to what extent even neurons of the rhombencephalic tegmentum have direct synaptic connections with the preganglionic motor neurons of the spinal cord. By analogy with the somatic motor system, it would seem possible and even likely that rhombencephalic neurons affect the final common path neurons of the visceral nervous system largely by way of spinal interneurons. The existence of such "visceral interneurons" (that is, propriospinal neurons having efferent connections with visceral motor neurons) in the spinal cord is all the more likely as it appears from recent evidence[26] (Petras, personal communication) that no dorsal root fibers terminate within the clusters of preganglionic visceral motor neurons, an observation suggesting that spinal autonomic reflex arcs may have no monosynaptic components. The results of a recent series of experiments in the cat (Merrill and Nauta, unpublished) in which the vagus nerve was cut central to the nodose ganglion suggest the same conclusion with respect to the parasympathetic vagus neurons: in all cases examined, the Fink-Heimer stain failed to reveal any fiber degeneration within the borders of the dorsal motor nucleus of the vagus. It must be stressed that such negative findings do not preclude the possibility that some at least of the primary sensory fibers synapse with dendrites extending beyond the cytoarchitectonic limits of the motoneuronal cell groups. Extremely laborious histological procedures would be required to identify such axodendritic relationships.[5] At the least, however, the evidence here mentioned suggests a marked difference between visceral and somatic motoneuronal cell groups in the sense that in the mammalian spinal cord at least, the latter are indeed invaded (and known to be synaptically contacted) by primary sensory fibers.

## SEQUENTIAL "LEVELS" IN THE CENTRAL VISCEROMOTOR SYSTEM

All of the sparse and largely circumstantial evidence reviewed above is compatible with a morphological notion of the visceral nervous system as organized essentially in the form of serially connected levels. In the terms of this formulation, the "ground floor" of the central visceromotor organization could be thought to consist of a local motor apparatus composed of preganglionic motor neurons and their associated interneurons. Descending upon this most distal level of the central visceromotor apparatus is a complex longitudinal conduction system, the synaptic fractionation of which allows three serially connected rostrocaudal subdivisions to be distinguished: hypothalamic, mesencephalic and rhombencephalic. Each of these large "blocks" in the central visceromotor system must be assumed to be a highly complex neural organization in its own right, encompassing numerous specific functional subsystems. It seems probable, moreover, that each is supplied by its own characteristic afferent input systems, besides receiving information from functional realms common to all levels. It is, for example, likely from Morest's [16] findings that the viscerosensory information received by the nucleus of the solitary tract via vagal and glossopharyngeal afferents is distributed primarily to the rhombencephalic level. Fibers of the so-called spinothalamic tract probably are distributed to both rhombencephalic and mesencephalic levels, and all levels may be affected by the neural efflux from one or another region of the neocortex.

## ASCENDING CONNECTIONS IN THE CENTRAL VISCEROMOTOR SYSTEM

The more or less local nature of some of these afferent relationships naturally does not imply confinement of the information in question to the level or levels upon which it impinges, for there is ample evidence that the various levels are interconnected by ascending fiber systems no less closely than by descending "visceromotor pathways." The anatomical evidence concerning such ascending connections is not entirely equivocal. The results of studies by fiber-degeneration methods [19] suggest that—much like the descending pathway —ascending impulse conduction (very largely at least) sequentially involves rhombencephalic, mesencephalic and hypothalamic levels. In particular, such experimental studies have failed to demonstrate ascending fiber systems that reach the hypothalamus from levels below the mesencephalon. Those that thus far have been demonstrated by fiber degeneration methods all appear to originate in the paramedian region of the mesencephalon as components of the mammillary peduncle and the dorsal longitudinal fasciculus of Schütz. Very recently, however, evidence of a quite substantial fiber system ascending to the hypothalamus through more lateral zones of the mesencephalic tegmentum has been encountered in this laboratory (Graybiel, unpublished); the origin of this connection has not yet been determined.

In conflict with the notion that the origin of ascending afferents to the hypothalamus is confined to the mesencephalon, one Golgi study at least [25] has revealed a rhombencephalic neuron having an ascending axon that extends with some of its ramifications into the hypothalamus. One possible explanation of this discrepancy would be that such long ascending afferents of the hypothalamus are relatively few and widely spaced and thus easily escape detection in anterograde-degeneration studies. Alternatively, the neurons in question could have chemical properties that render their axons resistant to

demonstration by any of the current degeneration techniques. Observations in studies by histofluorescence methods,[2] for example, have suggested the presence of norepinephrin-containing neurons in the rhombencephalon that project directly to the hypothalamus and even beyond it to the cerebral cortex.

Once again, it must be emphasized that most of the central visceromotor circuitry will have to remain a matter of inference as long as the neural components intercalated between the hypothalamus and the preganglionic motor neurons remain as largely unidentified as they are at present. The existence of functionally specific systems in the organization would seem a virtual certainty, yet the anatomical identification of the corresponding neural continua so far has been hampered severely by the circumstance that they all form part of the brain stem reticular formation.

## THE RETICULAR FORMATION

In brief, the reticular formation can be described as a massive but only vaguely delimited neural apparatus that extends throughout the length of the spinal cord and brain stem. It is composed of neurons having (a) long, rectilinear and sparsely arborizing dendrites subtending large and widely overlapping dendritic fields (a characteristic that has led to the term "isodendritic core" [24] as an alternative designation of the reticular formation); (b) heterogeneous afferent relationships; and (c) axons disposed so as to favor the formation of predominantly polysynaptic lines of conduction in both the ascending and descending direction. Neural tissue of this description is usually interpreted as "nonspecific," but it must be emphasized that this term can be applied with certainty only to the afferent characteristics of the

formation. Naturally, the degree of functional specificity of any neural mechanism is determined not only by its input-properties but also by its intrinsic organization and its efferent relationships. The somatic motor neurons, for example, must be assumed to have extremely diverse afferent connections ranging from dorsal root to corticospinal fibers yet can hardly be considered nonspecific from a more general point of view. It would therefore seem profitable to attempt a functional characterization of the reticular formation in terms of its effector characteristics rather than its afferent relationships, even though the latter, perhaps more than any other peculiarity, identify the reticular formation as the most widely "open" system of the brain (that is, open to the greatest diversity of afferent channels) and thus distinguish it from other, more nearly unimodal ("closed line") neural mechanisms.

It is obviously impossible to characterize in a simple phrase the effector properties of a neural apparatus as vast and complex as the isodendritic core of the brain stem. Nevertheless, a major general function of the system appears to express itself in phenomena of adaptive stability or "posture." In the somatic sphere this property is manifested as somatic posture or *stability in space*. The analogous visceral expression of this general function is characterized by Cannon's term "homeostasis," or stability of the internal milieu. Finally, at the level of the cerebrum, the reticular formation appears to affect neural mechanisms involved in behavioral responsiveness (*alertness*), a function that could likewise be viewed as a form of stability. Needless to say, the term, "stability," as used in this account does not necessarily imply rigid maintenance of a particular unalterable functional set. The familiar diurnal, seasonal and other periodic fluctuations in all of the functional realms here mentioned are in all likelihood attributable to variations, largely of an intrinsic (metabolic?) nature, in the functional bias of the reticular formation. In a sense, the reticular formation thereby assumes the significance of an internal adjustment system of the brain itself.

It is important to note that the hypothalamus in its structural organization exhibits all of the essential features attributed to the reticular formation.[14] In fact, from an anatomical point of view, the hypothalamus appears to be a rostral extension of the midbrain reticular formation, in particular of the central grey substance and the ventromedial region of the mesencephalic tegmentum. In turn, this diencephalic extension of the mesencephalic reticular formation continues forward to form the preoptic and septal regions (rostral to this level the mammalian brain at least is composed of cortical formations which are structurally so unique as to require separate classification). The hypothalamus thus appears as a major component of a continuum of reticular (isodendritic) neural tissue that extends the brain stem reticular formation forward beyond the midbrain as far as the septal region.

Similarity of general organizational tissue properties does not imply functional or anatomical sameness throughout. The hypothalamus may be continuous, and reciprocally connected, with the rhombencephalic reticular formation by way of mesencephalic links, but it is nonetheless true that it, and not the rhombencephalon, receives direct projections from a variety of limbic forebrain structures, whereas the rhombencephalic reticular formation but not the hypothalamus is converged upon by a variety of afferents that include direct projections from the cerebellum, the sensorimotor cortex, the superior colliculus, the spinal cord, the vestibular nuclei and the nucleus of the solitary tract. The efferent associations of the hypothalamus, moreover, include the important hypothalamohypophyseal connection, a relationship not shared by any other part of the brain.

However, it is perhaps the most distinctive peculiarity of the hypothalamus that it appears to be the only major expanse of the reticular core of the brain stem that can be identified almost categorically with the visceral and endocrine effector systems. Even though the hypothalamus has been shown to have some of its effects in the realm of somatic motor function,[11] it appears to lie outside the path of the great efferent fiber systems of the sensorimotor cortex, the corpus striatum and cerebellum. Below the hypothalamic level, this high degree of territorial privacy in the visceromotor system rapidly dwindles. At the level of the rhombencephalon in particular, nearly all subdivisions of the reticular formation are distribution zones of one or several fiber systems generally regarded as somatic in character. The medial, gigantocellular zone of the reticular formation of the medulla oblongata, for example, is such a distribution area for fibers from the cerebellar nucleus tecti, and the same region receives a substantial number of fibers from the "premotor" area 6 of the cerebral cortex. In the lateral, smaller-celled zone of the medullary reticular formation many fibers from the nucleus ruber and cortical area 4 terminate. It seems certain that there is, at this level at least, a considerable regional overlap between the somatic and visceral motor systems. Whether this overlap is really only regional or, conversely, extends to single neurons is a matter of conjecture. Observations such as those reported at this conference by Snider, however, suggest

that cerebellar impulse efflux has access to one or other link in the central visceromotor pathways, and a fairly extensive literature, reviewed by Kennard,[10] indicates that the same is true of the projections from the premotor cortex. As it appears that neither the cerebellum nor the premotor cortex has direct efferent connections with either the spinal cord or hypothalamus, it seems likely—though by no means certain—that such somatovisceral linkages are located at the level of either the rhombencephalon or mesencephalon, or both. In any event, the nature of these physiological observations indicates that the reticular formation, despite the heterogeneous afferentation of its constituent neurons, may subserve functions of marked qualitative and topographic specificity. In at least some of the functional aspects of the reticular formation the characteristic of afferent heterogeneity evidently serves purposes of convergent integration rather than diffuseness, an integration that appears to extend across the border between somatic and visceral effector categories.

## NEURAL ASSOCIATIONS OF THE HYPOTHALAMUS

The fiber connections of the hypothalamus have recently been reviewed in detail elsewhere [18] and only a brief summary of the known associations of the hypothalamus with other parts of the brain will here be given. The most fundamental conclusions that have been suggested by the results of fiber-degeneration studies are (a) that the hypothalamus has its most massive extrinsic connections with the tegmentum and central grey substance of the midbrain and with limbic structures of the cerebral hemisphere and (b) that with the certain exception of efferents from the olfactory cortex, and the possible exception of some fibers from the retina, no specific sensory

pathways have direct access to the hypothalamic mechanisms.

### Hypothalamic Connections with the Mesencephalon

The relationships of the hypothalamus with the mesencephalon were already mentioned earlier in this account. Direct hypothalamic efferents have been found distributed over a large region of the midbrain tegmentum as well as over the ventral half of the central grey substance. Reciprocating projections to the hypothalamus from the central grey substance and from the medial tegmental nuclei of Bechterew and Gudden appear to close a hypothalamomesencephalic circuit. This mesencephalohypothalamic projection prominently involves the lateral hypothalamic region—largely by way of fibers of the mammillary peduncle that bypass the mammillary body, ascend in the medial forebrain bundle and in part extend forward as far as the medial zone of the septum—but also (by ascending components of the dorsal longitudinal fasciculus of Schütz) the medial and periventricular hypothalamic zones.

The circuitous hypothalamomesencephalic connection here outlined is doubtless an important conductor of impulses to and from the hypothalamus. Its probable role in the conduction of descending discharge from the hypothalamus was discussed earlier in this chapter. It is no less likely involved in the transmission of neural impulses ascending to the hypothalamus from rhombencephalic and spinal levels. Numerous fibers of the spinothalamic tract, for example, terminate in the circumaqueductal central grey substance along with fibers ascending from the reticular formation of the medulla oblongata. Morest's [16] findings suggest a similar distribution for some fibers ascending from the nucleus of the solitary tract. As even the most painstaking studies by the aid of fiber-degeneration techniques have failed to disclose direct projections from either the spinal cord or the medulla oblongata (but see earlier discussion of this matter), it seems likely that most at least of the visceroceptive and nociceptive information received at rhombencephalic and spinal levels can reach the hypothalamus only by way of mesencephalic way stations. It is logical, in view of the anatomical evidence, to suggest the "limbic midbrain area" (central grey substance and nuclei of Bechterew and Gudden) represents

such an intermediary processing mechanism, but it appears likely that this paramedian midbrain region is not the only link in ascending conductions to the hypothalamus, especially in view of the recent evidence (mentioned earlier) of hypothalamic afferents that follow a more lateral trajectory through the mesencephalic tegmentum. Until further data become available, the present account of ascending conduction routes to the hypothalamus must therefore be considered provisional and incomplete.

### Connections of the Hypothalamus with Limbic Forebrain Structures

Most typical perhaps among the extrinsic associations of the hypothalamus are its strong and in part reciprocal connections with the limbic system. The term, "limbic system," is derived from the name *great limbic lobe* originally used by Broca to designate the marginal region (the "hem") of the pallial mantle that encircles the hilus of the mammalian cerebral hemisphere as a somewhat protuberant ring. This "lobe" harbors a marked variety of neural structures, some of which are paleocortical in organization (the olfactory cortex of the piriform lobe and the cortical fields composing the hippocampal formation), others neocortical (the cingulate or limbic cortex), yet others noncortical (the amygdaloid complex). The use of a collective term to denote such an heterogeneous array of neural organizations has been severely criticized; yet some justification for a categoric denomination may be found in the clinical experience that most or all of the components of Broca's limbic lobe appear to share the characteristic of having a low seizure threshold and in the anatomical evidence that all have close relationships with the septo-preoptico-hypothalamic continuum. The term, "limbic system," has the merit of being non-committal in the sense that it, unlike the earlier designation, *rhinencephalon,* does not allude to a dominant relationship to any particular sensory modality.

The most voluminous afferent connections of the hypothalamus with the limbic system originate from the hippocampal formation, the amygdaloid complex and the olfactory cortex. The hippocampo-hypothalamic connection is established by the fornix system. Approximately half of this massive fiber bundle distributes itself in the septal region which, in turn, originates a septohypothalamic fiber system that forms a large component of the medial forebrain bundle, a major longitudinal hypothalamic conduction system that extends caudally into the mesencephalon, largely by way of synaptically interposed neurons of the lateral hypothalamic region. The remaining half of the fornix system forms the compact columna fornicis which terminates chiefly in the anterior thalamic nuclei and in the mammillary body. From the latter structure arises the massive and compact bundle of Vicq d'Azyr which bifurcates into (a) the mammillothalamic tract distributed to the anterior thalamic nuclei and (b) the smaller mammillotegmental tract to the nuclei of Gudden in the paramedian region of the caudal midbrain. Both by direct fornix connections and by way of the mammillary body and mammillothalamic tract the hippocampus is linked to the anterior thalamic nuclei, a cell complex that in turn projects to the cingulate cortex, a cortical region thought to be connected by long association fibers to the parahippocampal cortices (presubiculum and entorhinal area) which have massive efferent connections with the hippocam-

pus. This complex circuitous sequence of connections unites the hippocampus, mammillary body, thalamus and cingulate cortex into the so-called Papez circuit. This circuit is only one—and certainly a very composite—instance of the reciprocity which exists in the relationships between the limbic system and the hypothalamic region. A less complex "circuit" of this general nature may be closed by fibers ascending from the lateral hypothalamic region to the septum which is known to project in part at least to the hippocampus and thus completes a neural sequence: hippocampus—septal region—lateral hypothalamic nucleus—septal region—hippocampus.

The major direct afferent connection of the hypothalamus with the amygdaloid complex is established by the stria terminalis, a fiber system that, unlike the hippocampal afferents, appears to involve mostly the medial hypothalamic zone and consequently does not (or only minimally) contribute to the medial forebrain bundle. In recent studies by the Fink-Heimer methods three circumscript areas of stria terminalis distribution have been identified—namely, one in the anterior hypothalamic region, a second one in the so-called cell-poor zone surrounding the ventromedial nucleus and a third in the region of the ventral premammillary nucleus. A second major amygdalohypothalamic conduction route may be contained in the so-called ventral amygdalofugal pathways, a conduction system spreading widely in the forebrain and including a component that terminates in the lateral hypothalamus. Although apparently avoiding the medial hypothalamic zones, there is physiological evidence [7] that this ventral conduction route secondarily converges upon the ventromedial hypothalamic nucleus. It must be emphasized that it is not certain that this pathway actually originates in the amygdaloid complex proper and not in the overlying periamygdaloid region of the olfactory cortex.[23] The amygdalohypothalamic relationship, much like the hippocampal connections with the hypothalamus, is to some extent at least reciprocal, for both the stria terminalis and the ventral amygdalofugal pathway contain fibers of the opposite polarity.

All of the structures composing the limbic system are closely associated with the hypothalamus, but the relationship may not invariably be a reciprocal one. The results of a recent experimental study, for example, have raised some doubts as to whether the cingulate cortex has any direct efferent connections with the hypothalamus.[6] Although undoubtedly implicated in hypothalamic circuitry by virtue of its afferent relationship with the anterior thalamic nuclei, the cingulate cortex, according to the results of this study in the rat, projects to subcortical structures not known to be associated with the hypothalamus. This observation suggests the possibility that the primary influence of the cingulate cortex is upon functional realms other than the visceromotor system.

### The Limbic Forebrain-Midbrain Circuit

The evidence summarized in the foregoing account implicates the hypothalamus in at least two complex neural circuits, one of which connects it reciprocally with major components of the limbic lobe, the other with the paramedian region of the mesencephalon. In a wider context, these two sets of associations could be thought to compose a large circuit that reciprocally links the limbic forebrain structures with the paramedian midbrain region. The hypothalamus—its lateral zone in particular—is synaptically intercalated in both the ascending and decending limbs of this larger "limbic forebrain-midbrain circuit."

### Specific Sensory Connections of the Hypothalamus

The only sensorium having anatomically identifiable direct connections with the hypothalamus is the olfactory. The connection is established by fibers that originate in the piriform cortex and olfactory tubercle, join the medial forebrain bundle and thus in all likelihood are distributed mainly in the lateral hypothalamic region.

There are several reports of direct projections to the hypothalamus from the retina. Thus far, however, the evidence for such a direct visual connection has been rather equivocal. The well-known influence of the diurnal light-cycle upon gonadotrophic mechanisms in many vertebrate species appears to require some form of neural communication from the retina to the medial hypothalamic region, yet such impulse conduction could conceivably be mediated by intercalated cell groups such as the medial terminal nucleus of the accessory optic tract (nucleus opticus tegmenti), a circumscript cell group in the ventral tegmental region immediately adjoining the hypothalamus.

### Connections with the Neocortex

Of all subdivisions of the neopallium, only the granular frontal cortex ("prefrontal cortex") is known to project directly to the hypothalamus. In the rhesus monkey, a major part of this substantial connection arises from the caudal half of the orbital surface, but frontohypothalamic fibers have been traced also from the convexity of the frontal lobe. It seems somewhat ironical that the granular frontal cortex on the basis of this projection to the hypothalamus could be considered part of the limbic system, a suggestion accentuated further by the massive direct associations of large areas of the prefrontal cortex with the cingulate cortex and presubiculum.

As the remainder of the neopallium appears to lack direct communication channels to the septohypothalamic continuum, it must be assumed that the neural efflux of the great sensory fields of the neocortex can affect the hypothalamic mechanism only indirectly by way of either (or both) higher-order connections of the neocortex with the limbic system, or neural links connecting the subcortical projections of these cortical fields with the hypothalamus. Recent findings in the monkey [21,22] suggest the prefrontal cortex, the inferior temporal region and the cingulate gyrus as major convergence areas in the sequential processing of sensory informations by the neocortex. All three of these regions have efferent connections with the limbic system: the prefrontal and cingulate fields by way of their close association with the hippocampal mechanism, the inferior temporal region in the form of a substantial projection to the amygdaloid complex. It is tempting in view of this evidence to consider the prefrontal, cingulate and inferior temporal regions as the origin of major links between the neocortex and limbic system, the prefrontal cortex moreover as the site of a neocortical mechanism directly affecting hypothalamic functions.

Elsewhere in this account, the hypothalamus was characterized as the only major subdivision of the brain stem that can be identified almost entirely with visceral (and the closely allied endocrine) effector mechanisms. This statement, however, requires some qualification, for the reason that clinical observations [14] and behavioral findings in animal experiments [27] suggest that the hypothalamus is implicated in the

neural mechanisms governing the organism's behavioral attitudes. Such mechanisms, needless to say, are extremely complex, for they must encompass processes that integrate information from both external and internal environments received over quite literally all peripheral sensory systems and, moreover, by way of nonneural chemical transmissions from the blood plasma itself (for example, blood glucose; target-gland hormones) to more or less selectively receptive central neurons. In mammalian species, this integration must commonly comprise neural codes resulting from elaborate information processing by the cerebral neocortex. It would be naive, therefore, to ascribe the neural substrata of behavioral motivation to any one particular level of the central nervous organization. Nonetheless, the distribution of the sites at which electrical stimulation or tissue destruction have been found to cause marked biases in the organism's behavioral responses to his environments appears to outline, however vaguely, a neural mechanism centrally involved in processes underlying "moods" and, hence, vital drive states. This mechanism extends from the limbic structures of the cerebral hemisphere caudalward to the septo-pre-optico-hypothalamic continuum and medial parts of the thalamus and beyond these into medial regions of the mesencephalon. In its apparent outline, this motivational mechanism largely corresponds to the complex system of neural connections reciprocally linking the limbic structures of the cerebral hemisphere with the paramedian mesencephalic region, a system of associations indicated in the foregoing account by the collective term, "limbic forebrain-midbrain circuit." The hypothalamus, preoptic region and septum are integral components of this mechanism in the sense that part at least of their neuronal population is intercalated in both the ascending and descending limbs of the limbic forebrain-midbrain circuit. The hypothalamus thus appears to be an essential link in the neural substratum of affect and motivation no less than it is an important component of the central visceral and endocrine effector mechanisms. In fact, in the light of this dual functional relationship it is possible to see the hypothalamus, together with the septum and the paramedian region of the mesencephalon, as forming that level of the brain at which the visceral and endocrine effector systems emerge as major efferent channels from a wider neural mechanism underlying the organism's affective involvement in his environments.

## REFERENCES


1. Alpers, B. J.: Personality and emotional disorders associated with hypothalamic lesions. *Res. Publ. Ass. Res. Nerv. Ment. Dis., 20:*725, 1940.

2. Andén, N-E., Dahlström, A., Fuxe, K., Larsson, K., Olson, L., and Ungerstedt, U.: Ascending monoamine neurons to the telencephalon and diencephalon. *Acta. Physiol. Scand., 67:*313, 1966.

3. Beattie, J., Brow, G. R., and Long, C. N. H.: Physiological and anatomical evidence for the existence of nerve tracts connecting the hypothalamus with spinal sympathetic centers. *Proc. Roy. Soc.* [Biol] *106:*253, 1930.

4. Bellone, G. B., and Terzian, H.: Autonomic nervous system and mental pathology. In Von Monakow, K. H. (Ed.): *The Autonomic Nervous System. Akt. Fragen Psychiat. Neurol., 4:*139, 1966.

5. Blackstad, T. W.: Electron microscopy of Golgi preparations for the study of neuronal relations. In Nauta, W. J. H., and Ebbesson, S. O. E. (Eds.): *Contemporary Research Methods in Neuroanatomy.* Berlin, Springer Verlag, 1970, pp. 186–216.

6. Domesick, V. B.: Projections from the cingulate cortex in the rat. *Brain Res., 12:*296, 1969.

7. Dreifuss, J. J., and Murphy, J. T.: Contrasting effects of two identified amygdaloid efferent pathways on single hypothalamic neurons. *J. Neurophysiol., 31:*237, 1968.

8. Guillery, R. W.: Degeneration in the hypothalamic connexions of the albino rat. *J. Anat., 91:*91, 1957.

9. Karplus, J. P., and Kreidl, A.: Gehirn und Sympathicus. II Ein Sympathicus-Zentrum in Zwischenhirn. *Arch. Ges. Physiol., 135:*401, 1910.

10. Kennard, M. A.: Autonomic function. In Bucy, P. C. (Ed.): *The Precentral Motor Cortex.* Urbana, University of Illinois Press, 1944, pp. 294–306.

11. Koella, W. P.: Control of skeletal motor activity, with emphasis on the role of diencephalic mechanisms. In Haymaker, W., Anderson, E., and Nauta, W. J. H. (Eds.): *The Hypothalamus.* Springfield, Charles C Thomas, 1969, pp. 645–658.

12. Kuypers, H. G. J. M., Fleming, W. R., and Farinholt, J. W.: Subcorticospinal projections in the rhesus monkey. *J. Comp. Neurol., 118:*107, 1962.

13. Magoun, H. W., Ranson, S. W., and Hetherington, A.: Descending connections for the hypothalamus. *Arch. Neurol. Psychiat., 39:*1127, 1938.

14. Millhouse, O. E.: A Golgi study of the descending medial forebrain bundle. *Brain Res., 15:*341, 1969.

15. Morest, D. K.: Connexions of the dorsal tegmental nucleus in the rat and rabbit. *J. Anat., 95:*229, 1961.

16. Morest, D. K.: Experimental study of the projections of the nucleus of the tractus solitarius and the area postrema in the cat. *J. Comp. Neurol., 130:*227, 1967.

17. Nauta, W. J. H.: Hippocampal projections and related neural pathways to the midbrain in the cat. *Brain, 81:*319, 1958.

18. Nauta, W. J. H., and Haymaker, W.: Hypothalamic nuclei and fiber connections. In Haymaker, W., Anderson, E., and Nauta, W. J. H. (Eds.): *The Hypothalamus.* Springfield, Charles C Thomas, 1969, pp. 136–209.

19. Nauta, W. J. H., and Kuypers, H. G. J. M.: Some ascending pathways in the brain stem reticular formation of the cat. In

Jasper, H. H., Proctor, L. D., Knighton, R. S., Noshay, W. C., and Costello, R. T. (Eds.): *Reticular Formation of the Brain.* Boston-Toronto, Little Brown & Co., 1958, pp. 3–30.

20. Nyberg-Hansen, R.: Functional organization of descending supraspinal fibre systems to the spinal cord. Anatomical observations and physiological correlations. *Ergebn. Anat. Entwicklungsgesch,* 39:2,3, 1966.

21. Pandya, D. N., Hallet, M., and Mukherjee, S. K.: Intra- and inter-hemispheric connections of the neocortical auditory system in the rhesus monkey. *Brain Res., 14:*49, 1969.

22. Pandya, D. N., and Kuypers, H. G. J. M.: Cortico-cortical connections in the rhesus monkey. *Brain Res., 13:*13, 1969.

23. Powell, T. P. S., Cowan, W. M., and Raisman, G.: The central olfactory connexions. *J. Anat., 99:*791, 1965.

24. Ramón-Moliner, E., and Nauta, W. J. H.: The isodendritic core of the brain stem. *J. Comp. Neurol., 126:*311, 1966.

25. Scheibel, M. E., and Scheibel, A. B.: Structural substrates for integrative patterns in the brain stem reticular core. In Jasper, H. H., Proctor, L. D., Kington, R. S., Moshay, W. C., and Costello, R. T. (Eds.): *Reticular Formation of the Brain.* Boston-Toronto, Little Brown, 1958, pp. 31–55.

26. Shriver, J. E., Stein, B. M., and Carpenter, M. B.: Central projections of spinal dorsal roots in the monkey. *Amer. J. Anat., 123:*27 (two parts), 1968.

27. Valenstein, E. S. (Ed.): Biology of drives. *Neurosciences Research Program Bulletin, 6 (No. 1),* 1968.

28. Taber, E.: The cytoarchitecture of the brainstem of the cat. *J. Comp. Neurol., 116:*27, 1961.


## DISCUSSION

**Dr. Schallek:** Dr. Nauta, you will recall that in 1953, Brady and you described the hyperirritability of rats with lesions in the septal area of the brain. I wonder if you would give us your current views as to what causes the irritability with such forebrain lesions?

**Dr. Nauta:** I have always been afraid of that question. The explanation at the time, much in line with the traditional notion of "higher centers" inhibiting "lower centers," was that the hypothalamic mechanism is released from some restraining telencephalic influence, and I am afraid that I cannot go beyond that tautological statement.

**Dr. Mogenson:** The septal region and the cingulate region have similar effects on somatic motor behavior in animals running down alleys and engaged in other motor behavior. Findings of this sort would very much support the point you are making about the cingulate region not being primarily involved in autonomic functions. But they do have one function in common, that of inhibiting ongoing somatomotor behavior.

**Dr. Nauta:** This idea could, perhaps, be checked by examining the effects of bilateral ablation of the cingulate cortex upon, respectively, operant and autonomic responses to an avoidance situation. It would gain in likelihood if such an ablation were to abolish or impair the operant response (for example, bar pressing to avoid a noxious stimulus) while leaving the autonomic responses to the warning signal unchanged.

**Dr. Mogenson:** Well the medial thalamic region seems to do just this, doesn't it?

**Dr. Nauta:** I would think so—yes.

**Dr. Reis:** I think there are experiments by Garson in which he made lesions in the amygdala in the "flight area" and showed that the animal was unable to respond by flight to a stimulus. It is very interesting in that only this particular aspect of behavior was blocked.

**Dr. Nauta:** Dr. MacLean has reported somewhat similar findings in a study in which he implanted crystals of carbachol in the hippocampus of cats. These animals did show signs of agitation as the warning signal came on, but they seemed confused and unable to release the appropriate avoidance response.

**Dr. MacLean:** To go back to the cingulate again, I thought that you had shown that the so-called dorsal fornix projected to quite a different part of the midbrain from the regular fornix. Cragg has some findings which would go along with this from stimulation of the dorsal fornix and getting blood pressure changes. Do the recent findings of the people in your laboratory agree? Are these observations counter to what you originally observed yourself as far as the dorsal fornix is concerned?

**Dr. Nauta:** We have lately become very uncertain to what extent the cingulate cortex contributes fibers to the fornix dorsalis. In Domesick's study, degeneration in the dorsal fornix appeared to be related to lesions of the indusium griseum rather than retrosplenial granular cortex. A much more detailed study of this matter seems necessary than has been done thus far. Incidentally, it is virtually certain that the fornix longus contains fibers originating in the CA-1 sector of Ammon's horn and likely that it represents a major connection of the fornix system with the paramedian zone of the mesencephalon.

**Dr. French:** The Scheibels identified fibers from the reticular nucleus of the thalamus as a very important feedback mechanism, presumably for exerting inhibitory control over the mechanism for behavioral events. I think that Reis' work is related to this phenomenon physiologically. Do you see any corollary of such a thalamic feedback control mechanism in the hypothalamic-limbic system? It seems to me that you did point out some similarities between the somatically oriented system and perhaps one more viscerally oriented. Could you comment on these similarities?

**Dr. Nauta:** The cingulate cortex projects quite massively to the reticular nucleus of the thalamus but, so it seems, do all or most cortical regions. It is unfortunate that the reticular nucleus is so inaccessible to lesion experiments. The nucleus forms, so to speak, a shell around the thalamus. Most of the thalamocortical projections penetrate it and so do many corticothalamic pathways. Any lesion to the nucleus would affect such traversing fiber systems as much as it would interfere with the caudally oriented connections arising in the reticular nucleus itself. The functional significance of the reticular nucleus will therefore probably prove difficult to identify. But your suggestion, if I have understood it correctly, Dr. French, can be answered affirmatively in the sense that of all the telencephalic structures considered "limbic," only the cingulate cortex appears to project to the reticular nucleus of the thalamus.

**Dr. French:** Then do I understand that the reticular nucleus may show similar effects upon hypothalamic structures through its connections with the cingulate system?

**Dr. Nauta:** The Scheibel's findings would suggest the possibility that the reticular nucleus can modulate neuronal activity patterns in the anterior thalamic nuclei and thus affect the mechanisms of the cingulate and parahippocampal cortices. If not directly, the latter regions could influence hypothalamic structures indirectly via the hippocampus and fornix system.

**Dr. Hockman:** I should like to ask if you know of any link between the red nucleus and the limbic system? In stimulating this nucleus, we have elicited a host of visceromotor responses.

**Dr. Nauta:** No, I don't know of any link

between the limbic circuits and the red nucleus. The closest I can come is the fields of Forel which adjoin the red nucleus on the rostral side and receive, among other afferents, a projection from the mammillary body—the so-called mammillo-subthalamic bundle. It is, however, unknown to what extent, if any, the fields of Forel project to the red nucleus. Despite our ignorance concerning limbic afferents to the red nucleus, your finding that electrical stimulation of the nucleus elicits visceral effects ought not to be surprising. By way of the rubrobulbar connection, the red nucleus could be thought to affect the rhombencephalic reticular formation. All systems having such links with the reticular formation could be expected to be potential modulators of one or the other central autonomic regulation mechanism (see, for example, Hoff and Green's [1932] report of cardiovascular changes elicited by stimulation of the motor cortex). Later, Dr. Snider will tell us of autonomic effects observed in the course of cerebellar stimulation. Some of these may be mediated by the red nucleus.

**Dr. Lacey:** What sort of autonomic effects would you expect from the red nucleus, for example, in a chronic animal?

**Dr. Hockman:** We haven't worked with chronic animals. We have stimulated cats that have been immobilized with either gallamine or succinylcholine, and we have observed fairly pronounced increases in blood pressure and alterations in cardiac rate and rhythm. I should expect to see similar responsiveness in the freely moving animal.

**Dr. Gloor:** Are there fibers going to the red nucleus which might explain this? Fibers that are really not connected to the nucleus at all but just going through it?

**Dr. Nauta:** I don't know of any that do. The red nucleus seems to be a fairly "closed" territory in the sense that the great conduction pathways ascending and descending through the midbrain tend to circumvent rather than traverse it.

**Dr. Lacey:** This is a very important question for me, and I would like to pursue it. You say that there is no communication from the red nucleus to the reticular system. Did I understand you to say that this is simply mimicking the activity of the reticular stimulation, or what reason did you give?

**Dr. Nauta:** I only said that the red nucleus has descending efferents that distribute themselves in the rhombencephalic reticular formation. I could imagine, seeing how widely the neurons of the autonomic nervous system are distributed over the reticular formation, that just about any such fiber system could have some synaptic connections with neurons forming part of central autonomic circuits. The picture of such relationships, seen from the anatomical point of view, would suggest diffuseness, but we should be careful not to equate anatomical diffuseness with functional nonspecificity. From the anatomist's point of view, the entire central autonomic nervous system, from the hypothalamus on down, appears to be a wide open neuronal system in the sense that it can be entered over a great variety of local, ascending and descending conduction pathways. We may have to interpret the openness of the system as a reflection of the multitude of environmental and internal (including "motivational" and "motor") changes that necessitate some or other visceral and endocrine adjustments.

**Dr. Livingston:** Is it because the indicators are too gross to identify its selectivity —that the sophistication of your end-point is not sufficient to discriminate? The discrimination must be there.

**Dr. Lacey:** And it is. Discrimination and

specificity become clear if the correct behavioral circumstances are used. I will talk about this point later, but I must point out that there are many data that argue against the notion of diffuseness and lack of specificity.

**Dr. Nauta:** It would be folly to raise anatomical arguments against physiological findings. Moreover, I cannot find anything particularly disconcerting in the contrast between functional specificity and apparent anatomical diffuseness. The anatomist is at a disadvantage in this issue: human perception cannot cope with visual complexities such as offered by the neuronal arrangements of the brain stem reticular formation. Only rarely can a neuronal concatenation be made to stand out selectively and be traced through. The physiologist, by contrast, can record visceral and endocrine effects and correlate these with particular, more or less controlled, inputs.

**Dr. Lacey:** But it can be played upon by a variety of influences that will produce rather specific patterns of activity.

**Dr. Livingston:** But it may be necessary to use a more focal end-point, which we are not really yet able to do.

**Dr. Lacey:** I hate to impose on Don Reis, but he has some beautiful fractionations of effects on the carotid sinus reflex, for example, in which he can independently influence the systolic and diastolic pressures and the heart rate. Can you talk about that, Don?

**Dr. Reis:** I want to dodge discussing my own work on this for a minute to comment on some rather impressive evidence for fractionation of autonomic effects of the sympathetic division of the autonomic nervous system. The first clear evidence is the work of Neal Miller and his associates at the Rockefeller University, showing that in operant conditioning one can get very impressive differentiation of sympathetic vasomotor activity. For example, under the appropriate experimental conditions, one can obtain slow changes limited to one limb or one organ system independently of mean blood pressure. Secondly, there are studies by Feigl and also by Ueno and associates showing that punctate stimulation of the hypothalamus can produce changes in blood flow restricted to one organ. By moving the stimulating electrode less than one millimeter away in the hypothalamus, a new pattern of organ flow can emerge. I think that the evidence, even at that level, is that there are highly differentiated organ blood-flow patterns represented in the brain. Now one wonders if this patterning of organ blood flow evoked by brain stimulation does not represent the visceral component of specific behavioral patterns organized within that same area of the brain. In other words, the question is whether or not such stimulation brings into play different kinds of behavior in which by only isolating blood flow, one can miss the whole pattern of the animal's activity. This sort of analysis has been explored by Folkow and Rubenstein. Thus there are several converging lines of evidence that there is a viscerotopic organization within the brain.

**Dr. Lacey:** As Karl Pribram has said, "Give me the freedom of seven neurons and I will take you from any one place to any other place in the nervous system."

**Dr. Nauta:** Could I say that certain somatic motor neurons are estimated to have something in the order of eight thousand synapses on their membrane. Obviously, this multitude represents a tremendous convergence of afferents. Despite the anatomist's confusion in confronting such a convergence, the motor neuron seems by no means confused: its participation in the range of possible movements is evidently very precisely defined. There is no a priori reason to think that matters should be fundamentally different in the reticular formation. I would like to emphasize once more that terms such as "diffuse" or "nonspecific" refer to the *afferent* side of neural organizations. The effector side simply demands specificity or there would be no free-ranging organism alive today. To give an example: the central mechanisms governing blood pressure must be very precisely organized, yet be accessible to the neural signals emanating from peripheral monitor regions such as the carotid sinus, as well as to impulses signaling changes in the animal's behavioral set. Here, too, there must be neurons involved that are converged upon from many sides. The anatomical appearance of such an organization could seem to be diffuse, yet its effector manifestations are obviously very specific.

**Dr. Lacey:** Not only the effector specificity, but a more global behavioral specificity.

**Dr. MacLean:** To go back more specifically to the anatomy, Walle, would you comment on the recent findings of Shute and Lewis on the two ascending systems—the dorsal and ventral tegmental systems that they claim they have demonstrated with Koelle's cholinesterase stain? Then, of course, they have gone from the septum on up to the hippocampus too and studied the afferent systems that way.

**Dr. Nauta:** Histochemical techniques, such as those demonstrating cholinesterase and the histofluorescence methods for the demonstration of neuronal systems characterized by monoamine content, have considerably enriched the neuroanatomical technology. Observations such as those of Shute and Lewis, and Fuxe and his colleagues, deserve serious attention. For one thing, their methods demonstrate chemical specificity of particular systems or subsystems and can be traced out in anatomically intact brains. These histochemical techniques do not require the production of tissue defects, and investigators using these techniques are thus absolved of some perennial doubts that beset the users of degeneration methods, in particular the uncertainty as to whether the lesions sampled an adequately large region of the brain. Naturally, the histochemical methods likewise have some blind spots. One major worry should be the question whether these techniques can be relied upon to show the synaptic fractionation of polyneuronal conduction systems; in other words, whether perhaps connections that appear to be direct may not in fact be interrupted by synaptic junctions. In time, such questions will, we hope, be answered.

**Dr. Reis:** I wonder if I can ask a question relating to your first comments on the function of the hypothalamus. When a cat is acutely stimulated, as Bard and Cannon showed, the animals have spontaneous outbursts of sham rage. Acute decerebration at the midcollicular level, as Sherrington showed, results in a quiet animal as long as he is unperturbed. However, while the midcollicular decerebrated cat does not move spontaneously, similar autonomic and somatic components of sham rage can be evoked by intense noxious stimulation. This has raised the question in the minds of some as to whether the hypothalamus functions more as an energizer of certain activities of lower brain centers rather than serving as a seat for integrating behavior. I wonder if you have any thoughts on this, Dr. Nauta?

**Dr. Nauta:** Bard's findings, more than any other Sherringtonian observations, suggest the important assumption that behaviors directed toward remote (that is, non-body contact) goals are largely if not entirely governed by the forebrain. More caudal levels of the brain, as well as the spinal cord, are obviously involved in such behaviors, but their participation would seem to be more nearly janitorial than strategic or "policy-making," in the usual sense of these words. If the forebrain should indeed be the only level at which behavioral itineraries are written out in neural code, it is likely also to encompass the neurological substratum of those central states that, on the basis of outward and subjective manifestations, are called "motivational" or "drive states." Along with many others, I interpret the physiological and anatomical evidence that has accumulated since Bard's observations were reported as strong suggestions that the limbic ring of the cerebral hemisphere together with its main brain stem correspondent, the septohypothalamic continuum, are central components of the sensorimotor substratum underlying such behavioral states. This substratum must be accessible to all forms of information from both internal and external environments and exquisitely sensitive to all those features of this information that signal the existence of physiological needs and social-environmental threats or opportunities. The visceral and endocrine manifestations of hypothalamic function in the intact animal seem forever to reflect the fluctuations of this "central state" which in turn could be seen as the end-product of more or less elaborately processed informations from the animal's "total world." To this extent we could indeed consider the hypothalamus as an "energizer" of the neural mechanisms in more caudal parts of the brain. This formulation, however, cannot accommodate the subtle modulations that we must assume the hypothalamus to be able to impose upon the visceral and endocrine effector systems. However, instead of delivering this lengthy disquisition I should perhaps have returned the question to you, Dr. Reis, and have asked if the "energizer" notion appeals to you as a physiologist.

**Dr. Reis:** I am not sure of it as a universal, but in this particular aspect of behavior it is extremely interesting because it does set the key for the spontaneous outburst. Without the hypothalamus you have all the machinery for this sham rage behavior, but you don't have the spontaneous display.

# III

# General Organization of the Central Nervous System

# HIPPOCAMPAL PROJECTIONS AND RELATED
# NEURAL PATHWAYS TO THE MID-BRAIN IN THE CAT

BY

WALLE J. H. NAUTA

(*Department of Neurophysiology, Walter Reed Army Institute of Research,
Washington, D.C.*)

RECENT studies in the rat (Guillery, 1956; Nauta, 1956) have confirmed much earlier descriptions of widespread distributions of the fornix system to the diencephalon and the rostral mid-brain regions. Besides the well-known massive hippocampal projections to the septal region (Ganser, 1882) and the mammillary body (Gudden, 1881), fornix components have been traced from the hippocampus to the preoptic region and the hypothalamus (Cajal, 1911), to the anterior nuclear complex and rostral intralaminar nuclei of the thalamus (Gudden, 1881; Vogt, 1898; Cajal, 1911), and to the rostral part of the central grey mid-brain substance (Edinger and Wallenberg, 1902).

The wide sphere of influence outlined by such distributions of the fornix system is undoubtedly further expanded by secondary neural pathways arising from the recipient structures of the direct hippocampal projection. Of such secondary pathways, those arising in the septal area (Nauta, 1956) and lateral hypothalamic region (Guillery, 1957) have recently been analysed in some detail in the rat by experimental methods. It is apparent from these studies that, in the rat, both structures project directly to the mid-brain by way of the medial fore-brain bundle, while an additional pathway of septal origin follows the stria medullaris to the medial nucleus of the habenular complex. The mesencephalic projections from both regions could be traced to an extensive medial region of the mid-brain, including Tsai's (1925) ventral tegmental area, those from the septum also to the voluminous raphe nucleus of Bechterew (n. centralis tegmenti superior) in the caudal half of the mid-brain. From the mammillary body, the mammillo-tegmental tract had already been traced experimentally (v. Valkenburg, 1912; Sanz Ibáñez, 1935) to Gudden's deep tegmental nucleus, a medial cell group closely associated with Bechterew's raphe nucleus. These findings appeared to indicate the existence of a convergence area for secondary hippocampal projections in a fairly circumscript medial zone of the mid-brain tegmentum. The present experiments were designed to subject the generality of this notion to further experimental test by tracing the mesencephalic projections arising in several further structures known to receive direct or indirect hippocampal pathways.

Needless to say, it is unlikely that fibre systems traced in this manner represent discharge channels of hippocampal impulses exclusively. It is known, for example, that the septal and preoptic regions receive projections not only from the hippocampus, but also from the amygdaloid complex (Gloor, 1955; Nauta and Valenstein, 1958). The medial fore-brain bundle, which serves as the main connexion of these regions with the mid-brain, must consequently be regarded as a common trajectory for descending projections of various origin. Most, if not all, such projections have their most cranial representation in palæocortical and subcortical structures in the basal and medial walls of the cerebral hemisphere, collectively designated by the term "limbic system" (for the historical derivation of this term *see* MacLean, 1952, 1954). The fibre systems to be discussed in the present study are thus extremely heterogeneous, anatomically as well as, no doubt, functionally. They are composed of direct projections from "limbic" structures, as well as the trans-synaptic subcortical extensions of such pathways. Where collective reference seems necessary, they will be referred to as "limbic pathways," in an attempt to obviate the use of the undoubtedly too restrictive terminology "rhinencephalic" or even "olfactory." However, this is done with the understanding that the term, in its present usage, is hardly less vague than designations such as "extrapyramidal."

A complication in the course of the present experiments was presented by the observation of some significant differences between mammalian species in regard to some of the primary and secondary hippocampal projections studied. Several of the previous findings in the rat proved not to be immediately applicable to the cat. Comparative data to be published elsewhere (Valenstein and Nauta, in preparation) suggest that such inter-specific differences consist of variations in the number of synaptic interruptions rather than the ultimate distribution of the pathways involved. Where relevant, such variations will be mentioned in the description of the findings.

## Material and Methods

The present report is based upon findings in approximately 20 cats in which small electrolytic lesions had been placed by the aid of a stereotactic instrument. The survival times varied between eight and fourteen days. Axon degeneration was demonstrated histologically in frozen sections by the Nauta and Gygax (1954) silver technique, and recorded in projection drawings of selected sections.

## OBSERVATIONS

(1) *Hippocampal projections.*—The direct hippocampal projections contained in the fornix system were found to be similar in many respects to those described in the rat by Guillery (1956) and Nauta (1956). As in the rat, fibres of hippocampal origin were traced via the massive precommissural fornix to the septal nuclei and preoptic region, and through the fornix column to certain rostral thalamic cell groups, to the lateral hypothalamic region, and to the mammillary body. However, direct hippocampal projections to the medial hypothalamic region, conspicuous in rodents (Cajal, 1911; Nauta, 1956) could not be identified with certainty in the cat. A further difference from the findings in the rat was noted in the number of fornix fibres distributing to the rostral region of the central grey mid-brain substance: whereas only few such fibres could be demonstrated in the rat, this fornix component appears to be quite massive in the cat. It is formed by numerous fibres which bypass the mammillary body and curve steeply dorsalward a short distance behind this structure in bundles which course along the mid-line, cross in part, and distribute to the ventral half of the central grey substance at the level of Darkschewitsch' nucleus and the rostral half of the oculomotor complex.

(2) *Projections from the septal nuclei.*—As in the rat, it proved to be difficult to obtain an accurate picture of the subcortical projections from the septal region, as lesions of the region invariably involve hippocampal projection pathways. However, it appears certain that, in the cat also, the medial part of the stria medullaris contains a fairly massive septal projection to the medial nucleus of the habenula. As in the rat, this pathway, long considered to be of hippocampal origin, was found degenerated only following certain lesions in the septal region; it was not affected in any case of hippocampal lesion. In the cat, this septo-habenular tract appears to originate largely in the caudal region of the supracommissural septum.

The precommissural fornix is joined in its septal trajectory by numerous fibres of septal origin. The resultant fibre group, which forms a large part of Zuckerkandl's "olfactory bundle of Ammon's horn" is one of the most massive roots of the medial fore-brain bundle. Whereas its hippocampal constituents appear not to distribute beyond the lateral preoptic region, septal fibres, in the cat as well as in the rat (Nauta, 1956), continue farther caudalward in the medial fore-brain bundle and distribute to the entire extent of the lateral hypothalamic region. In the rat, a considerable number of such septal fibres even extend into the mesencephalon, where they terminate in the ventral tegmental area (i.e. the region immediately ventral to the medial half of the red nucleus), and in Bechterew's superior central tegmental nucleus in the caudal half of the mid-brain. In the cat, however, such direct mesencephalic projections from the septal region appear to be much less well developed in both volume and extent. In the present experiments, only a few degenerated fibres could be traced from septal lesions via the medial fore-brain bundle to the ventral tegmental area; none were followed to Bechterew's nucleus. It is apparent from these findings that the septal projection in the cat does not extend much farther caudalward than do the direct hippocampal pathways.

(3) *Projections from the lateral preoptic and hypothalamic regions.*—It is evident from the foregoing observations that the lateral preoptic region is an important distribution area of hippocampal and septal projections, while many fibres of septal origin and a limited number of hippocampal fornix fibres in addition distribute to the lateral hypothalamic region. It therefore appeared of interest to determine the projections arising in this general region. It must be realized that the lateral preoptic and hypothalamic regions compose a continuous longitudinal cell group interstitial to the medial fore-brain bundle. The individual distributions of preoptic and hypothalamic efferents contained in the medial fore-brain bundle can therefore be identified only by careful comparison between the results of lesions located at various frontal levels. Such efferents can best be described by the aid of two cases.

A. *Lesion of the lateral preoptic region: Case CT* 124.—In this case, the lesion had been produced by a slender electrode introduced at an angle of about 45 degrees from the contralateral side and at the same time oriented slightly caudalward. It was localized in the lateral preoptic region coextensive with the rostral half of the optic chiasma. The caudal pole of the lesion is indicated in fig. 1. The fornix column was in no way involved in the defect. Only the fibre degeneration extending caudalward from the lesion will be described.

As shown by figs. 1–6, three main tracts of degenerating fibres can be followed from the lesion. The most voluminous of these is incorporated in the medial fore-brain bundle, while a second massive pathway follows the inferior thalamic peduncle, and a third group of degenerating fibres spreads laterally to the amygdaloid complex.

(*a*) The degeneration in the medial fore-brain bundle extends caudalward through the lateral zone of the hypothalamus and, beyond it, into the mid-brain. Profuse terminal degeneration accompanies the system throughout the lateral hypothalamic nucleus. A considerable number of degenerating axons, however, spreads medially from the bundle in a rather diffuse fashion and can be followed to the anterior and dorsomedial nuclei, some also to the ventromedial nucleus of the hypothalamus. Some of these medial offsets also distribute to the supraoptic nucleus (fig. 1), and a slightly larger number can be traced to the paraventricular hypothalamic nucleus (fig. 2).

Farther caudally it is evident that the mammillary body receives a considerable number of fibres from the medial fore-brain bundle (fig. 3).

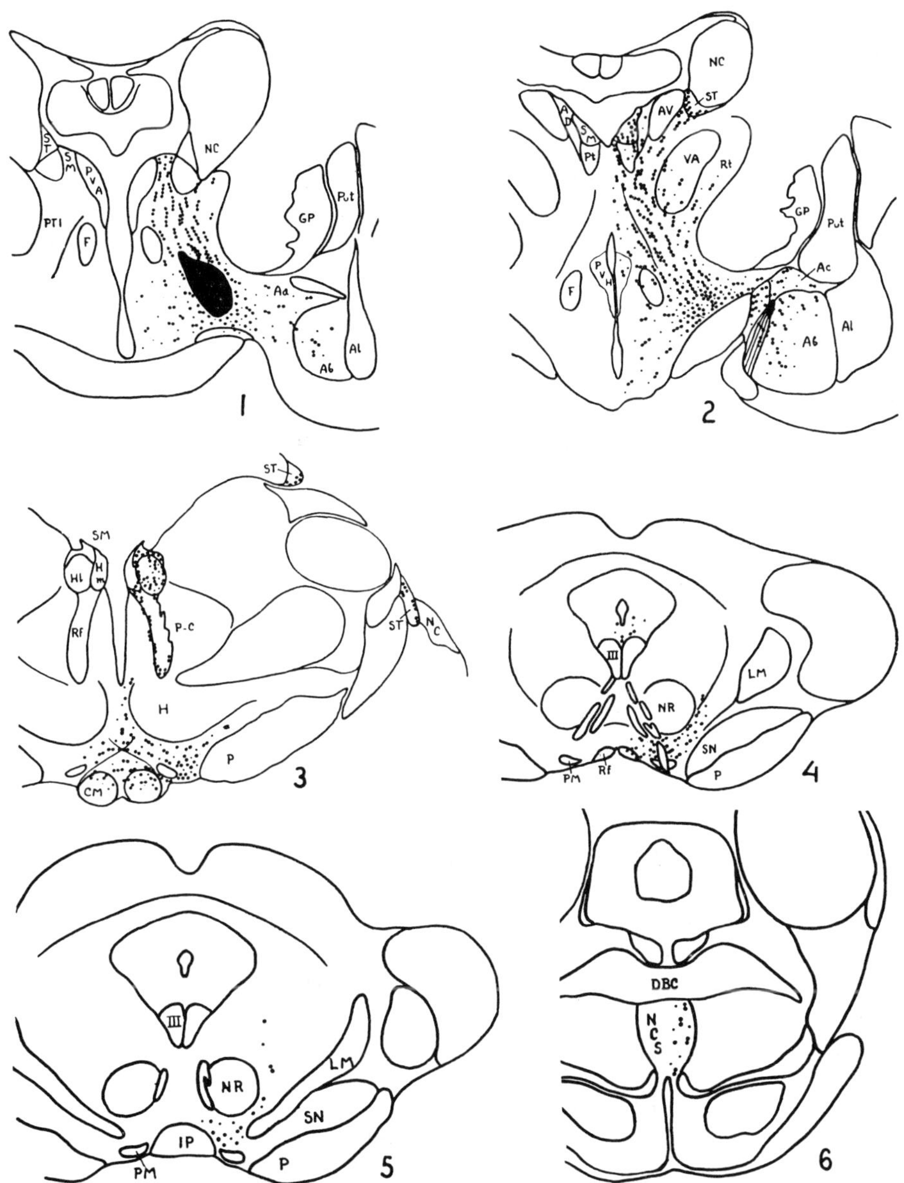

Figs. 1–6.—Fibre degeneration following a lesion of the lateral preoptic region (Case CT 124). The caudal part of the lesion is indicated in jet black in fig. 1. Coarse stipple indicates degenerating fibres of passage; fine stipple terminal degeneration. Abbreviations *see* p. 339.

Degenerating axons enter the structure from the lateral side and distribute rather diffusely in both medial and lateral cell groups. Some such fibres cross transversely into the contralateral medial mammillary nucleus.

Other degenerating fibres of the same group distribute to the supramam-millary region and to the extreme rostral part of the central grey mid-brain substance. It is of interest to note here that medial fore-brain bundle afferents to the mammillary body were already observed by Tello (1936) in young mouse embryos. According to Tello's findings, the medial fore-brain bundle precedes the fornix system in embryonic development as the major source of afferent supply to the mammillary body.

Sections caudal to the mammillary body show degenerating components of the medial fore-brain bundle in the ventral tegmental region of Tsai, adjoining the ventral aspect of the red nucleus (fig. 4). A moderate number of longer fibres of this group can be followed past the interpeduncular nucleus to the caudally adjacent raphe nucleus of Bechterew (n. centralis tegmenti superior, fig. 6) in which they terminate.

(*b*) A second massive system of degenerating fibres extends from the lesion in the dorsomedial direction, following the inferior thalamic peduncle. In the thalamus, this system traces a wide path, with offsets to the nuclei reticularis, ventralis anterior, parataenialis, and antero-dorsalis (fig. 2), as well as to the n. dorsomedialis. Numerous further degenerating fibres enter the stria medullaris (figs. 1, 2), while a smaller number extend laterally in the ventral fibre capsule of the anterior nuclear complex and join the stria terminalis (fig. 2). Those enclosed in the stria medullaris run caudalward to an area of profuse terminal degenera-tion which occupies most of the lateral habenular nucleus (fig. 3). Some fibres of the same group bypass the habenula and join the fasciculus retroflexus with which they reach the ventral surface of the brain-stem, where they appear to leave the bundle, entering the ventral tegmental area, and mingling with the medial fore-brain bundle components terminat-ing in the same region (fig. 4).

The degenerating fibres which join the stria terminalis follow the bundle in its semicircular course to the amygdaloid complex. The ultimate distribution of these elements cannot be determined accurately, as they blend with

(*c*) a third, rather diffuse group of degenerating axons which extends laterally from the preoptic lesion through the substantia innominata, and enters the amygdaloid complex from the medial side (figs. 1–2). These fibres, part of which travel caudally over some distance alongside the optic tract, distribute to the anterior, medial, central, and basal amygdaloid nuclei; none appear to reach the lateral nucleus of the complex (fig. 2).

B. *Lesion of the caudal part of the lateral hypothalamic region: Case CT 99.*—By contralateral approach, using an angle of approximately 45 degrees, a small electrolytic lesion was placed, immediately dorsal and slightly lateral to the mammillary body (fig. 7). The lesion and the electrode track had interrupted part of the supramammillary decussation, the com-mon stem ("pars principalis," Koelliker) of the mammillothalamic and mammillotegmental tracts, as well as a major part of the medial fore-brain bundle. A small part of the posterior hypothalamic nucleus was pierced by the electrode track. It is of importance to note that Forel's field H, the cerebral peduncle, and the subthalamic nucleus were in no way involved in the lesion.

Immediately caudal to the lesion, the medial fore-brain bundle appears as the central structure in a widespread field of fibre degeneration. It occupies its characteristic position ventral to the red nucleus (fig. 8). Spreading from the bundle in a laterodorsal direction are degenerating slender fascicles which distribute to the tegmental region adjoining the substantia nigra on the dorsal side. This lateral spread of the medial fore-brain bundle appears to be similar to, although more massive than that observed following lesions in the preoptic region. More caudally, however, a significant difference from the results of preoptic lesion is noted in the form of numerous degenerating axons which curve around the lateral side of the red nucleus (fig. 9) and collect in scattered fasciculi running caudalward immediately dorsal to the stratum lemnisci. Throughout its course along the ventral border of the mid-brain tegmentum, this group of degenerating fasciculi releases numerous fibres which sweep dorsalward through the tegmentum (fig. 9). Some such fibres appear merely to traverse the tegmentum *en route* to the central grey mid-brain substance; others, however, terminate diffusely in the central tegmental region. Farther caudally, slightly rostral to the frontal level of the trochlear nucleus, this entire lateral fibre system initiates a rather abrupt dorsal sweep through the tegmentum (fig. 10). On this dorso-caudal trajectory also, it is accompanied throughout by signs of diffuse tegmental termination, while additional fibres to the central grey mid-brain substance can be followed dorsally. Finally, at the level of the trochlear nucleus, the remaining fibres of the system collectively enter the central grey mid-brain substance (fig. 11), where they are lost among numerous further degenerated fibres distributing in the region.

The remaining part of the medial fore-brain bundle continues caudalward through the ventral tegmental area, immediately dorsal to the fasciculus retroflexus and the mammillary peduncle (figs. 8, 9). This medial group of degenerating fibres is more massive but otherwise comparable with that observed following preoptic lesion. As in the preceding case, it can be traced along the lateral aspect of the interpeduncular nucleus (fig. 9) and from here dorsalward along the median plane of the caudal mid-brain tegmentum. Its main termination appears to be in Bechterew's raphe nucleus (figs. 12, 25) and a narrow adjoining medial tegmental region. Unlike the preoptic projection, a number of fibres of the group continue beyond the raphe nucleus and enter the central grey substance near the mid-line (figs. 12, 13).

A further degenerated fibre system descending to the mid-brain consists of slender fascicles which extend from the lesion, first in the dorsomedial direction (fig. 8), then caudalward immediately ventral to the medial

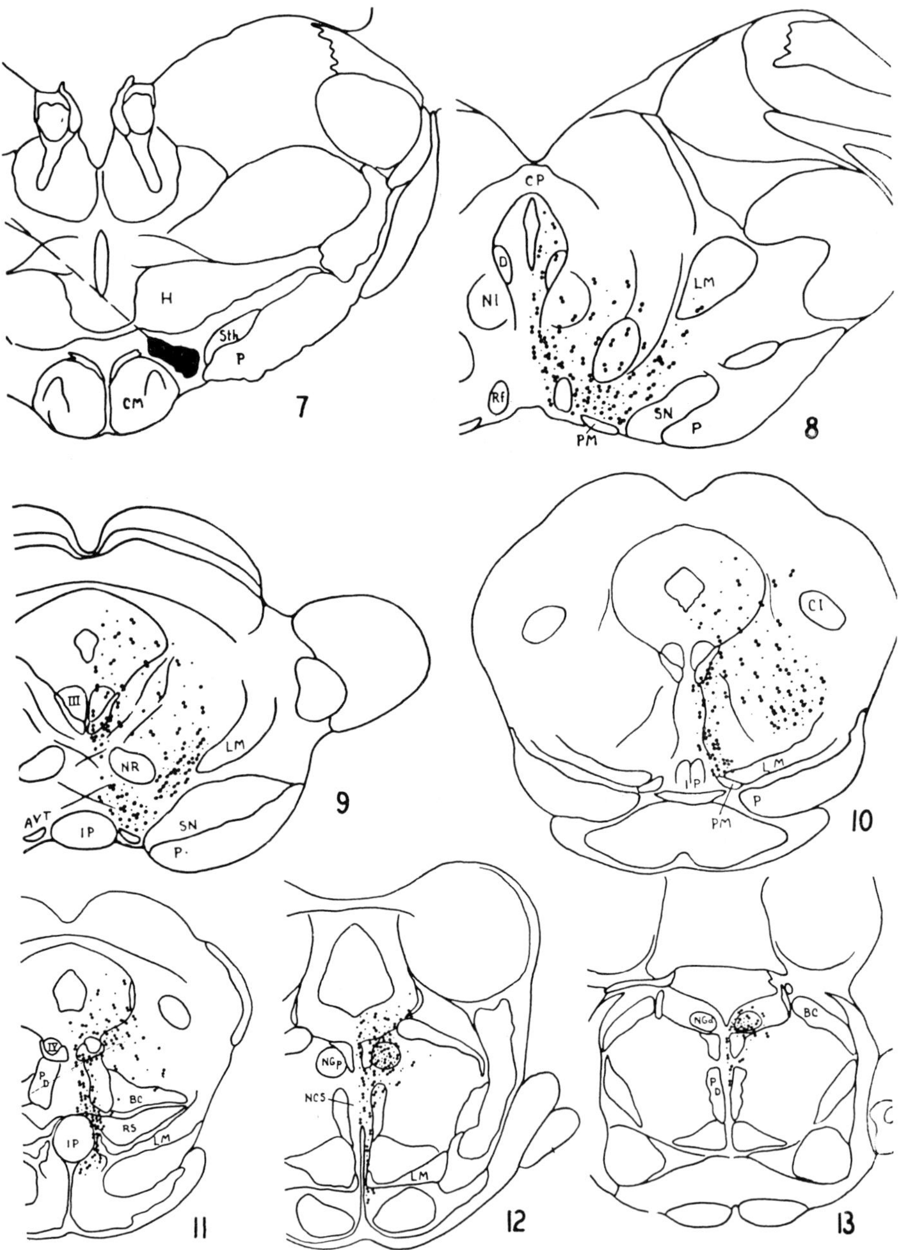



FIGS. 7–13.—Fibre degeneration observed following a small lesion in the caudal part of the lateral hypothalamus (Case CT 99). The lesion is indicated in jet black in fig. 7. Coarse dots: degenerating fibres of passage; fine stipple: terminal degeneration.

longitudinal fasciculus (figs. 9–11). Throughout its descending course the system is easily identifiable by its relative compactness. It appears to mingle to some extent with the ventral fibres of the medial longitudinal fasciculus. There can be little doubt that this degenerating fibre group represents the mammillo-tegmental tract: its appearance in the present case corresponds closely to Sanz Ibáñez (1935) observations in Marchi material. A large proportion of its fibres can be followed to a degenerating plexus of extremely fine axon ramifications in the deep ("ventral") tegmental nucleus of Gudden (figs. 12, 26), but many other fibres of the bundle enter the central grey substance and terminate in the medial part of Gudden's dorsal tegmental nucleus and its surroundings (fig. 13). Dense terminal degeneration in Gudden's tegmental nuclei as observed in the present case was consistently found lacking following hypothalamic lesions which had spared the mammillary body and the pars principalis of Vicq d'Azyr's bundle.

Finally, the slight involvement of the posterior hypothalamic nucleus noted above has caused the disintegration of a fairly considerable number of fibres descending in Schütz's dorsal longitudinal fasciculus. It is impossible to trace these fibres caudalward, due to their mingling with large numbers of degenerating fibres of different origin which enter the central grey substance at various frontal levels. However, as no fibre degeneration can be identified in the central grey substance one millimetre caudal to Gudden's dorsal tegmental nucleus, it seems doubtful that Schütz's fibre system contains elements descending directly from the hypothalamus beyond the isthmus region.

(4) *Projections from the habenula: Case VA 4.*—The foregoing cases have confirmed the existence of two major projections to the habenular complex. One such pathway originates in the supracommissural part of the septal region, joins the stria medullaris as a branch of the fornix bundle, and terminates in the medial habenular nucleus. The second (Koelliker, 1896) appears to arise from the lateral preoptic region as a component of the inferior thalamic peduncle; it also joins the stria, but its habenular termination is confined to the lateral habenular nucleus. Both pathways could be interpreted to represent indirect projections from the limbic system, in turn relayed to the mesencephalon via the habenulo-interpeduncular tract (fasciculus retroflexus of Meynert). Two cases of degeneration in the fasciculus retroflexus will be described next.

(*a*) In *Case VA* 4 a small lesion had been placed immediately ventral to the habenular complex, by contralateral and rostrocaudal approach (fig. 14). It had completely severed the fasciculus retroflexus just below its exit from the habenula. Also involved was a small part of the peri-ventricular grey substance ventral to the habenular complex. The posterior commissure was entirely intact.

The fasciculus retroflexus is completely degenerated and can easily

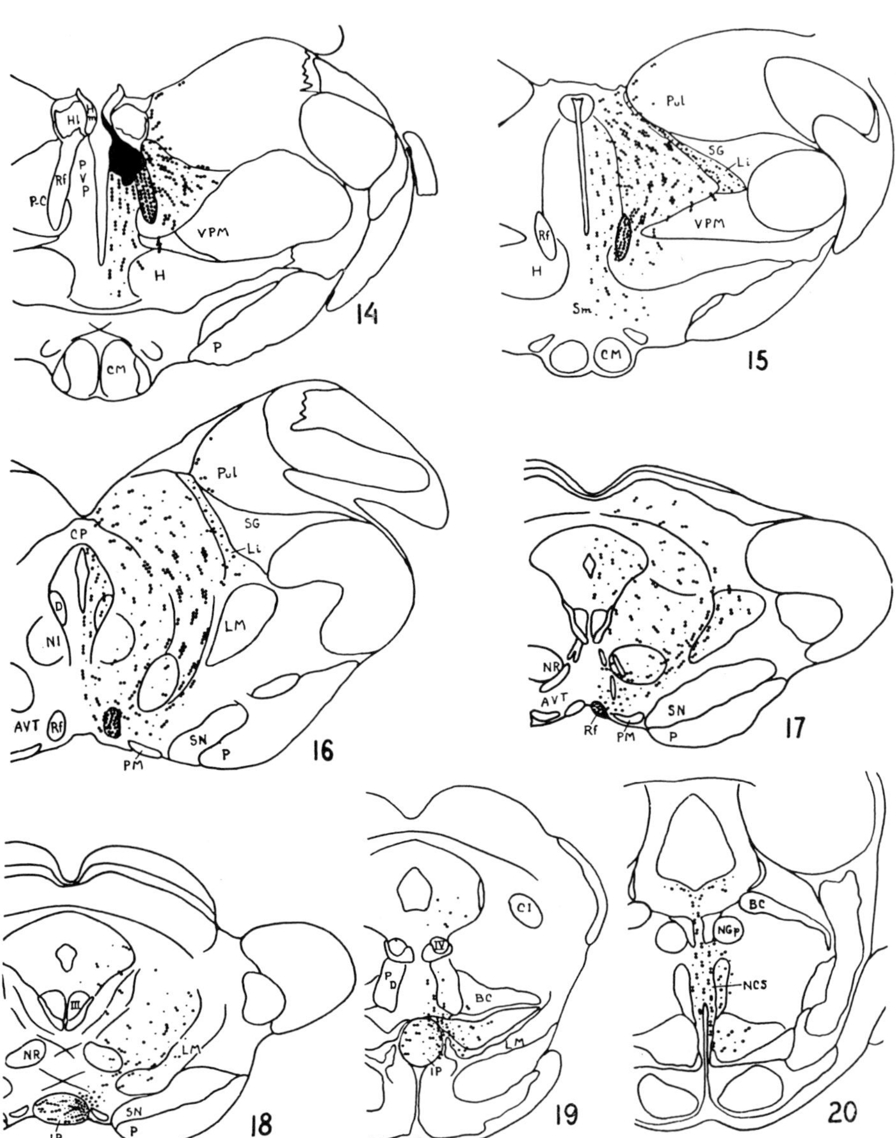

Figs. 14–20.—Fibre degeneration resulting from complete interruption of the fasciculus retroflexus (Case VA 4). Coarse dots: degenerating fibres of passage; fine stipple: terminal degeneration. Abbreviations *see* p. 339.

be followed in its characteristic course to the basal surface of the mesencephalon and thence caudalward. This case, however, demonstrates a greater complexity of the caudal habenular projections than would be recognized from a cursory examination of the fasciculus retroflexus in normal material: apart from the compact bundle of Meynert, a more diffuse system of arcuate fascicles, here degenerated, sweeps in broad curves lateralward from the lesion through the parafascicular nucleus—centre median complex (figs. 14, 15), then caudoventrally through the lateral tegmentum of the upper mesencephalon (fig. 16), and finally medialward underneath the red nucleus (figs. 17, 18). From here caudalward this more diffuse pathway of apparently habenular origin follows the compact fasciculus retroflexus on the lateral and dorsal sides. It is of great interest to note that this diffuse habenular pathway, throughout its arcuate trajectory, releases numerous fibres which distribute to the parafascicular nucleus—centre median complex and the adjacent nucleus limitans (figs. 14, 15), as well as to an extensive central and lateral tegmental region of the upper mid-brain (figs. 15–17). Some such fibres spread into the central grey substance where they are lost among fairly numerous degenerating axons interrupted by the periventricular lesion ventral to the habenula.

The termination of the compact fasciculus retroflexus in the interpeduncular nucleus is evident at levels represented by figs. 18 and 24. Slender fascicles of extremely fine degenerating fibres course medialward from the bundle, traverse the entire width of the interpeduncular nucleus, and turn back abruptly in a characteristic "figure 8" fashion as described by Cajal (1911). Degenerating terminal fibres are difficult to identify in this system of extremely fine axons; those that can be interpreted as such are more numerous on the side ipsilateral to the lesion.

Few if any fibres of the compact fasciculus retroflexus appear to continue caudalward beyond the interpeduncular nucleus. The more diffuse concomitant fibre group, however, by-passes the interpeduncular nucleus on the lateral side (figs. 18, 19) and, near the caudal pole of the nucleus, gradually curves dorsalward, spreading over Bechterew's nucleus centralis tegmenti superior and an adjacent narrow tegmental region outlined by the predorsal fasciculus (figs. 19, 20). The longest fibres of the group (fig. 20) extend dorsalward along the mid-line and enter the caudal part of the central grey substance, where their distribution appears to be restricted to a cell group described as "fountain nucleus" by Sheehan (1933).

(*b*) *Case CT* 112.—Among several additional cases of involvement of Meynert's bundle this case is of particular interest, as the lesion was entirely restricted to the medial habenular nucleus. In this instance, the compact fasciculus retroflexus showed massive degeneration, but the more diffuse concomitant lateral fibre system was entirely normal. This case suggests that the origin of the latter projection is confined to the lateral habenular nucleus.

DISCUSSION

The present study has served to emphasize the existence of various pathways by which the hippocampus, as well as other members of the limbic system, can transmit its influence to the mid-brain. In the cat as in the rat (*see* Nauta, 1956), only one such pathway seems to represent a direct hippocampal projection. It consists of a number of fornix fibres which by-pass the mammillary body and distribute to the rostral part of the central grey mid-brain substance. It is interesting that this direct hippocampo-mesencephalic projection is much more strongly developed in the cat than it is in the rat.

Far more massive than this direct projection are the indirect limbic connexions with the mid-brain. Such connexions, established by pathways arising in the septal region, lateral preoptic and hypothalamic regions, mammillary body, and habenula, follow the medial fore-brain bundle and the mammillo-tegmental tract as well as the route outlined by the stria medullaris and fasciculus retroflexus. The pathways in question are illustrated diagrammatically by fig. 21 A and B. As indicated by the present findings, direct hippocampal projections articulate with such secondary limbic pathways mainly in the septal and lateral preoptic regions and in the mammillary body, to a lesser extent also in the lateral hypothalamus. The amygdaloid complex, as indicated by Gloor's (1955) electrophysiological findings as well as by anatomical data (Nauta and Valenstein, 1958) projects to the septal and lateral preoptic regions and could there connect with the same secondary pathways[1].

It is of importance to note that the septal and lateral preoptic regions constitute a nodal area in the limbic projection channels. In this extensive region, primary hippocampal and amygdaloid projections appear to overlap at least to a certain extent, and from it, two separate pathways to the mid-brain originate: a ventral fibre system incorporated in the medial fore-brain bundle (fig. 21 A) and a more dorsal pathway which forms the stria medullaris, synapses in the habenula, and is continued by the fasciculus retroflexus (fig. 21 B). Of these two secondary fibre systems, the medial fore-brain bundle is the most difficult of analysis, for it contains long axons as well as numerous synaptic chains of shorter connexions. Of the many fibres of this bundle which enter the mid-brain, some have their origin in the lateral preoptic region (in the rat such fibres even originate in the septum), but a far larger number arise more caudally, i.e. in the lateral hypothalamus.

In spite of considerable diversity of origin and trajectory, all of the major limbic pathways to the mid-brain share certain characteristics in their caudal distribution. Somewhat schematically, all three: the medial fore-brain bundle, mammillo-tegmental tract, and fasciculus retroflexus, can be said to have a dual distribution in the mid-brain. Of each, a lateral component terminates in central and lateral tegmental

---

[1]Actually, the anatomical data have demonstrated that the distribution of the direct amygdalo-septal projection is restricted to the basal part of the septal region. It appears justifiable, however, to assume that secondary pathways exist which allow for a further spread of the amygdaloid projection to more dorsal septal areas.

regions (broken lines in fig. 21 A and B), while the remaining fibres distribute over a smaller or larger part of an extensive medial and paramedian mid-brain region to be discussed below. None of the pathways in question appears to extend beyond the level of the isthmus.

(1) *Lateral mesencephalic projections of the limbic system.*—The more lateral projections of the limbic system distribute widely in central and lateral tegmental regions, as follows:

(*a*) A prominent pathway of lateral distribution is established by a lateral component of the medial fore-brain bundle (broken lines in fig. 21 A).

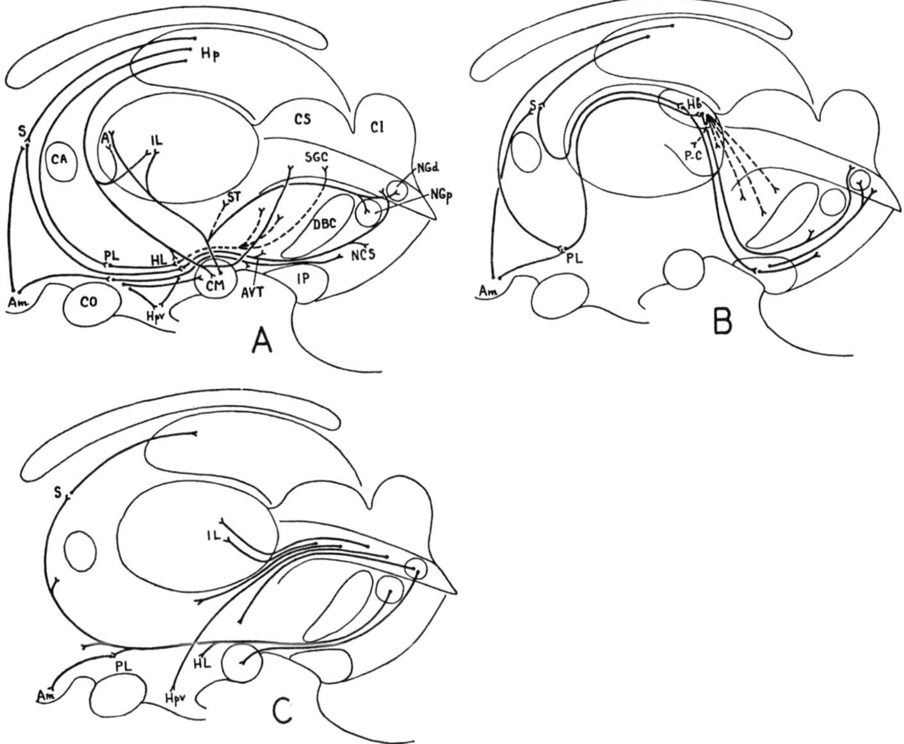

FIG. 21.—Diagrammatic representation of the connexions between some limbic fore-brain structures and the mid-brain, as described in the text. Broken lines indicate projections to central and lateral regions of the brain-stem tegmentum. Pathways distributing to medial tegmental regions ("limbic mid-brain area") are indicated by solid lines. A: limbic pathways incorporated in the fornix, medial fore-brain bundle, and mammillo-tegmental tract. B: limbic pathways following the stria medullaris and fasciculus retroflexus. C: ascending mesencephalic projections related to the limbic system. Abbreviations: A, anterior nuclear complex of thalamus; Am, amygdaloid complex; CA, anterior commissure; CO, optic chiasma; HL, lateral hypothalamic region; Hp, hippocampus; Hpv, medial and periventricular hypothalamic regions; IL, intralaminar thalamic nuclei; S, septal region; SGC, central grey substance, ST, subthalamic region. Other abbreviations *see* p. 339.

Significantly, it degenerates following lesions in the lateral hypothalamus, but lesions in the lateral preoptic region do not appreciably affect it. In its present description (*see* Case CT 99) it appears to correspond at least to some extent to a lateral part of the "basal olfactory tract" described by Wallenberg (1901) in the rabbit, and to a component of Bischoff's (1900) "fasciculus olfacto-mesencephalicus" of the hedgehog. It possibly also corresponds in part to the dorsal and posterior hypothalamo-tegmental tracts described by Crosby and Woodburne (1951) in the monkey.

This more lateral projection in all likelihood represents one of the main pathways by which the lateral hypothalamus transmits its influence, via synapses in the mid-brain reticular formation, to the visceral motor nuclei farther caudally in the neuraxis. Its wide spread in the mid-brain tegmentum could supply an explanation for the diffuseness of the trans-tegmental hypothalamic discharge demonstrated in physiological experiments (Magoun, Ranson and Hetherington, 1938; Ranson and Magoun, 1939). In addition, however, it must be realized that this lateral hypothalamic projection spreads over the tegmental region occupied by the diffuse and multineuronal reticulo-thalamic and reticulo-subthalamic pathways which are believed to form the anatomical substratum of Moruzzi and Magoun's (1949) diffuse cortical activating system (*see* Nauta and Kuypers, 1958). By connecting synaptically with the shorter, mesencephalic components of this widespread ascending brain-stem system, the hypothalamus, and by its mediation the limbic structures, could conceivably influence basic patterns of cortical activity related to states of wakefulness or sleep. The present anatomical findings would thus suggest a possible interpretation of the drowsy or sleep-like states accompanying inflammatory or traumatic lesions in or immediately behind the caudal hypothalamus (v. Economo, 1918; Ranson, 1939; Ranson and Magoun, 1939; Nauta, 1946; Collins, 1954, and others) as a result of interference with a phasic descending excitatory discharge from the hypothalamus, via lateral components of the medial fore-brain bundle articulating with ascending projections from the mid-brain reticular formation.

(*b*) A second pathway of somewhat comparable distribution to the central and lateral tegmental region appears to originate in the lateral nucleus of the habenula (broken lines in fig. 21 B). It consists of numerous arcuate fibres which spread over the parafascicular nucleus, centre median and nucleus limitans as well as farther caudalward over a broad area of the rostral mid-brain tegmentum (*see* the present Case VA 4). This projection does not extend as far caudally in the central and lateral mid-brain tegmentum as does the lateral hypothalamic projection described above.

(*c*) An additional lateral projection related to the limbic system is represented by a fairly substantial lateral offset from the mammillo-tegmental tract, distributing to Forel's field H and the zona incerta. It terminates among the complex conduction systems ascending and descending through the subthalamic region.

(*d*) Further potentially important connexions of the limbic system with the central and lateral tegmental regions are established by a massive and diffuse fibre system which originates in the central grey mid-brain substance and spreads radially over most of the tegmental cross section, as well as over the deep layers of the superior colliculus (Weisschedel's (1937) radiatio grisea tegmenti, *see below*). In considering the possible functional significance of this massive radial pathway, however, it must be borne in mind that the central grey substance receives afferents from many sources outside the limbic system (*see below*).

It is appropriate at this point to recall the somatic motor phenomena which have been elicited by stimulation of the amygdala and hippocampus (Kaada *et al.*, 1953, 1954; Magnus and Lammers, 1956; Votaw, 1957). It seems possible that such motor phenomena, like the visceral effects of stimulation of limbic fore-brain structures, are mediated by some or all of the above-mentioned limbic pathways to the subthalamus and central tegmental regions of the mid-brain reticular formation.

In the Australian marsupial, the phalanger, Adey *et al.* (1956) described a projection from the entorhinal area which follows the stria medullaris and, without synapsing in the habenula, distributes directly to dorsolateral regions of the mid-brain tegmentum. In the cat, a comparable pathway could not be identified in the present experiments. Two cases in which the stria medullaris was interrupted rostral to the habenula failed to show evidence of fibre degeneration in the lateral mid-brain tegmentum. However, this discrepancy may be due to interspecific variation which, as pointed out before, is quite commonly encountered in comparative studies of the limbic projection systems.

(2) *Limbic pathways distributing in medial mid-brain regions.*—Numerous fibres of the medial fore-brain bundle, mammillo-tegmental tract, and fasciculus retroflexus distribute to a medial tegmental region which has its rostral beginning in Tsai's ventral tegmental area, extends caudalward along the base of the mesencephalon where it includes the interpeduncular nucleus, and then arches dorsalward along the caudal aspect of the brachial decussation as the nucleus centralis tegmenti superior of Bechterew, which in turn becomes continuous with the central grey substance caudal to the trochlear nucleus (*see* fig. 21 A). The dorsal and deep tegmental nuclei of Gudden and the ventral part of the central grey substance must be included in the same mid-brain region.

The region in question largely lies outside the massive stream of the reticulo-thalamic and reticulo-subthalamic conduction systems, from which, however, it receives short connexions (Nauta and Kuypers, 1958). Since, excepting the central grey (*see below*), it appears to be related specifically to the limbic system, it seems justifiable to use the term "limbic mid-brain area" as a schematic designation of the region here discussed.

(*a*) In the region thus outlined, the medial fore-brain bundle distributes to the ventral tegmental area, Bechterew's tegmental nucleus and the caudal part of the central grey substance. The corresponding fibres originate in the lateral preoptic region (in the rat also in the septum (Nauta, 1956)), and especially in the lateral hypothalamus. In the cat, the contributions from the preoptic region appear not to extend farther caudalward than the rostral part of Bechterew's nucleus (Case CT 124). The present findings agree to some extent with Wallenberg's (1901) observation of medial fore-brain bundle termination in ventral and medial tegmental regions in the rabbit, with Tello's (1936) tracing of similar fibres to the regions of Gudden's nuclei in the mouse, and with Guillery's (1957) report of medial fore-brain bundle distribution to the ventromedial tegmentum and the caudal end of the central grey substance in the rat. In their typical position ventral to the red nucleus, the medially distributing components of the medial fore-brain bundle described above also correspond to the anterior hypothalamo-tegmental fibre system observed in the monkey by Crosby and Woodburne (1951) who, however, could not follow the system beyond the level of the oculomotor nerve exit in that species.

Although this medial component of the medial fore-brain bundle is quite massive, physiological findings (Magoun, Ranson and Hetherington, 1938) indicate that it does not represent the main pathway by which the hypothalamus exercises its influence upon the visceral motor system. Instead, such findings tend to emphasize the importance for such functions of more laterally placed transtegmental pathways, probably represented by the lateral component of the medial fore-brain bundle discussed above.

(*b*) The fasciculus retroflexus appears to have a dual mode of distribution in the limbic mid-brain area (fig. 21 B). Its most conspicuous, compact part terminates in the interpeduncular nucleus, where it articulates with Ganser's (1882) "tegmental bundle of the interpeduncular nucleus" (*see* Koelliker's (1896) figs. 635 and 643), a massive system of relatively fine fibres which extend from the nucleus caudodorsally along the median plane and distribute to Bechterew's nucleus and the caudal end of the central grey, probably including Gudden's dorsal tegmental nucleus. Additional fibres of the fasciculus retroflexus, however, form a massive and more diffuse lateral group which by-passes the interpeduncular nucleus and distributes directly to Bechterew's nucleus and the caudal part of the central grey substance. This fibre group undoubtedly corresponds to the "habenulo-tegmental tract" described by Bürgi and Bucher (1955) in the cat.

(*c*) Finally, the mammillo-tegmental tract was traced to both the dorsal and deep tegmental nuclei of Gudden. In contrast to Guillery's (1957) findings in the rat, the termination of this fibre system in the deep tegmental nucleus, although characterized by an extreme fineness of the terminal axon arborizations, is quite profuse in the cat. The mammillo-tegmental tract was originally traced to this nucleus by Gudden (1889), and later by Koelliker (1896), v. Valkenburg (1912), Sanz Ibáñez (1935), and others. It has not been possible in the present material to trace any fibres of the mammillo-tegmental tract caudalward into the pontine and medullary tegmentum as described by Tello (1936) and Sanz Ibáñez (1935).

*Dorsal longitudinal fasciculus and central grey substance.*—It would seem appropriate to include the dorsal longitudinal fasciculus of Schütz among the medial pathways related to the limbic system. For a discussion of the earlier findings in regard to this fibre system reference is made to Ingram's (1940) excellent review of hypothalamic anatomy. Although the prevailing direction of conduction appears to be ascending (Bucher and Bürgi, 1945; Johnson, 1953), the bundle contains descending fibres which appear to have a wide origin in medial and periventricular hypo-thalamic cell groups, distribute to the ventral half of the central grey substance, and are probably augmented by axons arising in the central grey proper. Undoubtedly, the dorsal tegmental nucleus of Gudden receives a number of such periventricular hypothalamic fibres, but the system rapidly dwindles caudal to this nucleus and no part of it seems to extend beyond the level of the abducens nucleus. It even appears doubtful from a study of normal material that Schütz's system is massively continued caudalward by fibres arising from the dorsal tegmental nucleus and its surroundings, as was suggested by Ariens-Kappers, Huber and Crosby (1936). Rather, experimental anatomical data suggest that Schütz's periventricular system and the dorsal longitudinal fasciculus essentially represent an association system of the medial hypothalamus and the central grey mid-brain substance.

As indicated by the present data, further limbic pathways distributing to the central grey are formed by: (*a*) hippocampal and possibly septal fibres to the rostral one-third of the structure; (*b*) fibres from the lateral hypo-thalamus via the lateral component of the medial fore-brain bundle to at least the rostral and middle thirds, and (*c*) a variety of further paths from the lateral hypothalamus, habenula, and interpeduncular nucleus to the caudal one-third. The central grey substance, however, also receives many afferents of non-limbic origin. Such afferents are, for example, contained in the "spinothalamic" tract and in the diffuse ascending reticular pathways (*see* Nauta and Kuypers, 1958), as well as in the brachium con-junctivum (Thomas *et al.*, 1956; Mehler, unpublished). It is important to note that, on the efferent side, the central grey connects with the hypo-thalamus (*see below*), and through a massive and diffuse system of radial

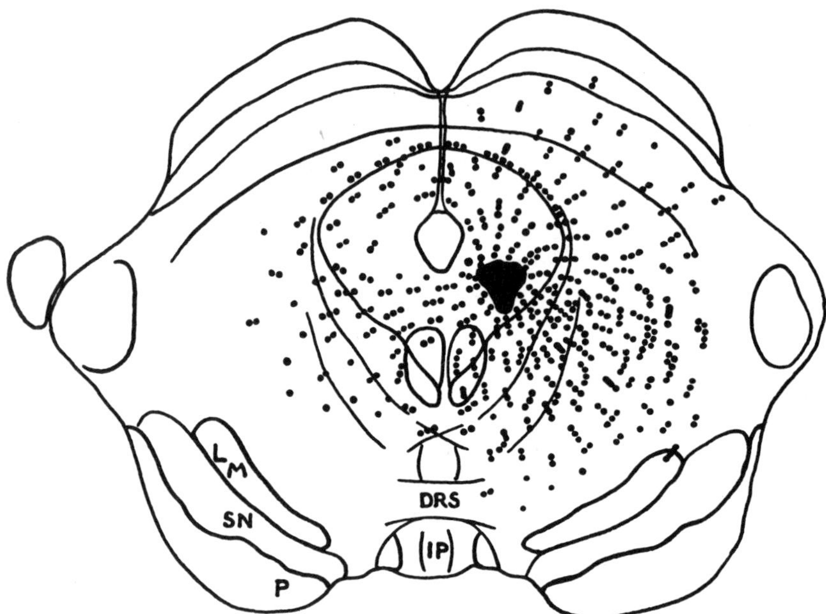



FIG. 22.—Degeneration of Weisschedel's "radiatio grisea tegmenti," in a case of sagittal-tubular lesion of the ventral half of the central grey substance (CT 116).

fibres (Weisschedel's (1937) "radiatio grisea tegmenti," *see* fig. 22) with almost the entire cross section of the mid-brain tegmentum and the superior colliculus. The latter fibre system appears to furnish a massive link between the limbic structures and the widespread ascending and descending tegmental conduction systems.

*Ascending pathways related to the limbic system.*—The close association of the medial mid-brain area with the limbic fore-brain structures is further emphasized by the distribution of ascending projections arising in the area. Such projections (*see* fig. 21 C) are represented by at least two fibre systems: (*a*) the ascending component of Schütz's dorsal longitudinal fasciculus which appears to originate throughout the length of the central grey substance inclusive of Gudden's dorsal tegmental nucleus, and (*b*) the system of the mammillary peduncle which arises from both the dorsal and ventral tegmental nuclei, possibly also from adjoining tegmental regions, and appears to receive further contributions from the ventral tegmental area (for a review of the pertinent literature *see* Ingram, 1940; Guillery, 1956; Nauta and Kuypers, 1958). The ascending component of Schütz's system ends in part in the intralaminar thalamic nuclei (Nauta and Kuypers, unpublished), but the majority distributes to periventricular and medial hypothalamic cell groups and to the dorsal hypothalamic area. The mammillary peduncle, on the other hand, connects massively with the mammillary body, but a large part of the bundle continues rostralward in the medial fore-brain bundle, distributing to the lateral hypothalamic and preoptic regions and to the medial septal nucleus (Guillery, 1956; Nauta and Kuypers, 1958), augmented by further ascending projections from the lateral hypothalamus (Guillery, 1957). The medial septal nucleus, in turn, has been shown to project back to the hippocampus (Daitz and Powell, 1954), while the present findings indicate the existence of a similar recurrent projection from the preoptic region to the amygdaloid complex (*see* the present Case CT 124).

*The limbic system—mid-brain circuit.*—Comparing these ascending fibre systems with the limbic projection pathways, a striking reciprocity appears in the connexions between the limbic fore-brain structures and the "limbic mid-brain area." Substantial escape pathways from this circuitous neural apparatus would seem to lead mainly in three directions: to the medial hypothalamic region, to the diffuse ascending and descending conduction systems of the mid-brain tegmentum (*see* diagram in fig. 23), and to the anterior and intralaminar nuclei of the thalamus.

Of these escape routes, those to the medial hypothalamus merit special emphasis, as this region (Hpv in fig. 23) appears to be the part of the

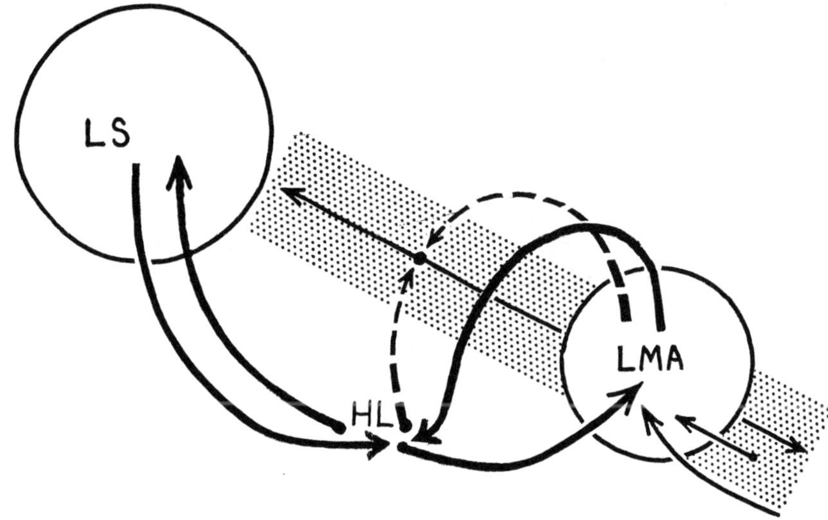


FIG. 23.—Simplified diagram of the limbic system—mid-brain circuit as described in the text. Broken arrows indicate collateral projections from the circuit to central and lateral regions of the mid-brain tegmentum (shaded): arrow 1 summarizes the lateral limbic pathways indicated by broken lines in fig. 21 A and B, arrow 2 represents Weisschedel's "radiatio grisea tegmenti" from the central grey substance. Abbreviations: LMA, "limbic mid-brain area"; LS, limbic fore-brain structures (hippocampus, septal region, amygdala); HL, lateral hypothalamic region; Hpv, medial and periventricular hypothalamic regions.

hypothalamus most directly involved in neural mechanisms controlling anterior hypophysial functions. As indicated in greatly simplified form in fig. 23, the medial hypothalamic region receives offsets from both the descending and ascending limbs of the limbic system—mid-brain circuit. Such afferents reach the region directly (from the limbic fore-brain structures via the stria terminalis and medial fornix components, from the limbic mid-brain area via Schütz's bundle), as well as indirectly over the lateral hypothalamic and preoptic regions. The supraoptic nucleus of the hypothalamus is probably connected with the same circuit via the lateral preoptic regions from which it receives short fibres (*see* Case CT 124). In view of these collateral connexions, it appears reasonable to assume that the limbic system—mid-brain circuit can influence hypothalamic mechanisms controlling anterior and posterior hypophysial as well as visceromotor functions. The reciprocity in the connexions between the limbic fore-brain structures and the medial mid-brain area tends to suggest a mutual engagement of both brain regions in mechanisms subserving visceral and endocrine homeostasis.

Although somewhat separated spatially from the major diffuse conduction pathways of the brain-stem, the mesencephalic pole of the circuit, especially its central grey substance component, nevertheless receives afferents ascending from the anterolateral fasciculus of the spinal cord, as well as offsets from the "mainstream" of ascending reticular conduction (Nauta and Kuypers, 1958). Such afferent connexions (*see* fig. 23) would seem to furnish a likely route whereby the limbic system—mid-brain circuit could be activated by impulses of sensory origin. The limbic mid-brain area could thus represent a major relay in the afferent pathways involved in, for example, the anterior pituitary response to physical stress. It is, however, not possible at the present time to preclude the existence of further pathways connecting the diffuse ascending reticular conduction pathways with the hypothalamus. Such additional connexions could conceivably originate along the subthalamic and intralaminar trajectories of the ascending reticular pathways; their existence, however, remains to be demonstrated.

Finally, it appears likely that, besides lemniscal and extralemniscal sensory impulses, certain neocortical mechanisms can affect the neural mechanisms discussed in the present paper. In spite of much speculation, the anatomical substratum for such neocortico-limbic interaction is almost entirely obscure; it would seem to hold a lively challenge for future investigation.

## SUMMARY

Direct and potential (i.e. transsynaptic) hippocampal projection pathways to the mid-brain were traced experimentally in the cat by the aid of the Nauta-Gygax silver technique. Conclusions are as follows:

(1) Direct hippocampo-mesencephalic projections appear to distribute only to the rostral part of the central grey substance.

(2) Indirect hippocampal pathways, in part interrupted by further relays, and probably also representing indirect projections of the amygdaloid complex, originate in the septum and in the lateral preoptic and hypothalamic regions, as well as in the mammillary body. Such pathways reach the mid-brain in three fibre systems, viz. the medial fore-brain bundle, the fasciculus retroflexus, and the mammillo-tegmental tract.

(3) Each of these three bundles has a dual distribution in the mid-brain: part of their fibres terminate in extensive central and lateral regions of the tegmentum, while the remaining component distributes to various subdivisions of a paramedian mid-brain region encompassing the ventral part of the central grey substance, Tsai's ventral tegmental region, interpeduncular nucleus, Bechterew's n. centralis tegmenti superior, and Gudden's dorsal and deep tegmental nuclei.

(4) The connexions with this medial ("limbic") mid-brain region are reciprocated by ascending pathways, part of which ultimately lead to the hippocampus and amygdaloid complex. Such ascending connexions appear to close a limbic system—mid-brain circuit, which in turn has extensive collateral connexions with the hypothalamus, as well as with the mesencephalic components of the diffuse ascending and descending reticular conduction systems.

ACKNOWLEDGMENTS

The author is greatly indebted to Mrs. Mildred H. Albrecht, Miss Ruth Fernstrom, Mrs. Hertha Middleton, and Mr. Gordon Fletcher for their technical assistance in this study.

ABBREVIATIONS IN FIGURES 1–20

Aa: anterior amygdaloid area; Ab: basal amygdaloid nucleus; Ac: central amygdaloid nucleus; AD: n. anterodorsalis thalami; Al: lateral amygdaloid nucleus; AV: n. anteroventralis thalami; AVT: ventral tegmental area of Tsai; BC: brachium conjunctivum; CI: capsula interna; CM: mammillary body; CP: posterior commissure; D: n. Darkschewitsch; DBC: decussation of brachium conjunctivum; DRS: decussation of rubrospinal tract; F: fornix; GP: globus pallidus; H: field H of Forel; Hl: n. lateralis habenulæ; Hm: n. medialis habenulæ; IP: n. interpeduncularis; Li: n. limitans thalami; LM: medial lemniscus; NC: n. caudatus; NCS: n. centralis tegmenti superior (Bechterew); NGd: n. tegmenti dorsalis (Gudden); NGp: n. tegmenti profundus (Gudden); NR: n. ruber; P: cerebral peduncle; P-C: parafascicular nucleus—centre median; PD: fasciculus predorsalis; PM: mammillary peduncle; Pt: n. paratenialis thalami; PTI: inferior thalamic peduncle; Pul: pulvinar thalami; Put: putamen; PVA: n. periventricularis anterior thalami; PVH: n. paraventricularis hypothalami; PVP: n. periventricularis posterior hypothalami; Rf: fasciculus retroflexus (Meynert); RS: rubrospinal tract; Rt: n. reticularis thalami; SG: n. suprageniculatus; Sm: supramammillary region; SM: stria medullaris; SN: substantia nigra; ST: stria terminalis; STh: n. subthalamicus; Va: n. ventralis anterior thalami; III: n. oculomotorius.

## REFERENCES


ADEY, W. R., MERRILLEES, N. C. R., and SUNDERLAND, S. (1956) *Brain*, **79**, 414.

ARIENS-KAPPERS, C. U., HUBER, G. C., and CROSBY, E. C. (1936) "The Comparative Anatomy of the Nervous System of Vertebrates, Including Man." New York. Vol. II.

BISCHOFF, E. (1900) *Anat. Anz.*, **18**, 348.

BUCHER, V., and BÜRGI, S. (1945) *Confin. neurol.*, **6**, 317.

BÜRGI, S., and BUCHER, V. M. (1955) *Dtsch. Z. Nervenheilk.*, **174**, 89.

CAJAL, RAMON Y S. (1911) "Histologie du Système nerveux de l'Homme et des Vertébrés." Paris, vol. II.

COLLINS, E. H. (1954) *J. comp. Neurol.*, **100**, 661.

CROSBY, E. C., and WOODBURNE, R. T. (1951) *J. comp. neurol.*, **94**, 1.

DAITZ, H. M., and POWELL, T. P. S. (1954) *J. Neurol. Neurosurg. Psychiat.*, **17**, 75.

ECONOMO, C. v (1918) "Die Encephalitis lethargica." Wien.

EDINGER, L., and WALLENBERG, A. (1902) *Arch. Psychiat. Nervenkr.*, **35**, 1.

GANSER, S. (1882) *Morph. Jb.*, **7**, 591.

GLOOR, P. (1955) *Electroenceph. clin. Neurophysiol.*, **7**, 223.

GUDDEN, B. (1881) *Arch. Psychiat. Nervenkr.*, **11**, 428.

—— (1889) "Gesammelte und hinterlassene Abhandlungen." Wiesbaden.

GUILLERY, R. W. (1956) *J. Anat., Lond.*, **90**, 350.

—— (1957) *J. Anat., Lond.*, **91**, 91.

INGRAM, W. R. (1940) *Res. Publ. Ass. nerv. ment. Dis.*, **20**, 195.

JOHNSON, F. H. (1953) *Anat. Rec.*, **115**, 327.

KAADA, B. R., ANDERSEN, P., and JANSEN, E. (1954) *Neurology*, **4**, 48.

——, JANSEN, J., Jr., and ANDERSEN, P. (1953) *Neurology*, **3**, 844.

KOELLIKER, A. (1896) "Handbuch der Gewebelehre des Menschen." 6th edition. Leipzig. Bd. II.

MACLEAN, P. D. (1952) *Electroenceph. clin. Neurophysiol.*, **4**, 407.

—— (1954) *J. Neurosurg.*, **11**, 29.

MAGNUS, O., and LAMMERS, H. J. (1956) *Folia psychiat. neerl.*, **59**, 555.

MAGOUN, H. W., RANSON, S. W., and HETHERINGTON, A. (1938) *Arch. Neurol. Psychiat., Chicago*, **39**, 1127.

MORUZZI, G., and MAGOUN, H. W. (1949) *Electroenceph. clin. Neurophysiol.*, **1**, 455.

NAUTA, W. J. H. (1946) *J. Neurophysiol.*, **9**, 285.

—— (1956) *J. comp. Neurol.*, **104**, 247.

——, and GYGAX, P. A. (1954) *Stain Technol.*, **29**, 91.

——, and KUYPERS, H. G. J. M. (1958). In the press.

——, and VALENSTEIN, E. (1958) *Anat. Rec.*

RANSON, S. W. (1939) *Arch. Neurol. Psychiat., Chicago*, **41**, 1.

——, and MAGOUN, H. W. (1939) *Ergebn. Physiol.*, **41**, 56.

SANZ IBÁÑEZ, J. (1935) *Trab. Lab. Invest. biol. Univ. Madr.*, **30**, 211.

SHEEHAN, D. (1933) *Arb. neurol. Inst. Univ., Wien*, **35**, 1.

TELLO, J. F. (1936) *Trab. Lab. Invest. biol. Univ. Madr.*, **31**, 77.

THOMAS, D. M., KAUFMAN, R. P., SPRAGUE, J. M., and CHAMBERS, W. W. (1956) *J. Anat., Lond.*, **90**, 371.

TSAI, C. (1925) *J. comp. Neurol.*, **39**, 217.

VAN VALKENBURG, C. T. (1912) "Caudal Connections of the Corpus Mammillare." *Proc. kon. ned. Akad. Wet.*, **4**, 1118.

VOGT, O. (1898) *C. R. Soc. Biol., Paris*, **50**, 206.

VOTAW, C. (1957) *Anat. Rec.*, **127**, 382.

WALLENBERG, A. (1901) *Anat. Anz.*, **20**, 175.

WEISSCHEDEL, E. (1937) *Arch. Psychiat. Nervenkr.*, **107**, 443.

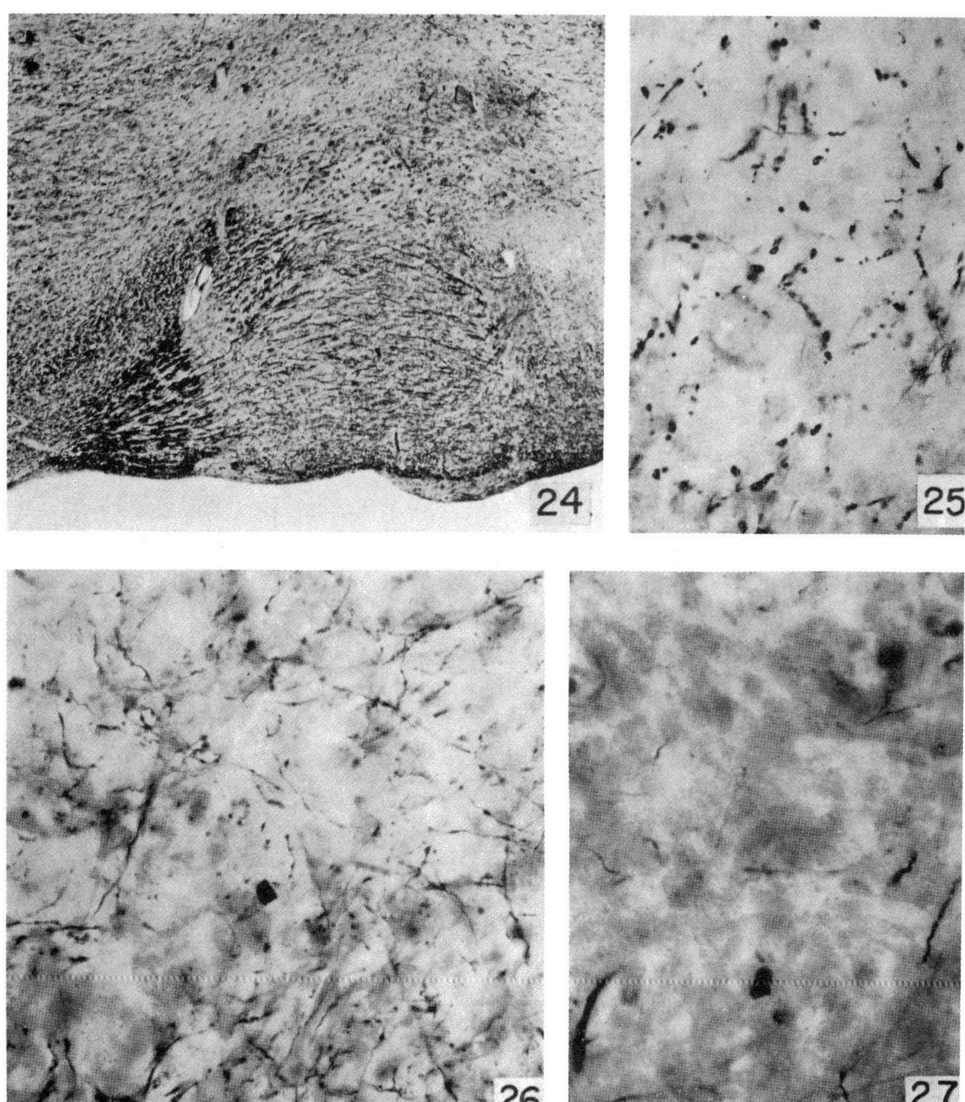

FIG. 24.—Unilateral degeneration of the fasciculus retroflexus following interruption of the bundle at its exit from the habenula (Case VA 4). Note the compact bundle releasing slender fascicles into n. interpeduncularis, and the more diffuse degenerating fibre group dorsal to it. ×45.

FIG. 25.—Axon degeneration in Bechterew's n. centralis tegmenti superior following caudolateral lesion of the hypothalamus (Case CT 99). ×600.

FIG. 26.—Degeneration of fine axons in Gudden's deep tegmental nucleus following ipsilateral lesion of the mammillo-tegmental tract (Case CT 99). ×600. Compare with fig. 27.

FIG. 27.—Absence of axon degeneration in Gudden's deep tegmental nucleus contralateral to lesion of mammillo-tegmental tract. Same case as fig. 26. ×600.

*To illustrate article by Walle J. H. Nauta.*

Reprinted from THE JOURNAL OF COMPARATIVE NEUROLOGY
Vol. 113, No. 3, December 1959

# A COMPARISON OF THE DISTRIBUTION OF THE FORNIX SYSTEM IN THE RAT, GUINEA PIG, CAT, AND MONKEY

ELLIOT S. VALENSTEIN AND WALLE J. H. NAUTA
*Departments of Experimental Psychology and Neurophysiology,
Walter Reed Army Institute of Research,
Washington, D. C.*

TWENTY-FIVE FIGURES

Detailed anatomical studies of the mammalian fornix system have, with few exceptions, been restricted to the lower mammals, viz., the mole, rabbit, and rodents. This is especially true of studies conducted by the aid of degeneration techniques, with the result that little is known of the mode of distribution of the fornix in higher mammalian forms.

The present study was suggested several years ago by incidental observations which indicated a marked difference in the volumes of the fornix connection with the central gray midbrain substance in the rat and cat respectively, and a further, though less striking dissimilarity between these two species in the thalamic distribution of certain fornix fibers. The monkey and guinea pig were included in this study for the purpose of further comparison.

## MATERIAL AND METHODS

Localized lesions were made in selected regions in the hippocampus, fimbria fornicis, fornix dorsalis, and septal region using 7 rats, 20 guinea pigs, 6 cats and 4 monkeys. Following a survival period of 6 to 12 days the animals were killed by an overdose of nembutal. Serial frozen sections of the brains were impregnated by the Nauta-Gygax ('54) silver technique. The localization of axon degeneration was determined microscopically and recorded on enlarged photographs or projection drawings of sections at representative levels.

OBSERVATIONS

In all of the species examined, maximal degeneration of the fornix system was found following lesions of the septal region. In this region the two major components of the fornix system (fimbria fornicis and fornix dorsalis) converge to form the compact fornix column as well as the more diffuse but no less massive precommissural fornix. Both groups of fornix fibers together with the contributions to the fornix system arising in the septal cell groups proper, can be completely interrupted by relatively small septal lesions. For this reason the distribution of the fornix in each of the species studied can be appreciated most readily from a description of the fiber degeneration following a septal lesion. The differential distribution of the various components of the system will, so far as significant data are available, be discussed thereafter.

I. Guinea pig

*1. Lesion of the septal region*

*a. Case GPS 3.* The lesion in this case, produced electrolytically by the aid of a stereotactic instrument, had destroyed a large part of the septum on the right side, and involved a narrow medial zone of the same structure on the left. Approaching the dorsal border of the anterior commissure the defect had completely severed the right fornix column. On the left side, a medial part of the fornix dorsalis had been interrupted in its ventral sweep through the septum. The lesion extended caudally into the ventral psalterium.

As illustrated by figures 1–8, the resultant fiber degeneration is similar in many respects to that observed following comparable lesions in the rat (Nauta, '56): numerous degenerating precommissural fibers follow Zuckerkandl's bundle into the medial forebrain bundle with which they are

141

distributed to the lateral preoptic and hypothalamic regions (figs. 1–4), while the degenerated main fornix bundle, en route to the mammillary body, sends a sizable fiber contingent dorsalwards which distributes itself to rostral thalamic cell groups (figs. 2–3). However, a more detailed examination of the thalamic fornix offset (Gudden's (1881) "upper uncrossed fornix bundle") reveals that the distribution of this fiber group does not correspond completely to the previous findings in the rat. Whereas most of the fibers in question terminate in the rat in the nucleus anterior ventralis, the main distribution area in the guinea pig appears to be an intralaminar cell region immediately ventral to the anterior nucleus (figs. 2–3). Only the caudal part of the nuclei anterior ventralis and anterior medialis appear to receive some fibers directly from the fornix (fig. 3). The intralaminar cell region in question appears to be a rostral extension of the nucleus centralis lateralis; locally it can hardly be delimited from the ventrally adjoining nuclei ventralis anterior and reticularis thalami. Moreover, in contrast to the findings in the rat, a few degenerating fornix fibers enter and terminate in a rostral part of the lateral thalamic complex (nucleus lateralis dorsalis, fig. 3).

In addition to these variations in the thalamic distribution of the fornix system, a striking difference from the findings in the rat is encountered in the number and distribution of the fornix fibers which bypass the mammillary body. This postmammillary fornix component is far more massive than it is in the rat. It decussates in large part in the dorsal capsule of the mammillary body and especially in the postmammillary fornix decussation (Ganser's (1881) posterior hypothalamic decussation, see figs. 6, 22–23). After this incomplete decussation, the postmammillary fornix divides into three components, one of which spreads dorsalward along the medial plane to the rostral part of the central gray midbrain substance and posterior hypothalamic nucleus (figs. 6–7), while a second fiber group, the largest of the three

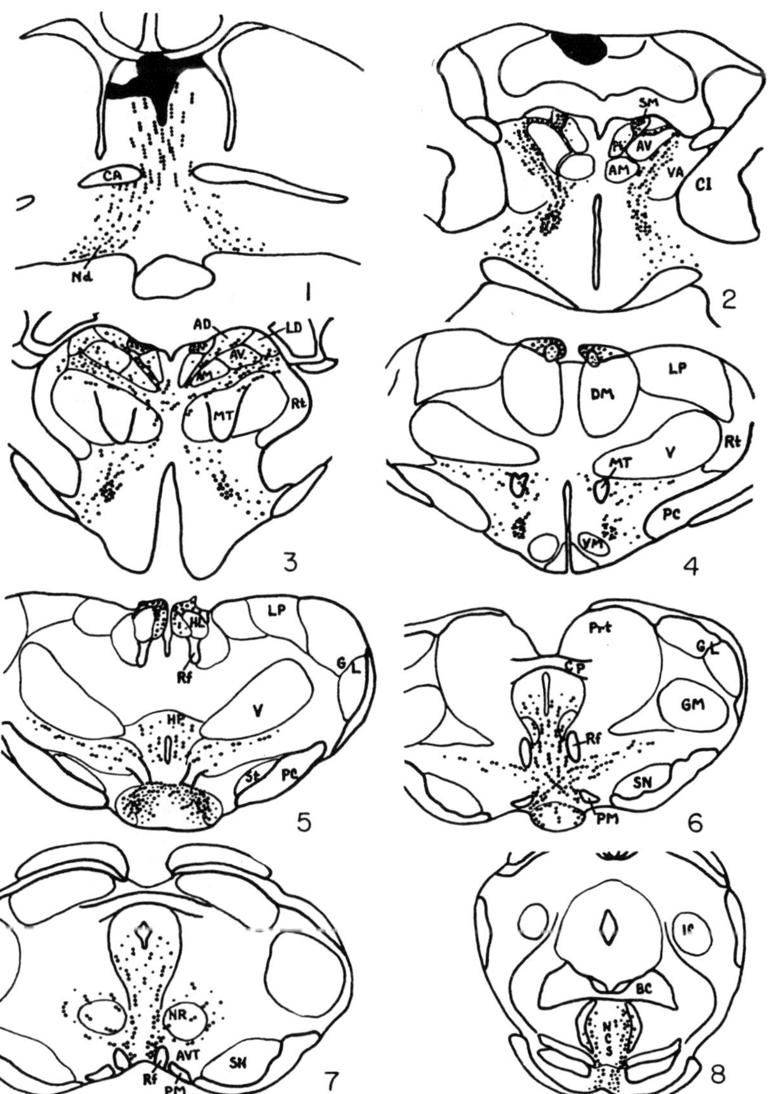

Figs. 1–8   Fiber degeneration following septal lesion in the guinea pig (case GPS 3).   Coarse dots indicate degenerating fibers of passage; fine stipple, degenerating terminal fibers.

components and apparently also the one containing the largest proportion of crossed fibers, continues caudally in a medial and ventral position, first medial to the fasciculus retroflexus (fig. 7), then dorsal to the interpeduncular nucleus. Caudal to the latter structure it curves dorsalward along the lateral side of Bechterew's nucleus centralis tegmenti superior, and appears to terminate in this nucleus as well as in the laterally adjoining medial tegmental region (figs. 8, 24–25). None of its fibers appears to reach the neighboring tegmental nuclei of Gudden or the caudal central gray substance.

Both of the above-mentioned postmammillary fornix bundles differ from similar fiber groups in the rat only by their much larger volume. A third mesencephalic component of the guinea pig fornix, however, has not been observed in the rat. It consists of a large number of fibers which, after partial crossing through the fornix decussation, spread laterally and distribute themselves widely to central regions of the mid-

---

*Abbreviations to figures*

| | |
|---|---|
| AD, n. anterior dorsalis thalami | NCS, n. centralis tegmenti superior |
| AM, n. anterior medialis thalami | (Bechterew) |
| AV, n. anterior ventralis thalami | Nd, diagonal band |
| AVT. ventral tegmental area | NR, n. ruber |
| BC, brachium conjunctivum | PC, cerebral peduncle |
| CA, anterior commissure | PM, mammillary peduncle |
| CI, internal capsule | Prt, pretectal region |
| CM, mammillary body | Pt, n. paratenialis thalami |
| CMd, centre median | Re, n. reuniens |
| CP, posterior commissure | RF, fasciculus retroflexus |
| DM, n. dorsomedialis thalami | Rt, n. reticularis thalami |
| GL, lateral geniculate body | SM, stria medullaris |
| GM, medial geniculate body | SN, substantia nigra |
| H, field H of Forel | ST, stria terminalis |
| Hl, n. lateralis habenulae | St, n. subthalamicus |
| Hm, n. medialis habenulae | Tol, olfactory tubercle |
| HP, n. posterior hypothalami | V, n. ventralis thalami |
| IC, inferior colliculus | VA, n. ventralis anterior thalami |
| LD, n. lateralis dorsalis thalami | VL, n. ventralis lateralis thalami |
| MFB, medial forebrain bundle | VM, n. ventromedialis hypothalami |
| MT, mammillo-thalamic tract | ZI, zona incerta |

brain tegmentum (fig. 6) and to a rostrally adjoining region of the subthalamic field H of Forel (figs. 4–5). The existence of this direct connection with more lateral tegmental areas represents the most striking difference between the fornix systems of the guinea pig and rat respectively.

In a previous study of the fornix system in the rat (Nauta, '56), fornix fibers to the caudomedial midbrain tegmentum (Bechterew's nucleus and surroundings) were believed to arise in the septal region. Several observations made in the present study, however, militate against this conclusion. A case in point is furnished by the findings in GPS 5, to be described next.

*b. Case GPS 5.* Stereotactic surgery had in this case produced an unusual lesion which destroyed parts of both the medial and lateral nuclei of the septum. Because of an oblique contralateral approach, the lesion left most of the fibers of the fornix dorsalis undisturbed. The fornix column was likewise intact, but some lateral fibers of the fimbria fornicis were interrupted by the caudal extension of the defect.

Only sparse degeneration is evident in the fornix column. Immediately behind the interventricular foramen, however, the dorsal aspect of the fornix bundle is crowded by numerous degenerating axons, some of which loop medially over the bundle and continue ventrally to form the medial cortico-hypothalamic fascicles. The distribution of this medial hypothalamic connection of the fornix system corresponds in detail to that observed in the rat (Nauta, '56). As in the latter species, its degeneration appears to be caused by interruption of lateral fibers of the fimbrial fornix component originating in distal parts of the hippocampal formation.

Very few fibers can be followed to the thalamus, except a fair number of degenerating elements which enter the stria medullaris and can be traced to the medial habenular nucleus.

The most massive fiber degeneration observed in this case follows Zuckerkandl's precommissural bundle into the nucleus of the diagonal band and the medial forebrain bundle. The degeneration in the latter fiber system is quite dense. It spreads diffusely over the lateral preoptic and hypothalamic cell groups, but few if any disintegrating fibers appear to extend into the ventral tegmental area (Papez's ('38) nucleus of the mammillary peduncle). The postmammillary fornix decussation is intact; no degenerating fibers appear in the central gray substance or any other mesencephalic structure.

It appears likely from the foregoing case that fornix fibers of mesencephalic distribution do not, as suggested previously, arise in the septal region. Instead, a comparison of cases GPS 3 and GPS 5 suggests that most at least of the postmammillary fornix components could be conveyed by the fornix dorsalis, the latter bundle being intact in GPS 5 and interrupted in GPS 3. This suggestion is substantiated by the findings made in 5 cases of selective interruption of the fornix dorsalis. One of these cases will be described next.

## 2. *Lesion of the fornix dorsalis*

*Case GPH 15.* By the use of an extremely fine electrode the fornix dorsalis was interrupted completely on the left side at the level of the ventral psalterium. There was only slight damage to the corresponding bundle on the left, and minimal involvement of the psalterium and its interstitial commissural nucleus.

A short distance rostral to the lesion the degenerated fornix dorsalis divides into a precommissural and a columnar component. The precommissural fibers bend sharply ventralward along the median plane, and there is evidence of terminal degeneration among the cells of the medial septal nucleus. However, only a few of the degenerating fibers continue ventralward beyond the horizontal level of the anterior commissure, and none can be followed into either the diagonal band or the medial forebrain bundle.

The degenerating fibers of the columnar component of the fornix dorsalis mingle diffusely with the normal fibers of the fimbria, as shown by the even scattering of degenerating axons amongst the more numerous normal fibers of the fornix bundle.

The fiber degeneration in the thalamus, although considerably less massive, corresponds in its distribution to that found following a septal lesion: degenerating fibers spread to rostral intralaminar and anterior nuclei, as well as to the medial habenular nucleus. The remaining degeneration in the fornix bundle extends to the mammillary body in the customary fashion, i.e., principally to the lateral part of the medial mammillary nucleus, whereas numerous degenerating fornix fibers bypass the mammillary body, and as in GPS 3, undergo a partial decussation to spread laterally to the subthalamic region and central midbrain tegmentum, dorsally to the central gray substance, and caudally to medial regions of the caudal mesencephalon. (see figs. 22–25).

Apart from the considerably smaller amount of degeneration in the diencephalon, and the virtual absence of degenerating fibers in the medial forebrain bundle, the degeneration in this case of a fornix dorsalis lesion is quite similar to that following most lesions of the septal region. This is especially true of the postmammillary fornix degeneration which is approximately equal in volume to that found in the cases with a large septal lesion (e.g., GPS 3).

Comparable fornix dorsalis lesions in three rats also produced degeneration of long postmammillary fornix fibers to the rostral part of the pontine protuberance and Bechterew's nucleus. Although entirely comparable to the postmammillary fornix degeneration in the guinea pig as regards distribution area, the amount of degeneration was, however, much smaller in the rat.

### 3. Lesions of the hippocampus: case GPH 3

A large part of the hippocampal formation had been removed in this case by suction following extensive ablation of the overlying cortex. The lesion was largely confined to the caudal half of the hippocampus.

The fornix degeneration in this animal corresponds almost entirely to the findings made in a previous study of the rat

(Nauta, '56). Degenerating precommissural fibers distribute to both the medial and lateral septal nuclei and to the nucleus of the diagonal band. A moderate number of these fibers follow the medial forebrain bundle to the lateral preoptic nucleus, but none appears to continue caudally to the lateral hypothalamic region.

The degeneration in the fornix column can be followed to (a) the periventricular zone of the anterior hypothalamus via the medial cortico-hypothalamic fascicles; (b) rostral intralaminar and anterior thalamic nuclei; (c) the mammillary body, and (d) the rostral part of the central gray substance. As in the rat, the last-mentioned hippocampal connection is established by only a small number of fibers. No other postmammillary fornix fibers are degenerated.

Finally, it is noteworthy that, in contrast to the findings in the rat, a few degenerating fibers can be followed from the fornix bundle via the stria medullaris to the medial habenular nucleus.

*4. Conclusions from the findings in the guinea pig*

The observations reported above indicate that direct hippocampal pathways in the guinea pig, although distributing widely in the diencephalon, do not extend farther caudalward than the rostral part of the central gray substance of the midbrain. Fornix fibers to more caudal mesencephalic regions appear to be supplied at least in large part by the fornix dorsalis. Almost all of the fornix connections with the lateral hypothalamic region are established by way of the medial forebrain bundle, by the intermediary of the septal and preoptic regions. Fornix fibers to rostral intralaminar and anterior thalamic nuclei are contained in both the fornix dorsalis and fimbria fornicis. Fornix connections with the medial habenular nucleus, conveyed by the stria medullaris, come principally from the septal region but also from the hippocampus and from the fornix dorsalis.

## II. Cat

### 1. Lesion of the septal region: case CF 13

The lesion, produced by means of a stereotaxic apparatus, was confined in this case largely to the supracommissural septum where it occupied a paramedian zone of approximately 1.5 mm in width on the left, with some slight involvement of the right side. The left fornix column had been almost completely destroyed. Extending ventralward to the anterior commissure, the lesion had also interrupted a large part of the pars precommissuralis striae terminalis.

As shown in figures 9–14, massive fiber degeneration follows the precommissural fornix, as well as the fornix column and stria terminalis. As in the two species of rodent, the degenerating precommissural fornix fibers can be followed to the diagonal band, and via the medial forebrain bundle to the entire extent of the lateral preoptic and hypothalamic regions. A small part of the medial forebrain bundle degeneration extends into the ventral tegmental area of the rostral midbrain.

Degenerating fibers of the compact fornix bundle are distributed to the thalamus, mammillary body and central gray substance. The findings in the thalamus correspond largely to those made in the guinea pig: thalamic fornix offset divides into (a) a fiber group which follows the stria medullaris to the medial habenular nucleus, and (b) a large fiber contingent which courses laterally and caudally through the internal medullary lamina. The latter fiber group, as in the guinea pig, but unlike the rat, terminates largely in the intralaminar cell region ventral to the anterior thalamic nucleus (fig. 11); only few of its fibers appear to ramify in the anteromedial and anteroventral thalamic nuclei. Many fibers of the same group enter the stratum zonale thalami by passing over the rostral aspect of the anterior nucleus (fig. 10) as well as around these cell groups (figs. 11–12). Some of these zonal fibers distribute to the anterodorsal nucleus, but

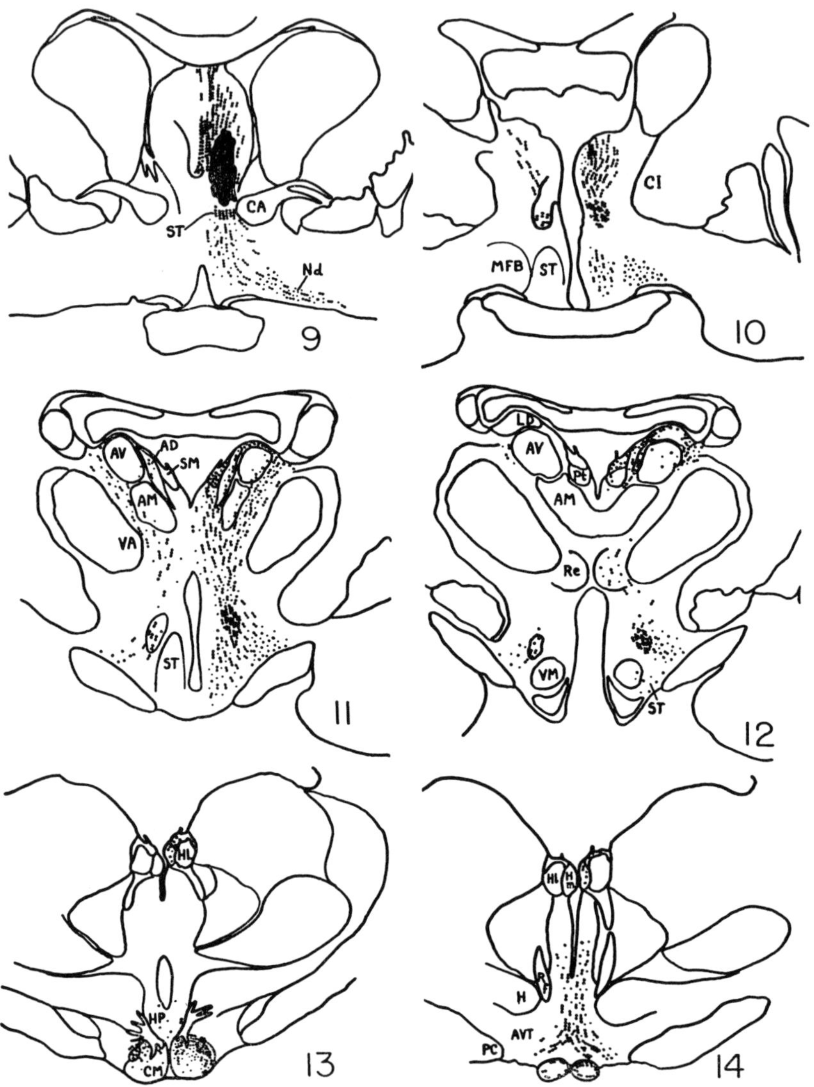

Figs. 9–14  Fiber degeneration observed in a case of septal lesion in the cat (case CF 13). In addition to the fornix system, the lesion has interrupted the precommissural component of the stria terminalis. Symbols as in figures 1–8

others continue caudally to end in the anterior part of the nucleus lateralis dorsalis (fig. 12).

Figure 13 shows the dense termination of the degenerating fornix in the mammillary body. Numerous fascicles of degenerating fibers bypass the mammillary body and turn sharply dorsalward immediately caudal to this structure. After a partial decussation (fig. 14) these fornix components are distributed to the rostral part of the central gray substance. In contrast to the findings in the guinea pig, however, no fibers of the postmammillary fornix appear to extend farther caudally in the mesencephalon; nor can any degenerating fibers be traced laterally to central tegmental regions or the subthalamic area.

### 2. Lesions of the hippocampus: case CF 5

In this case approximately the caudal half of the hippocampus had been extensively damaged by aspiration, following exposure of Ammon's horn by means of ablation of most of the posterior ectosylvian gyrus.

Massive fiber degeneration extends forward from the defect in the fimbria fornicis. The fornix dorsalis appears entirely normal. The distribution of the degenerating fibers is similar to that found in the foregoing case of septal lesion, except for a complete absence of degeneration in the stria medullaris and habenula. As in the guinea pig and rat, massive degeneration spreads over the entire septal region with the exception of the nucleus accumbens. Numerous degenerating precommissural fibers extend into the diagonal band and, via the medial forebrain bundle, into the lateral preoptic region. In contrast to the observations in the rat and guinea pig, an appreciable number of degenerating fibers is present in the lateral hypothalamic region, suggesting the existence of relatively long precommissural fornix fibers in the cat's medial forebrain bundle. Furthermore, the fiber degeneration in the lateral hypothalamic region appears to be augmented by short degenerating fascicles which enter the region at intervals from the fornix bundle proper. These direct connections of the hippocampus with the lateral hypothalamus are much less well-developed in the rat and guinea pig. On the other hand, there is no evidence in the cat of a direct hippocampal pathway to more medial zones of the hypothalamus. As pointed out above, the latter connection is established in the two rodent species by the quite prominent medial cortico-hypothalamic fascicles.

Postmammillary fornix degeneration is hardly less massive in this case than in the cases of septal lesion. After crossing primarily in the fornix decussation, its distribution appears to be limited to the rostral central gray substance. No degeneration can be traced to the caudal mesencephalic tegmentum.

*Conclusions from the findings in the cat.* It is evident from the two foregoing cases that in the cat no part of the fornix system extends caudalward beyond the most rostral mesencephalic levels. Most of these postmammillary fornix fibers — they are much more numerous than in the rat and comparable in number to those found in the guinea pig — are distributed to the anterior part of the central gray substance. Furthermore, direct hippocampal connections with the lateral hypothalamus are much more developed in the cat than in the two rodent species. By contrast, the cat appears to lack the prominent direct hippocampal connection with the medial hypothalamic zones which is found in the rodents. The thalamic distribution of the fornix system in the cat is comparable to that in the guinea pig, with this exception: in the cat, as in the rat, fornix fibers to the habenula appear to arise only in the septal region.

## III. Monkey

Lesions of the fornix were produced in 6 monkeys under direct vision. The corpus callosum was exposed by retraction of the left hemisphere, and divided by a longitudinal incision leading into the frontal horn of the opposite lateral ventricle. The fornix of that side was visualized by gentle retraction of the septum, and divided by ''hooking'' and avulsion at or a slight distance behind the interventricular foramen. In three of the animals the septal region of the same side was then destroyed by aspiration. Unfortunately, it was found that in the three cases in which no septal lesion was intended, the avulsion of the fornix behind Monroe's foramen had caused varying degrees of damage to the supracommis-

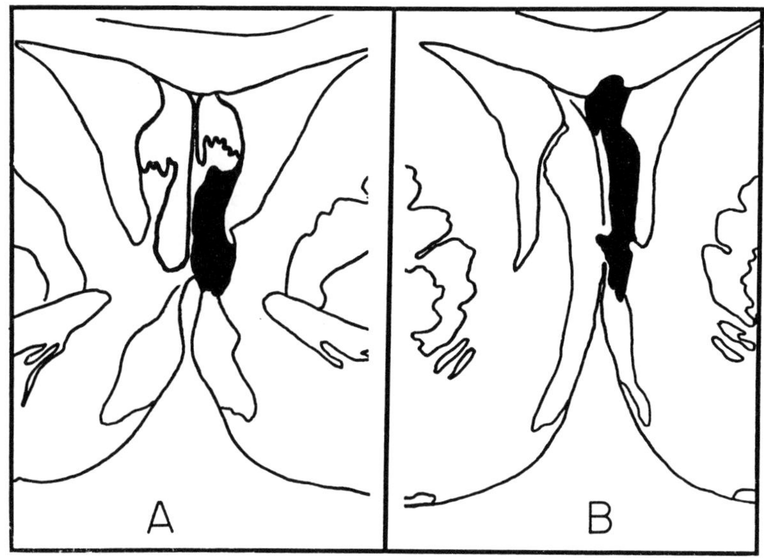


Fig. 15   Extent of the septal lesion in monkey MF 10.

sural septum. Therefore no uncomplicated case of a fornix lesion can be reported here.

*Lesion of the fornix and septal region: case MF 10.* As shown by figure 15, the lesion in this case completely divided the fornix column on the right side. On the same side, most of the supracommissural septum had been removed. In addition, a ventral extension of the septal lesion almost completely interrupted the stria terminalis at the point of its passage over the anterior commissure. There was some slight involvement of the medial part of the left fornix column.

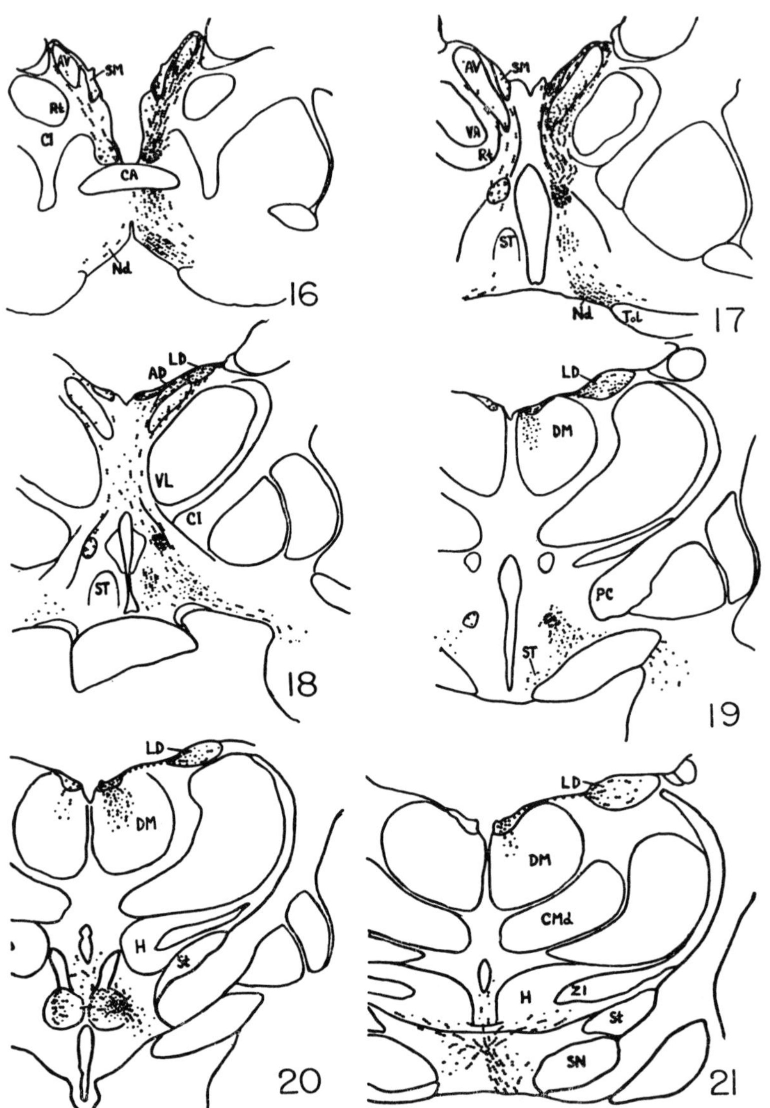


Figs. 16–21  Distribution of fiber degeneration in a case of septal lesion in the monkey (case MF 10). The extent of the lesion is shown in figure 15.

As illustrated by figures 16–21, the distribution of the fornix degeneration is comparable to that observed in other species. However, several noteworthy differences appear upon more detailed comparison. In the first place, there is a considerably greater development of the connection of the thalamic fornix bundle with the nucleus lateralis dorsalis thalami (figs. 18–21). Second, a connection not encountered in any of the other mammalian species is represented by degenerating fibers of the stria medullaris which terminate in the medial, magnocellular division of the dorsomedial nucleus (figs. 19–21). Finally, it is remarkable that the precommissural fornix degeneration which follows the diagonal band extends much farther laterally than was noted in other species. A small portion of these degenerating fibers actually enter the amygdaloid complex where their termination appears to be confined to the medial amygdaloid nucleus (fig. 19).

The distribution of postmammillary fornix fibers is much the same as that observed in the cat; after a partial decussation these fibers, which are considerable in number, distribute to the rostral part of the central gray substance (fig. 21); none can be traced farther caudalward in the midbrain. However, a small contingent of the postmammillary fornix fibers, after an apparently complete decussation, continue laterally and are distributed to a ventral region of Forel's tegmental field H in the subthalamic region (fig. 21). This subthalamic fornix component is quite comparable to the somewhat more massive lateral fornix connections encountered in the guinea pig.

DISCUSSION

Edinger and Wallenberg ('02) appear to have been the first to recognize interspecific variation as a possible source of disagreement between investigators of the fornix system. It is interesting that Edinger and Wallenberg based this suggestion upon their observation of rather marked dissimilarities in the postmammillary fornix connections in different

155

strains of rabbits. Later, Tello ('36) noted considerable variation in the same mesencephalic fornix connections when comparing his findings in the rat and mouse with those in the cat and guinea pig.

The present study has served to emphasize further the existence of interspecific variations in the fornix distribution. Such variations were noted not only in the mesencephalic connections, but also in the diencephalic distribution of the fornix system.

*The thalamic connections of the fornix.* The findings reported above indicate that the thalamic offset of the fornix bundle terminates massively in the anterior nuclear complex in the rat, and also to a lesser extent in the monkey, while the same fiber group in the cat and guinea pig appears to distribute largely to a rostral intralaminar cell group situated between the nuclei anterior and ventralis anterior thalami. In both of the last-mentioned species, however, at least some fornix fibers end directly in the anterior nucleus, while conversely in the rat and monkey a small contingent of thalamopetal fornix fibers connects with the rostral intralaminar zone.

A further point of variation is encountered in the monkey, in which a considerable number of fornix fibers terminate in the nucleus lateralis dorsalis, establishing a connection which is only just identifiable in the other species studied.

Although the functional significance of these interspecific differences must at present remain a matter of speculation, the massive fornix connection with the anterior thalamic nucleus in the rat could be interpreted as evidence of a more direct trans-thalamic influence of certain fornix impulses upon the mechanisms of the cingulate gyrus than would appear likely in the other species. The numerous fornix fibers of the monkey which terminate in the laterodorsal nucleus suggest more direct connections with the parietal association area adjoining the cingulate gyrus than appear to exist in either the rat, guinea pig, or cat. It is, however, of interest in this connection to note that in the rat the laterodorsal nucleus receives, instead of direct fornix fibers, a rather massive projection from the caudal granular region of the cingulate gyrus (Nauta, '53). This cingulate projection is also prominent in the cat and guinea pig, but has not been investigated in the monkey. In any case, with proper qualification it would seem justifiable to consider the most rostral part of the lateral thalamic nucleus to have marked afferent relations to limbic regions of the cerebral mantle.

Also in the monkey, a somewhat comparable connection in the fornix system is represented by fornix fibers which follow the stria medullaris and terminate in the medial subdivision of the dorsomedial thalamic nucleus. It seems likely that these fibers correspond to the septo-dorsomedial pathway recently identified in the cat by Guillery ('59). The absence of a corresponding degeneration in our cases of septal lesion in the cat suggests that this connection in the cat originates in ventral, i.e., subcommissural parts of the septal region not involved in any of the animals used in the present study. The pathway in question may be comparable to a massive projection of much the same distribution which arises in the amygdaloid complex in the monkey (Nauta and Valenstein, '58). It furnishes another example of limbic pathways articulating with "specific" thalamic cell groups, in this case one related to the orbitofrontal cortex.

Finally, it is interesting to note that the fornix connection with the medial habenular nucleus in the rat and cat appears to arise in the septal region exclusively, while in the guinea pig this connection has a wider origin which includes the hippocampus proper, and probably also the caudal part of the cingulate gyrus (see the discussion of the fornix dorsalis, below).

*Hypothalamic connections of the fornix system.* Of the direct cornu-hypothalamic connections, those with the lateral preoptic and hypothalamic regions appear to be the most consistent. A more detailed study of this connection, however, brings to light certain interesting variations in the organization of the pathways involved. In all 4 mammalian species here studied the precommissural fornix appears to constitute the main hippocampal contribution to the medial forebrain bundle and, hence, to the afferent fiber connections of the lateral preoptic and hypothalamic cell groups. In the rat and guinea pig, however, hardly any such precommissural fornix fibers extend caudally beyond the lateral preoptic region. In these same species extremely few fibers enter the lateral hypothalamus as premammillary offsets from the compact fornix bundle. It can be inferred that in the rat and guinea pig the hippocampal connection with the lateral hypothalamus is almost entirely indirect, with relays in both the septal and lateral preoptic cell groups (Nauta, '56). In the cat, by contrast, a considerable number of hippocampal fornix fibers distribute directly to the lateral hypothalamic region, both as precommissural fornix fibers accompanying the medial forebrain bundle and as premammillary ramifications of the compact fornix bundle. The organization of the corresponding pathways in the monkey could not be ascertained, as no cases of isolated lesion of the hippocampal fornix were available.

Fornix connections with the medial and periventricular zones of the hypothalamus exhibit more striking evidence of species difference. In both the rat and guinea pig, a fairly massive fiber system, the medial cortico-hypothalamic tract, connects the hippocampus directly with the periventricular zone of the preoptic and rostral hypothalamic regions; its longest elements appear to reach the arcuate nucleus of the infundibular region. According to previous findings in the rat (Nauta, '56) this fiber group arises exclusively in the caudal part of the hippocampal formation. It is interesting that no corresponding medial fornix connections could be identified in either the cat or monkey. It would thus appear that in both these higher mammalian forms impulses of hippocampal origin can reach the medial and periventricular hypothalamic regions only by way of synaptic relays in adjoining regions such as the lateral hypothalamic which in the cat has been shown to connect with the more medial hypothalamic zones by short transverse fibers (Nauta, '58). Such a lateral "re-routing" of the prominent direct medial hippocampo-hypothalamic connection of the rodent forms could explain in part at least the relatively large number of hippocampal fornix fibers to the lateral hypothalamic region in the cat. It seems likely, however, that in all species additional hippocampal connections with the medial hypothalamic zones are established by precommissural fornix fibers via relays in the septal region.

*Mammillary connections of the fornix.* This connection shows a remarkable similarity in the 4 species studied. In all, the fornix terminates most massively in the lateral part of the medial mammillary nucleus, and to a considerably lesser extent in the remaining parts of the mammillary body. The only noteworthy variation was observed in the cat, in which an appreciable number of fornix fibers distribute to the contralateral mammillary body.

*Postmammillary fornix connections.* In all species studied here a part of the fornix bundle bypasses the mammillary body and distributes to the rostral part of the central gray midbrain substance. In the guinea pig and monkey it distributes as well to the subthalamic region and adjoining central tegmental areas, and in the guinea pig and rat to medial regions of the caudal mesencephalic tegmentum, particularly the nucleus centralis tegmenti superior of Bechterew.

a. Fornix fibers passing via a postmammillary hemidecussation (Ganser's (1881) "decussatio hypothalamica posterior") to the central gray appear to have been observed for the first time in Marchi experiments in the rabbit by Edinger and Wallenberg ('02). It is of interest that these authors found the connection well-developed in one particular rabbit strain ("Riesenlapin"), but failed to demonstrate it in another type of rabbit, and in the mouse. Furthermore, Tello ('36) from an admirable study of normal material, concluded that the fornix in the mouse and rabbit terminates almost entirely in the mammillary body, but in the guinea pig and cat connects massively with the central gray substance.

The present findings agree well with Tello's observations. Although a partially decussating fornix contingent to the central gray substance was observed in all 4 species studied, such fibers seemed sparse in the rat (Nauta, '56), and numerous in the guinea pig, cat, and monkey. The distribution area of these fibers appears to be limited to the caudal part of the posterior hypothalamic nucleus and the ventral half of the central gray substance, and extends caudalward no farther than the level of the middle one-third of the oculomotor nucleus. The nucleus of Darkschewitsch seems to receive some such fibers in the cat, but no corresponding connection could be identified in any of the other species studied.

b. Fornix fibers to the subthalamic region were observed only in the guinea pig and monkey. This connection, quite conspicuous in the guinea pig but less striking in the monkey, is established by slender and discrete postmammillary fornix bundles which separate laterally from the group of fascicles destined for the central gray substance. In the monkey, the distribution of the fibers in question appears to be limited to the caudal subthalamic region proper, but the corresponding fascicles in the guinea pig spread more widely and also distribute to adjoining regions of the central mesencephalic tegmentum. Either connection could provide direct pathways for certain fornix impulses to the "mainstream" region of longitudinal intra-reticular conduction. No such direct lateral connection in the fornix system could be identified in either the rat or cat, but in the cat comparable bundles with a lateral distribution are issued by the mammillo-thalamic tract (Nauta, '58). Thus it would seem possible to conclude that although direct fornix fibers to the subthalamic region appear to be lacking in the cat, a comparable but indirect connection in this species is mediated by the mammillary body.

c. Fornix fibers which extend as far caudally as the level of the inferior colliculus were found only in the rat and guinea pig. This component of the postmammillary fornix, particularly massive in the guinea pig, constitutes a direct limbic projection to paramedian regions of the caudal mesencephalic tegmentum and pontine protuberance and to the lateral zone of Bechterew's nucleus centralis tegmenti superior. Although this direct fornix connection with the caudal midbrain tegmentum appears to be absent in the cat and monkey, a pathway with identical distribution in the cat originates in the lateral hypothalamic region as a component of the medial forebrain bundle (Nauta, '58). It thus appears that the direct and largely crossed fornix connection with the caudal mesencephalic tegmentum in the rat and guinea pig is supplanted in the cat by a largely uncrossed, indirect pathway relayed by the lateral hypothalamic region.[1] This observation provides a further example of marked and seemingly fundamental interspecific variation in the fornix connections, which upon closer analysis appears to concern the neuronal organization rather than the ultimate distribution of the pathways involved.

*Fornix dorsalis.* It is clear from the present study that the long fornix fibers which in the rodents extend to the caudal mesencephalic tegmentum do not degenerate following either large lesions of the hippocampus, or septal lesions sparing the fornix dorsalis. Such fibers do, however, disintegrate as a result of selective interruption of the fornix dorsalis. The conclusion is inescapable that fornix fibers to the caudal mesencephalon, instead of being of septal origin as suggested in an earlier paper (Nauta, '56), are components of the fornix dorsalis. This conclusion agrees to some extent with Sprague and Meyer's ('50) observation that in the rabbit degeneration of the postmammillary fornix appeared to be related to a lesion of the fornix dorsalis. However, according to the present study it does not apply to all postmammillary fornix connections, for in the guinea pig at least some

[1] A further polyneuronal fornix connection with the caudal mesencephalic region in question appears to involve relays in the septal and preoptic regions, and the habenular complex (Nauta, '58).

degenerating fornix fibers could be traced to the central gray substance from hippocampal defects which had not caused any degeneration in the fornix dorsalis (cf. the present case GPH 3), while the same connection in the cat appears to be even more independent of the fornix dorsalis (case CF 5).

The importance of distinguishing between the fornix dorsalis and fimbrial fornix fibers becomes evident when it is realized that the fornix dorsalis, in contrast to the fimbria, originates largely at least — according to Cajal ('11) entirely — from structures dorsal to the corpus callosum. Although Elliot Smith (1896) attributed to the fornix dorsalis an origin exclusively from the "dorsal limb of the hippocampus" (i.e., the induseum griseum and striae Lancisi), Ganser (1881), Koelliker (1896) and Cajal ('11) counted the white matter of the gyrus fornicatus among the sources of the fornix dorsalis. The latter notion is supported by the finding (unpublished) of degenerating fibers in the fornix dorsalis of the cat following small lesions located in either the rostral or caudal part of the gyrus cinguli and not involving the induseum griseum. If correct, it would imply a wider telencephalic origin of the fornix system than corresponds to the extent of the hippocampal formation even as outlined broadly by Elliot Smith.

It is interesting that the fornix dorsalis fibers, despite their apparent origin from a structurally very different region of the limbic telencephalon, mingle diffusely with the fimbrial fornix components and are not strikingly different from the latter in their terminal distribution. The exceptions have been noted that (a) the direct fornix connection with the caudal mesencephalon in the rat and guinea pig is established exclusively by the fornix dorsalis, whereas (b) precommissural components of the fornix dorsalis, in contrast to those of the fimbria, do not extend beyond the septal region and thus do not contribute to the medial forebrain bundle.

## SUMMARY

A comparative study of the rat, guinea pig, cat and monkey has revealed considerable variability in the distribution of the mammalian fornix system. Following section of the fornix in the septal region fiber degeneration can be traced in all 4 species to rostral thalamic cell groups and habenula, nucleus of the diagonal band, the preoptic region and hypothalamus, the mammillary body, and via postmammillary fornix fibers to the rostral part of the central gray midbrain substance. Interspecific variations can be summarized as follows.

1. The thalamic distribution of the fornix is primarily to the rostral part of the intralaminar complex in the cat and guinea pig, with only few fibers ending in the anterior nucleus. In the rat and monkey this ratio is reversed.

2. In all 4 species, fornix fibers to the hypothalamus terminate in the lateral preoptic and hypothalamic regions. Additional direct fornix connections with more medial hypothalamic zones appear to exist only in the rat and guinea pig.

3. Postmammillary fornix fibers in the guinea pig and monkey, but in neither the cat or rat, distribute laterally to the subthalamic region (tegmental field H of Forel).

4. Additional postmammillary fornix fibers to the medial region of the caudal midbrain (mainly to the tegmental nucleus of Bechterew) are found only in the two rodent species. This fornix connection, particularly well-developed in the guinea pig, appears to arise at least primarily in the gyrus fornicatus.

5. From a consideration of pathways synaptically related to the fornix system it is concluded that interspecific differences in the fornix connections consist of variations in the number of synaptic relays rather than in the ultimate distribution of the pathways involved.

## ACKNOWLEDGMENTS

It is a pleasure to acknowledge the unfailing technical assistance of Mrs. Mildred H. Albrecht, Mrs. Hertha Middleton, and Mr. Gordon V. Fletcher. The photographs were made by Miss Sally Craig.

## ADDENDUM

Since this paper went to the press, a recent paper by B. G. Cragg and L. H. Hamlyn (Histologic connections and electrical and autonomic responses evoked by stimulation of the dorsal fornix in the rabbit. Exp. Neurol., *1: 187–213, 1959*) came to our attention. Cragg and Hamlyn also described fornix dorsalis fibers arising from the cingulate cortex, but in addition such from the hippocampus ($CA_1$) and subiculum. According to their findings, the fornix dorsalis of the rabbit terminates in the mammillary body and cranial end of the pons, but not in the tegmentum of the lower midbrain.

## LITERATURE CITED

CAJAL, S. R. Y 1911 Histologie du système nerveux de l'homme et des vertébrés. Maloine, Paris.

EDINGER, L., AND A. WALLENBERG 1902 Untersuchungen über den Fornix und das Corpus mammillare. Arch. f. Psychiat., *35:* 1–21.

ELLIOT SMITH, G. 1896 The ''fornix superior.'' J. Anat. Physiol., *31:* 80–94.

GANSER, S. 1881 Vergleichend-anatomische Studien über das Gehirn des Maulwurfs. Morph. Jahrb., *7:* 591–725.

GUDDEN, B. V. 1881 Beitrag zur Kenntniss des Corpus mammillare und der sogenannten Schenkel der Fornix. Arch. f. Psychiat. Nervenkr., *11:* 428–452.

GUILLERY, R. W. 1959 Afferent fibers to the dorsomedial thalamic nucleus in the cat. J. Anat. Lond., *93:* 403–419.

KOELLIKER, A. V. 1896 Handbuch der Gewebelehre des Menschen, vol. II. Engelmann, Lpz.

NAUTA, W. J. H. 1953 Some projections of the medial wall of the hemisphere in the rat's brain (cortical areas 32 and 25, 24 and 29). Anat. Rec., *115:* 352. (Abstract.)

————— 1956 An experimental study of the fornix in the rat. J. Comp. Neur., *104:* 247–272.

————— 1958 Hippocampal projections and related neural pathways to the mid-brain in the cat. Brain, *81:* 319–340.

NAUTA, W. J. H., AND P. A. GYGAX 1954 Silver impregnation of degenerating axons in the central nervous system: a modified technic. Stain Technol., *29:* 91–93.

NAUTA, W. J. H., AND E. S. VALENSTEIN 1958 Some projections of the amygdaloid complex in the monkey. Anat. Rec., *130:* 346. (Abstract.)

PAPEZ, J. W. 1938 Thalamic connections in a hemidecorticate dog. J. Comp. Neur., *69*: 103–120.

SPRAGUE, J. M., AND M. MEYER 1950 An experimental study of the fornix in the rabbit. J. Anat. Lond., *84:* 354–368.

TELLO, J. F. 1936 Évolution, structure et connexions du corps mammillaire chez la souris blanche, avec des indications chez d'autres mammifères. Trav. Lab. Rech. Biol. Univ. Madrid, *31:* 77–142.


PLATE 1

EXPLANATION OF FIGURES

Photomicrographs of fiber degeneration in a case of unilateral transsection of the fornix dorsalis in the guinea pig (Case GPH 15).

22  Low power photograph of the postmammillary fornix decussation. $\times$ 30.

23  Detail of figure 22. $\times$ 750.

24  Degenerating postmammillary fornix fibers in the lateral zone of Bechterew's n. centralis tegmenti superior of the caudal mesencephalon. The degeneration is largely, but not entirely, contralateral with respect to the lesion. $\cdot$ $\times$ 30.

25  Detail of figure 24. $\times$ 750.

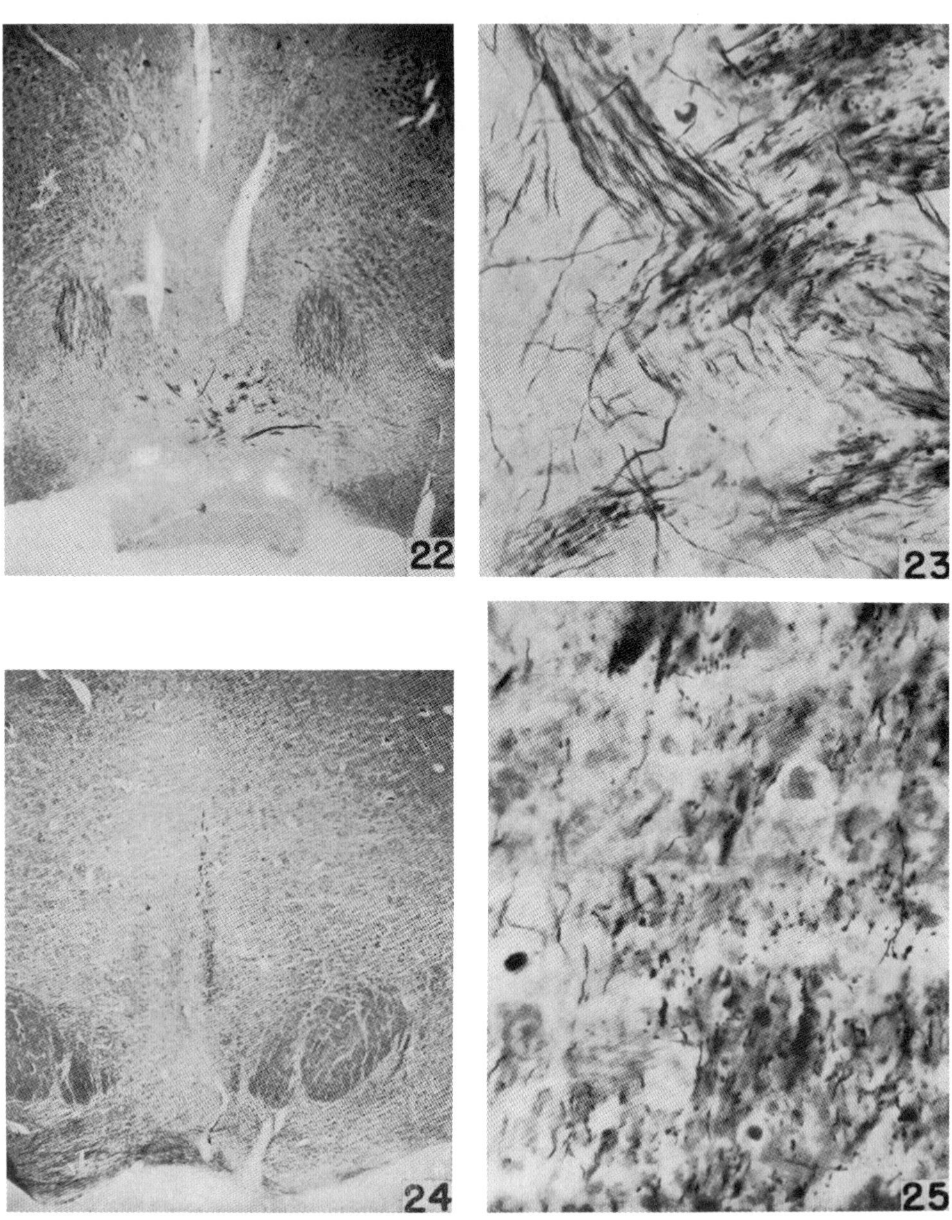

# ASCENDING AXON DEGENERATION FOLLOWING ANTERO-LATERAL CORDOTOMY. AN EXPERIMENTAL STUDY IN THE MONKEY

BY

WILLIAM R. MEHLER, MARTIN E. FEFERMAN[1] AND WALLE J. H. NAUTA

*Departments of Neurophysiology and Neurosurgery*
*Walter Reed Army Medical Center, Washington, D.C.*

"We know much less of the sensory path. Even in the spinal cord it is uncertain. The older theory (of Brown-Séquard), that sensation is chiefly conducted in the gray matter, is not disproved, but there is reason to believe that some sensation is conducted by the fibres of the lateral column in front of the pyramidal tract, and by fibres of the posterior columns. In the medulla and pons the path probably passes up in the posterior half, above the 'fillet,' perhaps chiefly in the curious network of fibres called 'the reticular formation'."

W. R. GOWERS, 1885
*Diagnosis of Diseases of the Brain and Spinal Cord.*

IN the same year in which Gowers wrote the prophetic statement quoted above, Bechterew on the basis of mylogenetic studies described two ascending fibre systems in the anterolateral funiculus of the spinal cord, two systems that were directly continuous into the reticular formation of the brain-stem. These two pathways consisted of a medial "ground bundle" of the lateral funiculus and a more lateral system of "the rest of the lateral funiculus," exclusive of the dorsal spinocerebellar tract of Flechsig (text-fig. 1).

The following year Gowers (1886) traced the more conspicuous of these lateral funicular fibres into the cerebellum as a system distinct from those observed by Flechsig in 1876. Even though uncertain of the polarity of conduction, Gowers reported that the sensory path continued to ascend beneath the corpora quadrigemina to higher centres.

Myelogenic studies on Amphibia and newborn cats led Edinger (1889, 1890) to report that fibres, crossing in the anterior grey commissure and ascending in the anterolateral *"Grundbündel,"* reached as far rostrally as the diencephalon. Edinger's "Tractus spino-bulbo-thalamicus" (1908) thus extended rostrally the spinal projection which Bechterew had traced to bulbar levels. The first experimental confirmation of Edinger's

---

[1]Present address: 315, Sherland Building, South Bend, Indiana.

crossed spinothalamic fibres appeared in Mott's (1895) investigations of ascending spinal tracts of the monkey. According to Edinger (1904), Boyce in 1897 concurred with similar experimental findings in the dog. In the same year von Sölder (1897) traced degenerating fibres into the external medullary lamina of the diencephalon in a clinical case with an extensive cervical spinal lesion. Like Gowers, von Sölder could not demonstrate "a direct irradiation . . ." of these fibres into the thalamic nuclei and concluded that these fibres must terminate in the hypothalamus along with those of the medial lemniscus. In 1898 both Quensel and Wersiloff identified the caudal region of the "nucleus externus thalami" as the terminus of the longest ascending spinal fibres in man. In the same year Rossolimo reported "ascending" spinal fibres which terminated in the putamen, while Choroschko (1909) described spinal fibres extending directly to the cortex. Quensel's and Wersiloff's findings in man, however, were substantially confirmed within the next decade by Amabilino (1901), Henneberg (1901), Thiele and Horsley (1901), Dydynski (1903), Marburg (1903), Collier and Buzzard (1903), Rothmann (1906), Petrén (1910) and Goldstein (1910). During the same period only a few confirmatory animal experiments appear in the literature (Wallenberg, 1899, and Kohnstamm, 1900a, in the rabbit; Probst, 1902, cat; Rothmann, 1903, dog; Vogt, 1909, monkey). In 1936, however, the extensive experiments of Clark showed the existence of a topically organized system of spinothalamic fibres. Clark's findings in the monkey were soon confirmed and extended to the chimpanzee (Walker, 1937, 1938).

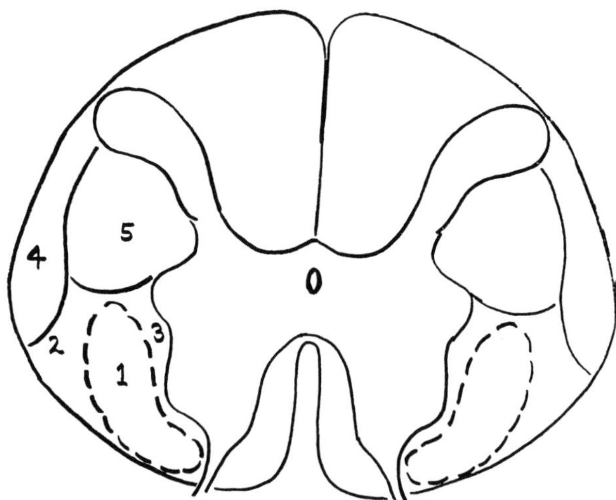



Fig. 1.—Spinal "Grundbündel," after Bechterew. 1, Ground bundles of the anterior and lateral funiculi; 2, the rest of the lateral funiculus; 3, lateral limiting fibre layer of the ventral grey column; 4, spinocerebellar systems; 5, lateral corticospinal tract.

Kuru (1949), in addition to confirming classical spinothalamic connexions in a number of clinical cordotomy cases, demonstrated an extensive system of fibres ascending through the medial region of the bulbar reticular formation. These fibres Kuru identified as the phylogenetically older ventral spinothalamic tract and noted that the majority of these fibres terminated in the bulbar reticular formation. Comparable observations in the monkey and cat by Morin *et al.* (1951) led to the identification of these fibres as a "spino-reticular tract," a term previously coined by Batten and Holmes in 1913.

Current interest in the bulbar reticular formation, renewed by the neurophysiological concepts of Moruzzi and Magoun (1949), was given further impetus by subsequent anatomical confirmation of spinoreticular systems. Employing recently modified silver impregnation techniques extensive spinal connexions with the lateral reticular nucleus were confirmed by Brodal (1949), while Bechterew's profuse, fine fibred, medical spinoreticular connexions were experimentally demonstrated by Johnson (1954). Johnson's findings in the cat were soon confirmed (Rossi and Brodal, 1957; Nauta and Kuypers, 1958) and extended to the primates (Mehler, Feferman and Nauta, 1956) and man (Bowsher, 1957). Comparable medial and lateral spinoreticular connexions, observed in a representative series of experimental mammals (Mehler, 1957), have demonstrated the phyletic constancy of the medial spinoreticular system, in contrast to a phylogenetic increase in the classical, long ascending pathways coursing more peripherally.

In view of such contrasts and the current interest in "reticular" pathways, a detailed report on the relationship of the phylogenetically older spinoreticular pathways to the classical, neospinal pathways in the monkey was initiated.

MATERIALS AND METHODS

Four young adult *Macaca rhesus* and 7 Cynomologus monkeys comprised the material studied. Spinal cordotomy was performed at cervical or thoracic levels on all of the animals. Six cervical cordotomies were performed ranging from simple lateral funicular interruption to complete anterior and lateral funicular destruction, with or without involvement of the contralateral anterior funiculus. In the five remaining cases middle or upper thoracic cordotomies were performed, the results of which varied in extent from lateral funicular interruption to complete hemisection. In all of the cordotomies some direct surgical and/or vascular involvement of the spinal grey matter was encountered upon examination of serial paraffin sections of the operated segments.

The animals were sacrificed nine to twelve days post-operatively by transcardial perfusion with isotonic saline and 10 per cent formalin following lethal doses of nembutal. The brains and spinal cords were stored in 10 per cent formalin for periods ranging from three months to one year.

The brains, with the exception of one cervical case prepared by the Swank-Davenport Marchi method, were stained by the more current modifications of the Nauta-Gygax (1954) selective silver impregnation technique following frozen serial sectioning in either the transverse, sagittal or horizontal plane. In all cases complementary series in either buffered thionin or cresylecht violet, and Weil or Sudan black were prepared to serve as controls. The selectively impregnated ascending degenerating spinal fibres in each case were microscopically charted on enlarged photographs or projection drawings of the silver impregnated sections.

The terminology used throughout the course of this report is based on Olszewski's[S] (1952) atlas of the thalamus of *Macaca mulatta* and upon Olszewski and Baxter's (1954) atlas of the human brain-stem.

## OBSERVATIONS

In order to avoid unnecessary repetition only one representative case of a high cervical anterolateral cordotomy will be described in detail. Although some dorsal root damage occurred in all cases, sometimes complicated by slight involvement of the dorsal funiculus, the resulting patterns of degeneration in the dorsal funicular nuclei will not be dealt with in detail. The findings in the monkey essentially confirm and extend Liu's (1956) findings in the cat that the dorsal roots of $C_1$ to $T_5$ distribute profusely to the nucleus cuneatus lateralis in the form of a spiral oblique lamella. Supplementary observations from other cases contrasting or confirming specific points will be identified accordingly.

### Spino-bulbar Connexions

Commencing at levels just caudal to the pyramidal decussation, ascending degenerating fibres of passage are seen dispersed throughout the white matter of the anterior and lateral funiculi. Excepting the medium-sized fibres of the fasciculus anterolateralis superficialis, the dorsal funiculi and the non-degenerated zone of the lateral cortical spinal tract, no individual tracts can be delineated in the massive picture of ascending degeneration. Dispersed in the anterolateral region and reaching as far medially as the central grey column, numerous small and medium-sized degenerating fibres are seen interspersed among normal fibres. Transverse or horizontal sections through this region show a gradual decrease in the most medial ascending degenerating fibres, a decrease in both number and general fibre calibre. In the anterior funiculus at $C_1$ degenerating fibres appear as far medial as the ventro-medial tip of the anterior grey column.

In general, all of these fibres, when followed cranially, show a progressive dorso-lateral shift. At all spinal levels rostral to the lesion large numbers of degenerating collaterals are seen coursing at right angles from the ascending degenerating fibres. These collaterals infiltrate the anterior grey column in the spinomedullary transition zone and terminate in the nuclei retroambigualis and supraspinalis and locally in the dorsal and ventral divisions of the nucleus medullæ oblongatae centralis. Thoracic cordotomy produces essentially the same degeneration pattern at this level, but the density of medially ascending fibres and collaterals is greatly reduced. The terminal degeneration areas, however, are more distinctly limited to the nucleus retroambigualis and to a limited ventromedial area of the subnucleus dorsalis, medullæ oblongatae centralis.

In their rostral course the ascending fibres of the anterolateral fasciculus, including the most anterior and medial fibres, are displaced dorsolaterally. This displacement appears to be caused first by the decussating pyramidal tracts, then by the initial formation of the medial lemniscus, and finally by the intervention of the caudal pole of the inferior olive. This displacement causes the majority of the fibres of the anterolateral fasciculus to pass through the nucleus medullæ oblongatæ lateralis (text-fig. 2B). All of the subdivisions of this nucleus exhibit extremely profuse pericellular terminal degeneration. Medially, in the pars ventralis of the nucleus medulæ oblongatæ centralis some terminal degeneration is seen as well as increasing numbers of degenerating fibres and collaterals ascending obliquely dorsomedially through this nucleus and into the medial half of its pars doraslis where some terminal degeneration is seen. At this level the displaced anterior funicular "ground bundle" fibres sweep dorsomedially over the caudal "tail" of the medial accessory olive. This subdivision of the inferior olivary complex exhibits profuse pericellular degeneration which, as in all other cases, is limited to its caudal one-third. Similar pericellular degeneration is, however, distributed throughout the lateral two-thirds of the dorsal accessory olivary lamina in all cases. (Plate I, fig. 7).

At the level of appearance of the dorsal accessory olive (text-fig. 2C) the majority of degenerating fibres in the apex of the fasciculus anterolateralis ascend dorsal to the inferior olivary complex. In the monkey the nucleus gigantocellularis appears almost concurrently with the dorsal olivary lamina, and in this nucleus numerous degenerating terminal axons, forming conspicuous peridendritic plexuses, are found (Plate CIX, fig. 8). The medial ascending fibres and collaterals continue rostrally in a dorsomedial direction in a zone occupying the whole ventromedial half of the medullary tegmentum. Medially, clusters of profuse terminal degeneration often outline groups of cells in the nucleus interfascicularis hypoglossi and the more orally situated nucleus of Roller.

In the thoracic cordotomy cases evidence suggestive of some topical order in the distribution of ascending spinal pathways was noted in the nucleus medullæ oblongatæ lateralis. In these cases profuse terminal degeneration was confined to the caudal and ventrolateral parts of this nucleus and absent in its rostral, subtrigeminal division. Terminal degeneration could not be demonstrated in Roller's nucleus following thoracic cordotomy. The nuclei interfascicularis hypoglossi and gigantocellularis, however, exhibited terminal degeneration which differed only in density from that observed in cases of high cervical cordotomy.

Sections at levels just rostral to the inferior olive pass through the nuclei paragigantocellularis lateralis and dorsalis, and through the more medially situated gigantocellularis, pars oralis and raphe pallidus subdivision of the substantia reticularis grisea, all of which receive ascending spinal fibres or their collaterals. The pronounced medial trend of the spinoreticular fibres is best seen at this level and at levels rostral to it as far as the oral pole of the facial nucleus. Dense terminal degeneration is confined to the nucleus gigantocellularis and the pars lateralis of the nucleus paragigantocellularis, while in the pars dorsalis of the latter nucleus and the nucleus raphe pallidus only minimal terminal degeneration is encountered.

In the thoracic cordotomy cases, little if any degeneration was found in the pars dorsalis of the nucleus paragigantocellularis, while degeneration in the pars lateralis and in the nucleus raphe pallidus was significantly reduced. By contrast, the density of terminal degeneration seen in the medial, gigantocellular reticular "core" was only moderately decreased in such cases.

Decussation of ascending spinal fibres is rarely observed in sections through the caudal half of the medulla oblongata. Rostral to the inferior olive, however, a significant number of degenerating fibres can be followed to the opposite side, some traversing the nucleus raphe pallidus and other, more numerous fibres, crossing through the bilaterally fused pars oralis of the gigantocellular nucleus. To date we have not been able to determine satisfactorily whether these decussating fibres merely terminate in the contralateral pars oralis or join ascending spinoreticular fibres of the opposite side in their course rostralwards.

Rostral to the oral pole of the inferior olive lies the facial nucleus. In such a position it is interposed directly in the path of the A–L fasciculus which, even in normal Weil preparations, it appears to divide into medial and lateral components. In silver preparations through this level such a division is discernible by the appearance of concentrations of degenerating fibres on both sides of the ventral aspect of the nucleus (text-fig. 2B). Significant terminal degeneration of ascending spinal fibres is observed in the facial nuclear complex only following cervical cordotomy and it is confined to the medial, ventromedial and intermediate subdivisions of the facial nucleus.

A dorsolateral shift in the remaining medial spinoreticular fascicles appears at the level of the rootlets of the abducens nerve. These fascicles are concentrated at this level in a zone running dorsomedially from the superior olivary complex to the rootlets of the abducens nerve. This zone encompasses the nucleus subcoeruleus ventralis and that portion of the nucleus pontis centralis caudalis which lies lateral to the abducens rootlets. Terminal degeneration is only moderately dense in the latter nucleus even in high cervical cases, and is almost absent

## KEY TO LETTERING

| | | | |
|---|---|---|---|
| AV | n. anterior ventralis | MoV | n. nervi trigemini motorius |
| BC | Brachium conjunctivum | MV | n. vestibularis medialis |
| BCi | Brachium colliculi inferioris | NC | n. caudatus |
| Ci | n. colliculi inferioris | ND | n. dentatus |
| Cl | n. centralis lateralis | NSpV | n. tractus spinalis trigemini |
| CM | n. centrum medianum | nT | n. trapezoidalis |
| Cnd | n. medullæ oblongata centralis, subnucleus dorsalis | Os | n. olivaris superior |
| | | P | Brachium pontis |
| Cns | n. centralis superior, subnucleus medialis | Pa | n. paraventricularis |
| | | PC | Commissura posterior |
| Cnv | n. medullæ oblongata centralis, subnucleus ventralis | Pc | n. parvocellularis |
| | | Pcn | n. paracentralis |
| Coe | n. locus cœruleus | Pf | n. parafascicularis |
| CP | Cerebral peduncle | Pgl | n. paragigantocellularis lateralis |
| CR | Corpus restiforme | Po.c. | n. pontis centralis caudalis |
| Csl | n. centralis superior lateralis | Po.o. | n. pontis centralis oralis |
| Cu | n. cuneatus | PpL | n. papillioformis |
| Cul | n. cuneatus lateralis | Pro.sl. | Processus griseum pontis supra-lemniscalis |
| DBC | Decussation of the brachium conjunctivum | Pro.t.L. | Processus tegmentosus lateralis (Ptl, fig. 6) |
| dOi | n. olivaris inferior accessorius dorsalis | Prp | n. præpositus hypoglossi |
| | | PrV | n. nervi trigemini sensibilis principalis |
| DPy | Pyramidal decussation | Pt | Regio pretectalis |
| G | n. gracilis | Pul | Pulvinar thalami |
| Gc | n. gigantocellularis | | |
| Gc.o. | n. gigantocellularis, pars oralis | R | n. reticularis thalami |
| GL | n. geniculatus lateralis | Ro | Roller's nucleus |
| GM | n. geniculatus medialis | | |
| GP | Globus pallidus | Sc.d. | n. subcœruleus, subnucleus dorsalis |
| H | Forel's field H | Sc.v. | n. subcœruleus, subnucleus ventralis |
| H1. | — field $H_1$ | | |
| H2 | — field $H_2$ | SN | substantiæ nigræ |
| Hb | Habenula | Sol | n. tractus solitarii |
| | | Sth | n. subthalamicus |
| If | Nucleus interfascicularis hypo-glossi | T | Corpus trapezoidalis |
| | | TSpV | Tractus nervi trigemini spinalis |
| LD | n. lateralis dorsalis | | |
| LP | n. lateralis posterior | V | Nervus trigeminus |
| LR | Lateral reticular n. | VA | n. ventralis anterior |
| LV | n. vestibularis lateralis | VH | Ventral grey column |
| MDdc | n. medialis dorsalis, pars denso-cellularis | VL | n. ventralis lateralis |
| | | vLL | n. lemnisci lateralis ventralis |
| MDmc | — pars magnocellularis | VPI | n. ventralis posterior inferior |
| MDmf | — pars multiformis | VPL | n. ventralis posterior lateralis |
| MDpv | — pars parvocellularis | VPM | n. ventralis posterior medialis |
| ML | Medial lemniscus | | |
| mOi | n. olivaris inferior accessorius medialis | ZI | Zona incerta |

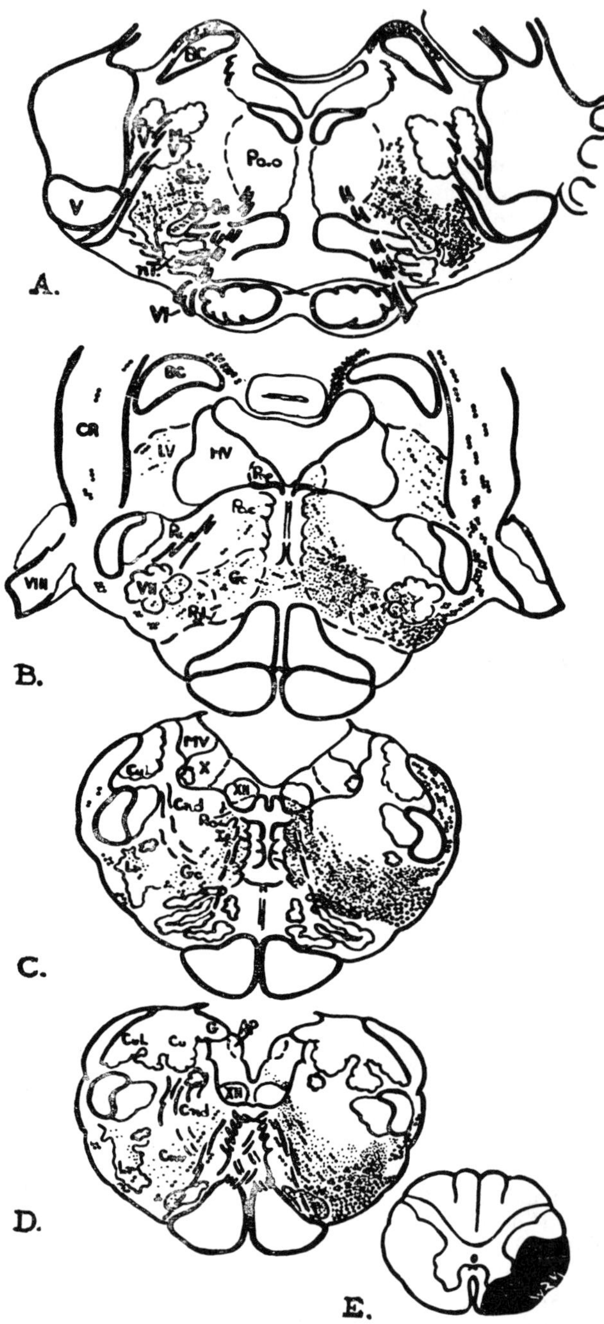

FIG. 2.—Spinobulbar distribution of ascending fibre degeneration following antero-lateral cordotomy at the upper cervical level (E., solid black, Case MST2).   Coarse stipple indicates degenerating fibres of passage; fine stipple, areas of terminal degeneration.   The density of symbols used in the illustrations represents only a relative picture of the histological findings.   However, an attempt has been made to approximate a ratio of 1 : 10 between the two quantities.

For abbreviations *see* p. 724.

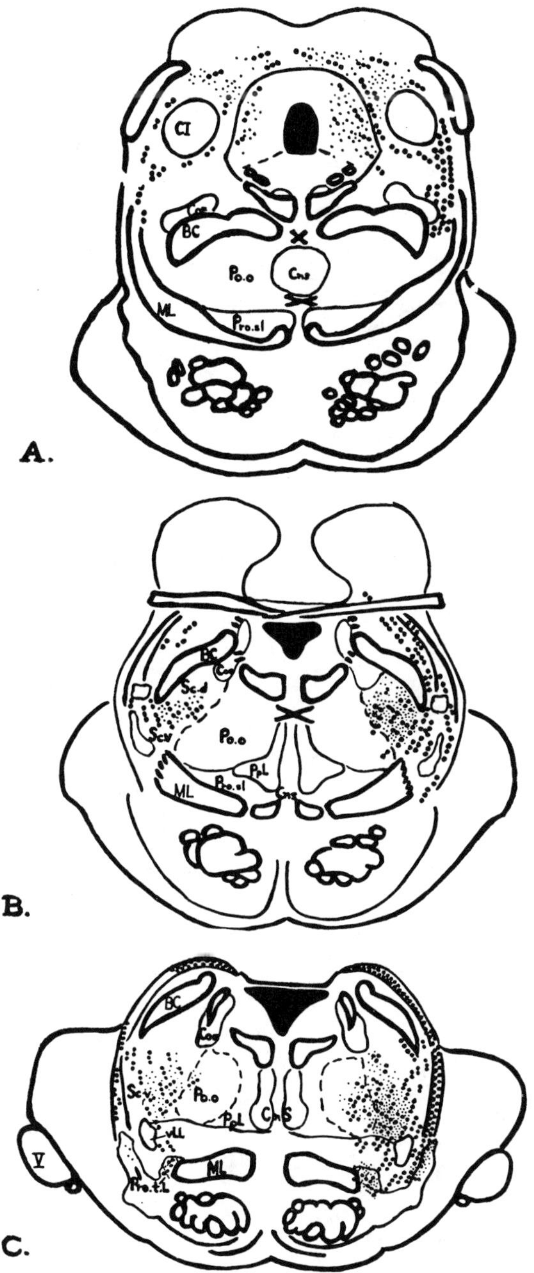

FIG. 3.—Distribution of degeneration in the pons and mesencephalon (Case MST2).

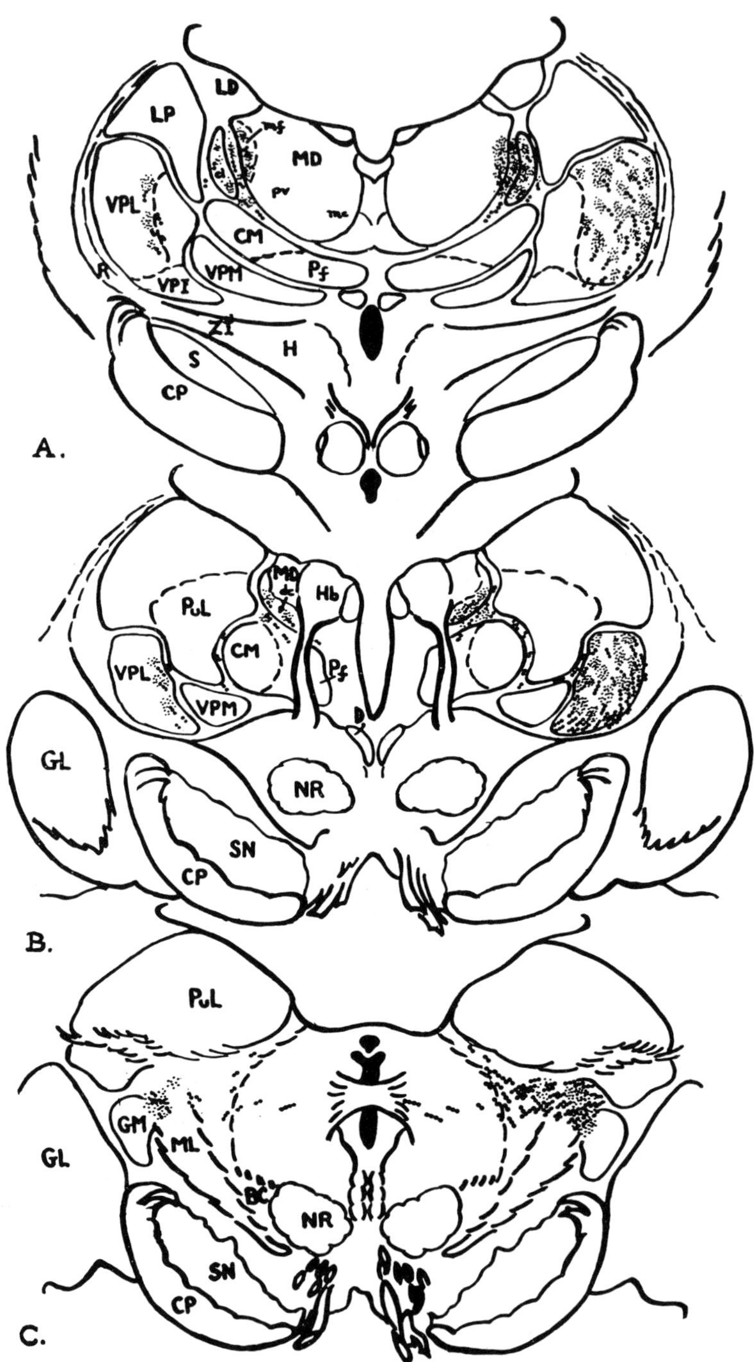



Fig. 4.—Spinothalamic distribution (Case MST2).

following thoracic cordotomy. Terminal degeneration is profuse in the nucleus subcœruleus ventralis in all cases of cervical cordotomy; and only slightly less massive following thoracic cordotomy.

Sections through the motor nucleus of the trigeminus (text-fig. 2A) present a picture of further depletion of the remaining medial spino-reticular components in their dorsolateral shift through the lateral pontine tegmentum. In several cases some terminal degeneration was seen in the lateral regions of the nucleus pontis centralis oralis, but in all cases the main field of terminal degeneration was limited to the sub-nucleus ventralis and the more orally situated subnucleus dorsalis of the subcœruleus complex (text-fig. 3C). In the region of the superior olivary complex the medial spino-reticular system appears separated from the fasciculus anterolateralis. This is due to the fact that the majority of the fibres of the A–L fasciculus are larger and that they present a more compact arrangement than those of the more medial, diffuse, fine-fibred spinoreticular system. In the cases of high cervical cordotomy this separation is more apparent than real but in the thoracic cases a true division can be observed showing medial and superficial systems separated by a zone of normal fibres. An abrupt dorsal displacement of the A–L fascicular fibres at this level further separates the fasciculus from the medially coursing spinoreticular system. It is the largest of these fibres that describes the classical dorsal curve of Gower's spinocerebellar tract around the ventral aspect of the root of the trigeminus.

In the monkey the nucleus processus griseum pontis tegmentosus lateralis appears beneath the root of the trigeminus as an interstitial nucleus of the A–L fasciculus, coextensive with the oral half of the superior olive. Profuse pericellular degeneration of collaterals from the fasciculus anterolateralis infiltrates this nucleus and delineates it from non-terminal pontine areas. Sagittal and horizontal sections through this nucleus best show its relationship with the A–L fasciculus, and in the thoracic cases with their more distinct topographic dichotomy of systems this relationship is especially clear (e.g. text-figs. 5B and 6).

Passing dorsalward, caudolateral to the lateral lemniscus, the fasciculus anterolateralis becomes flattened in a rostrocaudal direction on the lateral aspect of the ponto-mesencephalic isthmus. This flattening is due to the dorsocaudal reflection of the larger fibred Gowers' tract toward the cerebellum, while the medium and small fibre elements course dorsorostrally around the lateral side of the dorsal nucleus of the lateral lemniscus. At the oral pole of the nucleus these fibres become inextricably mixed with the remaining medial spinoreticular fibres which ascend on the medial aspect of the lateral lemniscus. Hence, from this level cranially, the ascending spinal fibres are grouped into a single pathway.

*Terminations in the Mesencephalon*

Transverse sections at levels through the trochlear nucleus show the remaining ascending degenerating spinal fibres in a limited zone ventrolateral to the inferior collicular nucleus (text-fig. 3A). From this zone fibres course in two directions around the inferior collicular nucleus to enter the nucleus intercollicularis. Some fibres course dorsomedially beneath the oral pole of the inferior collicular nucleus and penetrate the stratum album profundum of the superior colliculus. Fine fibres, or collaterals, from this group terminate in the subnucleus lateralis of the central grey substance of the mid-brain, while others following this course re-cross the stratum album profundum and terminate in the region of the nucleus intercollicularis. However, the majority of fibres to this nucleus pass around the rostrolateral aspect of the inferior collicular nucleus before turning dorso-medially and terminating. A small number of fibres following a similar course continue to terminate throughout superior collicular levels in the strata grisea profundum and medium. Some of these fibres decussate to the contralateral side through the tectal commissure before terminating. Other than a modetate decrease in the number of degenerating fibres of passage and some decrease in the density of terminal degeneration in the tectal areas cited, no significant differences from the above description were observed in cases of thoracic cordotomy.

In the rostral mesencephalon the classical spinothalamic fibres ascend in somewhat dispersed, small, loose fascicles in a narrow zone medial to the brachium of the inferior colliculus. At these levels the fascicles "cap" the medial lemniscus just caudal to their entrance into the diencephalon. Beginning at the caudal pole of the medial geniculate body increasing numbers of fine fibres are seen coursing dorsomedially away from the classical spinothalamic fascicles. These fine fibres appear to originate at right-angles to the classical fascicles throughout the levels of the meso-diencephalic junction, suggesting a "collateral" origin from the classical spinothalamic fibres. These fibres filter diffusely through the pretectal region and into the internal medullary lamina (text-figs. 5B and fig. 6).

In their course through the upper mesencephalon the fascicles of the classical pathway appear to issue few terminal fibres. The first signs of significant terminal arborization appear following their entrance into the pars magnocellularis of the nucleus geniculatus medialis. Terminal degeneration could be demonstrated in this nucleus in all cases (text-fig. 4C). The position of the lesser amount of terminal degeneration observed in cases of thoracic cordotomy is suggestive of topical arrangement within this connexion.

*Terminations in the Thalamus*

(*a*) *The nucleus ventralis thalami.*—At the caudal pole of the nucleus ventralis posterior lateralis (VPL) the degenerating fibres in the classical spinothalamic fascicles divide into a number of equal-sized daughter fibres. Small fascicles of fibres course dorsolaterally around the medial geniculate body and penetrate the pars caudalis of the VPL complex. While some of these degenerating fibres course through the medial region of the pars caudalis, the majority curve laterally and spread out on the inner surface of the external medullary lamina. Dense "bursts" of pericellular terminal degeneration are seen within the pars caudalis in relation to the more medially coursing degenerating fibres, while those fibres which course rostrally through the ventrolateral region of the VPL complex send their preterminal ramifications dorsomedially into the more lateral and oral regions of the VPL before terminating in similar pericellular "bursts" (text-fig. 4B and A). In the cases of cervical cordotomy such disseminated islands of highly concentrated pericellular terminal degeneration were seen throughout the VPL complex. Evidence confirming earlier reports of a topical order of distribution of the spino-thalamic fibres within the VPL was demonstrated in all cases of thoracic cordotomy. In such cases consistent patterns were observed in the limitation of terminal bursts to the lateral and oral regions of the nucleus.

*The Intralaminar Thalamic Nuclei*

The fine degenerating fibres issuing dorsomedially from the classical spino-thalamic fascicles throughout upper mesencephalic levels, course diffusely through the regio pretectalis and nucleus limitans and enter the internal medullary lamina. These fibres sweep over the dorsal and medial aspects of the nucleus centrum medianum and begin terminating among clusters of medium-sized and small cells throughout a caudal paralaminar zone. This zone comprises the dorsolateral region of the parafascicular nucleus, and ventral and lateral regions of the pars denso-cellularis and pars multiformis of the dorsomedial nucleus as outlined by Olszewski (1952). At the level of the oral pole of the CM (text-fig. 4A) increasing amounts of terminal degeneration are seen throughout the nucleus centralis lateralis. In this nucleus, and in the more caudal paralaminar zone, the extremely fine degenerating preterminal fibres are restricted to the clusters of medium-sized and small cells (Plate CX, fig. 13).

In addition to the similarity in cell sizes, a common rostrocaudal "intralaminar" polarity was consistently observed in these cells. Terminal degeneration continues to appear within the limits of the nucleus centralis lateralis rostral to the level of appearance of the nucleus paracentralis which partially overlaps the central lateral subdivision. No terminal degeneration, however, could be demonstrated in relation to the large multipolar cells included by Olszewski in the nucleus centralis lateralis, in the more reticulated nucleus paracentralis or in the dorsally situated nucleus centralis superior lateralis in any of the material examined.

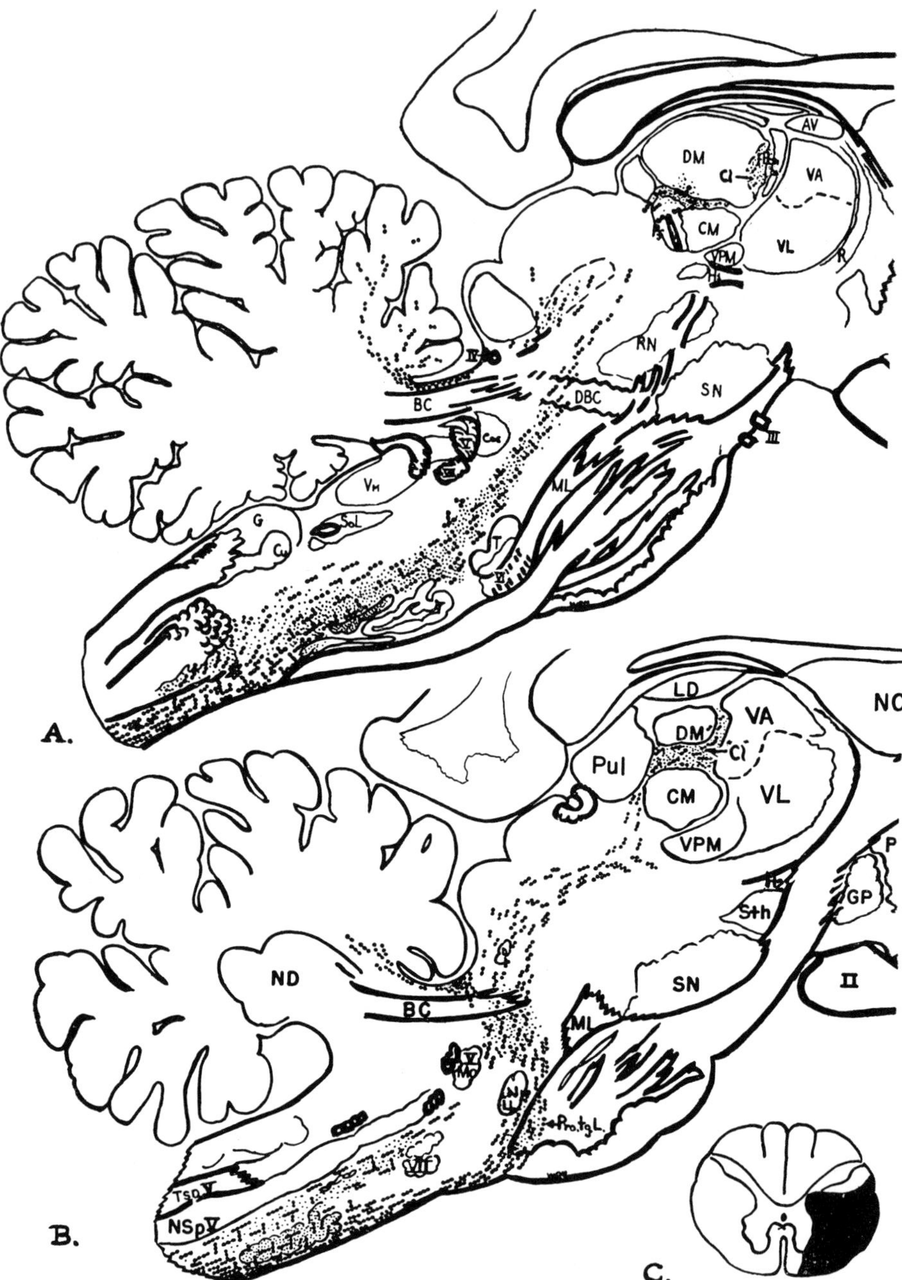


FIG. 5.—Sagittal series (Case MST5) illustrating the course of ascending fibre degeneration in a case of high cervical cordotomy (C, solid black). A, The medial spinoreticular system as seen in a medial sagittal section. B, A more lateral section showing mainly the more peripherally coursing "classical" spinal pathway. Terminal degeneration in the nucleus ventralis posterior lateralis lies lateral to the planes shown in these figures.

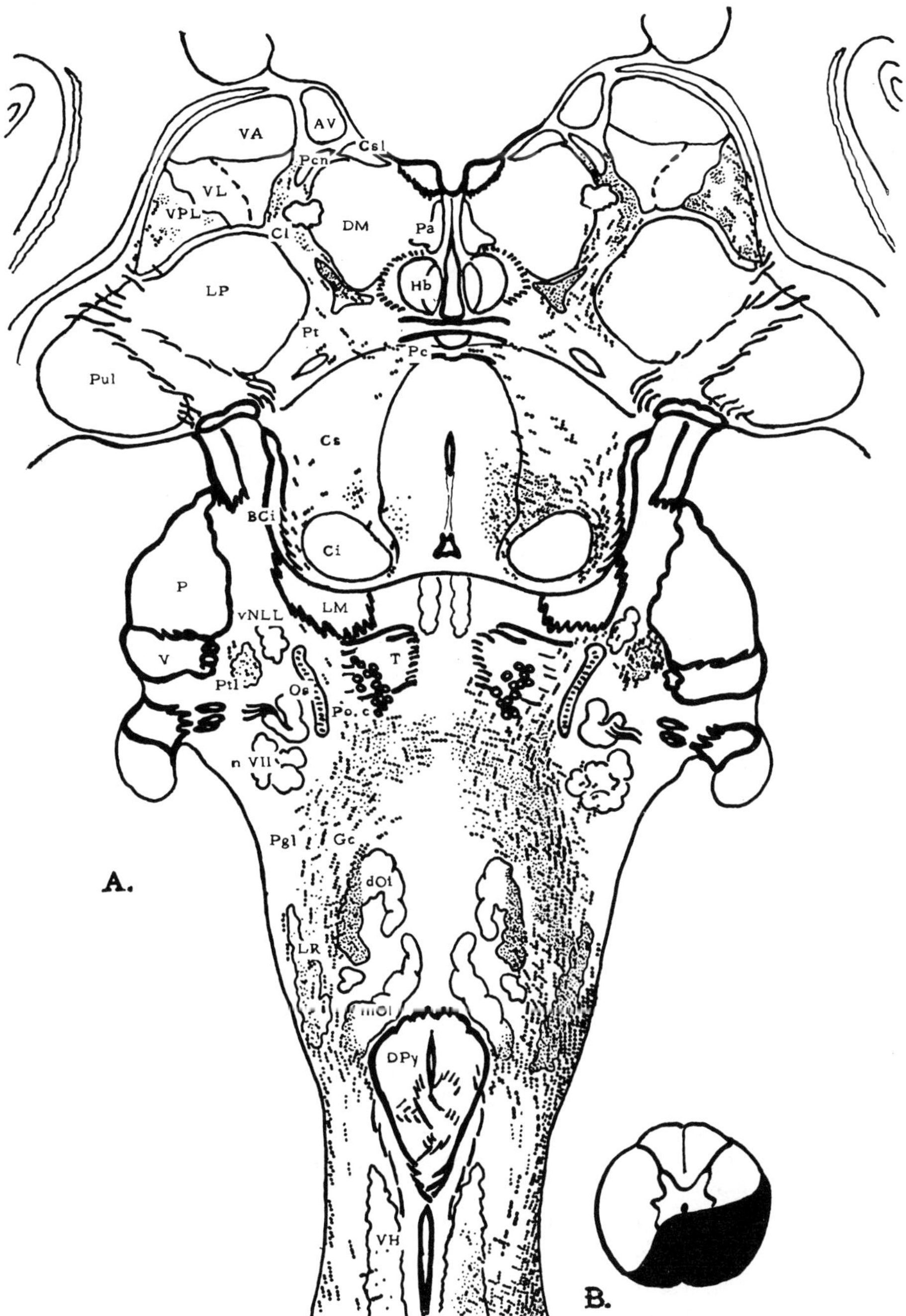


FIG. 6.—Horizontal composite (Case MST14) with a section through the dorsal thalamus and mid-brain superimposed upon a section through the pons and medulla oblongata. Lesion (B, solid black) level of Th. 6. The pars superficialis of the fasciculus anterolateralis passes out of the plane of section at the level of the lateral reticular nucleus (LR) but reappears at the level of the nucleus processus tegmentosus lateralis (Ptl). The "classical" spinothalamic component appears just lateral to the nucleus of the inferior colliculus (Ci) and again, terminating in "bursts" in the oral regions of the nucleus ventralis posterior lateralis thalami.

In contrast to the findings in the VPL reported above, no evidence of discrete somatotopic distribution of spinothalamic fibres could be seen in the non-specific thalamic nuclei.

## DISCUSSION

The results of the present study amply confirm previous descriptions of massive and widespread connexions of the anterolateral spinal funiculus with the brain-stem reticular formation. In addition to these numerous spinobulbar fibres, the anterolateral fasciculus contains axons which terminate in the superior colliculus and central grey substance of the midbrain, and other fibres to certain intralaminar thalamic cell groups while only a relatively small number of anterolateral fascicular fibres represent the classical spinothalamic tract which distributes directly to the posterior part of the ventral thalamic nucleus.

### Spinobulbar Fibre Systems

(1) *Spinoreticular connexions.*—In view of the recent interest in the brain-stem reticular formation it appears appropriate to point out that spinoreticular connexions had been observed before the turn of the century. This may be illustrated by a short historical survey of the concept of the central pain pathway.

(*a*) *Historical notes:* Bechterew (1885) described as "mediales Grundbündel" a system of relatively fine fibres in the anterolateral funiculus which, after perforating the remnants of the ventral horn, partially occupied the interolivary fibre layer and the area dorsal to the inferior olive. Bechterew made the important observation that the fibres of this system gradually decreased in number from the level of the hypoglossal nucleus rostralward to the level of the mesencephalic nucleus centralis superior, while a more lateral fibre system, the "rest of the lateral funiculus," traversed and distributed fibres to the lateral reticular nucleus of the medulla oblongata. In a subsequent description Bechterew (1900) subdivided the "ground bundle" into a medial and a lateral part, and stated that most of the medial fibres disappeared at the level of the "noyau réticulé de la calotte" (i.e. the papillioform nucleus of the present nomenclature), but that some continued into the "noyau central supérieur externe," probably identical to the nucleus pontis oralis.

In the classical spinothalamic studies cited in the introduction, it is interesting to note that many investigators recognized similar medial fibres ascending through the medullary reticular formation. Quensel (1898), for example, in his original description of the spinothalamic tract, observed such fibres but stated that he could not decide whether they actually terminated in the reticular formation. However, experimental studies in the rabbit (Kohnstamm, 1900a) in the cat (Probst, 1902), and in the dog and monkey (May, 1906), all confirmed the existence of spinal fibres which followed a medial trajectory and appeared to terminate in the bulbar reticular formation. The term "spinoreticular" was first applied to these connexions by Batten and Holmes (1913) who reported that these fibres, distinct from those distributing to the inferior olive and the lateral reticular nucleus, terminated gradually throughout the ventral part of the bulbar reticular formation as far rostralward as the "nucleus centralis inferior" (probably corresponding to the nuclei raphe pallidus and gigantocellularis pars oralis of the current terminology).

In the light of the modern concepts of the reticular formation it appears particularly interesting to reappraise the early notion of a *"centrum receptorium or sensorium"* within the bulbar reticular formation. This comprehensive term was introduced by Kohnstamm (Kohnstamm and Quensel, (1908*a*) to indicate bulbar reticular cell groups receiving spinal afferents. On the basis of previous experimental observations in the rabbit (1899, 1900*a* and *b*) Kohnstamm included in the centrum Jakobsohn's magnocellular nuclei as well as a group of nuclei which he labelled the "paralemniscal nuclei." These nuclei lie dorsal to the superior olive and medial to the lateral lemniscus, and from a comparison of Kohnstamm's figures appear to be homologous with the subcœruleus complex of the present study. By a study of retrograde cell reactions in the bulbar reticular formation following high mesencephalic lesions, Kohnstamm and Quensel (1908*a*, *b*, *c*) also demonstrated the existence of ascending pathways connecting the "centrum receptorium" with higher brain levels. It was Quensel (1907), however, who first suggested the reticulothalamic projections as a possible link in a multineuronal pathway for the conduction of sensations of pain and temperature to higher levels of the brain. In the absence of adequate physiological information on the reticular formation, Kohnstamm and Quensel's advanced notions did not receive proper recognition. In 1910, however, Petrén noted that relatively few spinal fibres reached the thalamus, and on this basis suggested that in considering the central pathway for pain attention should be paid to Kohnstamm and Quensel's concept.

The comparative-anatomical studies of Edinger, Koelliker, and the Herricks all furnished evidence of extensive spinoreticular connexions in lower vertebrates. The existence of homologous systems in mammals, although reported by such outstanding authorities as Bechterew, Koelliker and Cajal, was not generally accepted. It appears that their relegation to obscurity for over half a century has been due to the strong emphasis placed by clinicians and physiologists upon the phylogenetically newer, classical spinal pathways. For example, Henneberg (1901), in an argument similar to that of Gowers (1886), concluded that since decerebellation did not allay pain, then the more rostrally terminating spinothalamic fibres must be identified with the pain tract. The cumulative evidence for a localization of the central pain pathway in the anterolateral fasciculus led Spiller in 1905 to advocate its surgical interruption in cases of intractable pain and although many investigators realized that only a few ascending spinal fibers reached the thalamus, it appears that the pain tract became identified with the spinothalamic component, and the latter synonymous with the anterolateral fasciculus. As recently as 1932 Foerster and Gagel recognized the fact that the bulk of the anterolateral fasciculus ended in the bulbar reticular formation, but even then the interest aroused by such observations was overshadowed by the continued emphasis of clinical circles upon specific, well-defined pathways to the thalamus.

(b) *Recent findings in regard to spinobulbar pathways:* As pointed out above, evidence of spinal pathways distributing to the medullary and pontine reticular formation was first obtained by normal-anatomical studies, and later confirmed by experimental studies with the Marchi technique (Probst, 1902; May, 1906; Morin *et al.*, 1951, and others). However, more precise data regarding the terminal distribution of such spinobulbar pathways could be obtained only by the use of the silver techniques as recently modified for experimental purposes. Following Brodal's (1949) report of lateral reticular connexions, Johnson (1954) demonstrated that a large number of medial reticular nuclei also received ascending spinal fibres. These connexions, demonstrated in the cat, were soon confirmed in the monkey (Mehler *et al.*, 1956) and throughout the mammalian scale (Mehler, 1957; Rossi and Brodal, 1957; Bowsher, 1957). On the whole, the descriptions given by these authors satisfactorily agree with each other. Such differences as have appeared in these descriptions may be more apparent than real, and a matter of "atlas semantics" more than significant disagreement of observation. For example, Johnson's report appears to omit a number of reticular nuclei which were included by subsequent workers in the distribution area of spinal fibres. A comparison of terminologies, however, shows that Johnson and the Oslo group have projected their observations upon Meessen and Olszewski's (1949) description of the reticular formation of the rabbit, while most others have followed the nomenclature used in Olszewski and Baxter's (1954) analysis of the human brain-stem. Olszewski (1954), however, has pointed out the remarkable similarity and constant spatial relationship of reticular cell groups in the rabbit, cat, monkey and man, and Brodal (1957) has qualified many of the homologies between the two cytoarchitectonic studies. From a comparative-anatomical study of ascending spinal pathways (Mehler, 1957, to be published in more detail elsewhere) the homologous nature of the bulbar reticular nuclei receiving spinal afferents became quite apparent in all of the mammalian species examined. From such comparative observations it becomes increasingly clear that little if any actual disagreement exists between Johnson's report and, for instance, the present findings which, in order to avoid unnecessary repetition of the foregoing description, are here presented in tabulated form (Table I, overleaf).

It is an interesting question whether the spinal connexions to the bulbar reticular formation are established by collaterals from a peripherally situated spinothalamic pathway, or by an independent medial spinoreticular pathway of wide distribution. The progressive decrease in the number of ascending spinal fibres in their course through the medulla and pons, a decrease noted by Bechterew and others, appears especially to affect the finer medial fibres, a fact that in itself suggests the existence of a true medial spinoreticular fibre contingent in the antero-lateral fasciculus. Additional evidence in favour of this assumption, although likewise circumstantial, can be derived from the observation that in low mammalian forms, such as the opossum and rat, the classical spinothalamic pathway is of very small size, yet the spinoreticular connexion seems to be no less massive than in the primates where the classical pathway increases some tenfold over that observed in the lower forms. On the other hand, collateral branches of axons of the "long sensory pathways" have been demonstrated by the Golgi technique arborizing in the medullary reticular formation (Koelliker, 1896; Cajal, 1909). Thus, evidence exists that both direct spino-reticular fibres and collaterals of the "long sensory pathways" supply the bulbar reticular formation. Rossi and Brodal (1957) have pointed out that this is certainly the case for the ventral and lateral reticular nuclei and have stated their agreement with the present authors and others that the gigantocellular reticular "core" is predominantly supplied by a medially coursing spinoreticular system of fibres.

*Spino-olivary Connexions*

Historically the question of spinal connexions with the inferior olive is closely related to the problem of spinoreticular fibre systems. The connexions with the inferior olives, however, were further obscured by the lack of information on the polarity of conduction of the fibres involved. Helweg (1887) first reported these connexions as descending but in 1894 Bechterew voiced objection to Helweg's interpretation and stated that the fibres in question were ascending fibres from the cervical region that ended in the caudal part of the inferior olive. These conflicting reports gave rise to two schools of thought, one supporting Helweg's descending "Olivo-spinale Dreikantenbahn" and the other accepting Bechterew's findings of an ascending "Olivenbündel des Halsmarks."

In 1903 Dydynski followed fibres from a spinal lesion of the fifth thoracic level which corresponded to Bechterew's Olivenbündel in both course and termination. Comparable Marchi studies on cordotomized rabbits (Kohnstamm, 1900*a*), dogs and monkeys (May, 1906) also demonstrated the spinal origin of these connexions, but due to the inherent inability of the Marchi technique to stain the finer preterminal and terminal fibres of these connexions many contradictory interpretations are found in the literature concerning the terminal field of these ascending anterior funicular fibres.

In respect to the afferent connexions of the inferior olive, it was Cajal (1909) who first recognized that there were regional differences in the afferent connexions of this apparently homogenous nucleus. By means of the Golgi technique, Cajal demonstrated in the mouse that the spino-olivary connexions were collaterals of medially ascending anterolateral funicular fibres, collaterals which terminated in restricted areas of the inferior olive. Cajal suggested that these areas might be homologous to the human dorsal accessory olivary nucleus and a pars ventrolateralis of the main olive. With the exception of Gerebtzoff's (1939b) report of spino-paleo-olivary connexions in the rabbit, few authors recognized the brilliance of Cajal's observations until the problem was re-examined by improved silver techniques.

Using the Glees' technique Brodal, Walberg, and Blackstad (1950) extended Cajal's findings to the cat and confirmed the regional distribution of these connexions. Nauta-Gygax studies in the cat (Johnson, 1954; Nauta and Kuypers, 1958) and monkey (Mehler *et al.*, 1956) have all confirmed Cajal's findings. While not specifically mentioned in an initial report of comparative anatomical findings (Mehler, 1957), comparable spino-olivary connexions have been observed in all of the sub-primate species examined. In the species examined the degenerating terminal field selectively delimits homologous paleo-olivary regions from the phylogenetically more recent olivary subdivisions.

On the basis of the present study we can add little to the description of Brodal *et al.* (1950). We have confirmed in the monkey that the majority of the fibres which terminate in the paleo-olivary structures have their cells of origin at levels below the seventh thoracic segment and that their course through the upper thoracic and cervical levels is essentially restricted to the ipsilateral anterior funiculus.

It should be pointed out, however, that premature application of these findings to man should be avoided. While comparable spino-olivary connexions have been observed in several species of monkeys, as well as a representative series of sub-primate species, we have been unable to demonstrate satisfactorily homologous connexions in either the chimpanzee or man following complete anterolateral cordotomy.

While early Marchi studies of clinical cases (Dydynski, 1903; Collier and Buzzard, 1903; Goldstein, 1910; Batten and Holmes, 1913) report such findings, Bowsher's (1957) omission of mention of these connexions may stem from similar negative findings in cases of human cordotomy. At the present time one can only speculate as to whether this incongruity represents an actual interspecific difference in the spinal connexions with

TABLE I.—CORRELATIVE SUMMARY OF THE BRAIN-STEM NUCLEAR HOMOLOGUES IN MAN, MONKEY AND CAT IN WHICH TERMINAL DEGENERATION HAS BEEN DEMONSTRATED FOLLOWING ANTEROLATERAL CORDOTOMY

| Studies | Johnson, 1954 | Rossi and Brodal, 1957 | Bowsher, 1957 | | Mehler et al. | | | |
|---|---|---|---|---|---|---|---|---|
| Term: Olszewski and Baxter, 1954 | Meessen and Olszewski 1949 | Brodal, modificat. 1957 | 1954 Terminol. | (Abbr.) | 1954 Terminol. (C) | (Th) | Ipsi. | Contra. |
| N. med. obl. cent., dors. | +sN. ret. dors. | (−) [N. ret. parvi.] | + | Cn. d. | (+) | − | | |
| ——— vent. | +——— vent. | + [N. ret. vent.] | ++ | Cn. v. | ++ | + | + | (−) |
| N. med. obl. lat. sN. vent. | +N. lat. caud. magno | +N. ret. lat. | + | LR | +++ | + | +++ | (+) |
| ——— subtrigem. | +N. lat. oral. | (Brodal, 1949) | − | | + | (−) | + | (−)*? |
| N. interfasc. XII | ? [sN. ret. vent.] | + [paramed. ret. N.] (Brodal and Gogstad, 1957) | − | If | ++ | + | ++ | (−) |
| N. Roller | N.M. (same) | + (same) | N.M. | Ro. | (+) | − | (+) | (−) |
| N. raphe pallidus | N.M. [n. raphe. magn.] | + [N. raphe and N. ret. giganto.] | + | Ra. pa. | (+) | (−) | (+) | (−) |
| N. gigantocell. caud. | + (N. ret. giganto.) | same, no div. | ++ | Gc. | +++ | ++ | +++ | + |
| ——— oral. | | | + | Gc. o | ++ | + | ++ | + |
| N. paragiganto. dors. | + [N. ret. pont. caud.] | + [N. ret. pont. caud.] | + | Pg. d. | (+) | (−) | (+) | (−) |
| ——— lat. | + [N. "giganto, α"] | ["x" or N. ret. lat.] | + | Pgl | ++ | + | ++ | (−) |
| N. parvocellularis | (−) | (−) | (−) | Pc. | (−)* | − | (−)* | − |
| N. pont. cent., caud. | + [N. ret. pont. caud in PL. IX to XIII] | + (Same) | + | Po. C. | (+) | (−) | (+) | (−) |
| ——— oral | +[dorsomed. half of N. ret. pont. oral. in PL. XIV] | + (Same) | + | Po. o. | (−)* | − | (−)* | − |
| N. subcœruleus, vent. | +[pars, α N. ret. pont. caud., lat. half PL. XII and XIII] | + "N. cœruleus, α" or N. subcœruleus | + (no div.) | Sc. v. | ++ | + | ++ | + |
| ———, dors. | [N. ret. pont. oral, α PL. XIV, lat. half] | + „ | | Sc. d. | ++ | + | ++ | + |

| | | | | | | | | |
|---|---|---|---|---|---|---|---|---|
| Proc. gr. pont. teg. lat. | +"Spino-pont. system to lat. dors. N." | N.M. | N.M. | P. tg. L | ++ | + | ++ | (+) |
| Proc. gr. pont. supralem. | N.M. | N.M. | N.M. | Pro. sl | (+)* | − | (+) | (+) |
| Griseum pontis | "...no degenerating fibres issue from cortico-spinal tr." | "...spinocortical collaterals ..." N. pedunc., lat., vent. (Walberg and Brodal, 1953) | N.M. | Gr. p. | − | − | − | − |
| Gr. cent. pont. | N.M. | N.M. | + | Gr. c. p. | − | − | − | − |
| N. papillioformis | N.M. (Same) | +N. ret. teg. pont. of Bechterew (Walberg and Brodal, 1953) | N.M. | Ppl. | − | − | − | − |
| N. tr. sp. trigem. oral. | + (Same) | + (Rossi and Brodal, 1956) | N.M. | Sp. V. o. | (+)* | − | (+)* | − |
| N. teg. pedunculopont., sN. compactus (Kölliker) | N.M. | N M. | (−) | Tg. cm. | − | − | − | − |
| N. paralemniscalis | N.M. | N.M. | (−) | Pl. | − | − | − | − |
| N. intercollicularis | N.M. | N.M. | − | Icol. | + | (+) | + | (+) |
| N. interstitialis (Cajal) | N.M. | N.M. | (+) | Is. | − | − | − | − |
| Spino-solitary | N.M. | + (Rossi and Brodal, 1956) | N.M. | pars comm. only | (+) | (−) | (+) | (−) |
| Spinovestibular | + | +(Pompeiano and Brodal 1957) | N.M. | (−) | (+) | (−) | (+) | (−) |
| Spino-olivary N. dors. acc.<br>N. med. acc. | +<br>+ | +(Brodal *et al.*, 1950)<br>+ | N.M. | d0i<br>m0i | ++<br>+ | ++<br>+ | ++<br>+ | (−)<br>(−) |
| Spinofacial | + | − | N.M. | nVII | (+) | − | (+) | (−) |
| Spino-aqueductal | + | N.M. | + | sNLat. | ++ | + | ++ | (+) |
| Spinocortical | − | +(Brodal and Walberg, 1952) | N.M. | | − | − | − | − |

Key to symbols: −, no terminal degeneration.  (−), few, i.e., not entirely negative.  (+), some significant terminal degeneration.  + to +++, relative estimates of the density of terminal degeneration observed.  *, only in cases involving N. spinalis V.  [ ], included in N.M., not mentioned.

the inferior olivary complex, or a failure of current histological procedures to impregnate this extremely fine terminal plexus in the higher primates and man.

*Spinal fibres terminating in the facial nucleus* were reported as early as 1896 by Koelliker and were demonstrated experimentally by Wallenberg in 1899.   Excepting Cajal (1909) few investigators have promulgated such a connexion.   Johnson (1954) reported spinofacial connexions in the cat but Rossi and Brodal (1957) were unable to confirm this finding in their Glees' preparations.   In an initial report of the present findings (Mehler *et al.*, 1956) we confirmed the existence of these terminations in the monkey and indicated that they were restricted to the ventral subdivisions of the facial nucleus.   Accordingly, the present description merely reconfirms our initial conclusions, with the qualification that these fibres were demonstrated in the medial, ventromedial and intermediate subdivisions of the facial nucleus.   It should be emphasized, however, that in all cases in which a significant amount of terminal degeneration could be demonstrated in these subdivisions, the cordotomy had been performed at the level of the third cervical segment or rostral to this level.

In view of the circumstance that the spinal nucleus of the trigeminus extends for a variable distance into the cervical cord, the possibility must be considered that the fibres in question are in reality secondary trigemino-facial fibres.   However, control lesions of the tuberculum cinereum were found to cause terminal degeneration essentially restricted to the ipsilateral dorsal and lateral subdivisions of the facial nucleus. Hence, it seems logical to conclude that spinofacial fibres independent of trigemino-facial connexions indeed exist.[1]   Again, mention must be made of findings in sub-primates.   In the opossum and rat the density of these spinal connexions to homologus facial sub-nuclei is much greater than that observed in the monkey or even the cat.   We have interpreted such findings as further illustrations of interspecific changes in synaptology.   Whereas the previously mentioned spino-olivary findings suggest a complete withdrawal of a connexion, the phylogenetic changes in the density of the spinofacial connexions portends other evolutionary changes.   Similarly the discrepancies in the observations on the distribution and densities of spinovestibular connexions (text-fig. 2B) mentioned by Pompeiano and Brodal (1957) are clearly another example of interspecific changes in synaptology.   For the present, suffice it to say that we have found evidence of a constant phylogenetic decrease in both

[1]It is interesting to compare the differential mode of topical distribution of spinal and trigeminal afferents to the facial nucleus with the efferent connexions of the various subdivisions of the nucleus.   According to Papez (1927) the dorsal and lateral cell groups innervate facial muscles supplied by the sensory trigeminus, while the ventral and medial groups control phylogenetically old muscle units which receive their afferent supply essentially over the cervical plexus (Huber, 1930).

the distribution and the density of spinovestibular as well as spinofacial connexions.

(*c*) *Spinal pathways to the mesencephalon:* The studies of Clark (1936) in primates, in addition to demonstrating fibres running dorsomedially from the anterolateral fasciculus into the medullary reticular formation, confirmed Mott's (1892; 1895) demonstration of spino-tectal fibres in the monkey. Clark also appears to have been the first to report spinal fibres entering the central grey mid-brain substance. Both spino-tectal and spino-annular fibres were observed in the present study, as they had been in Johnson's (1954) original investigation with the Nauta-Gygax technique. Mehler (1957) has demonstrated the phyletic constancy of such spino-mesencephalic connexions throughout the mammalian scale, and Bowsher (1957) and Mehler and Nauta (unpublished) have confirmed these findings in man.

In the present study in the monkey, termination of spinal fibres in the mesencephalon was consistently found to be restricted to the deep layers of the superior colliculus and to the subnucleus lateralis of the central grey substance. The existence of spinal connexions with Flechsig's parabigeminal area (apparently synonymous with the rostral part of the nucleus of the lateral lemniscus) as reported by Monakow (1905) and more recently by Giok (1956) could not be confirmed; nor could termination of such fibres in the nuclei pedunculopontinus, paralemniscalis, or interstitialis of Cajal (Bowsher, 1957) be identified.

The functional significance of the spino-tectal and spino-annular connexions is at the present time a matter of conjecture, but it is interesting to note that behavioral reactions suggesting pain sensations have been observed during stimulation of the central grey substance (Magoun *et al.*, 1937) and tectum mesencephali (Spiegel *et al.*, 1954) in the cat. These observations, considered together with the primordial nature of the central grey substance and tectum mesencephali, suggest the existence of other possible links in a multisynaptic pathway for the transmission of nociceptive impulses. It is of interest to recall in this connexion Bechterew's (1900) conception of the central grey substance as "nodal point" in the network of conduction pathways, a region which exhibits multiple afferent and efferent connexions. Kohnstamm and Quensel (1908*a*) suggested the tectum as the final link in their "multi-neuronal pain pathway" to the cortex and Walker (1943) concluded that it might be the lowest level of integration for pain. For a discussion of some of the known connexions of the central grey substance the reader is referred to a recent study of the hippocampal projection pathways (Nauta, 1958).

The course of the classical spinothalamic tract through the upper mesencephalon medial to the brachium of the inferior colliculus was first demonstrated by Quensel (1898). A slight distance farther rostrally, the tract is found just medial to the medial geniculate body. At this transitional level between the mid-brain and diencephalon terminal degeneration was consistently observed in the so-called medial nucleus or pars magnocellularis of the medial geniculate body. This cell group, according to Clark (1932), cannot be regarded as part of the medial geniculate body, a conclusion borne out by the experiments of Rose *et al.* (1949, 1952) which demonstrated the non-auditory nature of the pars magnocellularis. A termination of part of the spinothalamic tract in the pars magnocellularis was already observed by Lewandowsky (1904) who labelled the cell group "parageniculate nucleus" and interpreted it as an outlying caudal part of the ventral thalamic nucleus. According to the present findings, this connexion of the spinothalamic tract is rather strikingly bilateral, which may explain Whitlock and Perl's (1957) recent observations of bilateral soma discharges in this nucleus following unilateral cutaneous stimulation in a preparation with only one anterolateral fasciculus intact.

(2) *Spinothalamic connexions.*—The spinothalamic component of the anterolateral fasciculus exhibits a dual distribution in the thalamus: the major portion of the pathway terminates in the nucleus ventralis posterior lateralis as classically described, while a more diffuse, fine fibred component distributes to certain intralaminar cell groups.

(*a*) *Spinal afferents to the principal thalamic nuclei:* Quensel's demonstration of spinal fibres terminating in "nucleus externus thalami" was amplied by the extensive primate studies of Clark (1936) and Walker (1937, 1938). Corroborating earlier reports on the course of spinotectal and spinothalamic fibres, these authors demonstrated that the spinothalamic fibres terminated in a somatotopic fashion within the nucleus ventralis posterior lateralis (VPL). Such somatotopic distribution of ascending fibre systems within the ventral nuclei of the thalamus, originally suggested by Wallenberg (1905), was conclusively confirmed by Chang and Ruch (1947). The findings of the present study are in agreement with the topical patterns demonstrated by these workers.

The classical spinothalamic fibres which enter the pars caudalis of the nucleus ventralis posterior lateralis have been described in this report as dividing into a number of equal-sized daughter fibres. While such division may only represent the first axon bifurcation in a complicated terminal arborization, a significant increase in the number of degenerating fibres can be demonstrated by comparing the number of fibres seen in caudal levels of the thalamus with the relatively few observed at mesencephalic levels. Fibres separating from the small fascicles of the daughter fibres can be traced into cell clusters of varying size. The profuse but sharply delimited "bursts" of pericellular degeneration (Plate CX, figs. 12 and 14) appear to correspond in localization with such cellular clusters. It is important to note, however, that while similar distribution patterns have been observed in the chimpanzee (Mehler, 1957) and man (Mehler and Nauta, unpublished), no comparable mode of parcellated distribution could be demonstrated in a study of the spinothalamic tract in the cat or other subprimates (Mehler, 1957; Nauta and Kuypers, 1958).

Studies on the thalamus (Rioch, 1929; Clark, 1932) have shown that subdivision of the thalamus is essentially the result of the development and elaboration of fibre systems relating the thalamus to the cerebral mantle. Similarly, the cell clusters observed in this study may only be due to further elaboration of such factors in the primates and man. The complicated nature of the terminal arborization of the lemniscal systems terminating in the ventral nuclei of the thalamus was first reported by Cajal (1911). In 1932 Clark noted that the terminal fibres of the fillet systems that enter the ventral thalamic nuclei tend to separate constituent cell groups into clusters and to give the whole nucleus a characteristic lobulated appearance. By contrast, it should be noted that following lesions of the posterior funicular nuclei (Mehler and Nauta, unpublished) the degenerating medial lemniscus, ascending without evidence of any termination caudal to the thalamus, was observed to enter the VPL *en masse* and terminate profusely throughout the nucleus in a terminal pattern which contrasts sharply with the parcellated mode of distribution described for the spinothalamic fibres. On this basis it appears justifiable to suggest that the classical spinothalamic tract overlaps in distribution with the medial lemniscus in "archipelago" fashion, i.e. in such a manner that only disseminated cell clusters of the nucleus ventralis posterior lateralis are contacted by both spinothalamic and medial lemniscal fibres. Such an arrangement introduces the possibility that functionally the classical spinothalamic component only contributes a quality of sharpness, an "epicritic" quality, to the appreciation of certain noxious stimuli, while the more diffuse ascending spinal fibres, which relay at levels caudal to the thalamus or in the more medially situated non-specific intralaminar cell groups, may represent the pathway for a more slowly conducting, "protopathic" component of the pain system.

(*b*) *Spinal afferents of the intralaminar thalamic nuclei:* Most authors recognize four cell groups in the thalamic grey intercalated in the internal medullary lamina: nucleus centralis medius, nucleus paracentralis, nucleus centralis lateralis, and a caudal complex composed of the centre median of Luys and the nucleus parafascicularis. It must be emphasized, however, that the accuracy with which the intralaminar distribution of spinal and other afferents can be described is limited by the circumstance that the intralaminar cell groups have rather vague boundaries with each other as well as with the surrounding specific thalamic nuclei. Due to the disparity of opinion among authors in subdividing these groups and, in order to avoid introducing still another terminology, we have attempted to adhere as closely as possible to the scheme of subdivision followed by Olszewski (1952). According to the present findings the intralaminar distribution of spinal fibres is limited to a region encompassing certain small-celled clusters contained within the dorsolateral region of Olszewski's borders of the nucleus parafascicularis, the pars densocellularis and multiformis of the nucleus medialis dorsalis, and throughout the nucleus centralis lateralis.[1] These nuclei, delimited by the termination of spinal projections, are the same para-, and intralaminar nuclei that we have observed to receive medially terminating brachium conjunctivum fibres following lesions of the dentate and interpositus nuclei in primates (Mehler, Vernier and Nauta, 1958). Such distribution essentially limits the significant terminal field of these spinal fibres to the nucleus centralis lateralis and to a number of caudally located cytologically similar cell clusters probably belonging to this nucleus. In order to clarify further the terminal nuclei in question, it should be noted that in none of the cases examined could terminal degeneration be demonstrated in the other intralaminar nuclear subdivisions recognized by Olszewski; viz. centralis superior lateralis, paracentralis or in the caudally situated centre median-parafascicular complex.

The problem of termination of spinal fibres in the centre median-parafascicular complex merits special attention. Spinal fibres terminating in the nucleus centrum medianum of the cat have been reported by Anderson and Berry (1956) but were expressly denied by other investigators studying the monkey, including the present authors. For a proper understanding of the problem, it is necessary to digress for a moment and deal briefly with the cytoarchitectonic aspect of the region in question.

It appears to the present authors that the region interpreted as "centre median" in atlases of the cat's diencephalon (Ingram *et al.*, 1932; Jimenez-Castellanos, 1949; Jasper and Ajmone-Marsan, 1954) includes for the most part the parafascicular nucleus and the clusters of cells representing the caudal elements of the nucleus centralis lateralis. Apart from these cell groups, the region in question also encompasses a small, almost acellular ventrolateral area which appears to approximate to the region originally indicated by Rioch (1931) as the possible homologue of the centre median in carnivores.

In the primates and man, however, it is this ventrolateral region of smaller, pale-staining cells which disproportionately increases in size in comparison to the more dorsomedially situated paralaminar cell groups. Olszewski (1952) includes most of these latter cell groups

[1]Olszewski (1952) describes three sizes of cells within the region he has delimited as the nucleus centralis lateralis: viz. large, medium and small. He points out that these small cells often form "small separate clusters" and that in both the pars densocellularis and multiformis of the nucleus medialis dorsalis cells similar to the small-celled component of the nucleus centralis lateralis are found.

within his paralaminar subdivisions of the dorsomedial nucleus while the Vogts (1941) and others have included them within the limits of the centrum medianum. Dekaban (1953) has followed the Vogts in this respect but pointed out that the dorsal and medial borders of the "centrum medianum" are the most indefinite and therefore were only tentatively indicated.

Following the nomenclature of Dekaban, Bowsher (1957) reported spinal fibres terminating in the nucleus centrum medianum of man. In the absence of a more specific localization of the terminal degeneration in Bowsher's report, it appears possible that his findings were restricted to the dorsomedial zone which has been interpreted by the present authors as a nuclear area containing cell clusters belonging to the central lateral nucleus rather than to the centre median proper.

Such an interpretation is also consistent with the topical distribution of intralaminar thalamostriate fibres as shown by Simma (1951) and Powell (1952) in man and recently confirmed in the monkey by Powell and Cowan (1956). These studies have shown that the more acellular ventrolateral "centre median" region projects to the putamen while the larger-celled paralaminar region and the nucleus centralis lateralis are essentially related to the caudate nucleus.

Spino-parafascicular connexions reported by Gerebtzoff (1939) in the rabbit could not be confirmed by Getz (1952) employing the Glees technique. In our initial report on the monkey (Mehler *et al.*, 1956) such connexions were not indicated but their existence was intimated in a generalized summary of phylogenetic intralaminar connexions (Mehler, 1957). This apparent contradiction stems from several factors. In the opossum and rat the nucleus which is the apparent cytological homologue of the nucleus parafascicularis (nomenclature: Bodian, 1939, opossum; Gurdjian, 1927, rat) does receive some degenerating fibres following high cervical cordotomy. However, as was previously pointed out, the nucleus caudalis of the trigeminus in lower forms consistently extends for variable distances into the upper cervical segments of the spinal cord, and possible involvement of this nucleus in high hemisections of the spinal cord could interrupt trigemino-parafascicular fibres. Gerebtzoff (1939a) concluded that this was essentially the case in the rabbit, and other authors (Papez and Rundles, 1937; Johnson, 1951) have reported trigemino-parafascicular connexions in other species. On the other hand, the undifferentiated nature of the sub-primate "CM-Pf" complex, with its inclusion of some central lateral cell clusters in its periphery, or within its "atlas" borders, could be the contributing factor in the identification of spinal connexions to this region. These problems are under examination and will be dealt with in a subsequent publication of comparative neuro-anatomical findings.

The possibility of spinal fibres by-passing the "CM-Pf" complex and coursing into the region of the internal medullary lamina was first recognized by Clark (1936). It was Clark's impression that the terminal field of these fibres was restricted to the more rostral intralaminar nuclei, especially the nucleus centralis lateralis. Getz similarly precluded the termination of spinal fibres in the CM-Pf nuclei and noted that the greatest density of degenerating boutons was observed in the nucleus centralis lateralis and nucleus paracentralis (nomenclature: Jimenez-Castellanos, 1949). Correcting for atlas semantics, our findings are essentially compatible with these observations and have been confirmed by Nauta and Kuypers (1958) in the cat. However, reports of terminal degeneration within the nucleus reticularis thalami (Getz, 1952; Bowsher, 1957) and nucleus lateralis, pars posterior (Getz, 1952), previously denied (Mehler, 1957), still cannot be confirmed in the monkey. While an occasional fibre of passage has been observed in the periphery of the external medullary lamina no evidence of terminal degeneration could be demonstrated in the nuclei intercalated in its superficial layers.

The fibres terminating in the intralaminar nuclei have been interpreted as "collaterals" of the long ascending pathways (i.e. spinothalamic). However, in studies of ascending spinal pathways in subprimates (Anderson and Berry, 1956; Mehler, 1957; Nauta and Kuypers, 1958) there is evidence that the fibres terminating in the intralaminar nuclei follow a medial tegmental trajectory independent of the classical spino-thalamic component. In view of these findings and disproportionate densities of terminal degeneration observed in the principal and intra-laminar thalamic nuclei, we have tentatively interpreted these medially terminating fibres in the primate to be homologous to the phylogenetically older system even though they become lateralized and inextricably mixed with the classical pathway in their course through the mesencephalon.

### The Problem of Fibres Decussating Above Spinal Levels

Spinal fibres decussating at medullary levels as illustrated in fig. 2B of the present study were previously reported by Long (1914) and Clark (1936). Although an occasional decussating fibre is seen at lower levels of the medulla the majority of these fibres have been observed at the level of the pars oralis of the gigantocellular nucleus. Due to the presence of some contralaterally ascending fibres in all of the cases examined, it was impossible to determine whether these fibres terminated in the contralateral pars oralis or continued rostrally to higher levels.

Similarly, other decussating fibres of spinal origin have consistently been observed in the tectal and posterior commissures. The fibres crossing in the posterior commissure (Probst, 1900; Chang and Ruch, 1947) however, deserve some further qualification since their place in the over-all pattern of ascending degeneration appears to have escaped recognition. These fibres, readily demonstrated in Marchi frontal sections, have tended to overemphasize the significance of the posterior commissure as a path for decussating spinal fibres. According to the present findings, rather than forming an independent decussation, the fibres in question merely represent the most rostral elements of a number of widely distributed decussating fibres which for the most part utilize the tectal commissures in their course rostralward. These rostrally decussating fibres appear to terminate in contralateral intralaminar nuclei, a conclusion supported by the following observations. In cases of unilateral cordotomy, almost equal amounts of terminal degeneration were observed in the intralaminar nuclei of both sides; while in the ventral nuclei, by far the greatest density of terminal degeneration was always found on the ipsilateral side. These differences are readily explained by the fact that the contralateral intralaminar nuclei, in addition to receiving some contralaterally ascending fibres, also receive a number of these rostrally decussating fibres.

### Summary and Conclusions

Fibres ascending in the anterior and lateral funiculi of the Simian spinal cord, particularly such contained in the deeper funicular strata (at cervical levels approximately corresponding to Bechterew's ground bundles), distribute massively to various cell groups in the medullary and pontine reticular formation (cf. Table I). Other, more superficially situated spinal fibres connect with the lateral region of the mesencephalic central grey matter and with the deeper layers of the superior colliculus. The relatively small remaining group of spinal fibres dichotomously terminates in certain para- and intralaminar nuclei and in the nucleus ventralis posterior lateralis of the thalamus. From concurrent studies on lower mammalian forms (Mehler, 1957) it appears that the spinal fibres of para- and intralaminar distribution represent a paleo-spinothalamic pathway distinct from the remaining classical spinothalamic fibres terminating in the ventral nucleus. Some phylogenetic trends in the distribution of spinal fibres to certain bulbar cell groups (e.g. facial motor nucleus, vestibular nuclei) have been discussed.

### Acknowledgments

We are greatly indebted to Dr. Jerzy Rose for his valuable criticism in preparation of the manuscript and to Mrs. Mildred H. Albrecht and Mr. Curtis King for their technical assistance,

### REFERENCES

AMABILINO, R. (1901) *Neurol. Zbl.*, **20**, 909.
ANDERSON, F. D., and BERRY, C. M. (1956) *Anat. Rec.*, **124**, 252.
BATTEN, F. E., and HOLMES, G. (1913) *Brain*, **35**, 259.
BECHTEREW, V. M. (1885) *Neurol. Zbl.*, **4**, 337.
—— (1894) *Neurol. Zbl.*, **13**, 433.
—— (1900) "Les voies de conduction du cerveau et de la moelle." Lyon.

BODIAN, D. (1939) *J. comp. Neurol.*, **71**, 259.

BOWSHER, D. (1957) *Brain*, **80**, 606.

BRODAL, A. (1949) *J. comp. Neurol.*, **91**, 259.

—— (1957) "The Reticular Formation of the Brain Stem." Edinburgh.

——, and GOGSTAD, A. C. (1957) *Acta Anat.*, **30**, 133.

——, and JANSEN, J. (1946) *J. comp. Neurol.*, **84**, 31.

——, WALBERG, F., and BLACKSTAD, T. (1950) *J. Neurophysiol.*, **13**, 431.

——, —— (1952) *Arch. Neurol. Psychiat. Chicago*, **68**, 755.

CHANG, H. T. and RUCH, T. C. (1947) *J. Anat., Lond.*, **81**, 150.

CHOROSCHKO, W. K. (1909) *Mschr. Psychiat. Neurol.*, **26**, 534.

CLARK, W. E. LE GROS (1932) *Brain*, **55**, 406.

—— (1936) *J. Anat., Lond.*, **71**, 7.

COLLIER, J., and BUZZARD, E. F. (1903) *Brain*, **26**, 559.

DEKABAN, A. (1953) *J. comp. Neurol.*, **99**, 639.

DYDYŃSKI, L. v (1903) *Neurol. Zbl.*, **22**, 898.

EDINGER, L. (1889) *Anat. Anz.*, **4**, 121.

—— (1890) *Dtsch. med. Wschr.*, **16**, 421.

—— (1904) "Vorlesungen über den Bau der nervösen Zentralorgane des Menschen und der Tiere." 7th edition. Leipzig, vol. I.

—— (1908) Vorlesungen über den Bau der nervösen Zentralorgane des Menschen und der Tiere. Leipzig, vol. II.

FLECHSIG, P. E. (1876) "Die Leitungsbahnen im Gehirn und Rückenmark des Menschen." Leipzig.

FOERSTER, O., and GAGEL, O. (1932) *Z. ges. Neurol. Psychiat.*, **138**, 1.

GEREBTZOFF, M. (1939a) *Cellule*, **48**, 91.

—— (1939b) *J. belge. Neurol. Psychiat.*, **39**, 719.

GETZ, B. (1952) *Acta Anat.*, **16**, 271.

GIOK, S. P. (1956) "Localization of Fiber Systems within the White Matter of the Medulla Oblongata and the Cervical Cord in Man." (Thesis) Leiden.

GOLDSTEIN, K. (1910) *Neurol. Zbl.*, **29**, 898.

GOWERS, W. R. (1885) "Diagnosis of Diseases of the Brain and of the Spinal Cord." New York.

—— (1886) *Lancet*, **1**, 1153.

GURDJIAN, E. S. (1927) *J. comp. Neurol.*, **43**, 1.

HELWEG, O. (1887) *Arch. Psychiat. Nervenkr.*, **19**, 104.

HENNEBERG, R. (1901) *Neurol. Zbl.*, **20**, 334.

HUBER, E. (1930) *Quar. Rev. Biol.*, **5**, 133.

INGRAM, W. R., HANNETT, F. I., and RANSON, S. W. (1932) *J. comp. Neurol.*, **55**, 333.

JASPER, H. H., and AJMONE-MARSAN, C. (1954) "A Stereotaxic Atlas of the Diencephalon of the Cat." Ottawa.

JIMENEZ-CASTELLANOS, J. (1949) *J. comp. Neurol.*, **91**, 307.

JOHNSON, F. H. (1951) *Anat. Rec.*, **109**, 309.

—— (1954) *Anat. Rec.*, **118**, 316.

KOELLIKER, A. (1896) "Handbuch der Gewebelehre des Menschen." 6th edition. Leipzig, vol. II.

KOHNSTAMM, O. (1899) *Arch. Psychiat. Nervenkr.*, **32**, 681.

—— (1900a) *Neurol. Zbl.*, **19**, 242.

—— (1900b) *Mschr. Psychiat. Neurol.*, **8**, 261.

——, and QUENSEL, F. (1908a) *Dtsch. Z. Nervenheilk.*, **36**, 182.

——, —— (1908b) *Neurol. Zbl.*, **27**, 242.

——, —— (1908c) *J. Psychol. Neurol., Lpz.*, **13**, 89.

KURU, M. (1949) "Sensory Paths in the Spinal Cord and Brain Stem of Man." Tokyo.

LEWANDOWSKY, M. (1904) Denkschr. Med. Naturw. Ges. (Jena), 10, *Neurobiol. Arb.*, Series II, 1, 63–150.

LIU, C.-N. (1956) *Arch. Neurol. Psychiat., Chicago,* **75,** 67.

LONG, E. (1914) *Nouv. Iconogr. Salpêt.,* **27,** 61.

MAGOUN, H. W., ATLAS, D., INGERSOLL, E. H., and RANSON, S. W. (1937) *J. Neurol. Psychopathol.,* **17,** 241.

MARBURG, O. (1903) *Mschr. Psychiat. Neurol.,* **13,** 486.

MAY, W. P. (1906) *Brain,* **29,** 742.

MEESSEN, H., and OLSZEWSKI, J. (1949) "A Cyto-architectonic Atlas of the Rhombencephalon of the Rabbit." Basel.

MEHLER, W. R. (1957) *Anat. Rec.,* **127,** 332.

——, FEFERMAN, M. E., and NAUTA, W. J. H. (1956) *Anat. Rec.,* **124,** 332.

——, VERNIER, V. G., and NAUTA, W. J. H. (1958) *Anat. Rec.,* **130,** 430.

MONAKOW, C. (1905) "Gehirnpathologie." 2nd ed. Wien.

MORIN, F., SCHWARTZ, H. G., and O'LEARY, J. L. (1951) *Acta psychiat. neurol. scand.,* **26,** 371.

MORUZZI, G., and MAGOUN, H. W. (1949) *Electroenceph. clin. Neurophysiol.,* **1,** 455.

MOTT, F. W. (1892) *Brain,* **15,** 215.

—— (1895) *Brain,* **18,** 1.

NAUTA, W. J. H., and GYGAX, P. A. (1954) *Stain Technol.,* **29,** 91.

——, and KUYPERS, H. G. J. M. (1958) In Jasper's "Reticular Formation of the Brain." Boston, p. 3.

—— (1958) *Brain,* **81,** 319.

OLSZEWSKI, J. (1952) "The Thalamus of the Macaca Mulatta." Basel.

—— (1954) *In* Council for International Organizations of Medical Sciences. "Brain Mechanisms and Consciousness," Oxford, p. 54.

——, and BAXTER, D. (1954) "The Cytoarchitecture of the Human Brain Stem." Basel.

PAPEZ, J. W. (1927) *J. comp. Neurol.,* **43,** 159.

——, and RUNDLES, W. (1937) *J. nerv. ment. Dis.,* **85,** 505.

PETRÉN, K. (1910) *Arch. Psychiat. Nervenkr.,* **47,** 495.

POMPEIANO, O., and BRODAL, A. (1957) *J. comp. Neurol.,* **108,** 353.

POWELL, T. P. S. (1952) *Brain,* **75,** 571.

——, and COWAN, W. M. (1956) *Brain,* **79,** 364.

PROBST, M. (1900) *Arch. Psychiat. Nervenkr.,* **33,** 1.

—— (1902) *Mschr. Psychiat. Neurol.,* **11,** 3.

QUENSEL, F. (1898) *Neurol. Zbl.,* **17,** 482.

—— (1907) *Neurol. Zbl.,* **26,** 1138.

RAMON Y CAJAL, S. (1909) "Histologie du Système Nerveux de l'Homme et des Vertébrés." Paris, vol. I.

—— (1911) "Histologie du Système Nerveux de l'Homme et des Vertébrés." Paris, vol. II.

RIOCH, D. M. (1929) *J. comp. Neurol.,* **49,** 1.

—— (1931) *J. Anat., Lond.,* **65,** 324.

ROSE, J. E., and GALAMBOS, R. (1952) *J. Neurophysiol.,* **15,** 343.

——, and WOOLSEY, C. N. (1949) *J. comp. Neurol.,* **91,** 441.

ROSSI, G. F., and BRODAL, A. (1956) *Confin. neurol.,* **16,** 321.

——, —— (1957) *Arch. Neurol. Psychiat., Chicago,* **78,** 439.

——, and ZANCHETTI, A. (1957) *Arch. ital. Biol.,* **95,** 199.

ROSSOLIMO, G. (1898) *Arch. Neurol., Paris,* **6,** 343.

ROTHMANN, M. (1903) *Neurol. Zbl.,* **22,** 744.

—— (1906) *Berl. klin. Wschr.,* **43,** 47.

SIMMA, K. (1951) *Mschr. Psychiat. Neurol.,* **122,** 32.

SOLDER, F. V. (1897) *Neurol. Zbl.*, **16**, 308.
SPIEGEL, E. A., KLETZKIN, M., and SZEKELY, E. G. (1954) *J. Neuropath. exp. Neurol.*, **13**, 212.
SPILLER, W. G. (1905) *J. nerv. ment. Dis.*, **32**, 318.
THIELE, F. H., and HORSLEY, V. (1901) *Brain*, **24**, 519.
THOMAS, A. (1897) *C. R. Soc. Biol., Paris*, **49**, 88.
VOGT, C. (1909) *J. Psychol. Neurol., Lpz.*, **12**, 285.
——, and VOGT, O. (1941) *J. Psychol. Neurol., Lpz.*, **50**, 32.
WALBERG, F., and BRODAL, A. (1953) *J. comp. Neur.*, **99**, 251.
WALKER, A. E. (1937) *Proc. Kon. Aka. Wetensch. Amst.*, **40**, 198–206.
—— (1938) "The Primate Thalamus." Chicago.
—— (1943) *Res. Publ. Ass. nerv. ment. Dis.*, **23**, 63.
WALLENBERG, A. (1899) *Neurol. Zbl.*, **18**, 829.
—— (1900) *Anat. Anz.*, **18**, 81.
—— (1905) *Anat. Anz.*, **26**, 145.
WERSILOFF, B. (1898) Quoted by Bechterew, 1900.
WHITLOCK, D. G., and PERL, E. R. (1957) *Anat. Rec.*, **127**, 388.

## LEGENDS FOR PLATES

### PLATE CIX

FIG. 7.—Spino-olivary connexions: Note profuse pericellular axon degeneration in the dorsal accessory olive and absence of degenerating fibres in the adjacent part of the main olive (lower left part of picture).

FIG. 8.—Spinoreticular connexions: Arrows indicate some of the fine degenerating axons in contact with dendrites of a large cell of the nucleus giganto-cellularis.

FIG. 9.—Composite photographs from a horizontal section showing ascending degeneration in the anterolateral fasciculus of the operated side. Note degenerating medial offsets, possibly collaterals of longitudinal axons, distributing to the ventral grey column.

FIG. 10.—Contral.: Composite photographs of comparable fields from the contra-lateral, unoperated side.

### PLATE CX

FIG. 11.—Spino-intralaminar connexions: Fine axon degeneration in the nucleus centralis lateralis thalami. This photograph was taken from the section used in the preparation of fig. 6. The field shown approximately corresponds to the area outlined in the Nissl section shown in fig. 13 (X, indicates a blood vessel appearing in both sections).

FIG. 12.—Nucleus ventralis posterior lateralis showing one of the disseminated clusters of pericellular degeneration following ipsilateral cordotomy at Th. 6 (same case as figs. 6 and 11).

FIG. 13.—Cresylecht violet section (horizontal) adjacent to the silver-stained section shown in fig. 11. Note the large, medium-sized and small cells in the region of the nucleus centralis lateralis (*see* footnote[1] p. 744).

FIG. 14.—High-power view of pericellular degeneration in the nucleus ventralis posterior lateralis.

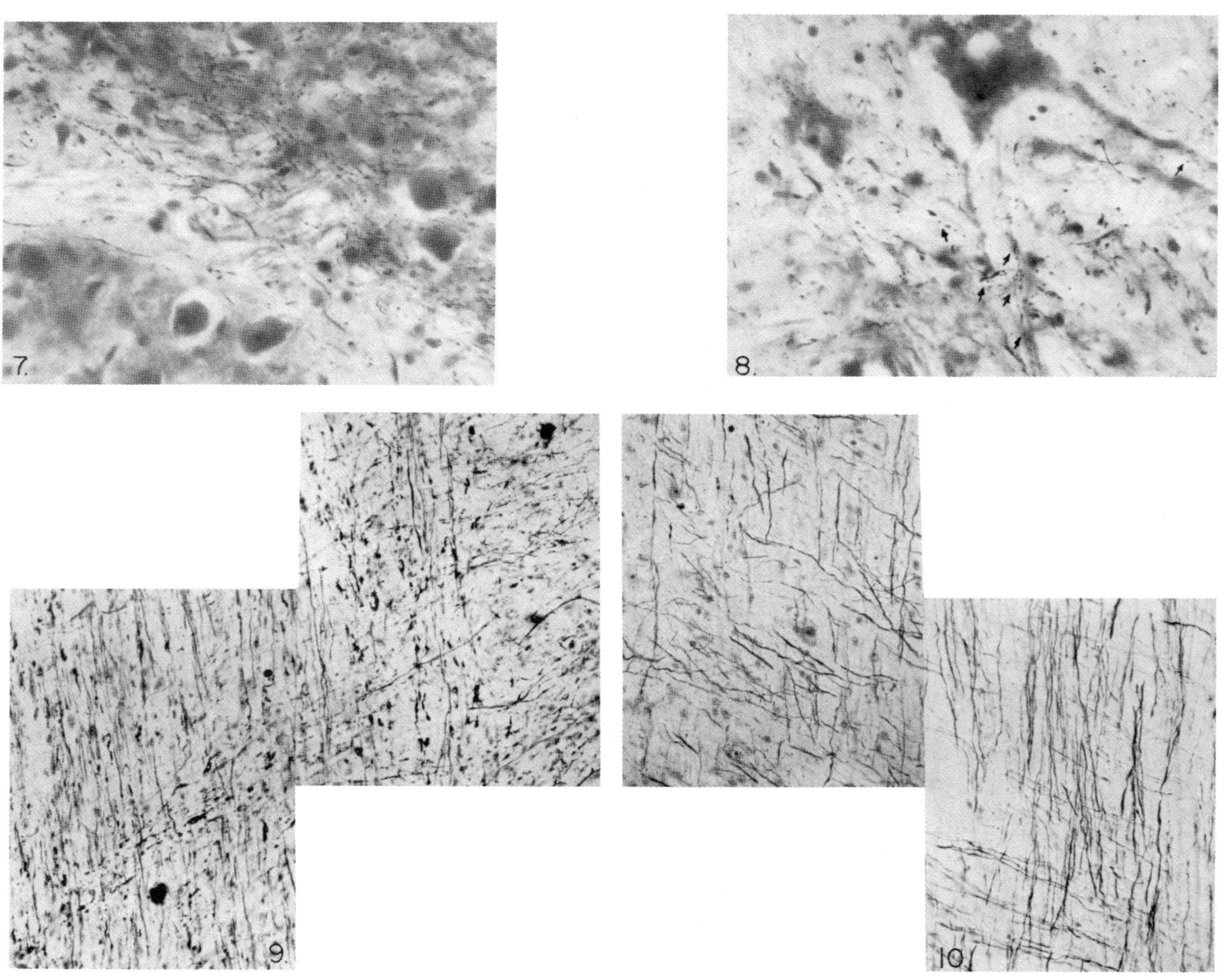

*To illustrate article by William R. Mehler, Martin E. Feferman and Walle J. H. Nauta.*

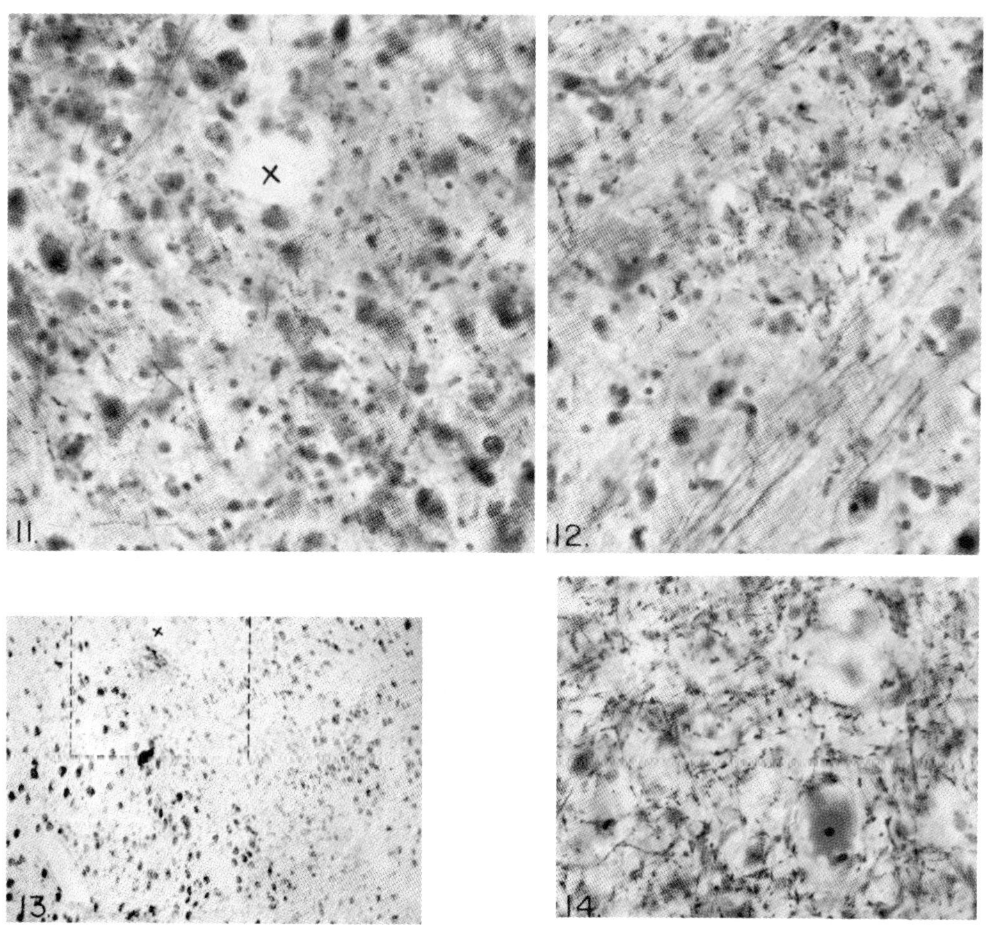

*To illustrate article by William R. Mehler, Martin E. Feferman, and Walle J. H. Nauta.*

[Reprinted from the JOURNAL OF ANATOMY
Vol. 95, Part 4, October 1961]

     

# FIBRE DEGENERATION FOLLOWING LESIONS OF THE AMYGDALOID COMPLEX IN THE MONKEY

By W. J. H. NAUTA

*Department of Neurophysiology, Walter Reed Army Institute of Research,
Washington, D.C.*

Of the amygdaloid projection pathways the stria terminalis has traditionally received strong emphasis in anatomical studies, and less attention has customarily been given other fibre systems originating in the amygdaloid complex. It is, however, known that the amygdala is connected with other basal telencephalic structures and with the diencephalon by a massive ventral fibre system which spreads forward and medially through the region underneath the lentiform nucleus. This fibre system appears to have been recognized first by Johnston (1923), who considered it to be an amygdalofugal component of his 'longitudinal association bundle'. Johnston limited his account of this projection system to the statement that such fibres beneath the globus pallidus join 'the general system of precommissural fibres passing up through the parolfactory area'. Later workers, on the basis of experimental-anatomical observations in the cat, described distributions of the ventral amygdalofugal fibre system to the preoptic region and hypothalamus (Lammers & Lohman, 1957; Hall, 1960), the bed nucleus of the stria terminalis (Fox, 1943), the caudate nucleus and the subcallosal gyrus (Lammers & Lohman, 1957).

In Marchi experiments in the monkey, Fox (1949) made the interesting observation that fibres of apparently the same general category join the inferior thalamic peduncle and terminate in the dorso-medial nucleus of the thalamus. In a recent study by the aid of the Nauta–Gygax silver technique the existence of a direct amygdalo-thalamic projection in the monkey was confirmed, whereas other components of the ventral amygdalofugal fibre system were traced to the substantia innominata, the lateral preoptic and hypothalamic regions, the basal septal region, olfactory tubercle, and rostral limbic cortex (Nauta & Valenstein, 1958). These findings have hitherto been published only in the form of an abstract. The present paper will serve to present a more detailed account of the pertinent observations.

## MATERIAL AND METHODS

This report is based largely on observations made in six young adult *Macaca mulatta* monkeys in which surgical lesions had been placed in the amygdaloid complex. Several further incidental observations made in other material will be mentioned in the Discussion.

The operative procedure was as follows. With the animal in deep pentobarbital anaesthesia, unilateral lesions of the amygdaloid complex were made, using the sub-frontal approach indicated by Scoville & Milner (1957). According to this procedure, a large frontal bone flap was turned, the dura opened widely, and the frontal lobe lifted from the orbital roof, with all exposed cortex thoroughly protected from mechanical injury by strips of cottonoid soaked in saline. Gentle elevation of the fronto-temporal junction exposed the rostral aspect of the amygdalo-piriform prominence, and a narrow suction tip, fashioned of a suitably curved 20-gauge injection cannula with the cutting tip ground off, was inserted into the amygdalo-piriform complex, using the internal carotid and middle cerebral arteries respectively as medial and dorsal landmarks. Following the aspiration of amygdaloid substance through the small puncture hole, the dura was carefully sutured and the bone flap replaced and fastened.

The animals were killed by an overdose of pentobarbital 9–12 days post-operatively. Fixation of the brain with formalin was initiated by perfusion and extended by storage in the fixative for 6–12 weeks. Frozen sections of the brains were stained following the Laidlaw modification of the Nauta–Gygax silver technique, and axon degeneration, as identified microscopically, was recorded in projection drawings of selected sections.

### OBSERVATIONS

As expected, marked variations were encountered in the localization of the lesions. In all cases, the suction tip had penetrated either the piriform cortex or the cortical amygdaloid nucleus, and the lesion extended laterally from this point of entry into at least the basal and accessory basal nuclei. The lateral nucleus was involved in four animals, the medial nucleus in one. In two cases the defect was found to encroach upon the temporal white matter covering the lateral aspect of the amygdala; one of these cases showed additional slight involvement of the ventral edge of the putamen. In all remaining cases the lesion was entirely confined to the amygdalo-piriform complex.

Further variations encountered in this series of experiments concerned the position of the lesion with respect to the dorso-ventral and rostro-caudal coordinates. As concerns the ventral amygdalofugal pathways, such variations appeared to affect the quantity rather than the distribution of the fibre degeneration, more massive degeneration apparently being related to greater involvement of the dorsal amygdaloid regions. The degeneration observed in the stria terminalis appeared to be more strikingly dependent on the localization of the lesion in regard to both quantity and distribution.

In the following account two of the six cases will be discussed in some detail.

### (1) *Case MA 3* (Text-fig. 1)

The lesion in this case was largely limited to the rostro-ventral quarter of the amygdaloid complex. It involved mainly the basal and accessory basal nuclei, and to a lesser extent the lateral nucleus. The suction tip had caused additional damage to the cortical nucleus and, more rostrally, to the piriform cortex. The lesion was separated from the temporal white matter adjoining the amygdaloid complex laterally by at least a millimeter of apparently normal tissue.

#### A. *Ventral amygdalofugal pathways*

It is evident even under low magnification that massive fibre degeneration extends from the lesion in the dorsal and rostral directions. Within the amygdaloid complex such degenerated fibres compose a complicated mazework (Text-fig. 1*b*) in which,

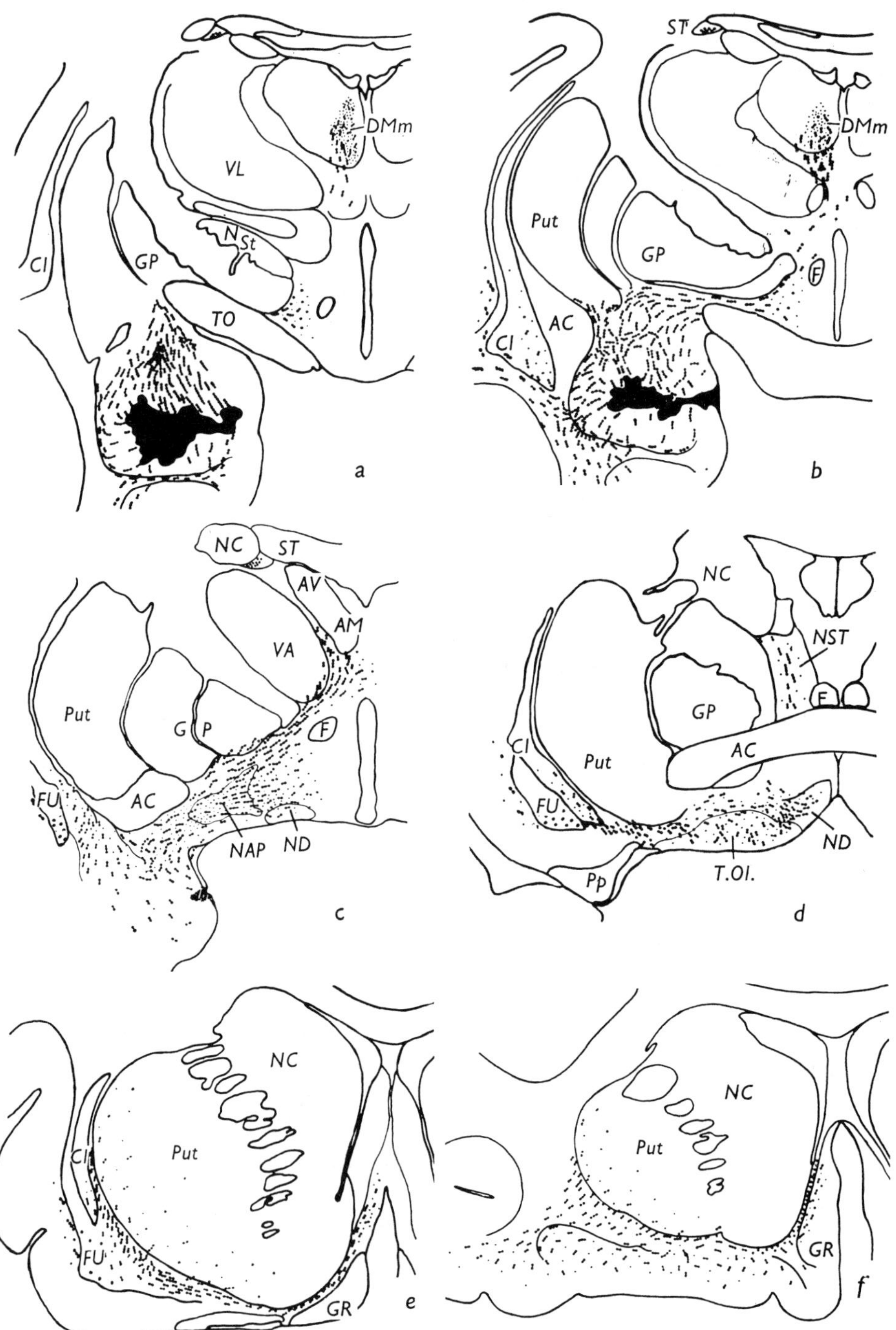



Text-fig. 1. Fibre degeneration observed in case MA 3. The amygdaloid lesion is indicated in jet black. Coarse dots indicate degenerating fibres of passage, fine stipple preterminal and terminal degeneration. Abbreviations: see p. 531.

          *W. J. H. Nauta*

however, at least one more condensed group of relatively fine fibres can be distinguished immediately dorsal to the basal amygdaloid nuclei (Text-fig. 1*a*). This fibre group is tentatively identified as the longitudinal association bundle, but it is noted that the bundle forms part of a more diffuse and widespread fibre system apparently originating mainly from the basolateral group of nuclei. It is of further interest to note the presence of numerous scattered degenerating axons of heavy calibre which, like the finer constituents of the longitudinal association bundle, in general follow dorsal and rostral trajectories through the amygdala.

*Substantia innominata.* From the level of the caudal border of the chiasma rostralward, large numbers of degenerating fibres curve medially and enter the sublenticular region often labelled substantia innominata. The latter region is pervaded by innumerable disintegrating axons of various calibres (Text-fig. 1*c*; Pl. 1*a*). In the ventral parts of the region fine fibres predominate, whereas coarse axons are more numerous in the dorsal zone adjoining the lentiform nucleus. The occurrence of fine degenerating pericellular fibres indicates that some amygdalofugal fibres actually terminate in the substantia innominata, both in the large-celled dorsal area known as the nucleus ansae peduncularis (Meynert's 'Basalganglion'; Ganser's nucleus basalis) and in the less well defined ventral zones which include a caudal extension of the nucleus of Broca's diagonal band. However, most of the fibres of the ventral amygdalofugal pathway only pass through the substantia innominata *en route* to more distant structures. Somewhat schematically, it can be said that such transit fibres are disposed in fan-tail fashion, with the caudal components oriented medially, the rostral ones rostrally in nearly sagittal planes.

*Preoptic region and hypothalamus.* The more caudal sublenticular transit fibres are distributed to the lateral preoptic and hypothalamic regions adjoining the substantia innominata (Text-fig. 1*a–c*). They terminate diffusely among cell groups scattered between the longitudinal fibres of the medial forebrain bundle. Only few such fibres accompany the medial forebrain bundle caudalward, and consequently little, if any, degeneration appears in the lateral hypothalamus at infundibular levels or farther caudally.

The most ventral of these amygdalo-hypothalamic fibres course immediately dorsal to the supraoptic nucleus, and a few scattered degenerating elements are seen to enter this cell group. There is, however, no convincing evidence for arborization of such fibres within the nucleus. Signs of fibre termination are, by contrast, abundant in the ventral hypothalamic zone immediately dorsal to the supraoptic nucleus.

*Thalamus.* The fibre degeneration extending to the lateral preoptic region is accompanied dorsally by a considerable number of degenerating coarse axons which join the inferior thalamic peduncle (Text-fig. 1*c*). Before gaining the peduncle most of these fibres follow dorsal paths in their medial course through the substantia innominata; some of the most dorsal ones even trace weaving trajectories through a ventral zone of the globus pallidus without, however, displaying signs of termination in that structure. In the peduncle the degenerating fibres form a rather scattered group (Pl. 1*b*) which, upon entering the thalamus, issues a few fibres medially to the nucleus reuniens (Text-fig. 1*c*) and laterally to the medial part of the nucleus reticularis thalami. The bundle then breaks up into a number of scattered fascicles which curve caudalward and follow the medial one-third of the internal medullary lamina to the medial, magnocellular part of the ipsilateral dorso-medial thalamic nucleus (Text-fig. 1*a*, *b*). In this cell group the constituent coarse axons terminate with profuse pericellular arborizations, as indicated by the extremely dense feltwork of disintegrating fine axons which fills the nucleus and sharply delimits it from its surroundings (Pl. 1*c*). A small number of amygdalo-thalamic fibres decussate in the internal medullary lamina and disperse in the medial part of the contralateral dorso-medial nucleus.

*Olfactory tubercle, septal region, gyrus subcallosus, rostral limbic cortex.* Rostrally the substantia innominata continues into the region of the substantia perforata anterior. This region is characterized by the appearance of the olfactory tubercle, a circumscript cortical formation flanked medially by the nucleus of Broca's diagonal band, and laterally by the prepiriform cortex (Text-fig. 1*d*). In its caudal half the tubercle is separated from the more dorsally situated lentiform nucleus by a rather loosely structured rostral extension of the nucleus ansae peduncularis (n. basalis of Ganser). More rostrally, however, the diagonal band curves dorsally into the septal region, and the nucleus ansae peduncularis tapers to a vague rostral limit, leaving the olfactory tubercle in immediate contact with the ventral aspect of the caudato-putaminal junction. Farther rostrally still the olfactory tubercle reaches its rostral boundary; from here forward the fundus striati is covered by the orbito-frontal cortex.

As shown by Text-fig. 1*d*, numerous fibres of the sublenticular amygdalofugal fibre system extend forward into the region of the olfactory tubercle. In the tubercle proper such fibres appear to terminate in the multiform as well as in the pyramidal cell layers. Other fibres are distributed to the nucleus ansae peduncularis deep to the tubercle, whereas densely packed fine degenerating axons follow a more medial path alongside and in the diagonal band. Undoubtedly many of these more medial fibres terminate in the diagonal nucleus, but this degeneration does not follow the nucleus over more than a short distance into the septal region (Text-fig. 1*e*), and consequently no axon degeneration is detectable in more dorsal parts of the septum.

Fibres of the same medial group that conveys amygdaloid efferents to the nucleus of the diagonal band extend forward beyond the septum, in the white matter of the gyrus rectus (Text-fig. 1*e*). Many of these fibres arborize in the grey matter of the gyrus subcallosus (Text-fig. 1*f*). Other fibres of the same group continue even farther forward, bend around the genu corporis callosi and become dispersed among the fibres of the fasciculus cinguli. Degenerating arborizations of these long amygdalofugal fibres are found scattered in the ventral region of approximately the rostral one-third of the gyrus cinguli.

*Temporal cortex, insula, putamen, claustrum, orbito-frontal cortex.* As shown by Text-fig. 1*e–f*, the aforementioned degeneration spreading to the septum and rostral regions of the gyrus fornicatus forms only a medial part of a widespread stratum of degenerating fibres that covers the ventral half of the putamen and nucleus accumbens. Approximately the lateral half of this degenerated fibre stratum appears to be made up of fibres closely related to the fasciculus uncinatus and the ventral margins of the outer capsules. It seems likely that these degenerating axons belong to a fibre system which takes a lateral exit from the amygdaloid complex, enters the white matter of the temporal lobe (Text-fig. 1 *b*) and spreads from here in various directions: (*a*) ventralward to a rostral part of the inferior temporal gyrus, (*b*) lateralward to rostral parts of the middle and superior temporal gyri, and (*c*) rostrally to the claustrum (Text-fig. 1 *b–e*), the ventral part of the insular cortex (Text-fig. 1 *b*, *d*, *e*), the lateral zone of the putamen (Text-fig. 1 *e–f*), and the caudal orbito-frontal cortex (Text-fig. 1 *f*). It must be remarked, however, that with the exception of the fibres to the inferior temporal gyrus and claustrum, a continuous tracing of the paths followed by the degeneration in question has not been possible. It is especially difficult to establish the connexion of the degeneration in the temporal white matter with that in the fasciculus uncinatus and in the base of the outer capsules. Hence, as will be pointed out more fully in the Discussion, the evidence in regard to some of the amygdalo-cortical connexions indicated above appears to be somewhat less than conclusive.

B. *Stria terminalis*

In this case only a small ventro-lateral part of the stria terminalis is degenerated. The fibres involved extend forward to the bed nucleus of the stria, in which cell groups all appear to terminate (Text-fig. 1 *d*). No strial degeneration can be traced ventralward past the anterior commissure.

(2) Case MA 11 (Text-figs. 2, 3)

This case is briefly described here for the supplementary information which it furnishes concerning the stria terminalis. The lesion in MA 11 was situated farther caudally in the amygdaloid complex than was the case in MA 3, and it extended into that part of the complex which extends caudally in the roof of the temporal horn of the lateral ventricle (Text-fig. 2). As in MA 3, the lesion involved mostly the basolateral cell groups, but it extended farther dorsally; also it had spared the rostral half of the amygdaloid complex, which was involved to a considerable extent in MA 3.

*Stria terminalis*

The stria terminalis shows massive degeneration. In contrast to its localization in case MA 3, the degeneration is densest in the dorso-medial half of the stria; it maintains this relative position throughout its course through the bed nucleus of the stria (Text-fig. 3 *a*). Few if any of the disintegrating fibres appear to end in the latter nucleus, and virtually all continue around the rostral and caudal aspects of the anterior commissure into the medial preoptic region and beyond it into the hypothalamus (Text-fig. 3 *b*, *c*). Most of these hypothalamic stria fibres appear to distribute to the anterior hypothalamic nucleus (Text-fig. 3 *b*; Pl. 1 *b*). In their course caudalward through this nucleus the degenerating fibres shift progressively farther ventrally, rapidly decreasing in number. In frontal sections involving the caudal one-third of the optic chiasma only a small number have remained. These strial components are here found scattered in regions ventral and somewhat lateral to the rostral pole of the ventro-medial nucleus (Text-fig. 3 *c*). Only a few isolated degenerating fibres can be traced some distance beyond the caudal border of the chiasma; their distribution area appears to be ventral to the ventro-medial nucleus and possibly involves a small lateral part of the arcuate nucleus. No degenerating fibres can be traced into the ventro-medial nucleus proper.

The only further component of the stria terminalis which is recognizable in this case appears in the form of a small number of degenerating fibres which join the dorsal stratum of the anterior commissure (Text-fig. 3*a*). Although these few fibres can be followed across the midline, it is not possible to establish their ultimate

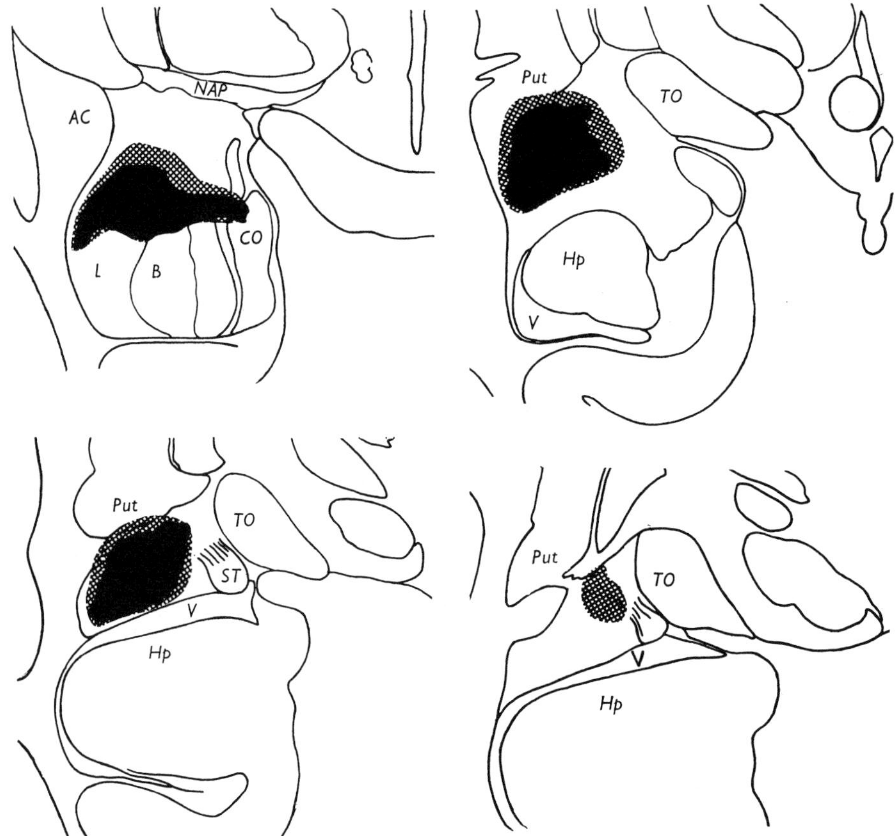


Text-fig. 2. Four drawings of transverse sections showing the extent of the amygdaloid lesions in case MA 11. Complete tissue loss is indicated in jet black, heavy gliosis with loss of cell bodies by cross-hatching. Abbreviations: see p. 531.

distribution. It is likely that this scant degeneration corresponds to the much more massive pars commissuralis striae terminalis described in the opossum by Johnston (1923), and in several subprimate mammalian forms by others.

*Ventral projection pathways*

The degenerating ventral amygdalofugal system charted in this case appears similar to that observed in MA 3, with the exception that the degeneration in several of the rostral components of the system is considerably less massive than in MA 3.

This is especially true of the degenerations traceable to the orbito-frontal cortex, putamen, claustrum, and olfactory tubercle. By contrast, the degeneration distributing to the nucleus of Broca's diagonal band, as well as to the gyrus subcallosus and rostral limbic cortex appears comparable in volume to that observed in **MA 3**.

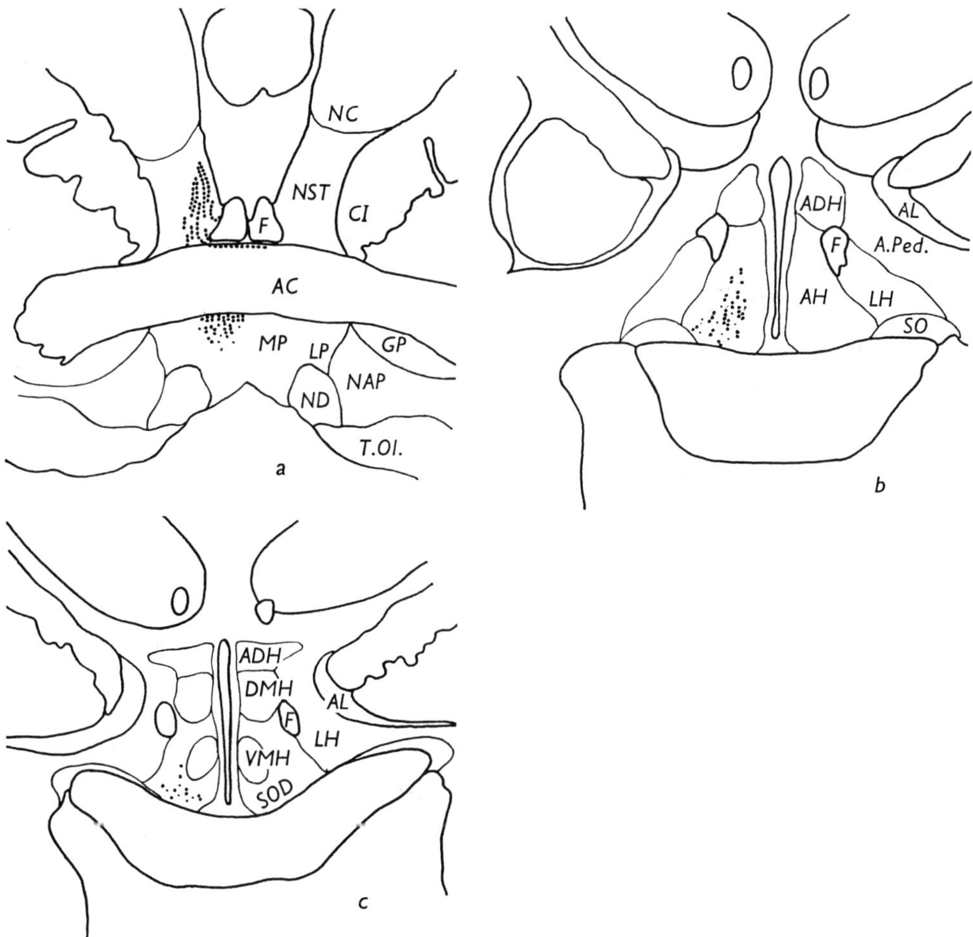


Text-fig. 3. Degeneration of the stria terminalis as observed in case MA 11.
Symbols as in Text-fig. 1. Abbreviations: see p. 531.

## DISCUSSION

The amygdaloid complex, as is well known, consists of several cell groups of various architecture, and it is unlikely that all these nuclei contribute in the same manner to the pathways described in the foregoing account. Unfortunately, even small surgical lesions in heterogeneous structures such as the amygdala destroy not only the cell bodies which fall within their boundaries, but also fibres of passage originating outside the area of the lesion proper. Only gross impressions of the mosaic of origin can therefore be expected from studies involving the surgical production of amygdaloid lesions. Apart from this restriction, the particular surgical procedure followed in the present study has the advantage of obviating the track-damage to other structures which unavoidably complicates stereotactic lesions and which would have vitiated several of the present observations by involving either the internal capsule or the temporal white matter lateral to the amygdala.

There remains, none the less, reason for caution in the interpretation of fibre degenerations following amygdaloid lesions: such lesions almost certainly involve transit projection fibres from the piriform cortex. The production of selective defects of the monkey's periamygdaloid cortex, which could have helped to clarify this point, has so far been unsuccessful in our hands. No attempt can consequently be made at this time to distinguish between paleocortical fibres passing through the amygdala, and projection fibres from the amygdaloid cell groups proper. Hence, although the term 'amygdalofugal pathways' is used here, the connexions in question are perhaps more cautiously interpreted as projections from the amygdalo-piriform complex as a whole.

The observations reported above have confirmed the existence of two main amygdalo-subcortical projection pathways in the macaque, namely the relatively compact stria terminalis, and a much more diffuse and widespread ventral amygdalo-fugal fibre system. Furthermore, the present findings suggest the existence of a third efferent pathway which connects the amygdala with several cortical regions.

As the literature pertaining to the fibre connexions of the amygdala has recently been reviewed succinctly by Gloor (1955, 1959), a detailed bibliographical survey would seem superfluous at this time. In the following account references will therefore be limited mostly to those previous experimental observations which seem immediately pertinent to the findings described in the foregoing account.

### A. *Ventral amygdalofugal pathways*

The present findings emphasize the great volume of the ventral amygdalo-subcortical pathways, as well as the multiplicity of their connexions. As pointed out in the description of case MA 3, this fibre system, corresponding in part at least to Johnston's (1923) longitudinal association bundle, emerges from the amygdala in the dorso-medial direction. It spreads medially and forward underneath the lentiform nucleus, distributing fibres to (*a*) basal forebrain structures, namely, the substantia innominata, the lateral preoptic and hypothalamic areas, the substantia perforata anterior including the olfactory tubercle, and the nucleus of the diagonal band; (*b*) rostral parts of the gyrus fornicatus: subcallosal gyrus and anterior cingulate cortex; and (*c*) the magnocellular element of the dorso-medial thalamic nucleus, via the ansa peduncularis.

*Amygdaloid projections to the septal-lateral preoptic-lateral hypothalamic region* can presumably serve to lead impulses of amygdaloid origin into the paths of both the medial forebrain bundle and stria medullaris-fasciculus retroflexus. The same would appear to hold for the amygdaloid pathways to the substantia innominata, for there is evidence that this structure also contributes a large number of fibres to the medial forebrain bundle (Mehler & Nauta, unpublished observations in the monkey). By virtue of these synaptic relationships with several sources of origin of the medial forebrain bundle and related fibre trajectories such as the stria medullaris, the ventral amygdalo-subcortical pathway appears comparable to the fornix system. Much like the latter, it appears to furnish a first link in a multi-synaptic chain of conduction which, partly through the intermediary of the lateral hypothalamus, reciprocally connects the limbic forebrain region with an extensive medial area of the midbrain tegmentum and central grey midbrain substance. As indicated previously (Nauta, 1958), this 'limbic system–midbrain circuit' is connected by massive escape pathways with the more laterally located central mesencephalic and subthalamic reticular formation, connexions by which the limbic forebrain structures could conceivably affect a diversity of reticular mechanisms in addition to the autonomic and endocrine functions represented in the midbrain and hypothalamus. It seems possible, for example, that such complex motor stereotypes as the licking, sniffing and chewing movements which have been observed in the course of amygdaloid stimulation experiments (see Gloor, 1959) were elicited by the medium of such indirect amygdalo-reticular connexions.

It is interesting to compare the foregoing anatomical considerations with the results of Gloor's (1955) electrophysiological study in the cat. The distribution of responses of shortest latency (less than 7 msec.) recorded by Gloor in the septum, preoptic region and anterior hypothalamus agrees well with the spread of the ventral amygdalo-subcortical system observed in the present study. Responses of longer latency, presumably denoting indirect connexions, were recorded from more caudal hypothalamic areas, as well as from a large expanse of mesencephalic reticular formation. Although Gloor's observations in general corroborate the anatomical evidence of widespread, if indirect, amygdalo-reticular pathways, it is only fair to point out that his results do not support the notion that such connexions are established mainly by the medial forebrain bundle or via the stria medullaris-fasciculus retroflexus trajectory. Widely scattered responses to amygdaloid stimulation were, for example, recorded from the mesencephalic tegmentum with delays considerably shorter than those registered in caudal regions of the lateral hypothalamus, i.e. along the more caudal part of the hypothalamic trajectory of the medial forebrain bundle. Actually, Gloor's electrophysiological evidence appears to point to extremely medially placed pathways, spreading caudalward through the periventricular hypothalamic region, as the more direct amygdalo mesencephalic conductors. Conceivably, one such pathway could be furnished by amygdalo-hypothalamic fibres articulating with descending components of Schütz's periventricular fibre system. However, even the relatively fast potentials recorded along this medial route fail to match the surprisingly short latency of some widespread responses obtained in Gloor's experiments from the rostral midbrain tegmentum. In commenting on these findings, Gloor himself suggests the possible existence of a direct or at most oligo-synaptic amygdalo-tegmental pathway by-passing the hypothalamus. If such a connexion indeed exists, it appears likely that it would follow the internal capsule and cerebral peduncle. At this time, however, one is forced to conclude that no data are available which could explain all the details of Gloor's observations in terms of known anatomical pathways.

*Amygdalo-hippocampal connexions.* The present study has failed to produce evidence of the direct amygdalo-hippocampal connexions which have been described on the basis of observations in normal material (Hilpert, 1928, and others). However, the existence of alternate pathways subserving amygdalo-hippocampal interaction is made likely by the conspicuous amygdaloid projection to the nucleus of the diagonal band, a cell group which is believed to project to the hippocampus either directly (Daitz & Powell, 1954) or via the presubiculum (Cragg & Hamlyn, 1957). The pathway in question probably conducts in both directions: the hippocampal formation is known to project to the entire septal region, and from the latter fibres can be followed along the diagonal band to the immediate vicinity of the amygdala, in the monkey even directly into the medial amygdaloid region (Valenstein & Nauta, 1959). Amygdalo-hippocampal conduction via the entorhinal area as suggested by Gloor (1955) seems conceivable also, although the evidence regarding this pathway is controversial. Adey & Meyer (1952) failed to obtain anatomical evidence of connexions from the amygdalo-piriform complex to the entorhinal area. On the other hand, the spread of strychnine spikes from the periamygdaloid cortex to caudal regions of the hippocampal gyrus, observed by Pribram & MacLean (1952), tends to support Gloor's suggestion.

*Amygdalo-thalamic connexions.* In agreement with Fox's (1949) observations in Marchi experiments in the monkey, a quite massive amygdaloid projection could be traced via the inferior thalamic peduncle to the dorso-medial nucleus of the thalamus. As degeneration was found in this pathway in all of the six cases of amygdaloid lesion, it is not possible to identify the contributing amygdaloid cell groups with certainty. It is, however, of interest that the lesions produced in the present study all involved the cortical nucleus, the accessory basal nucleus, and the basal nucleus proper. The heavy calibre of most of the constituent axons tends to suggest the basal and lateral nuclei as the most likely sources of origin of the amygdalo-thalamic connexion.

Although not specifically mentioned by Fox, it is clear from the present findings that the amygdalo-thalamic pathway terminates almost exclusively in the medial, magnocellular division of the dorso-medial nucleus. Some further sparse termination seems to take place in the rostral midline region and in the paracentral intralaminar nucleus. No fibres of the connexion could, however, be identified in the lateral part of the dorso-medial nucleus.

The detailed analysis of the dorso-medial thalamo-cortical projection by Pribram, Chow & Semmes (1948) has shown that the medial, magnocellular component of the dorso-medial nucleus projects specifically upon the orbito-frontal cortex. The present findings thus suggest that Fox's amygdalo-thalamic tract represents the first link in a major transthalamic amygdalo-orbito-frontal connexion. The extremely dense arborization of the amygdaloid projection fibres in the nucleus suggests furthermore that fibres of amygdaloid origin furnish the major afferent supply to the medial element of the dorso-medial thalamic nucleus. It must, however, be noted that other, apparently less massive pathways to the pars medialis of the dorso-medial nucleus, have been traced from the septal region (Guillery (1959) in the cat; Valenstein & Nauta (1959) in the monkey), and from the inferior temporal gyrus (Whitlock & Nauta (1956) in the monkey). The temporal cortical projection to the dorso-medial nucleus, like that from the amygdala, follows the inferior thalamic peduncle, but it is unlikely that it was involved in the present experiments as it courses lateral to the amygdala in the white matter of the temporal lobe.

It is noteworthy that in the cat no amygdalo-thalamic projections have been identified by either experimental-anatomical (Fox, 1943; Lammers & Lohman, 1957; Hall, 1960) or electrophysiological (Gloor, 1955) methods. However, thalamopetal fibre degeneration entirely comparable to the present findings in the monkey has been produced in the cat by lesions in the preoptic region (Nauta, 1958) and substantia innominata, structures that both receive numerous fibres from the amygdala. It thus appears that, despite the apparent absence of a direct connexion, an anatomical pathway for amygdalo-thalamic conduction is present in the cat also. Comparable interspecific variations in neuronal organization have been noted in various other connexions related to the limbic system (Valenstein & Nauta, 1959). It is tempting to speculate that such anatomical differences between species could reflect important functional variations in the neural mechanisms concerned.

Amygdaloid projections to the pulvinar as mentioned by Fox (1949) could not be identified in the present study. Projections apparently comparable to that observed by Fox have, however, been traced from a large extent of the temporal cortex (Whitlock & Nauta, 1956). As Fox's observations were reported only in the form of an abstract it is not possible to say to what extent degeneration of fibres to the pulvinar could have been caused in his experiments by surgical involvement of such cortico-thalamic connexions.

*Amygdalo-cortical connexions.* In several of the cases of amygdaloid lesion here studied axon degeneration could be traced in continuity from the lesion to (*a*) the gyrus subcallosus and rostral cingulate cortex, and (*b*) the rostral half of the inferior temporal gyrus. The former connexion, established by a moderate number of axons which accompany the fibre pathway to the nucleus of the diagonal band, would seem to correspond to the amygdaloid projection to the gyrus subcallosus observed in the cat by Lammers & Lohman (1957). As regards the fibres to the inferior temporal gyrus, there is little reason to suspect that their degeneration could have been caused by non-specific factors, for besides the uncus no part of the temporal lobe was actually touched during surgery. It would thus seem justified to accept the existence of a rather sparse and diffuse projection of the amygdaloid complex to the inferior temporal gyrus, a connexion which reciprocates an apparently somewhat more massive cortico-amygdaloid projection arising in the same general region of the temporal cortex (Whitlock & Nauta, 1956).

Less unequivocal are the present data regarding amygdaloid pathways to rostral parts of the middle and superior temporal gyri, to the ventral insular region, and to the caudal orbito-frontal cortex. In all of the present cases fibre degeneration was observed to spread to these cortical regions, apparently largely via the uncinate fasciculus, but it was impossible to trace it in continuity from the amygdaloid lesion. Naturally, this failure could have resulted from a peculiar (e.g. recurrent and diffuse) mode of junction of amygdalofugal fibres with components of the uncinate bundle. On the other hand, however, the possibility must be considered that the degeneration in question was caused by inadvertent damage to the orbito-frontal cortex inflicted during the surgical procedure. Such damage could have caused the degeneration of temporopetal fibres in the uncinate fasciculus, and could at the same time have mirrored the amygdaloid projection to the orbito-frontal cortex suggested by the present findings. The circumstance that microscopic evidence of punctate lesions in the orbito-frontal cortex could be found in only one of the cases does not entirely preclude this possibility.* On the other hand again, the suspicion of orbito-frontal lesion is contradicted on one important point: whereas surgically produced lesions of the orbito-frontal cortex are followed by conspicuous fibre degeneration in the internal capsule, capsular degeneration was absent in all but one of the present cases. It is difficult to conceive of orbito-frontal lesions which could have caused substantial degeneration of associated cortical efferents without concomitant disintegration of subcortical projection fibres. Furthermore, in the present experiments the amount of degeneration in the orbito-frontal cortex appeared to be dependent on the extent of damage inflicted to the rostral half of the amygdaloid complex. It was, for example, notably larger in case MA 3 than in MA 11 in which the lesion was confined to the caudal half of the complex. The surgical procedure being the same in all cases, this observation suggests the existence of a true amygdaloid projection to the orbito-frontal cortex, arising largely in the rostral half of the amygdaloid complex.

In conclusion, the weight of evidence appears to favour the actual existence of amygdaloid projections to (a) rostral parts of the middle and superior temporal gyri, (b) a ventral region of the insular cortex, and (c) a large extent of the caudal orbito-frontal cortex. Some reserve in accepting the present evidence of these amygdalo-cortical connexions remains necessary, especially because it has not been possible to trace the pathways in question in continuity. However, even if the presence of the direct amygdalo-cortical connexions in question be discounted there can be little doubt that the amygdaloid complex can influence the neural mechanisms of at least the orbito-frontal cortex through the intermediary of the dorso-medial thalamic nucleus.†

### B. *Stria terminalis*

For detailed normal anatomical descriptions of the various components of the stria terminalis in several subprimate forms especial reference is made to the publications of Johnston (1923), Berkelbach v. d. Sprenkel (1926), Humphrey (1936), and Ariens Kappers, Huber & Crosby (1936).

From a comparison of cases MA 3 and MA 11 it is apparent that the stria terminalis originates largely in the caudal half of the amygdaloid complex, a finding which agrees well with Fox's (1943) and Adey & Meyer's (1952) conclusions from previous experimental studies. The present observations suggest further that some stria fibres, originating in the rostral half of the complex, do not extend beyond the bed nucleus of the stria and hence do not contribute to the preoptic and hypothalamic components of the system. Conversely, the findings in MA 11 indicate that stria

* This statement is based on our experience that strong elevation of the occipitotemporal cortex in the cat, even if carried out extradurally, can cause massive intracortical and corticofugal fibre degeneration despite the absence of identifiable gross or histological lesions of the cortex.

† After this discussion was written, Drs J. Klingler and P. Gloor kindly sent us the typescript of a gross-anatomical study of temporal lobe connexions in man, performed by the aid of Klingler's dissection technique. In this study fibre tracts were dissected which extend between the amygdala on the one hand, the tip of the temporal lobe, the insula and the orbito-frontal cortex on the other hand. The pathways in question emerge from the lateral side of the amygdala and follow curved trajectories in close relationship to the uncinate fasciculus. Drs Klingler and Gloor point out that the gross dissection technique can offer little information regarding the polarity of fibre connexions. However, the appearance of the fibre tracts demonstrated by their analysis is in several respects consistent with the experimental evidence of amygdalo-cortical pathways discussed above.

fibres of more caudal origin by-pass the bed nucleus and make up the bulk of the preoptic and hypothalamic components.

This study has failed to confirm the existence of supracommissural fibres of the stria terminalis to the septal region. The possibility cannot be excluded that such fibres originate in amygdaloid regions not involved in the present experiments. In a case of complete surgical interruption of the stria several millimetres caudal to the anterior commissure (case MF 13), fibre degeneration was found in the septal region, but this finding was considered inconclusive for the reason that the fimbria fornicis was to some extent involved in the lesion.

The commissural component of the stria terminalis would seem to be of minimal volume in the monkey. Both in case MA 11 and in the case of stria terminalis section (MF 13) mentioned above, only a few degenerating fibres could be followed across the midline in the dorsal stratum of the anterior commissure. As in previous Marchi studies in the cat (Fox, 1943; Ban & Omukai, 1959), the termination of these commissural fibres could not be determined.

Stria terminalis fibres to the preoptic region and hypothalamus appear to form by far the largest component of the stria in the monkey. According to the present findings, such fibres distribute largely, if not exclusively, to the medial zone of the preoptic and hypothalamic regions. The stria terminalis differs in this respect from the ventral amygdalo-hypothalamic pathway which appears to connect primarily with more lateral preoptico-hypothalamic areas, and specifically with the region interstitial to the medial forebrain bundle. Within the medial zone most of the stria terminalis fibres appear to terminate among the cells of the medial preoptic and anterior hypothalamic nuclei. Only few fibres could be followed farther caudalward, and all of these appeared to terminate at and only slightly behind the caudal border of the optic chiasma, in the extreme ventral hypothalamic region containing the scattered cells of the so-called nucleus supraopticus diffusus (Rioch, Wislocki & O'Leary, 1940). Some of the longest stria fibres may end in contact with the arcuate nucleus of the infundibulum.

From an experimental study by the Glees technique in the monkey, Adey & Meyer (1052) concluded that amygdalo hypothalamic fibres arc distributed in large part to the ventro-medial hypothalamic nucleus. The same study furthermore indicated a virtually symmetrical bilateral distribution of the amygdalo-hypothalamic projection. The present observations differ from these conclusions in major respects. Naturally, in comparing the present findings with those of Adey & Meyer the possibility must be considered that certain amygdalofugal fibre contingents had escaped degeneration in all of the present cases of amygdaloid lesion. However, incidental findings made in three further cases likewise failed to confirm the existence of bilateral amygdaloid projections to the ventro-medial hypothalamic nucleus. In one of these supplementary cases (MF 13 mentioned before in this Discussion) complete unilateral interruption of the stria terminalis resulted in exclusively ipsilateral degeneration of hypothalamic stria fibres in a distribution comparable to that found in MA 11. Two other cases, in which the lateral half of the substantia innominata had been extensively damaged in an unsuccessful attempt to produce stereotactic lesions of the globus pallidus, again showed absence of any but ipsilateral hypothalamic degeneration; in neither case was fibre degeneration observed in the ventro-medial nucleus. The lesions in the two latter cases had undoubtedly severed most of the ventral amygdalo-hypothalamic fibres in their sublenticular passage medialward. Thus, in the present experiments neither a variety of amygdaloid lesions nor the massive interruption of the two known amygdalo-hypothalamic pathways was followed by contralateral hypothalamic fibre degeneration or degeneration of preterminal fibres in the ventro-medial nucleus of either side. Even when allowance is made for the greater ease with which details of terminal degeneration can be identified with the Glees method than by the Nauta–Gygax technique (Bowsher, Brodal & Walberg, 1960), the very small number of the degenerated fibres of passage which in the present study could be followed into the medial hypothalamic zone caudal to the optic chiasma appears to contradict the existence of a significant direct amygdaloid projection to the ventro-medial hypothalamic nucleus.

## SUMMARY

The fibre degenerations resulting from lesions in the amygdaloid complex in. the monkey were studied by means of the Nauta–Gygax technique. The results confirm the existence of two major amygdalo-subcortical fibre systems, namely, a relatively diffuse ventral amygdalofugal pathway, and the compact stria terminalis.

1. The ventral amygdalofugal pathway, apparently the most massive amygdaloid projection system, spreads medially and forward ventral to the lentiform nucleus and connects with the substantia innominata, lateral preoptic and hypothalamic regions, nucleus of Broca's diagonal band, and the olfactory tubercle. A prominent further component of the system by-passes the preoptic region and follows the inferior thalamic peduncle to terminate in the medial, magnocellular division of the dorso-medial thalamic nucleus. Furthermore, the ventral amygdalofugal pathway contains an amygdalo-cortical component which accompanies the pathway to the nucleus of Broca's diagonal band and terminates in rostral parts of the gyrus fornicatus (gyrus subcallosus and rostral cingulate cortex).

2. The stria terminalis originates mostly in the caudal half of the amygdaloid complex. Fibres arising most rostrally in the complex appear to terminate largely in the bed nucleus of the stria terminalis. Other fibres of more caudal origin form a prominent preoptico-hypothalamic component distributing fibres to the medial preoptic nucleus, anterior hypothalamic nucleus, and the region of the nucleus supraopticus diffusus. No stria terminalis fibres could be followed to the ventro-medial hypothalamic nucleus. Only ipsilateral amygdalo-hypothalamic fibres could be identified.

3. Evidence was obtained of an additional amygdalofugal fibre system which emerges through the lateral and ventral sides of the amygdala and distributes fibres to rostral parts of the superior, middle and inferior temporal gyri, ventral insular cortex, claustrum, rostral putamen, and caudal orbito-frontal cortex. As most of these connexions could not be followed in continuity, the present evidence of their existence cannot by itself be considered conclusive.

It is a pleasure to acknowledge the valuable technical assistance of Mrs M. H. Albrecht and Messrs. Gordon Fletcher and Curtis King. The photographs were

530                 *W. J. H. Nauta*

made by Miss Sally Craig. The author also wishes to thank Drs J. Klingler and P. Gloor of the Montreal Neurological Institute for their kind communication of important data prior to publication.



## REFERENCES

ADEY, W. R. & MEYER, M. (1952). Hippocampal and hypothalamic connexions of the temporal lobe in the monkey. *Brain*, **75**, 358–383.

ARIENS KAPPERS, C. U., HUBER, G. C. & CROSBY, E. C. (1936). *The Comparative Anatomy of the Nervous System of Vertebrates, including Man*, vol. II. New York: MacMillan.

BAN, T. & OMUKAI, F. (1959). Experimental studies on the fiber connections of the amygdaloid nuclei in the rabbit. *J. comp. Neurol.* **113**, 245–279.

BERKELBACH V. D. SPRENKEL, H. (1926). Stria terminalis and amygdala in the brain of the opossum (*Didelphys virginiana*). *J. comp. Neurol.* **42**, 211–254.

BOWSHER, D., BRODAL, A. & WALBERG, F. (1960). The relative values of the Marchi method and some silver impregnation techniques. A critical survey. *Brain*, **83**, 150–160.

CRAGG, B. G. & HAMLYN, L. H. (1957). Some commissural and septal connexions of the hippocampus in the rabbit. A combined histological and electrical study. *J. Physiol.* **135**, 460–485.

DAITZ, H. M. & POWELL, T. P. S. (1954). Studies of the connexions of the fornix system. *J. Neurol.* **17**, 75–82.

FOX, C. A. (1943). The stria terminalis, longitudinal association bundle and precommissural fornix fibers in the cat. *J. comp. Neurol.* **79**, 277–295.

FOX, C. A. (1949). Amygdalo-thalamic connections in *Macaca mulatta*. *Anat. Rec.* **103**, no. 2, 537–538 (abstract).

GLOOR, P. (1955). Electrophysiological studies on the connections of the amygdaloid nucleus in the cat. Part I. The neuronal organization of the amygdaloid projection system. *Electroenceph. clin. Neurophysiol.* **7**, 223–242.

GLOOR, P. (1959). Amygdala. In *Handbook of Physiology*, Section I: Neurophysiology, chapter LVII. Baltimore: Williams and Wilkins.

GUILLERY, R. W. (1959). Afferent fibres to the dorsomedial thalamic nucleus in the cat. *J. Anat., Lond.*, **93**, 403–419.

HALL, E. (1960). Efferent pathways of the lateral and basal nuclei of the amygdala in the cat. *Anat. Rec.* **136**, no. 2, 205 (abstract).

HILPERT, P. (1928). Der Mandelkern des Menschen. I. Cytoarchitektonik und Faserverbindungen. *J. Psychol. Neurol., Lpz.*, **36**, 44–73.

HUMPHREY, T. (1936). The telencephalon of the bat. I. The non-cortical nuclear masses and certain pertinent fibre connections. *J. comp. Neurol.* **65**, 603–711.

JOHNSTON, J. B. (1923). Further contributions to the study of the evolution of the forebrain. *J. comp. Neurol.* **35**, 337–481.

LAMMERS, H. J. & LOHMAN, A. H. M. (1957). Experimenteel anatomisch onderzoek naar de verbindingen van piriforme cortex en amygdalakernen bij de kat. *Ned. Tijdschr. Geneesk.* **101**, 1–2.

NAUTA, W. J. H. (1958). Hippocampal projections and related neural pathways to the mid-brain in the cat. *Brain*, **81**, 319–340.

NAUTA, W. J. H. & VALENSTEIN, E. S. (1958). Some projections of the amygdaloid complex in the monkey. *Anat. Rec.* **130**, no. 2, 346 (abstract).

PRIBRAM, K. H., CHOW, K. L. & SEMMES, J. (1948). Limit and organization of the cortical projection from the medial thalamic nucleus in monkey. *J. comp. Neurol.* **98**, 433–448.

PRIBRAM, K. H. & MACLEAN, P. D. (1952). Neuronographic analysis of medial and basal cerebral cortex. II. Monkey. *J. Neurophysiol.* **16**, 324–340.

RIOCH, D. McK., WISLOCKI, G. B. & O'LEARY, J. L. (1940). A precis of preoptic, hypothalamic and hypophysial terminology with atlas. *Res. Publ. Ass. nerv. ment. Dis.* **20**, 3–30.

SCOVILLE, W. B. & MILNER, B. (1957). Loss of recent memory after bilateral hippocampal lesions. *J. Neurol.* **20**, 11–21.

VALENSTEIN, E. S. & NAUTA, W. J. H. (1959). A comparison of the distribution of the fornix system in the rat, guinea pig, cat, and monkey. *J. comp. Neurol.* **113**, 337–363.

WHITLOCK, D. B. & NAUTA, W. J. H. (1956). Subcortical projections from the temporal neocortex in *Macaca mulatta*. *J. comp. Neurol.* **106**, 183–212.

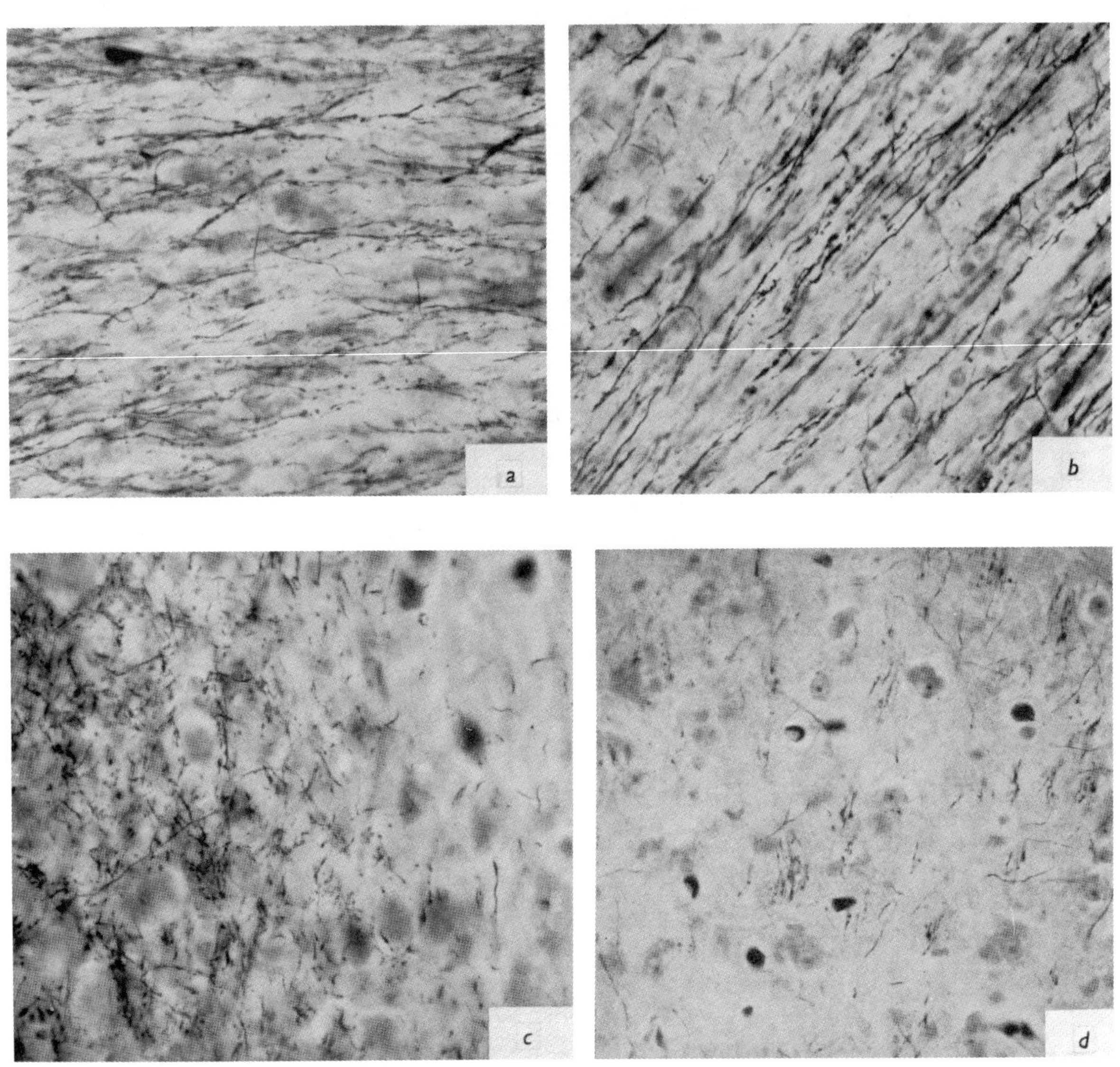

NAUTA—AMYGDALOID PROJECTIONS IN THE MONKEY

(*Facing p.* 531)

## EXPLANATION OF PLATE

Photomicrographs ( × 230) of fibre degeneration following lesion of the amygdaloid complex.

*a*, degenerating axons of various calibre traversing the substantia innominata. Case MA 3.

*b*, degenerating coarse axons in the inferior thalamic peduncle. Case MA 3.

*c*, dense pericellular and intercellular axon degeneration in a circumscript region (pars medialis) of the n. dorsomedialis thalami. The nucleus periventricularis anterior thalami appears near the right margin of the picture. Case MA 3.

*d*, degenerating fascicles of the stria terminalis passing through the anterior hypothalamic nucleus. Case MA 11.

## KEY TO ABBREVIATIONS USED IN FIGURES

*AC*, anterior commissure; *ADH*, area dorsalis hypothalami; *AH*, nucleus anterior hypothalami; *AL*, ansa lenticularis; *AM*, nucleus anterior medialis thalami; *A.Ped.*, ansa peduncularis; *AV*, nucleus anterior ventralis thalami; *B*, basal amygdaloid nucleus; *CI*, capsula interna; *Cl*, claustrum; *DMH*, nucleus dorsomedialis hypothalami; *CO*, cortical amygdaloid nucleus; *DMm*, nucleus dorsomedialis thalami, pars medialis; *F*, fornix; *FU*, fasciculus uncinatus, *GP*, globus pallidus; *GR*, gyrus rectus; *Hp*, hippocampus; *L*, lateral amygdaloid nucleus; *LH*, nucleus lateralis hypothalami; *LP*, nucleus preopticus lateralis; *MP*, nucleus preopticus medialis; *NAP*, nucleus ansae peduncularis; *NC*, nucleus caudatus; *ND*, nucleus of the diagonal band of Broca; *NST*, nucleus striae terminalis; *NSt*, nucleus subthalamicus; *Pp*, cortex prepiriformis; *Put*, putamen; *SO*, nucleus supraopticus; *SOD*, nucleus supraopticus diffusus; *ST*, stria terminalis; *TO*, tractus opticus; *T.Ol.*, tuberculum olfactorium; *V*, lateral ventricle; *VA*, nucleus ventralis anterior thalami; *VL*, nucleus ventralis lateralis thalami; *VMH*, nucleus ventromedialis hypothalami.

# NEURAL ASSOCIATIONS OF THE AMYGDALOID COMPLEX IN THE MONKEY

BY

WALLE J. H. NAUTA

*Department of Neurophysiology*
*Walter Reed Army Institute of Research*
*Washington, D.C.*

THIS paper is the result of a series of experiments performed in an attempt to determine the wider neural organizations of which the amygdaloid complex forms a part. As a first step toward this aim a study was made of the fibre degenerations resulting from lesions in the amygdaloid complex in the monkey (Nauta and Valenstein, 1958; Nauta, 1961). Some of the results of that study are summarized in diagrammatic form in fig. 14.

Of particular importance to the subject of the present study is the existence in the monkey of a well-developed, direct pathway from the amygdaloid complex to the dorsomedial nucleus of the thalamus. This direct amygdalo-thalamic connexion, reported for the first time by Fox (1949), leaves the amygdaloid complex as a thick-fibred component of the ventral amygdalo-subcortical pathway, joins the inferior thalamic peduncle and terminates almost exclusively in the medial, magnocellular subdivision of the dorsomedial thalamic nucleus. It appears to distribute no fibres to the larger lateral part of the nucleus.

The virtually selective termination of amygdalo-thalamic fibres in the medial subdivision of the dorsomedial nucleus would seem to hold a clue in regard to the particular thalamo-cortical mechanism affected by impulse transmission over the amygdalo-thalamic pathway. The results of several experimental studies, most recently those of Pribram, Chow and Semmes (1953) indicate that the medial, magnocellular part of the dorsomedial nucleus projects to at least a caudal region of the orbito-frontal cortex, in contrast to the lateral subdivision of the nucleus which is largely related to cortical areas on the convexity of the frontal lobe. It would thus seem justified to interpret Fox's amygdalo-thalamic bundle as the first link in a transthalamic pathway from the amygdaloid complex to the orbitofrontal cortex. Judging from the extremely high density of pericellular axon degeneration in the medial part of the dorsomedial thalamic nucleus following amygdaloid lesions, the amygdala must represent a major potential influence upon the thalamo-orbitofrontal mechanisms.

As the anatomical evidence thus far available clearly indicated that the amygdaloid complex, the magnocellular part of the dorsomedial thalamic nucleus, and the caudal orbitofrontal cortex are serially and in that order connected, it was thought to be of interest to investigate the possible existence of further connexions between the three structures. The present experiments were performed with this objective in mind.

MATERIAL AND METHODS

Seven young adult monkeys (*Macaca mulatta*) were used in this study. In three of the animals small electrolytic lesions were placed in the medial part of the dorsomedial thalamic nucleus. In the remaining four animals the caudal orbitofrontal region identified as the main recipient of the thalamo-orbitofrontal projection was to various extents decorticated by suction. After a postoperative survival period of 10 to 14 days the animals were killed by an overdose of nembutal. The observations to be described below were made on frozen sections of the formalin-fixed brains, stained by the Nauta-Gygax (1954) and Albrecht-Fernstrom (1959) techniques of silver impregnation for degenerated axons.

OBSERVATIONS

(1) *Lesions of the medial, magnocellular subdivision of the dorsomedial thalamic nucleus.*—Substantially similar results were found in all three animals of this group. The following case is described in some detail.

*Case MA-c.*—The lesion in this animal had been produced by stereotaxic electrolysis, using a slender, tapered electrode introduced at an angle of 20 degrees through the contralateral hemisphere in a successful attempt to by-pass the stria medullaris. Avoidance of the stria was considered of some importance for the reason that it has recently been found to contain rostrally directed fibres distributed, among others, to anterodorsal regions of the hypothalamus (Massopust, 1960, personal communication). The resulting defect involved the medial, magnocellular cell group of the dorsomedial nucleus bilaterally, as well as the narrow paramedian and small-celled nucleus periventricularis thalami which separates the nuclei of both sides.

As indicated in figs. 1 and 2, degenerating fascicles spread from the lesion rostralward in the internal medullary lamina. A short distance behind the rostral pole of the dorsomedial nucleus the degenerating fibre group begins to join the inferior thalamic peduncle, with which it turns ventralward (figs. 2 and 3). In this initial part of their ventral course the more caudal fibres of the group traverse medial parts of the nuclei ventralis lateralis and ventralis anterior, apparently without termination in these cell groups. Slightly farther rostrally, however, there is evidence of pre-terminal and terminal degeneration in the most medial part of the nucleus reticularis thalami (fig. 3). Degenerating fibres passing through this region

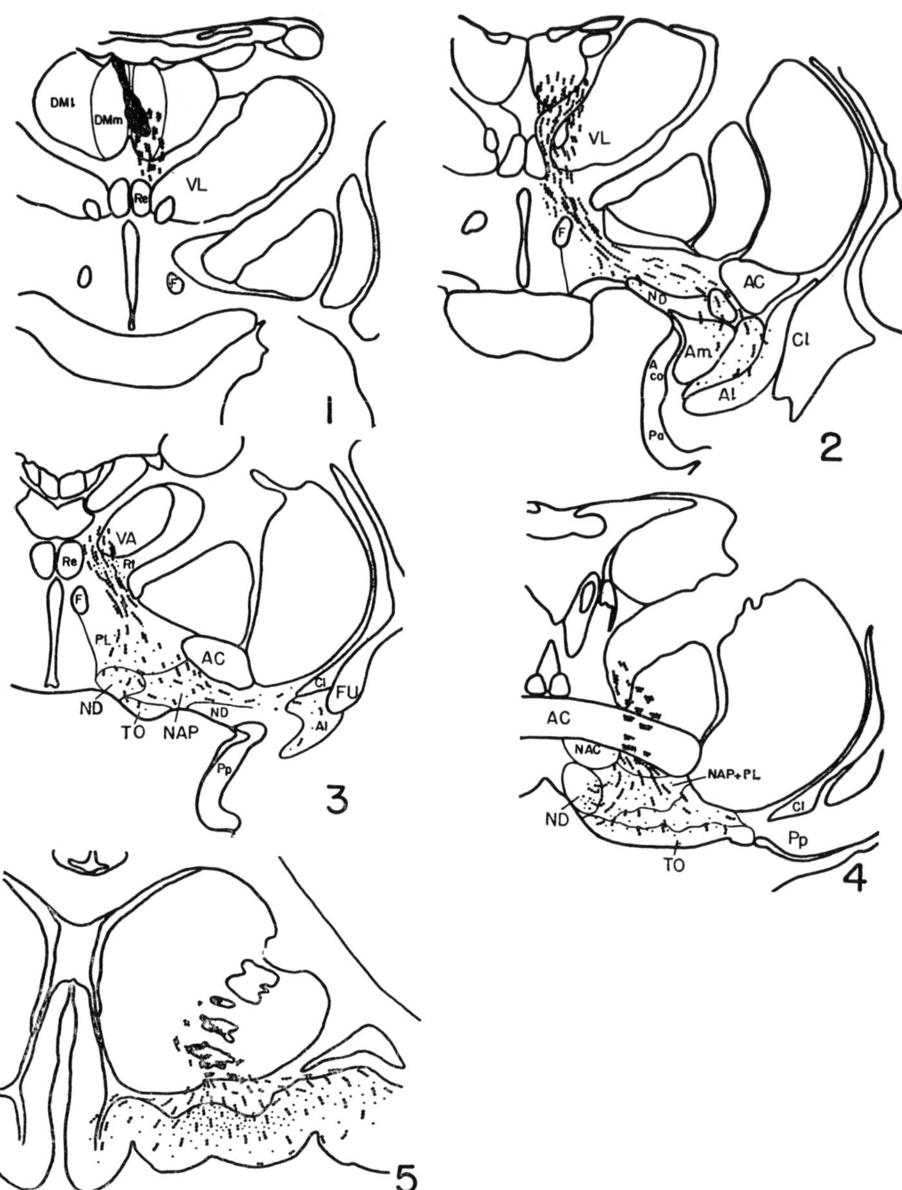



Fig. 1–5.—Fibre degeneration observed in case MA-c. The lesion is indicated in jet black in fig. 1. Coarse dots indicate degenerating fibres of passage, fine stipple degenerating terminal and preterminal fibres.    Abbreviations on page 519.

continue ventrally in the inferior thalamic peduncle and disperse in the dorsal hypothalamic area (fig. 2), lateral preoptic and hypothalamic regions, nucleus of the diagonal band, substantia innominata (figs. 2–4),

33                                                                    

substantia perforata anterior, and olfactory tubercle (figs. 3 and 4). A small number of the degenerating peduncular fibres course laterally through the substantia innominata and enter the amygdala near its rostral pole. The distribution of these thalamo-amygdaloid fibres appears to involve mostly the rostral pole of the basal amygdaloid nucleus, and to a lesser extent the medial, lateral and central nuclei (figs. 2 and 3). The fibre degeneration in the nucleus of the diagonal band extends in the rostral direction and can be followed as far forward as the basal part of the septal region. No components of this group can be identified in the supracommissural septum.

The thalamofugal fibres described in the foregoing account form part of the large extracapsular component of the inferior thalamic peduncle. The latter contains additional fibres which enter the medial edge of the internal capsule, and contribute to the anterior thalamic radiation. In the present case MA-c numerous fibres of this capsular component of the peduncle are degenerated. They can be followed rostrally in discrete fascicles, some of which perforate the anterior commissure, and fan out over a large extent of the caudal orbitofrontal cortex (fig. 5). Intracortical degeneration is dense in the caudal orbital gyrus (fig. 20, Plate XXIX), and considerably sparser in the adjoining gyri rectus and orbitalis lateralis.

The connexions here described are summarized in diagrammatic form in fig. 15.

(2) *Lesions of the caudal orbitofrontal cortex.*—The present observations in cases of orbitofrontal ablation suggest that the subcortical projections of the orbitofrontal cortex are largely composed of fibres of extremely small calibre. In the degenerated state such fine axons can be traced only with difficulty, especially when the terminal arborization takes place in regions traversed by numerous thick fibres. Of the 4 cases here studied case MA 13, which afforded the most detailed observations, is described below. The 3 remaining cases provided essentially confirmatory evidence.

*Case MA 13.*—As shown by fig. 7 (insert) the lesion found in this case straddled the lateral orbital sulcus and involved part of Walker's (1940b) area 13 which occupies most of the caudal orbitofrontal gyrus, as well as a small part of the lateral orbital gyrus. It fell entirely within the distribution limits of the thalamo-orbitofrontal degeneration observed in the foregoing case MA-c.

From the area of the lesion, fibre degeneration spreads caudalward in two prominent streams (figs. 6–11). One of these follows the uncinate fasciculus and basal parts of the outer capsules, distributes numerous fibres to the ventral two-thirds of the claustrum, and spends itself in the rostral part of the superior, middle, and inferior temporal gyri (not illustrated). Some of the sparse fibre degeneration spreads from the extreme capsule over the anteroventral part of the insular cortex (figs. 7–9). No component of this pathway appears to connect with the amygdaloid complex or piriform cortex.

A more medial system of degenerating fibres represents the more strictly subcortical projections from the lesion.  As shown by the sequence of figs. 6–8, two main fibre contingents can be distinquished in this system: a dorsal group that immediately enters the most ventral bundles of the internal capsule, and a more ventral group which extends caudalward through the basal forebrain region underneath the lentiform nucleus. Some fibres leave the dorsal, i.e. capsular, bundles and become dispersed in the adjacent rostral parts of the caudate nucleus and putamen (fig. 6). The most ventral of these capsular fascicles perforate the anterior commissure (fig. 7).

The ventral, sublenticular contingent of degenerating fibres passes through the substantia perforata anterior, distributing some fibres to the ventrolateral margin of the putamen (figs. 6 and 7), to the nucleus ansae peduncularis and to the lateral preoptic region (figs. 8, 19, Plate XXIX). Somewhat farther caudally (fig. 9) a moderate number of degenerating fibres, apparently components of the same group, are present in the lateral hypothalamic region.

At frontal levels involving the preoptic region, prominent bundles of degenerating axons are found in the ansa peduncularis (inferior thalamic peduncle), ventral to the medial edge of the internal capsule (figs. 8, 18, Plate XXVIII).   It appears likely that these bundles consist of fibres of the sublenticular corticofugal pathway which have swung dorsally and are about to mingle with the degenerating fibre group in the medial edge of the internal capsule.  It is possible, however, that at least some fibres course in the opposite direction, i.e. from the internal capsule ventralward, augmenting the corticofugal fibre degeneration in the sublenticular and hypothalamic regions.  Some fibres of the bundles in question course through the medial part of the external pallidal segment, and there are clear signs of fibre termination restricted to this region of the globus pallidus (fig. 8).

From the area of degeneration centering around the medial edge of the internal capsule large numbers of degenerating fibres stream dorsalward. Some of these fibres terminate in the medial part of the reticular thalamic nucleus, whereas others continue across this region to the extreme medial part of the nucleus ventralis lateralis (fig. 9), to the nucleus paracentralis (fig. 9), and to the medial element of the dorsomedial nucleus (figs. 9–11). The fibre degeneration in the last-mentioned cell group is considerably less massive than that observed after lesions of the amygdaloid complex. This is true not only in the present case but also in cases with more extensive ablation of orbitofrontal cortex.

Finally, a considerable number of degenerating fibres spread from the degenerated medial capsular fibre group into the subthalamic region (figs. 10 and 11), and beyond it into the central tegmental area of the

rostral mesencephalon (figs. 12 and 13).  In the subthalamus the terminal
axon degeneration is mostly confined to the regions defined by Mettler
(1945) as the subthalamoprerubral and ventrolateral tegmental fields.  The
rather sparse mesencephalic tegmental degeneration apparently represents
longer elements of the same fibre group which extend caudolaterally and
terminate in the extreme lateral zone of the parvicellular subdivision of

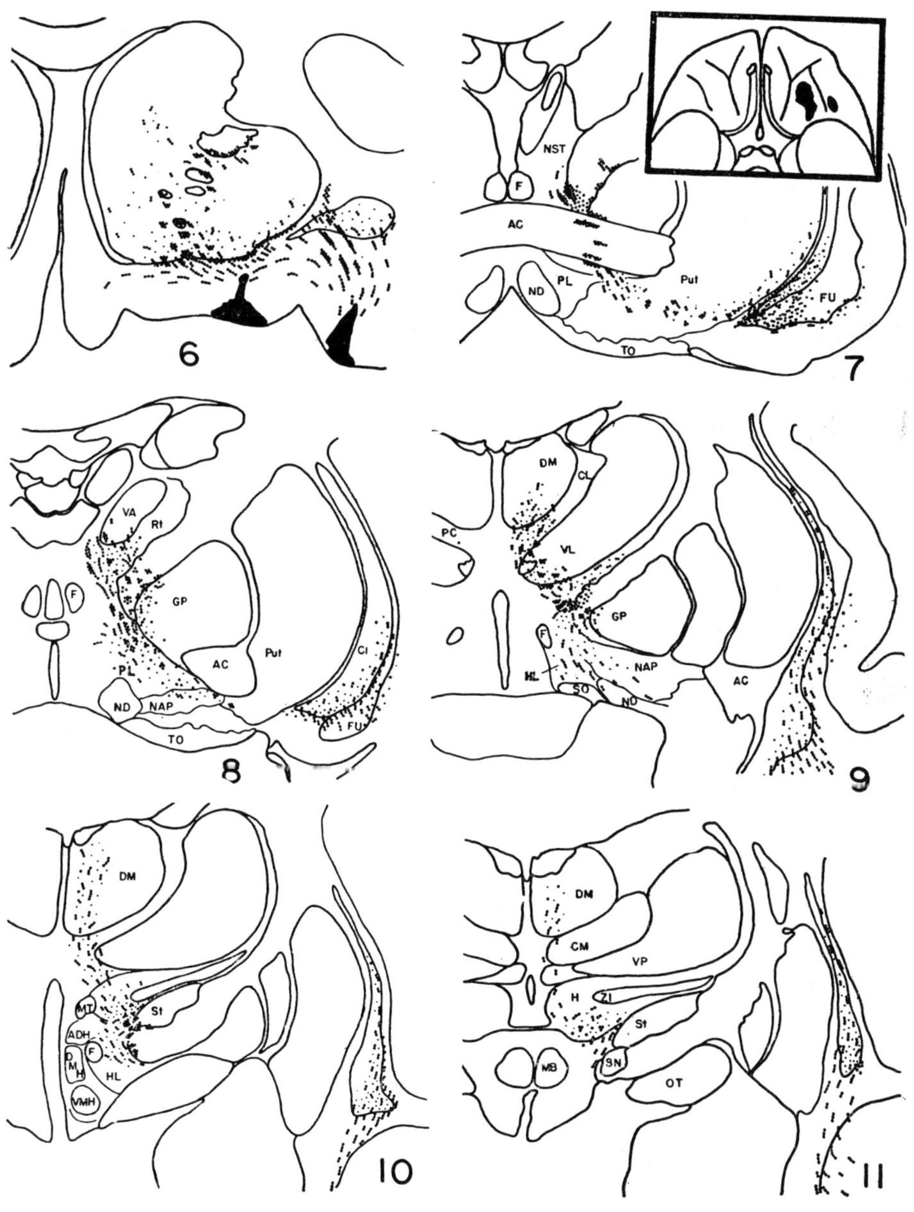

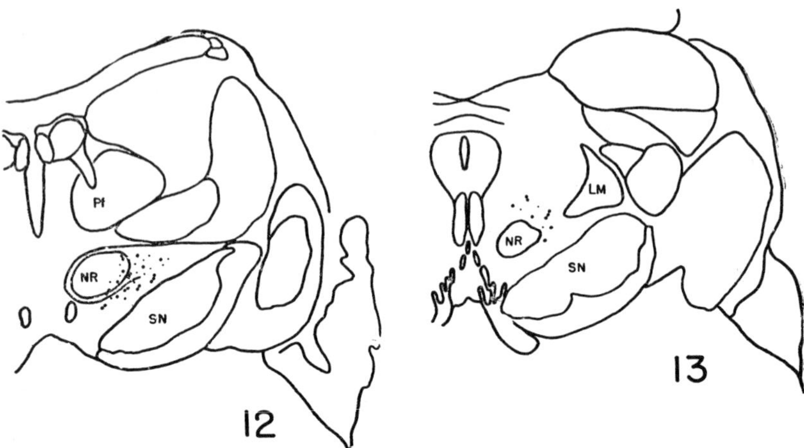


Figs. 6–13.—Fibre degeneration in case MA 13.  The lesion of the orbital surface of the frontal lobe is indicated in fig. 7 (insert).  Symbols and abbreviations as in figs. 1–5.

the red nucleus, as well as in the tegmental area ventral and lateral to the nucleus.

The degenerated pathways here described are summarized in diagrammatic form in fig. 16.

DISCUSSION

A. *Projections from the medial, magnocellular component of the dorsomedial thalamic nucleus.*—The findings reported above agree with earlier evidence, obtained by the method of retrograde degeneration, that the medial part of the dorsomedial nucleus projects to the orbital surface of the frontal lobe (Walker, 1940a; Freeman and Watts, 1947; Pribram, Chow and Semmes, 1953; Ström-Olsen and Northfield, 1955).  The present experiments give no reliable information regarding the total extent of the cortical projection area, for no case of complete destruction of the thalamic nucleus was studied.  However, in good accord with the conclusions of Pribram *et al.*, degeneration fibres appeared to end in greatest density in the caudal orbitofrontal gyrus.  It is noteworthy that the area of intracortical fibre degeneration included the most caudal part of this gyrus, ablation of which alone caused no apparent thalamic cell loss in Pribram's experiments

In addition to its orbitofrontal projection, the magnocellular part of the dorsomedial nucleus evidently emits fibres of subcortical distribution. Like the cortical projection bundle, such fibres leave the thalamus with the inferior thalamic peduncle, but they by-pass the medial edge of the internal capsule ("extracapsular thalamic peduncle," *see* Klingler and Gloor, 1960) and become part of the ansa peduncularis, a massive and heterogeneous

226

system of generally arciform fibres occupying the basal forebrain region medial and ventral to the lentiform nucleus.[1] Papez (1942) suspected that part of these fibres formed a projection from the dorsomedial thalamic nucleus to the globus pallidus, but no corresponding distribution of degenerating fibres could be identified in the present study. Instead, the thalamofugal fibres in question were traced to the most lateral zone of the anterior hypothalamus and preoptic region, the substantia innominata, the olfactory tubercle and other cell groups of the anterior perforated substance, the nucleus of Broca's diagonal band, and rostral parts of the amygdaloid complex. The last-mentioned connexion is of particular interest, as it evidently reciprocates Fox's amygdalo-thalamic pathway. It must be noted that it appears to be very much less massive than the latter connexion, and that it is questionable whether its intra-amygdaloid distribution overlaps at all with the origin of the amygdalo-thalamic projection. It is, however, possible and even likely that the thalamo-amygdaloid connexion is augmented by fibres relaying in basal forebrain structures such as the lateral preoptic region and substantia innominata.

It is remarkable that no fibre degeneration could be traced from the thalamic lesion to medial regions of the hypothalamus. Thalamo-hypothalamic connexions in the periventricular fibre system have been suggested on the basis of observations in normal material (Roussy and Mosinger, 1934), and were described recently by Showers (1958) from findings in Marchi experiments. The only direct hypothalamic projection which could be traced from the dorsomedial lesion in the present study followed the inferior thalamic peduncle and appeared to end exclusively in the lateral hypothalamic region.

No evidence was found of efferents from the dorsomedial nucleus to the caudate nucleus as observed by Showers (1958). Such thalamo-caudate fibres could conceivably arise in more lateral parts of the monkey's dorsomedial nucleus. In an earlier study in the cat, no thalamo-caudate fibres could be traced from massive lesions in the dorsomedial nucleus, although certain lesions in the intralaminar region of the thalamus did cause the degeneration of some such fibres (Nauta and Whitlock, 1954).

B. *Projections from the caudal orbitofrontal cortex.*—(*a*) *Orbitofronto-temporal connexions.*—The present experiments have confirmed previous experimental anatomical (Showers, 1958) and neuronographical evidence (Pribram and MacLean, 1953) of a massive associational pathway between the caudal orbitofrontal cortex and the rostral part of the superior, middle and inferior temporal gyri. This pathway largely follows the uncinate

[1]For an account of the various nomenclatures which have been used in descriptions of the sublenticular region reference is made to Papez and Aronson (1934). A very useful description of the major fibre systems related to the region in question is found in Klingler and Gloor's (1960) dissection study of the human temporal region.

fasciculus, and distributes numerous fibres to the ventral two-thirds of the claustrum, as well as some to the ventral half of the insular cortex. In contrast to the two earlier studies, the present findings furnished no evidence of direct connexions from the orbitofrontal cortex to the hippo-campal gyrus.

(*b*) *Subcortical projections of the caudal orbitofrontal cortex.*—The subcortical projections traced in the present study follow a dual trajectory: whereas part of the corticofugal fibres travel caudally in the medial edge of the internal capsule, others follow a wide extracapsular path which leads through basal forebrain regions and appears to be related to, if not actually part of, the general system of the medial forebrain bundle. It is noteworthy that a similar dual mode of transit is encountered in the projections of the cingulate gyrus, which follow in part the internal capsule and cerebral peduncle, in other part the dorsal fornix. As the internal capsule-cerebral peduncle is composed of fibres related to the neocortex, the fornix and medial forebrain bundle of projections to and from paleocortical formations and associated subcortical structures, the dual path followed by the projections from the cingulate and caudal orbitofrontal cortices could be interpreted as further evidence of an intermediate status of the two last-mentioned cortical regions (Rose's (1927) mesocortex).

According to the present findings, the projections from the caudal orbitofrontal cortex have a wide distribution which involves parts of the basal ganglia, thalamus, subthalamus, preoptic region and hypothalamus, and the mid-brain tegmentum.

The projection to the medial component of the dorsomedial thalamic nucleus, even though apparently of rather modest volume, is of some interest in the context of the present study, for it clearly represents a reciprocal connexion to the more massive thalamo-orbitofrontal pathway. Other fibres of the orbitofronto-thalamic projection are distributed to the non-specific reticular and paracentral thalamic nuclei.

Evidence of orbitofrontal projections to the hypothalamus has been obtained in the monkey by anatomical methods as well as by strychnine-neuronography. Ward and McCulloch (1947) recorded spike potentials from the posterior hypothalamic area folowing strychninization of the caudal orbitofrontal cortex. In a similar study Sachs, Brendler and Fulton (1949) obtained strychnine spikes especially when the recording electrodes straddled the medial fore-brain bundle or, alternatively, Forel's field H, whereas firing was also observed in the lateral preoptic area. Using the Glees silver method Clark and Meyer (1950) and Wall, Glees and Fulton (1951) found considerable terminal degeneration in the ventromedial hypothalamic nucleus following unilateral surgical lesion of Walker's area

13. In Clark and Meyer's study such terminal degeneration was observed bilaterally.

The results of the present study further support the notion of orbito-frontohypothalamic projections. In contrast to the findings in the Glees studies, however, fibre degeneration appeared to be confined to the lateral preoptic and hypothalamic regions ipsilateral to the lesion, and no degenerating axon arborization or transit fibres could be identified in the medial and periventricular hypothalamic cell groups. The lateral hypothalamic fibre degeneration appeared to reach its caudal limit a slight distance rostral to the level of the mammillary body. In its more caudal extent it was immediately continuous with an area of fibre degeneration in the subthalamo-prerubral region. The preoptic, hypothalamic, and subthalamic distribution of orbitofrontal projections suggested by the present findings matches to a remarkable extent Sachs, Brendler and Fulton's (1949) identification of the lateral preoptic region, medial fore-brain bundle and field H as the areas of most active response to orbito-frontal stimulation. Owing to the circumstance that the area of fibre degeneration straddled the vague hypothalamo-subthalamic boundary, the trajectory of the orbitofronto-hypothalamic projection could not be identified with certainty. It seems likely that the fibres to the lateral preoptic region follow the medial fore-brain bundle, but those ending in the more caudal part of the lateral hypothalamic region could well include fibres travelling in the medial edge of the cerebral peduncle.

The orbitofrontal projection to the rostral mid-brain tegmentum forms a caudal continuation of the connexion with the subthalamic region. The fibres of this pathway most likely travel in the medial part of the cerebral peduncle. The area of their apparent termination—lateral and dorsal to the red nucleus—coincides at least in part with that of certain lateral components of the medial fore-brain bundle (*see* Nauta, 1958; Valenstein and Nauta, 1959). It hardly needs emphasis, however, that the tegmental region under consideration has numerous additional afferent connexions.

In a recent electrophysiological study in the cat Newman and Wolstencroft (1959), stimulating the orbital gyrus, recorded responses of extremely short latency from the pontine and medullary reticular formation. The present experiments have provided no evidence in support of the existence of the direct orbitofronto-bulbar projections indicated by Newman and Wolstencroft's findings. It must be remarked, however, that the homology of the cat's orbital gyrus with the monkey's caudal orbitofrontal cortex appears at this time far from certain.

CONCLUSIONS

The findings discussed in the foregoing account suggest that the amygdaloid complex forms part of a widespread neural apparatus which also includes the magnocellular component of the dorsomedial thalamic

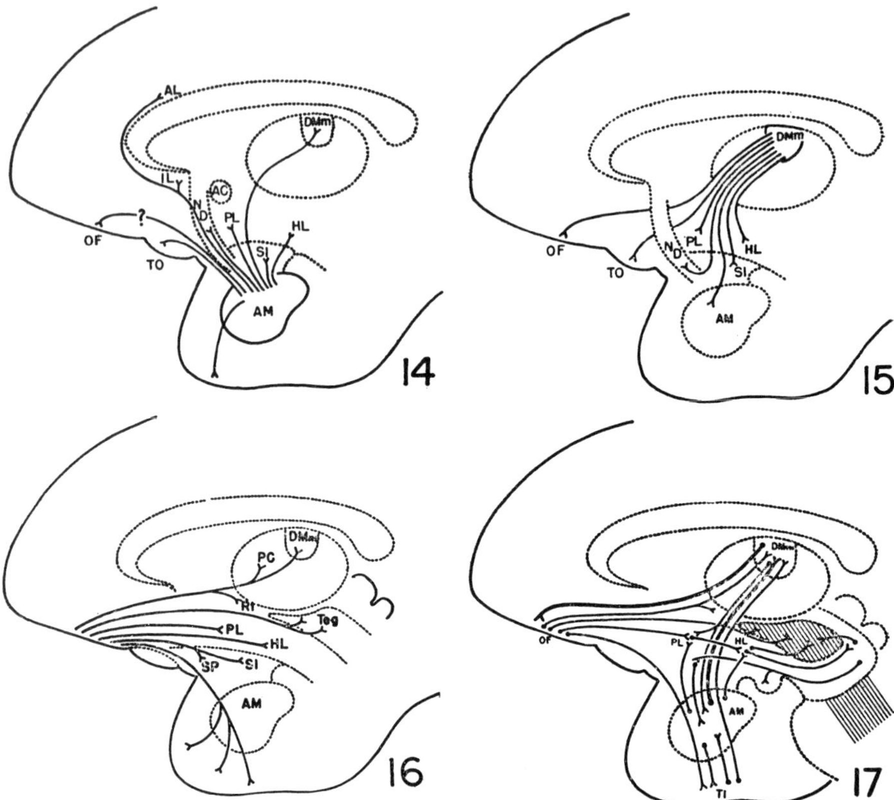

FIG. 14.—Summary diagram of projections from the amygdaloid complex (from Nauta and Valenstein, 1958, and Nauta, 1961). The stria terminalis is not included in the diagram. With the exception of the projections to the temporal and orbitofrontal cortex, the indicated connexions are components of the so-called ventral amygdalofugal pathway.

FIG. 15.—Summary of efferent connexions of the medial component of the dorso-medial thalamic nucleus, as described in the text.

FIG. 16.—Summary diagram of the projections from the caudal orbitofrontal cortex described in the text.

FIG. 17.—Diagram of neural pathways associated with the amygdaloid complex. The figure is largely a composite of figs. 14–16, but also incorporates some previous findings. The pathways to and from the mesencephalon are indicated according to Nauta (1958); those from the lower temporal cortex according to Whitlock and Nauta (1956). The slash marks outline central regions of the brain-stem reticular formation extending into the subthalamic tegmental fields.

nucleus, and at least the caudal part of the orbitofrontal cortex approximately corresponding to Walker's area 13 and v. Bonin and Bailey's area FF. The fibre connexions between the three structures in question can be summarized as follows (*see also* fig. 17): the amygdaloid complex emits a considerable fibre bundle to the thalamic nucleus, which in turn projects massively to the caudal orbitofrontal cortex. There are reciprocal connexions to both of these systems, but the direct thalamo-amygdaloid and orbitofronto-thalamic pathways appear to be much less massive than their respective opposites. No direct connexion from the orbitofrontal cortex to the amygdala could be demonstrated, but the existence of indirect pathways relaying in the preoptic region and substantia innominata appears likely. In addition, there is good evidence that the orbitofrontal cortex is associated by way of the uncinate fasciculus with the rostral part of the temporal cortex, and at least the inferior temporal gyrus has been shown to project directly to the amygdaloid complex (Whitlock and Nauta, 1956). It should, of course, be added that the potential neural complexity introduced by the interpolation of a cortical structure in a neuronal chain is such that it hardly appears justifiable to interpret these data as evidence of an orbitofronto-temporo-amygdaloid "conduction pathway." Nevertheless, the available evidence strongly suggests that the neural mechanisms of the amygdaloid complex can be directly affected by those of an extensive orbitofronto-temporal region of cortex (v. Bonin and Bailey's (1947) areas FF and TG), a region in turn easily accessible to impulses of amygdaloid origin.

The connexions set forth above are indicated schematically in fig. 17, which also illustrates some of the subcortical pathways most likely related to the amygdalo-thalamo-orbitofrontal mechanism. In regard to the latter's efferent connexions, it is noteworthy that fibres from the amygdaloid complex as well as from the medial part of the dorsomedial thalamic nucleus and from the caudal orbitofrontal cortex converge upon the basal fore-brain region composed of the lateral preoptico-hypothalamic region and the adjoining substantiæ innominata and perforata anterior. In this region, so far as can be judged from some of its known efferent connexions, the projections of the amygdalo-thalamo-orbitofrontal mechanism could articulate with at least two major descending fibre systems, viz. the medial fore-brain bundle and the trajectory outlined by the stria medullaris, habenula, and fasciculus retroflexus. According to the results of an earlier study (Nauta, 1958), a considerable part of both these conduction systems is distributed to a paramedian region of the mid-brain tegmentum and central grey substance ("limbic mid-brain area"), whereas the remaining fibres spread widely over more lateral parts of the mesencephalic reticular formation. There is reason to believe that both pathways form a series of first links in a massive system of multineuronal connexions involved in part in visceral and endocrine functions, in part in mechanisms of general somatic and cortical activation, and, finally, in the neural events associated with "mood" and "psychical attitude." It is interesting that the conduction pathways of medial mesencephalic distribution are reciprocated by ascending connexions to the hypothalamus, preoptic region and septal area. These ascending pathways undoubtedly serve as conductors of impulses from the "limbic mid-brain area" to the septo-preoptico-hypothalamic complex, and could, by the latter's mediation, also represent a mesencephalic projection to the amygdala and hippocampal formation. The circuitous connexions in question have been reviewed in more detail elsewhere (Nauta, 1958); in the present diagram of fig. 17 only a few of the known pathways in this general category are indicated. Specifically, the diagram omits the neural links with the medial and periventricular hypothalamic cell groups presumably in part at least directly involved in hypothalamo-pituitary mechanisms. Such pathways of medial hypo-thalamic distribution are known to arise in the amygdala (stria terminalis), lateral hypothalamic region (Nauta, 1958), and central grey substance (Schütz's periventricular system).

None of the pathways mentioned earlier in this discussion represents a direct connexion from the amygdalo-thalamo-orbitofronto-temporal mechanism to the mesencephalic reticular formation. There is no com-plete agreement regarding direct amygdalo-mesencephalic projections. Whereas such direct connexions were suggested by the results of Gloor's (1955) electrophysiological analysis in the cat, as well as by Klingler and Gloor's (1960) findings in a dissection study in the human brain, no direct amygdalo-mesencephalic pathway could be identified in Nauta and Valenstein's (1958) experiments in the monkey. However, even if the amygdala should lack direct mesencephalic associations, there is little doubt that the larger amygdalo-thalamo-orbitofronto-temporal organi-zation of which it forms part is directly connected with the mesencephalic and subthalamic tegmentum, viz. by pathways originating in the orbito-frontal (cf. the present case MA 13) and rostral temporal cortex (*see* Whitlock and Nauta, 1956; Klingler and Gloor, 1960). It thus seems likely that the amygdalo-thalamo-orbitofronto-temporal organization can exercise a direct influence upon mechanisms of the rostral mesencephalic tegmentum. The previously discussed indirect pathways to the mid-brain could conceivably serve to amplify this influence, at the same time in-volving hypothalamic mechanisms. Needless to say, anatomical findings such as here discussed cannot be expected to furnish more than vague indications in regard to function. They do, however, strongly suggest that the amygdaloid complex, together with its associated fore-brain structures, can profoundly modulate neural mechanisms underlying most if not all major categories of central nervous function.

In the form suggested above, the efferent conduction systems of the amygdalo-thalamo-orbitofrontal organization are quite comparable to those of the hippocampal formation. It appears certain, for example, that impulses from both sources have access to common longitudinal conduc-tion systems such as the medial fore-brain bundle. This circumstance strongly suggests that the hippocampal formation and amygdaloid complex are engaged in the same general functional mechanisms, even though this joint involvement could conceivably be characterized in part by a mutual antagonism.

Finally, besides the indicated similarity of subcortical projection pathways, the present findings suggest an interesting analogy in the wider neural associations of the amygdaloid complex and hippocampus respectively. Both of these limbic structures give rise to projections, by way of a thalamic nucleus, to an adjoining cortical region which in turn is connected with the limbic structure under consideration. In more specific terms, the circuit: hippocampus—anterior thalamic nucleus—gyrus fornicatus— hippocampus appears to find a parallel in the organization: amygdala— dorsomedial thalamic nucleus—orbitofronto-temporal cortex—amygdala.

## SUMMARY

An attempt was made to determine the neural mechanisms associated with the amygdaloid complex in the monkey. The following conclusions are drawn.

(1) The medial, magnocellular subdivision of the dorsomedial thalamic nucleus, known to receive fibres from the amygdaloid complex, projects via the extracapsular component of the inferior thalamic peduncle to the lateral preoptic and hypothalamic regions, substantia innominata, rostral pole of the amygdaloid complex, nucleus of Broca's diagonal band, and olfactory tubercle. Numerous additional fibres from the nucleus follow the internal capsule to the caudal orbitofrontal gyrus.

(2) The caudal orbitofrontal cortex projects to the medial subdivision or the dorsomedial thalamic nucleus and to the thalamic nuclei paracentralis and reticularis. Further efferents are distributed to the lateral preoptic and hypothalamic regions, the substantiæ innominata and perforata anterior, and to the subthalamic and rostral mesencephalic tegmentum. The caudal orbitofrontal cortex has additional efferent connexions with anterior regions of the temporal cortex, and with anteroventral parts of the insular cortex.

(3) These findings, in combination with previous data, suggest the notion of an amygdalo-thalamo-orbitofronto-temporal organization with multiple discharge pathways to the hypothalamus, subthalamus, and mesencephalic reticular formation.

## ACKNOWLEDGMENTS

It is a pleasure to acknowledge the valuable technical assistance of Mrs. Mildred H. Albrecht and Messrs. Gordon Fletcher and Curtis King. The photographs were made by Miss Sally Craig.

## Abbreviations in Figures 1–17

AC: anterior commissure; Aco: cortical amygdaloid nucleus; ADH: dorsal hypothalamic area; AL: anterior limbic cortex; Al: lateral amygdaloid nucleus; AM: amygdaloid complex; Am: medial amygdaloid nucleus; CL: n. centralis lateralis; Cl: claustrum; CM: centre median; DM: dorsomedial thalamic nucleus; DMH: dorsomedial hypothalamic nucleus; DMl: lateral subdivision of dorsomedial thalamic nucleus; DMm: medial subdivision of dorsomedial thalamic nucleus; F: fornix; FU: fasciculus uncinatus; GP: globus pallidus; H: Forel's field H; HL: lateral hypothalamic region; IL: subcallosal gyrus; LM: medial lemniscus; MB: mammillary body; MT: mammillo-thalamic tract; NAC: bed nucleus of anterior commissure; NAP: n. ansæ peduncularis; ND: n. diagonalis Brocæ; NR: red nucleus; NST: bed nucleus of stria terminalis; OF: orbitofrontal cortex; OT: optic tract; PC: n. paracentralis thalami; Pf: n. parafascicularis; PL: lateral preoptic region; Pp: prepiriform cortex; Put: putamen; Re: n. reuniens; Rt: n. reticularis thalami; SI: substantia innominata; SN: substantia nigra; SO: supraoptic nucleus; ST: subthalamic nucleus; Teg: tegmental reticular formation of mesencephalon and subthalamus; TI: inferior temporal gyrus; TO: olfactory tubercle; VA: n. ventralis anterior thalami; VL: n. ventralis lateralis thalami; VMH: n. ventromedialis hypothalami; ZI: zona incerta.

## REFERENCES


ALBRECHT, M. H., and FERNSTROM, R. C. (1959) *Stain Tech.*, **34**, 91.

BONIN, G. V., and BAILEY, P. (1947) "The Neocortex of *Macaca mulatta*." Urbana, Ill.

CLARK, W. E. LE GROS, and MEYER, M. (1950) *Brit. med. Bull.*, **6**, 341.

FOX, C. A. (1949) *Anat. Rec.*, **103**, 537.

FREEMAN, W., and WATTS, J. W. (1947) *J. comp. Neurol.*, **86**, 65.

GLOOR, P. (1955) *Electroenceph. clin. Neurophysiol.*, **7**, 223.

KLINGLER, J., and GLOOR, P. (1960) *J. comp. Neurol.*, **115**, 333.

METTLER, F. A. (1945) *J. comp. Neurol.*, **82**, 169.

NAUTA, W. J. H. (1958) *Brain*, **81**, 319.

—— (1961) *J. Anat., Lond.* (In the Press.)

——, and GYGAX, P. A. (1954) *Stain Tech.*, **29**, 91.

——, and VALENSTEIN, E. S. (1958) *Anat. Rec.*, **130**, 346.

——, and WHITLOCK, D. G. (1954) *In:* Council for International Organizations of Medical Sciences. Brain Mechanisms and Consciousness. Oxford. p. 81.

NEWMAN, P. P., and WOLSTENCROFT, J. H. (1959) *J. Neurophysiol.*, **22**, 516.

PAPEZ, J. W. (1942) *Res. Publ. Ass. nerv. ment. Dis.*, **21**, 21.

——, and ARONSON, L. R. (1934) *Arch. Neurol. Psychiat., Chicago*, **32**, 1.

PRIBRAM, K. H., CHOW, K. L., and SEMMES, J. (1953) *J. comp. Neurol.*, **98**, 433.

——, and MACLEAN, P. D. (1953) *J. Neurophysiol.*, **16**, 324.

ROSE, M. (1927) *J. Psychol. Neurol., Lpz.*, **35**, 65.

ROUSSY, G., and MOSINGER, M. (1934) *Rev. neurol.*, **41**, 848.

SACHS, E. JR., BRENDLER, S. J., and FULTON, J. F. (1949) *Brain*, **72**, 227.

SHOWERS, M. J. C. (1958) *J. comp. Neurol.*, **109**, 261.

STRÖM-OLSEN, R., and NORTHFIELD, D. W. C. (1955) *Lancet*, **1**, 986.

VALENSTEIN, E. S., and NAUTA, W. J. H. (1959) *J. comp. Neurol.*, **113**, 337.

WALKER, A. E. (1940*a*) *J. comp. Neurol.*, **73**, 87.

——, (1940*b*) *J. comp. Neurol.*, **73**, 59.

WALL, P. D., GLEES, P., and FULTON, J. F. (1951) *Brain*, **74**, 66.

WARD, A. A. JR., and MCCULLOCH, W. S. (1947) *J. Neurophysiol.*, **10**, 309.

WHITLOCK, D. G., and NAUTA, W. J. H. (1956) *J. comp. Neurol.*, **106**, 183.

## LEGENDS FOR PLATES

### PLATE XXVIII

FIG. 18.—Degeneration of thin fibres in the inferior thalamic peduncle following lesion of the caudal orbitofrontal cortex (case MA 13). Note the presence of numerous impregnated normal fibres, mostly of larger calibre. × 380.

### PLATE XXIX

FIG. 19.—Fibre degeneration in the lateral preoptic region at a level corresponding approximately to that of fig. 8. Case MA 13. × 380.

FIG. 20.—Fibre degeneration in the caudal orbitofrontal cortex following lesion of the medial part of the dorsomedial thalamic nucleus. Case MA-c. × 380.

———

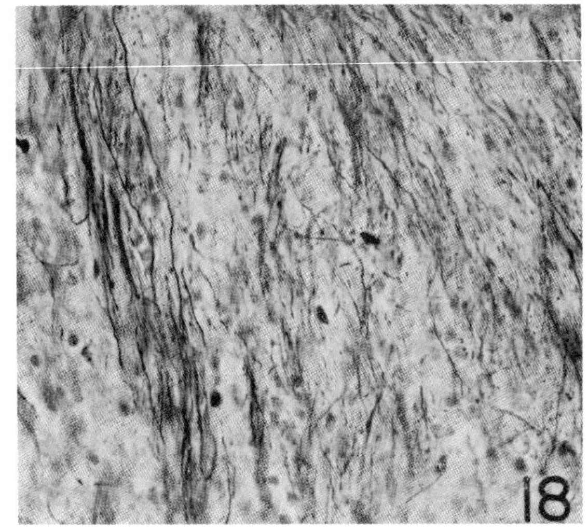



*To illustrate article by Walle J. H. Nauta.*

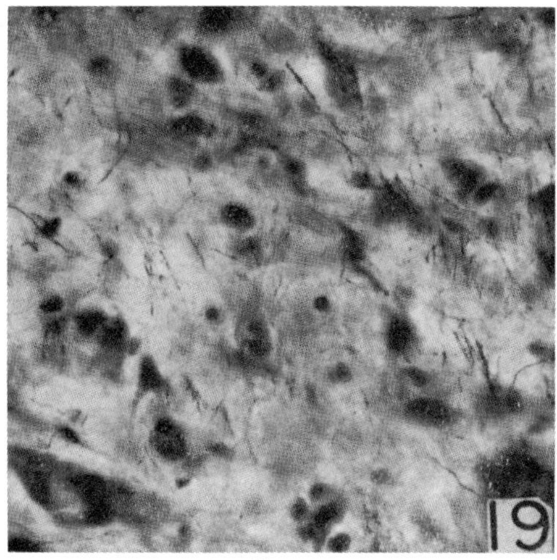

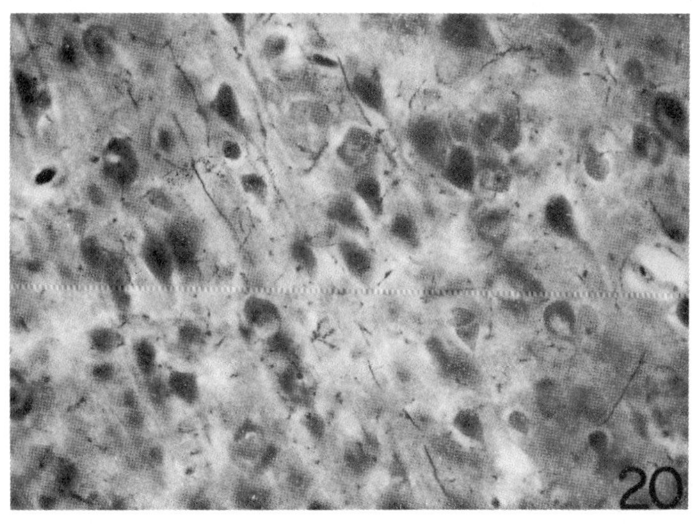

*To illustrate article by Walle J. H. Nauta.*

Reprinted from THE JOURNAL OF COMPARATIVE NEUROLOGY
Vol. 126, No. 3, March 1966

# The Isodendritic Core of the Brain Stem

E. RAMÓN-MOLINER AND W. J. H. NAUTA
*Université Laval, École de Médecine, Québec, Canada, and
Massachusetts Institute of Technology, Department of
Psychology, Cambridge, Massachusetts* [1]

*ABSTRACT*    According to their degree of morphological specialization, the cell populations of the brain stem may be classified into three groups: isodendritic, allodendritic and idiodendritic. The isodendritic neurons, or generalized neurons, are the most frequently encountered. If one discards those isodendritic centers that by common definition are sensory or motor, an *isodendritic core* is left which displays very little histological variation throughout the whole extent of the brain stem. This core corresponds, with certain restrictions, to the regions that are usually regarded as reticular formation. It constitutes a *continuum of overlapping dendritic fields* that extends from the spinal cord to the diencephalon. In view of the similarities that exist between the histology of the isodendritic core and the relatively disorganized nervous system of the lower vertebrates, it is postulated that it represents a pool of pluripotential neurons which in the course of phylogeny have remained relatively undifferentiated and in charge of processing afferent signals of very heterogeneous origin. By contrast, the allodendritic and idiodendritic centers can be regarded as relatively specialized centers from the point of view of their dendritic morphology, connections and functions. Attention is paid to the fact that the diffuse characteristics of the isodendritic core do not necessarily entail ill-defined physiological properties.

This paper is the result of an attempt to apply recent findings concerning dendritic patterns of the brain stem (Mannen, '60; Ramón-Moliner, '62a, b, '63; Leontovich and Zhukova, '63) to the demarcation of a vast neuronal territory largely corresponding to those regions usually regarded as reticular formation (Allen, '32; Brodal, '57; Brodal and Rossi, '55; French et al., '52, '53; Magoun, '52, '54; Moruzzi and Magoun, '49; Nauta and Kuypers, '58; Pitts et al., '39; Ramón y Cajal, '09a, b; Scheibel and Scheibel, '58; Scheibel et al., '55; Starzl and Magoun, '51; Starzl et al., '51a, b, to cite only some of the most representative communications).

Obviously, the concept of a reticular formation would lose in significance if its presumptive extent were to be amplified excessively. The term would become entirely meaningless if as a result of future research it came to be identified with the nervous system as a whole. If the concept is to survive at all, regions must be identified which lack "reticular" properties, or to which such properties apply only in a nonsignificant manner.

It is also imperative to avoid the use of circular definitions. We must realize, for example, that one can accept neither the amalgamation of a number of physiological properties on the basis of an *assumed* unitary morphology nor, conversely, a herding of anatomical regions on the basis of an *assumed* common physiological role. In other words, one must avoid the vicious circle whereby the physiologist aggregates heterogeneous functions in the assumption that they share a common anatomical feature, while the anatomist assembles miscellaneous territories because he assumes that they share some common physiological role.

Any demarcation of the reticular formation as a structure of unitary character should meet the following requirements; (1) There must be at least one outstanding property, either positive or negative, attributable to all its territories; (2) this property must constitute an actual peculiarity (in other words it must apply to those territories and only to them); (3) such a property may be a morphological feature, a physiological role, or both. It can also

[1] This study was initiated under contract DA-49-007-MD-1005 of the U. S. Army Research and Development Command with the University of Maryland Medical School, and continued under the support of the Medical Research Council of Canada (MA-1550).

be a coincidence of characteristics, provided some definite significance can be attributed to that coincidence.

In a previous paper one of us (Ramón-Moliner, '62a) submitted a classification of nerve cells based on the sharp contrast between, on the one hand, cell regions exhibiting "generalized" dendritic features, and on the other hand, certain well-defined cell clusters such as the cochlear nuclei, which display "specialized" dendritic characteristics. Mannen's ('60) suggested classification of the brain stem nuclei into "noyaux ouverts" and "noyaux fermés" was based largely on similar criteria. Leontovich and Zhukova ('63) likewise used the generalized dendritic configuration as a basis for their recent tentative delineation of the reticular formation.

It is indeed tempting on the basis of these recent studies to regard the generalized dendritic pattern as an identifying characteristic of the reticular formation. The present study, however, has produced evidence imposing certain restrictions upon this notion. In the following account the question will be examined in how far dendritic peculiarities, alone or in combination with other histological properties, can nevertheless be regarded as valid criteria in the conceptuation of the reticular formation. Before this can be done, certain new terms must be introduced. Two of these — "isodendritic regions" and "isodendritic core" — will be used here with reference to a vast region which extends throughout the brain stem and spinal cord, forming a matrix in which other cell groups with more specialized dendritic features lie embedded. The other two terms, "allodendritic" and "idiodendritic" refer to specialized dendritic patterns the details of which will be the subject of further communications. These terms are believed to reflect three degrees of morphological differentiation apparently corresponding to three main stages in the evolution of brain stem nuclei.

## MATERIALS AND METHODS

Brains of cats, sharks (Squalus acanthias) and lampreys (Petromyzon marinus) were impregnated in a Golgi-Cox solution to which sodium tungstate was added (Ramón-Moliner, '58) embedded in cel-

loidin and alkalinized in alcohol-ammonia (Van der Loos, '56). Some of the sections were counterstained with the procedure recommended by Van der Loos, slightly modified according to Ramón-Moliner et al. ('64). Many of the observations and conclusions reported in this paper were complemented by the study of Golgi-Cox stained material of a variety of other animals (monkey, rat, mouse).

### OBSERVATIONS

#### 1. Isodendritic, allodendritic and idiodendritic neurons

A careful examination of figures 3 and 4 reveals similarities as well as contrasts. Figure 3 shows a neuron found in the region corresponding to the lateral portion of Taber's ('61) nucleus prepositus hypoglossi, whereas the neurons shown in figure 4 belong to the nucleus vestibularis interstitialis (Ramón y Cajal, '09a, vol. I, fig. 323). The neuron in figure 3 shares several characteristics with the cells appearing in figure 4; all of the cells shown have ovoidal cell bodies and the number of dendritic trunks is approximately the same. One could expect that all of these cells, had they been stained by the Nissl method, would have looked similar. Another common feature is that in both locations the cells are interposed between heavily myelinated bundles of passing fibers. When, however, attention is paid to the configuration and disposition of the dendrites, a clear contrast appears. The dendrites of the neuron in figure 3 are relatively rectilinear, show little tendency to branch and, when they do branch, the resulting segments are, as a rule, longer than those from which they take origin. By contrast, the dendrites of the neurons in figure 4 follow an irregular or twisted course and show a much greater tendency to divide into segments of second and higher order; the resulting dendritc appearance could be described as "wavy," as "tufted," or as a combination of both.

It is also important to note the relationship of the dendrites to the heavily myelinated transit fiber bundles. In one case (fig. 3) dendrites mingle freely with such fiber bundles, penetrating the latter's interstices, whereas in the other case (fig.

4) they seem to avoid such intermingling. Mannen ('60), the first to emphasize this differential relationship, considered the free intermingling of dendrites and axonal bundles a characteristic of the regions denoted by his term "noyaux ouverts." In his "noyaux fermés" no such mingling was observed.

By and large, the most common type of neuron found in the brain stem is the one represented by figure 3. The size of the perikaryon, the extent of the dendritic field, and the number of dendrites per individual cell may show considerable variation (figs. 9–12). Nevertheless, in most cases it is impossible to use these variable features as distinguishing characteristics of any given cell group, for throughout the length of the brain stem large and small neurons of the same general type, but with varying dendritic richness, are frequently found side by side (figs. 7 and 8).

In contrast with the widely distributed cell type exemplified by figure 3, other neurons display dendritic arborizations of a less common pattern. Figure 4 gives an example of such forms. As pointed out in earlier communications (Ramón-Moliner, '62a, b) such more special neurons are usually confined to relatively well-defined territories. Some of these dendritic patterns are so peculiar that the corresponding cell group can be identified on the basis of only one successfully impregnated neuron.

In the earlier communications it was suggested that specialization of dendrites in the brain stem can occur following two main trends which, in extreme cases, result in either "wavy" or "tufted" dendrites. However, since "wavy" and "tufted" dendrites constitute only extreme cases, and since intermediate forms are more commonly encountered, a more differentiated nomenclature will be suggested here (fig. 1). Those neurons characterized by the presence of generalized dendrites (lower left corner) will be named "isodendritic" (*isos:* similar, unchanging, uniform) and those that deviate from this prototype, "allodendritic" (allos: different) or "idiodendritic" (*idios:* peculiar) depending on the degree of differentiation. Figures 9–12 show individual examples of isodendritic

neurons, and figures 7 and 8 present somewhat wider views of regions populated by these neurons. Figures 5 and 6 illustrate two examples of allodendritic cell populations with dendrites of intermediate configuration. Figure 5, taken from the inferior colliculus, shows relatively specialized dendrites with a tendency toward "tuftedness" resembling that found in certain thalamic nuclei. Figure 6 represents the pontine nuclei and shows neurons with dendrites likewise relatively specialized, but showing a greater tendency toward "waviness" reminiscent of the patterns found in the inferior olivary nucleus (fig. 1, top left).

### 2. *General cytological features of the isodendritic regions of the brain stem*

Figures 9 through 12 show nerve cells sampled from a large number of territories in the brain stem of the cat. Despite considerable differences in size and over-all dendritic richness these neurons share so many characteristics that they can be regarded as variations of one and the same neuronal type. In general, the perikaryon of these cells is polygonal or triangular, more rarely round or oval. As a rule, the larger the size, the more closely the cell approximates the polygonal or triangular shape. The largest neurons (fig. 9) are found in the nucleus of Deiters, the nucleus gigantocellularis of the medulla oblongata and the masticatory nucleus. Polygonal cell bodies are also observed, although more rarely in medium and small-sized neurons (figs. 10 and 11). In the smaller neurons the cell body is usually more ovoidal, and often fusiform (fig. 12). Many of these "spindles" actually represent flattened neurons which in photographs appear elongated. Upon careful observation, it becomes apparent that the term "lens-shaped" would describe such cells more accurately. Generally, whenever the dendrites are markedly oriented, the cell body follows the same trend. The converse is not true: an extremely elongated or flattened cell body is not necessarily associated with any particular dendritic orientation.

With exceptions, the specialized dendritic patterns appear to be more com-

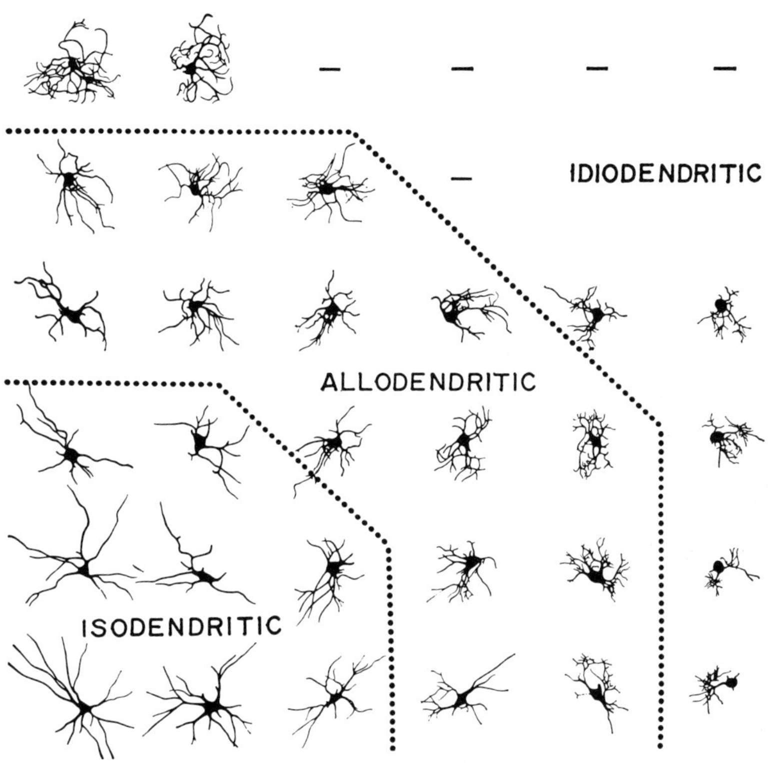


Fig. 1    Varieties of dendritic patterns in the brain stem of the cat. In the lower left corner the generalized or isodendritic patterns are found. The allodendritic neurons represent a step towards specialization. The idiodendritic neurons represent the highest degree of specialization which results in either a pronounced degree of waviness (upper left) or tuftedness (lower right) of their dendrites.

monly associated with round or ovoid cell bodies, whereas the generalized dendritic patterns illustrated in figures 7–12 are usually found in cells having polygonal cell bodies. Occasionally, the size and distribution of the Nissl substance can serve as an index of the dendritic disposition: a generalized dendritic configuration is frequently associated with irregularly distributed Nissl bodies, varying in size and extending well into the dendritic trunks. By contrast, territories containing large neurons with chromatic granules of relatively even size, preferably arranged in the periphery of the cell body and leaving a "halo" around the nucleus, often have dendrites of specialized configurations. With decreasing size of the cell body this rule becomes less reliable, small neurons show-

ing no evident correlation between dendritic and Nissl substance patterns.

The characteristic features of the nonspecialized dendritic expansions can be observed in figures 7 and 8. It can be seen that the dendrites are relatively rectilinear, show a marked tendency to taper as they extend from the cell body, and have relatively little tendency to branch. It is difficult to study the branching rule in view of the fact that so many dendrites soon reach the cut surface of the section. It nevertheless appears that, in general, the distal segments are longer than the proximal ones, a characteristic noted in certain neurons of the cerebral cortex by Sholl ('53) and regarded by one of us (Ramón-Moliner, '62a, '63) as a criterion permitting a distinction of "reticular" and motor neurons in the brain stem from most secondary, and certain precerebellar, neurons.

In general, the dendrites of the brain stem, whether generalized or specialized, exhibit only a small number of "spines" or "gemmules." Current Russian authors appear to have attributed much importance to this fact (Leontovich and Zhukova, '63). It is certainly true that no neurons are found in the brain stem comparable in this regard to, for example, the pyramidal cells of the cerebral cortex, the cells of the caudate nucleus, or the Purkinje cells of the cerebellum. If the dendritic spines, as seems likely (Gray, '59), manifest the presence of a special variety of synapse, one must conclude that such synapses are relatively infrequent in the brain stem.

Cytological polymorphism is another striking feature in figures 7 and 8. In the motor nuclei, this polymorphism is less pronounced than it is in the remaining isodendritic regions, a characteristic which is occasionally useful in attempts to identify a group of motor neurons in Golgi stained material. At this time, however, there is no reliable criterion by which a large cell of the nucleus gigantocellularis of the medulla oblongata could be distinguished from, for example, a neuron of the masticatory nucleus.

It is not possible to analyze in statistical terms the distribution of the various sizes of generalized neurons throughout the brain stem, in view of the fact that the Golgi method can easily give a very distorted sample. Figure 2 may, nevertheless, convey an approximate idea of the apparent overall frequency distribution of variation in cell size and richness of the dendritic ramification.

### 3. The intermingling of dendrites and passing fibers

When attention is paid to the sharp contrast between the free intermingling of dendrites and passing fibers which is shown in figure 3 on the one hand, and the absence of such mixing in the case of figure 4 on the other, one may wonder if these two different relationships could reflect the rules of economy of space, or represent some unreported interaction between nerve cells and passing fibers in the "open" nuclei which, in the case of the

"closed" nuclei is conspicuously avoided. The majority of the territories usually included in the brain stem reticular formation exhibit the characteristic of "hodophilic" dendrites, in contrast to, for example, the inferior olive and pontine nuclei in which the dendrites are rarely found to extend into fiber bundles and could thus be described as "hodophobic."

Mannen ('60) pointed out that an extensive mingling of dendrites and passing fibers often cause the boundaries of cell territories to become very indistinct. His term "open nuclei" does not adequately convey this characteristic and a change to "open regions" might therefore be more appropriate. In any event, Mannen's observation has furnished a criterion of unquestionable interest, as it permits a clear-cut morphological distinction between two types of central nervous tissue, one of which, the "open" variety, appears to correspond largely to that region which here is referred to as "generalized"or "isodendritic." It should be pointed out, however, that a close mingling of passing fibers and dendrites is by no means a peculiar feature of the core of the lower brain stem. The deeper layers of the cerebral cortex, the reticular nucleus of the thalamus, the subthalamus, the zona incerta, certain intralaminar nuclei of the thalamus, the substantia innominata, etc., are examples of the same type of histological arrangement. Leontovich and Zhukova ('63) have attributed to this extensive territory dendritic peculiarities similar to those of the reticular formation of the mesencephalon, pons and medulla oblongata. On the basis of a cursory survey of our Golgi material we are inclined to agree with this point of view.[2]

### 4. The continuum of overlapping dendritic fields

The geometrical features displayed by the neurons populating the isodendritic or "open" territories suggest the conclusion that there exists in the brain stem a vast neural territory, extending from the me-

[2] If confirmed, it might represent a morphological corollary to the functional similarities that have been postulated to exist between the reticular formation and the non-specific thalamic projection system (Morison and Dempsey, '42; Jasper, '49 and '54; Starzl et al., '51a,b, '52; Nauta and Whitlock, '54; and others).

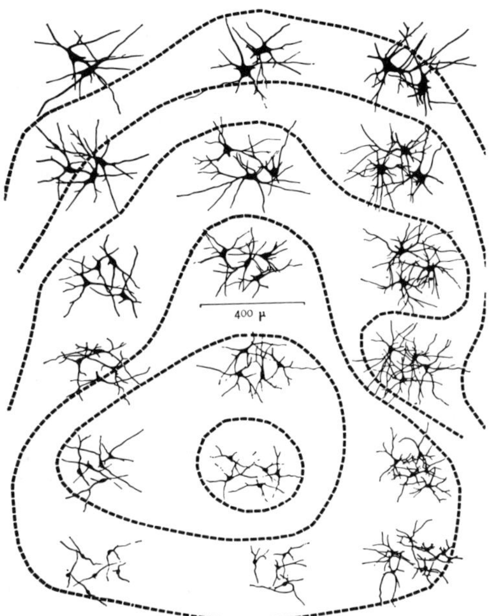



Fig. 2 A graphic representation of the apparent frequency distribution of the isodendritic neurons in the brain stem of the cat. The nerve cells have been grouped according to the variables of size and over-all dendritic richness. In view of the many sources of error involved in the sampling of neurons stained with the Golgi method, this diagram is only of relative value. The contour lines encircle the various neuronal groups like the geodesic lines of a map. The most frequently encountered neurons are shown at the "summit" (middle column, second group from the bottom).

dulla oblongata to the mesencephalon (and probably even further), in which no sharp territorial boundaries can be drawn. This peculiarity naturally does not preclude the existence of regional differences in morphology, presumably corresponding to functional "subsets." The Nissl picture affords at least some measure of regional parcellation (Meesen and Olszewski, '49; Olszewski and Baxter, '54). However, a glance at a suitable Golgi preparation shows that in this extensive territory no imaginary line, or plane, or surface of any sort, can be drawn that does not intersect dendrites of relatively widely separated nerve cells. We can therefore speak of a *dendritic continuum* that pervades much of the volume of the brain stem.

It must be emphasized that this continuum, despite its large extent, does not encompass all the neurons of the brain stem tegmentum. In a coarse analogy, it could be compared to the connective tissue that supports well outlined glandular acini. For obvious reasons this analogy cannot be carried too far. The dorsal and ventral tegmental nuclei of Gudden, the papillioform nucleus of the tegmentum (Olszewski and Baxter, '54), and part of the so-called lateral reticular nucleus of the medulla oblongata, to cite only a few cell groups customarily subsumed under the term reticular formation, are examples of allodendritic nuclei which cannot be included in the more diffuse dendritic continuum. In the latter, the recognition of "gradients" or gradual structural transitions would seem at this stage more objectively founded than any sharp linear demarcation of nuclei and subnuclei.

The average dendritic length found in this continuum is well over 300 μ (see figs. 7 and 8). Thus, the average neuron easily covers a distance of half a millimeter. Consequently, unless we are dealing with one of the relatively scarce "closed" nuclei, any arbitrary borderline will cut structures covering a band at least half a millimeter wide. Many striking examples of such dendritic overlapping were shown by Mannen ('60). Our Golgi-stained material has disclosed neurons of the raphe nuclei which emit dendrites extending bilaterally as far as the magnocellular portion of the trigeminal nucleus. Within the latter, neurons are found that have dendrites which, coursing transversely in opposite directions, reach the subnucleus gelatinosus as well as the neighborhood of the raphe nuclei. The dendrites of the dorsal motor nucleus of the vagus overlap with those of the solitary tract (Mannen, '60) and the latter nucleus has dendrites entering the area postrema (Morest, '60). In fact, one could group all these nuclei, together with the central grey substance, into a single neural apparatus in which the processes of motor, sensory, and internuncial neurons mingle closely.

The phenomenon of dendritic overlapping is difficult to describe. It has little intuitive appeal and turns the spatial notions of neighborhood and distance into meaningless terms. Thus, if we define the "degree of neighborhood" between two neurons in terms of the shortest distance between their cell membranes, two nerve cells could be identified as relatively "close neighbors" even if their perikarya were situated far away from each other. Conversely, two neurons with adjacent somata may show so little overlapping of their dendritic fields that in actuality they could be regarded as mutual "strangers." It is evident that examples of the first case must abound in the isodendritic regions (figs. 7 and 8).

A problem arises when one considers the daily necessity of reporting the exact location of electrodes, lesions, fields of axonal distribution, etc. The unitary concept of the reticular formation has often tempted physiologists and anatomists alike to be less than optimally accurate in such reports. For practical reasons, the consistent use of any parcellating nomenclature can still be justified even if we know that the phenomenon of dendritic overlapping may lessen its objective value. In the present paper, for example, the nomenclature proposed by Olszewski and Baxter ('54) in the human brain stem and that of Taber ('61) in the brain stem of the cat, have been used to describe the location of the numerous nerve cells displayed in figures 9–12. One could, instead, have given the stereotaxic coordinates of the regions in question. This practice, however, would turn any exchange of ideas between research workers into an insurmountable task.

### DISCUSSION

#### 1. Dendritic patterns and phylogeny

The phylogenetic process by which nerve cell groups and fiber systems have gradually evolved from a primordial diffuse net into well-established pathways and other anatomical entities began to be understood after the works of Herrick ('20, '48), and Herrick and Bishop ('58). The conclusions reached by these authors on the basis of their comparative neuroanatomical studies correspond in part to Coghill's ('29) concept of the embryological development of the amphibian nervous system. Herrick and Bishop have pointed out that well-defined neural pathways are sparse in lower vertebrates and that, consequently, a primordial diffuse net must have preceded the stage of at least partially more distinct organization found in higher forms. It is difficult to imagine an evolutive process that would lead to the sudden appearance of well-outlined nuclei and well-defined pathways, and the notion that a more diffuse arrangement of nerve cells and fibers must have preceded that stage seems both logical and consistent with objective data. The term "phylogenetic segregation" will be used here to denote this assumed course of events, for it identifies a notion applicable to many aspects of the evolution of the nervous system. In fact, numerous cell groups which in mammals are characterized by a high degree of spatial confinement display a much more diffuse architecture in lower vertebrates. A sharp demarcation between the white and grey matters of the spinal cord, for example, is never completely achieved in vertebrates below the reptilian level. In such lower forms (fig. 16) the dendrites of the spinal cord motor neurons invade the interstices of the funiculi in a manner recalling the characteristics of the "reticular cells" in higher vertebrates (fig. 3). Several other regions which in mammals are organized in well-defined layers, (olfactory lobe, cerebellum, etc.), are more diffusely structured in lower forms. Apparently, the process of increasingly sharp functional and morphological specification leaves a considerable volume of the more diffuse structured primordial organization unaffected even in the highest mammals. The histological features of the mammalian reticular formation are such that it is difficult not to regard it as the most likely representative of that primordial net. Mammalian reticular neurons have, for example, dendritic configurations resembling those of the average neuron of the shark more than they do those of the more sharply defined mammalian allodendritic or idiodendritic cell groups. Figure 14 shows two neurons from the rhombencephalic tegmentum of the shark, one spindle-shaped and the other with radiating and relatively rectilinear dendrites, both of which resemble those of the reticular nucleus centralis pontis oralis of the cat (fig. 13). Although the similarity may not be complete, it is obviously greater than that between, for example, the isodendritic mammalian cells shown in figure 13 on the one hand, and the likewise mammalian but allodendritic neurons of figure 6, on the other. One significant difference between the neurons of the mammalian nervous system and the average neuron of the brain stem of lower forms seems to lie in the number of dendrites per cell. It is generally less in the latter forms: spindle-shaped cells are relatively numerous in the brain stem of the shark (fig. 15). In this regard, the nervous system of lower vertebrates is comparable to those zones of the mammalian brain stem and diencephalon immediately surrounding the central canal and the ventricles. The shark neurons shown in figure 15, for example, are very similar to those found in larger numbers in the hypothalamus and mesencephalic central gray substance of mammals. Most other regions of the mammalian reticular formation also show this slender type of neuron (figs. 12 and 13), but here it is mixed in various proportions with other elements of richer dendritic ramification. However, if the geometrical configuration of the dendrites has any specific functional significance (Rall, '59, '60, '61, '62), the conclusion suggests itself that especially the periventricular regions of the mammalian brain stem contain neurons which have retained in a relatively unchanged manner certain functional properties already present in the nervous system of fishes.

## 2. The concept of the reticular formation

At present a generally accepted definition of the term, reticular formation, is lacking and in most cases one is forced to reconstruct each author's point of view by the way in which he uses it. When reference is made to the reticular formation several connotations are possible. The term can be used: (1) as if it corresponded to a definite anatomical entity with features found nowhere else in the nervous system; (2) as an admission of ignorance and as a way of avoiding the task of analyzing the various cell groups or nuclei that may compose the region, and, (3) as an ill-defined anatomical concept corresponding to a well-defined or at least useful physiological concept.

1. Among anatomists the term is probably never meant to denote a well-defined anatomical entity, but precisely this notion appears to prevail among investigators whose studies rarely, if ever, involve direct anatomical observations. Some classical illustrations display the reticular formation as a well-delimited cylinder placed inside a ghost of the brain stem. Although useful in discussions between neuroanatomists and neurophysiologists, such visual aids easily become misleading in a wider interdisciplinary exchange. Today, in the course of a discussion on the reticular formation some questions are commonly asked which are indicative of this state of affairs. To cite only a few: "Is it possible to stain in a selective manner the cells of the reticular formation?"; "Is there a reticular formation in the nervous system of the fly?"; "When studied with the electron microscope, what are the specific features of the reticular formation?"; "How many nuclei does the reticular formation contain?"

It is important to recall that the term "reticular formation" in its original usage did not refer to any peculiar feature of the local neuropil or to any other histological characteristic. In fact, not one single histological feature can be attributed to the reticular formation that cannot be found also in other regions of the nervous system. If one should decide to regard certain admixtures of myelinated nerve fibers and gray matter as the distinguishing characteristic, then the pontine nuclei could well be included in the reticular formation. If one chooses the fact that the reticular formation contains cells with axons dividing into ascending and descending branches (Ramón y Cajal, '09b; Scheibel and Scheibel, '58; Valverde, '61), then the neurons of the dorsal root ganglia could also be included in the concept. If the generalized dendritic pattern were considered the most distinctive feature, then the motor neurons would qualify as components of the reticular formation. The classical findings of Moruzzi and Magoun ('49); Moruzzi ('54); and Scheibel et al. ('55) would suggest the heterogeneous nature of the afferent supply to its constituent cells as the most typical of all its attributes, but in that case, the deep layers of the superior colliculus, the territories commonly described as vestibular nuclei, the motor nuclei, and other cell aggregates should be included in the reticular formation.

2. In view of the above considerations one may be tempted to the other extreme and conclude that the use of the term, "reticular formation," should be discontinued. It has been suggested (Olszewski and Baxter, '54) or implied (Allen, '32) that as a result of future investigations the term, reticular formation, may eventually disappear from our vocabulary and that an inventory of nuclei and subnuclei might be substituted for it. This point of view implies that, even if no regional subdivisions are apparent, "a priori" they must always be assumed to exist. We believe that this implicit assumption is by no means incontestable. In fact, the opposite suggestion would seem equally or even more reasonable if one takes into account the phenomenon of dendritic overlapping (Herrick, '48). It is true that the presence of dendritic overlapping does not necessarily prove the morphological or functional equivalence of neighboring regions, but it can be used to question the dogmatic value of many proposed parcellations. This means that when extensive dendritic overlapping exists between two adjacent territories, there is some reason to regard them as one single region until the contrary is indicated. A nucleus or center can be identified on the basis of any one of a variety of distinctive properties: specific functional effects following stimulation or lesion; histochemical properties; pigment content; specific afferent or efferent relationships; or, simply evident cytological individuality. On the basis of any one of these criteria, individual circumlineation would seem justified even if the Golgi method should reveal extensive dendritic overlapping. However, if this overlapping is present and none of the above-mentioned criteria applies, it would seem defensible to regard any number of territories as components of one single region.

3. In its reference to a physiological identifiable neural apparatus, the term reticular formation is more conservative and may be used as a compromise between the above-mentioned extremes. This no-tion, probably the most commonly accepted, would seem to offer no serious implications. It is impossible to remain aloof of the fruitful developments engendered by the awareness among neurophysiologists that there exists in the mammalian brain a neuronal system of "reticular" organization. Nevertheless, it should be borne in mind that this notion was originally an anatomical one. It is, therefore, somewhat paradoxical that the reticular formation should now be regarded as an ill-defined anatomical concept corresponding to a useful physiological notion, and hence as a concept submitted by the anatomist at the request of the physiologist. We believe that this point of view is no less contestable than any of its previously discussed alternatives. We have seen that every one of the features that have been attributed to the reticular formation can be identified, at least singly, in other regions of the brain and that, consequently, when considered individually, such features cannot provide a suitable basis for a definition. However, their *aggregation* is significant, since the reticular formation is one of the few regions of the brain, if not the only one, where most or all of the above listed characteristics are found together. In addition, this extensive territory displays in Golgi stained material a relatively uniform histological appearance:

a. Cytological polymorphism: large and small neurons are found side by side and their individual dendritic trees show a wide variation in overall richness of ramification.

b. Generalized dendrites: long, radiate, relatively rectilinear processes branching in such a manner that the distal segments are, as a rule, longer than the proximal ones.

c. Apparent lack of distinctive regional dendritic characteristics.

d. Considerable degree of dendritic overlapping: as a result of the generally great length attained by the dendrites, and their rectilinear course, a continuum of overlapping dendritic fields is formed that appears to extend throughout the length of the neuraxis.

e. Free intermingling of dendrites and passing myelinated and unmyelinated fiber bundles: in this regard the dendrites of the isodendritic core differ radically from those of most other cell regions.

This coincidence of individually non-distinctive morphological features suggests itself as a peculiarity justifying a unitary concept even in the absence of any physiological data. It is true that the reticular formation is "diffuse" and its boundaries "ill-defined," but we believe that the *concept* can nevertheless be precise: all of the above-mentioned attributes taken together describe a type of neural tissue that is present throughout the length of the core of the brain stem. If the same coincidence of attributes should be identified in any other region of the brain, this would simply mean that a similar neural formation exists elsewhere.

### 3. The isodendritic core

Even if the foregoing remarks are accepted, the question whether or not a given cell group or region belongs to the reticular formation is not always answered easily. The term has often been used to include such cell groups as the lateral reticular nucleus of the medulla and the nucleus papilliformis (pterygoid nucleus of the tegmentum) which, at least in part, do not display generalized dendrites. On the other hand, the literature on the reticular formation has attained such un-assimilable proportions that it is at this time practically impossible to revoke many of the connotations that have been imposed on the term. For this reason we submit the term *isodendritic core* as an alternative name referring to most of the territories usually regarded as reticular formation. The main advantage of the term is that the inclusion of a given cell territory in the family of isodendritic regions can be based on purely descriptive grounds: it is always relatively easy to determine whether a particular cell group displays generalized or specialized dendritic patterns. It is thus possible to state, for example, that the pars medialis of the spinal trigeminal nucleus, as customarily outlined, is isodendritic whereas the so-called lateral reticular nucleus of the medulla, at least in part, is not. As is well known, the term "reticular substance" in its original usage referred to a particular interspersion of gray and white matter. Apparently intuitively, the early anato-

mists excluded the pontine nuclei from the reticular substance, even though a comparable intermingling also exists in these nuclei. This restriction agrees remarkably well with the fact that the cells of the pontine nuclei differ considerably from the standard isodendritic pattern, and with the circumstance that these nuclei appear to have rather well-defined connections. The same holds for the "lateral reticular nucleus" of the medulla oblongata, a cell group which has often figured in the argument against a unitary character of the reticular formation because of its rather circumscript fiber connections. It happens that, on the basis of the present criteria, this cell group cannot be entirely included in the isodendritic territory for many of its neurons show a relatively specialized dendritic pattern.

On the other hand, it should be emphasized that the isodendritic pattern is also encountered in regions which were never considered part of the reticular formation, such as the vestibular nuclei, motor nuclei, superior colliculus, deeper layers of certain sensory nuclei, and probably other regions. In other words, the two concepts "reticular formation" and "isodendritic regions" overlap but do not coincide. However if only a few restrictions are imposed, it is possible to attain a nearly perfect equivalence. Thus, if certain isodendritic regions, such as the so-called vestibular nuclei and the moto-neuronal groups are excluded, a vast territory remains which extends from the spinal cord to the diencephalon without outstanding histological changes (Leontovich and Zhukova, '63).

### 4. Biological significance of the isodendritic neurons

At one time the reticular formation could be defined as "what was left over once the other nuclei had become established" (Allen, '32). The present point of view is that this definition is far from being merely a facetious one, and that the considerable extent attributed to the reticular formation by Leontovich and Zhukova ('63) is at least compatible with general biological principles.

Leontovich and Zhukova ('63) have illustrated "reticular" neurons evidently corresponding to the isodendritic type of neuron found in Mannen's ('60) "noyaux ouverts" and described as "generalized neuron" by Bodian ('62) and Ramón-Moliner ('62a). Such neurons, according to Leontovich and Zhukova, are not confined to the brain stem and can be found even in the cerebral cortex. From a general survey of the dendritic patterns throughout the central nervous system it was suggested (Ramón-Moliner, '62a) that this generalized or isodendritic type of neuron prevails in brain regions characterized by afferent connections of heterogeneous origin. By contrast, many of the so-called secondary and tertiary sensory neurons were found to display more specialized ("tufted" or "wavy") dendrites (Ramón-Moliner, '62a), an observation also reported by Leontovich and Zhukova. However, in the brain stem there are, at least, two conspicuous exceptions to this rule: the nucleus of the solitary tract and the vestibular nuclei which are populated by isodendritic neurons comparable to those of the adjacent reticular formation. In the nucleus of Deiters not even the narrow band which is known to receive primary vestibular fibers (Brodal, '60; Brodal et al., '62) was found to contain more specialized neurons. In this regard, the vestibular area and the nucleus of the solitary tract could be compared to the nucleus proprius of the spinal cord. In all three cases, we seem to be confronted with "secondary sensory" neurons which in their dendritic patterns resemble more closely the "reticular" neurons than they are similar to the secondary neurons of other afferent systems.

The prevalence of generalized, i.e., isodendritic, neurons in some of the known secondary cell groups, a category of nuclei generally characterized by more specialized allodendritic or idiodendritic types, would appear to suggest that morphological specialization of neurons is conditioned by one or more peculiarities of their afferent relationships. Obviously, an adequate analysis of this suggestion could be made only if it were possible to ascertain the origin, distribution and functional modality of all fiber systems terminating in each secondary sensory cell group. In the absence of such detailed information, only one of several possibilities can be examined here, viz. the earlier suggestion

(Ramón-Moliner, '62a) that specialized dendrites could be related to a high degree of homogeneity of the "input." A particularly suggestive argument in its favor is furnished by an observation reported by Kuypers et al. ('61). According to these workers, those subdivisions of the nuclei cuneatus and gracilis which receive both dorsal root fibers and corticofugal projections are characterized by dendritic patterns less specialized than those found in the cell nest region of the nuclei where only dorsal root fibers are known to terminate. The isodendritic characteristics of the nucleus proprius in the dorsal horn of the spinal cord could be related to a similar overlapping of dorsal root fibers and corticofugal projections. The vestibular nuclei appear to receive afferents from a variety of sources (Brodal et al., '62) and thus might have to be regarded as a subdivision of the isodendritic core which differs from the rest of this core only in that it receives primary sensory fibers from the vestibular ganglia. The isodendritic features of motor neurons might likewise be explained by highly heterogeneous afferent relationships. All these considerations suggest that generalized, isodendritic neurons are characteristic of cell regions not completely (or sufficiently) "monopolized" by a particular specific afferent system.

Attractive as this notion may appear to be, there are reasons to view it with reserve. Observations in several brain stem nuclei strongly suggest that homogeneity of input or, at least, a strongly dominant relationship to one particular afferent fiber system, cannot be the only factor operative in the coining of specialized dendritic configurations. For example, the pars compacta of the nucleus tegmenti pedunculopontinus (Olszewski and Baxter, '54) in the caudal mesencephalic tegmentum receives an extremely dense plexus of afferent fibers from the globus pallidus (Nauta and Mehler, '66), and yet appears to contain only isodendritic neurons. Conversely, several cell groups known to project selectively to the cerebellar cortex (inferior olive, lateral reticular nucleus) display allodendritic and even idiodentritic neurons, despite the presence of both spinal and cortical fibers among their known afferent systems. These are, no doubt, conflicting facts that prevent the formulation of a general law correlating dendritic configuration and type of afferent supply. One is thus forced to conclude that although homogeneity of input appears to play an important part, other factors can likewise be involved in the development of morphologically specialized dendrites. In the case of the specific precerebellar nuclei, a particular common projection target may have determined such specialization. In the case of the medial nucleus of the habenula, the determining factor could be some specific role in the control of endocrine functions (see Szentagothai et al., '62), and one could suspect that other cell groups might acquire allodendritic characteristics by virtue of other functional specializations, the nature of which cannot be precisely stated at this time.

The above-mentioned controversies could easily lead to the conclusion that no general functional significance can be attributed to a lack of morphological specialization of dendrites. Upon closer analysis, however, at least one general corollary to dendritic specialization can be suspected, namely, a reduction in the number of functional potentialities. Many examples can be drawn from various fields of biology to show that lack of specialization is often a manifestation of pluripotentiality. If, in the particular case of neurons, heterogeneity of connections were regarded as a particular example of multiplicity of potential attributes, it would be possible to account for the apparent rule that generalized dendritic patterns are associated with diversified connections, afferent, efferent, or both. The "dendritic specificity gradient" expressed by the terms, isodendritic, allodendritic and idiodendritic, could in that case be considered to reflect a gradual decrease in the number of potential connections and hence, of functional properties. The paucity of allodendritic and idiodendritic neurons in the brain stem of lower vertebrates suggests that, from a phylogenetic point of view, the isodendritic core is a primordial entity. It appears to be a pool of pluripotential neurons which in the course of evolution remained diffusely distributed throughout the brain stem, retaining generalized mor-

phological features. The usually more circumscript allodendritic and idiodendritic cell groups may have become segregated from this isodendritic pool, some as a result of a gradually increasing dominance of one, or at most a very few, among the original afferent systems, others possibly as a consequence of increasing specificity in their efferent relationships, and yet others by factors of an unexplained nature. *This postulated course of events should be regarded as a phylogenetic process and not as an ontogenetic one.* In all likelihood, it represents the outcome of a lengthy interaction between mutations and natural selection which eventually has become perpetuated in the genetic code of the individual. The allodendritic and idiodendritic configurations appear very early in embryonary life, long before the corresponding cell groups are invaded by afferent fibers, i.e., long before any specific functional properties can be implicated. This means that, strictly speaking, the dendritic patterns are genetically determined. In the final analysis, however, the development of any genetic code cannot be separated from the history of the species and from the past requirements imposed by the environment. For this reason, we feel that there is essentially little conceptual difference between the physiological "modifiability of the neuron," as postulated by Weiss ('60) and what could be described as "modifiability by adaptive evolution."

### 5. Physiological considerations

The question arises whether it is possible to assign some form of "general physiological property" to the isodendritic neuronal apparatus. This amounts to asking whether a single term could be proposed which would encompass the large variety of functions that have been attributed to the reticular formation, functions that range from the control of muscle tone to the maintenance of consciousness (French et al., '52, '53; Magoun, '52; and others), and include visceral mechanisms apparently mediated by anatomical territories overlapping the same regions that are involved in either muscle tone, or consciousness, or both (Pitts, Magoun, and Ranson, '39; Bonvallet et al., '55; Tang, '55; Baxter and Olszewski,

'55; Dell, '58; Harris, '58; Morest and Sutin, '61; and others).

It is difficult to propose a term encompassing this multiplicity of functions without crossing the borderline between neurobiology and philosophy. One could, for example, regard consciousness as a manifestation of the "tone of the brain" and muscle tone as an expression of some central excitatory state not very different in principle from that of the "activated cortex." In order to account for other physiological properties ascribed to the same neural apparatus, one could propose the term "homeostasis" or "maintenance of posture," or suggest a mechanism controlling the general "degree of responsiveness" of the organism. Speculative as they may appear, all these terms may nonetheless refer to a common physiological property, at present vague and elusive, but not without promise of sharper definition. On the basis of purely anatomical observations it would seem imprudent and presumptuous to go beyond these general suggestions.

One important prejudice to avoid is that it is a priori impossible to assign differentiated types of activity to a neural apparatus characterized by a high degree of dendritic overlapping or other "diffuse" characteristics. In the first place, it is by no means certain that such "diffuse" characteristics reflect lack of spatial organization. The diffuse appearance may be due simply to a failure of our three-dimensional intuition of space to grasp any organization which does not manifest itself in the form of clearcut "clusters" or "layers" of nerve cells. On the other hand, it could be asked whether the widespread overlapping of dendritic and axonal fields could not represent the very material basis for what we usually call "integrative functions of the nervous system." It is possible to defend the thought that what "makes the nervous system work" is the fact that its cellular elements do not meet the spatial restrictions imposed on the cells of other organs. The receptivity of nervous matter must often involve the repeated use of the same nervous elements by signals of very heterogeneous origin. The integrative aspects of nervous activity must forcibly result from a constant combination and recombination of signals, and it is difficult to imagine

how a nervous system devoid of some kind of overlapping circuitry could perform all these operations. Following a related line of thought, it could be said that if the "higher nervous activities" were to be associated with a high degree of neurohistological organization, then the cerebellar cortex, with its orderly arrangement of neurons, should be regarded as the most likely candidate for such "higher functions." Instead, it appears to be comparable to a servomechanism, highly useful for the execution of movements but not essential for the conduct of intelligent, goal-directed behavior in a more general sense. By contrast, the cerebral cortex is justifiably regarded as indispensable for the performance of highly complex patterns of behavior, and the spatial distribution of its neurons is far from being an "orderly" one. To regard the cortex as a variety of reticular formation (Bishop, '58a) may become a controversial issue. However, there are undoubtedly more points of histological similarity between the cerebral cortex and the reticular formation than between the reticular formation and the cerebellar cortex. In any case, physiological observations such as those reported by Bard and Rioch ('37) suggest that the diffuse structural characteristics of the isodendric brain stem apparatus are compatible with functional mechanisms of considerable complexity.

One cannot overlook another possibility whereby both "diffuse" and "specific" functions could be performed by the same neuron or group of neurons. This could be the case if radically different properties were attributed to axo-dendritic and axo-somatic contacts. One could, for example, assume that the cell body and proximal parts of the dendrites are involved in the collection of relatively specific messages, well localized and probably coded in terms of an all-or-none type of response, whereas the more distal dendritic branches could be the recipients of information regarding the overall degree of activity in neighboring or more remote territories. On purely speculative grounds, one could also invoke some kind of non-synaptic influence or "field effect" exerted by the powerful bundles of passing fibers on the dendrites of these neurons which, as the Scheibels ('58) and

Valverde ('61) have shown, are preferentially perpendicular to such passing fibers. One may also wonder to what extent one can extrapolate to the central nervous system the phenomenon of ephaptic transmission (Jasper and Monnier, '38; Arvanitaki, '42a, b; Grundfest, '59). In any case, even if only the orthodox view that neurons act upon each other exclusively by means of synaptic contacts is accepted, there is always the possibility that axodendritic and axosomatic synapses might have different physiological properties. There appears to be some evidence that the dendritic membrane, in contrast to the somatic and axonal membranes, is capable of graded responses which propagate in a decremental manner and do not show an evident refractory period (Bishop, '56; 58b; Clare and Bishop, '55). According to Rall ('62), the mathematical analysis of the physiological properties of dendrites leads to the theoretical prediction of a functional difference between axo-dendritic and axo-somatic synapses: the former would be expected to dominate slow adjustments of the background excitation level, while the axosomatic synapses would be best suited for the triggering of neuronal discharges.

All of the above theories attribute to the dendrites some sort of tonic role determining the degree of "readiness" of soma and axon to originate the discharge. This notion must affect any physiological concept concerning the generalized neurons pervading the core of the brain stem with their overlapping dendritic fields. It could, for example, be suggested that, while the axosomatic synapses may provide specific messages, the axo-dendritic contacts could serve the purpose of keeping the neurons at appropriate levels of readiness to respond to specific information. In this manner, a compromise appears possible whereby the principle of "mass action" postulated by Lashley ('31) can be made compatible with the extensive list of neural structures known to have specific functions.

It has been pointed out (Herrick and Bishop, '58; Nauta, '63) that the great internuncial net must combine the fundamentally different properties of *analysis* and *integration*. One may now take one further step and attribute the segregation of the allodendritic and the idiodendritic cell groups to a progressive increase in *input discrimination* a possibility that would account for the sharp contrast between generalized and specialized dendritic patterns and their respective degree of cytological overlapping. The lesser degree of dendritic overlapping would appear to suggest allodendritic and idiodendritic cell groups as the more logical instrument for discriminative functions. By contrast, the considerable degree of dendritic overlapping in the isodendritic core would suggest that the neurons of this apparatus are less likely to be recipients of specialized types of input, although they may offer a more optimal substratum for integrative functions.

## ACKNOWLEDGMENTS

The authors wish to thank Mlle. A. Turcotte for technical contribution and Mlle. M. Lebelle for secretarial help. We are particularly indebted to Miss Cynthia J. Dunnan for her valuable criticism, editing advice and secretarial aid.

## LITERATURE CITED

Allen, W. F. 1932 Formatio reticularis and reticulospinal tract, their visceral functions and possible relationships to tonicity and tonic contractions. J. Wash. Acad. Scien., 22: 490–495.

Arvanitaki, A. 1942a Effects evoked in an axon by the activity of a contiguous one. J. Neurophysiol., 5: 89–108.

——— 1942b Interactions électriques entre deux cellules nerveuses contiguës. Arch. Int. Physiol., 52: 381–407.

Bard, P., and D. M. Rioch 1937 A study of four cats deprived of neocortex and additional portions of the forebrain. Bull. Johns Hopks. Hosp., 60: 73–148.

Baxter, D. W., and J. Olszewski 1955 Respiratory responses evoked by electrical stimulation of pons and mesencephalon. J. Neurophysiol., 18: 276–287.

Bishop, G. H. 1956 Natural history of the nerve impulse. Physiol. Rev., 36: 376–399.

——— 1958a The place of cortex in a reticular system. In: Reticular Formation of the Brain. Ed. by H. H. Jasper, L. D. Proctor, R. S. Knighton, W. C. Noshay, R. T. Costello. Little Brown and Co., Boston. Chapt. XX, 413–421.

——— 1958b The dendrite: receptive pole of the neuron. E. E. G. Clin. Neurophysiol., suppl. 10: 12–21.

Bodian, D. 1962 The generalized vertebrate neuron. Sci., 137: 323–326.

Bonvallet, M., A. Hugelin and P. Dell 1955 Sensibilité comparée du système réticulé activateur ascendant et du centre respiratoire au gaz du sang et à l'adrénaline. J. Physiol. Paris, 47: 651–654.

Brodal, A. 1957 The Reticular Formation of the Brain Stem. Anatomical Aspects and Functional Correlations. Oliver & Boyd, Edinburgh.

———— 1960 Fiber connections of the vestibular nuclei. In: G. L. Rasmussen and W. F. Windle (EDS.), Neural Mechanisms of the Auditory and Vestibular Systems. Charles C Thomas, Springfield, Ill.

Brodal, A., O. Pompeiano and F. Walberg 1962 The vestibular Connections and Functional Correlations. The Henderson Trust Lectures. Oliver and Boyd, Edinburgh.

Brodal, A., and G. Rossi 1955 Ascending fibers in brain stem reticular formation of cat. A.M.A. Arch. Neurol. and Psychiat., 74: 68–87.

Clare, M. H., and G. H. Bishop 1955 Facilitation and recruitment in dendrites. E.E.G. Clin. Neurophysiol., 7: 486–489.

Coghill, G. E. 1929 (reprinted 1964) Anatomy and the Problem of Behavior. Hafner, New York and London.

Dell, P. 1958 Humoral effects on the brain stem reticular formation. In: Reticular Formation of the Brain. Ed. by H. Jasper, L. D. Proctor, R. S. Knighton, W. C. Noshay, R. T. Costello. Little and Brown, Boston, Chapt. XVIII, 365–378.

French, J. D., F. K. von Amerongen and H. W. Magoun 1952 An activating system in brain stem of monkey. A.M.A. Arch. Neurol. Psychiat., 68: 577–589.

French, J. D., M. Verzeano and H. W. Magoun 1953 An extralemniscal sensory system in the brain. A.M.A. Arch. Neurol. Psychiat., 69: 505–518.

Gray, E. G. 1959 Electron microscopy of synaptic contacts on dendrite spines of the cerebral cortex. Nature, 183: 1592–1593.

Grundfest, H. 1959 Synaptic and ephaptic transmission. In: Handbook of Physiology, Vol. I. Neurophysiology. Ed. by J. Field, Williams and Wilkins, Baltimore, p. 147–194.

Harris, G. V. 1958 The reticular formation, stress and endocrine activity. In: Reticular Formation of the Brain. Ed. by H. H. Jasper, L. D. Proctor, R. S. Knighton, W. C. Noshay, R. T. Costello. Little and Brown, Boston. Chapt. IX, 207–218.

Herrick, C. Judson 1920 Irreversible differentiation and orthogenesis. Sci., suppl. 51: 621–625.

———— 1948 The brain of the tiger salamander. University of Chicago Press, Chicago, Ill.

Herrick, C. Judson and G. H. Bishop 1958 A comparative survey of the spinal lemniscus systems. In: Reticular Formation of the Brain. Ed. by H. H. Jasper, L. D. Proctor, R. S. Knighton, W. C. Noshay, R. T. Costello. Little and Brown, Boston. Chapt. XVII, 353–359.

Jasper, H. H. 1949 Diffuse projection systems: integration action of the thalamic reticular system. E E. G. and Clin. Neurophysiol., 1: 405–420.

———— 1954 Functional properties of the thalamic reticular system. In: Brain Mechanisms and Consciousness. Ed. by J. F. Delafresnaye. Blackwell Scient. Pub., Oxford, 374–401.

Jasper, H., and A. M. Monnier 1938 Transmission of excitation between excised non-myelinated nerves. An artificial synapse. J. Cell. and Comp. Physiol., 11: 259–277.

Kuypers, H. G. J. M., A. L. Hoffman and R. M. Beasley 1961 Distribution of cortical 'feedback' fibers in the nuclei cuneatus and gracilis. Proc. Soc. Exptl. Biol. Med., 108: 634–637.

Lashley, K. S. 1931 Mass action in cerebral function. Sci., 73: 245–254.

Leontovich, T. A., and G. P. Zhukova 1963 The specificity of the neuronal structure and topography of the reticular formation in the brain and spinal cord of carnivora. J. Comp. Neur., 121: 347–381.

Magoun, H. W. 1952 An ascending reticular activating system in the brain stem. A.M.A. Arch. Neurol. Psychiat., 67: 145–154.

———— 1954 The ascending reticular system and wakefulness. In: Brain Mechanisms and Consciousness. Ed. by J. F. Delafresnaye. Blackwell Scient. Pub., Oxford, 1–20.

Mannen, H. 1960 "Noyau fermé" et "noyau ouvert." Contribution à l'étude cytoarchitectonique du tronc cérébral envisagée du point de vue du mode d'arborisation dendritique. Arch. Ital. Biol., 98: 330–350.

Meesen, H., and J. Olszewski 1949 A cytoarchitectonic Atlas of the Rhombencephalon of the Rabbit. Karger, Basel and New York.

Morest, K. 1960 A study of the area postrema with Golgi methods. Am. J. Anat., 107: 291–303.

Morest, K., and J. Sutin 1961 Ascending pathways from an osmotically sensitive region of the medulla oblongata. Exper. Neurol., 4: 413–423.

Morison, R. S., and E. W. Dempsey 1942 A study of the thalamo-cortical relations. Amer. J. Physiol., 135: 281–292.

Moruzzi, G. 1954 The physiological properties of the brain stem reticular formation. In: Brain Mechanisms and Consciousness. Ed. J. F. Delafresnaye. Blackwell Scient. Publ. Oxford, 21–53.

Moruzzi, G., and H. W. Magoun 1949 Brain stem reticular formation and activation of the EEG. EEG. and Clin. Neurophysiol., 1: 455–473.

Nauta, W. J. H. 1963 Central nervous organization and endocrine motor systems: In: Advances in Neuroendocrinology. Ed. by A. V. Nalbandov. University of Illinois Press, Urbana, 5–20.

Nauta, W. J. H., and H. G. J. M. Kuypers 1958 Some ascending pathways in the brainstem reticular formation. In: Reticular Formation of the Brain. Ed. by H. H. Jasper, L. D. Proctor, R. S. Knighton, W. C. Noshay, R. T. Costello. Little and Brown, Boston, Chapt. I, 3–28.

Nauta, W. J. H., and W. R. Mehler 1966 Projections of the lentiform nucleus in the monkey. Brain Research, 1: 3–42.

Nauta, W. J. H., and D. G. Whitlock 1954 Anatomical analysis of the non-specific thalamic projection system. In: Brain Mechanisms and Consciousness. Ed. by J. F. Delafresnaye. Blackwell Scient. Pub., Oxford, 81–104.

Olszewski, J., and D. Baxter 1954 Cytoarchitecture of the human brain stem. S. Karger, New York and Basel.

Pitts, R. F., H. W. Magoun and S. W. Ranson 1939 Localization of the medullary respiratory centers in the cat. Am. J. Physiol., 126: 673–689.

Rall, W. 1959 Branching dendritic trees and motoneuron membrane resistivity. Exper. Neurol., 1: 491–527.

———— 1960 Membrane potential transients and membrane time constant of motoneurons. Exper. Neurol., 2: 503–532.

———— 1961 Theory of physiological properties of dendrites. Ann. N. Y. Acad. Sci., 96: 1071–1092.

———— 1962 Electrophysiology of a dendritic neuron model. Biophys. Journ. 2: suppl.: 145–167.

Ramón y Cajal, S. 1909a Histologie du Système Nerveux de l'homme et des vertébrés. Volumes I and II. Maloine, Paris, Reprinted: Consejo Superior de Investigaciones Cientificas, 1952, Madrid.

———— 1909b Contribucion al estudio de los ganglios de la substancia reticular del bulbo. Trab. Lab. Invest. Biol., Madrid., 7: 259–284.

Ramón-Moliner, E. 1958 A tungstate modification of the Golgi-Cox method. Stain Technol., 33: 19–29.

———— 1962a An attempt at classifying nerve cells on the basis of their dendritic patterns. J. Comp. Neur., 119: 211–227.

———— 1962b The distribution of non-specific dendritic patterns in the brain stem (abstract). Anat. Rec., 142: 270.

———— 1963 Dendritic patterns of the cat's brain stem. (Demonstration at the 76th Amer. Meet. of Anatomists). Anat. Rec., 145: 366.

Ramón-Moliner, E., M. Vane and G. V. Fletcher 1964 Basic dye counterstaining of sections impregnated by the Golgi-Cox method. Stain Technol., 39: 65–70.

Scheibel, M. D., and A. B. Scheibel 1958 Structural substrates for integrative patterns in the brain stem reticular core. In: Reticular Formation of the Brain. Ed. by H. H. Jasper, L. D. Proctor, R. S. Knighton, W. C. Noshay, R. T. Costello. Little and Brown, Boston, Chap. II, 31–53.

Scheibel, M. E., A. B. Scheibel, A. Mollica and G. Moruzzi 1955 Convergence and interaction of afferent impulses on single units of the reticular formation. J. Neurophysiol., 18: 309–331.

Sholl, D. A. 1953 Dendritic organization in the neurons of the visual and motor cortices of the cat. J. Anat. Lond., 87: 387–406.

Starzl, T. E., and H. W. Magoun 1951 Organization of the diffuse thalamic projection system. J. Neurophysiol., 14: 133–146.

Starzl, T. E., C. W. Taylor and H. W. Magoun 1951a Ascending conduction in the reticular activating system with special reference to the diencephalon. J. Neurophysiol., 14: 461–477.

———— 1951b Collateral afferent excitation of the reticular formation of the brain stem. J. Neurophysiol., 14: 479–496.

Starzl, T. E., and D. G. Whitlock 1952 Diffuse thalamic projection system in the monkey. J. Neurophysiol., 15: 449–468.

Szentágothai, J., B. Flerko, B. Mess and B. Halasz 1962 Hypothalamic control of the Anterior Pituitary. Akademi Kiado, Budapest.

Taber, E. 1961 The cytoarchitecture of the brain stem of the cat. I. Brain stem nuclei. J. Comp. Neur., 116: 27–69.

Tang, P. C. 1955 Levels of brain stem and diencephalon controlling micturition reflex. J. Neurophysiol., 18: 583–595.

Valverde, F. 1961 Reticular formation of the pons and medulla oblongata. A Golgi study. J. Comp. Neur., 116: 71–100.

Van der Loos, H. 1956 Une combinaison de deux vieilles méthodes histologiques pour le système nerveux central. Monatschr. f. Psychiat. u. Neurol., 132: 330–334.

Weiss, P. 1960 Modifiablity of the neuron. A.M.A. Arch. Neurol., 2: 595–599.

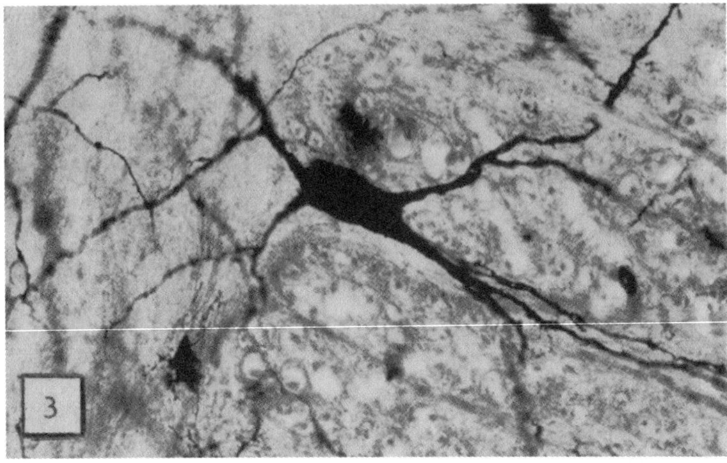

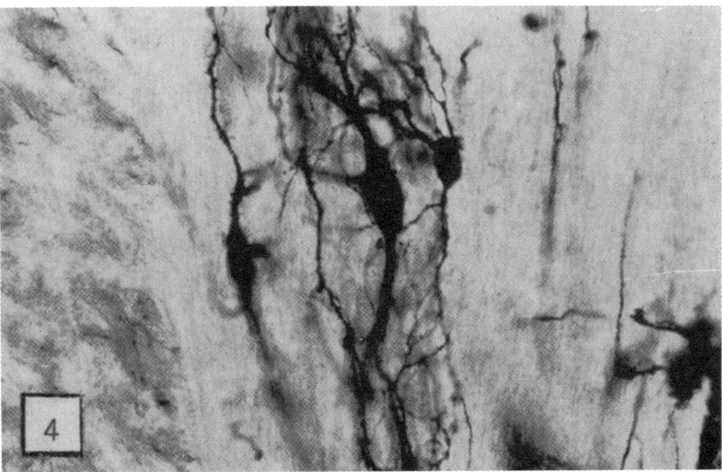

**EXPLANATION OF FIGURES**

3 A neuron from the region of the nucleus prepositus hypoglossi. Note the generalized dendritic pattern and the free intermingling of its dendrites with the transversally cut fibers of the medial longitudinal fasciculus ($\times$ 300).

4 Nucleus vestibularis interstitialis. (See Ramón y Cajal, '09a, vol. I, fig. 323.) Note the highly differentiated dendritic patterns and the slight tendency of the dendrites to mingle with passing fibers. In spite of its name, this cell group has probably little relationship with the vestibular nuclei properly speaking. The latter are characterized by generalized neurons ($\times$ 300).

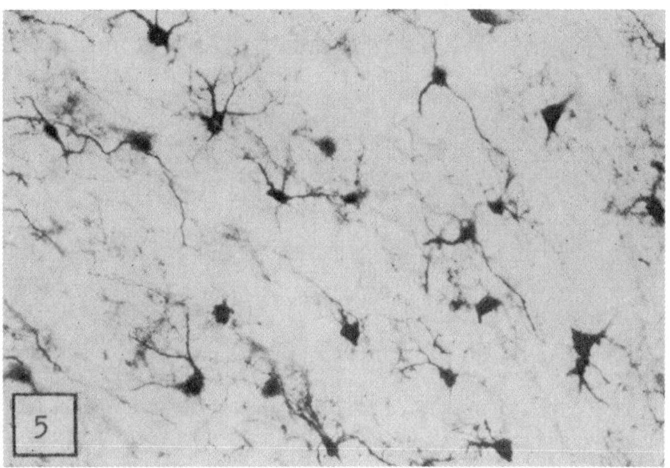

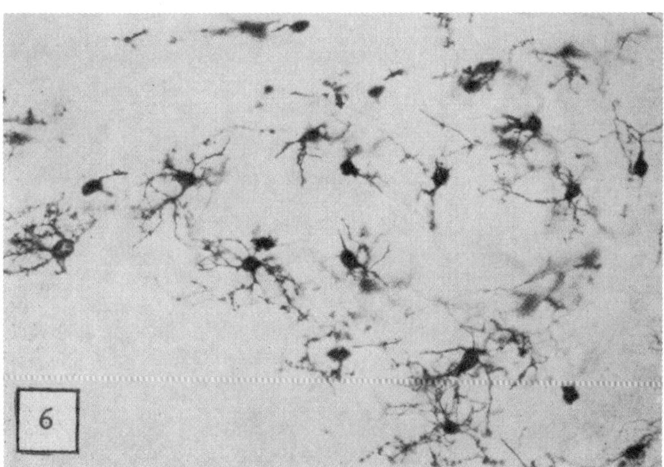

**EXPLANATION OF FIGURES**

5  Allodendritic neurons of the inferior colliculus. Note the dendritic tufts and the relatively small size of the over-all dendritic fields ($\times$ 100).

similar to those found in the inferior olivary nucleus ($\times$ 100).
6  Allodendritic neurons of the pontine nuclei. Note their tendency to have wavy dendrites

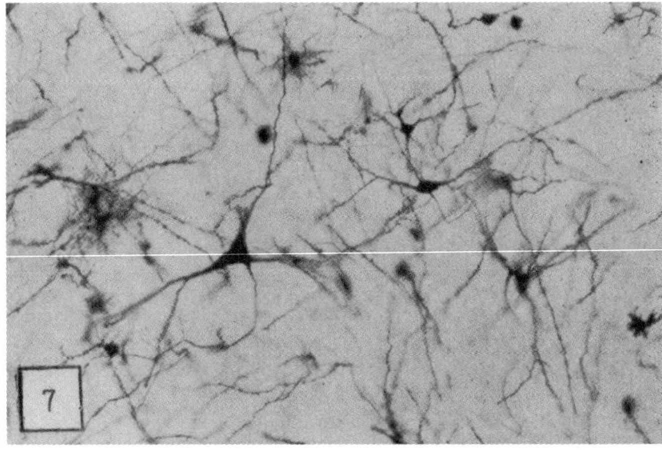

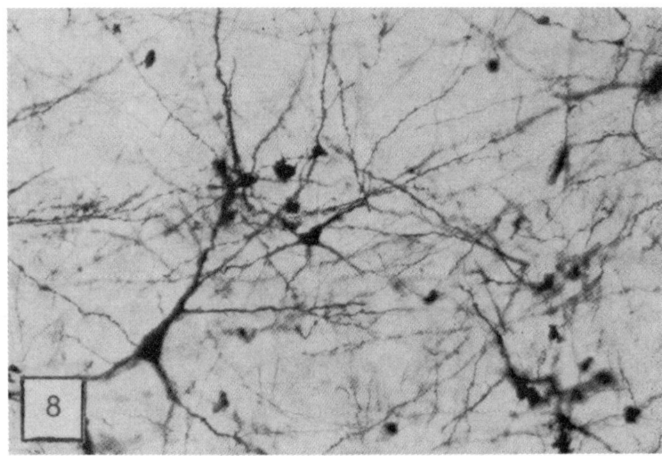

**EXPLANATION OF FIGURES**

7 Isodendritic neurons from the ventral tegmental area of Tsai ($\times$ 100).

8 Isodendritic neurons from the area centralis tegmenti pontis ($\times$ 100).

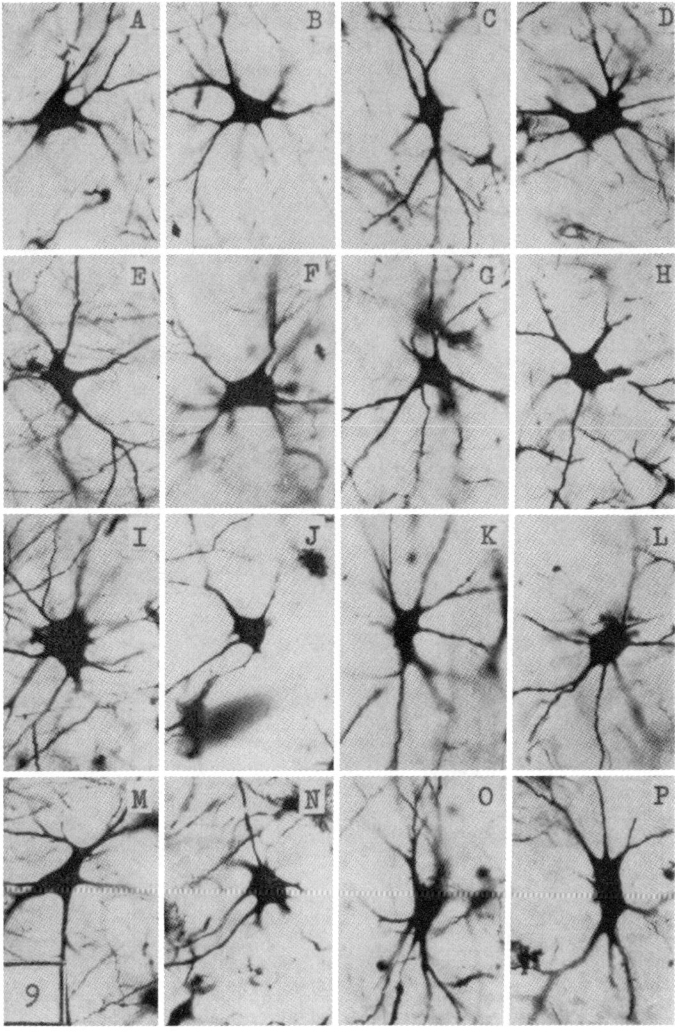

**EXPLANATION OF FIGURE**

9 Large isodendritic neurons sampled from a variety of regions of the cat's brain stem and assembled to demonstrate their similarity. A, B, and C, lower layers of the superior colliculus; D, E, and F, area centralis pontis oralis; G, H, and I, area gigantocellularis medullae oblongatae; J, nucleus oculomotorius; K, nucleus nervi abducentis; L, nucleus nervi trigemini motorius; M, and N, nucleus nervi facialis; O, nucleus vestibularis lateralis; P, nucleus vestibularis superior (× 100).

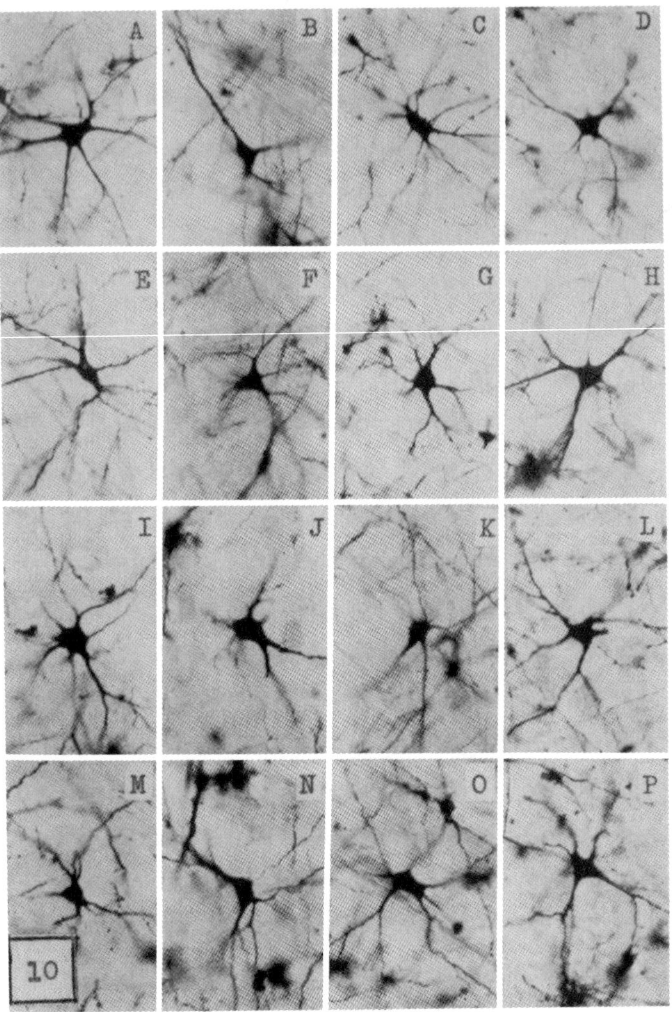

**EXPLANATION OF FIGURE**

10   Medium sized isodendritic neurons sampled from a variety of regions of the cat's brain stem and assembled to demonstrate their similarity. A, and B, area cuneiformis tegmenti; C, area subcuneiformis tegmenti; D, region of the nucleus coeruleus (subcoeruleus?); E, F, and G, area paralemniscalis dorsalis, H, superior colliculus (stratum lemnisci?); I, nucleus oculomotorius; J, nucleus nervi trochlearis; K, substantia nigra compacta; L, and M, area centralis medullae oblongatae; N, area parvicellularis medullae oblongatae; O, area medullae oblongatae centralis ("subnucleus" ventralis?); P, area paramediana medullae oblongatae (× 100).

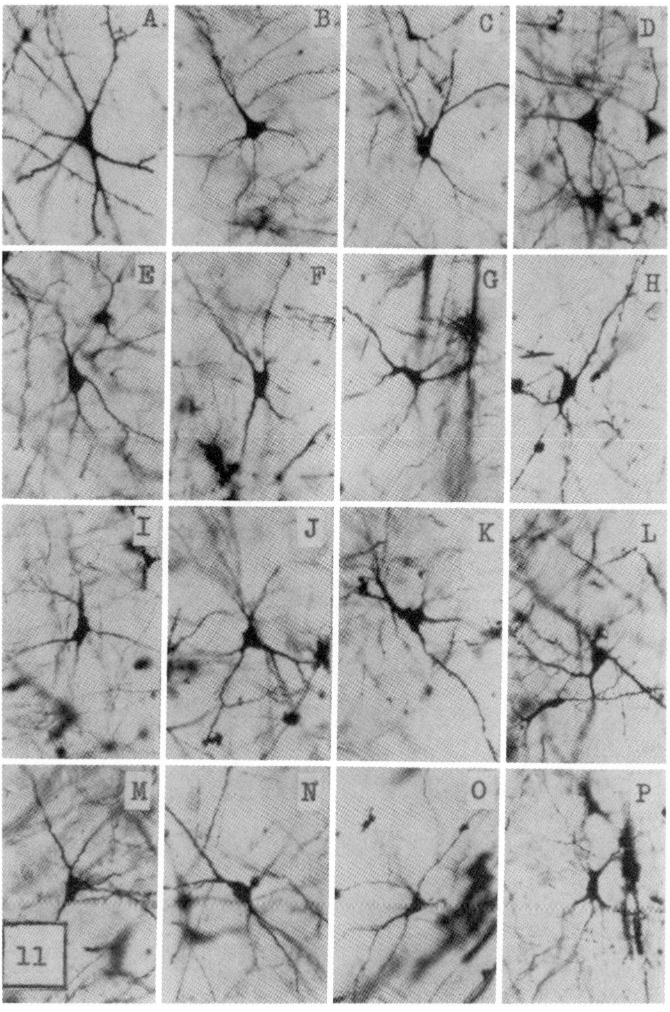

**EXPLANATION OF FIGURE**

11  Medium sized isodendritic neurons sampled from a variety of regions of the cat's brain stem. A, superior colliculus (stratum lemnisci?); B, area paralemniscalis tegmenti (cuneiformis?); C, substantia nigra lateralis; D, and E, area centralis superior tegmenti; F, and G, area tegmenti pontis centralis; H, and I, area parvicellularis medullae oblongata; J, and K, area gigantocellularis; L, and M, area paragigantocellularis dorsalis; N, and O, region of the nucleus raphae magnus; P, region of the nucleus prepositus hypoglossii ($\times$ 100).

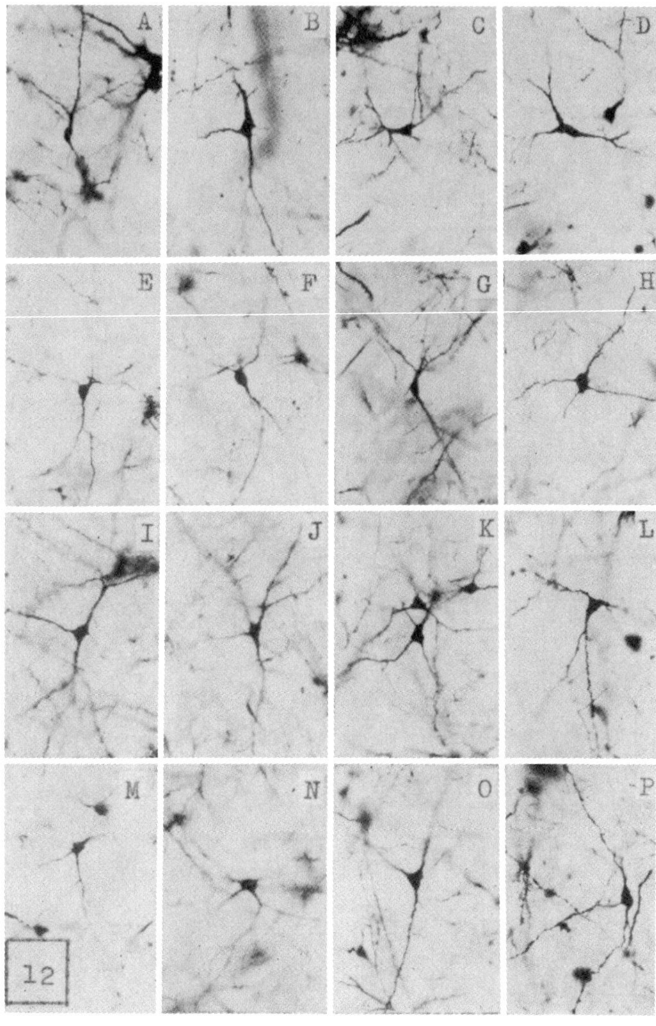

**EXPLANATION OF FIGURE**

12 Small isodendritic neurons sampled from a variety of regions of the cat's brain stem and assembled to demonstrate their similarity. A, area centralis medullae oblongatae; B, area gigantocellularis; C, and D, area parvocellularis; E, area peripeduncularis tegmenti; F, region of the "nucleus" sagulum; G, and H, substantia nigra lateralis; I, region of the "nucleus" interstitialis tegmenti; J, area paralemniscalis; K, and L, area centralis superior tegmenti; M, "nucleus" tegmenti pedunculopontinus, par oralis; N, and O, area subcuneiformis tegmenti; P, area parvocellularis medullae oblongatae ($\times$ 100).

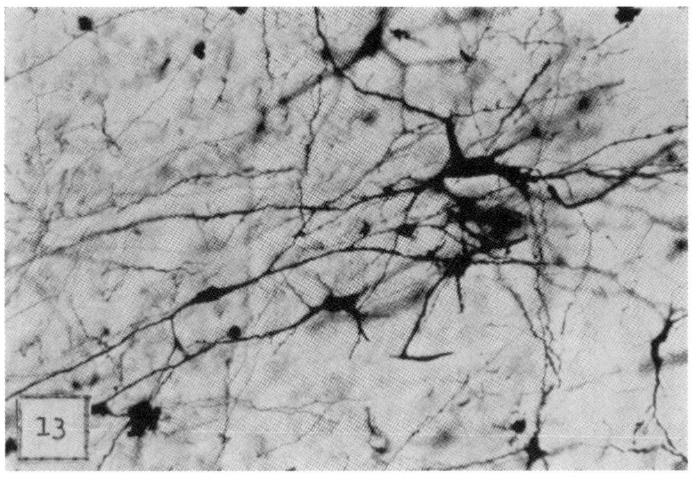

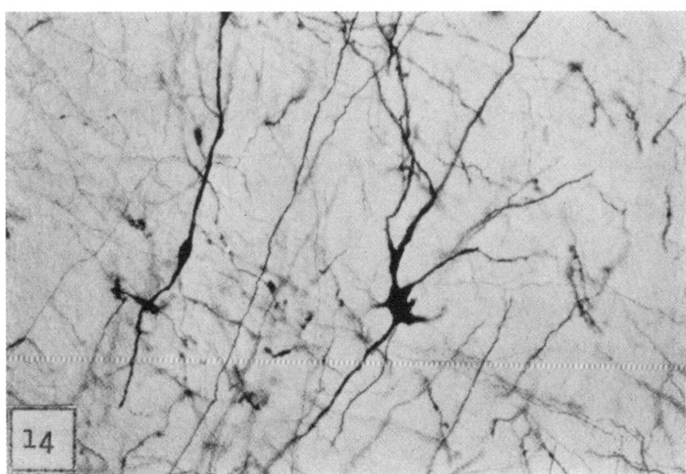

### EXPLANATION OF FIGURES

13  Area centralis tegmenti pontis of cat. Note that the generalized neurons present in this region resemble those of the tegmentum of the shark (fig. 14), while differing considerably from those of the allodendritic centers of the brain stem of the cat in figures 5 and 6 ($\times$ 100).

14  Tegmentum of the shark (Squalus acanthias). The coexistence of spindle-shaped and radiate neurons is a phenomenon found both in the tegmentum of the lower vertebrates and in the reticular formation (isodendritic core) of mammals ($\times$ 100).

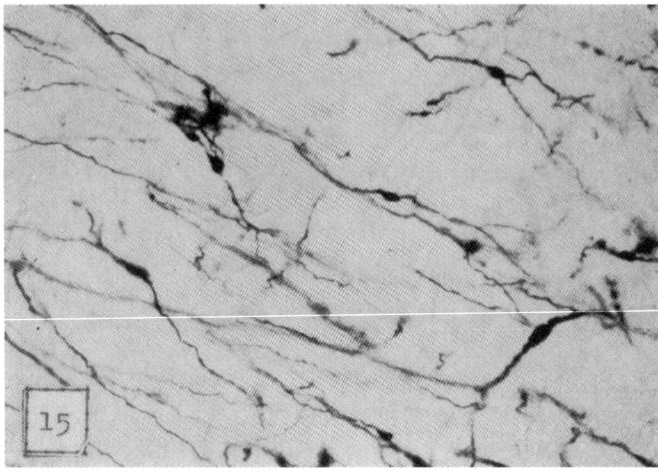

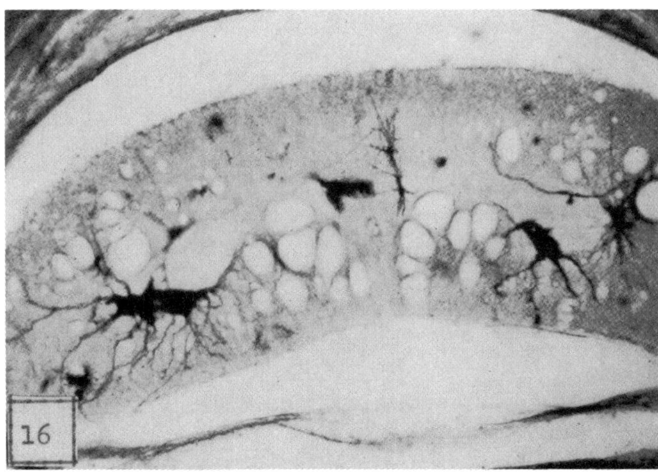

**EXPLANATION OF FIGURES**

15  Tegmentum of the shark (Squalus acanthias). These spindle-shaped neurons are relatively more numerous in lower vertebrates than in mammals. In the latter they are prevalent in the central gray substance and in other periventricular regions ($\times$ 100).

16  Transverse section of the spinal cord of lamprey (Petromyzon marinus). Note the free intermingling of the motoneuronal dendrites with the longitudinal fiber bundles, as well as the lack of sharp demarcation between gray and white matters. There is a certain similarity between these neurons and the ones shown in figure 3 ($\times$ 100).

Reprinted from
*Brain Research*
Elsevier Publishing Company, Amsterdam
Printed in the Netherlands

# THE HYPOTHALAMIC DISTRIBUTION OF THE STRIA TERMINALIS IN THE RAT

LENNART HEIMER AND WALLE J. H. NAUTA

*Department of Psychology, Massachusetts Institute of Technology, Cambridge, Mass. 02139 (U.S.A.)*

(Accepted October 20th, 1968)

INTRODUCTION

Despite numerous painstaking studies of normal brain material, and a considerable amount of experimental work by the use of anterograde degeneration methods, the hypothalamic distribution of the stria terminalis has remained a controversial subject. In their fundamental descriptive studies, Johnston[18], Berkelbach van der Sprenkel[3], Gurdjian[12], Humphrey[16] and others could identify two major contingents of amygdalo-hypothalamic stria fibers (Johnston's fourth and second subdivisions, synonymous with, respectively, the precommissural or supracommissural, and preoptic or postcommissural stria components of later workers), but the generally small fiber caliber precluded an accurate tracing of these fiber groups through the intricate hypothalamic fiber plexus. Even Cajal's[5] Golgi studies led no farther than the statement (Vol. II, p. 723) that the stria ('voie de projection de l'écorce temporale') distributes itself to the ventral hypothalamic region ('région sous-thalamique la plus inférieure') where its axons ramify in the suprachiasmatic region in an elongated cell territory ventral and caudal to the anterior commissure, the 'noyau interstitiel de la voie de projection de l'écorce temporale'. Elsewhere (Vol. II, p. 479), Cajal suggests the 'noyaux supérieur ou postérieur du tuber' (probably the dorsomedial nucleus and the ventral premammillary nucleus) as likely sites of stria fiber termination. Krieg[20], likewise working with the Golgi method, stated that the preoptic component disperses its fibers in the preoptic region, whereas the precommissural component becomes lost between and in the anterior hypothalamic area and the nucleus paraventricularis anterior. Krieg also notes some precommissural fibers extending into the infundibular region; although he cautiously avoided naming any particular cell group, his Fig. 18 suggests that he considered the ventromedial nucleus to be a likely recipient of such long stria fibers. Valverde[23] from a more recent Golgi study in the cat and rat describes the postcommissural stria fibers as spreading out underneath the anterior commissure, with only a few fibers continuing caudally in the medial forebrain bundle. The precommissural component, according to Valverde, passes caudally among many other fibers coursing sagittally through the lateral hypothalamic region.

Valverde's description can be interpreted as reinstating the view that the stria terminalis incorporates itself to a considerable extent into the medial forebrain bundle. This notion was expressed earlier by Johnston[18] and Berkelbach van der Sprenkel[3], but was questioned by Gurdjian[12] who described both the pre- and postcommissural stria components as passing along the medial side of the medial forebrain bundle, readily distinguishable from the latter in normal silver-impregnated material.

Normal descriptive methods other than the Golgi, while identifying the major subdivisions of the bundle and their general relationship to the preoptic area and hypothalamus, have likewise failed to clarify the details of the hypothalamic stria distribution. Most authors using such methods—notably the fully-impregnating silver techniques of Cajal—have therefore expressed themselves with much reserve, with the exception of Kappers *et al.*[19] who state (Vol. II, p. 1181): '... stria terminalis fibers, arising from the amygdaloid regions of the hemisphere, distribute to all of the major hypothalamic areas as far caudal as the premammillary nuclei, except the nucleus ovoideus, the nucleus filiformis, the nucleus tangentialis, the nucleus supraopticus and the periventricular grey'.

Experimental studies by Wallerian degeneration methods have led to views of the hypothalamic stria distribution no less conflicting than those derived from the study of normal material. The evidence from a Marchi study in the cat reported by Fox[10] suggested a distribution confined to the rostral half of the hypothalamus. By the use of the Glees method, however, Adey and Meyer[1] found degenerating boutons terminaux in the ventromedial hypothalamic nucleus of monkeys with unilateral surgical lesions of the amygdaloid region; the most remarkable aspect of their report is that this terminal degeneration appeared bilaterally in the nucleus. The interpretation of Adey and Meyer's findings was later placed in doubt by Cowan and Powell's[6] observations which suggested the presence of a Glees-positive material in the hypothalamus of normal monkeys, and thus raised the possibility of a normal phenomenon simulating axon degeneration. In later studies by the Nauta-Gygax method, no evidence of axon degeneration was found in the ventromedial nucleus of either side in the monkey[21] and cat[22]; the findings in both studies were interpreted as suggesting that the stria terminalis distributes its fibers largely to hypothalamic levels rostral to the ventromedial nucleus. By more recent observations in the rat, Cowan *et al.*[7] were led to a similar conclusion, but it is of interest that, in describing their case A8, these authors note the presence of some fine degenerative axon fragments along the lateral borders of the ventromedial and dorsomedial hypothalamic nuclei.

The study reported below was prompted by the recent development of a silver impregnation technique[9] affording a more nearly quantitative visualization of degenerating axon arborizations and their synaptic terminals than could be obtained by the use of the earlier Nauta–Gygax process. In sections successfully impregnated by this method, the synaptic fields of several degenerating fiber systems at least have been identified by the appearance of more or less densely massed black-impregnated spherular structures in the respective terminal regions. In case of doubt as to the true nature of such in themselves rather non-distinctive black corpuscles, electron microscopic examination of the region in question can be and has been used[14,15] to verify the presence of degenerating synaptic endstructures, for degenerating boutons terminaux can be identified electron microscopically independent of previous silver impregnation[2]. It was hoped that such combined light and electron microscopic analysis might disclose the synaptic relationships of the stria terminalis with the hypothalamus in more detail than appears to have been attainable in earlier studies.

MATERIALS AND METHODS

This report is based upon observations in 12 rats. In 6 of these, the stria terminalis was cut unilaterally under direct vision following exposure of the bundle by partial removal of overlying neocortex, hippocampus and fimbria fornicis. In 4 animals, extensive lesions were placed in the amygdalo-piriform region by suction under direct vision.

All but 2 of the animals were killed 3–6 days postoperatively by an overdose of pentobarbital, and perfused transcardially with 10% formalin. Following further fixation in 10% formalin for 1–6 weeks, the brains were immersed in 30% sucrose for 3 days, and thereafter sectioned on a freezing microtome at 25 $\mu$. Alternating series of sections were stained for degenerating axons by a uranyl–nitrate modification of the Nauta–Gygax method, for degenerating synaptic endstructures by the Fink–Heimer[9] procedures, and for cell bodies with cresylecht-violet.

The 2 remaining animals were used in an attempt to verify in the electron microscope some of the conclusions drawn from the light microscopic observations. These animals were sacrificed 3 and 5 days, respectively, after stria terminalis section by being perfused under surgical anesthesia with a glutaraldehyde–formaldehyde solution buffered at pH 7.2 (ref. 24). The brains were removed 1 h after perfusion, stored overnight in the same fixative, and cut frontally in slices of approximately 1 mm thickness. The slices were then trimmed down to the borders of the hypothalamic region, and the fairly large blocks thus obtained were postfixed in 2% osmium tetroxide in phosphate buffer for 2 h, dehydrated in ethanol, and embedded in an Epon–Araldite mixture. In order to identify the fields of terminal degeneration to be examined, and thus ensure optimal sampling for electron microscopy, semi-thin sections (5–7 $\mu$) taken at appropriate levels were stained by a relatively simple method allowing the silver impregnation of degenerating axon terminals in Epon–Araldite-embedded material[13]. The fields of terminal degeneration thus identified were carefully plotted in projection drawings of the sections, using blood vessels and other incidental profiles as landmarks. Once this was done, the face of the Epon–Araldite block was trimmed down to the area of immediate interest.

OBSERVATIONS

Either stria terminalis section or extensive removal of the amygdalo-piriform complex caused massive degeneration of hypothalamic stria terminalis fibers. The latter surgical procedure in addition caused widespread degeneration of the so-called ventral amygdalofugal pathway[7,11,21], part of which is distributed to the hypothalamus.

Our findings with respect to this shorter, transversally oriented fiber system from the amygdalo-piriform complex consistently indicated that its distribution does not, or only minimally, overlap that of the hypothalamic stria terminalis components. Unlike the latter, it appears to confine its termination to the lateral hypothalamic region; at any rate, none of the present cases provided evidence of ventral amygdalofugal fibers extending through the medial forebrain bundle into more medial hypothalamic zones as suggested by Szentágothai *et al.*[22] and Johnson[17].

A more likely risk of confusion arises in connection with the medial cortico-hypothalamic tract[12], Cajal's 'faisceau ammonique au tuber cinéréum'. Experimental evidence indicates that this component of the fornix system—striking in various rodents and the rabbit, but apparently absent or less well-developed in the cat and monkey—extends forward in the fimbria fornicis in or near the latter's free margin. It detaches itself from the fornix column below the interventricular foramen in compact fascicles which extend toward the infundibulum, following a periventricular path immediately medial to the supracommissural component of the stria terminalis, and locally nearly indistinguishable from the latter's most medial fascicles. As a result of the unavoidable involvement of the fimbria fornicis, all cases of stria terminalis section were complicated by a complete degeneration of the medial cortico-hypothalamic tract. Sporadic degenerating fibers, however, appeared in the fascicles even in the cases of amygdalectomy, presumably as a result of damage of the temporal tip of the hippocampus.

The following account of the hypothalamic stria terminalis distribution is based upon a careful subtractive comparison of the degeneration patterns elicited by each of the two different surgical procedures described above.

*Bed nucleus of the stria terminalis*

The most proximal and perhaps most massive termination of stria terminalis fibers takes place in the bundle's bed nucleus. In Nauta–Gygax sections this relationship is apparent by a feltwork of fragmented fine fibers extending throughout the nucleus. Sections impregnated by the Fink–Heimer procedures, by contrast, show the nucleus densely filled with small argyrophilic corpuscles (Fig. 1A–C), similar in appearance to those identified electron microscopically elsewhere in the brain as silver-impregnated degenerating preterminal axon branches and boutons terminaux[14].

Although decussating or commissural stria components have been left outside the scope of this paper, it is of interest to note the presence of some terminal degeneration also in the contralateral bed nucleus of the stria terminalis (Fig. 1B). This relatively sparse contralateral degeneration appeared in approximately equal quantity following either amygdalectomy or stria terminalis section, a finding suggesting that it is not attributable to a direct involvement of the ipsilateral bed nucleus in the sectioning of the stria. It therefore seems likely that it reflects the existence of stria fibers originating in the amygdala and distributed to the contralateral bed nucleus by way of the caudal part of the anterior commissure.

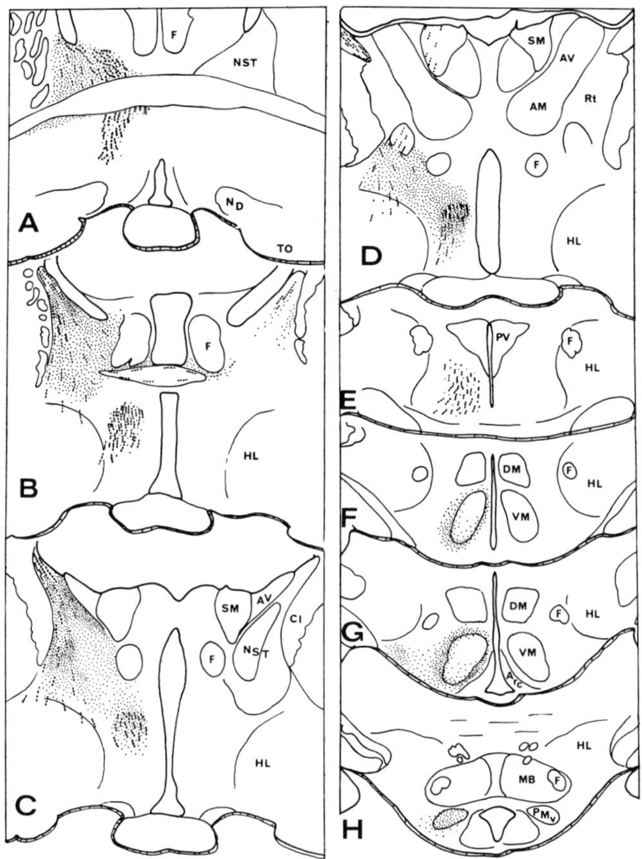


Fig. 1. Shown in this plate are the common features of the degeneration patterns elicited by amygdal-ectomy and stria terminalis section, respectively. Abbreviations: AM: anteromedial thalamic nucleus; Arc: arcuate nucleus; AV: anteroventral thalamic nucleus; CI: internal capsule; DM: dorsomedial nucleus; F: fornix; HL: lateral hypothalamic region; MB: mammillary body; ND: nucleus of diagonal band; NST: bed nucleus of stria terminalis; PMv: ventral premammillary nucleus: Rt: reticular nucleus of thalamus; TO: olfactory tubercle; VM: ventromedial nucleus; SM: stria medullaris; PV: paraventricular nucleus

*Pars preoptica (postcommissural component of the stria terminalis)*

Immediately caudal to the anterior commissure (Fig. 1B, C) the dense field of terminal degeneration filling the bed nucleus of the stria terminalis is seen to expand in the ventromedial and caudal direction into the anterior hypothalamus, extending between the medial forebrain bundle ventrally and laterally, the fornix dorsally (Fig. 1C, D). In its medial extent the field reaches near the third ventricle, where it is traversed by the degenerating fascicles of the precommissural stria component. The sharply demarcated region in question is largely confined to dorsal and medial hypothalamic regions and appears not, or only minimally, to involve the more poly-morphous cell region of the lateral hypothalamus traversed by the medial forebrain

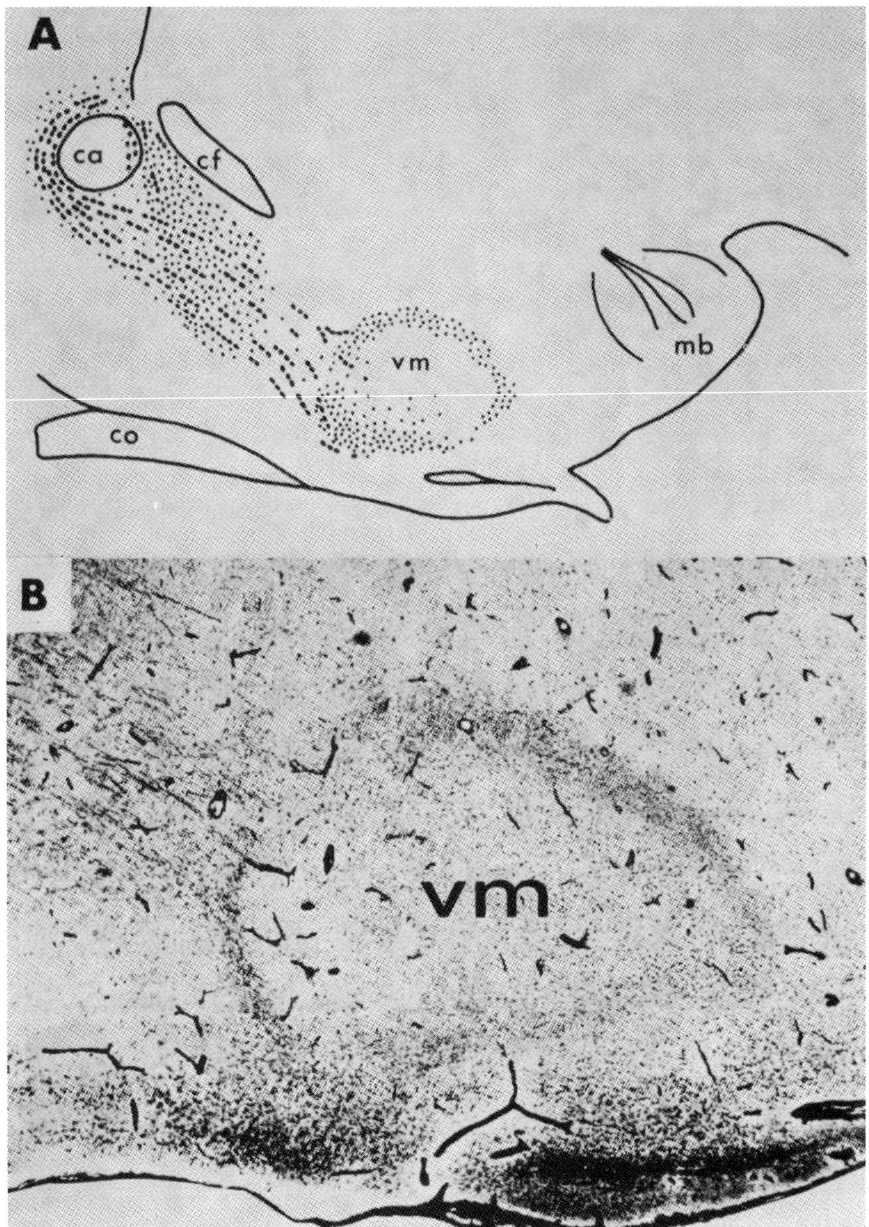


Fig. 2. The upper half of this figure (A) shows the course and terminal distribution of stria terminalis fibers in the hypothalamus, charted from a single paramedian section. B, Low-power photograph of the tuberal region in a nearby sagittal section, showing the zone of dense terminal degeneration surrounding the ventromedial nucleus. Fink–Heimer stain, procedure 2. Abbreviations: ca: anterior commissure; cf: fornix column; co: optic chiasm; mb: mammillary body; vm: ventromedial nucleus.

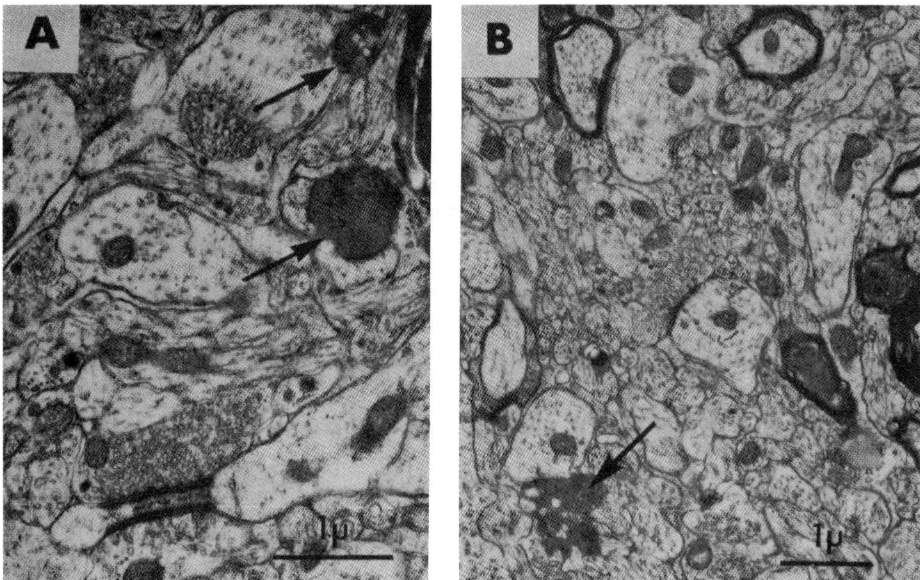

Fig. 3. Two electron micrographs showing degenerating boutons terminaux (arrows) in contact with dendrites in the rostral half of the anterior hypothalamic nucleus of a rat sacrificed 5 days after stria terminalis section. The location corresponds to the medial part of the area of terminal degeneration shown in Fig. 1D. The degenerating thin myelinated axons appearing near the right margin of frame B probably are constituents of the precommissural stria component traversing the region.

bundle. Its rather sharp caudal boundary lies slightly rostral to the level of the caudal border of the optic chiasm (Fig. 2).

That part at least of the argyrophilic corpuscles characterizing the region are indeed degenerating axon terminals was verified by the finding of degenerating boutons terminaux in electron micrographs (Fig. 3).

The topographic characteristics of this extensive field of stria fiber termination, and especially its uninterrupted continuity with the bed nucleus of the stria behind the anterior commissure, leave little doubt that it corresponds to the postcommissural (preoptic) component of the stria terminalis. It is remarkable that no indication of fiber degeneration was found in the region in sections impregnated by the Nauta–Gygax method. In Fink–Heimer sections, few degenerating fibers of passage could be recognized, although terminal degeneration appeared with clarity. Only in the lateral part of the field could a number of degenerating postcommissural fibers be identified. Some of these lateral—and apparently somewhat thicker—fibers were seen to enter the lateral hypothalamic zone in which they disappeared among the fascicles of the medial forebrain bundle (Fig. 1B–D).

*Pars precommissuralis (supracommissuralis)*

This fiber group, the most striking of the hypothalamic stria components, could be followed with comparative ease throughout its anterior hypothalamic trajectory.

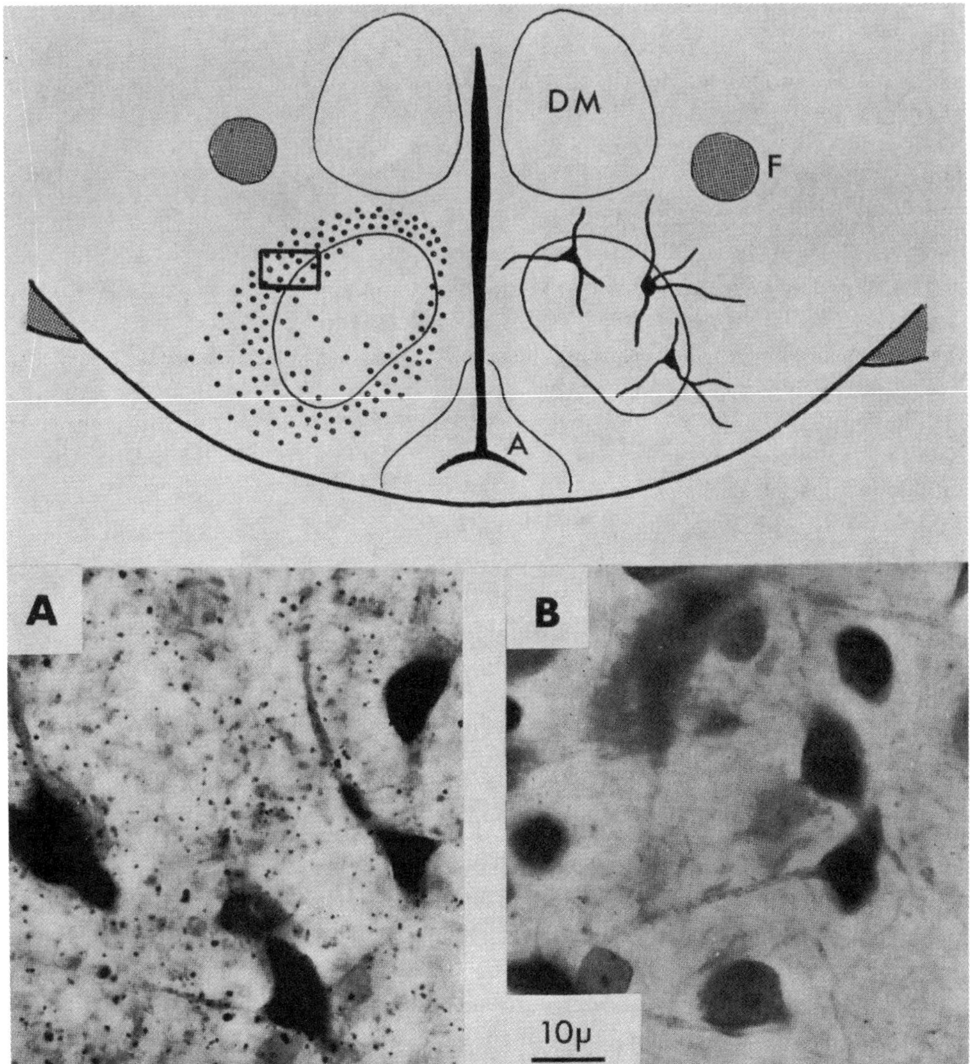


Fig. 4. Photograph A shows terminal degeneration near the lateral border of the ventromedial nucleus in a rat sacrificed 4 days after ipsilateral amygdalectomy. Approximate position of the field shown is indicated by rectangular frame in upper drawing. B shows the corresponding field on the opposite side. Fink–Heimer stain, procedure 1.

Apparently composed largely of the more medial fibers of the stria terminalis (Fig. 1A), it curves caudalward over the rostral aspect of the anterior commissure and follows a nearly sagittal course through the medial hypothalamic zone, organized in slender discrete fascicles. In the initial part of its course, corresponding to the medial preoptic region (Fig. 1A, B; Fig. 2) only sparse evidence of fiber termination appears among the fascicles of the supracommissural stria component. Slightly farther caudally, however, the fiber group passes through the rostral half of the anterior hypothalamic nucleus where it traverses the dense field of terminal degeneration interpreted in the foregoing description as representing the postcommissural stria component (Fig. 1C, D). It is impossible to ascertain from the material available whether any fibers of the supracommissural component actually terminate in this region. In their further course, leading through the caudal half of the anterior hypothalamic nucleus, the degenerating fascicles once again appear unaccompanied by any considerable amount of terminal degeneration (Fig. 1E).

Beyond this level, *i.e.* upon entering the tuberal region, the supracommissural component abruptly loses its fascicular composition and disperses its fibers into a fairly sharply defined zone of dense terminal degeneration. This zone surrounds the ventromedial nucleus and appears largely to correspond in location to the cell-poor zone separating the nucleus from adjacent cell groups in Nissl material. The shell-like disposition of this synaptic field surrounding the ventromedial nucleus is clearly visible both in frontal (Fig. 1F, G) and sagittal sections (Fig. 2); its great density is apparent from the low-power photograph shown in Fig. 2. As indicated in the same figures, a notable inward expansion of the field of terminal degeneration across the cytoarchitectonic limits of the ventromedial nucleus appears to involve only ventral and lateral regions of the nucleus; the density of this intranuclear axon degeneration nowhere approximates that of the field surrounding the nucleus. Fig. 1G, moreover, illustrates a pronounced lateral spread of the terminal degeneration along the ventral surface of the tuberal region. This wing-like expansion appeared to be more massive in the cases of stria terminalis section than following amygdalectomy, so that the possibility must be considered that it represents, in part at least, fibers of extra-amygdalar origin such as the medial cortico-hypothalamic tract or other components of the fornix system.

Fig. 4A illustrates the appearance of the terminal degeneration in silver sections under somewhat higher magnification. That many of the argyrophilic corpuscles shown in this figure indeed represent degenerating synaptic endstructures is documented by the electron micrographs showing several degenerating boutons terminaux in contact with dendritic profiles in animals sacrificed 3 and 5 days after ipsilateral stria terminalis section (Fig. 5).

The most caudal area of terminal degeneration identified in the present study corresponds to the ventral premammillary nucleus (Fig. 1H), and appears to be a caudal extension of the tuberal field of degenerating axon terminals. In contrast to the latter's largely perinuclear disposition, however, terminal degeneration at this more caudal level is densest within the cytoarchitectonic limits of the ventral premammillary nucleus.

No evidence of contralateral hypothalamic connections of the stria terminalis appeared in this study.

DISCUSSION

The findings here reported indicate that the hypothalamic termination of stria terminalis fibers is confined largely to three fairly circumscript regions. The most rostral of these, apparently the major termination site of the postcommissural stria component, extends from the bed nucleus of the stria terminalis ventrally and medial-

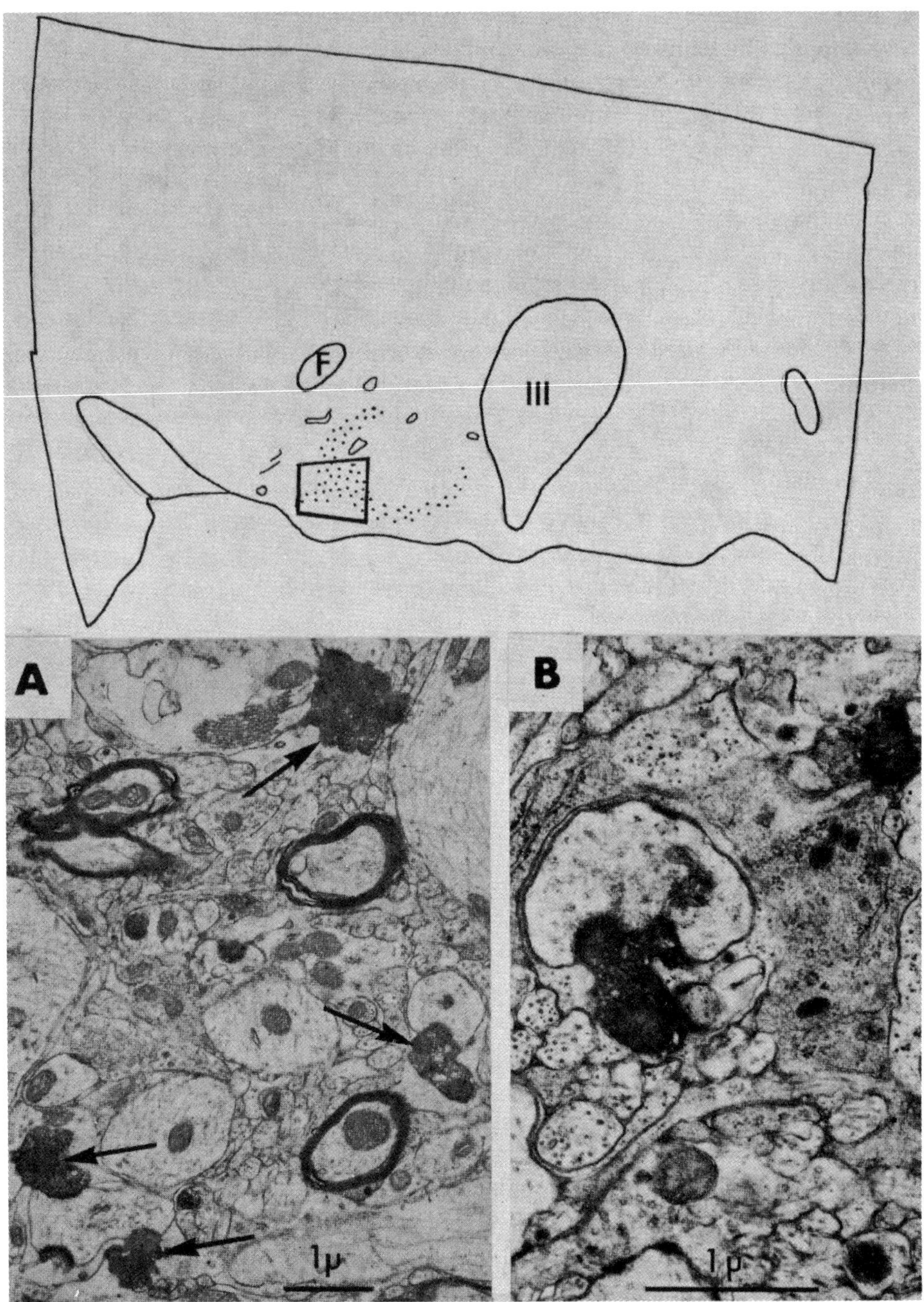



Fig. 5. Two electron micrographs showing degeneration of boutons terminaux (arrows) in synaptic contact with dendrites in the circum-ventromedial region, 3 days (A) and 5 days (B) after stria terminalis section. Upper drawing represents a silver-impregnated semi-thin section taken from an Epon–Araldite-embedded block and used for light microscopic charting of the field of terminal degeneration[13]. Electron micrograph B was made from a nearby ultrathin section cut after the block had been trimmed down to the small region outlined in heavy line in upper drawing.

ward into the rostral half of the anterior hypothalamic nucleus. The precommissural stria component, by contrast, appears to have its major synaptic distribution in two regions, *viz.* (1) the cell-poor zone surrounding the ventromedial nucleus, and, (2) farther caudally, in the ventral premammillary nucleus.

These observations fail to confirm earlier claims of a significant contribution of the stria terminalis to the medial forebrain bundle[3,18,23]. In considering this controversy it should, of course, be kept in mind that the medial forebrain bundle, instead of being a sharply defined, compact fiber group in the customary sense, is a loosely textured fiber system interspersed with grey matter and only vaguely delimited from adjoining groups of sagittally oriented intrinsic and extrinsic fibers. The denotation of the term, medial forebrain bundle, therefore cannot be made rigid. However, if the term is restricted to the prominent system of longitudinal fiber fascicles traversing the lateral preoptico-hypothalamic nucleus, it can be stated that the present study revealed no more than a few postcommissural stria terminalis fibers which entered the medial forebrain bundle. In agreement with Gurdjian's[12] description, the great majority of stria fibers were found to follow a course well medial to the latter fiber system, and their major terminal distribution areas nowhere appeared to spread within the cyto-architectonic limits of the lateral hypothalamic nucleus. It cannot be concluded, however, that the stria can affect the lateral hypothalamic region directly only by way of an apparently few postcommissural fibers which join the medial forebrain bundle. Suitable Golgi preparations reveal that numerous neurons of the hypothalamus have very long dendrites, and it is not difficult, particularly at more rostral levels, to find examples of lateral hypothalamic neurons with one or more dendrites protruding medially into areas containing synaptic terminations of the stria terminalis. Whether such extraneous dendrites indeed engage in synaptic contacts with stria fibers is a question that can be approached only by such exacting methods as the electron microscopic analysis of Golgi-impregnated experimental material introduced by Blackstad[4]. Pending such more definitive studies, it should be kept in mind that the stria terminalis may have axo-dendritic contacts with hypothalamic neurons whose cell bodies are located well outside the generally sharp limits of its synaptic fields.

The most striking of the latter areas, from a topographic point of view, is the dense zone of stria terminalis endings surrounding the ventromedial nucleus. Although much less densely populated than the ventromedial nucleus proper, and hence often described as a cell-poor zone, the region in question contains a considerable number of neuronal perikarya. More striking, however, is the large number of dendrites radiating into the zone, mostly from the ventromedial nucleus but to a lesser extent also from neighboring cell groups such as the arcuate and dorsomedial nuclei (Fig. 6). It appears quite likely from these relationships that stria terminalis fibers penetrating into the tuberal region terminate predominantly on outlying dendrites of ventromedial neurons, in smaller quantity also within the limits of the nucleus—and, hence, possibly in part at least in synaptic contact with cell bodies—and on dendrites protruding from adjoining tuberal cell groups. The present evidence suggesting the ventromedial nucleus as the dominant recipient of stria terminalis fibers distributed to the tuberal region is compatible with the recent observation of Dreifuss *et al.*[8] that slow potentials

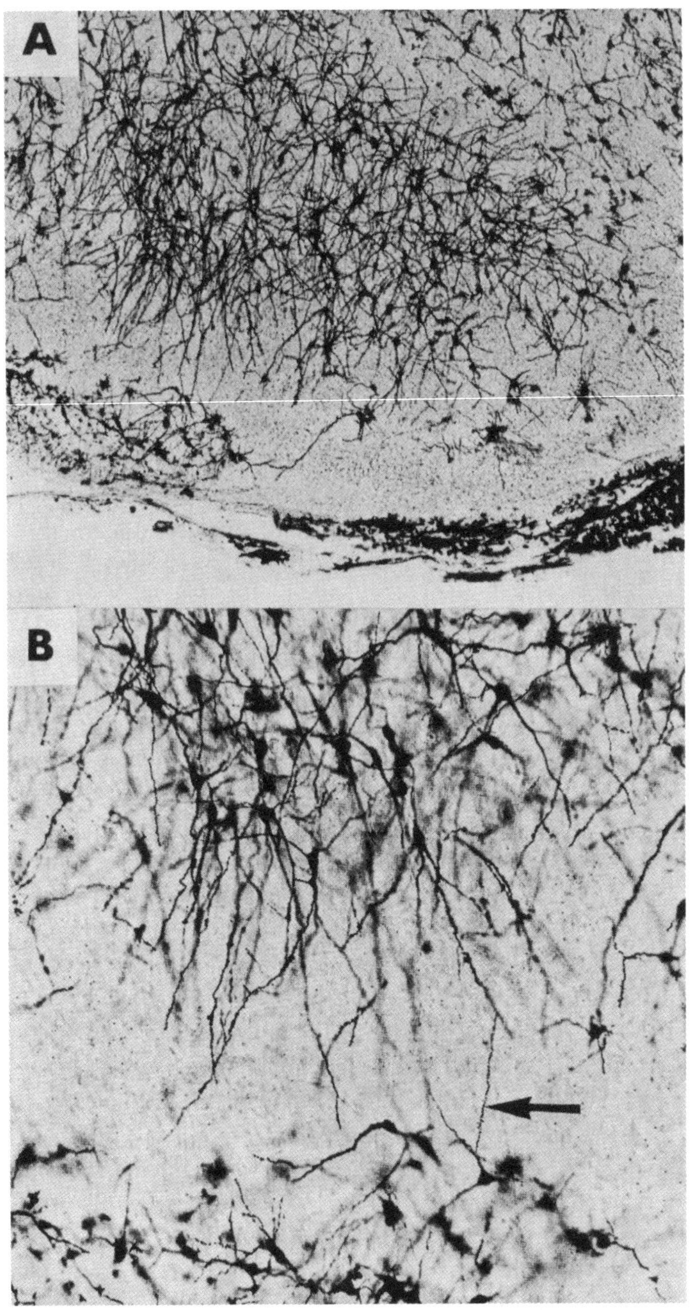

Fig. 6. A, The ventromedial nucleus as appearing in a sagittal section through the tuberal region. Note the large number of dendrites radiating from the nucleus into the surrounding zone. Golgi–Cox method. B, Detail of A at higher magnification. Near lower margin are cells of the arcuate nucleus, one of which (arrow) shows a dorsally oriented dendrite which extends into the clear zone between the two cell groups.

and unit responses evoked in the region by impulses traveling over the stria terminalis are virtually confined to the area of the ventromedial nucleus. The same study revealed that these responses differ fundamentally from those elicited in the same region by stimulation of the ventral amygdalofugal pathway, a connection presumably involving a synaptic relay in the lateral hypothalamic region. An analysis of the differences, if any, in the mode of termination of these two major amygdalofugal conduction systems converging upon the ventromedial nucleus would be of obvious interest.

## SUMMARY

The hypothalamic distribution of the stria terminalis was studied by the Fink–Heimer methods and electron microscopy in rats sacrificed 3–6 days after unilateral amygdalectomy or stria terminalis section. Each of the two major hypothalamic stria components was found to be distributed to fairly sharply defined regions, as follows: (1) The postcommissural component terminates massively in the rostral half of the anterior hypothalamic nucleus; an apparently small number of its fibers join the medial forebrain bundle. (2) The supracommissural component terminates predominantly in a dense synaptic zone surrounding the ventromedial nucleus and containing a massive plexus of dendrites protruding from this nucleus, in smaller quantity also in lateral and ventral parts of the ventromedial nucleus proper and in the ventral premammillary nucleus. The prevalence in the hypothalamus of neurons with long dendrites suggests that the stria terminalis may have synaptic contacts not only with perikarya and dendrites of neurons lying within its synaptic fields, but also with dendrites of neurons whose cell bodies lie well outside these fields.

## ACKNOWLEDGEMENTS

This study was supported by U.S. Public Health Service Research Grants NB 06542 and NB 08022, and by Career Development Award MH 40516 from the National Institute of Mental Health to L.H.

The authors gratefully acknowledge the technical assistance of Mrs. Anneliese Hanf and Mr. Harry Greenlaw.

## REFERENCES

1 ADEY, W. R., AND MEYER, M., Hippocampal and hypothalamic connexions of the temporal lobe in the monkey, *Brain*, 75 (1952) 358–383.
2 ALKSNE, J. F., BLACKSTAD, TH. W., WALBERG, F., AND WHITE, L. E., Electron microscopy of axon degeneration: a valuable tool in experimental neuroanatomy, *Ergebn. Anat. Entwickl.-Gesch.*, 39, Heft 1 (1966) 6–31.
3 BERKELBACH VAN DER SPRENKEL, H. J., Stria terminalis and amygdala in the brain of the opossum (*Didelphys virginiana*), *J. comp. Neurol.*, 42 (1926) 211–254.
4 BLACKSTAD, TH. W., Mapping of experimental axon degeneration by electron microscopy of Golgi preparations, *Z. Zellforsch.*, 67 (1965) 819–834.
5 CAJAL, S. RAMÓNY , *Histologie du Système Nerveux de l'Homme et des Vertébrés, Tôme II*, Maloines, Paris, 1911.

6 COWAN, M. W., AND POWELL, T. P. S., A note on terminal degeneration in the hypothalamus, *J. Anat. (Lond.)*, 90 (1956) 188–192.

7 COWAN, W. M., RAISMAN, G., AND POWELL, T. P. S., The connexions of the amygdala, *J. Neurol. Neurosurg. Psychiat.*, 28 (1965) 137–151.

8 DREIFUSS, J. J., MURPHY, J. T., AND GLOOR, P., Contrasting effects of two identified amygdaloid pathways on single hypothalamic neurons, *J. Neurophysiol.*, 31 (1968) 237–248.

9 FINK, R. P., AND HEIMER, L., Two methods for selective silver impregnation of degenerating axons and their synaptic endings in the central nervous system, *Brain Research*, 4 (1967) 369–374.

10 FOX, C. A., The stria terminalis, longitudinal association bundle and precommissural fornix in the cat, *J. comp. Neurol.*, 79 (1943) 277–295.

11 GLOOR, P., Electrophysiological studies on the connections of the amygdaloid nucleus in the cat. I. The neuronal organization of the amygdaloid projection system, *Electroenceph. clin. Neurophysiol.*, 7 (1955) 223–242.

12 GURDJIAN, E. S., Olfactory connections in the albino rat, with special reference to the stria medullaris and the anterior commissure, *J. comp. Neurol.*, 38 (1925) 127–163.

13 HEIMER, L., Silver impregnation of degenerating axons and their terminals in Epon–Araldite sections, *Brain Research*, 12 (1969) 246–249.

14 HEIMER, L., AND PETERS, A., An electron microscope study of a silver stain for degenerating boutons, *Brain Research*, 8 (1968) 337–346.

15 HEIMER, L., AND WALL, P. D., The dorsal root distribution to the substantia gelatinosa of the rat with a note on the distribution in the cat, *Exp. Brain Res.*, 6 (1968) 89–99.

16 HUMPHREY, T., The telencephalon of the bat. The noncortical nuclear masses and certain pertinent fiber connections, *J. comp. Neurol.*, 65 (1936) 603–711.

17 JOHNSON, T. N., An experimental study of the fornix and hypothalamo-tegmental tracts in the cat, *J. comp. Neurol.*, 125 (1965) 29–39.

18 JOHNSTON, J. B., Further contributions to the study of the evolution of the forebrain, *J. comp. Neurol.*, 35 (1923) 337–481.

19 KAPPERS, C. U. A., HUBER, C. C., AND CROSBY, E. C., *The Comparative Anatomy of the Nervous System of Vertebrates, Including Man*, MacMillan, New York, 1936.

20 KRIEG, W. J. S., The hypothalamus of the albino rat, *J. comp. Neurol.*, 55 (1932) 19–89.

21 NAUTA, W. J. H., Fibre degeneration following lesions of the amygdaloid complex in the monkey, *J. Anat. (Lond.)*, 95 (1961) 515–531.

22 SZENTÁGOTHAI, J., FLERKÓ, B., MESS, B., AND HALÁSZ, B., *Hypothalamic Control of the Anterior Pituitary*, Akad. Kiadó, Budapest, 1962 19–105.

23 VALVERDE, F., *Studies on the Piriform Lobe*, Harvard Univ. Press, Cambridge, Mass., 1965, 128 pp.

24 VAUGHN, J. E., AND PETERS, A., Aldehyde fixation of nerve fibers, *J. Anat. (Lond.)*, 100 (1966) 687

Reprinted from THE JOURNAL OF COMPARATIVE NEUROLOGY
Vol. 173, No. 1, May 1, 1977  © The Wistar Institute Press 1977

# Afferent Connections of the Habenular Nuclei in the Rat. A Horseradish Peroxidase Study, with a Note on the Fiber-of-Passage Problem[1]

MILES HERKENHAM[2] AND WALLE J. H. NAUTA
*Department of Psychology, Massachusetts Institute of Technology, Cambridge, Massachusetts 02139*

ABSTRACT    The afferent connections of the habenular complex in the rat were examined by injecting horseradish peroxidase (HRP) into discrete portions of the habenular nuclei by microelectrophoresis.

1. HRP deposits confined to the lateral half of the lateral habenular nucleus labeled a multitude of cells in the entopeduncular nucleus. Numerous labeled cells also appeared in such cases in the lateral hypothalamus, indicating that the lateral habenular nucleus is a major convergence point of projections from these otherwise apparently quite separate cell regions. Moderate-to-small numbers of labeled cells were also found in the nuclei of the diagonal band, substantia innominata, lateral preoptic area and, more caudally, in the ventral tegmental area, the region of the mesencephalic raphe, and the central gray substance.

2. HRP injected into the medial part of the lateral habenular nucleus labeled cells in the same regions, but more in the diagonal band and fewer in the entopeduncular nucleus than were labeled by more lateral injections. The contrast suggests that the projections from the basal forebrain and entopeduncular nucleus to the lateral habenular nucleus are somewhat topographically organized.

3. Injections of the medial habenular nucleus labeled an abundance of cells in the posterior parts of the supracommissural septum, but also a small number of cells in the diagonal band and mesencephalic raphe.

4. HRP injected into the stria medullaris labeled cells in all of the afore-mentioned areas and, in addition, cells in several olfactory structures, confirming that HRP may be taken up by fibers of passage and label their cells of origin, and suggesting that olfactory structures contribute fibers to the stria medullaris that do not terminate in the habenula.

The habenular nuclei together with their associated fiber systems, the stria medullaris and fasciculus retroflexus (habenulo-interpeduncular tract), compose a conspicuous link between fore- and midbrain (Nauta, '58). However, until recently little was known about the sources of habenular afferents, largely because of the long-standing lack of adequate retrograde-tracing methods. Early claims based on the study of normal material placed the origin of habenulo-petal fibers in a host of limbic and olfactory forebrain structures (Gurdjian, '25; Marburg, '44; Ariëns Kappers et al., '60). The results of later studies by an-terograde fiber-degeneration methods likewise suggested that habenular afferents originate in various limbic and olfactory components of the forebrain and midbrain tegmentum, in particular the septum (Nauta, '56; Guillery, '59; Cragg, '61; Raisman, '66; Powell, '68), olfactory tubercle and piriform cortex (Kusama and Hagino, '61; Ban and Zyo, '62; Powell et al., '65; Price and Powell, '71), lateral preoptic and hypothalamic areas (Nauta, '58; Cragg, '61;

---

[1] Supported by PHS Grants NS 06542 and MH 25515 and NIH Fellowship NS 02101.
[2] Present address: Laboratory of Neurophysiology, National Institute of Mental Health, Bethesda, Maryland 20014.

Wolf and Sutin, '66; Mizuno et al., '69; Huang and Mogenson, '72), amygdala (Laursen, '55; Cragg, '61; Kusama and Hagino, '61), anterior thalamus (Cragg, '61) and the paramedian "limbic midbrain area" (Nauta, '58; Nauta and Kuypers, '58), in particular the median and dorsal raphe nuclei (Conrad et al., '74; Bobillier et al., '75; Pierce et al., '76) and the interpeduncular nucleus (Cragg, '61; Massopust and Thompson, '62; Mitchell, '63; Smaha and Kaelber, '73).

It must be added that not all of these source-identifications have remained uncontested. For example, the notion of a close association of the habenular with the olfactory system (Herrick, '48; Crosby et al., '62) has been challenged by the results of an electron-microscopical analysis which suggested that the stria-medullaris fibers that degenerate following lesions in the olfactory tubercle or piriform cortex merely pass through the rostral part of the habenula, and actually terminate in the subjacent mediodorsal nucleus (Heimer, '72).

On the other hand, a recent study by the autoradiographic fiber-tracing method has demonstrated a previously rarely suspected but quite substantial afferent connection of the habenula with the globus pallidus (Nauta, '74). This finding makes a second modification necessary in the existing concepts of habenular circuitry: evidently, the habenular complex receives inputs of both striatal and limbic origin, and therefore comes to appear as one of the few brain regions in which these two major parallel systems are known to converge.

The present study by the method of retrograde cell-marking with horseradish-peroxidase (HRP) (Kristensson and Olsson, '71; LaVail and LaVail, '72; LaVail et al., '73) was undertaken in an attempt to obtain more accurate and direct identification of the various sources of habenular afferents. Such information is more difficult, and often impossible, to obtain by anterograde tracing methods, especially in the case of afferrent fibers originating from cytoarchi-

tecturally heterogeneous regions such as the basal forebrain.

## MATERIALS AND METHODS

Small unilateral deposits of HRP were placed in either the medial or lateral habenular nucleus in 15 adult albino rats of both sexes (Charles River Laboratories). In order to confine the retrograde marker to each one of these small structures the injections were made by microelectrophoresis (Graybiel and Devor, '74) from glass micropipettes (internal tip diameter 10-20 $\mu$m) filled with a 13% solution of HRP (Sigma, type VI) in Tris-HCl buffer at pH 8.6. The driving current was delivered by a constant-current source (Midgard Electronics, Watertown, Massachusetts at a pulse rate of seven seconds on, seven seconds off, for a total duration of 12 to 20 minutes. The "on" pulse was a constant positive 2 $\mu$A current passed through the solution via a silver wire. The placement of the pipette tip was guided either by the coordinates supplied by König and Klippel's ('63) stereotaxic atlas, or in five cases, by direct vision with the aid of a surgical microscope after aspiration of the overlying cortex and hippocampus.

Twenty-four hours postoperatively (3 days in one case) the rats were deeply anaesthetized and perfused through the heart with physiological saline followed immediately by a freshly prepared solution of 1.5% paraformaldehyde, 2% glutaraldehyde and 5% sucrose in 0.1 M phosphate buffer of pH 7.4, cooled to 4°C. The brains were removed, placed in the same fixative for 10 to 12 hours, stored overnight in 0.1 M phosphate buffer containing 10% sucrose, and sectioned on a freezing microtome at 50 $\mu$m in a frontal plane corresponding to that of König and Klippel's atlas. The sections were collected serially into six compartments containing Tris-buffer and each compartment was then immediately processed according to the method of Graham and Karnovsky ('66). The sections from every compartment were mounted on gelatin-coated slides and counterstained with thionin. Each section was microscopically examined for the presence and location of HRP-positive neuronal perikarya, using both conventional and darkfield illumination. Labelled cells were counted in all sections. The location of the injection site and that of labeled neurons in representative afferent regions were charted onto projection drawings of selected sections. These charts provided the illustrations presented in the text.

RESULTS

The habenular nuclei proved to be targets remarkably suitable for small, localized microelectrophoretic label injection. Especially at more caudal levels, the accumulation of fibers that form the fasciculus retroflexus provides an effective ventral boundary favoring the confinement of HRP within the borders of the habenula. Since at such levels the habenular complex is also largest, the most successfully restricted injections were located in the caudal one-third of the habenula. Usually there was little or no invasion of HRP into the adjacent mediodorsal nucleus even though many of the injection sites were rather long and spindle-shaped, a consequence of extending the initial pipette penetration beyond the calculated depth in order to insure that the pial-glial barrier over the stria medullaris was pierced. Probably owing to the small tip-diameter, inadvertent spillage of HRP into the cortex and hippocampus along the pipette track rarely presented a problem in this study; moreover, it naturally was entirely obviated in the five cases in which the overlying structures had been removed by aspiration.

## 1. Lateral habenular nucleus

Ten animals received injections in the lateral habenular nucleus, but one case particularly clearly illustrates the sources of afferents to this structure. The spindle-shaped injection in this case (Hb-14) was located in the lateral and caudal corner of the nucleus (fig. 1A). The area of densest deposit was approximately 250 $\mu$m in diameter; a much lighter brown halo of HRP

reaction product filled the lateral half of the nucleus. The track left in the cortex and hippocampus showed no significant traces of HRP.

Numerus HRP-positive cells appeared in this case in pallidal and other regions of the basal forebrain, including the nuclei of the diagonal band and of the substantia innominata, the lateral preoptic and hypothalamic areas and, in especially large numbers, the entopeduncular nucleus (fig. 2). The majority of these cells were located ipsilateral to the side of the injection, but some cells were also found in each of the corresponding contralateral regions.

Proceeding in the rostrocaudal direction, the first HRP-marked cells were found scattered throughout the areas that comprise the magnocellular basal forebrain (Divac, '75). These cells were typically large and multipolar or spindle-shaped. Such cells appeared sporadically in the nucleus of the diagonal band, and in somewhat larger number in the region immediately ventral to the temporal limb of the anterior commissure; Heimer and Wilson ('75) have identified this component of the substantia innominata, whose large multipolar and pear-shaped cells are placed lateral to the subcommissural part of the bed nucleus of the stria terminalis, as a rostroventral extension of the globus pallidus (ventral pallidum, *vp* in fig. 2a). Marked cells became more numerous at rostral levels of the preoptic area (figs. 2b,c). At such levels HRP-positive cells were scattered across the lateral preoptic area and the ventral pallidum immediately beneath the anterior commissure. A few additional marked cells were located in the caudal part of an oval-shaped grouping of large, rounded, deeply staining multipolar cells within the substantia innominata (Cragg, '61). This cell group has been interpreted as part of the nucleus of the horizontal limb of the diagonal band by Price and Powell ('70a) and was called magnocellular preoptic nucleus by Swanson ('76), but its lateral position has suggested to some that it may correspond to the nucleus basalis (nb) of primates (Nauta

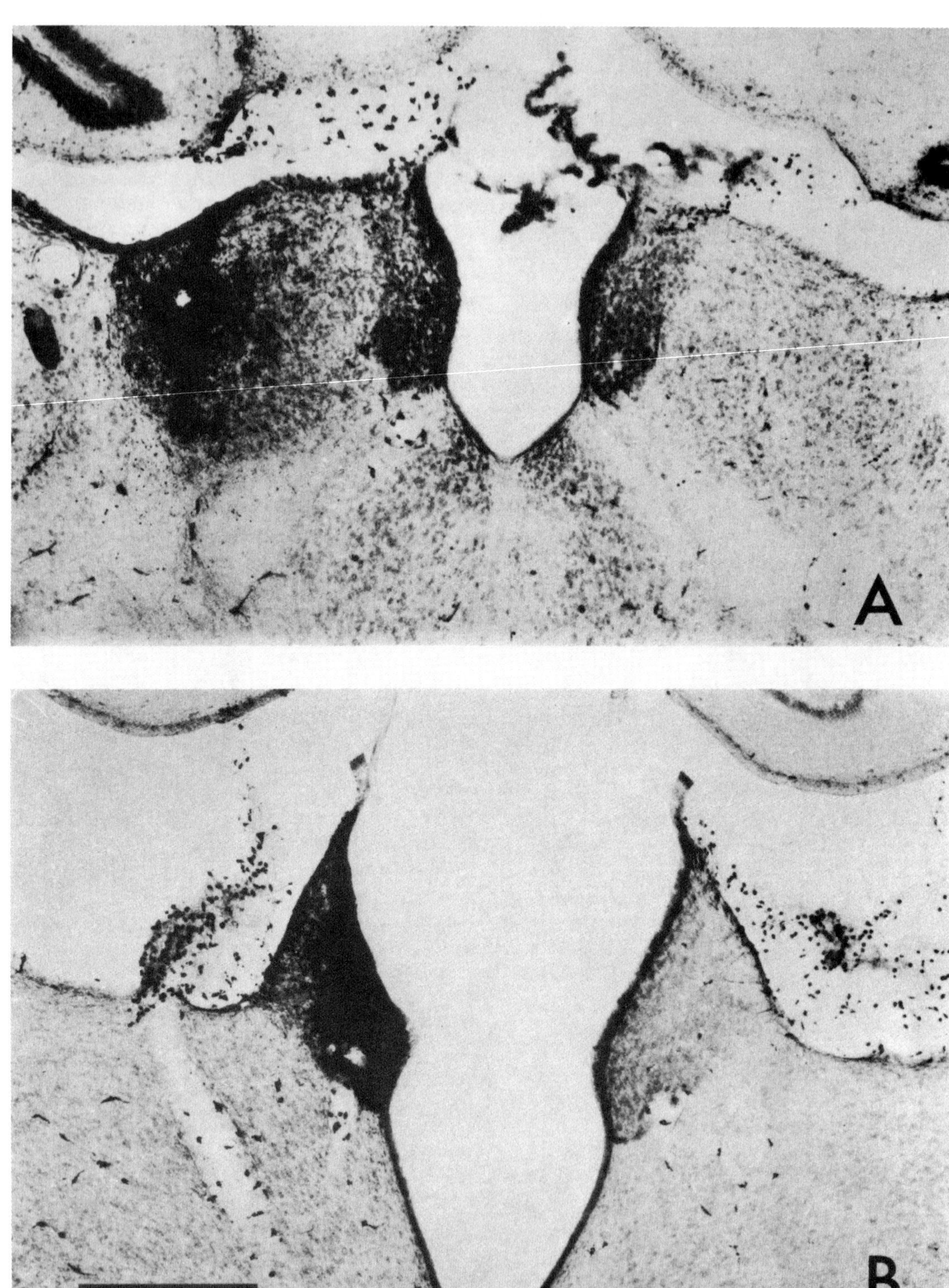

Figure 1

and Haymaker, '69; see also Jones et al., '76).

By far the largest number of HRP-marked cells in this case appeared in the entopeduncular nucleus and adjacent lateral hypothalamic regions (figs. 2d-f). The number of marked cells in the lateral hypothalamus reached a maximum in areas lying in the path of the inferior thalamic peduncle (fig. 2d), and from here caudalward declined gradually to approach zero at levels caudal to the entopeduncular nucleus. Thus, labeled cells in this case were confined largely to the rostral half of the lateral hypothalamus. It is interesting to note that in practically every section at this general level marked cells were found to surround the fornix bundle (figs. 2d-f, 3); often a dendrite's beaded shadow could be seen to enter the fornix bundle, and occasionally a marked cell actually lay within the bundle.

The number of labeled hypothalamic neurons was exceeded only by the multitude of labeled cells in the adjacent entopeduncular nucleus (figs. 2d-f). Judged by the present findings, this medial segment of the globus pallidus constitutes the major single source of afferents to the lateral habenular nucleus. Heavily and lightly granulated multipolar and spindle-shaped cells were massed throughout its extent. In the region of densest labeling (figs. 2e, 3) virtually every entopeduncular cell was marked; sectors lying rostral and caudal to that level contained proportionately fewer marked cells.

Additional, but generally rather sparse, HRP-positive cells were found in the midbrain. Most were ipsilateral to the

injection. Some such cells appeared in a circumscript region medial to the fasciculus retroflexus in the ventral tegmental area, close to the midline and just rostral to the interpeduncular nucleus. The median raphe nucleus, or nucleus centralis superior, likewise contained marked cells in a fairly random distribution in and at varying distance from the median plane, at levels extending from the caudal tip of the interpeduncular nucleus caudally to the frontal plane involving the dorsal tegmental nucleus. At these most caudal levels a few labelled cells were found in the cell group called nucleus tegmenti dorsalis lateralis, by Palkovits and Jacobowitz ('74), in the central gray substance of the isthmus region.

## 2. Medial habenular nucleus

In case Hb-15, HRP was deposited in the caudal end of the habenular complex, where the medial habenula is unaccompanied by its lateral counterpart. Figure 1B shows the center of this injection. The dark-brown reaction product completely filled the nucleus at this level and partially invaded the habenular commissure. Only a very faint HRP reaction appeared at more rostral levels containing the lateral habenula, and it thus seems likely that this injection was entirely restricted to the medial habenular nucleus. Moreover, the spillage of HRP along the penetration track was negligible.

Relatively few brain regions contained HRP-marked neurons in this case. At rostral levels such cells were found in the nucleus of the diagonal band, mainly in the horizontal limb (figs. 4a,b). However, no marked cells appeared elsewhere in the basal forebrain with the exception of the caudal end of the cell group here called nucleus basalis which contained a few heavily marked cells bilaterally (figs. 4d,e). A few additional, very faintly labeled cells were found in the lateral hypothalamus (fig. 4f).

In contrast to these sporadic findings in the basal forebrain region, a very large number of HRP-positive cells appeared in

Fig. 1(A) Lightfield photomicrograph of a frontal section at the center of the HRP injection site in case Hb-14. The spindle-shaped injection is bounded laterally and ventrally by white matter. Dorsally the stria medullaris contains marker and medially a diffused reaction product fills the lateral half of the nucleus. There was no significant spread of HRP into other structures. (B) Photomicrograph of the center of the injection site in case Hb-15. The reaction product completely fills the medial habenula and invades the overlying fibers that farther caudally enter the habenular commissure. The bar represents 0.5 mm in both photographs.

the nuclei septofimbrialis and triangularis of the supra- and postcommissural septum (figs. 4c-f). These small, spindle-shaped neurons were fewer in number and generally showed fainter labeling in the more rostral septofimbrial nucleus than in the nucleus triangularis septi in which most of the numerous marked cells appeared quite brightly labeled in the darkfield (fig. 5).

In this case, contrary to those in which HRP had been deposited in the lateral habenular nucleus, no more than sporadic labeled cells could be found in the paramedian region of the midbrain; a few more appeared at even more caudal levels, in the pontine raphe nucleus.

### 3. *Topography of habenular afferents*

In the cases described up to this point HRP had been deposited in fringe regions of the habenular complex: the first was localized to the lateral one-third of the caudal part of the lateral habenular nucleus, whereas the second filled the most caudal region of the medial habenula. Several other cases in which different parts of these nuclei were injected provide, first, an essential corroboration of the identified sources of afferents and, second, a suggestion that some of the afferent connections of the habenular nuclei are topographically organized.

### a. *Case Hb-5*

The injection in case HB-5 was quite similar to that in case HB-14 with respect to its size, shape and position near the caudal border of the lateral habenular nucleus. However, in contrast to the latter case this deposit was confined to the medial half of the lateral nucleus (fig. 6). The structures containing marked cells were the same in the two cases, but some notable differences appear in the quantitative distribution of the labeled cells. First, the nucleus of the diagonal band contained many more labeled cells in this case of more medially placed HRP injection. Second, the cells in the basal forebrain areas were generally placed more medially than were those marked by the more lateral injection: in the lateral preoptic area they were typically located close to the border of the medial preoptic area or dorsally, close to the border of the lateral septal nucleus and the bed-nucleus of the stria terminalis.

### b. *Case Hb-4*

A more rostral region of the lateral habenular nucleus and the stria medullaris was injected in case Hb-4 (fig. 6). In this case the aspiration of structures had caused considerable distortion and some expansion of the habenular complex on the injected side. The injection was larger than any of those previously described and, at its center point, slightly invaded the mediodorsal nucleus.

The cell labeling in this case was quite comparable to that caused by the lateral habenular injections described above, with the following exceptions. Moderately-to-

---

Abbreviations

| | |
|---|---|
| ac, Anterior commissure | ne, Entopeduncular nucleus |
| apl, Lateral preoptic area | nrt, Nucleus reticularis thalami |
| apm, Medial preoptic area | nsf, Nucleus septofimbrialis |
| cem, Nucleus centralis medialis | nst, Nucleus triangularis septi |
| cl, Nucleus centralis lateralis | ot, Olfactory tubercle |
| cp, Cerebral peduncle | pc, Nucleus paracentralis |
| f, Fornix | pf, Parafascicular nucleus |
| fr, Fasciculus retroflexus | pvp, Posterior periventricular nucleus |
| gp, Globus pallidus | sc, Suprachiasmatic nucleus |
| ic, Internal capsule | sm, Stria medullaris |
| lh, Lateral hypothalamic region | so, Supraoptic nucleus |
| md, Mediodorsal nucleus | st, Stria terminalis |
| mt, Mammillothalamic tract | v, Ventral nucleus |
| nb, Nucleus basalis | vm, Ventromedial nucleus |
| ndb, Nucleus of the diagonal band | vp, Ventral pallidum |

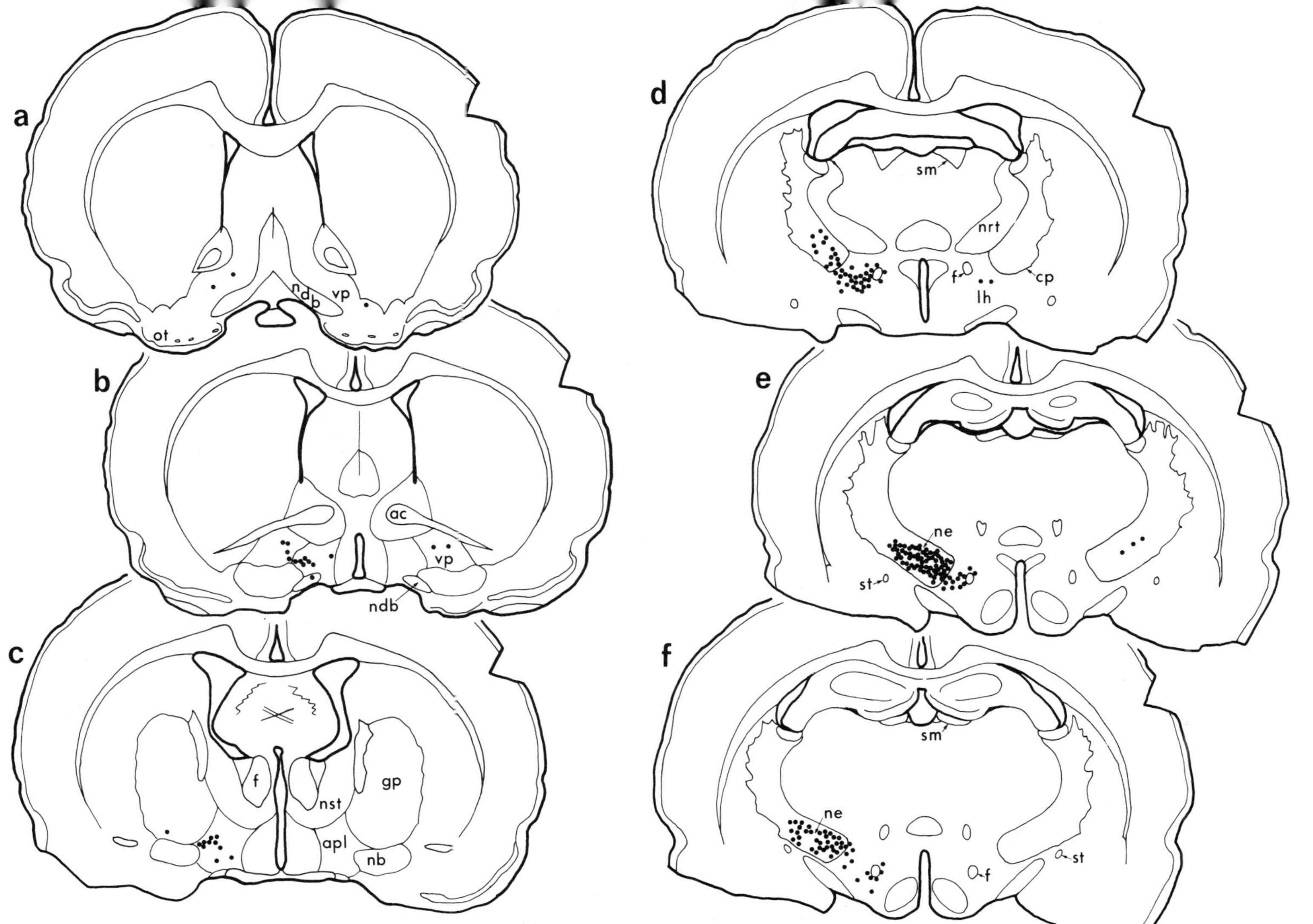


Fig. 2  Chartings of selected frontal sections in case Hb-14. Black dots indicate HRP-positive cells. Descriptions of each level are given in the text.

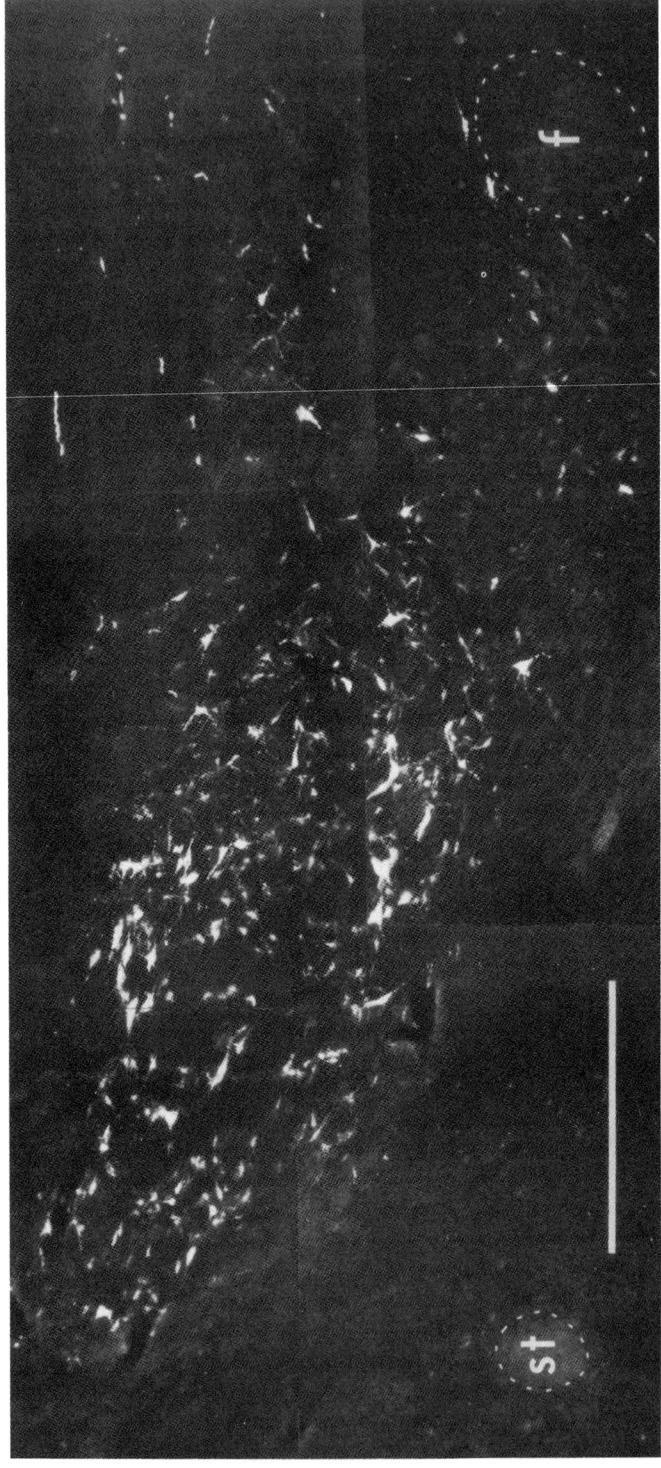

Fig. 3 A darkfield microphotomontage of the entopeduncular nucleus within the interstices of the cerebral peduncle and the adjacent lateral hypothalamus in case Hb-14. Same section as was charted in figure 2e. Intensely HRP-granulated cells are obvious in the picture, but more faintly labeled cells are also discernible. The stria terminalis and fornix are outlined; note the proximity of labeled cells to the latter. Bar represents 0.5 mm.

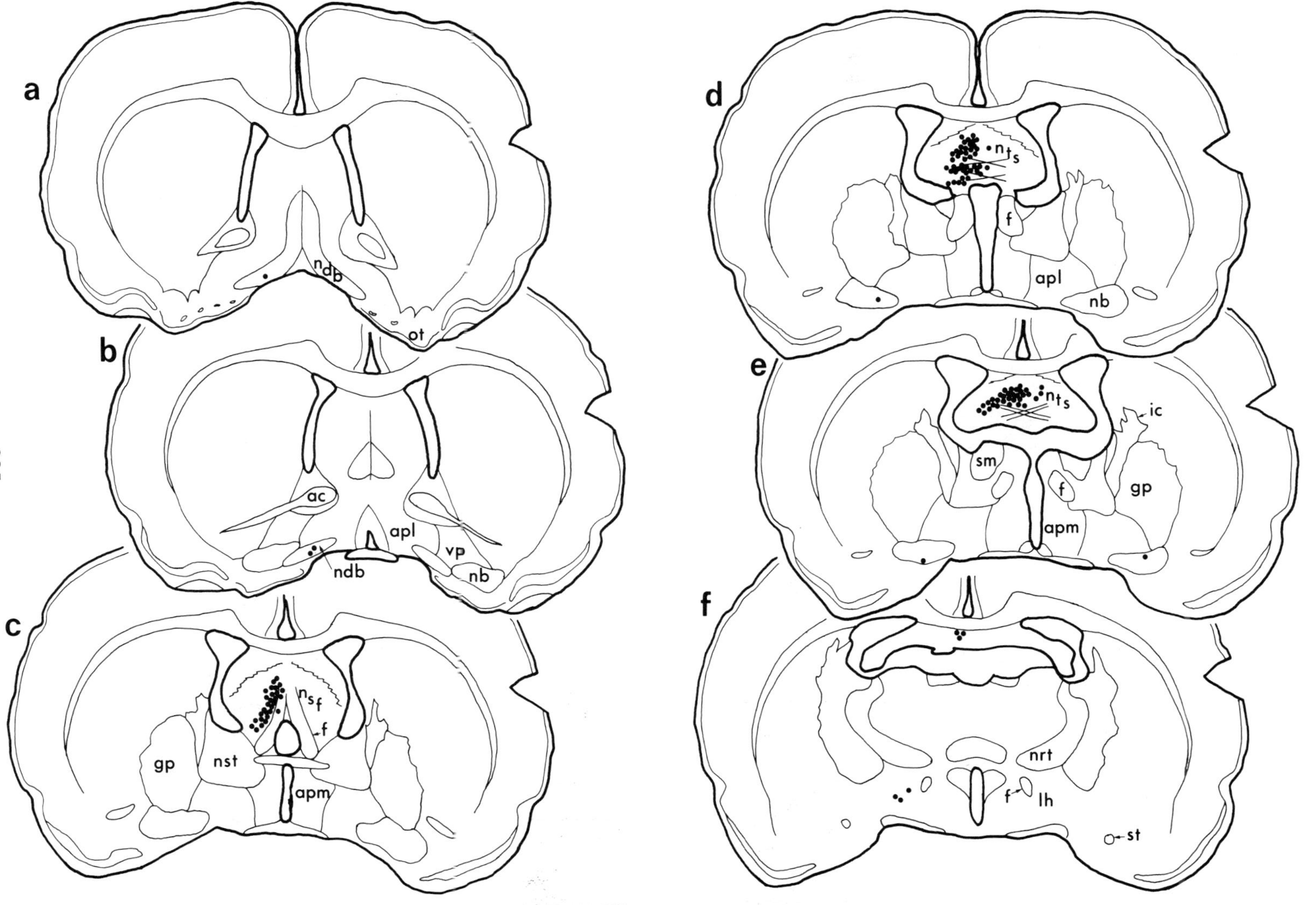


Fig. 4 Chartings of selected frontal sections in case Hb-15. Black dots indicate HRP-positive cells.

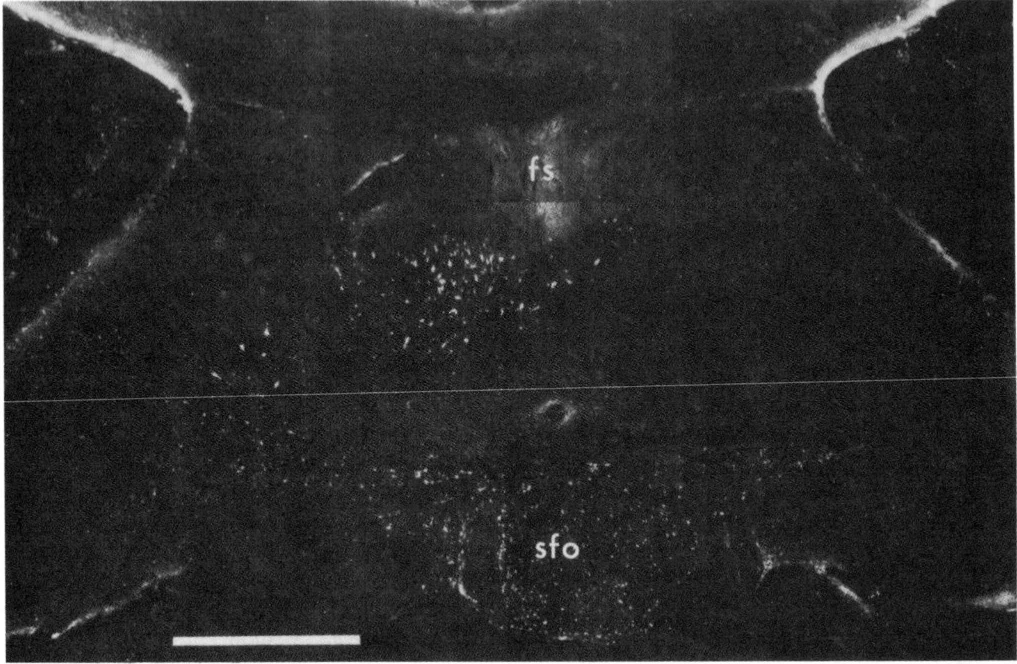


Fig. 5   A darkfield microphotomontage of the region of the ventral hippocampal commissure and embedded nucleus triangularis septi. Same section as was charted in figure 4e. A field of small, brightly HRP-marked cells is visible ventrolateral to the fornix superior (fs). Light speckles at the bottom of the picture represent endogenous peroxidase activity in the subfornical organ (sfo). Bar measures 0.5 mm.

intensely labeled large multipolar cells were found in the deep, polymorphous layers of the piriform cortex and olfactory tubercle. The most rostral of these cells appeared at levels immediately behind the anterior olfactory nucleus; still more were scattered throughout the deep stratum of the olfactory tubercle adjoining the nucleus of the diagonal band. Occasional marked cells occupied the deep paleocortical stratum dorsal to the lateral olfactory tract. Caudally, some of these labeled cells invaded the anterior amygdaloid nucleus. At more caudal levels a few labeled cells were found again in the deep layers of the piriform cortex but very few appeared in the subjacent major amygdaloid nuclei.

It is interesting to note that in this case many cells were found labeled bilaterally in the nucleus basalis, especially in caudal parts of the nucleus. The number of these labeled cells was much greater in this case than in cases of more caudally located injection of the lateral habenular nucleus. Elsewhere the pattern of labeling was not significantly different from that observed in either case Hb-5 or Hb-14.

### c. Case Hb-13

Finally, the medial habenular nucleus was injected at a level rostral to that of the HRP deposit in case Hb-15. In case Hb-13 (fig. 6) HRP was placed in the medial habenular nucleus and the medial edge of the lateral habenular nucleus by an angled stereotaxic approach. Although some HRP had leaked into the cortex at the penetration point, no significant HRP spillage had occurred in the hippocampus.

In this case, as in Hb-15, the resultant cell labeling was restricted to relatively few nuclei. As in Hb-15, by far the greatest number of labeled neurons were found in the postcommissural septum but, unlike

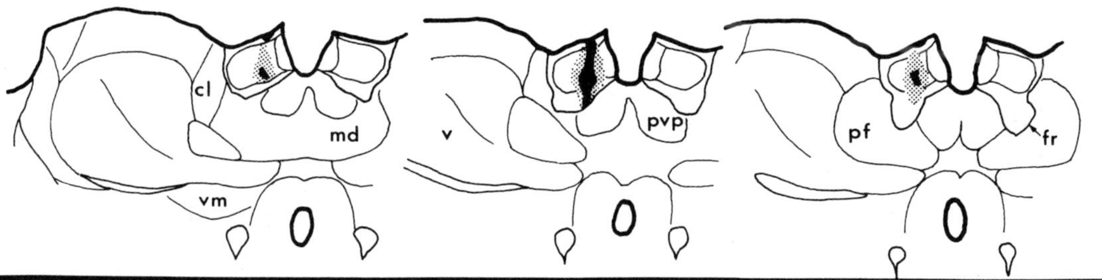


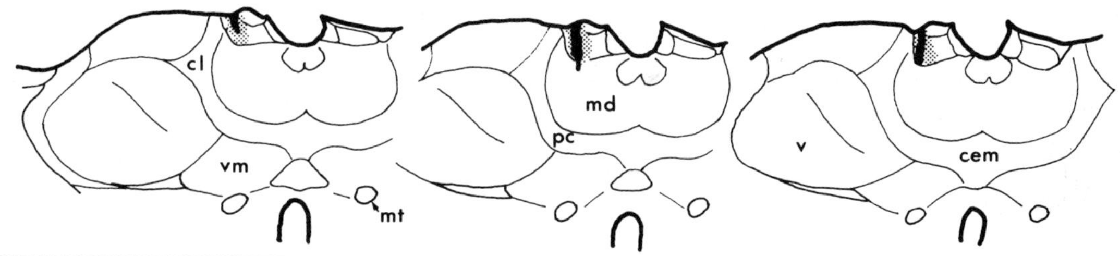


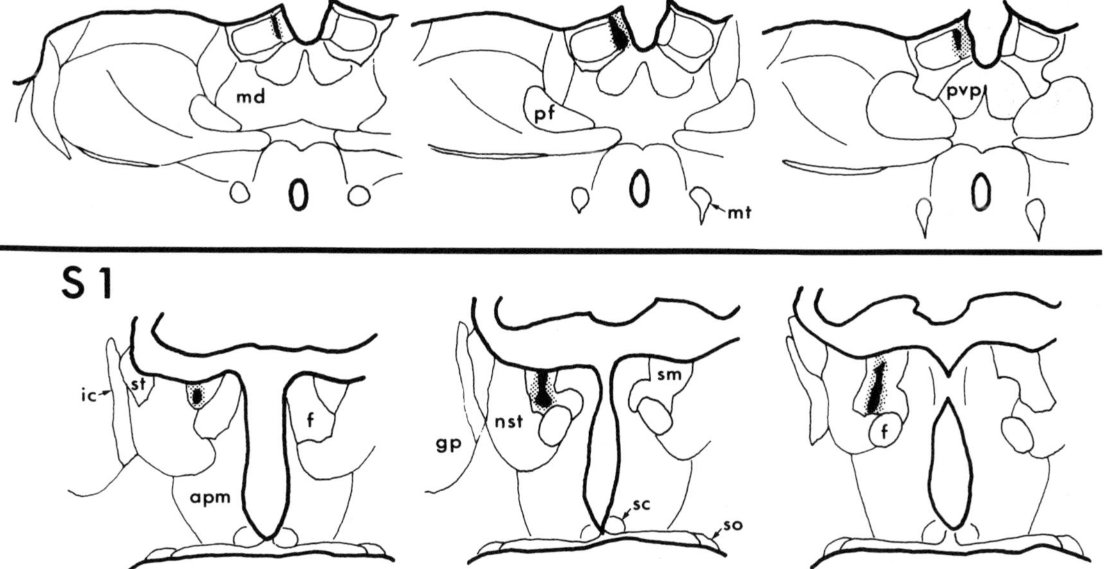


Fig. 6 Chartings of HRP-injection site in four cases discussed in the text. In each case the center of the injection site and levels 100 $\mu$m rostral and caudal to it are shown. The bar at lower left represents 1 mm.

Hb-15, the vast majority of these cells was concentrated in the ventral part of the septofimbrial nucleus surrounding the ventral third of the column fornicis. Virtually every cell seemed labeled in that region on the side of the injection, and a few such cells also appeared in the contralateral homologous region. Additional labeled cells were found in this case in the nucleus of the diagonal band, nucleus basalis, lateral preoptic and hypothalamic areas, mesencephalic raphe nuclei, and nucleus tegmenti dorsalis lateralis.

Finally, in this case sporadic labeled neurons were found in anteromedial parts of the nucleus accumbens, in the medial septum, and scattered throughout ventral parts of the lateral septal nucleus. It would thus seem that some fibers at least pass from these areas either to the medial habenular nucleus or to a medial part of the lateral habenular nucleus.

## 4. Injections of the stria medullaris: case S-1

At the rostral pole of the thalamus the various components of the stria medullaris converge from the rostral and ventral sides to form a thick fiber bundle. The injection in case S-1 was centered at this point—nowhere did HRP invade adjacent gray matter (fig. 6), and virtually no HRP had been spilled along the pipette track.

In this case a profusion of axons evenly filled with brown HRP-reaction product could be traced both caudalward from the injection site to the habenula and rostrally into the medial and lateral corticohabenular tracts (nomenclature after Gurdjian, '25). In addition to these solidly-filled fibers, numerous cells were found that exhibited the characteristic signs of retrograde HRP-labeling of perikaryon and dendrites. It is remarkable that, in contrast

TABLE 1

*Locations and amounts of HRP-positive cells in the cases described*

| | Location of injection site | | | | | |
|---|---|---|---|---|---|---|
| Afferent structure | Lateral part of lateral habenula Hb-14 | Medial part of lateral habenula Hb-5 | Caudal medial habenula Hb-15 | Rostral medial habenula Hb-13 | Rostral lateral habenula Hb-4 | Stria medullaris S-1 |
| Nuc. of diagonal band | 0-+ | ++ | + | ++ | ++ | ++ |
| Lateral septum & accumbens | 0 | 0-+ | 0 | + | 0-+ | + |
| Ventral pallidum | ++ | ++ | 0 | 0 | ++ | ++ |
| Olfactory tubercle | 0 | 0 | 0 | 0 | ++ | + |
| Piriform cortex | 0 | 0 | 0 | 0 | + | + |
| Nucleus basalis | 0-+ | + | 0-+ | + | ++ | + |
| Lateral preoptic area | ++ | ++ | 0 | ++ | ++ | ++ |
| Amygdala | 0 | 0 | 0 | 0 | + | + |
| Entopeduncular nucleus | +++++ | +++ | 0 | 0 | +++ | +++ |
| Lateral hypothalamus | +++ | +++ | 0-+ | ++ | +++ | +++ |
| Septofimbrial nuc. | 0 | ++ | +++ | ++++ | + | — — |
| Nuc. triangularis septi | 0 | ++ | ++++ | +++ | + | — — |
| Ventral tegmental area | ++ | 0-+ | 0 | 0 | ++ | ++ |
| Median raphe | + | + | 0-+ | +· | ++ | ++ |
| Dorsal raphe | 0 | 0 | 0 | 0 | 0 | 0-+ |
| Ventral central gray | 0-+ | 0-+ | 0 | 0-+ | 0-+ | 0-+ |

0-+, an occasional cell per section.
  +, less than three cells per section.
  ++, less than 20 cells per section.
  +++, less than 100 cells per section.
  ++++, less than 200 cells per section.
+++++, more than 200 cells per section.
  — —, solidly-filled cells

to the foregoing cases of habenula injection, cell-labeling in this case of stria-medullaris injection was uniformly heavy and, in many cells, characterized by a coarser granulation. Many of these labeled cells were located, as in the foregoing cases, in the basomedial forebrain and in the paramedian zone of the mesencephalon, more specifically in the septofimbrial nucleus, the nucleus of the diagonal band and the adjacent basal region of the lateral septal nucleus, the ventral pallidum, the lateral preopticohyothalamic region, the entopeduncular nucleus, the ventral tegmental area and the nucleus centrelis tegmenti superior (median raphe nucleus); sporadic labeled cells also appeared in the dorsal raphe nucleus and in the nucleus tegmenti dorsalis lateralis.

It is interesting to note that only in the ventral part of the septofimbrial nucleus, close to the injection site in the stria medullaris, many neuronal perikarya appeared evenly filled with HRP-reaction product; some of these cells were clearly continuous with an axon equally evenly marked. Of particular importance is that, as in case Hb-4 of rostral-habenula injection but unlike the cases of HRP deposits confined to caudal parts of the habenula, additional labeled cells appeared in the deep, polymorph layer of both the olfactory tubercle and piriform cortex. The significance of this finding will be discussed below.

A summary of cell labeling in this and the previously described cases is given in table 1.

### DISCUSSION

#### Technical considerations

The use of the histochemical marker horseradish-peroxidase for the anatomical identification of afferent-sources must still be considered a new technique with peculiarities that are not yet completely understood. It seems likely that, in some fiber systems at least, HRP is taken up exclusively by axon terminals and not by intact axons (LaVail et al., '73; Nauta et al., '74), but there is good evidence that axons severed or otherwise injured by the injection procedure readily take up and retrogradely transport HRP (Kristensson and Olsson, '74; DeVito et al., '74; Furstman et al., '75; Halperin and LaVail, '75; Kuypers and Maisky, '75). In such instances the labeled perikaryon usually has a granulated appearance identical to that seen after transport from axon terminals, but it may also show a more intensive granular reaction or an even-brown background stain (DeVito et al., '74; Halperin and LaVail, '75; Colman et al., '76).

The many cells labeled after an electrophoretic injection confined to the stria medullaris in the present case S-1 clearly indicate that even the use of fine-tipped micropipettes does not prevent uptake of HRP by numerous fibers-of-passage. Although light-microscopic examination of the injection site revealed no evidence of tissue damage, it seems likely that the enzyme had been taken up by injured axons. Axons had been evenly filled with HRP in both the cellulifugal and cellulipetal direction; according to earlier reports, such bidirectional Golgi-like impregnations are attributable to a diffusion of the enzyme within injured axons (LaVail and LaVail, '74; Kirstensson and Olsson, '74; Scalia and Colman, '74; Bunt et al., '75; Colman et al., '76). Moreover, some of the retrogradely filled axons could be traced back directly to their nearby cells of origin in the septofimbrial nucleus. These cells likewise were evenly filled with brown HRP-reaction product, suggesting that their axons had been injured and their perikarya filled by passive cytoplasmic diffusion (Jones and Leavitt, '74; Turner and Harris, '74; Halperin and LaVail, '75; Colman et al., '76; Parent, '76). By contrast, labeled cells located farther than about 2 mm from the injection site exclusively showed the characteristic granular HRP reaction that represents the active and rapid transport of the enzyme over discrete intracellular channels (LaVail and LaVail, '74; Nauta et al., '74). Parametric studies would be required to determine whether there is a limit to the distance over which passive intraaxonal diffusion of HRP will take place and cause solid cell-filling.

It seems clear from all these observations that the electrophoretic delivery technique causes damage to axons. Since the small pipette tips (10-20 $\mu$m) could hardly be thought to have injured directly the very large number of fibers impregnated in case S-1, some non-mechanical form of fiber damage seems more likely. Possibly the high voltages often needed to counteract high impedances at the pipette tip can cause electrolytic or thermal lesions. Alternatively, some form of chemical damage could be thought to occur, especially since this technique can produce extremely concentrated deposits of chemical compounds. However this may be, it seems clear that the microelectrophoretic technique, although it has a great advantage over all other delivery methods by permitting very small, concentrated deposits of marker substance, is no less likely to label fibers of passage and therefore does not reduce the need for cautious interpretation of cell labeling resulting from HRP deposits in brain structures traversed by transit fibers.

### Afferents of the lateral habenular nucleus

In view of the problems of interpretation discussed above it is important to note that, with the few exceptions noted below, the findings here reported are compatible with evidence obtained by anterograde (largely autoradiographic) fiber-tracing methods.

### Afferents from the entopeduncular nucleus

The most striking result of the present study confirms earlier autoradiographic evidence (H. J. W. Nauta, '74) that the entopeduncular nucleus—the internal segment of the globus pallidus of non-primate mammals—is a major source of afferents to the lateral habenular nucleus. Pallidal cells were labeled by all lateral habenular injections but in largest number by injections in the lateral half of the nucleus. In one such

case (Hb-14) it seemed that in a large part of the entopeduncular nucleus virtually every neuron was HRP-positive (fig. 3). This observation raises the possibility that the pallidohabenular connection, instead of being established by a separate sub-population of entopeduncular neurons, is made up largely at least by collaterals of the fibers composing the ansa lenticularis. The same possibility has been suggested on the basis of recent electrophysiological evidence (Filion et al., '76).

The present findings also suggest that the pallidohabenular projection is somewhat topographically organized. In the cat the projection terminates mainly in the lateral half of the lateral habenular nucleus (Nauta, '74), and this appears to be true also in the rat, as by far the largest number of entopeduncular cells were labeled by the most lateral injections. However, while a somewhat more medial injection (case Hb-5, fig. 6) labeled many fewer cells in the entopeduncular nucleus (table 1), these cells were confined to its medial district. It would thus seem that the medio-lateral dimension of the entopeduncular nucleus is represented in a compressed fashion in the lateral part of the lateral habenular nucleus. Finally, the results of an injection confined to the stria medullaris (case S-1) indicate that the several divergent fiber paths that together compose the pallidohabenular connection (Nauta, '74) likewise represent different parts of the entopeduncular nucleus: the most rostral of these paths joins the stria medullaris at the rostral pole of the thalamus, and the labeling pattern in case S-1 suggests that it consists of fibers from the most rostral portion of the entopeduncular nucleus. In which part of the lateral habenular nucleus these fibers terminate is, however, unknown.

### Preoptic and hypothalamic afferents

According to the present findings a second major afferent connection of the lateral habenular nucleus arises from a lateral preoptico-hypothalamic continuum stretching from the most anterior part of the preoptic region caudalward to the mid-hypothalamic level. The pattern of cell labeling in the lateral preoptic region suggests an orderly fiber arrangement such that cells in the lateral part of the lateral preoptic region project to the lateral part (fig. 2), more medially placed preoptic cells to the medial part of the lateral habenular nucleus. This finding agrees well with a recent autoradiographic demonstration of the preopticohabenular connection (Swanson, '76).

The largest number of labeled cells in the preoptico-hypothalamic continuum appeared in the anterior part of the lateral hypothalamic region. In all cases of lateral habenula injection these cells formed a diagonally oriented cluster extending from above the supraoptic nucleus to the perifornical region; at more caudal levels they bridged the distance between the entopeduncular nucleus and the fornix (figs. 2d-f). The more anterior of these labeled cells lay in the path of the convergent lateral olfactohabenular and lateral corticohabenular tracts of Gurdjian ('25), root bundles of the stria medullaris that issue from the medial forebrain bundle and were interpreted by Cajal ('11) as collateral offsets of the latter bundle. Because of the intricate commingling of cells and fibers in the lateral preoptic and hypothalamic region no unequivocal evidence that these fibers actually originate in this region could be obtained by fiber-degeneration studies (Guillery, '57; Nauta, '58; Zyo et al., '63; Wolf and Sutin, '66; Huang and Mogenson, '72); however, the connection has been suggested strongly by observations in Golgi material (Millhouse, '69), in cases of periventricular isotope-injection apparently encroaching on the lateral hypothalamus (Conrad and Pfaff, '76b), and in stimulation and recording experiments (Mok and Mogenson, '74). Some further remarks concerning the hypothalamohabenular connection will be made below in the section of this discussion dealing with stria-medullaris fibers by-passing the habenula.

## Afferents from the septal region

Previous findings have suggested certain septal nuclei (Guillery, '59; Powell, '68) and the nucleus accumbens (Powell and Leman, '76) as the origin of fibers to the lateral habenular nucleus. In the present study, small-to-moderate numbers of cells in the ventrolateral region of the septum, in the adjacent medial zone of the nucleus accumbens, and in the nucleus of the diagonal band were labeled by HRP deposits in the most medial part of the lateral habenular nucleus (table 1, cases Hb-5 and Hb-13). No cell labeling in these regions resulted from the more laterally placed habenula injection (case Hb-14), and injections of the medial habenular nucleus labeled cells only in the supracommissural septum and in the nucleus of the diagonal band. These findings suggest that a medial part of the nucleus accumbens and an adjacent region of the lateral septal nucleus project to a medial part of the lateral habenular nucleus and, furthermore, that the nucleus of the diagonal band projects to both habenular nuclei. The latter evidence confirms Domesick's ('76) and Conrad and Pfaff's ('76a) autoradiographic data; at present the nucleus of the diagonal band appears to be the only prosencephalic cell group that is known to project to both habenular nuclei.

It is interesting to note a probable alternate, indirect route by which the nucleus accumbens septi is connected with the lateral habenular nucleus: recent studies by anterograde methods (Wilson, '72; Heimer and Wilson, '75) have shown that the nucleus accumbens projects massively to the subcommissural part of the globus pallidus ("ventral pallidum" of Heimer and Wilson ('75), labeled *vp* in the present fig. 1), and in the present study all HRP injections of the lateral habenular nucleus labeled some cells of this pallidal cell group.

## Afferents of the medial habenular nucleus

The results of this study confirm previous reports by showing that the postcommissural septum comprises the major source of afferents to the medial habenular nucleus (Nauta, '56; Cragg, '61; Ban and Zyo, '62; Raisman, '66; Powell, '68). The patterns of cell-labeling in cases Hb-13 and 15 suggest that the rostral part of the medial habenular nucleus receives afferents mainly from the septofimbrial nucleus, whereas more caudal parts of the nucleus receive fibers from the nucleus triangularis septi. It appears that most or even all neurons of these septal nuclei project to the habenula, for in each case the areas of densest cell labeling contained virtually no unmarked neurons. Septo-habenular fibers for a short span follow the descending fornix column and below the interventricular foramen separate from this bundle to join the stria medullaris. The present pattern of cell-labeling is consistent with earlier conclusions that no structures caudal to the septum contribute to this path (Nauta, '56; Raisman, '66).

Injections restricted to the medial habenular nucleus also labeled cells in the nucleus of the diagonal band and in the median raphe nucleus of the midbrain (table 1). Since cells in these areas were labeled also by injections confined to the lateral habenular nucleus, they at present appear to be the only afferent sources shared by both habenular nuclei. Autoradiographic findings have shown that fibers from the nucleus of the diagonal band terminate in the medial part of the lateral nucleus and more densely in the medial nucleus (Domesick, '76). The projection from the median raphe is distributed preferentially to the medial nucleus according to one study (Conrad et al., '74) but was traced only to the lateral nucleus in another (Bobillier et al., '75). That a raphe projection to the habenula indeed exists and is at least in part serotonergic is suggested by the presence of serotonin axon terminals in the medial habenular nucleus (Björklund et al., '72).

Injections of the medial habenular nucleus also consistently labeled a few cells in the nucleus basalis and lateral hypothala-

mus. The intense labeling of sporadic cells in the former cell group most likely resulted from HRP-uptake by fibers decussating in the habenular commissure (Price and Powell, '70b), while the very faint labeling of lateral hypothalamic cells may have been caused by diffusion of the enzyme into the lateral habenular nucleus or by some uptake by decussating fibers. Only one publication has reported evidence of projections from these areas to the medial habenular nucleus (Zyo et al., '63). Reports of afferents to this nucleus from the medial and lateral nuclei of the precommissural septum (Zyo et al., '63; Powell, '68), the interpeduncular nucleus (Mitchell, '63) and the dorsal raphe nucleus (Pierce et al., '76) receive no support from the present findings.

The medial habenular nucleus has been found also to receive an adrenergic innervation from the superior cervical ganglion (Björklund et al., '72). This connection was not studied in the present material.

### Olfactory afferents

The habenula was once considered to be an olfactory way-station, on the basis of reports of afferent connections with olfactory forebrain structures, afferent fibers conveyed by the medial forebrain bundle (see Ariëns Kappers et al., '60, for a review). This view was buttressed by several reports of anterograde degeneration in the stria medullaris and lateral habenula following lesions of the piriform cortex, olfactory tubercle and amygdala (Laursen, '55; Cragg, '61; Kusama and Hagino, '61; Ban and Zyo, '62; Cowan et al., '65; Price and Powell, '71). However, in most of these studies the lesions were large enough to encroach on adjacent structures with habenular connections.

Careful examination of Fink-Heimer material (Leonard and Scott, '71; Heimer, '72) together with electron microscopic identification of degenerating axon terminals (Heimer, '72) showed that some and perhaps all of the structures that receive direct projections from the olfactory bulb project via the stria medullaris not to the habenula but to the subjacent mediodorsal nucleus. According to Heimer, fibers from the piriform cortex and olfactory tubercle follow a route from the medial forebrain bundle into the stria medullaris, but at an anterior-habenular level abruptly turn ventrally through the anterior part of the lateral habenular nucleus to terminate in a central sector of the mediodorsal nucleus. In the present study, cells in the piriform cortex and olfactory tubercle were labeled by HRP injections in either the stria medullaris (case S-1) or the attenuated rostral part of the habenula (case Hb-4) but not by more caudal habenular injections, a finding compatible with Heimer's more detailed and direct evidence that the fibers in question do not terminate in the habenula.

The present findings suggest a similar conclusion with respect to amygdalohabenular connections. A recent autoradiographic study (Krettek and Price, '74) has confirmed earlier evidence of a projection from the amygdala to the mediodorsal nucleus. The course of some of these fibers must be similar to that of the stria medullaris fibers originating from the piriform and periamygdaloid cortex, for, like these cortical structures, the amygdala contained labeled cells only in cases S-1 and Hb-4 (table 1). It thus appears that amygdalofugal fibers in the stria medullaris do not extend to the habenula. It is interesting to note that a recent electrophysiological study likewise failed to demonstrate a direct amygdalohabenular connection (Mok and Mogenson, '74).

The case of stria medullaris injection (S-1) incidentally failed also to confirm the anastomosis between the stria terminalis and stria medullaris described by Johnston ('23) as stria terminalis component 5. The HRP deposit in this case had caused solid-filling of axons in both the retrograde and anterograde direction for several millimeters from the injection site. Many of these fibers followed a ventral route corresponding to the lateral corticohabenular tract thought by Gurdjian ('25) to originate from olfac-

tory cortical structures; a few followed a more lateral route in the stratum zonale to a position closely medial to the stria terminalis, but none appeared anywhere within the stria terminalis itself. Several previous experimental studies have tended to confirm the existence of a link between the two striae (Cragg, '61; Leonard and Scott, '71), but a recent degeneration study of the stria terminalis by the sensitive cupric-silver method produced no evidence of Johnston's inter-strial anastomosis (De-Olmos and Ingram, '72).

### Stria medullaris fibers by-passing the habenula

The complexity of the stria medullaris has been only further underscored by the recent evidence that it contains fibers of olfactory and amygdalar origin that do not extend to the habenula. Earlier studies already had demonstrated that the bundle also contains some caudally directed fibers that by-pass the habenula and continue alongside the fasciculus retroflexus to ventromedial regions of the mesencephalic tegmentum (Nauta, '58; Cragg, '61).

Naturally, in the course of the present experiments the question arose whether the presence of these transit fibers in or around the habenula might lead to an erroneous identification of one or the other basal forebrain region as a source of habenular afferents. That such misidentification is unlikely is indicated by the outcome of a careful examination of a large autoradiographic case material collected in this laboratory. In this material descending stria-medullaris fibers by-passing the habenula as originally observed in fiber-degeneration studies in the cat (Nauta, '58) were found labeled only in four cases of tritiated-leucine-and-proline injection confined to the lateral preoptico-hypothalamic region between the frontal level of the anterior commissure and that of the anterior pole of the subthalamic nucleus. In these same cases, however, it was clear that a much larger number of labeled fibers actually terminated in the lateral habenular nucleus. No stria-medullaris fibers bypassing the habenula were found labeled in any case in which the radioactive amino-acid deposit was localized to the entopeduncular nucleus, ventral pallidum, nucleus accumbens, nucleus of the diagonal band, or nucleus basalis. These findings indicate that the moderate number of stria-medullaris fibers that bypass the habenula in transit to ventromedial parts of the mesencephalon originate from the same lateral preoptico-hypothalamic region which also gives rise to a far greater number of stria-medullaris fibers that terminate in the habenula. It thus appears possible to conclude that most, at least, of the lateral preoptico-hypothalamic neurons labeled in the present experiments actually project to the lateral habenular nucleus. It must be noted, however, that neither the retrograde nor the available autoradiographic evidence can answer the question as to whether habenular-by-pass fibers and fibers actually terminating in the habenula arise from the same or from different individual preoptico-hypothalamic neurons. In view of the latter possibility it must be considered likely that some of the preoptico-hypothalamic cells labeled by HRP deposits in the lateral habenular nucleus are hypothalamotegmental rather than hypothalamohabenular neurons.

*Commissural by-pass fibers*

There is evidence suggesting that the so-called habenular commissure is not a commissure but, rather, a decussation of fibers traveling in the stria medullaris. In the present study, HRP deposited in the habenula never labeled cells in any other part of the habenula or in the contralateral habenular complex; however, any given habenula injection labeled at least a few cells in the contralateral counterpart of each of the ipsilateral forebrain regions in which labeled cells appeared. This finding is consistent with anterograde evidence that some fibers in the stria medullaris decussate in the habenular commissure and terminate in the lateral nucleus of the contralateral habenular complex (Cragg, '61;

Price and Powell, '70b; Nauta, '74; Swanson, '76).

Two publications have reported evidence that the magnocellular nucleus of the substantia innominata here labeled nucleus basalis—but interpreted as part of the horizontal limb of the nucleus of the diagonal band by Price and Powell ('71)—sends fibers to the lateral habenula on both sides, as well as to its own contralateral counterpart by way of the habenular commissure (Price and Powell, '70b; Swanson, '76). Consistent with these reports is the pattern of cell labeling in the basal nucleus observed in the present study: in all cases of habenula injection in which labeled cells appeared on the contralateral side their number was largest in the nucleus basalis; moreover, in the only case in which the HRP deposit involved the medial habenular nucleus as well as the habenular commissure (Hb-15) some of the labeled cells in both the ipsi- and contralateral nucleus basalis exhibited the same extremely dense HRP granulation that marked many cells in the basal forebrain in the case of stria medullaris injection (S-1). On the basis of the arguments put forward in the description of case S-1 it seems likely that these heavily-labeled cells had incorporated the enzyme through a damaged segment of their axon.

*Ascending habenular afferents*

HRP deposited in the lateral habenular nucleus labeled small-to-moderate numbers of neurons in the following structures: 1. a medial part of the ventral tegmental area closely surrounding the fasciculus retroflexus immediately rostral to the interpeduncular nucleus, 2. the median raphe nucleus (nucleus centralis superior), and 3. ventral regions of the central gray substance at the isthmus rhombencephali. Injections in the medial habenular nucleus caused labeling of only a small number of cells in the median and pontine raphe nuclei. Most of the labeled cells were located ipsilateral to the habenular injection.

Not all of these locations match the origin of ascending habenulopetal fibers suggested by anterograde fiber-tracing studies. Some such studies have suggested that the interpeduncular nucleus projects to the habenula (Massopust and Thompson, '62; Mitchell, '63; Smaha and Kaelber, '73), but it is probable that the lesions in these studies encroached upon adjoining areas, in particular the ventral tegmental area which supplies dopaminergic habenular afferents (Kizer et al., '76), or involved fibers ascending from the median raphe nucleus (Conrad et al., '74; Bobillier et al., '75), some of which must correspond to the habenula's serotonergic afferents (Björklund et al., '72; Kuhar et al., '72). Findings in other studies have suggested habenulopetal fibers arising from the dorsal raphe nucleus (Conrad et al., '74; Bobillier et al., '75; Pierce et al., '76) but the lesions and autoradiographic injections in all cases exceeded the boundaries of the narrow raphe nucleus and, therefore, may have involved cells in adjacent parts of the central gray matter.

It is remarkable that in the present study only the stria-medullaris injection labeled neurons in the dorsal raphe nucleus (table 1). The fact that this injection also marked some cells in all of the other above-mentioned mesencephalic nuclei supports earlier conclusions that the stria medullaris contains ascending fibers that by-pass the habenular complex (Cragg, '61; Massopust and Thompson, '61; Mitchell, '63; Smaha and Kaelber, '73), but it is somewhat surprising that no fibers originating in the dorsal raphe nucleus appeared to take up label from HRP deposits in the habenula. However, in view of the experience that the HRP technique may fail to label otherwise well-documented connections (Nauta et al., '74), this negative result should be interpreted with caution.

Another negative outcome reported here is the failure of any habenula injection to label cells in the dorsal part of the central gray substance, the region from which Hamilton ('73) traced degenerating fibers to the lateral habenular nucleus. It is possible that the HRP method is unable to demonstrate this connection. Alternatively,

since the trajectory of ascending fibers from more caudal regions is unknown, it is possible that Hamilton's lesions interrupted fibers originating in the nucleus tegmenti dorsalis lateralis, the cell group (Palkovits and Jacobowitz, '74) which contained most of the HRP-labeled cells found in the central gray matter in the present study. A reinvestigation of the pathway in question with autoradiographic techniques might settle this issue.

### General comments

One of the most remarkable characteristics of the habenular complex is that it consists, as pointed out by Cajal ('11), of two cytologically quite dissimilar cell groups between which no axonal connections appear to exist. The two habenular nuclei thus, paradoxically, appear as mutually isolated processing stations in a dorsal conduction route between fore- and midbrain. Much like their respective cytoarchitectures, their principal afferent connections are quite different: whereas the lateral habenular nucleus receives its major afferent supplies from the entopeduncular nucleus and the lateral preoptico-hypothalamic region, the afferents of the medial habenular nucleus appear to originate very largely in the postcommissural septum. Nonetheless, the two habenular cell groups have at least two afferent relationships in common: they share afferents from the nucleus of the diagonal band of Broca and from the mesencephalic nucleus centralis superior (median raphe nucleus). Judged by their volume, however, these two projections must be classified as subsidiary rather than principal habenular afferents.

The question whether the habenular complex represents a cross-roads of pallidal and limbic pathways has been discussed by H. J. W. Nauta ('74), whose account of the problem can now be supplemented by more conclusive evidence (discussed above) that the lateral habenular nucleus, in addition to its massive pallidal input, indeed receives afferents from the lateral preoptico-hypothalamic region and from the nucleus of the diagonal band. Hence, it would now appear that at least in the lateral habenular nucleus substantial pallidal and limbic pathways converge, whereas the medial habenular nucleus, by contrast, composes a separate habenular compartment receiving limbic (more specifically, septal) but no pallidal projections.

Obviously, the foregoing inferences are open to question. Even though recent studies have virtually ascertained a hypothalamic projection to the lateral habenular nucleus, the nature of the cells giving rise to this projection is unknown. In particular, it could be asked if perhaps these cells are actually pallidal neurons lying outside the indistinct medial border of the entopeduncular nucleus, a possibility suggested by the arrangement of the labeled cells shown in figure 2d-f. This question may require electronmicroscopic examination of lateral preoptico-hypothalamic cells previously labeled by an HRP deposit in the lateral habenular nucleus. However, whether or not such cells should be found to exhibit the ultrastructural characteristics of pallidal neurons described by Fox et al. ('74), the fact would remain that they, unlike the cells of the entopeduncular nucleus, lie embedded among the fibers of the medial forebrain bundle and are therefore likely to be accessible primarily to impulses originating from limbic structures rather than from the caudatoputamen. The circumstance that the medial forebrain bundle recently has been shown to contain numerous fibers descending from the striatal nucleus accumbens septi (Swanson and Cowan, '75; Conrad and Pfaff, '76a; Powell and Leman, '76) would not invalidate this argument, for the reason that the nucleus accumbens, unlike the larger remainder of the striatum, is known to receive its major telencephalic afferents from the hippocampus (Sprague and Myer, '50; Raisman et al., '66), septum (Powell and Leman, '76) and amygdala (DeOlmos, '72) rather than from the neocortex.

A clue to the significance of the apparent differential distribution of habenular afferents may eventually emerge from the mode in which the habenulo-mesencephalic connection is organized. The outcome of an earlier fiber-degeneration study (Nauta, '58) has suggested that both medial and lateral habenular nucleus project to the region of the mesencephalic midline by way of the fasciculus retroflexus, but the projection from the medial nucleus appeared from that study to be limited to the interpeduncular nucleus, whereas that from the lateral habenular nucleus could be traced beyond it to the nucleus centralis superior (median raphe nucleus) and the fountain nucleus of Sheehan (dorsal raphe nucleus). Further details may become apparent from a current autoradiographic re-examination of the habenulo-mesencephalic connection, but the data available already leave little doubt that the habenular complex is a major source of afferents to the mesencephalic raphe region. They also suggest that by these afferents the habenula differentially distributes information, a. from the post-commissural septal nuclei, processed by its medial nucleus, and b. from the pallidum and lateral preoptico-hypothalamic region, integrated partially at least by its lateral nucleus.

## LITERATURE CITED

Ariëns Kappers, C. U., G. C. Huber and E. C. Crosby 1960 The Comparative Anatomy of the Nervous System of Vertebrates, Including Man. Hafner, New York. Vol. I, pp. 1101-1103.

Ban, T., and K. Zyo 1962 Experimental studies on the fiber connections of the rhinencephalon. I. Albino rat. Med. J. Osaka Univ., 12: 385-424.

Björklund, A., Ch. Owman and K. A. West 1972 Peripheral sympathetic innervation and serotonin cells in the habenular region of the rat brain. Z. Zellforsch., 127: 570-579.

Bobillier, P., F. Petitjean, D. Salvert, M. Ligier and S. Seguin 1975 Differential projections of the nucleus raphe dorsalis and nucleus raphe centralis as revealed by autoradiography. Brain Res., 85: 205-210.

Bunt, A. H., A. E. Hendrickson, J. S. Lund, R. D. Lund and A. F. Fuchs 1975 Monkey retinal ganglion cells: morphometric analysis and tracing of axonal projections with a consideration of the peroxidase technique. J. Comp. Neur., 164: 265-286.

Cajal, S. Ramón y 1911 Histologie du Système Nerveaux de l'Homme et des Vertébrés. Maloine, Paris. Vol. II, pp. 415-426.

Colman, D. R., F. Scalia and E. Cabrales 1976 Light and electron microscopic observations on the anterograde transport of horseradish peroxidase in the optic pathway in the mouse and rat. Brain Res., 102: 156-163.

Conrad, L. C. A., C. M. Leonard and D. W. Pfaff 1974 Connections of the median and dorsal raphe nuclei in the rat: an autoradiographic and degeneration study. J. Comp. Neur., 165: 179-206.

Conrad, L. C. A., and D. W. Pfaff 1976a Autoradiographic tracing of nucleus accumbens efferents in the rat. Brain Res., 113: 589-596.

——— 1976b Efferents from medial basal forebrain and hypothalamus in the rat. II. An autoradiographic study of the anterior hypothalamus. J. Comp. Neur., 169: 221-262.

Cowan, W. M., G. Raisman and T. P. S. Powell 1965 The connexions of the amygdala. J. Neur. Neurosurg. Psychiat., 28: 137-151.

Cragg, B. G. 1961 The connections of the habenula in the rabbit. Exp. Neur., 3: 388-409.

Crosby, E. C., T. Humphrey and E. W. Lauer 1962 Correlative Anatomy of the Nervous System. Macmillan, New York, pp. 271-273.

DeOlmos, J. S. 1972 The amygdaloid projection field in the rat as studied with the cupric-silver method. In: Neurobiology of the Amygdala. B. E. Eleftheriou, ed. Plenum, New York, pp. 145-204.

DeOlmos, J. S., and W. R. Ingram 1972 The projection field of the stria terminalis in the rat brain. An experimental study. J. Comp. Neur., 146: 303-334.

DeVito, J. L., K. W. Clausing and O. A. Smith 1974 Uptake and transport of horseradish peroxidase by cut end of the vagus nerve. Brain Res., 82: 269-271.

Divac, I. 1975 Magnocellular nuclei of the basal forebrain project to neocortex, brain stem, and olfactory bulb. Review of some functional correlates. Brain Res., 93: 385-398.

Domesick, V. B. 1976 Projections of the nucleus of the diagonal band of Broca in the rat. Anat. Rec., 184: 391-392 (Abstract).

Filion, M., C. Harnois and G. Guano 1976 Electrophysiological study of the distribution of axonal branches of individual entopeduncular neurons in the cat. Neurosci. Abstr., 2: 63 (Abstract).

Fox, C. A., A. N. Andrade, I. J. LuQui and J. Rafols 1974 The primate globus pallidus: a Golgi and electron microscopic study. J. für Hirnfforschung, 15: 75-93.

Furstman, L., S. Saporta and L. Kruger 1975 Retrograde axonal transport of horseradish peroxidase in sensory nerves and ganglion cells of the rat. Brain Res., 84: 320-324.

Graham, R. C., Jr., and M. J. Karnnovsky 1966 The early stages of absorption of injected horseradish peroxidase in the proximal tubules of mouse kidney: ultrastructural cytochemistry by a new technique. J. Histochem. Cytochem., 14: 291-302.

Graybiel, A. M., and M. Devor 1974 A microelectrophoretic delivery technique for use with horseradish peroxidase. Brain Res., 68: 167-173.

Guillery, R. W. 1957 Degeneration in the hypothalamic connexions of the albino rat. J. Anat. (London), 91: 91-115.

——— 1959 Afferent fibers to the dorso-medial thalamic nucleus in the cat. J. Ant. (London), 93: 403-419.

Gurdjian, E. S. 1925 Olfactory connections of the albino rat, with special reference to the stria medullaris and the anterior commissure. J. Comp. Neur., 38: 127-163.

Halperin, J. J., and J. H. LaVail 1975 A study of the dynamics of retrograde transport and accumulation of horseradish peroxidase in injured neurons. Brain Res., 100: 253-269.

Hamilton, B. L. 1973 Projections of the nuclei of the periaqueductal gray matter in the cat. J. Comp. Neur., 152: 45-58.

Heimer, L. 1972 The olfactory connections of the diencephalon in the rat. Brain, Behav. and Evol., 6: 484-523.

Heimer, L., and R. D. Wilson 1975 The subcortical projections of the allocortex: similarities in the neural associations of the hippocampus, the piriform cortex, and the neocortex. In: Golgi Centennial Symposium. Proceedings. M. Santini, ed. Raven, New York, pp. 177-193.

Herrick, C. J. 1948 The Brain of the Tiger Salamander. Univ. Chicago Press, Chicago, pp. 247-264.

Huang, Y. H., and G. J. Mogenson 1972 Neural pathways mediating drinking and feeding in rats. Exp. Neur., 37: 269-286.

Johnston, J. B. 1923 Further contributions to the study of the evolution of the forebrain. J. Comp. Neur., 35: 337-481.

Jones, E. G., H. Burton, C. B. Saper and L. W. Swanson 1976 Midbrain, diencephalic and cortical relationships of the basal nucleus of Meynert and associated structures in primates. J. Comp. Neur., 167: 385-420.

Jones, E. G., and R. Y. Leavitt 1974 Retrograde axonal transport and the demonstration of nonspecific projections to the cerebral cortex and striatum from thalamic intralaminar nuclei in the rat, cat and monkey. J. Comp. Neur., 154: 349-377.

Kizer, J. S., M. Palkovits and M. J. Brownstein 1976 The projections of the A8, A9 and A10 dopaminergic cell bodies: evidence for a nigral-hypothalamic-median eminence dopaminergic pathway. Brain Res., 108: 363-370.

König, J. F. R., and R. A. Klippel 1963 The Rat Brain. Krieger, Huntington, New York.

Krettek, J. E., and J. L. Price 1974 A direct input from the amygdala to the thalamus and the cerebral cortex. Brain Res., 67: 169-174.

Kristensson, K., and Y. Olsson 1971 Retrograde axonal transport of protein. Brain Res., 29: 363-365.

——— 1974 Retrograde transport of horseradish peroxidase in transected axons. 1. Time relationships between transport and induction of chromatolysis. Brain Res., 79: 101-109.

Kuhar, M. J., G. K. Aghajanian and R. H. Roth 1972 Tryptophan hydroxylase activity and synaptosomal uptake of serotonin in discrete brain regions after midbrain raphe lesions: correlations with serotonin levels and histochemical fluorescence. Brain Res., 44: 165-176.

Kusama, T., and N. Hagino 1961 Medial forebrain bundle and stria medullaris in rabbits. Folia psychiat. neurol. jap., 15: 229-245.

Kuypers, H. G. J. M., and V. A. Maisky 1975 Retrograde axonal transport of horseradish peroxidase from spinal cord to brain stem cell groups in the cat. Neurosci. Letters, 1: 9-14.

Laursen, A. M. 1955 An experimental study of pathways from the basal ganglia. J. Comp. Neur., 102: 1-25.

LaVail, J. H., and M. M. LaVail 1972 Retrograde axonal transport in the central nervous system. Science, 176: 1416-1418.

——— 1974 The retrograde intraaxonal transport of horseradish peroxidase in the chick visual system: a light and electron microscopic study. J. Comp. Neur., 157: 303-358.

LaVail, J. H., K. R. Winston and A. Tish 1973 A method based on retrograde intraaxonal transport of protein for identification of cell bodies of origin of axons terminating within the CNS. Brain Res., 58: 470-477.

Leonard, C. M., and J. W. Scott 1971 Origin and distribution of the amygdalofugal pathways in the rat: an experimental neuroanatomical study. J. Comp. Neur., 141: 313-330.

Marburg, O. 1944 The structure and fiber connections of the human habenula. J. Comp. Neur., 80: 211-234.

Massopust, L. C., Jr., and R. Thompson 1962 A new interpedunculo-diencephalic pathway in rats and cats. J. Comp. Neur., 118: 97-105.

Millhouse, O. E. 1969 S Golgi study of the descending medial forebrain bundle. Brain Res., 15: 341-363.

Mitchell, R. 1963 Connections of the habenula and of the interpeduncular nucleus in the cat. J. Comp. Neur., 121: 441-457.

Mizuno, N., C. D. Clemente and E. K. Sauerland 1969 Fiber projections from rostral basal forebrain structures in the cat. Exp. Neur., 25: 220-237.

Mok, A. C. S., and G. J. Mogenson 1974 Effects of electrical stimulation of the lateral hypothalamus, hippocampus, amygdala and olfactory bulb on unit activity of the lateral habenular nucleus in the rat. Brain Res., 77: 417-429.

Nauta, H. J. W. 1974 Evidence of a pallidohabenular pathway in the cat. J. Comp. Neur., 156: 19-28.

Nauta, H. J. W., M. B. Pritz and R. J. Lasek 1974 Afferents to the rat caudoputamen studied with horseradish peroxidase. An evaluation of a retrograde neuroanatomical research method. Brain Res., 67: 219-238.

Nauta, W. J. H. 1956 An experimental study of the fornix system in the rat. J. Comp. Neur., 104: 247-271.

——— 1958 Hippocampal projections and related neural pathways to the mid-brain in the cat. Brain, 81: 319-341.

Nauta, W. J. H. and W. Haymaker 1969 Hypothalamic nuclei and fiber connections. In: The Hypothalamus. W. Haymaker, E. Anderson and W. J. H. Nauta, eds. Thomas, Springfield, Ill., pp. 136-209.

Nauta, W. J. H., and H. G. J. M. Kuypers 1958 Some ascending pathways in the brain stem reticular formation. In: Reticular Formation of the Brain. H. H. Jasper, L. D. Proctor, R. S. Knighton, W. C. Noshay and R. T. Costello, eds. Little-Brown, Boston, pp. 3-30.

Palkovits, M., and D. M. Jacobowitz 1974 Topographic atlas of catecholamine and acetylcholinesterase-containing neurons in the rat brain. II. Hindbrain (Mesencephalon, rhombencephalon). J. Comp. Neur., 157: 29-42.

Parent, A. 1976 Striatal afferent connections in the turtle (Chrysemys picta) as revealed by retrograde axonal transport of horseradish peroxidase. Brain Res., 108: 25-36.

Pierce, E. T., W. W. Foote and J. A. Hobson 1976 The efferent connection of the nucleus raphe dorsalis. Brain Res., 107: 137-144.

Powell, E. W. 1968 Septohabenular connections in the rat, cat and monkey. J. Comp. Neur., 134: 145-150.

Powell, E. W. and R. B. Leman 1976 Connections of the nucleus accumbens. Brain Res., 105: 389-403.

Powell, T. P. S., W. M. Cowan and G. Raisman 1965 The central olfactory connections. J. Anat (London), 99: 791-813.

Price, J. L. and T. P. S. Powell 1970a An experimental study of the origin and the course of the centrifugal fibers to the olfactory bulb in the rat. J. Anat. (London), 107: 215-237.

——— 1970b The afferent connexions of the nucleus of the horizontal limb of the diagonal band. J. Anat. (London), 107: 239-256.

——— 1971 Certain observations on the olfactory pathway. J. Anat. (London), 110: 105-126.

Raisman, G. 1966 The connexions of the septum. Brain, 89: 317-348.

Raisman, G., W. M. Cowan and T. P. S. Powell 1966 An experimental analysis of the efferent projections of the hippocampus. Brain, 89: 83-108.

Scalia, F., and D. R. Colman 1974 Aspects of the central projection of the optic nerve in the frog as revealed by anterograde migration of horseradish peroxidase. Brain Res., 79: 496-504.

Smaha, L. A., and W. W. Kaelber 1973 Efferent fiber projections of the habenula and the interpeduncular nucleus. An experimental study in the opossum and cat. Exp. Brain Res., 16: 291-308.

Sprague, J. M., and M. Meyer 1950 An experimental study of the fornix in the rabbit. J. Anat. (London), 84: 354-368.

Swanson, L. W. 1976 An autoradiographic study of the efferent connections of the preoptic region in the rat. J. Comp. Neur., 167: 227-256.

Swanson, L. W., and W. M. Cowan 1975 A note on the connections and development of the nucleus accumbens. Brain Res., 92: 324-330.

Turner, P. T., and A. B. Harris 1974 Ultrastructure of

exogenous peroxidase in cerebral cortex. Brain Res., 74: 305-326.

Wilson, R. D. 1972 The neural associations of nucleus accumbens septi. Masters thesis, M.I.T., Cambridge, Mass.

Wolf, G., and J. Sutin 1966 Fiber degeneration after lateral hypothalamic lesions in the rat. J. Comp. Neur., 127: 137-156.

Zyo, K., T. Ôki and T. Ban 1963 Experimental studies on the medial forebrain bundle, medial longitudinal fasciculus and supraoptic decussations in the rabbit. Med. J. Osaka Univ., 13: 193-239.

Reprinted from THE JOURNAL OF COMPARATIVE NEUROLOGY
Vol. 187, No. 1, September 1, 1979  © The Wistar Institute Press 1979

# Efferent Connections of the Habenular Nuclei in the Rat [1]

MILES HERKENHAM [2] AND WALLE J. H. NAUTA
*Department of Psychology, Massachusetts Institute of Technology,
Cambridge, Massachusetts 02139*

*ABSTRACT*    The efferent connections of the medial (MHb) and lateral
(LHb) habenular nuclei in the rat were demonstrated autoradiographically following small injections of tritiated amino acids localized within various parts
of the habenular complex. Comparison of individual cases led to the following
conclusions.

MHb efferents form the core portion of the fasciculus retroflexus and pass to
the interpeduncular nucleus (IP) in which they terminate in a topographic pattern that reflects 90° rotations such that dorsal MHb projects to lateral IP, medial MHb to ventral, and lateral MHb to dorsal IP. Most MHb fibers cross in
the interpeduncular nucleus in the "figure 8" pattern described by Cajal, and
terminate throughout the width of IP with only moderate preference for the ipsilateral side. However, the most dorsal part of MHb projects almost exclusively to the most lateral IP zone in a cluster pattern that is particularly dense on
the ipsilateral side. The MHb appears to have no other significant projections,
but very sparse MHb fibers may pass to the supracommissural septum and to
the median raphe nucleus.

Except for some fibers passing ventrally into the mediodorsal nucleus, all of
the LHb efferents enter the fasciculus retroflexus and compose the mantle portion of the bundle. No LHb projections follow the stria medullaris. In the ventral tegmental area LHb efferents become organized into groups that disperse
in several directions: (a) Rostrally directed fibers follow the medial forebrain
bundle to the lateral, posterior and dorsomedial hypothalamic nuclei, ventromedial thalamic nucleus, lateral preoptic area, substantia innominata and ventrolateral septum. (b) Fibers turning laterally distribute to the substantia
nigra, pars compacta (SNC); a small number continue through SNC to adjacent
tegmentum. (c) The largest contingent of LHb efferents passes dorsocaudally
into paramedian midbrain regions including median and dorsal raphe nuclei,
and to adjacent tegmental reticular formation. Sparse additional LHb projections pass to the pretectal area, superior colliculus, nucleus reticularis tegmenti pontis, parabrachial nuclei and locus coeruleus. No LHb projections appear to
involve the interpeduncular nucleus. All of these connections are in varying
degree bilateral, with decussations in the supramammillary region, ventral tegmental area and median raphe nucleus.

On the basis of differential afferent and efferent connections, the LHb can be
divided into a medial (M-LHb) and a lateral (L-LHb) portion. The M-LHb, receiving most of its afferents from limbic regions and only few from globus
pallidus, projects mainly to the raphe nuclei, while L-LHb, afferented mainly
by globus pallidus and in lesser degree by the limbic forebrain, projects predominantly to a large region of reticular formation alongside the median raphe
nucleus. Both M-LHb and L-LHb, however, project to SNC. The reported data
are discussed in correlation with recent histochemical findings.

---

[1] These data first appeared in abstract form in Anat. Rec. ('77) *187:* 603.
[2] Present address: Laboratory of Neurophysiology, National Institute of Mental Health, Bethesda, Maryland 20205.

One of the most remarkable features of the habenula consists in the generally caudal orientation of its fiber connections. Although traditionally considered part of the thalamus, the habenular nuclei lack the prominent ascending projections entrained by all other thalamic cell groups, and instead project caudalward by way of the fasciculus retroflexus to a paramedian midbrain region that includes the interpeduncular nucleus and the mesencephalic raphe nuclei (Nauta, '58; Akagi and Powell, '68). Likewise in contrast with other thalamic cell groups, the habenular nuclei receive mainly descending afferents, conveyed by the stria medullaris from a variety of forebrain structures, in particular the supracommissural septum, the nucleus of the diagonal band, the lateral preoptic and hypothalamic region, and the internal segment of the globus pallidus (Herkenham and Nauta, '77). Since all but the last mentioned of these sources of habenular afferents lie in the path also of other conduction routes leading from the limbic forebrain to the paramedian zone of the midbrain, the dorsal pathway comprising the stria medullaris, habenular nuclei and fasciculus retroflexus has been interpreted as a dorsal parallel to a more ventral, transhypothalamic pathway formed largely by descending components of the medial forebrain bundle (Nauta, '58). The similarity of habenular and hypothalamic connections appears reflected in functional similarities between the two structures: various experimental manipulations of the habenula have been reported to affect autonomic (Kabat, '36; Cragg, '61b; Lengvari et al., '70) and endocrine control (Szentágothai et al., '62; Ford, '68; Motta et al., '68) as well as aspects of sexual (Zouhar and de Groot, '63; de Groot, '65; Modianos et al., '74), consummatory (Donovick et al., '69; Cooper and Van Hoesen, '72) and defensive behavior (Reinert, '64; Rausch and Long, '74).

On the other hand, the habenular nuclei have some connections that distinguish them from other diencephalic structures associated with the limbic system. First, they constitute the major source of afferents to the interpeduncular nucleus (Lenn, '76) and the mesencephalic raphe nuclei (Aghajanian and Wang, '77). Second, a long standing suggestion that the habenula receives fibers from the ansa lenticularis was recently confirmed by the demonstration, in the cat and rat, of a substantial projection from the entopeduncular nucleus (the non-primate homologue of the internal pallidal segment) to the lateral part of the lateral habenular nucleus (L-LHb) (Nauta, '74; Herkenham and Nauta, '77; Larsen and McBride, '79); as emphasized by Nauta ('74), this finding identifies the lateral habenular nucleus as one of the few forebrain structures in which efferent channels of corpus striatum and limbic system are known to converge. The medial habenular nucleus (MHb), by contrast, appears to receive afferents almost exclusively from the supracommissural septum, an afferent connection that apparently does not involve the LHb at all (Herkenham and Nauta, '77). Considered together, the recent data suggest a subdivision of the habenular complex into three compartments, each characterized by a single dominant afferent system: (1) the medial nucleus (MHb), dominated by afferents from the supracommissural septum, (2) a medial part of the lateral nucleus (M-LHb), dominated by afferents from the basal forebrain, and (3) a lateral division of the lateral nucleus (L-LHb) dominated by afferents from the internal pallidal segment.

The present study was prompted in part by the question whether the suggested tripartite nature of the habenular complex might find expression in a particular patterning of the habenular projection to the paramedian midbrain. Previous studies of the efferent connections of the habenular complex by fiber degeneration methods have been unable to settle this question, mainly as a consequence of the fiber-of-passage problem and the small size of the habenular nuclei.

MATERIALS AND METHODS

In each of 16 adult albino rats of both sexes (Charles River Laboratories) a small deposit of an equal part mixture of tritiated proline and leucine (proline, leucine and lysine in two cases) was placed in the habenular complex. The injections were made microelectrophoretically as described by Graybiel and Devor ('74) from a glass micropipette (internal diameter 10-20 $\mu$m at the tip) filled with 0.01 M acetic acid containing the isotopes at a concentration of 20 $\mu$Ci/$\mu$l, the driving force being supplied by a 0.5-1 $\mu$A positive current delivered by a constant current source (Midgard Electronics, Newton, Massachusetts) at a pulse rate of seven seconds on, seven seconds off, for a duration of eight to ten minutes. The rats were sacrificed 1 to 13 days postoperatively by an overdose of anesthetic, and their formaldehyde perfused brains, postfixed in formaldehyde for one to two weeks, were embedded in albumin-gelatin and sectioned on a freezing microtome at 25 μm in the frontal or sagittal plane. A series composed of every third section was mounted on subbed glass slides which were then coated with Kodak NTB-2 emulsion at 40-42°C, dried slowly, and stored in light- and moisture-proofed boxes at −15°C for 4 to 20 weeks. After being developed in Kodak D-19 at 16-17°C the sections were counterstained with cresylechtviolet, coverslipped, and examined microscopically under both brightfield and darkfield illumination for radioactively labeled nerve fibers.

## RESULTS

### Projections of the medial habenular nucleus

Of five cases in which the isotope was almost completely confined to the MHb, one (Hb-33) is particularly informative and will therefore be described in some detail. At the focus of the injection in case Hb-33 (figs. 1A, 2C) isotope completely fills the cross section of the nucleus at approximately the middle of the latter's longitudinal extent, and spreads slightly across the lateral border into the LHb, as well as into the thalamic periventricular nucleus. Nearly all of the labeled fibers leave the nucleus in a ventrolateral direction and assemble to form the compact core of the fasciculus retroflexus (figs. 1B, 2D), in which

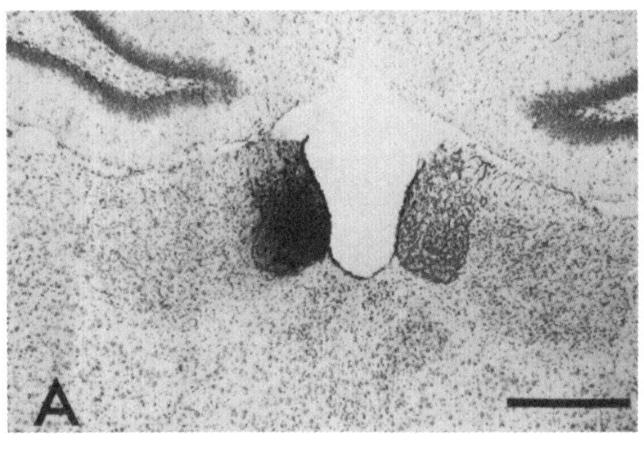
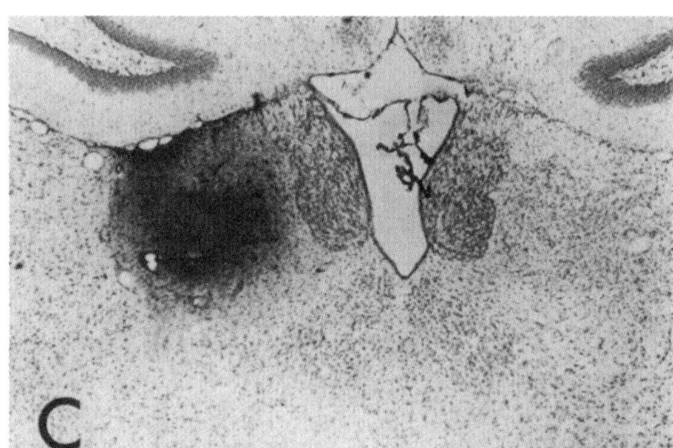
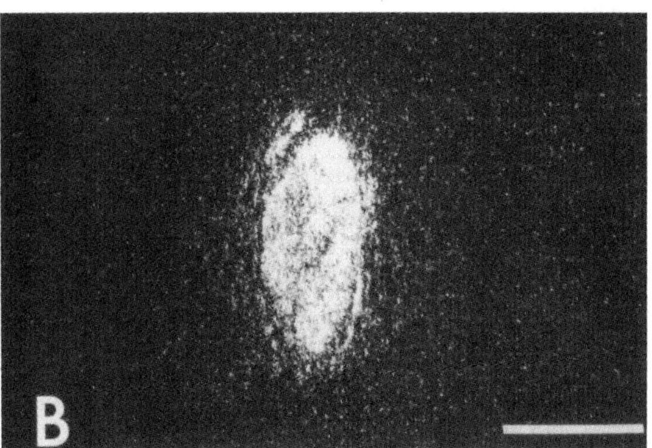
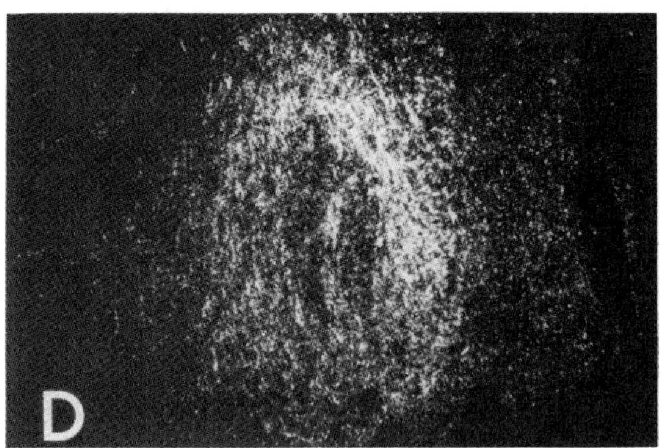

Fig. 1A   Case Hb-33. Brightfield microphotograph of center of isotope injection into the MHb located at A 4,110 μm of König and Klippel ('63). Twelve-week exposure. This section was charted in figure 2D. Black bar measures 0.5 mm for figures 1A and 1C.

B   Darkfield microphotograph of labeled MHb fibers within the core of the fasciculus retroflexus in Hb-33, at the level of the anterior ventral tegmental area.

C   Case Hb-12. Brightfield microphotograph of center of isotope injection into the LHb, located at A 3,990 μm. Fourteen-week exposure. This section was charted in figure 3G.

D   Darkfield microphotograph of labeled LHb fibers at the periphery of the fasciculus retroflexus in Hb-12, in the anterior ventral tegmental area.

they extend to the interpeduncular nucleus (figs. 2E, 13).

Although a complete charting of the label distribution in this case (fig. 2) suggests a variety of further MHb efferents, it seems likely from a critical comparison with other cases that the medial habenular nucleus has few if any projections to structures other than the interpeduncular nucleus. Sparse labeling lines the third ventricle (figs. 2A-D) and the aqueduct of Sylvius (figs. 2E-G), but since such labeling is found in all cases in which isotope is injected into or near the ventricular space, it probably indicates no more than a diffusion of the label into periventricular regions from the cerebrospinal fluid. Labeled fibers passing rostrally into the basal forebrain and bed nucleus of the stria terminalis (figs. 2A,B) can be seen in other cases (Ce-18, not shown) to arise from the thalamic periventricular nucleus, and are absent when the isotope does not invade this nucleus (for example, case Hb-

31, fig. 11). A very small contingent of labeled fibers can be traced rostrally in the stria medullaris to the supracommissural septal region containing the nuclei septofimbrialis and triangularis septi (figs. 2A,B); although no such fibers are labeled in any of the remaining four cases of MHb injection, their distribution suggests that they may represent a sparse reciprocation of the massive projection from the supracommissural septal region to the MHb (Herkenham and Nauta, '77). Alternatively, however, the label may have been transported retrogradely (see below). The sparse fiber labeling in the hypothalamus (fig. 2C) is indistinguishable from that seen in cases of isotope injection confined to the lateral habenular nucleus (fig. 3), and could have resulted from the slight spread of isotope beyond the lateral border of the MHb. The diffuse fiber labeling in the paramedian region of the midbrain tegmentum (figs. 2F-H) likewise closely resembles the pattern of LHb projections, and

---

*Abbreviations*

a, accumbens nucleus
AC, anterior commissure
ah, anterior hypothalamus
am, anteromedial nucleus
apl, lateral preoptic area
apm, medial preoptic area
AS, aqueduct of Sylvius
av, anteroventral nucleus
avt, ventral tegmental area
b, basal nucleus (nucleus of horizontal
    limb of diagonal band)
BC, brachium conjunctivum
CA, anterior commissure
cg, central gray substance
CP, cerebral peduncle
DBC, decussation of brachium
    conjunctivum
dh, dorsomedial hypothalamus
dr, dorsal raphe nucleus
DT, tegmental decussation
dt, dorsal tegmental nucleus
    (of Gudden)
F, fornix
FN, facial nerve (VII)
FR, faciculus retroflexus
ge, gelatinosus nucleus
gp, globus pallidus
H, field H of Forel
hi, hippocampal formation
hl, lateral habenula
hm, medial habenula
IC, internal capsule
ic, inferior colliculus
ip, interpeduncular nucleus
lc, locus coeruleus
lh, lateral hypothalamic region
LM, medial lemniscus
mb, mammillary bodies
md, mediodorsal nucleus
MLF, medial longitudinal fasciculus

MP, mammillary peduncle
mr, median raphe nucleus
MT, mammillothalamic tract
mV, mesencephalic nucleus of the
    trigeminus (V)
o, oculomotor nucleus (III)
ON, oculomotor nerve (III)
OT, optic tract (II)
ot, olfactory tubercle
P, pyramidal tract
p, pons
pb, parabrachial nucleus
pf, parafascicular nucleus
ph, posterior hypothalamus
pv, thalamic periventricular nucleus
pvp, posterior periventricular nucleus
r, red nucleus
re, reuniens nucleus
RF, reticular formation
rt, reticular thalamic nucleus
sc, superior colliculus
sf, septofimbrialis nucleus
sl, lateral septum
SM, stria medullaris
snc, substantia nigra, pars compacta
snr, substantia nigra, pars reticulata
ST, stria terminalis
st, nucleus of stria terminalis
sum, supramammillary region
tdl, tegmenti dorsalis lateralis
    nucleus
tp, reticularis tegmenti pontis nucleus
ts, triangularis septinucleus
v, ventral thalamic complex
vm, ventromedial nucleus
vmh, ventromedial hypothalamic nucleus
vp, ventral pallidum
vt, ventral tegmental nucleus
    (of Gudden)
zi, zona incerta

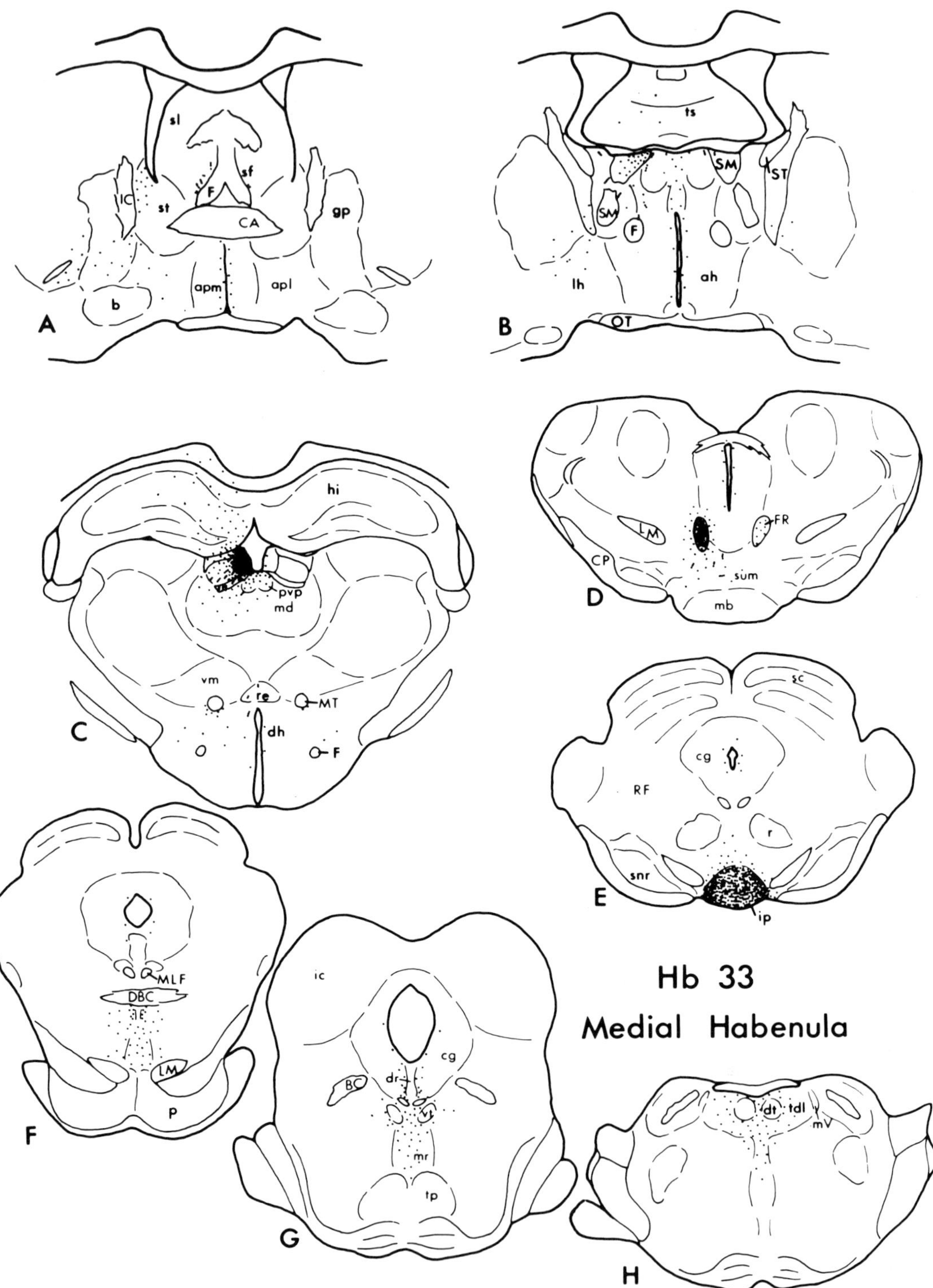



Fig. 2   Projection chartings of a rostro-caudal series of frontal sections in case Hb33. As in all chartings, the area of the injection (level C) is shown in solid black and the autoradiographically labeled axons are represented as short dotted lines, or as grouped dots when the labeled path appears in cross-section. The course of the labeled paths and the areas of axon termination are described in the text.

could be attributed to the same source. However, in cases in which the LHb is not at all involved in the injection (for example, case Hb-25, fig. 14), very sparse labeling can be traced caudally into paramedian tegmental regions but not into the lateral hypothalamus. Therefore, it is possible that the MHb maintains a sparse projection to midline tegmental regions, in particular to the median raphe nucleus.

The habenulo-interpeduncular projection

The foregoing findings indicate that, except for possible sparse projections to the supracommissural septum and to the paramedian zone of the midbrain tegmentum, the MHb projects exclusively to the interpeduncular nucleus. The habenulo-interpeduncular tract forms the central core of the fasciculus retroflexus; as will be described later, fibers from the lateral habenular nucleus surround this core (compare figs. 1B,D). The habenulo-interpeduncular tract enters the interpeduncular nucleus laterally at its rostral end and disperses into fascicles that extend across the nucleus in a horizontal plane. The autoradiographic appearance of the interpeduncular nucleus in case Hb-33 (fig. 13) accords well with Cajal's ('11) original observation that many habenulo-interpeduncular axons traverse the nucleus two or three times in a shoelace or "figure 8" manner before terminating.

Injections localized to different sectors of the MHb reveal some details of the habenulo-interpeduncular projection pattern. Injections placed in dorsal parts of MHb (case Hb-31, fig. 11) label axons that pass ventrolaterally through the lateral habenular nucleus (fig. 11) and into the habenulo-interpeduncular tract. As the labeled fibers enter the interpeduncular nucleus, some cross horizontally in the rostral sector, but many retain a lateral position in the nucleus on the ipsilateral side. In the caudal half of the nucleus a distinct dense patch of terminal label marks a lateral strip of the nucleus ipsilaterally (fig. 12); the same lateral zone is more sparsely labeled on the contralateral side. Injections placed medially (fig. 14) in the MHb (cases Hb-10 and Hb-25) label axons that are distributed to a ventral stratum of the interpeduncular nucleus (fig. 15), whereas more lateral injections (fig. 7A) of the MHb (case Hb-34) label projections to more dorsal strata of the nucleus (fig. 16). The topography of the habenulo-interpedun-

cular projection reflects a compound 90° rotation of the MHb with respect to the interpeduncular nucleus, associating dorsal MHb with lateral, medial MHb with ventral, and lateral MHb with dorsal parts of the interpeduncular nucleus. Consistent with this topographic relationship, isotope injections of the periventricular thalamic nucleus that spread into ventral parts of MHb (cases Ce-15 and Ce-18, not shown) label projections to all of the interpeduncular nucleus except for the lateral wings that receive habenular efferents from the dorsal portion of the MHb. The rostro-caudal coordinate of the projection was not systematically studied.

### Projections of the lateral habenular nucleus

Case Hb-12. Large injection

The injection in this case has its center at the level shown in figure 1C. The isotope is contained within the LHb laterally and ventrally by a thin fiber stratum that provides a natural barrier, but medially and dorsally label has been incorporated by some cells in the MHb and hippocampus, respectively. Additional label is spilled along the penetration track in the cortex. Longitudinal diffusion of label from the injection focus, rather more extensive in this case than is typical of iontophoretic injections, has resulted in labeling of LHb neurons extending over the caudal two-thirds of the nucleus. Whereas the label is contained by the habenula's fiber capsule at the injection center (figs. 1C, 3G), it diffuses into the thalamic nuclei mediodorsalis and centralis lateralis at more rostral levels (figs 3F, 17). The charts in figure 3 illustrate nearly the complete pattern of labeling found in this case, including the labeling resulting from the MHb involvement, but do not show the sparse labeling in the internal capsule, striatum and cortex that is accounted for by the escape of label to the nuclei mediodorsalis and centralis lateralis.

By far the greatest numbers of labeled LHb fibers directly enter the fasciculus retroflexus, in which they become arranged in a manner complementary to that of the MHb fibers: whereas the latter compose the compact core of the bundle (fig. 1B), the more loosely arranged LHb fibers surround this core in a mantle-like fashion (fig. 1D). A small number of labeled LHb fibers enter the fasciculus retroflexus only after describing a wide lateral detour through the ventral thalamic nucleus (fig. 3H) without apparent termination en route. Labeling of additional fibers spreading to rostral parts of the central gray substance and adjacent tectal and tegmental regions (figs. 3H,I) is sparse in this case but more prominent in a case of rostral LHb injection (case Hb-11, not shown). The only further labeled fibers pass ventrally from the injection site into lateral parts of the mediodorsal nucleus of both sides (figs. 3D-G, 17). The ipsilateral stria medullaris is heavily labeled (figs. 3C-E), but, as will be argued later, it is not certain that this represents an ascending LHb projection.

As the fasciculus retroflexus nears the interpeduncular nucleus most of its labeled fibers accumulate on the bundle's dorsomedial side (fig. 1D). At this point, they begin to disperse in various directions (figs. 5C, 6B) to form, (a) a large component passing caudally over the dorsal and lateral sides of the interpeduncular nucleus, then curving dorsalward into the paramedian tegmentum, ventral parts of the central gray substance, and superior colliculus, (b) a lateral projection into and through the substantia nigra, pars compacta, and (c) a rostral projection that ascends in the medial forebrain bundle.

Two main peculiarities of the projections from the lateral habenular nucleus must be emphasized. First, all projections are strongly bilateral and of only slightly less volume on the contralateral than on the ipsilateral side. Second, with only a single exception to be mentioned below, their mode of radioactive labeling fails to indicate any particular structure or region as a distinct target site; instead, labeled fibers traced along any particular route arrive at a given region and merely dwindle in number, suggesting a form of termination without profuse preterminal axon arborization. Therefore, target regions can be identified only by careful examination of the grain density along the pathways described.

*Caudal projections*

Most of the labeled fibers in the fasciculus retroflexus descend over the dorsolateral margin of the interpeduncular nucleus (figs. 3I,J). A moderate labeling of the nucleus in this case probably reflects the marginal involvement of the medial habenular nucleus noted above, since isotope deposits entirely restricted to the LHb (cases Hb-16 and Hb-20) do not label any

projection to the interpeduncular nucleus (figs. 5A,B, 6). Some fibers curve dorsally along the midline and pass through the nucleus raphis linearis and between the oculomotor nuclei into the ventral part of the central gray substance (fig. 3J). A small number of labeled fibers continue from here laterally to the lateral one-third of the intermediate and deep collicular layers (figs. 3J,K).

Most of the labeled LHb efferents continue caudally in a paramedian position to enter the nucleus centralis superior of Bechterew (median raphe nucleus). However, additional fiber labeling densely marks a tegmental field laterally adjacent to the raphe nucleus, in the interstices of the decussated brachium conjunctivum, especially at levels between those charted in J and K of figure 3. This terminal field will be described in more detail below. Throughout the length of the midbrain descending fibers cross, to follow similar courses on the contralateral side. Thus, labeled fibers bilaterally descend into a lateral zone of the nucleus centralis superior (figs. 3K-M) and from this position either turn medially into more central parts of the nucleus or continue dorsolaterally into the tegmental reticular formation adjoining the brachium conjunctivum proximal to its decussation (figs. 3K,L, 4). The most dorsal fibers enter the ventral region of the central gray substance, which at these levels contains the dorsal raphe nucleus. It is remarkable that labeling in the central gray substance is more dense along the sides of the dorsal raphe nucleus than within the nucleus itself (figs. 3K,L, 19). This labeling pattern, suggesting a projection that encapsulates rather than pervades a nucleus, is encountered once more, and in more pronounced form, in the regions containing the ventral and dorsal tegmental nuclei of Gudden (figs. 3L-O, 4, 20). Finally, at the level of figure 3K (also fig. 4) paired thin lines of grain appear near the midline in the nucleus reticularis tegmenti pontis.

The pattern of labeling remains essentially unchanged at more caudal levels. Here likewise, labeled fibers mark the raphe and ventral parts of the central gray substance, as well as adjacent regions of the reticular formation. Dorsally, labeled fibers are prominently concentrated in the median plane between left and right medial longitudinal fasciculus and tegmental nuclei of Gudden (figs. 3L-N, 4). Labeling surrounding the ventral

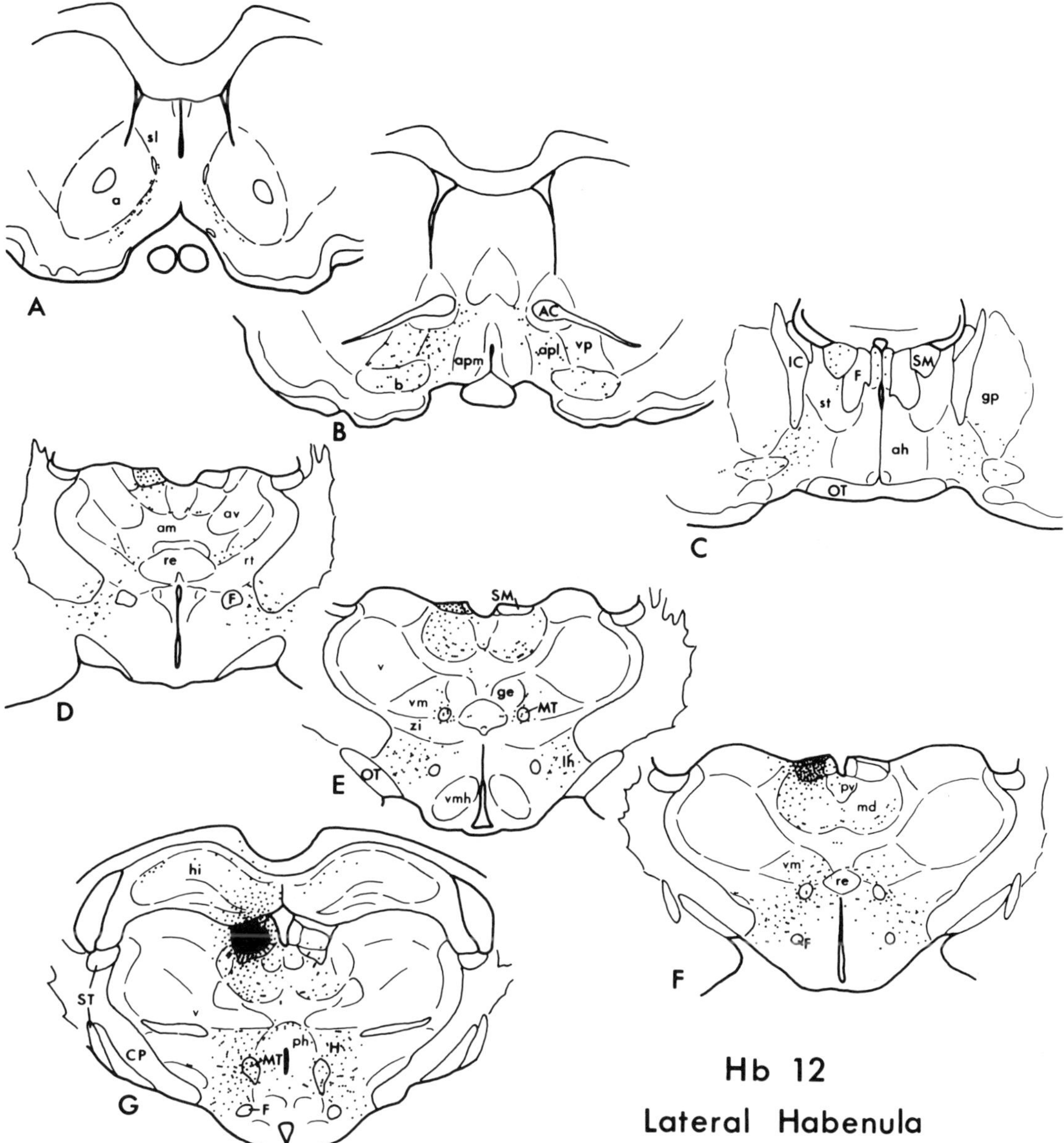


Fig. 3   Projection drawings of selected frontal sections in case Hb-12. Dots represent fiber pathways and terminal zones of the LHb projection.

tegmental nucleus is more dense than the sparse labeling within the nucleus (figs. 3L, 20). At the level of the isthmus, the lateral expansion of label encapsulates the dorsal tegmental nucleus and rather evenly fills the nucleus tegmenti dorsalis lateralis (figs. 3M-O, 4). The sectors of the dorsal tegmental nucleus defined by Morest ('61) as partes anterior, centralis and posterior receive no labeled fibers at all, and the pars ventromedialis is only sparsely labeled (fig. 4). In the most caudal part of the central gray substance labeling extends laterally to, but not across, the medial borders of the mesencephalic nucleus of the

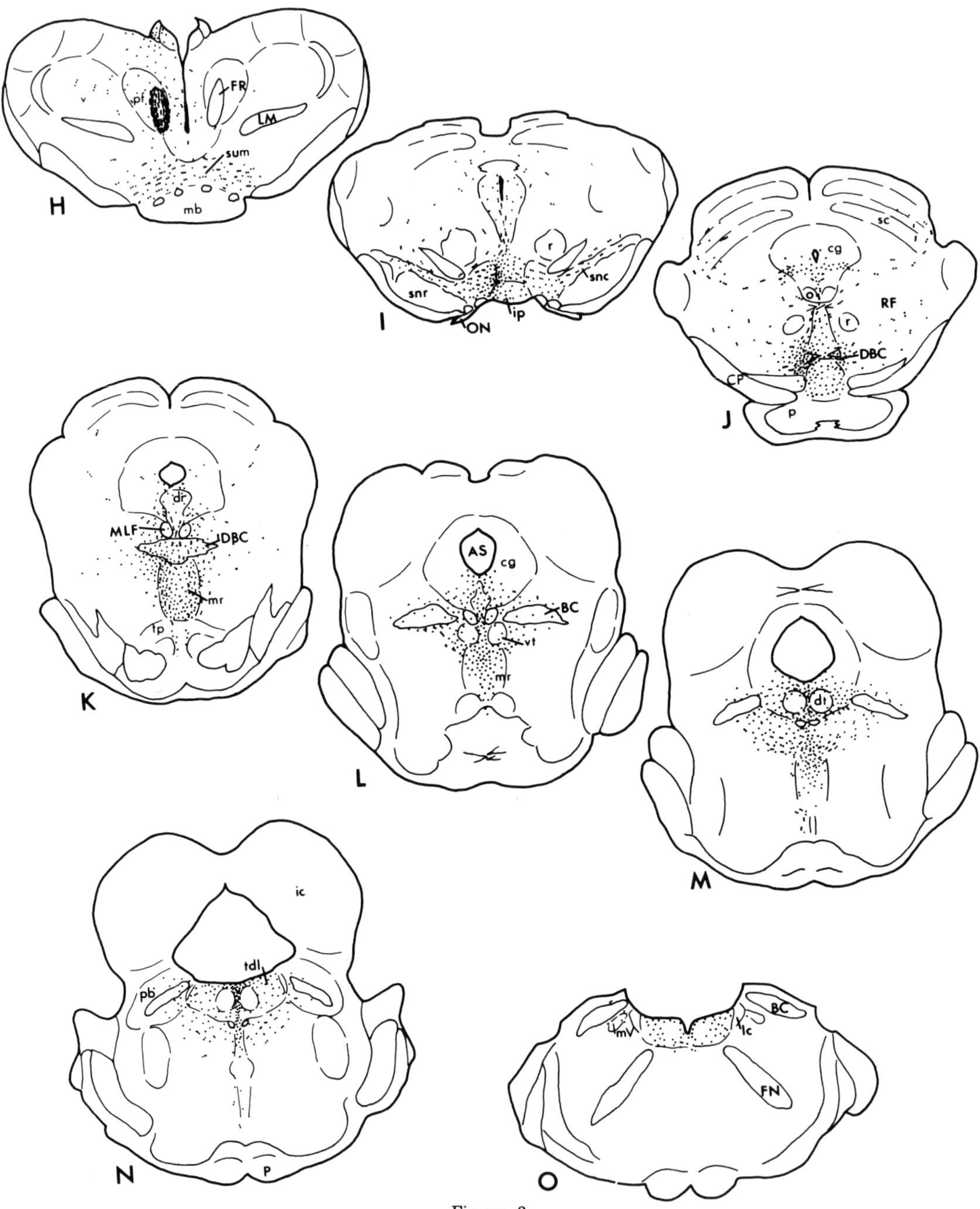

Figure 3

trigeminus and locus coeruleus (fig. 30). At these same levels there is some labeling of the parabrachial nuclei, but it is not clear whether the corresponding fibers enter the re-gion through the central gray substances or by a more ventral route.

A noteworthy feature of the fiber labeling in cases of injection involving most of the

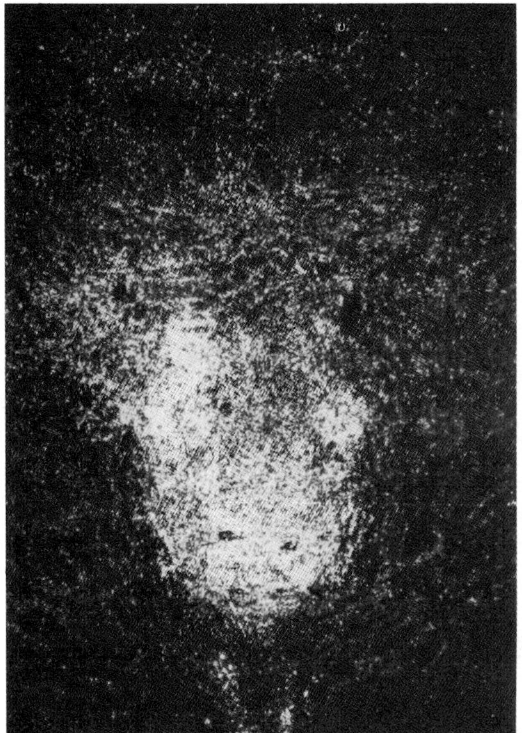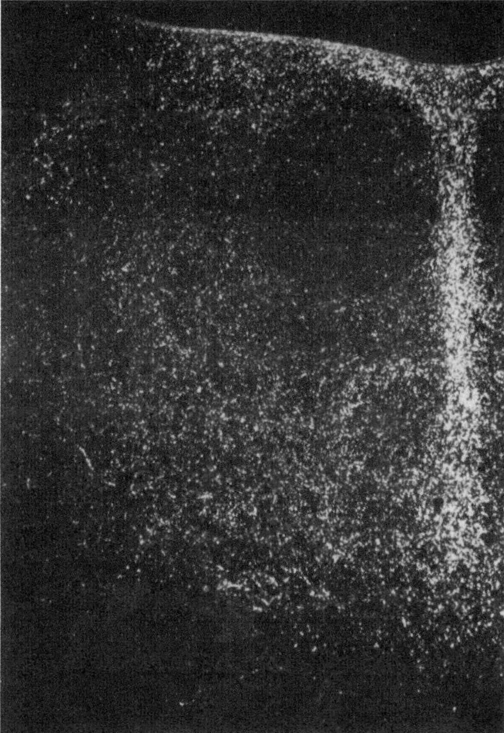

Fig. 4   LHb projections in darkfield photomicrographs.
Left.   Median raphe region at the level charted in figure 3K. In the surrounding neuropil the MLF can be distinguished as dark spots dorsally. The DBC occupies the immediately subjacent region that is characterized by intercalated labeled fibers. A strong projection to the ventrolateral interstices of the DBC appears on the left side. The area of greatest labeling is the mr, while two labeled streaks at the bottom represent the projection to the medial tp.
Right.   Dorsal tegmental nucleus region at the level charted in figure 3N. The dark unlabeled area is the dtn, within which the speckled pattern represents light reflected from glial and neuronal somata. Labeling is strongest on the midline, but it diffusely invades both the central gray and dorsal tegmental regions obscuring their common boundary.

width of LHb is that it spreads outside the lateral boundaries of the median raphe and central gray substance, and involves in equal or greater volume the adjacent reticular formation. Labeled fibers spread dorsolaterally through the dorsomedial quadrant of the reticular formation, by and large following the brachium conjunctivum caudalward (figs. 3K-N). Many of these fibers terminate in the interstices of the brachium, while others either terminate in peribrachial regions or turn medially again to enter the central gray substance. The dorsal flow of fibers in the reticular formation contributes to the encapsulation of the ventral tegmental nucleus and medial longitudinal fasciculus. Other fibers appear to flow in the opposite direction: out from the central gray into the region dorsal to the brachium conjunctivum. Most caudally, the labeled fibers occupy the medial half of the parabrachial nuclei (fig. 3N).

*Lateral projections*

Labeled LHb efferents that turn laterally from the fasciculus retroflexus in the ventral tegmental area pass into the substantia nigra, pars compacta, and the zone that lies between the pars compacta and the medial lemniscus (figs. 3I, 18). A small number of these fibers continue past the nigral region and turn dorsomedially into the mesencephalic reticular formation (figs. 3I,J), in a trajectory identical to that followed by efferents from the nucleus accumbens and lateral hypothalamus (Nauta et al., '78). Like the latter projections, these fibers follow a course toward the central gray substance, but they are lost among the more numerous LHb efferents passing caudodorsally through the tegmentum.

*Rostral projections*

Some labeled LHb fibers of the fasciculus retroflexus turn rostrally and, partly crossing over in the supramammillary decussation (fig. 3H), ascend bilaterally in the medial forebrain bundle. A few of these labeled fibers turn medially into the posterior (fig. 3G) and dorsomedial (fig. 3F) hypothalamic nuclei. Others aggregate at the ventral edge of the mammillothalamic tract and follow this bundle to the ventromedial thalamic nucleus, apparently terminating for the most part in the latter's ventromedial quarter (figs. 3D-F). The labeled fibers that remain in the medial forebrain bundle continue their rostral course in gradually decreasing number, distributing to the lateral hypothalamic and preoptic regions, and to the substantia innominata (fig. 3B). The longest of these ascending fibers extends into the basolateral septal zone lying between the nucleus accumbens and the diagonal band of Broca (fig. 3A). At slightly more rostral levels, where the nucleus accumbens and the nucleus of the diagonal band are virtually contiguous, the rostral remnants of the projection mark the border between the two cell groups but do not give rise to an identifiable terminal field in either nucleus.

It is important to note that none of the ascending LHb projections are carried by the stria medullaris. This conclusion is based upon several observations. First, labeling in the stria (figs. 3C-E) is strictly ipsilateral to the injection, whereas the forebrain labeling is strongly bilateral. Second, the density of grains marking the stria medullaris rapidly declines in the rostral direction, and falls to background levels at the rostral pole of the thalamus; accordingly, no labeled fibers can be traced in that division of the tract that interconnects the stria medullaris and the medial forebrain bundle. Third, the grains marking the stria medullaris are not aligned so as to suggest the presence of labeled axons. This is particularly striking in sagittal sections, as will be described below.

*Contralateral projections*

Habenular projections are universally bilateral. The major decussation lies in the ventral tegmental area where the fasciculus retroflexus divides into its various branches. Curiously, the habenular commissure appears to contain no crossed habenular efferents at all. The most rostral crossing of habenular efferents takes place in the supramammillary region (fig. 3H) and apparently involves largely or even exclusively the ascending projection from the lateral habenular nucleus. Numerous other fibers decussate over the dorsal border of the interpeduncular nucleus. Grains in the median raphe nucleus are often aligned in horizontal rows, somewhat like those marking the MHb fibers in the interpeduncular nucleus; it has not been possible in the present study to determine whether this grain orientation represents fibers passing through the raphe and continuing a contralateral course, or fibers establishing synaptic contacts within the median raphe nucleus.

In most regions the ratio of ipsilateral to contralateral grain density is about 3:2. A notable exception is the distribution of LHb efferents in the medial reticular formation (figs. 3J,K, 9). These fibers form the densest terminal field in the path of the LHb projection. It is bilateral, but the density of the ipsilateral field is several times higher than that of its contralateral counterpart.

The lateral habenular projection in sagittal sections

As LHb efferents generally extend caudally in a paramedian position, most of the pathway can be surveyed in a small number of closely spaced sagittal sections, and such a survey allows a somewhat wider view of the relationships described above.

Figure 5 illustrates the findings in a case, Hb-20, with a small injection (fig. 5D) confined to the lateral half of LHb. Only sparse and diffuse grain marks the M-LHb (fig. 5C) and MHb (fig. 5B). This case clearly illustrates the labeling of fibers composing the mantle portion of the fasciculus retroflexus (figs. 5C, 6B), the dissociation of this fiber assembly near the rostral pole of the interpeduncular nucleus into caudally and rostrally oriented components (fig. 6B), and the difficulty of determining whether any such fibers actually terminate in the ventral tegmental area (fig. 6C, showing diffuse grains among the labeled fascicles passing caudally over the interpeduncular nucleus).

It must be noted that in this case only few of the labeled fasciculus retroflexus fibers that

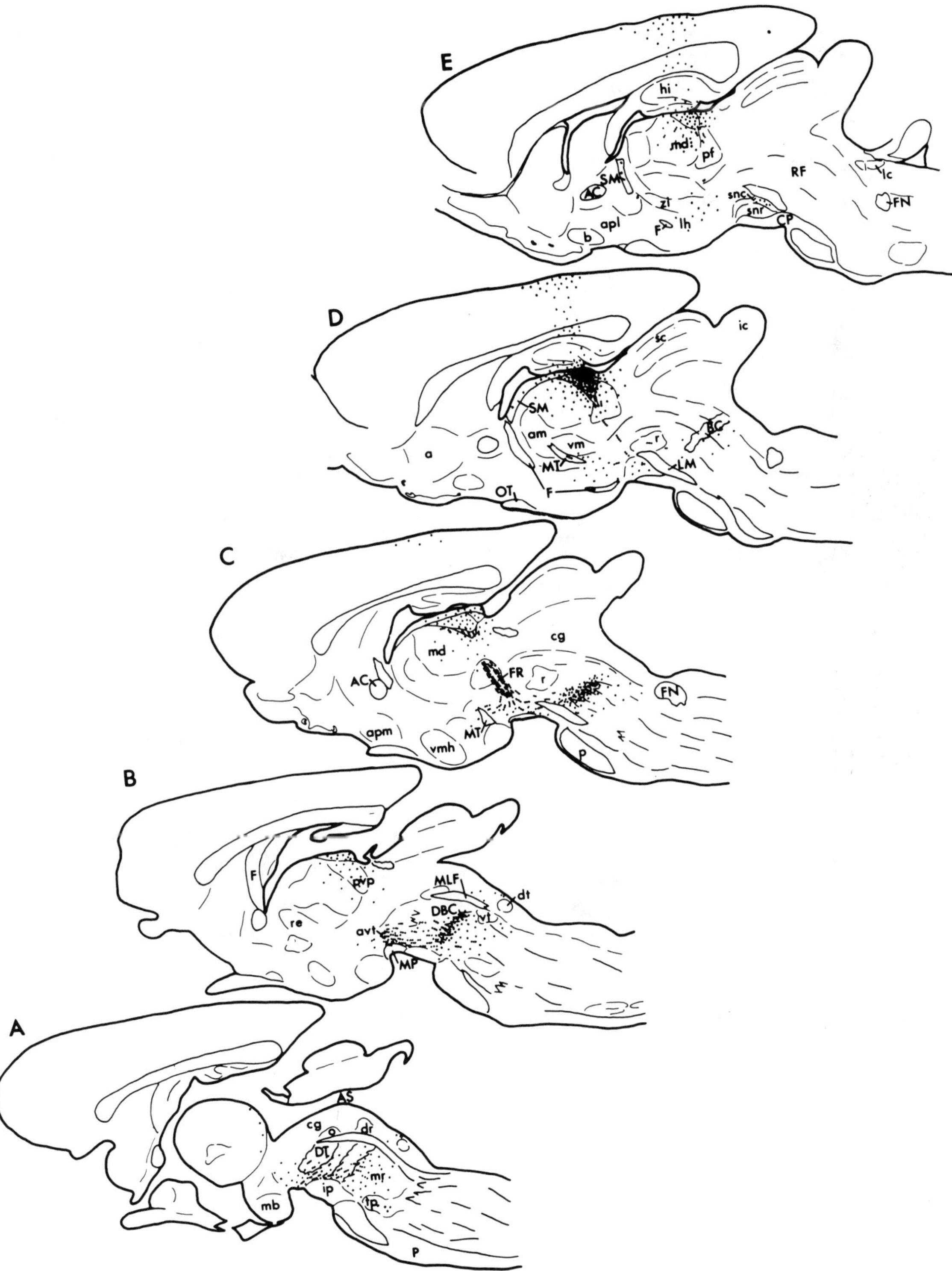

Fig. 5  Projection drawings of a series of closely spaced midsagittal and parasagittal sections in case Hb-20 of L-LHb injection. Level A is median, E is most lateral.

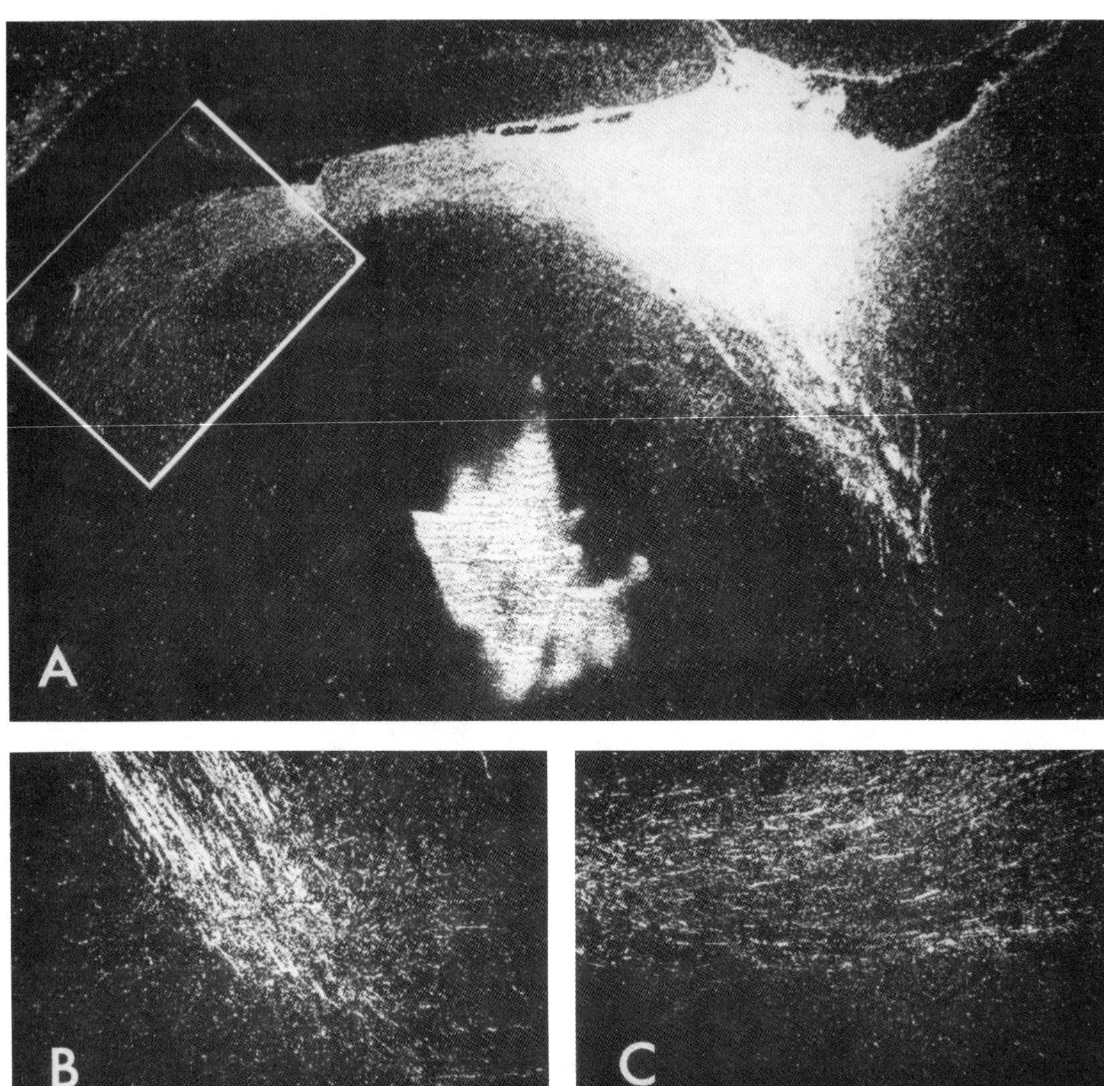

Fig. 6  Darkfield photomicrographs of LHb projections in case Hb-20; sagittal sections.
  A  Thalamic portion of section charted in figure 5D. Inset at upper left is a photomicrograph of stria medullaris region taken at same magnification as remaining photograph. The inset shows decreasing grain density from upper right to lower left. Larger aligned speckles represent the rows of glial cells within the SM.
  B  Fasciculus retroflexus at level charted in figure 5C. Note fiber flow both rostrally (left) and caudally.
  C  Habenulo-tegmental fibers passing dorsal to the interpeduncular nucleus at the level charted in figure 5B. Note that the interpeduncular nucleus is virtually free of grains.

continue caudalward are distributed to the median raphe nucleus and central gray substance. Instead, by far the most appear to terminate in a medial tegmental zone situated immediately lateral to the median raphe nucleus, and traversed by the fibers of the brachium conjunctivum (figs. 5B,C). This finding illustrates one of the main topo-graphic characteristics of the LHb projection that will be discussed below.

*Stria medullaris labeling*

The darkfield picture of grain distribution in the two major fiber tracts associated with the habenula (fig. 6A) shows that while both are labeled, the grains marking the fasciculus retroflexus are aligned with fibers and fiber fascicles, whereas the stria medullaris is marked by evenly distributed grains not aligned with the fiber flow. Moreover, the labeling in the fasciculus retroflexus retains a constant density throughout its course, while in the stria medullaris it evenly diminishes to background levels in the rostral direction. It seems doubtful from the lack of destination that the labeling marks an efferent pathway. Künzle ('77) reported retrograde transport of proline in the axons of some fiber systems, and the question arises whether the diffuse labeling of the major bundle of habenular afferents could represent such cellulipetal label transport. On the other hand, the diffuse grain pattern might suggest an extra-axonal movement of the label. Supporting the latter view, but by no means proving it correct, is (a) the fact that much sparser labeling of the stria medullaris is found in case Hb-13 which is almost identical to Hb-20 except that the animal survived the injection for 13 days instead of 1 day, and (b) the absence of any labeled cell bodies in the forebrain regions known to contribute to the stria medullaris.

### Differential projections of the medial and lateral segments of the lateral habenular nucleus

The descending projections of the LHb, as demonstrated by the foregoing cases Hb-12

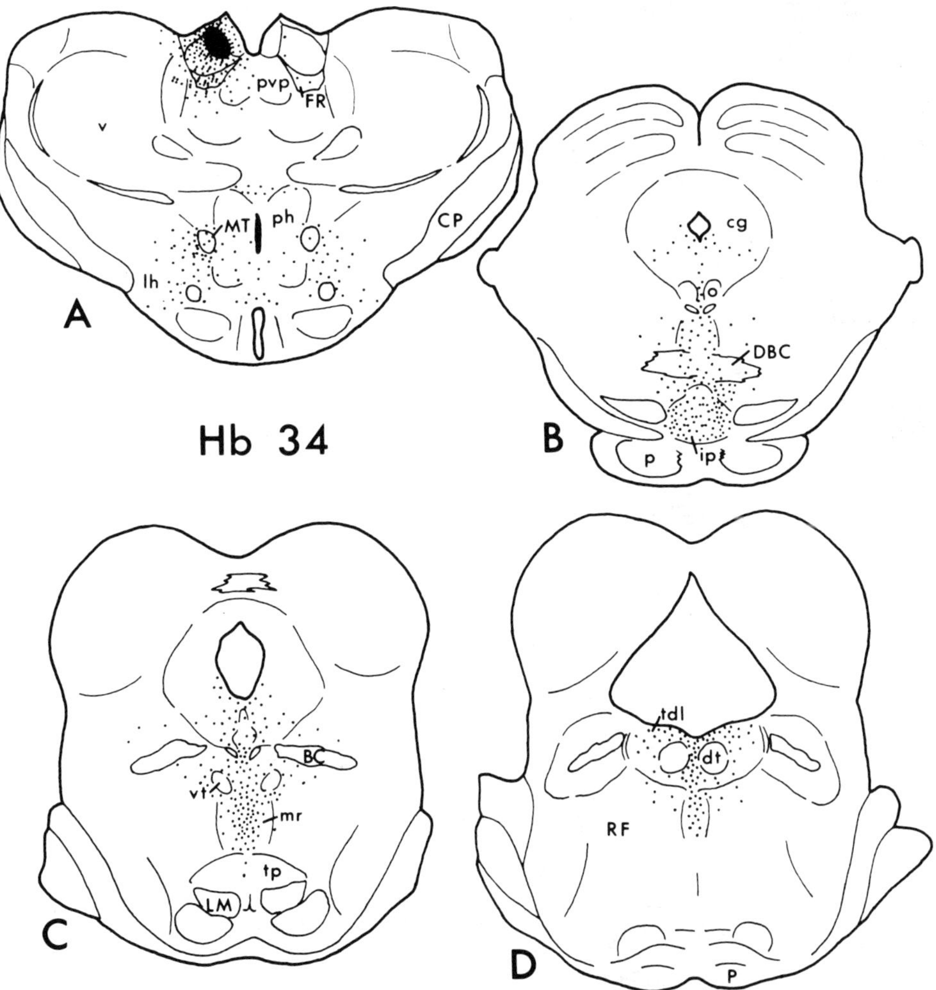


Fig. 7   Charts of sections in case Hb-34. Injection is within the M-LHb and the lateral edge of the MHb.

and Hb-20, innervate not only the median raphe nucleus and central gray substance, but also a tegmental zone adjoining the lateral sides of these midline structures. From a comparison of case Hb-34 (fig. 7) with Hb-16 (fig. 8) it is evident that the projections to the median raphe nucleus and central gray substance originate from the medial half (M-LHb), those to the laterally adjacent tegmental region from the lateral half (L-LHb) of the lateral habenular nucleus.

The small injection confined to the L-LHb in case Hb-16 (fig. 8) has labeled only sporadic fibers to the mesencephalic midline region; the descending fibers labeled in this case are very largely distributed to a medial tegmental region that adjoins the median raphe nucleus and is traversed by the decussated fibers of the brachium conjunctivum (figs. 8B,C). This connection is bilateral, but unlike all other habenular projections it is several times more massive ipsilaterally than contralaterally. Furthermore, it is the only one among the habenular projections labeled in this study that exhibits a striking increase in grain density in its terminal field (fig. 9), suggesting a more profuse and concentrated preterminal fiber arborization.

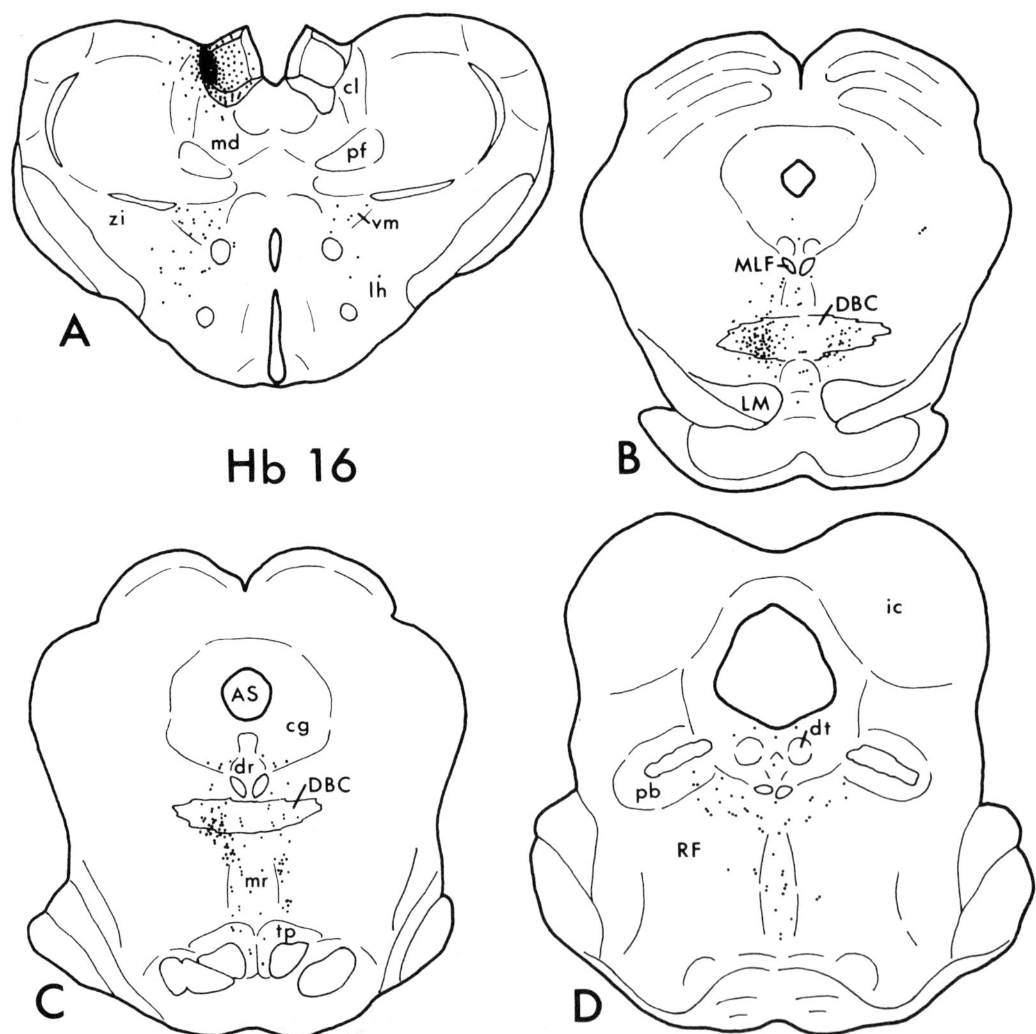


Fig. 8  Charts of sections in Hb-16, a case of L-LHb injection. Compare projections with figure 7.

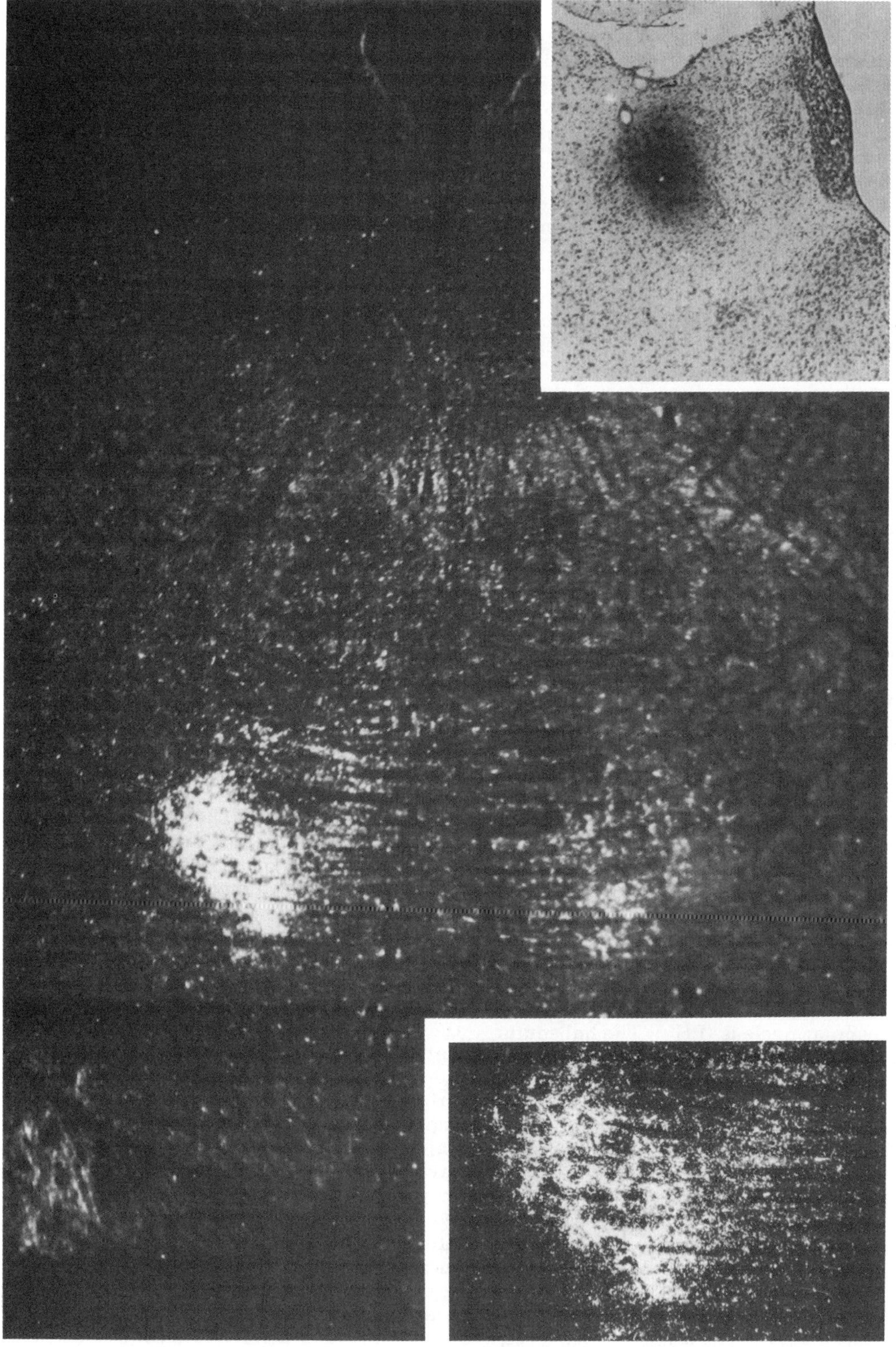

Figure  9

As illustrated by figure 7, the distribution of fibers labeled by a small injection centered in the M-LHb is nearly complementary to that of the L-LHb efferents. Fiber labeling is considerable in the midline structures and very sparse in more lateral parts of the tegmentum. The labeling of the interpeduncular nucleus (fig. 7B) in this case probably reflects some involvement of the medial habenular nucleus in the uptake of label.

Outside the paramedian tegmentum the M-LHb and L-LHb share several common targets but also show some important differences. Both project to the substantia nigra, pars compacta, and to the midline zone of the nucleus reticularis tegmenti pontis. In neither case could a projection to the superior colliculus be identified. Both regions of the LHb project rostrally into the lateral hypothalamus and ventromedial thalamic nucleus, but apparently only the M-LHb projects to the medially placed posterior and dorsomedial hypothalamic nuclei (fig. 7A). Furthermore, whereas M-LHb fibers ascend in the medial forebrain bundle as far rostrally as the lateral preoptic area and the ventrolateral septal region at the medial margin of the nucleus accumbens, the corresponding L-LHb projection appears not to extend beyond infundibular levels (figs. 8A, 5D). Finally, fibers to the lateral part of the mediodorsal nucleus are labeled only in the M-LHb case.

### DISCUSSION

#### *Organization and afferentation of the habenular complex*

The present study has demonstrated differential projections from three compartments within the habenular complex — the medial habenular nucleus (MHb) and medial (M-LHb) and lateral (L-LHb) parts of the lateral habenular nucleus (LHb). The habenular complex is conventionally subdivided on morphological grounds into two compartments, a medial and a lateral nucleus. The MHb in the rat is made up of a homogeneous population of darkly staining, small, round or piriform,

very tightly packed neurons with small dendritic fields (Cajal, '11; Iwahori, '77; Tokunaga and Otani, '78). The LHb contains a more heterogeneous population of cells, most of which are of medium size, are more loosely packed, and have dendritic fields that may span distances of several hundred microns (Cajal, '11; Iwahori, '77). Although the two nuclei abut each other, Golgi studies have shown little or no connection between them: some MHb axons pass through the LHb (fig. 11) but do not issue collaterals, and MHb dendrites do not invade LHb territory (Cajal, '11; Iwahori, '77). A few of the most medial LHb neurons, however, have dendrites protruding into the MHb (Iwahori, '77).

The afferent fibers to the habenula maintain a strict medio-lateral arrangement within the stria medullaris. Fundamentally different groups of neurons innervate each of the two habenular nuclei (Herkenham and Nauta, '77; Iwahori, '77), and the present findings indicate that the latter in turn project to separate brainstem nuclei. The afferent relationships of the medial habenular nucleus are unique: each of the medial stria medullaris fibers that enter the nucleus breaks up into a dense preterminal arbor of a diameter closely corresponding to that of the dendritic fields of MHb neurons (Cajal, '11; Iwahori, '77). This arrangement may provide a basis for a point-to-point transmission of impulses originating mainly in the nuclei of the supracommissural septum (Nauta, '56; Raisman, '66; Herkenham and Nauta, '77) and mediated by the MHb to the interpeduncular nucleus.

More laterally placed stria medullaris fibers issue a succession of long collaterals into the lateral habenular nucleus as they pass caudalward over the nucleus (Iwahori, '77). Since these collaterals give off only short branches, and their parent fibers in the stria medullaris maintain a sagittal orientation throughout, it seems likely that the ramifications of each stria fiber are confined to a fairly narrow, sagittally oriented slab of LHb tissue within which, however, they synapse with neurons having relatively wide dendritic fields. The arrangement of tissue elements described by Iwahori ('77) favors the notion of a medial-to-lateral segmentation of the lateral habenular nucleus, the functional significance of which is determined by the medial-to-lateral arrangement of functionally specified categories of habenular afferents in the stria medullaris, as well as by the differential distribution

Fig. 9 Photographs of L-LHb projections. Inset at upper right shows the injection site in case Hb-16, charted in figure 8A. Darkfield photograph shows the tegmental level charted in figure 8B. The exposure reveals light areas of tegmental neuropil and the darker central gray substance. The cerebral aqueduct appears at the top and the pontine nuclei at the bottom. The terminal fields appear in the DBC. The inset at the lower right is a higher power photomicrograph of the ipsilateral terminal field. Note the absence of grains elsewhere.

of the efferents arising from the respective segments.

Some evidence of such a medio-lateral zoning of the LHb is provided by the cytoarchitecture of the nucleus. In transverse Nissl stained sections at least three fairly distinct zones can be distinguished in the LHb of the rat: (1) a medial zone of closely spaced cells of small and medium size, (2) an intermediate zone of somewhat more loosely scattered cells mostly of medium size, and (3) a lateral zone of widely scattered medium and large cells.

The notion of habenular compartmentalization is further supported by recent histochemical observations. Both substance P and choline acetyl transferase, the enzyme catalyzing the synthesis of acetylcholine (ACh), have been localized to MHb perikarya (Hökfelt et al., '75; Hattori et al., '77). According to other reports, nerve endings containing substance P are contained within the M-LHb (Cuello et al., '78), and acetylcholinesterase, the degradative enzyme for ACh, is distributed throughout the LHb (Parent and Butcher, '76). A crescent shaped band rich in opiate receptors lies at the junction of the MHb and M-LHb (Pert et al., '76; Atweh and Huhar, '77). A GABAergic path from the entopeduncular nucleus to the LHb has been described (Gottesfeld et al., '77; Nagy et al., '78), as well as fibers from the septum to the MHb that are positive for the releasing factor of luteinizing hormone (Barry, '78; Silverman and Krey, '78) and from the suprachiasmatic nucleus to the M-LHb that are positive for vasopressin and neurophysin (Buijs, '78; Sofroniew and Weindl, '78). Finally, monoamines are found in the habenula, but studies of their distribution within the habenular compartments are incomplete. Autoradiographic findings in the rat suggest that the dopamine content of the habenula (Kizer et al., '76) is attributable to a projection from the ventral tegmental area distributed largely to the L-LHb (Beckstead et al., '79), whereas habenular serotonin (Kuhar et al., '72) corresponds to a projection from the median and dorsal raphe nuclei distributed largely to the M-LHb (unpublished findings in this laboratory). Histofluorescence studies have shown an adrenergic innervation of the MHb and M-LHb by the monoamine cell group A11 in the central gray substance (Lindvall et al., '74); a cluster of catecholamine terminals in the MHb, on the other hand has been reported to represent fibers originating both in the CNS

and in the superior cervical ganglion (Björklund et al., '72; Lindvall et al., '74).

Considered together, these data are compatible with a subdivision of the habenula into at least three different parts. The habenula has long been considered to be associated with limbic circuitry, and this notion is borne out by studies of its afferent connections. However, whereas the MHb and M-LHb indeed appear to receive by far the most of their afferents from medial and basal forebrain structures associated with the limbic system, the L-LHb receives its most massive afferent projection from the entopeduncular nucleus, the non-primate homologue of the internal pallidal segment of primate forms and, like the latter, the principal origin of long afferent connections of the corpus striatum. For this reason the L-LHb could schematically be thought of as the "pallidal" part, the M-LHb as the "limbic" part of the lateral habenular nucleus. It must be emphasized, however, that this offers little more than a caricature of the afferent relationships of the LHb. Findings in a study by the HRP method (Herkenham and Nauta, '77) suggest that limbic (i.e., basal forebrain) afferents, although terminating principally in the M-LHb, are distributed throughout the width of the lateral habenular nucleus. Conversely, afferents of pallidal origin terminate in great density in the L-LHb but in smaller numbers also in more medial parts of the LHb. Together with the considerable width of the dendritic fields of LHb neurons, this evidence indicates a relatively high degree of confluence of the two categories of afferents, and supports a view of the lateral habenular nucleus as a site of convergence for efferent channels of the limbic system and corpus striatum (Nauta, '74).

### Habenular efferents

The present results confirm some of the main findings reported from earlier attempts to trace habenular efferents by fiber degeneration methods, but conflict with several further conclusions drawn from such earlier studies. The conflicts appear attributable in large part to the familiar shortcomings of the lesion method. Considering the course of numerous MHb efferents across the LHb (fig. 11) it is obvious that the fiber degeneration following a lesion of the LHb would almost unavoidably lead to an erroneous identification of LHb as a source of habenulo-interpeduncular fibers. Moreover, habenular lesions would interrupt the well documented fibers (Cragg, '61a; Mitchell, '63; Iwahori, '77) that ascend from the midbrain along the fasciculus retroflexus and bridge over to the stria medullaris without terminating in the habenula. Similar risks of misinterpretation are introduced by: (1) other ascending stria medullaris fibers that originate in contralateral basal forebrain structures and have crossed in the habenular commissure (Yamadori, '69; Price and Powell, '70), and (2) basal forebrain efferents that follow the stria medullaris caudalward, bypass the habenula, and extend along the fasciculus retroflexus to the midbrain tegmentum (Nauta, '58; Cragg, '61a), as well as to the tectum (Powell and Hoelle, '67). For these reasons no attempt will be made to account for the discrepancies between the present findings and those reported from lesion studies. The autoradiographic method has in general permitted more detailed fiber chartings than could be achieved by the use of degeneration techniques and even more important, it does not label fibers of passage. Moreover, with the microelectrophoretic delivery method isotope injections can be confined to very small target areas.

Medial habenular nucleus

The present results show that the MHb projects almost exclusively to the interpeduncular nucleus. Moreover, the results of variously placed isotope injections of the habenular complex, as well as the retrograde labeling of neurons in the MHb but not LHb by horseradish peroxidase injected into the interpeduncular nucleus (Marchand et al., '78; Ahlenius and Nauta, '79), clearly indicate the MHb as the only source of habenulo-interpeduncular fibers. The projection appears not to be reciprocated: horseradish peroxidase injections of the habenula label no cells in the interpeduncular nucleus (Herkenham and Nauta, '77), and tritiated leucine injected into the interpeduncular nucleus labels no fibers to the habenula (Stofer and Edwards, '78; Ahlenius and Nauta, '79).

Habenulo-interpeduncular fibers compose the central core of the fasciculus retroflexus, whereas the mantle of the bundle is formed by fibers from the lateral habenular nucleus. Herrick ('48) noted this same concentric arrangement in the fasciculus retroflexus of the tiger salamander and described the core of the bundle as consisting of very thin, unmyelinated axons. In Golgi material of rodent brains, likewise, MHb efferents appear thin (Cajal, '11; Iwahori, '77), but in the rat numerous fibers throughout the cross section of the fasciculus retroflexus can be stained by the Weil and Klüver-Barrera techniques down to the bundle's entry into the interpeduncular nucleus, suggesting that a majority at least of MHb efferents are myelinated in that species. Barker (1899, footnote to p. 777) notes that in the human only the core portion of the fasciculus retroflexus is unmyelinated at birth.

The present autoradiographic evidence suggesting the medial habenular nucleus as the sole source of habenulo-interpeduncular fibers agrees well with the combined histochemical evidence that (1) the habenulo-interpeduncular connection is cholinergic (Kataoka et al., '73; Kuhar et al., '75; Leranth et al., '75), (2) choline acetyltransferase in the habenula is localized to cell bodies in the MHb (Hattori et al., '77), and (3) in the fasciculus retroflexus, fibers positive for acetylcholinesterase occupy the core but not the mantle portion of the bundle (Lewis and Shute, '67).

The habenulo-interpeduncular tract is positive also for substance P (Hong et al., '76; Mroz et al., '76; Cuello et al., '78). Since cell bodies immunoreactive for substance P are likewise localized to the MHb (Hökfelt et al., '75; Cuello et al., '78), and electron microscopic studies of the interpeduncular nucleus have revealed essentially one kind of habenulo-interpeduncular synapse (Leranth et al., '75; Lenn, '76; Hattori et al., '77), it has been suggested that substance P and ACh may reside in the same habenular neurons (Brownstein et al., '76). However, Cuello et al. ('78) recently argued that the two neurotransmitters are confined to separate habenular and interpeduncular compartments, and reported evidence that perikarya positive for substance P are located only in the dorsal part of the MHb. In one of the present experiments this region was injected with tracer label (fig. 11) and found to project selectively to the most lateral part of the interpeduncular nucleus (fig. 12). The region labeled in figure 12 is virtually identical to that of substance P positive fibers outlined (and, incidentally, mislabeled "ventral tegmental area") in the report of Cuello et al. ('78). It thus appears that among the several distinct compartments of the interpeduncular nucleus (Stofer and Edwards, '78) only the most lateral one receives habenular substance P efferents; the present findings support the conclusion of Cuello et al. that these substance P fibers originate from the dorsal part of the medial habenular nucleus. By inference, it seems likely that the MHb projection innervating more medial parts of the interpeduncular nucleus is largely or entirely cholinergic.

Another noteworthy topographic feature of the habenulo-interpeduncular connection is that, as noted, the lateral part of MHb projects to the dorsal part of the interpeduncular nucleus (fig. 16), suggesting a selective connection between two restricted areas that each contain a high concentration of opiate receptors (Atweh and Kuhar, '77).

### Lateral habenular nucleus

In the foregoing account evidence of cytoarchitectural and histochemical compartmentalization of this nucleus was discussed. Because of the small size of the LHb only two subdivisions, roughly corresponding to the medial (M-LHb) and lateral (L-LHb) halves of the nucleus, could be investigated individually in the present study. A more differentiated picture of the pattern of origin of habenular efferents might have resulted if a better resolution of the transverse dimension of LHb had been achieved. For example, it was not possible in the present study to determine separately the projections of a discrete zone at the medial border of the nucleus, a zone probably corresponding in part to the small-celled portion of LHb and singled out in histochemical studies for its high content of opiate receptors (Pert et al., '76; Atweh and Kuhar, '77) as well as substance P (Cuello et al., '78), norepinephrine (Lindvall et al., '74), vasopressin and neurophysin (Buijs, '78; Sofroniew and Weindl, '78). From an HRP study Pasquier et al. ('77) reported evidence suggesting that this narrow, small-celled zone contributes to the projection from M-LHb to the dorsal raphe nucleus; whether it shares in other M-LHb projections cannot be stated at present.

Together with previous findings the present data lead to the conclusion that the M-LHb, receiving afferents mainly from limbic structures at the base of the forebrain and in smaller number also from the entopeduncular nucleus and the raphe nuclei (Herkenham and Nauta, '77), projects predominantly to the median raphe nucleus, dorsal raphe nucleus and adjoining parts of the central gray substance, medial and lateral hypothalamic regions, lateral preoptic area, and ventral parts of the septum. By contrast, the L-LHb, pro-

jected upon mainly by the entopeduncular nucleus (Nauta, '74) and in lesser volume by the basal forebrain (Herkenham and Nauta, '77), and ventral tegmental area (Beckstead et al., '79), projects mainly to a large region of the midbrain reticular formation lying immediately lateral to the tegmental raphe zone and extending caudally into dorsal parts of the pontomesencephalic tegmentum including the parabrachial nuclei.

### Habenular projections to the hypothalamus

Judged by their respective afferent and efferent connections the M-LHb, the predominantly "limbic" part of the lateral habenular nucleus, receives afferents primarily from basal forebrain structures associated with the limbic system, and in turn projects to paramedian regions of diencephalon and midbrain, including the hypothalamus and the raphe nuclei. Its projection to the lateral preoptico-hypothalamic region could be thought to reciprocate, in part at least, the prominent afferent connection of the M-LHb with this region. This connection, as well as the previously unreported projections to the posterior and dorsomedial hypothalamic nuclei, suggests an anatomical corollary of various viscero-endocrine functions imputed to the habenula on the basis of physiological observations. Such observations indicate that the habenular complex modulates in particular thyroid (Szentágothai et al., '62; Ford, '68) and gonadal activity (Motta et al., '68; Modianos et al., '74).

The possibility that the medial habenular nucleus, likewise, is involved in neuroendocrine functions is suggested by the specialized relationship of this nucleus with the adjacent ventricular wall, by its dense capillary bed (Kumar and Kumar, '75; Tokunaga and Otani, '78) and by its high permeability to certain components of the cerebrospinal fluid (Ribak and Peters, '75). Moreover, an involvement of the MHb in autonomic functions has been suggested by findings in behavioral studies (Zouhar and de Groot, '63; de Groot, '65; Lengvari et al., '70). It is not entirely clear, however, by what pathways the MHb could affect the hypothalamo-hypophysial axis. Since the efferent connections of the MHb appear to be limited almost entirely to the interpeduncular nucleus, and the latter has no known direct projection to the hypothalamus, any link with the hypothalamus would have to originate beyond the interpeduncular nucleus, i.e., in the median raphe nucleus or central gray substance. However, the possibility of an alternative pathway leading from the MHb to the hypothalamus by way of the M-LHb cannot be ruled out entirely: negative Golgi findings by Cajal ('11) and Iwahori ('77) notwithstanding, evidence from a recent immunohistofluorescence study (Cuello et al., '78) suggests that the substance P pathway from the MHb includes a link to the M-LHb. Owing to the close mutual proximity of the MHb and M-LHb, the present study cannot help to resolve this controversy.

Habenular connections with monoamine
    cell groups

As discussed above, the habenular complex receives norepinephrine fibers in the MHb and M-LHb, while serotonin and dopamine afferents seem likely to terminate in, respectively, the M-LHb and L-LHb. The present findings suggest that the habenula may reciprocate these afferent monoamine connections at least to some extent: both M-LHb and L-LHb project to the pars compacta of the substantia nigra and probably also to the ventral tegmental area, and the M-LHb has substantial additional projections to the median and dorsal raphe nuclei as well as to the nucleus tegmenti dorsalis lateralis bordering the locus coeruleus on the medial side.

The LHb may be the source of the largest single projection to the raphe nuclei (Aghajanian and Wang, '77), and has been reported to exert a powerful inhibitory effect on the serotonin neurons of at least the dorsal raphe nucleus (Wang and Aghajanian, '78). Wang and Aghajanian, ('78) have reported evidence that this inhibition is mediated directly by GAGAergic habenular efferents, but excitatory responses recorded elsewhere in the brainstem by Mok and Mogenson ('74) suggest a more complex inhibitory pathway. However this may be, Wang and Aghajanian's observations clearly identify the lateral habenular nucleus as a prominent component of the neural mechanism modulating the activity of the serotonin system. According to the present findings the medial half of the nucleus (M-LHb) directly projects to the raphe nuclei, but the medial habenular nucleus (MHb) almost certainly can affect the raphe nuclei more indirectly by way of the interpeduncular nucleus and Ganser's (1881) "radiation of the ganglion interpedunculare."[3] Since the great majority of afferents to the MHb and M-LHb

originate from medial and basal forebrain structures associated with the limbic system, it seems likely that functional modulation of the serotonin nuclei by the habenula would largely reflect fluctuations in the functional state of the limbic system. It is interesting in this connection that a second major afferent system, descending to the raphe nuclei from the preoptic region and hypothalamus by way of the medial forebrain bundle (Nauta, '58; Conrad and Pfaff, '76a,b; Swanson, '76; Nauta and Domesick, '78) also is likely to be controlled primarily by the limbic system. Habenular projections to dopamine cell groups A9 and A10, by contrast, appear to originate exclusively from the lateral habenular nucleus (both M-LHb and L-LHb) and hence could convey information from the corpus striatum as well as from the limbic system, either separately or in integrated form. Of particular interest is the likelihood that the projection from the L-LHb to these nigral components closes a nigro-striato-pallido-habenulo-nigral circuit involved primarily in extrapyramidal function but entered upon at the habenular level by various limbic channels. The efferent connections of the habenula with the raphe nuclei and substantia nigra are summarized in schematic form by figure 10.

Habenular projections to the mesencephalic
    reticular formation and non-specific
    thalamic nuclei

The main projection of the L-LHb, the "pallidal" part of the lateral habenular nucleus, is to the tegmental reticular formation immediately adjoining the median raphe nucleus and central gray substance. The densest termination of the projection involves a reticular region embedded among the fibers of the brachium conjunctivum (fig. 9), and the question should be raised whether this region is perhaps related to (or even part of) the nucleus tegmenti pedunculopontinus pars compacta (TPC), a cell group traversed by the brachium at more caudal levels (corresponding approximately to figs. 3L,M) and known to receive projections from the globus pallidus (Nauta and Mehler, '66) and the substantia nigra pars reticulata (Beckstead et al.,

---

[3] This fiber system, noted first by Ganser (1882) in the mole and mouse, was illustrated most clearly by Koelliker (1896, see in particular his figs. 635 and 643) in the rabbit as a bilateral sagittal plate of thin, dorsally and caudally oriented fibers traversing the median raphe nucleus and extending between, through and around the two medial longitudinal fasciculi into the caudal part of the central gray substance.

MILES HERKENHAM AND WALLE J. H. NAUTA

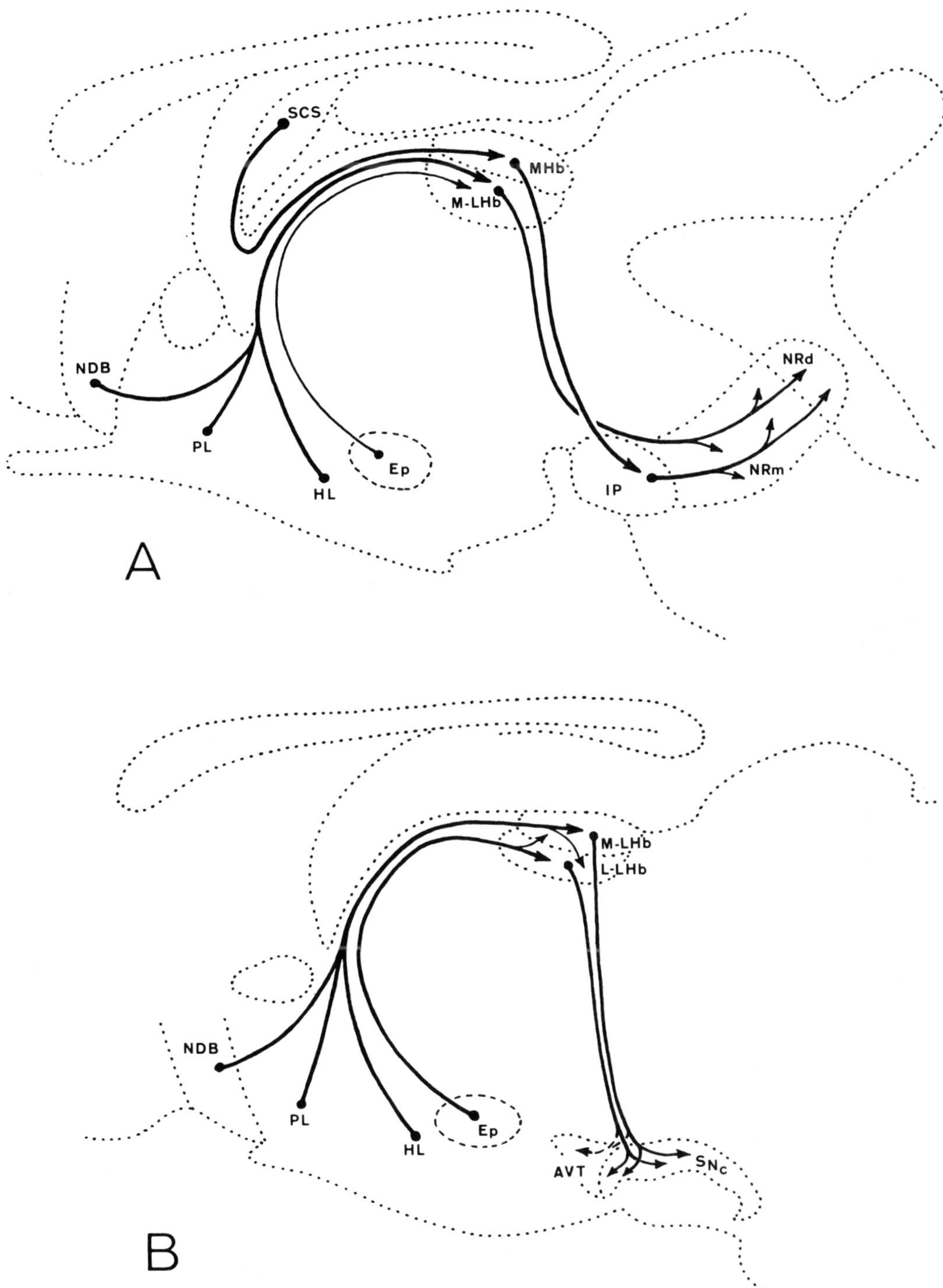



Fig. 10  Trans-habenular conduction pathways involving the raphe nuclei (A) and substantia nigra, pars compacta (B) as described in the text. Abbreviations for this figure: Ep, entopeduncular nucleus; HL, lateral hypothalamic region; Ip, interpeduncular nucleus; L-LHb, lateral subdivision of lateral habenular nucleus; MHb, medial habenular nucleus; M-LHb, medial subdivision of lateral habenular nucleus; NDB, nucleus of the diagonal band of Broca; NRD, dorsal raphe nucleus; NRm, median raphe nucleus; PL, lateral preoptic region; SCS, supracommissural septum.

'79). A comparison of the two regions, however, yields little support for that notion. Whereas the TPC is somewhat distinct from its surroundings by its relatively large, dark staining cell bodies, the more anterior tegmental region receiving the L-LHb projection lacks such characteristics and actually has a rather sparse neuron population. A clue to the functional significance of the region in question and, by inference, of the habenulo-tegmental projections may eventually come from a study of its efferent connections. Unpublished autoradiographic findings in this laboratory suggest that the region projects bilaterally rostralward to the ventral tegmental area and substantia nigra pars compacta, to the hypothalamus, subthalamus, and medial thalamic structures including in particular the LHb, periventricular, intralaminar, mediodorsal and ventromedial nuclei. The most rostral fibers pass through the lateral hypothalamus into the lateral preoptic area, substantia innominata, septum, and medial parts of the striatum including the nucleus accumbens. The last projection resembles that of the dopamine cell groups A8 and A10, and it is possible that some of the widely scattered caudal elements of these cell groups lie in the injected reticular zone (see, for example, Nauta and Domesick, '78). A major diffuse bilateral descending projection distributes itself to the mesencephalic and pontine reticular formation, to ventral parts of the central gray substance, raphe nuclei, parabrachial nuclei, locus coeruleus (rather strongly) and a ventromedial region of the medullary reticular formation.

Aside from the wide distribution of these projections, it is noteworthy that many of the ascending fibers follow the pathway of the electrophysiologically defined reticular activating system to the subthalamus and nonspecific thalamic nuclei (Starzl et al., '51; Robertson et al., '73). These thalamic cell groups receive a considerable variety of additional afferents and in turn project widely to the cerebral cortex (Herkenham, '78, '79). The present findings suggest that the L-LHb can affect these diffuse corticipetal mechanisms both by way of direct projections to the ventromedial thalamic nucleus and more indirectly via its strong projection to the mesencephalic reticular formation. Since the remaining habenular nuclei have substantial efferent connections with the likewise widely distributed serotonin system, it could be sus-

pected that in fact the entire habenular complex is involved in a great variety of general and more specific central integrative mechanisms.

## ACKNOWLEDGMENTS

This study was made possible by U. S. Public Health Service Grant MH 25515, National Science Foundation Grant BNS76-81227, and NIH Postdoctoral Fellowship NS 02101 to M. H.

## LITERATURE CITED

Aghajanian, G. K., and R. Y. Wang 1977 Habenular and other midbrain raphe afferents demonstrated by a modified retrograde tracing technique. Brain Res., *122:* 229-242.

Ahlenius, S., and W. J. H. Nauta (1979, in preparation) The connections of the interpeduncular nucleus in the rat: an experimental study using the techniques of autoradiography and horseradish peroxidase.

Akagi, K., and E. W. Powell 1968 Differential projections of habenular nuclei. J. Comp. Neur., *132:* 263-274.

Atweh, S. F., and M. J. Kuhar 1977 Autoradiographic localization of opiate receptors in rat brain. II. The brain stem. Brain Res., *129:* 1-12.

Barker, L. F. 1899 The Nervous System and its Constituent Neurons. New York, D. Appleton & Co.

Barry, J. 1978 Septo-epithalamo-habenular LRF-reactive neurons in monkeys. Brain Res., *151:* 183-187.

Beckstead, R. M., V. B. Domesick and W. J. H. Nauta 1979 Efferent connections of the substantia nigra and ventral tegmental area in the rat. Brain Res., in press.

Björklund, A., Ch. Owman and K. A. West 1972 Peripheral sympathetic innervation and serotonin cells in the habenular region of the rat brain. Z. Zellforsch., *127:* 570-579.

Brownstein, M. J., E. A. Mroz, J. S. Kizer, M. Palkovits and S. E. Leeman 1976 Regional distribution of substance P in the brain of the rat. Brain Res., *116:* 299-305.

Buijs, R. M. 1978 Intra- and extrahypothalamic vasopressin and oxytocin pathways in the cat. Cell Tiss. Res., *192:* 432-435.

Cajal, S. Ramón y 1911 Histologie du Système Nerveux de l'Homme et des Vertébrés. Maloine, Paris, Vol. II.

Conrad, L. C. A., and D. W. Pfaff 1976a Efferents from medial basal forebrain and hypothalamus in the rat. I. An autoradiographic study of the medial preoptic area. J. Comp. Neurol., *169:* 185-220.

———— 1976b Efferents from medial basal forebrain and hypothalamus in the rat. II. An autoradiographic study of the anterior hypothalamus. J. Comp. Neur., *169:* 221-262.

Cooper, W. E., and G. W. Van Hoesen 1972 Stria medullaris-habenular lesions and gnawing behavior in rats. J. Comp. Physiol. Psychol., *79:* 151-155.

Cragg, B. G. 1961a The connections of the habenula in the rabbit. Exp. Neur., *3:* 388-409.

———— 1961b The role of the habenula in the respiratory response of the rabbit to warmth or to restraint. Exp. Neur., *4:* 115-133.

Cuello, A. C., P. E. Emson, G. Paxinos and T. Jessell 1978 Substance P containing and cholinergic projections from the habenula. Brain Res., *149:* 413-429.

de Groot, J. 1965 The influence of limbic structures on pituitary functions related to reproduction. In: Sex and Behavior. F. A. Beach, ed. Wiley, New York, pp. 496-511.

Donovick, P. J., R. G. Burright, J. Kaplan and N. Rosenstreich 1969 Habenular lesions, water consumption, and palatability of fluids, in the rat. Physiol. Beh., *4:* 45-47.

Ford, D. H. 1968 Central nervous system-thyroid interrelationships. Brain Res., *7:* 329-349.

Ganser, S. 1882 Vergleichen-anatomische Studien über das Gehirn des Maulwurfs. Morphol. Jahrb., *7:* 591-725.

Gottesfeld, Z., V. J. Massari, E. A. Muth and D. M. Jacobowitz 1977 Stria medullaris: a possible pathway containing GABAergic afferents to the lateral habenula. Brain Res., *130:* 184-189.

Graybiel, A. M., and M. Devor 1974 A microeletrophoretic delivery technique for use with horseradish peroxidase. Brain Res., *68:* 167-173.

Hattori, T., E. G. McGeer, V. K. Singh and P. L. McGeer 1977 Cholinergic synapse of the interpeduncular nucleus. Exp. Neur., *55:* 666-679.

Herkenham, M. 1978 Intralaminar and parafascicular efferents to the striatum and cortex in the rat; an autoradiographic study. Anat. Rec., *190:* 420 (Abstract).

———— 1979 The afferent and efferent connections of the ventromedial thalamic nucleus in the rat. J. Comp. Neur., *183:* 487-518.

Herkenham, M., and W. J. H. Nauta 1977 Afferent connections of the habenular nuclei in the rat. A horseradish peroxidase study, with a note on the fiber-of-passage problem. J. Comp. Neur., *173:* 123-146.

Herrick, C. J. 1948 The habenula and its connections. In: The Brain of the Tiger Salamander. U. of Chicago Press, Chicago, pp. 247-264.

Hökfelt, T., J. O. Keller, G. Nilsson and W. Pernow 1975 Substance P localization in the central nervous system and in some primary sensory neurons. Science, *190:* 889-890.

Hong, J. S., E. Costa and H.-Y. T. Yang 1976 Effects of habenular lesions on the substance P content of various brain regions. Brain Res., *118:* 523-525.

Iwahori, N. 1977 A golgi study on the habenular nucleus of the cat. J. Comp. Neur., *171:* 319-344.

Kabat, H. 1936 Electrical stimulation of points in the forebrain and midbrain: the resultant alterations in respiration. J. Comp. Neur., *64:* 187-208.

Kataoka, K., Y. Nakamura and R. Hassler 1973 Habenulointerpeduncular tract: a possible cholinergic neuron in rat brain. Brain Res., *62:* 264-267.

Kizer, J. S., M. Palkovits and M. J. Brownstein 1976 The projections of the A8, A9 and A10 dopaminergic cell bodies: Evidence for a nigral-hypothalamic-median eminence dopaminergic pathway. Brain Res., *108:* 363-370.

Koelliker, A. von 1896 Handbuch der Gewebelehre. Vol. 2. W. Engelmann, Leipzig.

König, J. F. R., and R. A. Klippel 1963 The Rat Brain. Krieger, Huntington.

Kuhar, M. J., G. K. Aghajanian and R. H. Roth 1972 Tryptophan hydroxylase activity and synaptosomal uptake of serotonin in discrete brain regions after midbrain raphe lesions: correlations with serotonin levels and histochemical fluorescence. Brain Res., *44:* 165-176.

Kuhar, M., R. N. Dehaven, H. I. Yamamura, H. Rommelspacher and V. R. Simon 1975 Further evidence for cholinergic habenulo-interpeduncular neurons: pharmacologic and functional characteristics. Brain Res., *97:* 265-275.

Kumar, K., and T. C. Anand Kumar 1975 The habenular ependyma: a neuroendocrine component of the epithalamus in the rhesus monkey. In: Anatomical Neuroendocrinology. W. E. Stumpf and L. D. Grant, eds. Karger, Basel, pp. 40-51.

Künzle, H. 1977 Evidence for selective axon-terminal uptake and retrograde transport of label in cortico- and rubrospinal systems after injection of $^3$H-proline. Exp. Brain Res., *28:* 125-132.

Larsen, K. D., and R. L. McBride 1979 The organization of feline entopeduncular nucleus projections: anatomical studies. J. Comp. Neur., *184:* 293-308.

Lengvari, I., K. Koves and B. Halász 1970 The medial habenular nucleus and the control of salt and water balance. Acta Biol. Acad. Sci. (Hungary), *21:* 75-83.

Lenn, N. J. 1976 Synapses in the interpeduncular nucleus: electron microscopy of normal and habenula lesioned rats. J. Comp. Neur., *166:* 73-100.

Leranth, C. S., M. J. Browstein, L. Zaborsky, Z. S. Jaranyi and M. Palkovitz 1975 Morphological and biochemical changes in the rat interpeduncular nucleus following the transection of the habenulo-interpeduncular tract. Brain Res., *99:* 124-128.

Lewis, P. R., and C. C. D. Shute 1967 The cholinergic limbic system: projections to hippocampal formation, medial cortex, nuclei of the ascending cholinergic reticular system, and the subfornical organ and supro-optic crest. Brain, *90:* 521-540.

Lindvall, O., A. Björklund, A. Nobin and U. Stenevi 1974 The adrenergic innervation of the rat thalamus as revealed by the glyoxylic acid fluorescence method. J. Comp. Neur., *154:* 317-348.

Marchand, R., J. N. Riley and R. Y. Moore 1978 Afferents to the interpeduncular nucleus of the rat. Neurosci. Abst., *4:* 224 (Abstract).

Mitchell, R. 1963 Connections of the habenula and of the interpeduncular nucleus in the cat. J. Comp. Neur., *121:* 441-457.

Modianos, D. T., J. C. Hitt and J. Flexman 1974 Habenular lesions produce decrements in feminine, but not masculine, sexual behavior in rats. Behavioral Biol., *10:* 75-87.

Mok, A. C. S., and G. J. Mogenson 1974 Effects of electrical stimulation of the lateral habenular nucleus and lateral hypothalamus on unit activity in the upper brain stem. Brain Res., *78:* 425-435.

Morest, D. K. 1961 Connexions of the dorsal tegmental nucleus in rat and rabbit. J. Anat. (London), *95:* 229-249.

Motta, M., F. Franschini, G. Giuliani and L. Martini 1968 The central nervous system, estrogen and puberty. Endocrinol., *83:* 1101-1107.

Mroz, E. A., M. J. Brownstein and S. E. Leeman 1976 Evidence for substance P in the habenulo-interpeduncular tract. Brain Res., *113:* 597-599.

Nagy, J. I., D. A. Carter, J. Lehman and H. C. Fibiger 1978 Evidence for a GABA-containing projection from the entopeduncular nucleus to the lateral habenula in the rat. Brain Res., *145:* 360-364.

Nauta, W. J. H. 1956 An experimental study of the fornix system in the rat. J. Comp. Neur., *104:* 247-271.

———— 1958 Hippocampal projections and related neural pathways to the mid-brain in the cat. Brain, *81:* 319-341.

Nauta, H. J. W. 1974 Evidence of a pallidohabenular pathway in the cat. J. Comp. Neur., *156:* 19-28.

Nauta, W. J. H., and V. B. Domesick 1978 Crossroads of limbic and striatal circuitry: hypothalamo-nigral connections. In: Limbic Mechanisms. K. E. Livingston and O. Hornykiewicz, eds. Plenum, pp. 75-93.

Nauta, W. J. H., and W. R. Mehler 1966 Projections of the lentiform nucleus in the monkey. Brain Res., *1:* 3-42.

Nauta, W. J. H., G. P. Smith, R. L. M. Faull and V. B. Domesick 1978 Efferent connections and nigral afferents of the nucleus accumbens septi. Neurosci., *3:* 385-401.

Parent, A., and L. L. Butcher 1976 Organization and morphologies of acetylcholinesterase-containing neurons in the thalamus and hypothalamus of the rat. J. Comp. Neur., *170:* 205-226.

Pasquier, D. A., T. L. Kemper, W. B. Forbes and P. J. Morgane 1977 Dorsal raphe, substantia nigra and locus coeruleus: interconnections with each other and the neostriatum. Brain Res. Bull., *2:* 323-339.

Pert, C. B., M. J. Kuhar and S. H. Snyder 1976 Opiate receptor: autoradiographic localization in rat brain. Proc. Natl. Acad. Sci., *73:* 3729-3733.

Powell, E. W., and D. F. Hoelle 1967 Septotectal projections in the cat. Exp. Neur., *18:* 177-183.

Price, J. L., and T. P. S. Powell 1970 The afferent connexions of the nucleus of the horizontal limb of the diagonal band. J. Anat., *107:* 230-256.

Raisman, G. 1966 The connexions of the septum. Brain, *89:* 317-348.

Rausch, L. J., and C. J. Long 1974 Habenular lesions and avoidance learning deficits in albino rats. Physiol. Psychol., *2:* 352-365.

Reinert, H. 1964 Defence reaction from the habenular nuclei, stria medullaris and fasciculus retroflexus. J. Physiol., *170:* 28-29.

Ribak, C. E., and A. Peters 1975 An autoradiographic study of the projections from the lateral geniculate body of the rat. Brain Res., *92:* 341-368.

Robertson, R. T., G. S. Lynch and R. F. Thompson 1973 Diencephalic distributions of ascending reticular systems. Brain Res., *55:* 309-322.

Silverman, A. J., and L. C. Krey 1978 The luteinizing hormone-releasing (LH-RH) neuronal networks of the guinea pig brain. I. Intra- and extra-hypothalamic projections. Brain Res., *157:* 233-246.

Sofroniew, M. V., and A. Weindl 1978 Projections from the parvocellular vasopressin- and neurophysin-containing neurons of the suprachiasmatic nucleus. Am. J. Anat., *153:* 391-430.

Starzl, T. E., C. W. Taylor and H. W. Magoun 1951 Ascending conduction in reticular activating system, with special reference to the diencephalon. J. Neurophysiol., *14:* 461-477.

Stofer, W. D., and S. B. Edwards 1978 Organization and efferent projections of the interpeduncular complex in the cat. Neurosci. Abst., *4:* 228 (Abstract).

Swanson, L. W. 1976 An autoradiographic study of the efferent connections of the preoptic region in the rat. J. Comp. Neur., *167:* 227-256.

Szentágothai, J., B. Flerkó, B. Mess and B. Halász 1962 Hypothalamic Control of the Anterior Pituitary. Publ. House Hung. Acad. Sci., Budapest, pp. 143-173.

Tokunaga, A., and K. Otani 1978 Fine structure of the medial habenular nucleus in the rat. Brain Res., *150:* 600-606.

Wang, R. Y., and G. K. Aghajanian 1977 Physiological evidence for habenula as major link between forebrain and midbrain raphe. Science, *197:* 89-91.

Yamadori, T. 1969 Efferent fibers of the habenula and stria medullaris thalami in rats. Exp. Neur., *25:* 541-558.

Zouhar, R. L., and J. de Groot 1963 Effects of limbic brain lesions on aspects of reproduction in female rats. Anat. Rec., *145:* 358.

11  Darkfield microphotograph of injection site and transported label in case Hb-31, dorsal MHb injection. Bar measures 0.5 mm for all figures in this plate.

12  Appearance of grains in the lateral parts of the caudal interpeduncular nucleus in Hb-31, dorsal MHb injection. This part of the nucleus is bounded laterally by the medial lemniscus.

13  "Shoelace" pattern of MHb projection to the interpeduncular nucleus in case Hb-33. Injection shown in figure 1A.

14  Bilateral uptake of label in the medial edge of the MHb and in the periventricular nucleus. Case Hb-25. Pipette may have been situated in the ventricle during this injection. Labeled fibers aggregate ventrolaterally in the habenulo-interpeduncular tract.

15  Distribution of transported label in the ventral portion of the interpeduncular nucleus after injection of the medial part of the MHb, case HB-25, above.

16  Distribution of label in the dorsal portion of the interpeduncular nucleus after injection of the lateral part of the MHb, case Hb-34. Injection site charted in figure 7A.

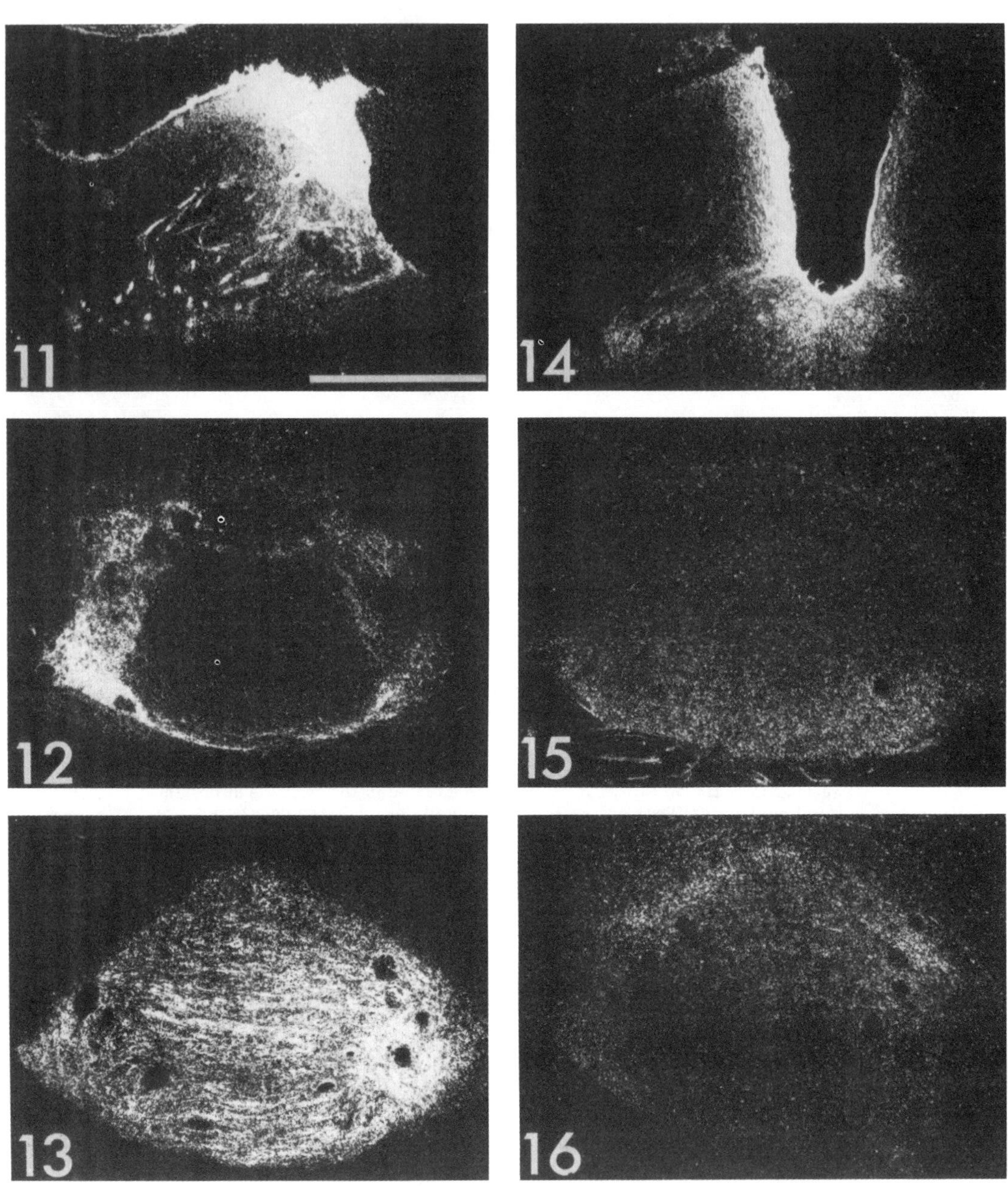

17 Darkfield microphotomontage of thalamus and hypothalamus in case Hb-12, injection of LHb. This level is charted in figure 3F. The rostral periphery of the injection site appears in the LHb at the top of the picture. Below, labeling fills the lateral part of the mediodorsal nucleus. The unlabeled intralaminar nuclei separate the mediodorsal projection from the labeled input to the ventromedial nucleus. A dense accumulation of label marks the ventrolateral side of the mammillothalamic tract. Labeling of medial and lateral hypothalamus is visible. The third ventricle is barely visible along the lowermost right edge. The lowest photograph of this composite is printed at a lighter exposure to facilitate the visualization of the fornix and ventral hypothalamic territory.

18 The innervation of the substantia nigra, pars compacta in case Hb-12. Labeling is most dense in the ventral tegmental area on the right. Labeled fibers are bounded dorsally by the medial lemniscus and ventrally by the nigral pars reticulata.

19 The innervation of the dorsal raphe nucleus and the adjacent ventral central gray substance in case Hb-12. The boundaries of the dr appear in ghost-like fashion by virtue of the relatively more dense labeling at its periphery. The paired MLF appears as dark regions at the bottom.

20 The bilateral encapsulation of the ventral tegmental nucleus in Hb-11, a case of large rostral LHb injection. The midline is at the center of the picture, the paired MLF appears near the top as dark regions.

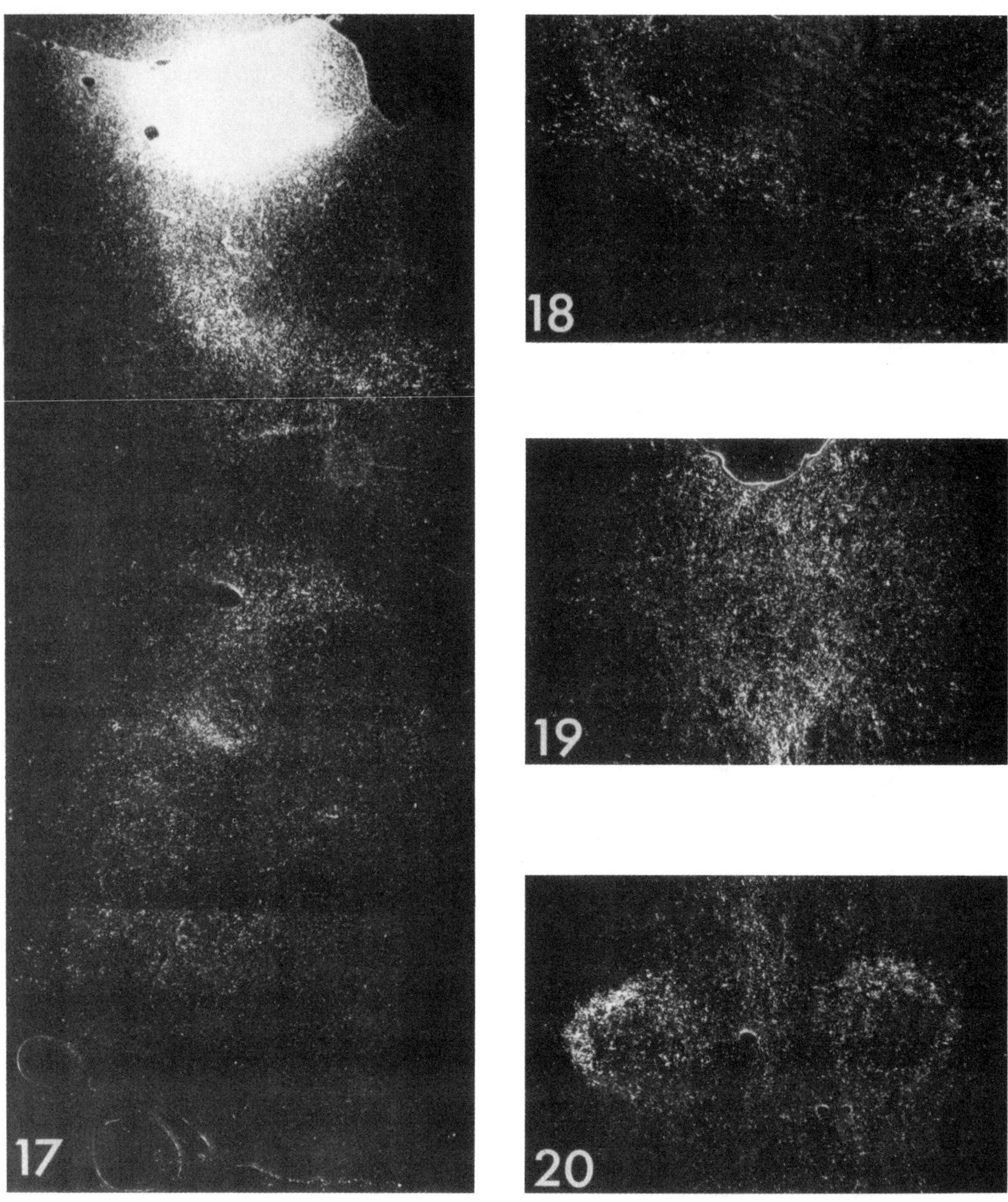

Published 1981 by Elsevier North Holland, Inc.
PSYCHIATRY AND THE BIOLOGY OF THE HUMAN BRAIN:
A SYMPOSIUM DEDICATED TO SEYMOUR S. KETY/Matthysse, ed.

# Ramifications of the Limbic System

## Walle J.H. Nauta and Valerie B. Domesick

*Department of Psychology and Brain Science, Massachusetts Institute of Technology, Cambridge, Massachusetts; Mailman Research Center, McLean Hospital, Belmont, Massachusetts*

## Dedication

We want to express our pleasure at the chance of contributing to this volume honoring Seymour Kety. Some of our fellow-participants in this conference can boast of having been his students. Neither of us is so fortunate, but we want to emphasize that the privilege of an association with him extends well beyond the circle of those he trained. We salute Seymour Kety for his vast accomplishments as a scientist, as a teacher, and as a promotor of our common purpose to learn more about the brain and its disorders. And last, not least, we honor him for his sterling qualities as a colleague and friend.

## Introduction

Even in its most restrictive connotation, the term *limbic system* refers to the hippocampal formation and amygdala, together with the gyrus fornicatus (gyrus cinguli, retrosplenial cortex, and parahippocampal gyrus) which is associated with the hippocampal formation by both direct and indirect fiber connections. It is a logical question to ask how such very dissimilar telencephalic structures came to be integrated into a unitary concept.

The term limbic system (MacLean, 1952) has its origin in the name "grand lobe limbique" coined a century ago by Broca (1878) to denote the somewhat distinct region of cortex nearest the margin (limbus) of the cortical mantle, a region which in the mammalian brain fully encircles the hilus of the cerebral hemisphere. Broca's term referred to a detail of cerebral surface anatomy that comprised the hippocampus and gyrus fornicatus, but did not explicitly include the deep-lying amygdala. Neither was the amygdala mentioned in the classical paper in which Papez (1937) stressed the interconnectedness of hippocampus and gyrus fornicatus and suggested that this cortical aggregate serves as a neural substrate of emotion. The amygdala became recognized as a limbic structure only when MacLean (1952) emphasized that it, like the hippocampal formation, is closely associated with the hypothalamus, and therefore should be considered to belong in a class with the hippocampus and the latter's superstructure, the gyrus fornicatus. For this category of telencephalic structures, MacLean introduced the name *limbic system*, adding to it in his earlier publications the physiological sub-title, "*visceral brain*." On the basis of MacLean's unifying concept, the limbic system became in essence defined as a heterogeneous group of basal and medial telencephalic structures together composing that part of the cerebral hemisphere which is most directly associated with the hypothalamus.

The main purpose of the present review is to explore the question, how far into the brain stem is it possible to trace the ramifications of the limbic system? An earlier examination of this question (Nauta, 1958) in the light of MacLean's then recent concept led to the notion that the limbic telencephalon is reciprocally connected with an uninterrupted continuum of subcortical grey matter that begins with the septum, continues from there caudalward over the preoptic region and hypothalamus, and extends beyond the latter over a paramedian zone of the mesencephalon that reaches caudally as far as the isthmus rhombencephali. The mesencephalic part of this continuum comprises the ventral tegmental area, the ventral half of the central grey substance (including the cell group now named dorsal raphe nucleus but then known as the fountain nucleus of Sheehan), the nucleus centralis tegmenti superior of Bechterew (now better known as the medial raphe nucleus), the interpeduncular nucleus, and the dorsal and ventral tegmental nuclei of Gudden. To be sure, in the three species in which these studies were done (rat, cat, and monkey), few if any direct projections from either hippocampus or amygdala could be traced caudalward beyond the septum, preoptic region, and hypothalamus. The paramedian zone of the mesencephalon nonetheless became included in the subcortical limbic continuum as "limbic midbrain area" when in the guinea pig it was found to receive a substantial contingent of fornix fibers (Valenstein and Nauta, 1959), and in the other three species turned out to be a common distribution area for direct and indirect projections descending from the septum, preoptic region, and hypothalamus. Such projections follow three fiber systems: (a) the medial forebrain bundle, (b) the route composed of stria medullaris and fasciculus retroflexus, and (c) the mammillotegmental tract. Moreover, the paramedian zone of the mesencephalon was found (Nauta and Kuypers, 1958) to be a main origin of projections ascending in the medial forebrain bundle and presumably linking up with known projections from the anterior hypothalamus and septum back to, respectively, the amygdala and the hippocampal formation. The combined evidence suggested a reciprocal connection ("limbic forebrain-midbrain circuit") between the amygdala and hippocampus at the rostral end and a paramedian zone of the midbrain at the caudal end, a circuit in which the continuum of the septum, preoptic region and hypothalamus, as well as the habenular complex, occupy intermediate positions. The construct as formulated more than 20 years ago is illustrated schematically in Figure 1.

In the past 15 years or so, the initial and quite schematic notion of a "limbic forebrain-midbrain circuit" has become much expanded and differentiated, especially as a result of studies by the aid of several new techniques, in particular the autoradiographic fiber-tracing method (Cowan et al., 1972) and the method of retrograde labelling of cell bodies with horseradish-peroxidase (Kristensson et al., 1971; LaVail et al., 1973). Some of the main modifications that must now be made in the original anatomical construct are indicated in

**Figure 1.** The "limbic forebrain-midbrain circuit" as initially proposed. (A) Descending pathways following medial forebrain bundle and mammillotegmental tract to the paramedian zone of the midbrain, and their suggested links with hippocampal and amygdalofugal projections. More laterally distributed components of medial forebrain bundle and mammillotegmental tract are indicated in broken lines. (B) Descending conduction lines following stria medullaris and fasciculus retroflexus. (C) Some ascending return links from the paramedian midbrain region to limbic forebrain structures; all of the lines ascending to hippocampus and amygdala were thought to be interrupted in the hypothalamo-preoptico-septal continuum.

SOURCE: *Reproduced from Brain, Vol. 81, 1958, by courtesy of MacMillan & Co., Ltd., London.*

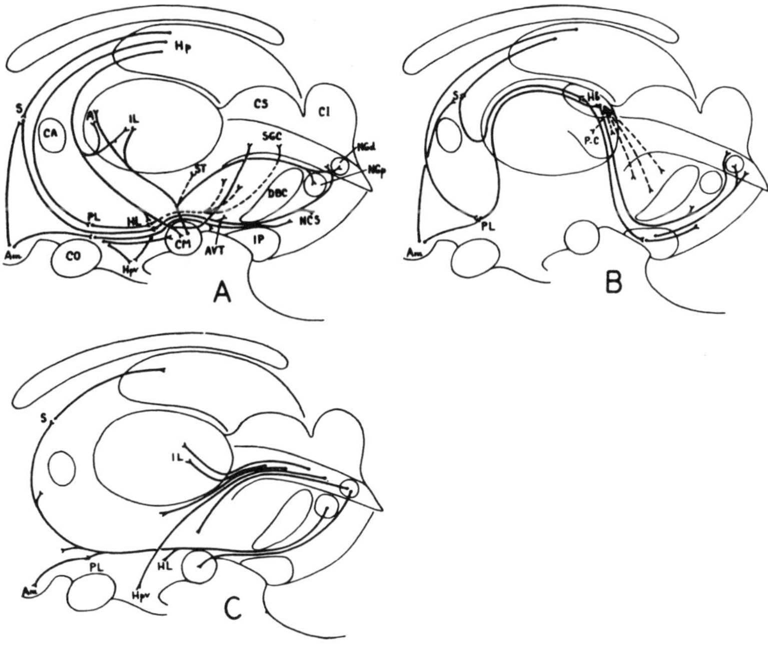

the following review. The survey will be illustrated by schematic drawings based on a standardized, sagittally oriented view of fore- and midbrain (Figure 2). For practical reasons, this ground figure represents the primate brain, but it must be emphasized that the more recently identified components of the circuitry illustrated were discovered for the most part in non-primate species, the rat in particular, and thus far have not been corroborated in primates. To prevent visual overcrowding, it was necessary to limit structure identification by lettering in the circuit diagrams; for a complete listing of labels reference is made to Figure 2.

**Figure 2.** Ground-figure to serve as base for circuits illustrated in Figures 3–9. Abbreviations: A: nucleus anterior thalami; Am: amygdala; CC: corpus callosum; CGS: central grey substance; C-P: caudatoputamen; EA: entorhinal area; GC: gyrus cinguli; GPi: globus pallidus, internal segment; Hbl: lateral habenular nucleus; Hbm: medial habenular nucleus; Hp: hippocampal formation; Ht: hypothalamus; IP: interpeduncular nucleus; lc: locus coeruleus; MB: mammillary body; MD: nucleus mediodorsalis thalami, Nacc: nucleus accumbens septi; NdB: nucleus of the diagonal band of Broca; NST: bed-nucleus of stria terminalis; OF: orbitofrontal cortex; OT: olfactory tubercle; PH: parahippocampal gyrus; Po: preoptic area; Pt: nucleus parataenialis thalami; Rd: dorsal raphe nucleus; Re: nucleus reuniens thalami; Rm: median raphe nucleus; Rs: retrosplenial cortex; SI: substantia innominata; td: dorsal tegmental nucleus of Gudden; tdl: nucleus tegmenti dorsalis lateralis; Ti: inferior temporal cortex.

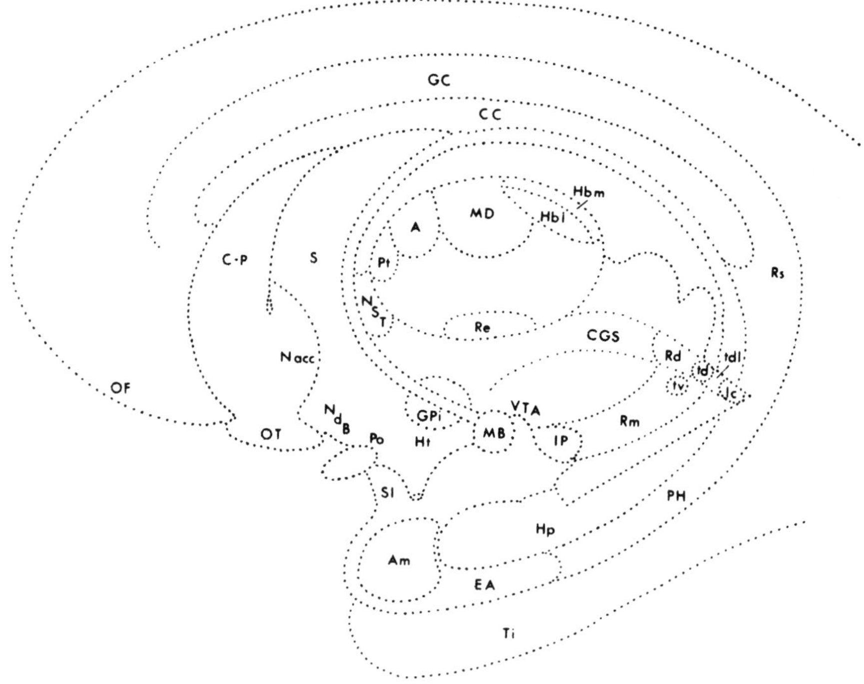

# Descending Components of the "Limbic Forebrain-Midbrain Circuit"

## Medial Forebrain Bundle

This complex longitudinal fiber system composes the basal of the two main limbic pathways leading to the mesencephalon, the more dorsal one being the stria medullaris. Rather than being a circumscript, compact fiber bundle, the medial forebrain bundle is a vaguely delineated fiber system that is interspersed throughout its extent with grey matter, in an arrangement similar to that of the brain stem reticular formation. Thus, the lateral preoptic and lateral hypothalamic nuclei can be interpreted as interstitial nuclei of the medial forebrain bundle (Millhouse, 1969), receiving fibers—or collateral axon branches—from the bundle, as well as adding fibers to it in either the caudal or rostral direction, or both. As a consequence of this complex grey-and-white composition, it is extraordinarily difficult to follow individual conduction lines through the length of the medial forebrain bundle, and impossible to express the true complexity of the system in a survey diagram. However, since the medial forebrain bundle has been found to contain also a considerable number of fibers that span much or even all of the distance between rostral forebrain levels and the pontomesencephalic transition, some statements about the origin and distribution of the system as a whole can be made with some confidence.

As indicated schematically in Figure 3, descending components of the medial forebrain bundle originate from a great variety of medial and basal forebrain structures. Most prominent among the sources of such fibers are the septal region and the lateral preoptico-hypothalamic continuum. Further descending fibers are known to originate in the olfactory tubercle and piriform cortex (Lundberg, 1960, 1962); substantia innominata, in particular the basal nucleus of Meynert (Jones et al., 1976); nucleus accumbens (Swanson and Cowan, 1975; Conrad and Pfaff, 1976c; Powell and Leman, 1976; Nauta et al., 1978); amygdala (Krettek and Price, 1978; Hopkins and Holstege, 1978); bed nucleus of the stria terminalis (Swanson and Cowan, 1979), and orbitofrontal cortex (Nauta, 1962; Beckstead, 1979). The medial forebrain bundle appears fully formed and most readily delineated in its sagittal passage through the lateral hypothalamic zone. Throughout this part of its course, however, the bundle receives numerous further contributions not only from the lateral hypothalamic nucleus but also from more medial hypothalamic cell groups (Conrad and Pfaff, 1976a,b; Saper et al., 1976; Krieger et al., 1979).

Upon entering the ventral tegmental area, the medial forebrain bundle begins to spread out widely in the lateral direction. From here on caudalward, it can be subdivided somewhat arbitrarily into (a) a medial division that maintains throughout its mesencephalic extent the sagittal orientation that characterized the hypothalamic trajectory of the system, and (b) a large lateral division composed of fibers that become deflected lateralward as they descend. It is mainly the medial subdivision that establishes the connection of the limbic

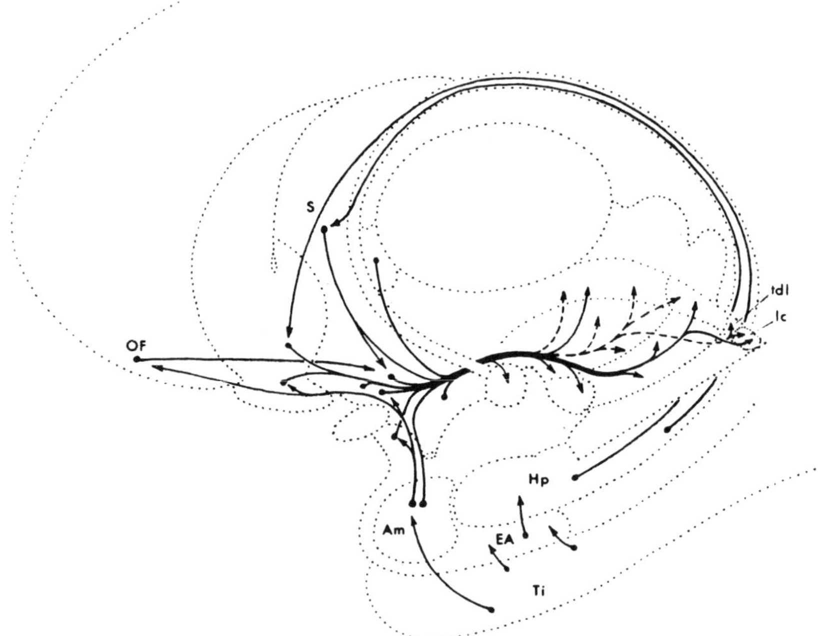



**Figure 3.** Diagrammatic representation of the descending components of the medial forebrain bundle. Broken lines represent bundle's lateral division, distributed to nigral complex, central regions of the mesencephalic reticular formation, peripeduncular nucleus, and parabrachial region. Not indicated are those lateral-division fibers that arise from the hypothalamus and central amygdaloid nucleus and continue caudalward to the nucleus of the solitary tract, dorsal motor nucleus of the vagus, and lateral horn of the spinal grey matter (see text). Also included in the diagram are projections from the antero-inferior temporal cortex (Ti) to amygdala (Am) and—indirectly, by way of the entorhinal area (EA)—to hippocampus. For identification of structures left un-labelled see Figure 2.

forebrain structures with the paramedian zone of the midbrain ("limbic midbrain area"). As schematically indicated in Figure 3 by continuous lines, its distribution involves the medial part of the ventral tegmental area, the inter-peduncular nucleus, median raphe nucleus, and dorsal raphe nucleus along with other ventral regions of the central grey substance including the nucleus tegmenti dorsalis lateralis and the locus coeruleus (Nauta, 1958; Conrad and Pfaff, 1976a,b; Swanson, 1976; Domesick, 1976).

The lateral subdivision of the medial forebrain bundle (indicated in Figure 3 by broken lines), by contrast, spreads widely in the lateral direction as it descends through the gradually narrowing ventral tegmental zone that lies between the medial lemniscus dorsally and the pars compacta of the substantia nigra ventrally. In this part of its course, it forms a fairly compact stratum of laterally and caudally oriented fibers, scattered among which lie the neurons that compose the lateral part of the dorsally outlying nigral dopamine cell group A10 of Dahlström and Fuxe (1964). (The larger part of this dopamine cell group occupies a more medial region of the ventral tegmental area extending from the substantia nigra to the midline, and lies embedded among the more medial fibers of the medial forebrain bundle (Nauta and Domesick, 1978; Nauta, 1979)). Throughout its course beneath the medial lemniscus, the lateral division of the medial forebrain bundle issues fibers in three main directions: (a) sparse fibers are deflected ventrally to be distributed to the pars compacta and the dorsal one-third of the pars reticulata of the substantia nigra (Swanson, 1976; Nauta and Domesick, 1978), (b) others extend laterally to the peripeduncular nucleus, a region of medium-sized cells occupying most of the interval between the dorsolateral margin of the substantia nigra and the medial geniculate body, and bordered medially by the medial lemniscus, and (c) a substantial fiber contingent curves sharply in the dorsomedial direction around the lateral margin of the medial lemniscus and along the lateral side of the red nucleus. These last-mentioned fibers traverse a wide central region of the midbrain tegmentum; most seem to terminate in this part of the midbrain reticular formation but some longer fibers extend into the ventral half of the central grey substance, a region also invaded by fibers of the medial subdivision.

The remaining fibers of the lateral division accumulate into a substantial fiber bundle that occupies a ventrolateral position in the midbrain tegmentum near the caudal pole of the substantia nigra. Embedded in it at this level is a second outlying nigral cell aggregate, Dahlström and Fuxe's (1964) dopamine cell group A8 (Nauta and Domesick, 1978). At this point in its caudal course, the lateral division of the medial forebrain bundle initiates a dorsomedial curve that leads it through a wide region of the caudal midbrain tegmentum, in particular the cuneiform and parabrachial nuclei. The longest of its fibers extend through the parabrachial nuclei into the caudal part of the central grey substance where they appear to converge with fibers of the medial division. From its gradual decrease in volume, it can be assumed that the lateral division of the medial forebrain bundle throughout this caudal part of its course distributes fibers to most or all of the tegmental regions straddling its path.

Fiber degeneration techniques have failed to reveal fibers of the medial forebrain bundle descending beyond the level of the locus coeruleus. Studies by anterograde and retrograde labelling methods, however, have demonstrated a direct hypothalamic projection to the nucleus of the solitary tract and the dorsal motor nucleus of the vagus in the medulla oblongata, as well as to the lateral horn of the spinal cord (Saper et al., 1976). Findings in retrograde-labelling experiments have indicated medial hypothalamic cell groups, in particular the paraventricular and dorsomedial nucleus, as the main origin of this direct hypothalamo-autonomic connection (Kuypers and Maisky, 1975; Saper et al., 1976). Hopkins and Holstege (1978), however, found the hypothalamic projection to the nucleus of the solitary tract and dorsal motor nucleus of the vagus accompanied by numerous fibers from the central amygdaloid nucleus (see below).

*Hippocampal and Amygdaloid Contributions to the Medial Forebrain Bundle*

Next to be considered is the question, at which points the efferents from the telencephalic limbic structures (hippocampus, amygdala, gyrus fornicatus, and orbitofrontal cortex) lead into the medial forebrain bundle. The hippocampal formation projects by way of the precommissural fornix to all parts of the septum and to the nucleus accumbens (Figure 3). Its massive postcommissural fornix projection to the mammillary body, by contrast, forms part of a separate, though closely related line that leads over the lateral mammillary nucleus to both the dorsal and the ventral tegmental nucleus of Gudden by way of the mammillotegmental tract (Figure 7). The amygdalofugal lines are more complex, for the amygdala projects into the medial forebrain bundle (or, in some instances, to origins of the latter's descending fibers) by way of two fiber systems: the stria terminalis and the ventral amygdalofugal pathway. The stria terminalis (not illustrated) distributes itself to its bed nucleus, to the nucleus accumbens (Olmos, 1972), and to medial preoptic and hypothalamic cell groups, prominently including the ventromedial hypothalamic nucleus (Heimer and Nauta, 1969; Olmos and Ingram, 1972; Krettek and Price, 1978). Since these medial regions are now known to contribute substantially to the medial forebrain bundle (Conrad and Pfaff, 1976a,b; Saper et al., 1976; Krieger et al., 1979), they can provisionally be interpreted *in the present context* as intermediaries in a medial amygdalohypothalamic conduction line leading, in part at least, into the larger limbic forebrain-midbrain channel. The amygdala has, however, additional and more direct access to this larger channel by way of the ventral amygdalofugal pathway, a complex fiber system intercalated in which as a bed-nucleus lie the cell groups composing the substantia innominata (a relationship more evident in primates than in carnivores and rodents). The ventral amygdalofugal pathway contains numerous fibers that reciprocally connect the amygdala with the septo-preoptic-hypothalamic region (Nauta, 1961; Krettek and Price, 1978). In a comprehensive and detailed autoradiographic study in the rat, Krettek and Price found that it also contains fibers from the central amygdaloid nucleus that project beyond the hypothalamus to the midbrain. Similar findings in the cat were reported by Hopkins and Holstege (1978). The chartings in both studies indicate that this direct amygdalo-mesencephalic projection follows the lateral division of the medial forebrain bundle, and terminate mainly in the parabrachial region. Hopkins and Holstege (1978) emphasize, however, that the projection has further substantial distributions that involve central regions of the mesencephalic tegmentum and the central grey substance, and that part of its fibers continue beyond the isthmus to the lateral reticular formation of the pons and medulla oblongata, the nucleus of the solitary tract and the dorsal motor nucleus of the vagus. The projection appears to be accompanied in its more rostral mesen-cephalic course by fibers to the peripeduncular nucleus from the basal nucleus of Meynert in the substantia innominata (Jones et al., 1976) and from the ventromedial hypothalamic nucleus (Saper et al., 1976; Krieger et al., 1979).

Other components of the ventral amygdalofugal pathway are distributed to the orbitofrontal cortex (Nauta, 1961, in the monkey), olfactory tubercle (Nauta, 1961; Krettek and Price, 1978) and nucleus accumbens (Krettek and Price, 1978), all of which have been reported to contribute fibers to the medial forebrain bundle. The fiber system also includes projections from the amygdala to the subcallosal region of the anterior cingulate cortex (Lammers and Lohman, 1957; Krettek and Price, 1977), as well as a return projection to the amygdala from the ventromedial nucleus of the hypothalamus (Krieger et al., 1979).

### Stria Medullaris and Fasciculus Retroflexus

These two circumscript fiber bundles together compose an alternate pathway from the rostral diencephalon to the paramedian zone of the midbrain. In contrast to the medial forebrain bundle, this conduction route leads over the dorsum of the thalamus, and it is the habenular complex, rather than the lateral preoptico-hypothalamic nucleus, that straddles its course.

The stria medullaris was originally thought to arise exclusively from medial and basal forebrain structures also giving origin to fibers of the medial forebrain bundle. More recently, however, it has been shown to include additional fibers that arise from the internal segment of the globus pallidus (Nauta, 1974). It thus appears that the habenular complex receives input not only of limbic but also of extrapyramidal origin. Some of the main anatomical features of the stria medullaris and its trans-synaptic extension, the fasciculus retroflexus, are shown diagrammatically in Figure 4.

By way of summary, the stria medullaris-fasciculus retroflexus route from the mediobasal forebrain to the midbrain can be described as comprising two distinct conduction lines. One of these, originating from the dorsal septal region—hence presumably transmitting largely hippocampal outflow leads over the medial habenular nucleus to the interpeduncular nucleus and thence to the raphe nuclei. A second, more diverse line originates in more ventral parts of the septum as well as in the lateral preoptico-hypothalamic region, and leads over the lateral habenular nucleus to the raphe nuclei, with side branches (a) ascending in the medial forebrain bundle to the septo-preoptico-hypothalamic continuum, (b) spreading laterally to the ventral tegmental area and the substantia nigra pars compacta, and (c) descending into the medial mesencephalic reticular formation. Leading into this second conduction line at the level of the lateral habenular nucleus is a substantial extrapyramidal pathway originating in the internal pallidal segment. It is important to note that the two lines appear to remain largely if not entirely separate in their passage through the habenular complex, and thus may converge only in the

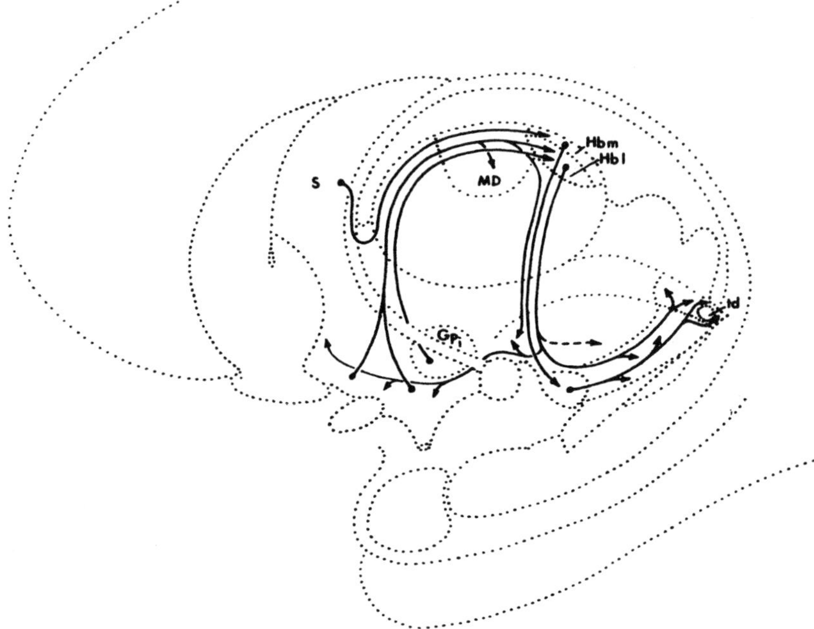

**Figure 4.** Conduction pathways involving the habenular complex, as described in the text. Broken line indicates projections from lateral habenular nucleus (Hbl) to structures lateral to the paramedian "limbic midbrain area," in particular substantia nigra, pars compacta, and tegmental reticular formation. For identification of structures not labelled in this drawing see Figure 2.

region of the raphe nuclei. Cajal (1911) was the first to emphasize the apparent absence of both axonal connections and dendritic overlap between the medial and the lateral habenular nucleus (for further references see Herkenham and Nauta, 1979).

## Ascending Components of the "Limbic Forebrain-Midbrain Circuit"

A detailed accounting of the various fiber systems ascending from the paramedian "limbic midbrain area" to the forebrain has thus far remained elusive, largely because only few of the likely sources of such fibers have been explored by appropriate combinations of anterograde and retrograde tracing methods. Important data, however, have come from studies by the histofluorescence method. This method provided the first evidence that several such projections, in part at least, affect the hippocampus, amygdala, and gyrus fornicatus directly rather than exclusively indirectly by way of the septo-preoptico-hypothalamic continuum as previously believed (Nauta, 1958; Nauta and Haymaker, 1969).

### *Projections Ascending from the Ventral Tegmental Area (VTA)*

Projections ascending from VTA are diagrammatically illustrated in Figure 5. In histofluorescence studies Andén et al. (1966) and Ungerstedt (1971) identified dopamine cell group A10 in the VTA as the source of the dopamine fibers innervating the nucleus accumbens and olfactory tubercle. Ungerstedt added the observation that A10 projects also to the central nucleus of the amygdala, the bed nucleus of the stria terminalis and the septum; apparently on the strength of these additional connections (and perhaps the consideration that A10 projects to that part of the striatum that receives projections also from the hippocampal formation and amygdala), Ungerstedt introduced the term *meso-limbic system* to denote the ascending A10 projection. Additional support for the notion that A10 is closely associated with the limbic system came from later evidence of A10 projections to the anteromedial (i.e., frontocingulate) cortex (Fuxe et al., 1974; Lindvall et al., 1974; Beckstead, 1976) as well as to the entorhinal area (Lindvall and Björklund, 1974).

The ascending VTA projections first identified in histofluorescence studies by virtue of their content in dopamine fibers have subsequently been demonstrated also by anterograde (autoradiographic) and retrograde labelling methods (Domesick et al., 1976; Fallon and Moore, 1978; Beckstead et al., 1979).

**Figure 5.** Projections ascending from the ventral tegmental area, as described in the text. Broken line indicates descending projections to tegmental reticular formation and locus coeruleus. For identification of unlabelled structures see Figure 2.

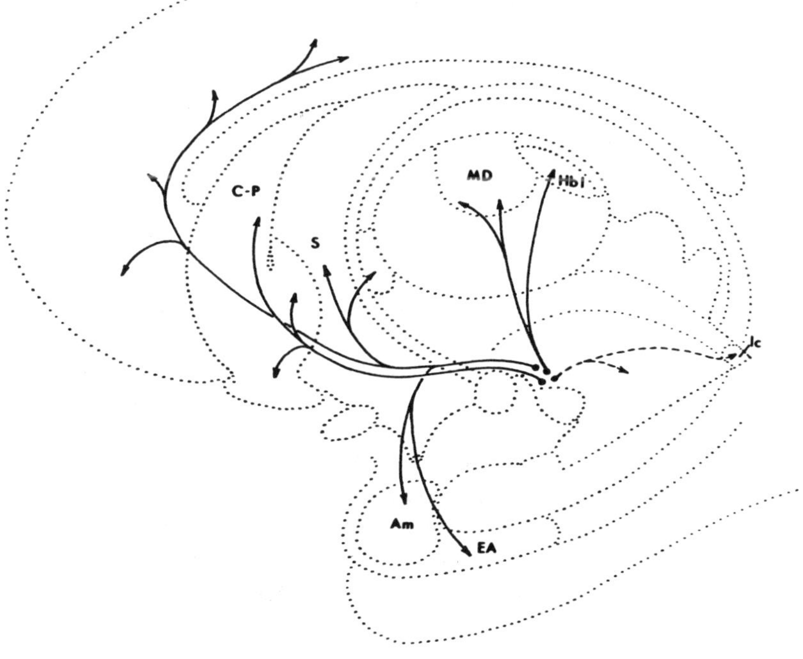

These methods have demonstrated additional VTA projections to the thalamus, in particular the lateral habenular nucleus and a circumferential zone of the mediodorsal nucleus (Herkenham and Nauta, 1977; Beckstead et al., 1979), as well as (broken lines in Figure 5) to the midbrain reticular formation (Domesick et al., 1976; Beckstead et al., 1979), dorsal raphe nucleus, and locus coeruleus (Simon et al., 1979). Since the methods used to demonstrate these additional VTA projections can supply no information whatever about the chemical characteristics of the neurons they label, it cannot *a priori* be concluded that VTA projections to striatal regions outside the nucleus accumbens, to the midbrain, and to the thalamus are dopaminergic.

### Projections Ascending from the Raphe Nuclei

A projection ascending the medial forebrain bundle from the paramedian zone of the caudal midbrain—the location of the major mesencephalic raphe nuclei—to the hypothalamus, preoptic area and septum was noted first in fiber-degeneration studies (Nauta and Kuypers, 1958), but it was the monoamine-histofluorescence method that supplied the first evidence that the projection is composed in part at least of serotonin fibers whose distribution extends beyond the septo-preoptico-hypothalamic continuum to the limbic structures of the cerebral hemisphere (Dahlström and Fuxe, 1964). Figure 6, in which the connections of the mesencephalic raphe nuclei with the forebrain are schematically indicated, is based upon these histochemical findings as well as upon later studies by anterograde labelling methods (Conrad et al., 1974; Bobillier et al., 1975; Moore et al., 1978).

### Non-Monoaminergic Projections Ascending from the Limbic Midbrain Area

It is still an open question whether the projections ascending from the ventral tegmental area and from the raphe nuclei are composed of monoamine fibers entirely or only in part. The difficulty is, that the histofluorescence method demonstrates exclusively monoamine fibers, whereas the autoradiographic technique—the most sensitive anterograde fiber-tracing method currently available—allows no distinction to be made between monoamine and non-monoamine fibers. Nonetheless, it is virtually certain that the ascending limb of the limbic forebrain-midbrain circuit is not exclusively monoaminergic. A projection ascending the medial forebrain bundle to the septum and hippocampus has recently been traced from the interpeduncular nucleus in the rat and cat (Stofer and Edwards, 1978; Baisden et al., 1979; Riley et al., 1979). The interpeduncular nucleus is not known to contain monoamine neurons. Neither is the dorsal tegmental nucleus of Gudden, from which a projection ascends in the mammillary peduncle to the lateral mammillary nucleus; beside fibers reciprocating the mammillotegmental connection, this ascending projection includes fibers that bypass the mammillary body and continue forward in the medial forebrain bundle as far rostrally as the medial septal nucleus

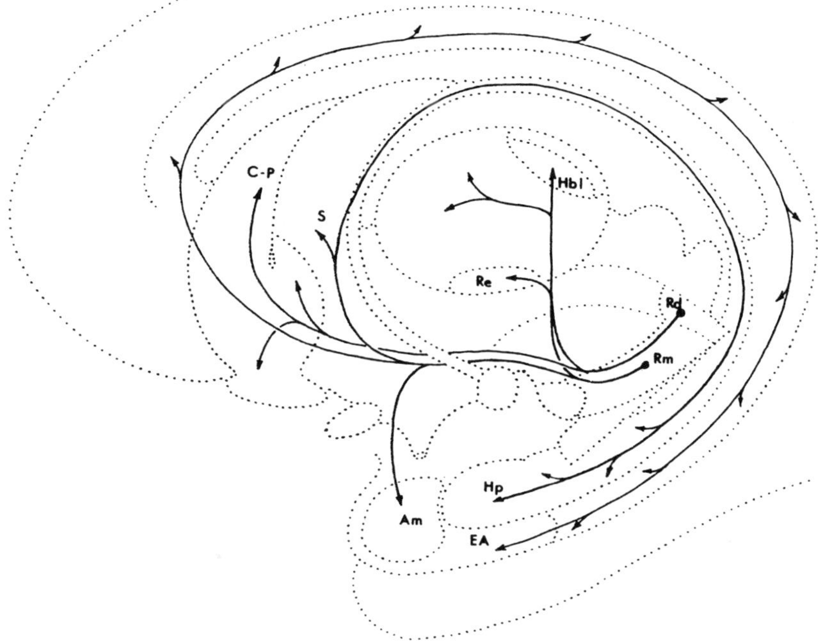


**Figure 6.** Projections ascending from the median (Rm) and dorsal (Rd) raphe nuclei of the midbrain, as described in the text. For identification of structures not labelled in this figure see Figure 2.

(Morest, 1961). Some of these fibers terminate in the supra-mammillary region, which projects in turn to both the hippocampus (Segal and Landis, 1974) and amygdala (Pretorius et al., 1979). Furthermore, Riley et al. (1979) have reported evidence of a direct projection from the nucleus tegmenti dorsalis lateralis to the hippocampus; this connection likewise seems unlikely to originate from monoamine neurons.

Finally, non-monoaminergic fibers are likely to be contained also in the dorsal longitudinal fasciculus of Schütz. In fiber-degeneration experiments (Nauta and Kuypers, 1958; Morest, 1961; Chi, 1970), ascending components of this fine-fibered system have been traced to posterior and dorsal regions of the periventricular and medial hypothalamic zones, but thus far neither the origin nor the distribution of the dorsal longitudinal fasciculus appear to have been determined by the aid of present-day labelling methods.

*Projections to the Limbic Forebrain from Lateral Brain Stem Structures*

Recently, evidence has been reported of projections to the limbic forebrain that arise not from the paramedian "limbic midbrain area" but, instead, from more lateral parts of the pontine and mesencephalic tegmentum, in particular the parabrachial region and the peripeduncular nucleus. The parabrachial region of the isthmus was found by Norgren (1976) to project not only to the thalamic gustatory nucleus but also to the central amygdaloid nucleus and the bed nucleus of the stria terminalis. According to Koh and Ricardo (1978), the two last-mentioned projections arise primarily from a dorsolateral part of the parabrachial region that receives secondary-sensory afferents mainly from the caudal, predominantly visceroceptive division of the nucleus of the solitary tract, and only more sparsely from the rostral, predominantly gustatory part of the nucleus. McBride and Sutin (1977) have reported evidence that a more anterior part of the parabrachial region projects directly to the ventromedial hypothalamic nucleus.

The peripeduncular nucleus has only recently become identified as a source of fibers ascending to the limbic forebrain. In an autoradiographic study, Jones et al. (1976) traced fibers from this lateral tegmental structure to the amygdala, and Pretorius et al. (1979) reported retrograde cell labelling in the peripeduncular by horseradish peroxidase injected into the amygdala.

*Rhombencephalic Projections to the Limbic Forebrain*

The first unequivocal evidence of a direct projection to the limbic forebrain from levels below the ponto-mesencephalic transition was reported by Ricardo and Koh (1978) in the form of a projection from the caudal third of the nucleus of the solitary tract to a remarkable diversity of mediobasal forebrain structures: the bed nucleus of the stria terminalis, the paraventricular, dorsomedial and arcuate nuclei of the hypothalamus, and the medial preoptic nucleus. This direct solitario-hypothalamic connection could be viewed as a sensory counterpart of sorts to the direct hypothalamo-autonomic projection demonstrated by Saper et al. (1976).

## Limbic Circuits Involving the Thalamus

Several limbic connections associated with the thalamus have been schematically indicated in Figure 7. Most familiar are those composing the so-called Papez circuit: hippocampal projections—originating from the subiculum hippocampi rather than from the CA fields (Swanson and Cowan, 1977)—pass to the anterior thalamic nucleus both directly (Gudden, 1881; Nauta, 1956) and by way of the mammillary body and mammillothalamic tract; the anterior thalamic nucleus projects, largely by way of the fasciculus cinguli, to the cingulate cortex, retrosplenial cortex and presubiculum (Domesick, 1973); the presubiculum in turn projects to the entorhinal area (Shipley, 1974) from which the massive perforant pathway to the hippocampus originates. A somewhat comparable circuit involving the amygdala is established by a massive projection from the amygdala to the medial division of the mediodorsal thalamic nucleus (Fox, 1949; Nauta, 1961, 1962; Krettek and Price, 1977) which in turn projects to the orbitofrontal cortex; the latter is reciprocally

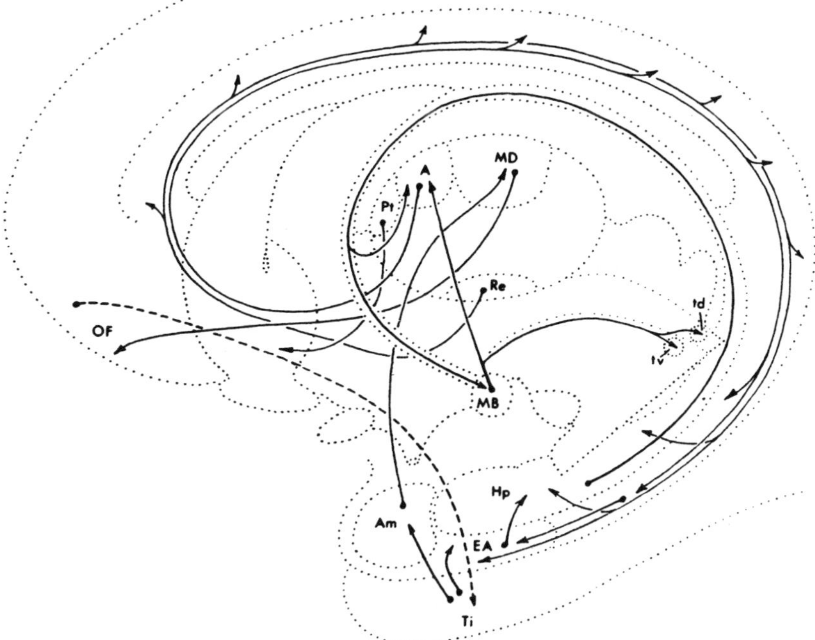



**Figure 7.** Limbic circuitry associated with the thalamus, as discussed in the text.

associated with the inferior temporal cortex (broken line in Figure 7), from which arise projections to the amygdala; an alternative orbitofronto-amygdaloid connection (not illustrated) is likely to lead by way of the medial forebrain bundle over the preoptic area and hypothalamus.

Two further thalamic connections are indicated in Figure 7. One of these is a substantial projection from the nucleus reuniens by way of the fasciculus cinguli to the full extent of the gyrus fornicatus, including the entorhinal area. At the level of the retrosplenial cortex this projection spreads into the hippocampus where it terminates profusely and selectively in the stratum lacunosummoleculare of field CA 1 (Herkenham, 1978). The projection is of interest especially as it establishes the only direct thalamohippocampal connection known at present. The information conveyed by it is as uncertain as are the afferent connections of the nucleus reuniens; of the latter little is known beyond the suggestion that they include fibers from the septum, subiculum hippocampi, anteromedial cortex, and mesencephalic raphe nuclei (Herkenham, 1978). Last to be mentioned is a projection to the nucleus accumbens from the parataenial nucleus of the thalamus (Cowan and Powell, 1955; Swanson and Cowan, 1975); this thalamic nucleus has been reported to receive a sparse reciprocating projection from the nucleus accumbens (Conrad and Pfaff, 1976c; Nauta et al., 1978), as well as afferents from the nucleus of the diagonal band (Domesick, 1976) and from the ventromedial hypothalamic nucleus (Krieger et al., 1979).

*Cross-Links of Limbic System and Corpus Striatum*

Several of the connections mentioned in the foregoing account can be viewed as efferent limbic channels bridging over to the striatum either directly or by way of structures known to project to the striatum. Direct limbico-striatal projections arise from the hippocampal formation, amygdala, and cingulate cortex. Those from hippocampus and amygdala together appear to involve an anteroventral part of the striatum that includes, but is considerably larger than, the nucleus accumbens as conventionally demarcated from the large remainder of the striatum (Krettek and Price, 1978; for a brief discussion of the problem of delimiting the nucleus accumbens cf. Nauta et al., 1978). The cingulate cortex projects to a narrow mediodorsal zone of the striatum; only its most anteroventral, infralimbic region has an additional and quite substantial projection to the nucleus accumbens (Beckstead, 1979).

An even larger part of the striatum—if not indeed its entire expanse—could be affected by the limbic system more indirectly, by way of the substantia nigra and the dorsal raphe nucleus. The entire nigral dopamine complex, including the outlying cell groups A10 and A8, lies within the distribution area of preoptic and hypothalamic efferents descending in the medial forebrain bundle (Nauta and Domesick, 1978), and cell groups A10 and A9 receive additional limbic afferents by way of the lateral habenular nucleus (Herkenham and Nauta, 1979). The raphe nuclei, likewise, receive inputs from limbic forebrain structures by way of both the medial forebrain bundle and fasciculus retroflexus.

Connections oriented in the opposite direction—from the striatum to the limbic system—appear to be more limited in number. The only direct striato-limbic connection thus far reported is established by a rather sparse projection from the nucleus accumbens to the lateral amygdaloid nucleus (Nauta et al., 1978). More indirect channels of striato-limbic communication, however, are potentially numerous. From the nucleus accumbens, fibers have been traced not only to the substantia nigra (Swanson and Cowan, 1975) but also to the lateral septal nucleus, bed-nucleus of the stria terminalis, various hypothalamic cell groups, and ventral tegmental area (Conrad and Pfaff, 1976c; Nauta et al., 1978). Longer accumbens efferents have been reported to follow the mesencephalic course of the medial forebrain bundle and to be distributed much like the larger number of fibers descending from the septo-preoptico-hypothalamic continuum. This striato-mesencephalic projection originates not only from the nucleus accumbens as conventionally defined but also from more lateral parts of the ventral striatal region (Nauta et al., 1978). It seems certain to be distributed in part at least to the ventral tegmental area and other paramedian regions from which projections ascend to limbic forebrain structures.

A somewhat similar, although more restricted confluence of efferent limbic channels with the circuitry of the extrapyramidal system takes place in the lateral habenular nucleus (Nauta, 1974; Herkenham and Nauta, 1977, 1979). In this instance it is not the striatum but the pallidum (more specifically its internal segment) that gives rise to an extrapyramidal cross-bridge into the circuitry of the limbic system.

## The "Limbic Forebrain-Midbrain Circuit"

It is plain from the foregoing survey that recent anatomical analyses have added much detail to the previously elaborated picture of subcortical limbic connections. None of these newer findings, however, appears to contradict the initial impression (Nauta, 1958; Nauta and Haymaker, 1969) that these connections in part compose a circuit reciprocally linking the limbic structures of the cerebral hemisphere with a continuum of subcortical grey matter extending from the olfactory tubercle and septal region caudalward over the preoptic region, the hypothalamus, and the paramedian zone of the midbrain. The most important revision of the original construct is dictated by the evidence that the ascending limb of the circuit is composed in part at least of fibers, especially of the monoamine variety, that extend directly to the limbic structures of the cerebral hemisphere. In the initial report, the ascending return loop of the circuit was assumed to be quantitatively interrupted in the septo-preoptico-hypothalamic continuum.

## A Second "Limbic Forebrain-Midbrain Circuit"

A further and more fundamental modification of the original notion of a reciprocal limbico-mesencephalic connection has become necessary, for data now available indicate the existence of a second limbic forebrain-midbrain circuit not identified in the earlier fiber-degeneration studies. This circuit involves *lateral* tegmental cell groups rather than the paramedian "limbic midbrain area," and appears to be part of the lateral rather than medial division of the medial forebrain bundle (Figure 8). Its descending limb is composed of (a) fibers from the central nucleus of the amygdala to the parabrachial region (Krettek and Price, 1978), and (b) fibers to the peripeduncular nucleus originating from the basal nucleus of Meynert in the substantia innominata (Jones et al., 1976) as well as from the ventromedial hypothalamic nucleus (Saper et al., 1976; Krieger et al., 1979). The ascending limb of this lateral circuit includes (a) fibers from the parabrachial region to the central nucleus of the amygdala, bed nucleus of the stria terminalis (Koh and Ricardo, 1978), and ventromedial hypothalamic nucleus (McBride and Sutin, 1977) and (b), a projection from the peripeduncular nucleus to the amygdala (Jones et al., 1976; Pretorius et al., 1979) and possibly to other limbic forebrain structures. It is interesting that this circuit at its rostral end involves the amygdala as well as several structures known to be closely associated with the amygdala: the

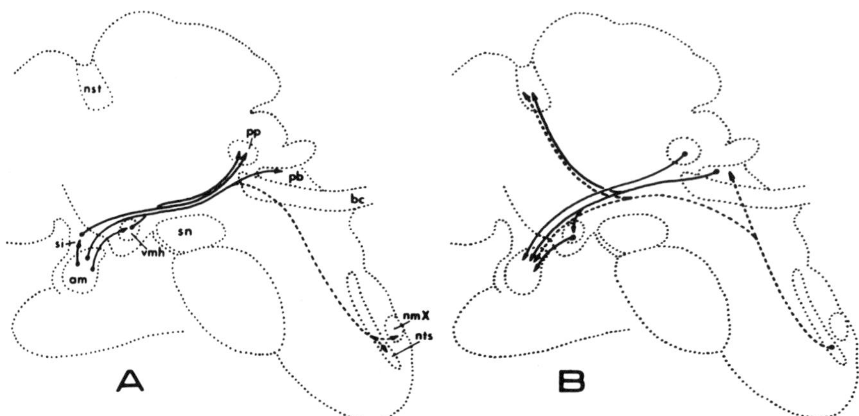


**Figure 8.** The lateral limbic forebrain-midbrain circuit as described in the text. (A) Descending components, leading from central nucleus of the amygdala (am) to the parabrachial region (pb), and from ventromedial hypothalamic nucleus (vmh) and substantia innominata (si) to the peripeduncular nucleus (pp). The direct amygdalofugal projection to dorsal motor nucleus of the vagus (nmX) and nucleus of the solitary tract (nts) (Hopkins and Holstege, 1978) is indicated in broken line. Other labels: bc: brachium conjunctivum; nst: bed-nucleus of stria terminalis; sn: substantia nigra. (B) Ascending components of the lateral circuit. Broken lines indicate projections from nucleus of the solitary tract to parabrachial region, central amygdaloid nucleus, and bed-nucleus of stria terminalis (Ricardo and Koh, 1978).

basal nucleus of Meynert, the bed nucleus of the stria terminalis, and the ventromedial hypothalamic nucleus. Of further interest is that the dorsolateral part of the parabrachial region, the central amygdaloid nucleus, and the bed nucleus of the stria terminalis have all been found to receive direct projections from the caudal part of the nucleus of the solitary tract (Ricardo and Koh, 1978). This appears to suggest that the lateral limbic forebrain-midbrain circuit is closely involved in the monitoring of neural inputs from viscera innervated by the vagus and glossopharyngeal nerves. The close proximity of visceral-afferent to gustatory-afferent channels in the circuit, not only at the level of the nucleus of the solitary tract but also in the parabrachial region (Koh and Ricardo, 1978), raises the question whether the lateral limbic forebrain-midbrain circuit might be involved in the visceral and motivational mechanisms underlying the selection and intake of food.

## The Action Radius of the Limbic System

It must be emphasized that the connection of the limbic system with the ponto-mesencephalic tegmentum is not defined by the forebrain-midbrain circuits mentioned above. A large contingent of descending medial-forebrain-bundle fibers is distributed to the wide expanse of central midbrain reticular formation that lies between the medial and lateral circuit. The available data indicate that this limbic projection to the central midbrain tegmentum originates largely in the preoptic region and hypothalamus (Nauta, 1958; Conrad and Pfaff, 1976a, b; Swanson, 1976; Saper et al., 1979) and receives lesser contributions from the central amygdaloid nucleus (Hopkins and Holstege, 1978); nucleus accumbens (Nauta et al., 1978) and bed nucleus of the stria terminalis (Swanson and Cowan, 1979); other basal forebrain structures may eventually be found also to contribute to it. A similar, although more restricted distribution to tegmental regions lateral to the "limbic midbrain area" has been found to characterize those fibers of the fasciculus retroflexus that originate from the lateral half of the lateral habenular nucleus (Herkenham and Nauta, 1979). It thus seems that the limbic forebrain can convey its influence to a wide cross-sectional area of the midbrain reticular formation by way of both the medial forebrain bundle and the conduction routes involving the lateral habenular nucleus. Both these descending pathways presumably synapse in the midbrain with the upper links of a generally polysynaptic descending reticular pathway ultimately affecting, among other structures, preganglionic visceral motor neurons of the brain stem and spinal cord. This putative transreticular supranuclear component of the central visceromotor system is paralleled by the above-mentioned direct hypothalamo-and-amygdalo-autonomic projections reported by Saper et al. (1976) and Hopkins and Holstege (1978), respectively.

These connections doubtless form part of the anatomical substratum for the visceromotor functions of the limbic system. However, by such wide access routes to the pontomesencephalic reticular formation the limbic system could be thought to modulate also the effects of the reticular formation upon the forebrain, effects exerted largely, it seems, over mesencephalic pathways ascending to the subthalamic region and intralaminar thalamic nuclei. But recent findings have suggested a further and shorter path by which the limbic system could affect the forebrain, in this instance the cerebral cortex in particular. In retrograde-labelling experiments, Divac (1975) and Kievit and Kuypers (1975) have demonstrated a projection originating from large neurons in the nucleus of the diagonal band, preoptico-hypothalamic regions and substantia innominata, and spreading to most or all of the neocortex. Judging from their very wide distribution, both this fiber system of basal-forebrain origin and the pathways ascending from the pontomesencephalic reticular formation seem likely to affect the functional state of the cortex as a whole, rather than that of any cortical area in particular. However, in the monkey, anatomical evidence has been found of several limbico-neocortical connections that have a more limited distribution, and therefore seem likely to have a more selective function. The most convincingly documented among these originate from the amygdala and are directed at the frontal and inferior temporal cortex. One is composed of fibers from the amygdala directly to the inferior temporal region (Nauta, 1961; Krettek and Price, 1974); a second one is a direct projection from the amygdala to the dorsal half of the convexity of the frontal cortex (Jacobson and Trojanowski, 1975), whereas a third one is a dual amygdaloorbitofrontal projection, in part direct (Nauta, 1961; Porrino and Goldman, 1979) (Figure 3), in part indirect by way of the mediodorsal nucleus of the thalamus (Figure 7). In addition to these well-documented direct amygdalocortical projections, direct pathways to temporal and frontal neocortex have recently been reported to originate from the subiculum hippocampi as well as from the parasubiculum and entorhinal area (Schwerdtfeger, 1979). The collective distribution of these more circumscript amygdalocortical and hippocampocortical projections is of considerable interest, for it appears to correspond largely to those cortical regions which are most heavily converged upon by the chains of cortico-cortical connections that lead away from each of the primary sensory areas of the cortex (Jones and Powell, 1970). From various parts of these same frontotemporal convergence regions projections, direct or indirect, have been traced in return to the amygdala (Whitlock and Nauta, 1956; Jones and Powell, 1970; Herzog and Van Hoesen, 1976) and also to the entorhinal area (Van Hoesen and Pandya, 1975; Van Hoesen et al., 1975) (Figure 3). The amygdalocortical connections mentioned above could thus be viewed as limbico-cortical projections reciprocating the flow of multisensory information from the cortex to the limbic system. In the light of these connections, the limbic system could be viewed as a neural mechanism that not only monitors the sensory processes of the cerebral cortex, but can also reach out to intervene in these processes. It could thus be suspected, even on the basis of no more than anatomical data, that the functional set of the limbic system affects not only, as widely acknowledged, the organism's visceral and endocrine functions and its motivational state, but also the sensory and associative mechanisms involved in its perceptions and ideational processes.

ACKNOWLEDGMENTS
This survey is based in part on studies made possible by USPHS Grants NB 06542 and MH 25515, and by NSF Grant 76-81227.

# References

Andén, N.-E., Dahlström, A., Fuxe, K., Larsson, K., Olson, L., and Ungerstedt, U.: Ascending monoamine neurons to the telencephalon and diencephalon. *Acta Physiol. Scand.* 67:313–326, 1966.

Baisden, R.H., Hoover, D.B., and Cowie, R.J.: Retrograde demonstration of hippocampal afferents from the interpeduncular and reuniens nuclei. *Neurosci. Letters* 13:105–109, 1979.

Beckstead, R.M.: Convergent thalamic and mesencephalic projections to the anterior medial cortex in the rat. *J. Comp. Neur.* 166:403–416, 1976.

Beckstead, R.M.: An autoradiographic examination of cortico-cortical and subcortical projections of the mediodorsal-projection (prefrontal) cortex in the rat. *J. Comp. Neur.* 184:43–62, 1979.

Beckstead, R.M., Domesick, V.B., and Nauta, W.J.H.: Efferent connections of the substantia nigra and ventral tegmental area in the rat. *Brain Res.* 175:191–217, 1979.

Bobillier, P., Petitjean, F., Salvert, D., Ligier, M., and Seguin, S.: Differential projections of the nucleus raphe dorsalis and nucleus raphe centralis as revealed by autoradiography. *Brain Res.* 85:205–210, 1975.

Broca, P.: Anatomie comparée des circonvolutions cérébrales. Le grand lobe limbique et la scissure limbique dans la série des mammifères. *Rev. d'Anthrop.* Ser. 2, 1:285–498, 1878.

Cajal, S.R.: Histologie du Système Nerveux de l'Homme et des Vertébrés, Vol. II. Maloine, Paris, 1911.

Chi, C. G.: An experimental silver study of the ascending projections of the central grey substance and adjacent tegmentum in the rat with observations in the cat. *J. Comp. Neur.* 139:259–272, 1970.

Conrad, L.C.A., Leonard, C.M., and Pfaff, D.: Connections of the median and dorsal raphe nuclei in the rat: an autoradiographic and degeneration study. *J. Comp. Neur.* 156:179–205, 1974.

Conrad, L.C.A. and Pfaff, D.W.: Efferents from medial basal forebrain and hypothalamus in the rat. I. An autoradiographic study of the medial preoptic area. *J. Comp. Neur.* 169:185–220, 1976a.

Conrad, L.C.A. and Pfaff, D.W.: Efferents from medial basal forebrain and hypothalamus in the rat. II. An autoradiographic study of the anterior hypothalamus. *J. Comp. Neur.* 169:221–262, 1976b.

Conrad, L.C.A. and Pfaff, D.W.: Autoradiographic tracing of nucleus accumbens efferents in the rat. *Brain Res.* 113:589–596, 1976c.

Cowan, W.M., Gottlieb, D.I., Hendrickson, A.E., Price, J.L., and Woolsey, T.A.: The autoradiographic demonstration of axonal connections in the central nervous system. *Brain Res.* 37:21–51, 1972.

Cowan, W.M. and Powell, T.P.S.: The projections of the midline and intralaminar nuclei of the thalamus of the rabbit. *J. Neur. Neurosurg. Psychiat.* 18:266–279, 1955.

Dahlström, A. and Fuxe, K.: Evidence for the existence of monoamine-containing neurons in the central nervous system. I. Demonstration of monoamines in the cell bodies of brain stem neurons. *Acta Physiol. Scand. Suppl. 232*, 62:1–55, 1964.

Divac, I.: Magnocellular nuclei of the basal forebrain project to neocortex, brain stem, and olfactory bulb. Review of some functional correlates. *Brain Res.* 93:385–398, 1975.

Domesick, V.B.: Thalamic projections in the cingulum bindle to the parahippocampal cortex of the rat. (Abstr.) *Anat. Rec.* 175:308, 1973.

Domesick, V.B.: Projections of the nucleus of the diagonal band of Broca in the rat. (Abstr.) *Anat. Rec.* 184:391–392, 1976.

Domesick, V.B., Beckstead, R.M., and Nauta, W.J.H.: Some ascending and descending projections of the substantia nigra and ventral tegmental area in the rat. (Abstr.) *Neuroscience Abstr.* II, Part 1, 61, 1976.

Fallon, J.H. and Moore, R.Y.: Catecholamine innervation of the basal forebrain. IV. Topography of the dopamine projection to the basal forebrain and neostriatum. *J. Comp. Neurol.* 180:545–580, 1978.

Fox, C. A.: Amygdalo-thalamic connections in Macaca mulatta (Abstr.) *Anat. Rec.* 103:537–538, 1949.

Fuxe, K., Hökfelt, T., Johansson, O., Jonsson, G., Lidbrink, P., and Ljungdahl, A.: The origin of the dopamine nerve terminals in limbic and frontal cortex. Evidence for meso-cortical dopamine neurons. *Brain Res.* 82:349–355, 1974.

Gudden, B., von: Beitrag zur Kenntniss des Corpus mammillare und der sogenannten Schenkel des Fornix. *Arch. f. Psychiat. Nervenkr.* 11:428–452, 1881.

Heimer, L. and Nauta, W.J.H.: The hypothalamic distribution of the stria terminalis in the rat. *Brain Res.* 13:284–297, 1969.

Herkenham, M.: The connections of the nucleus reuniens thalami: evidence for a direct thalamo-hippocampal pathway in the rat. *J. Comp. Neurol.* 177:589-610, 1978.

Herkenham, M. and Nauta, W.J.H.: Afferent connections of the habenular nuclei in the rat. A horseradish peroxidase study, with a note on the fiber-of-passage problem. *J. Comp. Neurol.* 173:123-146, 1977.

Herkenham, M. and Nauta, W.J.H.: Efferent connections of the habenular nuclei in the rat. *J. Comp. Neurol.* 187:19-48, 1979.

Herzog, A.G. and Van Hoesen, G.W.: Temporal neocortical afferent connections to the amygdala in the Rhesus monkey. *Brain Res.* 115:57-69, 1976.

Hopkins, D.A. and Holstege, G.: Amygdaloid projections to the mesencephalon, pons and medulla oblongata in the cat. *Exp. Brain Res.* 32:529-547, 1978.

Jacobson, S. and Trojanowski, J.Q.: Amygdaloid projections to prefrontal granular cortex in Rhesus monkey demonstrated with horseradish peroxidase. *Brain Res.* 100:132-139, 1975.

Jones, B.E. and Moore, R.Y.: Ascending projections of the locus coeruleus in the rat. II. Autoradiographic analysis. *Brain Res.* 127:23-53, 1977.

Jones, E.G., Burton, H., Saper, C.B., and Swanson, L.W.: Midbrain, diencephalic, and cortical relationships of the basal nucleus of Meynert and associated structures in primates. *J. Comp. Neurol.* 167:385-420, 1976.

Jones, E.G. and Powell, T.P.S.: An anatomical study of converging sensory pathways within the cerebral cortex of the monkey. *Brain* 93:793-820, 1970.

Kievit, J. and Kuypers, H.G.J.M.: Subcortical afferents to the frontal lobe in the Rhesus monkey studied by means of retrograde horseradish peroxidase transport. *Brain Res.* 85:261-266, 1975.

Koh, E.T. and Ricardo, J.A.: Afferents and efferents of the parabrachial region in the rat: evidence for parallel ascending gustatory versus viscerorceptive systems arising from the nucleus of the solitary tract. (Abstr.) *Anat. Rec.* 190, No. 2:449, 1978.

Krettek, J.E. and Price, J.L.: A direct input from the amygdala to the thalamus and the cerebral cortex. *Brain Res.* 67:169-174, 1974.

Krettek, J.E. and Price, J.L.: Projections from the amygdaloid complex to the cerebral cortex and thalamus in the rat and cat. *J. Comp. Neurol.* 172:687-722, 1977.

Krettek, J.E. and Price, J.L.: Amygdaloid projections to subcortical structures within the basal forebrain and brain stem in the rat and cat. *J. Comp. Neurol.* 178:225-254, 1978.

Krieger, M.S., Conrad, L.C.A., and Pfaff, D.W.: An autoradiographic study of the afferent connections of the ventromedial nucleus of the hypothalamus. *J. Comp. Neurol.* 183:785-816, 1979.

Kristensson, K., Olsson, Y., and Sjöstrand, J.: Axonal uptake and retrograde transport of exogenous proteins in the hypoglossal nerve. *Brain Res.* 32:399-406, 1971.

Kuypers, H.G.J.M. and Maisky, V.A.: Retrograde axonal transport of horseradish peroxidase from spinal cord to brain stem cell groups in the cat. *Neurosci. Letters* 1:9-14, 1975.

Lammers, H.J. and Lohman, A.H.M.: Experimenteel anatomisch onderzoek naar de verbindingen van piriforme cortex en amygdalakernen bij de kat. *Nederl. Tijdschr. Geneesk.* 101, no. 13:1-2, 1957.

Lavail, J.H., and Winston, K.R., and Tish, A.: A method based on retrograde intraaxonal transport of protein for identification of cell bodies of origin of axons terminating within the CNS. *Brain Res.* 58:470-477, 1973.

Lindvall, O. and Björklund, A.: The organization of the ascending catecholamine neuron systems in the rat brain as determined by the glyoxylic acid fluoresence method. *Acta Physiol. Scand., Suppl.* 412:1-48, 1974.

Lindvall, O., Björklund, A., Moore, R.Y., and Stenevi, U.: Mesencephalic dopamine neurons projecting to neocortex. *Brain Res.* 81:325-331, 1974.

Lundberg, P.O.: Cortico-hypothalamic connections in the rabbit. *Acta Physiol. Scand., Suppl.* 171:1–80, 1960.

Lundberg, P.O.: The nuclei gemini. Two hitherto undescribed nerve cell collections in the hypothalamus of the rabbit. *J. Comp. Neurol.* 119:311–316, 1962.

MacLean, P.D.: Some psychiatric implications of physiological studies on frontotemporal portion of limbic system (visceral brain). *E.E.G. & Clin. Neurophysiol. J.* 4:407–418, 1952.

McBride, R.L. and Sutin, J.: Amygdaloid and pontine projections to the ventromedial nucleus of the hypothalamus. *J. Comp. Neurol.* 174:377–396, 1977.

Millhouse, O.E.: A Golgi study of the descending MFB. *Brain Res.* 15:341–363, 1969.

Moore, R.Y., Halaris, A.E., and Jones, B.E.: Serotonin neurons of the midbrain raphe: ascending projections. *J. Comp. Neurol.* 180:417–438, 1978.

Morest, D.K.: Connections of the dorsal tegmental nucleus in rat and rabbit. *J. Anat. (Lond.)* 95:229–246, 1961.

Nauta, H.J.W.: Evidence of a pallidohabenular pathway in the cat. *J. Comp. Neurol.* 156:19–28, 1974.

Nauta, W.J.H.: An experimental study of the fornix system in the rat. *J. Comp. Neurol.* 104:247–271, 1956.

Nauta, W.J.H.: Hippocampal projections and related neural pathways to the midbrain in the cat. *Brain* 81:319–340, 1958.

Nauta, W.J.H.: Fiber degeneration following lesions of the amygdaloid complex in the monkey. *J. Anat. (Lond.)* 95:515–531, 1961.

Nauta, W.J.H.: Neural associations of the amygdaloid complex in the monkey. *Brain* 85:505–520, 1962.

Nauta, W.J.H.: Expanding borders of the limbic system concept. In Rassmussen, T. and Marino, R. (eds.), *Functional Neurosurgery.* Raven Press, New York, pp. 7–23, 1979.

Nauta, W.J.H. and Domesick, V.B.: Crossroads of limbic and striatal circuitry: hypothalamo-nigral connections. In K.E. Livingston and O. Hornykiewicz (eds.), *Limbic Mechanisms*, Plenum Publishing Corporation, New York & London, pp. 75–93, 1978.

Nauta, W.J.H. and Haymaker, W.: Hypothalamic nuclei and fiber connections. In Haymaker, W. et al. (eds.), *The Hypothalamus*, C.C. Thomas Publisher, Springfield, Ill., pp. 136–209, 1969.

Nauta, W.J.H. and Kuypers, H.G.J.M.: Some ascending pathways in the brain stem reticular formation. In *Reticular Formation of the Brain*, H.H. Jasper et al. (eds.), Little, Brown and Co., Boston & Toronto, pp. 3–30, 1958.

Nauta, W.J.H., Smith, G.P., Faull, R.L.M., and Domesick, V.B.: Efferent connections and nigral afferents of the nucleus accumbens in the rat. *Neuroscience* 3:385–401, 1978.

Norgren, R.: Taste pathways to hypothalamus and amygdala. *J. Comp. Neur.* 166:17–30, 1976.

Olmos, J.S. de: The amygdaloid projection field in the rat studied with the cupric-silver method. In B.E. Eleftheriou (ed.), *The Neurobiology of the Amygdala.* Plenum Press, New York & London, pp. 145–204, 1972.

Olmos, J.S. de and Ingram, W.R.: The projection field of the stria terminalis in the rat brain. An experimental study. *J. Comp. Neur.* 146:303–333, 1972.

Papez, J.W.: A proposed mechanism of emotion. *Arch. Neurol. Psychiat.*, Chicago 38:725–743, 1937.

Porrino, L.J. and Goldman, P.S.: Selective distribution of projections from amygdala to prefrontal cortex in Rhesus monkey. *Soc. Neurosci. Abstr.* 5:280, 1979.

Powell, E.W. and Leman, R.B.: Connections of the nucleus accumbens. *Brain Res.* 105:389–403, 1976.

Pretorius, J.K., Phelan, K.D., and Mehler, W.R. Afferent connections of the amygdala in rat. (Abstr.) Anat. Rec. 193:657, 1979.

Ricardo, J.A. and Koh, E.T.: Anatomical evidence of direct projections from the nucleus of the solitary tract to the hypothalamus, amygdala, and other forebrain structures in the rat. *Brain Res.* 153:1–26, 1978.

Riley, J.N., Marchand, E.R., and Moore, R.Y.: Diencephalic and brain stem afferents to the hippocampal formation of the rat. *Soc. Neurosci. Abstr.* Vol. 5:p. 281, 1979.

Saper, C.B., Loewy, A.D., Swanson, L.W., and Cowan, W.M.: Direct hypothalamo-autonomic connections. *Brain Res.* 177:305–312, 1976.

Saper, C.B., Swanson, L.W., and Cowan, W.M.: The efferent connections of the ventromedial nucleus of the hypothalamus of the rat. *J. Comp. Neurol.* 169:409–422, 1976.

Saper, C.B., Swanson, L.W., and Cowan, W.M.: An autoradiographic study of the efferent connections of the lateral hypothalamic area in the rat. *J. Comp. Neurol.* 183:689–706, 1979.

Schwerdtfeger, W.K.: Direct efferent and afferent connections of the hippocampus with the neocortex in the marmoset monkey. *Am. J. Anat.* 156:77–82, 1979.

Segal, M. and Landis, S.: Afferents to the hippocampus of the rat studied with the method of retrograde transport of horseradish peroxidase. *Brain Res.* 78:1–15, 1974.

Shipley, M.T.: Presubiculum afferents to the entorhinal area and the Papez circuit. *Brain Res.* 67:162–168, 1974.

Simon, H., Le Moal, M., Stinus, L., and Calas, A.: Anatomical relationships between the ventral mesencephalic tegmentum—A10 region and the locus coeruleus as demonstrated by anterograde and retrograde tracing techniques. *J. Neural Transmission* 44:77–86, 1979.

Stofer, W.B. and Edwards, S.B.: Organization and efferent projections of the interpeduncular complex in the cat. *Soc. Neurosci. Abstr.* 4:228, 1978.

Swanson, L.W.: An autoradiographic study of the efferent connections of the preoptic region in the rat. *J. Comp. Neurol.* 167:227–256, 1976.

Swanson, L.W. and Cowan, W.M.: A note on the connections and development of the nucleus accumbens. *Brain Res.* 92:324–330, 1975.

Swanson, L.W. and Cowan, W.M.: An autoradiographic study of the organization of the efferent connections of the hippocampal formation in the rat. *J. Comp. Neurol.* 172:49–84, 1977.

Swanson, L.W. and Cowan, W.M.: The connections of the septal region in the rat. *J. Comp. Neurol.* 186:621–656, 1979.

Ungerstedt, U.: Stereotaxic mapping of the monoamine pathways in the rat brain. *Acta Physiol. Scand.* 197, Suppl. 367:1–48, 1971.

Valenstein, E.S. and Nauta, W.J.H.: A comparison of the distribution of the fornix system in the rat, guinea pig, cat, and monkey. *J. Comp. Neurol.* 113:337–363, 1959.

Van Hoesen, G.W. and Pandya, D.N.: Some connections of the entorhinal (area 28) and perirhinal (area 35) cortices of the Rhesus monkey. I. Temporal lobe afferents. *Brain Res.* 95:1–24, 1975.

Van Hoesen, G.W., Pandya, D.N., and Butters, N.: Some connections of the entorhinal (area 28) and perirhinal (area 35) cortices of the Rhesus monkey. II. Frontal lobe afferents. *Brain Res.* 95:25–38, 1975.

Whitlock, D.G. and Nauta, W.J.H.: Subcortical projections from the temporal neocortex in Macaca mulatta. *J. Comp. Neurol.* 106:183–212, 1956.

# Cytoarchitecture, Fiber Connections, and Some Histochemical Aspects of the Interpeduncular Nucleus in the Rat

H.J. GROENEWEGEN, S. AHLENIUS, S.N. HABER, N.W. KOWALL, AND W.J.H. NAUTA
Department of Psychology and Brain Science, Massachusetts Institute of Technology, Cambridge, Massachusetts 02139, and the Mailman Research Center, McLean Hospital, Belmont, Massachusetts 02178

## ABSTRACT

The organization of afferent and efferent connections of the interpeduncular nucleus (IP) has been examined in correlation with its subnuclear parcellation by using anterograde and retrograde tracing techniques. Based on Nissl, myelin, and acetylcholinesterase staining five paired and three unpaired IP subnuclei are distinguished. The unpaired division includes the rostral subnucleus (IP-R), the apical subnucleus (IP-A), and the central subnucleus (IP-C). The subnuclei represented bilaterally are the paramedian dorsal medial (IP-DM) and intermediate subnuclei (IP-I) and the laterally placed rostral lateral (IP-RL), dorsal lateral (IP-DL), and lateral subnuclei (IP-L). Immunohistochemical techniques showed cell bodies and fibers and terminals immunoreactive for substance P, leu-enkephalin, met-enkephalin, or serotonin to be differentially distributed over the different IP subnuclei. Substance P–positive perikarya were found in IP-R, enkephalin neurons in IP-R, IP-A, and the caudodorsal part of IP-C, and serotonin-containing cell bodies in IP-A and the caudal part of IP-L.

Efferent IP projections were studied both by injecting tritiated leucine in IP and by injecting HRP or WGA-HRP in the presumed termination areas. The results indicate that the major outflow of IP is directed caudalward to the median and dorsal raphe nuclei and the caudal part of the central gray substance, i.e., the dorsal tegmental region. The projection appears to terminate mainly in the raphe nuclei, around the ventral and dorsal tegmental nuclei of Gudden, and in the dorsolateral tegmental nucleus. The descending projection to the dorsal tegmental region originates in virtually all IP subnuclei, but the main contribution comes from IP-R and the lateral subnuclei IP-RL, IP-DL, and IP-L. Sparser projections to the dorsal tegmental region originate in IP-C and IP-I, whereas the contribution of IP-A is only minimal. The projections from IP-R are mainly ipsilateral and those from IP-DM are mainly contralateral. IP fibers to the median and dorsal raphe nuclei originate predominantly in IP-R and IP-DM, and to a lesser extent in IP-C, IP-I, IP-RL, and IP-DL.

A much smaller contingent of IP fibers ascends to diencephalic and telencephalic regions. A relatively minor projection, stemming from IP-RL and IP-DL, reaches the lateral part of the mediodorsal nucleus, the nucleus

Accepted January 18, 1986.

Address reprint requests to Dr. H.J. Groenewegen at his present address, Department of Anatomy and Embryology, Faculty of Medicine, Vrije Universiteit, Van der Boechorststraat 7, 1081 BT Amsterdam, The Netherlands.

S. Ahlenius' present address is the Astra Pharmaceutical Company, Research and Development, Department of Pharmacology, S-15185 Sodertalje, Sweden.

S.N. Haber's present address is the Department of Anatomy, University of Rochester, Medical School, 601 Elmwood Ave., Rochester, NY 14642.

Parts of this study have been reported elsewhere in abstract (Ahlenius and Nauta, '80; Groenewegen et al., '83a,b).

gelatinosus, and some midline thalamic nuclei. These IP fibers follow either the habenulo-interpeduncular pathway or the mammillothalamic tract. Another group of ascending IP fibers follows the medial forebrain bundle, providing offsets to the supramammillary region, the medial and lateral hypothalamus, the nucleus of the diagonal band, and the lateral septum. A few of these fibers continue in the fornix to the cornu Ammonis and fascia dentata of the hippocampus. Retrograde tracing experiments revealed that this part of the ascending IP system stems from neurons in IP-A and the caudal part of IP-L.

Regarding the afferent connections of the interpeduncular nucleus, in the present study most emphasis has been placed on the anterograde tracing of these projections. The nucleus derives its main input from the medial habenular nucleus and the dorsal tegmental region. The habenula projects in moderate-to-heavy volume to all IP subnuclei, with the exception of IP-A, which receives a relatively sparse habenular input. Similarly, afferents from the dorsal tegmental region reach the entire IP, but this projection exhibits a more complex differential distribution. It most massively involves the contralateral IP-R and IP-DM, and IP-DL and IP-L bilaterally. Less prominent are its distributions to IP-A, IP-C, IP-I, and IP-RL. More rostral parts of the central gray substance, including the dorsal raphe nucleus, project heavily to the contralateral IP-DM, and to IP-RL and IP-DL bilaterally. The median raphe nucleus sends a strong projection to IP-C and IP-I and a minor projection to IP-L. Injections in the parabigeminal region labeled a very restricted but distinct projection to a dorsal part of IP-RL bilaterally. Within IP itself indications were found of reciprocal connections between IP-A and IP-R. Autoradiographic tracing of descending fibers to IP from the nucleus of the diagonal band and the lateral hypothalamus revealed only very sparse projections, mainly distributed to IP-C, IP-I, and IP-L.

The present study demonstrates that the inputs and outputs of the various subnuclei of IP are interrelated in a most complex way. Nevertheless, the massive projection from the medial habenular nucleus to IP and the close reciprocal connection of IP with the raphe nuclei and the dorsal tegmental nuclei suggests an important modulatory role of IP in the transmission of impulses descending from the limbic forebrain via the medial habenular nucleus to the sources of serotoninergic and other mesencephalic projections affecting most if not all principal subdivisions of the cerebral hemisphere.

Key words: interpeduncular circuitry, substance P, enkephalin, serotonin, acetylcholinesterase, limbic system

---

The interpeduncular nucleus constitutes an important link in the descending pathways of the limbic system–midbrain circuit (Nauta, '58; Herkenham and Nauta, '79; Nauta and Domesick, '81). Limbic forebrain structures such as the septal region and the preoptic area are connected through the medial habenular nucleus to the interpeduncular nucleus (Wang and Aghajanian, '77; Herkenham and Nauta '77, '79; Marchand et al., '80), which itself projects prominently to the region of the mesencephalic raphe nuclei. The median and dorsal raphe nuclei in turn affect all of the principal subdivisions of the forebrain by way of widespread serotoninergic projections that directly involve such diverse structures as the hippocampus, the amygdala, the striatum, and the cerebral cortex (Conrad et al., '74; Bobillier et al., '75; Azmitia and Segal, '78; Steinbusch, '81).

A descending projection to the mesencephalic raphe, first described by Ganser in 1882, doubtless represents the most substantial and well-defined efferent connection of the interpeduncular nucleus. Much less is known about other efferent connections of the nucleus. Older studies by the aid of fiber-degeneration methods have yielded some impression of these less prominent projections (Massopust and Thompson, '62; Mitchell, '63; Smaha and Kaelber, '73), but a more detailed analysis by such methods has been impeded by the small size of the nucleus and its position between large ascending and descending fiber tracts. Both degeneration experiments and more recent anterograde (Stofer and Edwards, '78) and retrograde tracer studies have indicated that the interpeduncular nucleus, in addition to its major descending projections, gives rise to ascending pathways as well. Thus, retrogradely labeled cells have been found in the interpeduncular nucleus following HRP injections into the hippocampus (Baisden et al., '79; Wyss et al., '79; Riley and Moore, '81), the mediodorsal thalamic nucleus (Velayos and Reinoso-Suarez, '82; Groenewegen and Nauta, '82), and the lateral hypothalamus (Stofer and Edwards, '78).

The nucleus interpeduncularis consists of several cytoarchitectonically distinct cell groups (Ives, '71; Hamill and Lenn, '81, '84; Hemmendinger and Moore, '84). Moreover, specific parts of the interpeduncular nucleus can be distinguished from other parts by a higher or lower content in substances such as acetylcholinesterase and one or the other of several peptides or transmitter substances, either in fibers and terminals or in cell bodies (Ljungdahl et al., '78; Sar et al., '78; Cuello and Kanazawa, '78; Steinbusch, '81; Finley et al., '81; Hemmendinger and Moore, '82, '84; Hamill et al., '84). Little is known, however, of the specific input-output relationships of such individual parts of the nucleus.

In the present study the autoradiographic fiber-tracing method was used to trace the efferent connections of the interpeduncular nucleus in the rat. Subsequently, the retrograde-tracing techniques were employed to corroborate the results of these and previously reported anterograde-tracing experiments and to determine the pathways in question with respect to the cytoarchitectonically and histochemically specified subdivisions of the interpeduncular nucleus.

The results of these experiments provided a rather detailed picture of the interpeduncular afferents, and in view of this it seemed worthwhile to restudy a number of afferent fiber systems of the interpeduncular nucleus with anterograde-tracing techniques in an attempt to correlate the afferent and efferent relationships of the nucleus.

## MATERIALS AND METHODS

The present report is based on observations and experiments in female rats of the Charles River CD strain, weighing 250–500 g. In a number of brains one or a combination of the following stains was used to study the normal structure and histochemistry of the nucleus interpeduncularis (IP). The cytoarchitecture was studied in material stained with cresyl violet and by silver impregnation of Nissl substance according to Merker's technique ('83). For the staining of the myelin pattern the Loyez method was used. The distribution of acetylcholinesterase (AChE) activity was determined with the aid of Geneser-Jensen and Blackstad's ('71) method. All procedures were carried out on aldehyde-fixed tissue cut on a freezing microtome at 20, 30, or 40 $\mu$m.

### Immunohistochemistry

For the localization of enkephalin-like, serotonin-like, and substance P–like immunoreactivity nine rats were deeply anesthetized and subsequently perfused with 4% paraformaldehyde in a 5% solution of sucrose in 0.1 M phosphate buffer of pH 7.2 and cooled to 4°C. Three of the animals had been pretreated 24–48 hours prior to perfusion with an intraventricular injection of 5 $\mu$l colchicine (15 $\mu$g/$\mu$l), and three other animals received an injection of 1 $\mu$l colchicine (15 $\mu$g/$\mu$l) near the interpeduncular nucleus to enhance staining of cell bodies. The brains were removed, postfixed for 90 minutes, transferred to increasing sucrose gradients (10%, 15%, 20%), and finally stored in 30% sucrose in phosphate buffer. The brains were frozen and cut on a freezing microtome at 20 or 30 $\mu$m.

Antisera to leu-enkephalin were kindly provided by Dr. R. Miller, University of Chicago (Miller et al., '78; Miller and Pickel, '80). Antibodies to met-enkephalin, serotonin, and substance P were generously provided by Dr. Robert P. Elde, University of Minnesota. Antisera to met-enkephalin (Haber and Elde, '82) and substance P (Ho and DePalatis, '80) were generated in response to immunogens produced by linking the peptides to bovine thyroglobulin and keyhole limpet hemocyanin, respectively, by means of glutaraldehyde coupling. Antisera against serotonin (5HT) were generated by linking the monoamine to bovine serum albumin by formaldehyde coupling (Maley and Elde, '82). Absorption controls for met-enkephalin, 5HT, and substance P were prepared by incubating each antiserum overnight with synthetic met-enkephalin, 5HT, or substance P (10 $\mu$g/ml diluted antiserum). The peroxidase-antiperoxidase (PAP) technique of Sternberger ('79) was used throughout the present study. The sections were incubated overnight at 4°C with the primary antisera diluted 1/500 with 0.3% Triton X-100 in phosphate-buffered saline (hereafter called PBS/T). After rinsing, the tissue was incubated in goat antirabbit IgG (1/100 in PBS/T) for 1 hour at room temperature and then rinsed and incubated in rabbit PAP (1/500 in PBS/T) for 1 hour at 22°C. After rinsing with Tris (0.05 M, pH 7.6) the tissue was incubated in 3,3'-diaminobenzidine tetrahydrochloride (0.5 mg/ml in Tris buffer containing 0.01% $H_2O_2$) for approximately 7 minutes at 22°C. The sections were rinsed, dehydrated, and coverslipped with Permount. Alternate sections were stained with cresyl violet or processed for enkephalin-like, 5HT-like, and substance P–like immunoreactivity, respectively.

### Tracing experiments

For the purpose of anterograde tracing of the projections from the nucleus interpeduncularis (IP) iontophoretic injections of tritiated leucine were placed in 18 animals in the nucleus or its immediate vicinity. Prior to surgery the animals were deeply anesthetized by intraperitoneal injection of 0.4 ml/100 g body weight Chloropent (a mixture of chloralhydrate and pentobarbital, Fort Dodge Laboratories). The following stereotaxic approach was followed: after determining the desired anteroposterior level of the injection, the occipital pole of the right hemisphere was removed by aspiration in order to expose the superior colliculus. A glass micropipette, measuring 15–20 $\mu$m ID at the tip, was introduced through the superior colliculus at an inward angle of 20° and lowered to the intended depth. The pipette contained a solution of L-(3.4.5-$^3$H) leucine (specific activity 110 Ci/$\mu$mol, New England Nuclear) in 0.01 N acetic acid (20 $\mu$Ci/$\mu$l). Ejection of the tritiated leucine was accomplished by the passage of a positive current of 0.2–0.5 $\mu$A, pulsed at a rate of 7 seconds on/7 seconds off, for 5–15 minutes. Upon completion of the injection, the pipette was left in situ for an additional 30 minutes.

Six to 8 days after surgery the animals were given an overdose of Chloropent and were perfused transcardially with saline, followed by formol-saline. The brains were postfixed for at least 7 days and subsequently transferred to 30% sucrose-formalin. Sections 30 $\mu$m thick were cut on a freezing microtome, mounted, defatted, and coated with Kodak NTB2 or NTB3 emulsion. Following exposure times of 8–16 weeks at −20°C, the emulsion was developed in Kodak D 19 at 16°C, and the sections were counterstained with cresyl violet.

In a second group of six animals, for the purpose of tracing both the afferents and efferents of IP, an injection of either free horseradish peroxidase (HRP Sigma, type VI) or the conjugate of wheat germ agglutinin with HRP (WGA-HRP, Sigma, *Tritium vulgaris*, peroxidase type VI) was placed in the IP. In these experiments a stereotaxic approach similar to that used in the autoradiographic experiments was followed. Injections were placed electrophoretically from a glass micropipette with an internal tip diameter of 15–35 $\mu$m. The free HRP was used in a 13% solution and the WGA-HRP in a 10% solution Tris-buffered at pH 8.6. A constant positive current of 2.0–3.0 $\mu$A was passed through the pipette for 5–15 minutes, and the pipette was left in situ for an additional 15–30 minutes. After

*Abbreviations*

| | | | | |
|---|---|---|---|---|
| AC | nucleus accumbens | | IP-DL | dorsal lateral subnucleus of IP |
| ac | anterior commissure | | IP-DM | dorsal medial subnucleus of IP |
| bc | branchium conjunctivum | | IP-I | intermediate subnucleus of IP |
| CA | cornu Ammonis | | IP-L | lateral subnucleus of IP |
| ca | commissura anterior | | IP-R | rostral subnucleus of IP |
| cc | corpus callosum | | IP-RL | rostral lateral subnucleus of IP |
| CL | nucleus centralis lateralis | | LC | locus coeruleus |
| CM | corpus mammillare | | LH | lateral hypothalamus |
| CP | caudate-putamen | | lHb | lateral habenula |
| DB | nucleus of the diagonal band of Broca | | lm | lemniscus medialis |
| DLTN | nucleus tegmenti dorsalis lateralis | | LS | nucleus septalis lateralis |
| DRN | nucleus raphe dorsalis | | LV | lateral ventricle |
| DTR | dorsal tegmental region | | MD | nucleus mediodorsalis thalami |
| DTN | nucleus dorsalis tegmenti of Gudden | | mfb | medial forebrain bundle |
| Ec | entorhinal cortex | | mHb | medial habenula |
| f | fornix | | MRN | nucleus raphe medialis |
| FD | fascia dentata | | MS | nucleus septalis medialis |
| flm | fasciculus longitudinalis medialis | | mt | tractus mammillothalamicus |
| G | nucleus gelatinosus thalami | | Pb | parabigeminal nucleus |
| GP | globus pallidus | | pc | pedunculus cerebri |
| Hab | habenula | | R | nucleus ruber |
| Hip | hippocamᵖus | | Re | nucleus reuniens thalami |
| hp | tractus habenulo-interpeduncularis | | Rh | nucleus rhomboidalis thalami |
| Hyp | hypothalamus | | Se | septum |
| IC | inferior colliculus | | SN | substantia nigra |
| IF | nucleus interfascicularis | | S | subiculum |
| IP | nucleus interpeduncularis | | VM | nucleus ventralis medialis thalami |
| IP-A | apical subnucleus of IP | | VTA | ventral tegmental area of Tsai |
| IP-C | central subnucleus of IP | | VTN | nucleus ventralis tegmenti of Gudden |

a survival of 48–72 hours, the animals were once again deeply anesthetized and perfused through the heart with a solution of 3% sucrose, 1.25% glutaraldehyde, and 0.5% paraformaldehyde in 0.1 M phosphate buffer at pH 7.4. The brains were stored overnight in a 20% solution of sucrose in phosphate buffer and sectioned on a freezing microtome at 50 $\mu$m. In experiments in which free HRP was used as a tracer, alternating series of sections were processed for the diaminobenzidine (DAB, Graham and Karnovsky, '66) and tetramethylbenzidine (TMB) reactions (Mesulam, '78). In some cases an additional series of sections was processed for the benzidine dihydrochloride method (BDHC, Mesulam, '76). The DAB-treated sections were counterstained with cresyl violet, the BDHC and TMB material with neutral red, and the remaining half coverslipped without counterstaining. The DAB and BDHC material was examined in bright- and darkfield illumination and the TMB-stained sections were examined in brightfield and polarized light (Illing and Wässle, '79; Hess and Schneider, '81).

In a third group of experiments smaller or larger injections of HRP or WGA-HRP were placed in the various structures in which anterogradely transported label had been found in the first two groups of experiments. Iontophoretic injections were placed in two animals in the lateral hypothalamus, and in three rats in the septum. In another three rats pressure injections of 0.2 $\mu$l of 30% HRP were placed in the hippocampus. In ten animals an iontophoretic injection was placed in the region of the dorsal tegmental nucleus and in three animals in the median raphe nucleus. Both of these mesencephalic structures were reached after exposure of the floor of the anterior recess of the fourth ventricle by aspiration of the anterior lobe of the cerebellum.

Included in this last group of experiments are a number of cases separately reported elsewhere (Groenewegen and Nauta, '82) in which the retrograde tracer had been injected into the mediodorsal thalamic nucleus. Several of these injections had been placed under direct visual guidance following aspiration of the overlying cortex, corpus callosum, and dorsal hippocampus. Further surgical and histological procedures for this group of HRP experiments were similar in detail to those described above for the second group of experiments.

In a fourth and last group of experiments, anterograde tracers were injected in regions known to project to the IP (our results; Marchand et al., '80; Flumerfelt et al., '81). Injections of tritiated amino acids or WGA-HRP were placed in the caudal central gray substance (16 rats), the median raphe nucleus (two rats), the habenular complex (three rats), the lateral hypothalamus (four rats), the nucleus of the diagonal band (four rats), and the parabigeminal region (two rats). The experimental procedures used for this group of experiments were similar to those described above.

## RESULTS
### Cyto- and myeloarchitecture

The interpeduncular nucleus (IP) can be subdivided into several subnuclei with different architectonic characteristics. As will be described below, these subnuclei differ among each other also by histochemical and hodological criteria. The delineation of the various IP subnuclei proposed in the present paper largely agrees with recent cytoarchitectural descriptions (Hamill and Lenn, '84; Hemmendinger and Moore, '84). In order to facilitate comparison of our results with those of other studies, the compromise nomenclature recently proposed by Lenn and Hamill ('84) has been adopted.

The IP is here considered to be made up of three unpaired midline nuclei, two bilateral paramedian nuclei, and two lateral groups, each consisting of three subnuclei. The border between the median/paramedian and lateral groups roughly coincides with a row of paramedian blood vessels, extending from the ventral surface dorsally into the mes-

356

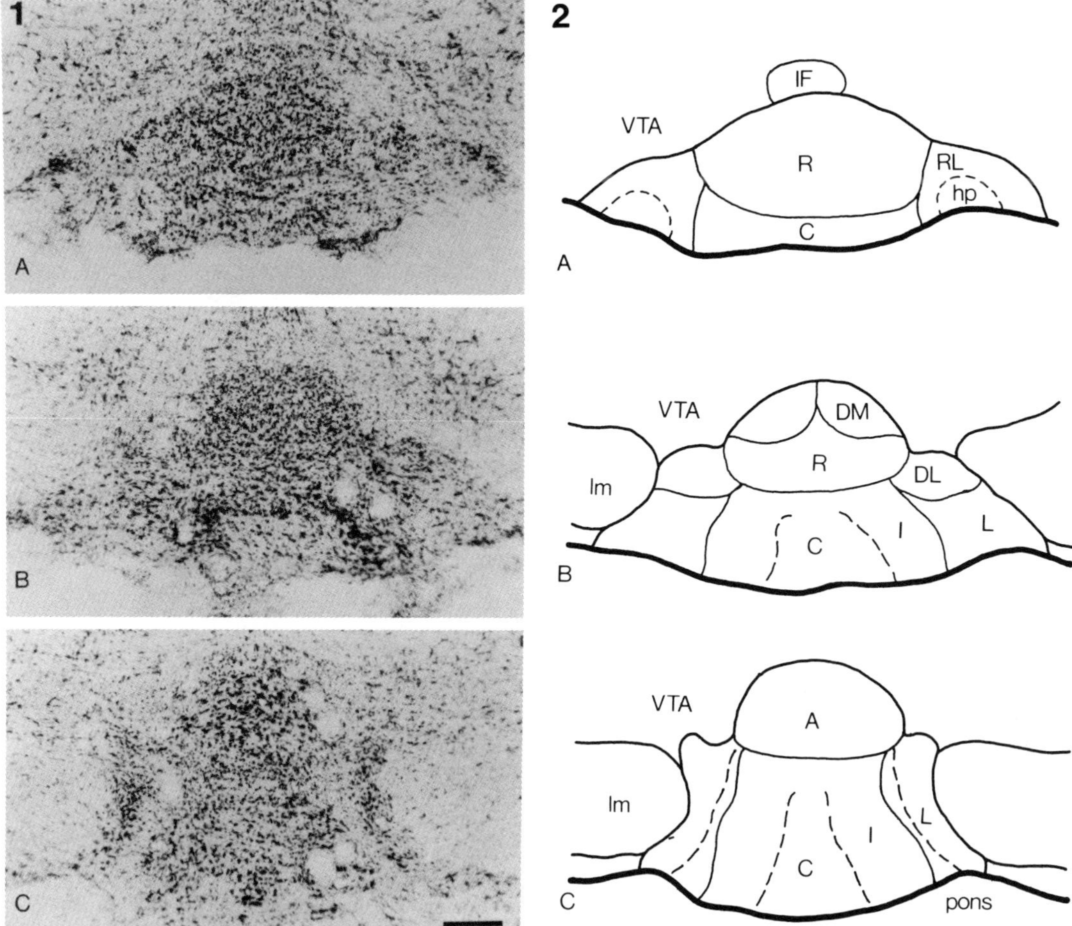



Fig. 1. Photomicrographs of Nissl-stained frontal sections through the nucleus interpeduncularis and adjacent regions in the rat. A. Rostral section. B. Midstrocaudal level. C. Caudal section. Scale bar represents 200 μm.

Fig. 2. Drawings of frontal sections of IP as illustrated in Figure 1, to indicate the boundaries of the various subnuclei.

encephalic raphe on either side of the midplane. This collection of blood vessels is most conspicuous in the caudal half of the IP but can usually be recognized also at more rostral levels (e.g., Figs. 1, 4). In sections stained with the Loyez method for myelin the IP contrasts sharply with all adjacent structures, except the interfascicular nucleus, in its low myelin content (Fig. 3). Differences in concentration and orientation of myelinated fibers support the proposed subdivision of the IP based on cytoarchitectonic criteria (Figs. 1–3).

The most prominent representative of the median group of the IP is the rostral subnucleus (IP-R; Figs. 1A, 2A). This unpaired cell group consists of small cells lightly stained in Nissl preparations and rather closely spaced. At the rostral pole the IP-R occupies almost the entire dorsoventral extent of the IP; more caudally it becomes progressively confined to the dorsal half of the nucleus (Figs. 1A,B, 2A,B). At a level approximately halfway through the length of the IP, the IP-R is replaced—or possibly augmented—by a pair of subnuclei only vaguely delineated from each other in the midplane and composed of numerous very small, closely spaced cells (Figs. 1B, 2B). In sections stained for myelin this pair of paramedian subnuclei can be distinguished from the IP-R by its slightly higher content in myelinated fibers (Fig. 3A,B). This striking bilateral parvicellular cell group, here to be named the dorsal medial subnucleus (IP-DM), extends caudalward over only a short distance, to be replaced in the caudal one-third of the IP by a similarly

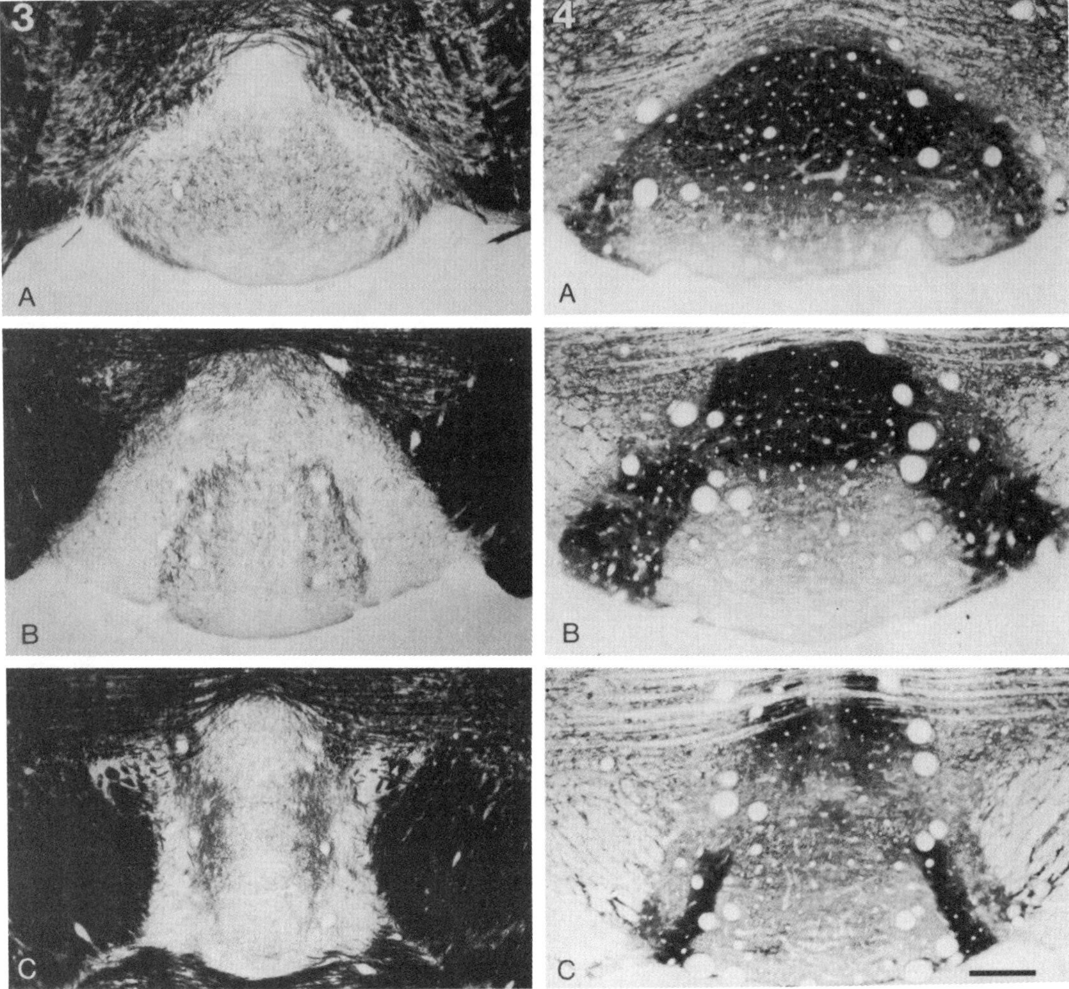

Fig. 3. Photomicrographs of Loyez-stained frontal sections through IP. Levels A–C are approximately at the same longitudes as those illustrated in Figures 1 and 2. Scale bar in Figure 4C applies also to Figure 3.

Fig. 4. Photomicrographs of frontal sections through IP, stained for acetylcholinesterase activity. Levels A–C are comparable to those in Figures 1 and 2. Scale bar represents 200 $\mu$m.

shaped but distinctly large-celled region, the apical subnucleus (IP-A) (Figs. 1C, 2C). The latter is composed mainly of large, dark-staining, spindle-shaped neurons, many of which are oriented in the horizontal plane (Fig. 9A). At the caudal pole of the IP the IP-A cannot be sharply demarcated from the dorsally adjacent raphe region, which contains numerous neurons of similar size, shape, and Nissl-staining properties.

Together, the three foregoing cell groups (IP-R, IP-DM, and IP-A) compose approximately the dorsal half of the median zone of the IP. The ventral half of this zone is occupied by the bilateral intermediate (IP-I) and the unpaired central subnucleus (IP-C). These subnuclei are most

prominent in the caudal two-thirds of the IP, where they can be demarcated from adjoining cell groups also in myelin preparations, as the IP-I contains a rather high concentration of myelinated fibers (Fig. 3B,C). Characteristic of the IP-C and IP-I in Nissl preparations is the uneven, clustered distribution of the small and medium-sized neurons. In the rostral half of the IP-C such clustering appears in the form of several transversely oriented cell rows separated by cell-poor zones (Fig. 1A). Caudally this horizontal pattern gradually changes into a less orderly one (Fig. 1B), and in the caudal half of the IP most of the cells are aggregated into three irregular, vertically oriented columns, the median one of which identifies the centrally located IP-C and is on

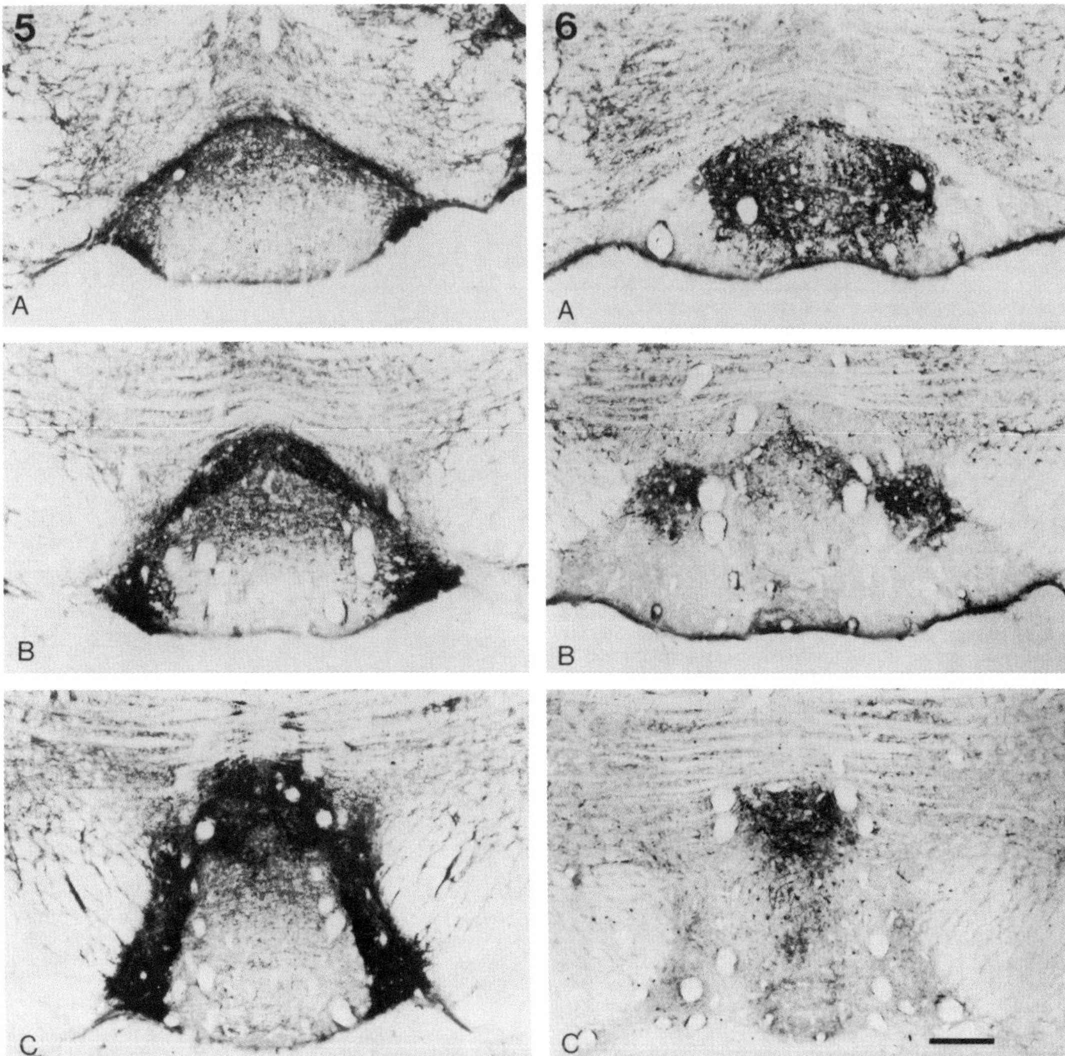

Fig. 5. Photomicrographs of substance P–like immunoreactivity in frontal sections of IP. Note the staining of neuronal cell bodies in IP-R (A,B). For structure identification compare with Figures 1 and 2. For scale bar see Figure 6C.

Fig. 6. Photomicrographs of leu-enkephalin-like immunoreactivity in frontal sections of IP. Note the staining of cell bodies in IP-R, IP-A, and the caudodorsal part of IP-C. Leves A–C are comparable to those in Figures 1 and 2. Scale bar represents 200 $\mu$m.

either side separated by a relatively cell-poor zone from the laterally adjoining intermediate subnucleus (IP-I; Figs. 1C, 2C).

The lateral part of the interpeduncular nucleus is subdivided into three subnuclei. The rostral lateral subnucleus (IP-RL) is a somewhat triangular region forming the lateral tip of the IP, and it consists of a rather cell-poor ventrome-dial part and a smaller lateral zone containing most of the small and medium-sized neurons that populate the IP-RL (Fig. 1A). It seems likely that the cell-poor zone of the IP-RL corresponds to the habenulo-interpeduncular tract of the fasciculus retroflexus, a correspondence also suggested by the arrangement of fibers (Fig. 3A) and the very high acetylcholinesterase content of this region (see below). At

359

middle longitudes of the IP the separate dorsal lateral subnucleus (IP-DL) consists of rather closely spaced, lightly stained, ovoidal neurons, slightly smaller than the darker-staining neurons in the ventrally adjacent lateral subnucleus (IP-L; Figs. 1B, 2B). Caudally, this latter subnucleus constitutes a sickle-shaped region just medial to the medial lemniscus. Most of the neurons in the IP-L are somewhat larger and more closely spaced than those of the IP-RL, but in the IP-L likewise the cells tend to form clusters separated from each other by cell-poor intervals (Fig. 1C).

## Acetylcholinesterase activity

The pattern of distribution of acetylcholinesterase (AChE) activity is, with a few minor exceptions, remarkably similar to that of substance P–like immunoreactivity (see below). The habenulo-interpeduncular tract in the rostral IP displays a strong acetylcholinesterase activity (Fig. 4A,B). Moderate-to-high activity of acetylcholinesterase is observed in the IP-RL and IP-DL. Staining over the laterally located cell clusters of the IP-L, by contrast, is only weak to moderate, but it is dense in the medial, cell-poor zone of the subnucleus (Fig. 4C). The rostral part of the IP-R shows a strong AChE reaction, but more caudally the reactivity of this nucleus wanes. In the caudal two-thirds of the IP, AChE staining in the IP-DM and IP-A is generally somewhat darker than in the IP-R (Fig. 4B,C). The lowest AChE activity is shown by the IP-C and IP-I; in the rostral part of the two subnuclei the AChE reaction is virtually negative, but more caudally the three vertically oriented cell columns show a faintly positive reaction (Fig. 4C). Two heavily stained paramedian fiber bundles leave the caudal pole of the IP in a dorsocaudal direction on either side of the midplane (Fig. 8A). After traversing the median raphe nucleus these fiber bundles enter the central gray substance by passing through the narrow interval between the left and the right medial longitudinal fasciculus. There can be no doubt that this discrete fiber system, earlier noted for its strong AChE reactivity (Wilson and Watson, '80), and according to the present findings also containing substance P–positive fibers (Fig. 8B), corresponds to the interpedunculotegmental tract of Ganser (1882).

## Distribution of substance P–like, enkephalin-like, and serotonin-like immunoreactivity

*Substance P.* Numerous substance P–positive cell bodies are present in the IP, all of them located in the IP-R, throughout the rostrocaudal extent of this cell group (Fig. 5). Staining of cell bodies is evident in both colchicine-treated (Fig. 9F) and nontreated animals, but in the latter the staining is considerably weaker. Substance P–like immunoreactivity of fibers—and presumably terminals—in the IP appears to be very specific. In the rostral half of the IP substance P–reactive fibers are relatively sparse and almost entirely confined to a thin layer marking the dorsal border of IP (Figs. 5A, 9F). However, the habenulo-interpeduncular tract, entering the IP's rostral pole in an extreme ventrolateral position, contains a substance P–positive fiber bundle that is slender rostrally but expands considerably at middle levels (Fig. 5B), and farther caudally almost completely enmeshes the IP-L in a dense fiber plexus. The central part of the IP-L exhibits a somewhat lower substance P-immunoreactivity (Fig. 5C). In the rostral part of the IP-R, except for processes of the local substance P–positive neurons, very few substance P–positive fibers are seen (Fig. 9F), while the most caudal part of the IP-R con-

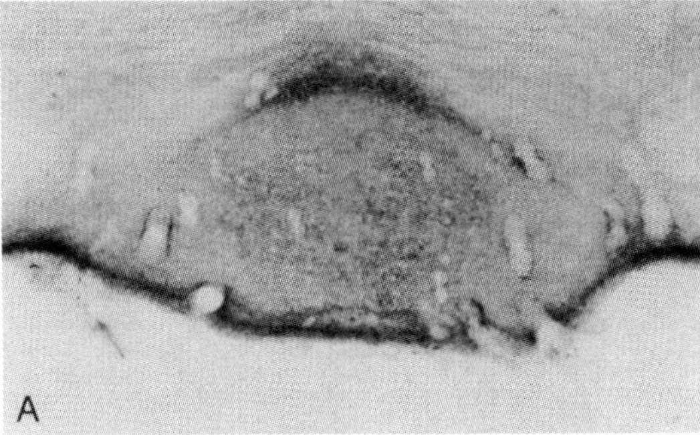
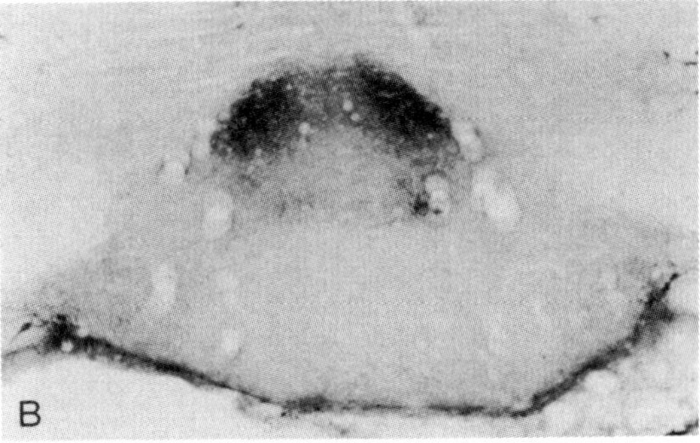
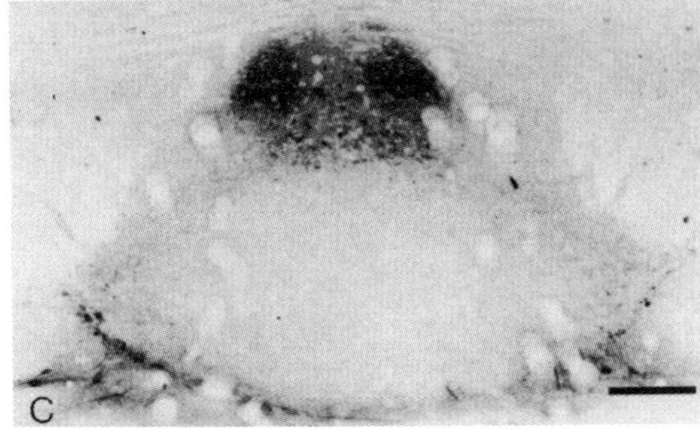

Fig. 7. Photomicrographs of frontal sections through IP to illustrate the distribution of serotonin-like immunoreactivity. Note the staining of cell bodies in IP-A (C) and IP-L (B,C). A–C are approximately at the same longitudes as in Figures 1 and 2. Scale bar represents 200 μm.

tains a moderate number of such fibers, as also do the IP-DM and IP-A. In contrast to their surrounding subnuclei, the IP-C and IP-I display virtually no substance P–like immunoreactivity (Fig. 5B,C). Leaving the caudal pole of the IP, a paired substance P–positive fiber bundle courses dorsocaudally in a paramedian position (Fig. 8B). These fibers appear to be part of the interpedunculotegmental tract of Ganser (1882), a fiber group also noted for its high acetylcholinesterase content.

*Enkephalin.* The distribution patterns of met-enkephalin and leu-enkephalin immunoreactivity over the various

360

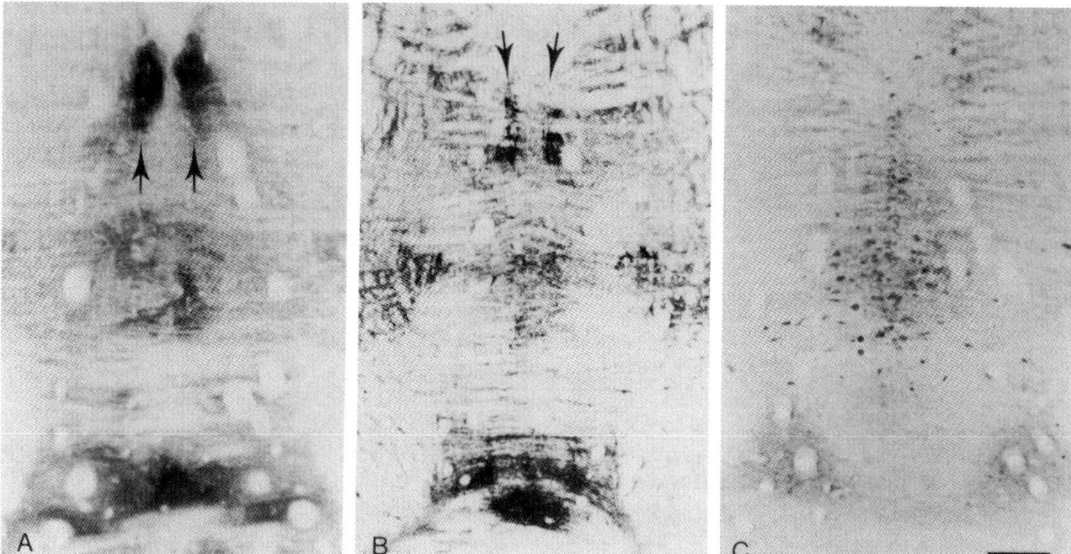

Fig. 8. Frontal sections through a level just caudal to IP. A. Acetylcholinesterase activity. Note the dark-stained interpedunculotegmental tract (arrows). B. Substance P–like immunoreactivity. Here, too, note the positive staining in the interpedunculotegmental tract (arrows). C. Serotonin-like immunoreactivity. The population of positive cell bodies in the midline, here located in the median raphe nucleus, is continuous with the group of positive cells in IP-A shown in Figures 7C and 9B. Scale bar represents 200 μm.

subnuclei of the IP appear to be very similar, although with the antisera here used staining for leu-enkephalin is generally stronger than for met-enkephalin. Enkephalin-positive cells in the IP, stainable even without colchicine pretreatment, are most numerous in the rostral subnucleus IP-R, but are also found in the IP-A and in the caudal part of the IP-C (Figs. 6, 9D,E). The IP-R contains also a dense enkephalin-positive fiber plexus that extends laterally into the dorsal lateral subnucleus IP-DL and ventrally into the central subnucleus IP-C (Fig. 6). Less dense fiber staining marks the IP-RL, IP-A, and IP-I, whereas the dorsal medial subnucleus IP-DM stands out by its conspicuous enkephalin negativity (Fig. 6B).

*Serotonin.* Serotonin-like immunoreactivity of neuronal somata is found only in the caudal one-third of the IP. By far the most numerous 5-HT-positive cell bodies are located in the IP-A (Figs. 7C, 9B,C), a population of neurons which at the caudal confines of the IP cannot be sharply demarcated from the 5-HT-positive neurons in the median raphe nucleus (Fig. 8C), the B8 cell group of Dahlström and Fuxe ('64). In addition, 5-HT-positive cells are present in the caudal half of the IP-L, concentrated for the most part in the latter's ventral and lateral periphery (Fig. 7B,C). These positive IP neurons have the same morphology as the scattered 5-HT-immunoreactive cells lying between the fascicles of the medial lemniscus, and belonging to the B9 cell group of Dahlström and Fuxe ('64). The densest plexus of 5-HT-positive fibers and/or terminals pervades the IP-DM (Figs. 7B,C,19C), while the adjacent subnucleus IP-A contains a lesser number of such fibers. Only a moderate-to-low density of fiber or terminal staining is seen in the IP-L

and IP-R, but a dense stratum of 5-HT-positive fibers caps the rostrodorsal border of the latter subnucleus (Fig. 7A). The subnuclei IP-C and IP-I are almost free of serotonin-like immunoreactivity (Fig. 7B,C).

### Anterograde tracing experiments

The autoradiographic material here examined consists of 18 cases of iontophoretic injection of tritiated leucine in or directly adjacent to the IP. Representative injection sites are collectively, and somewhat schematically, represented in Figure 10, in which only the last two digits of the experiment numbers are used to identify individual cases. One of the experiments, case *RRIP-15*, with a large injection apparently confined to the IP, provided the most comprehensive picture of ascending and descending IP projections, and therefore is described here first in some detail.

*Case RRIP-15.* The center of the injection site in this case is located at and about the midrostrocaudal level of the IP and involves primarily the IP-R and IP-DM. Small parts of the adjacent subnuclei of the IP, particularly the IP-C, IP-I, IP-A, and IP-DL, are included in the center of the injection (Figs. 10, 12A). The periphery of the injection site includes the rest of the IP except for its most rostral and lateral parts.

*Descending projections.* Most of the labeled fibers leave the IP dorsally and caudally, through and lateral to the interpedunculotegmental tract (of Ganser, 1882; Fig. 11H). In the median raphe nucleus labeling appears in a linear pattern over single fibers as well as in a more random pattern suggestive of labeled terminals (Figs. 11I, 12D). Dorsal and lateral to this nucleus the compact, heavily

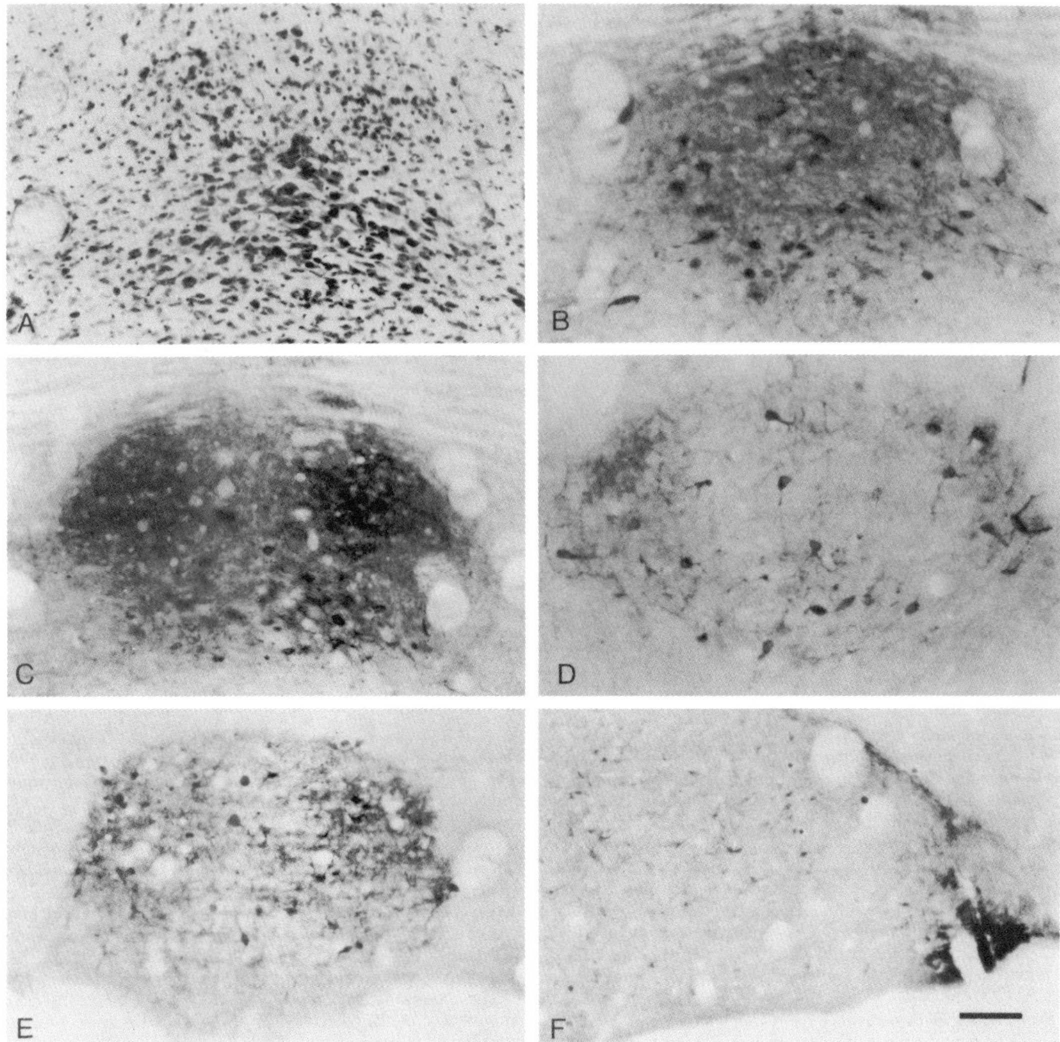

Fig. 9. Photomicrographs of morphological details of IP. A. Nissl-stained section through IP-A. Note the predominant orientation of the large cells in more or less horizontal planes. Parts of IP-DM and IP-R are located, respectively, dorsal and ventral to the magnocellular subnucleus IP-A. B. Serotonin-like immunoreactivity in cell bodies and fiber elements in the caudal part of IP-A. C. Serotonin-like immunoreactivity in cells and neuropil of IP-A (ventral) and in fiber elements in IP-DM (dorsal). Compare Figure 7C. D. Met-enkephalin-like immunoreactivity in IP-R at the midstrocaudal level of IP. E. Met-enkephalin-like immunoreactivity in the rostral part of IP-R. F. Substance P–like immunoreactivity in the rostral IP. Note the staining of cell bodies in IP-R and the dense fiber staining in the habenulo-interpeduncular tract (lower right corner). Scale bar represents 50 µm.

labeled paramedian fiber bundles skirt the margins of the ventral tegmental nucleus of Gudden and subsequently pass between the two medial longitudinal fasciculi to fan out dorsolaterally over the caudal part of the dorsal raphe nucleus (Figs. 11J, 12D,E). The ventral tegmental nucleus of Gudden is surrounded by labeled fibers, but only few pass through the nucleus (Fig. 12E). Immediately caudal to the ventral tegmental nucleus a large number of fibers deviate lateralward and curve dorsally to enter the central gray substance. In the caudal part of the central gray substance heavy labeling appears along the midline, in and around the caudal extension of the dorsal raphe nucleus (Figs. 11K, 12B). Furthermore, extremely heavy labeling surrounds the caudal half of the dorsal tegmental

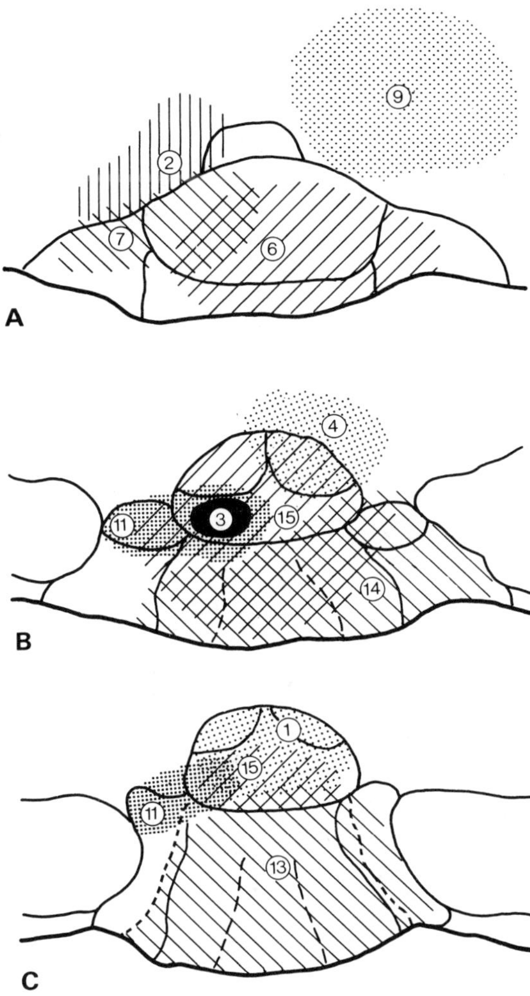

**A**

**B**

**C**

Fig. 10. Chartings of the injection site of tritiated leucine in 11 cases. Only case *RRIP-11* is represented at two levels (D,C), all other injection sites are outlined at only one level, which in some cases means that the injection site is not illustrated in its full rostrocaudal extent. Note that only the last two digits of the case designations have been used in the figure. The three levels represented in this figure are the same as those shown in Figures 1A–C and 2A–C.

nucleus of Gudden. The rostral two-thirds of the nucleus proper contains only very sparse labeling, but its caudal one-third—in which the cells are relatively widely spaced—is densely labeled (Figs. 11L, 12C). The nucleus tegmenti dorsalis lateralis (Figs. 11K, 12B) is moderately labeled over its entire rostrocaudal extent but includes a few small areas of higher grain concentration. Labeling in the central gray substance extends for some distance caudal to the dorsal tegmental nucleus in rapidly diminishing inten-

sity and reaches its caudal limit near the rostral margin of the genu nervi facialis.

*Ascending projections.* The rostral pole of the IP is marked by a high concentration of label, but only sparse grain appears in the ventral tegmental area immediately dorsal and lateral to the IP. Rostral to the IP labeled fibers are loosely arranged around the fasciculus retroflexus (Fig. 11F). At the level of the mesodiencephalic transition moderately dense fiber labeling fills the supramammillary region but the mammillary body itself is entirely free from label (Fig. 11E). From the caudal diencephalic border forward, labeled fibers follow three different paths. A number of fibers ascend surrounding the fasciculus retroflexus and can be followed to the caudal pole of the habenula (Fig. 11D). Very sparse labeling marks the medial half of the lateral habenular nucleus and continues rostrally into the stria medullaris, gradually scattering itself over the lateral part of the mediodorsal thalamic nucleus (Fig. 11C).

A second group of rather sparse labeled fibers courses alongside the mammillothalamic tract (Fig. 11D). Most of these fibers appear to terminate in the rostral part of the nucleus gelatinosus thalami (Fig. 11C). Sparser labeling appears lateral to the tract in the ventromedial thalamic nucleus. A few labeled fibers travel farther rostrally with the mammillothalamic tract, but can be followed only over a short distance. It is possible that they join the sparsely labeled fibers entering the mediodorsal nucleus from the overlying stria medullaris.

A third ascending component of the IP fibers follows the medial forebrain bundle. In case *RRIP-15* this transhypothalamic fiber group appears as rather loosely arranged labeled fibers diffusely distributed over the cross section of the lateral hypothalamic area (Fig. 11B–D). Some labeling is seen around the fornix, but very little is detectable in the medial zone of the hypothalamus. The same pattern of diffuse labeling is found in the lateral preoptic area and the nucleus of the diagonal band of Broca (Fig. 11A). A somewhat higher concentration of grains, however, marks the medial septal nucleus and the ventral part of the lateral septal nucleus (Fig. 11A). A few labeled fibers can be seen to enter the fornix, but these can be followed only over a short distance into the lateral part of the fimbria. Despite this difficulty in tracing interpeduncular efferents beyond the septum in continuity, this case provides evidence that some are distributed to the temporal quarter, at least, of the hippocampus. In this restricted hippocampal region labeling clearly above background appears in the stratum moleculare and hilus of the fascia dentata, as well as in the stratum radiatum of field CA3 (Fig. 11E,F).

## Differential efferent projections of interpeduncular subnuclei

Owing to the small size of the IP and, *a fortiori*, of its constituent subnuclei, it is nearly impossible to confine injections of tritiated amino acids to any particular subnucleus. In most cases more than two subnuclei of the IP are included in the injection site, and it is therefore not possible to discern more than a rough topographic organization in the labeled IP projections. A more differentiated topographic picture emerges when this autoradiographic material is studied in conjunction with the cases of retrograde cell labeling.

*Descending projections.* A topographic organization is more difficult to recognize in the descending than in the

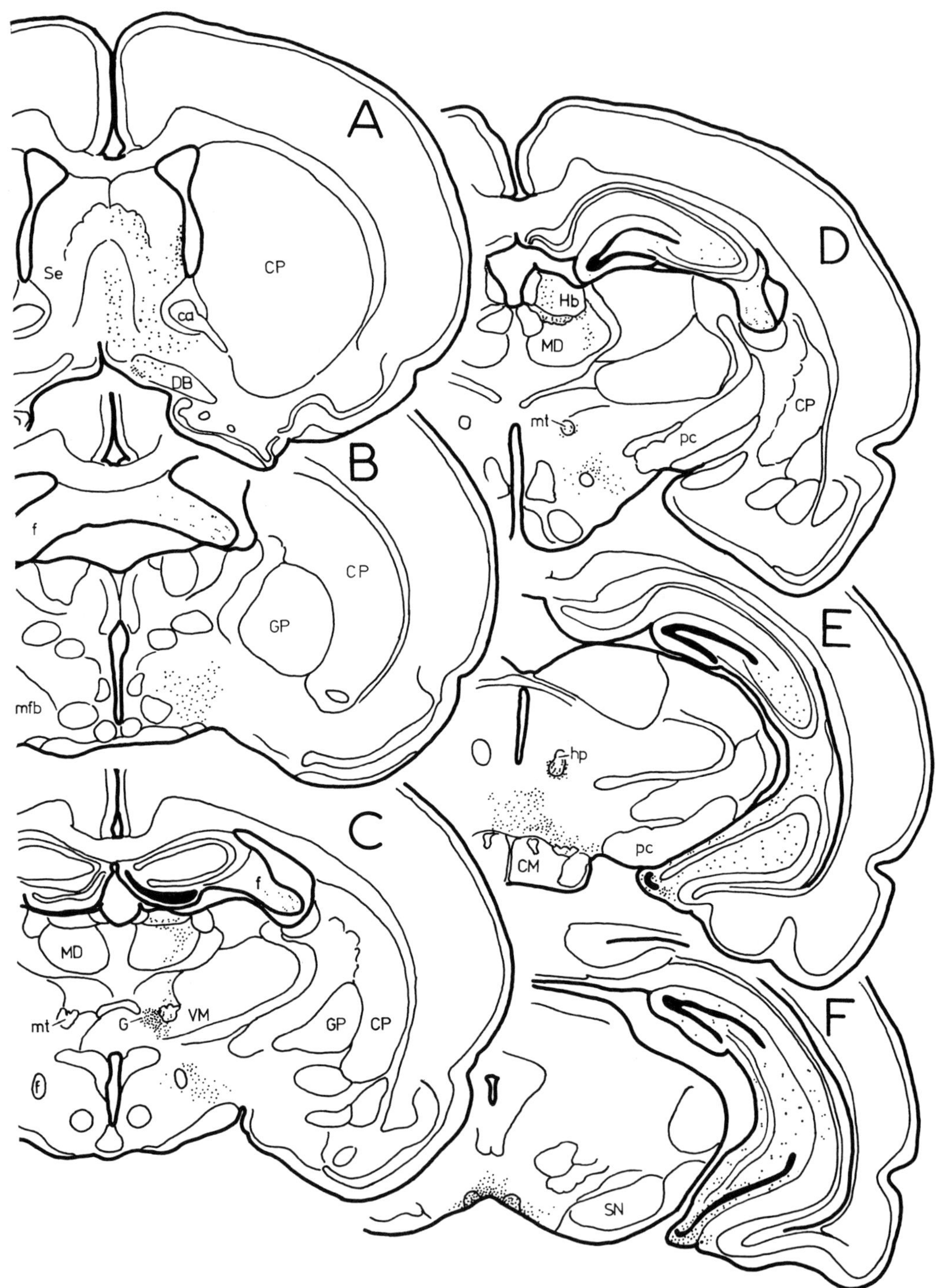


Fig. 11. Chartings of the fiber labeling resulting from an injection of tritiated leucine at midstrocaudal levels of IP (solid black area in G; pipette approach is indicated). A–F. Distribution of ascending IP efferents. H–L. Course and termination of descending IP fibers. Case *RRIP-15*.

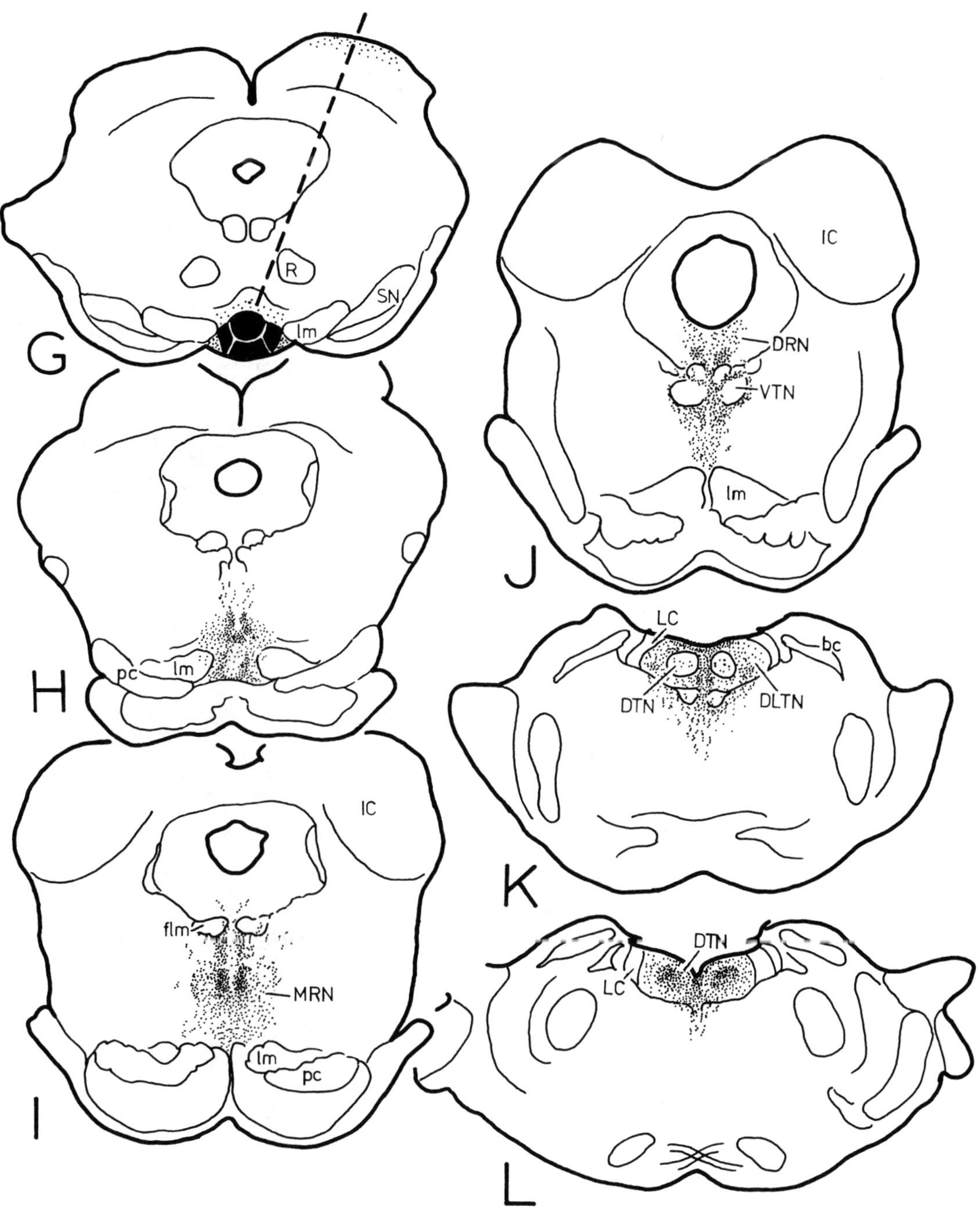



Figure 11 (cont'd)

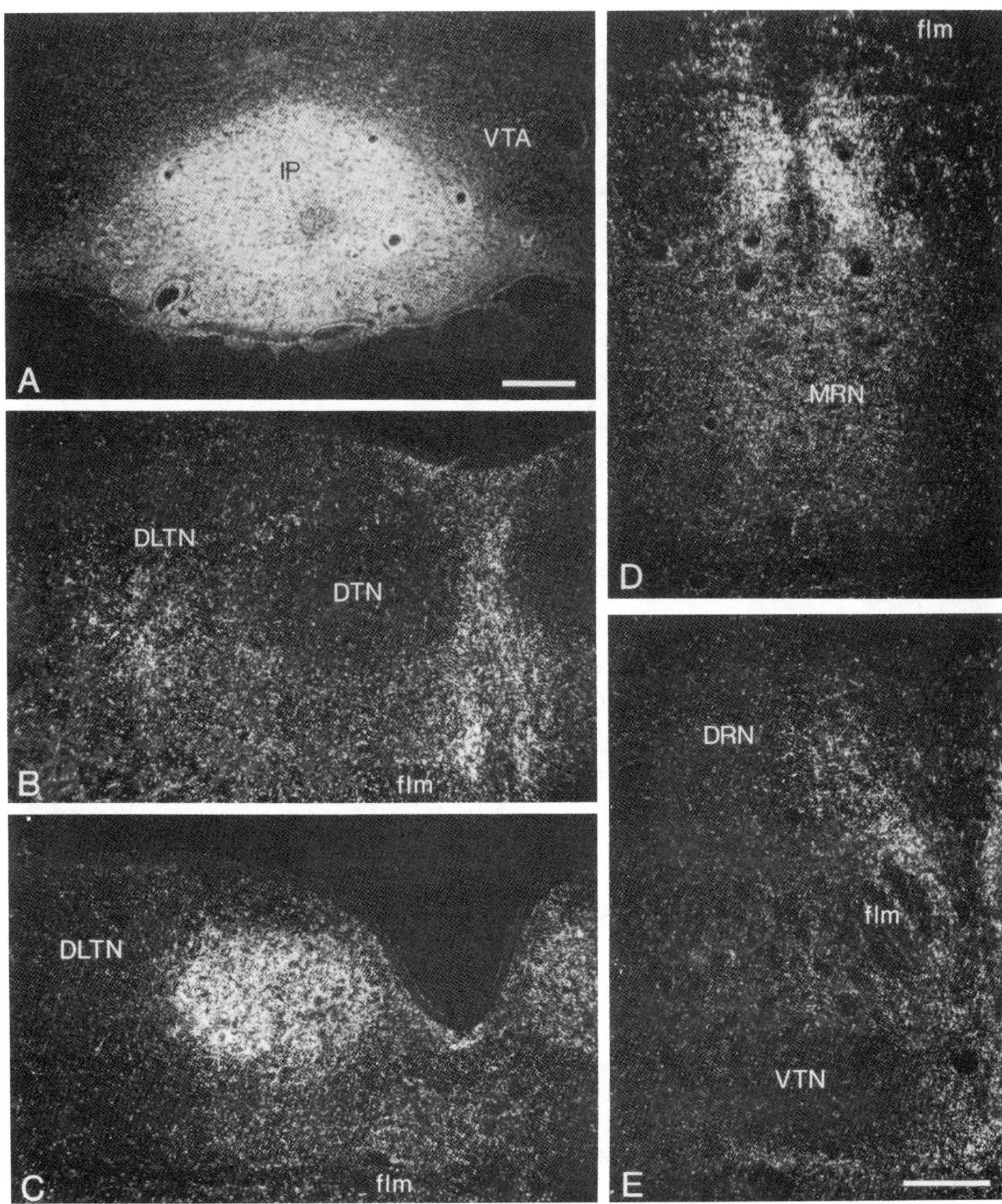



Fig. 12. Darkfield photographs of injection site (A) and descending fibers in case *RRIP-15*. Scale bar represents 200 μm. B. Labeling in the caudal central gray substance. At this relatively rostral level most of the labeling marks the midline structures and surrounds the sparsely labeled dorsal tegmental nucleus in the center. Note the patchy labeling in the dorsolateral tegmental nucleus. C. Strong labeling in the caudal one-third of the dorsal tegmental nucleus; the surrounding central gray matter shows only moderate labeling. D. The appearance of the labeling in the median raphe nucleus. The two patches of heavy labeling at the top represent the interpedunculotegmental tract of Ganser. A more diffuse pattern of grain distribution is seen in the median raphe nucleus proper. E. The innervation of the caudal part of the dorsal raphe nucleus and the encapsulation of the ventral tegmental nucleus of Gudden. The midline is at the right-hand margin of the field shown. Scale bar represents 200 μm.

ascending projections of the IP (see below). All isotope injections that included parts of the IP were found to result in anterograde label transport to the median and dorsal raphe nuclei and to the caudal part of the central gray substance. Differences in the labeling of the IP efferents among different autoradiographic experiments are mainly quantitative and do not suggest a clear-cut topographic arrangement. To illustrate the general nature of these differences a few experiments will be compared.

Injections of tritiated leucine in the IP, with a preference for one-half of the IP over the other, result in asymmetric fiber labeling in the median raphe nucleus and the dorsolateral tegmental nucleus. In experiment *RRIP-14* the injection involves the IP-C, IP-I, and parts of the IP-R and IP-A of both sides, as well as the IP-DL and IP-L of one side only (Fig. 10B). Heavy labeling appears in the median raphe nucleus with a clear predominance of the side ipsilateral to the involvement of IP-DL and IP-L. Projections to the dorsal raphe nucleus appear to be almost equal on either side of the midline. The distribution of labeled fibers over the caudal central gray substance is similar in pattern to that noted in *RRIP-15* (cf. Figs. 11J–L, 12B,C), but in *RRIP-14* a marked asymmetry appears, in the sense that the labeling directly surrounding the rostral two-thirds of the dorsal tegmental nucleus of Gudden, and especially the labeling of the dorsolateral tegmental nucleus, is much more massive on the side contralateral to the main involvement of the IP. By contrast, at more caudal levels of the central gray substance, the dense labeling marking the caudal one-third of the dorsal tegmental nucleus of Gudden appears to be equal on both sides.

Experiment *RRIP-11* (Fig. 10B,C), in which the injection site is largely unilateral in the IP-R and IP-A and involves the IP-DL only on the same side, confirms the alternating distribution of the IP fibers suggested in *RRIP-14*. In the median raphe nucleus labeling is heaviest on the side ipsilateral and in the dorsolateral tegmental nucleus contralateral to the main involvement of the IP. Contrary to cases *RRIP-14* and *-15*, in *RRIP-11* the projection from the IP to the caudal pole of the dorsal tegmental nucleus has remained unlabeled, and this is the case also in several other instances of injection restricted to the caudal half of the IP (*RRIP-1, -12,* and *-13*; Fig. 10C).

Most injections of retrograde tracers in the raphe nuclei and the caudal central gray substance label cells in almost all subnuclei of the IP. It must be stressed that even small injections in the caudal central gray tend to involve more than one of the cell groups recognized in this region and thus are likely to affect afferents of multiple rather than single cell groups. Moreover, injections in the median raphe nucleus are very likely to interrupt IP efferents in passage to more caudal termination sites. Despite these difficulties the retrograde tracing experiments can reveal at least some of the topographic features of the descending IP projection.

Following relatively large but unilateral injections in the caudal central gray substance a remarkable pattern of ipsi- and contralateral cell labeling appears in the IP. In experiment *RHI-81* the WGA-HRP injection covers a large part of the caudal central gray substance, including the dorsal tegmental nucleus of Gudden, the dorsolateral tegmental nucleus, and parts of the midline structures (Fig. 14A). In the IP both anterograde and retrograde labeling is found. In the rostral one-third of the IP retrograde cell labeling is predominantly ipsilateral to the injection site and confined

to the IP-R and IP-RL. Only sporadic labeled cells appear in the same subnuclei of the contralateral side. In the middle one-third of the IP an entirely different distribution of labeled cells is found: most are here located in the IP-DM and IP-DL contralateral to the injection site (Fig. 14A). At this middle level labeling is still almost entirely ipsilateral in the IP-R, and scattered labeled cells are found ipsilaterally in the IP-DM and IP-L. Labeling of neurons in the IP-C and IP-I starts at the midlevel of the IP but is more pronounced farther caudally (Fig. 14A). In the caudal one-third cell labeling in the IP-L is bilateral. At this level labeled neurons also appear along the dorsal and lateral margins of the IP, immediately dorsal and medial to the medial margin of the medial lemniscus.

The pattern of retrograde labeling described above was noted also in three other cases of HRP (*RHI-40* and *-52*; Fig. 16A,B) or WGA-HRP (*RHI-72*) injection in the caudal central gray substance. A slightly different distribution of the cells is found in cases of injection covering a smaller part of the caudal central gray. In case *RHI-49* a unilateral HRP injection is centered in the caudal half of the dorsal tegmental nucleus and extends ventrally into the medial longitudinal fasciculus (Fig. 14C). Retrogradely labeled cells are confined largely to the rostroventral part of the ipsilateral IP-R (Fig. 14C). The remaining few labeled neurons are located in the contralateral IP-DM and IP-L and bilaterally in the caudal part of the IP-I (not shown).

In contrast, small injections centered in more rostral parts of the dorsal tegmental region (cases *RHI-31* and *-73*) label relatively few cells in the IP-I and IP-C, and in the rostral ipsilateral IP-R. In these cases more labeling is found in the contralateral IP-DM and bilaterally in the IP-L (Fig. 14D). A small injection located in the caudal part of the dorsal raphe nucleus and extending into the medial longitudinal fasciculus and ventral tegmental nucleus of Gudden (*RHI-44*) labeled IP neurons located for the most part in the ipsilateral IP-L. In this case, only very few cells were labeled in the IP-R, IP-C, and IP-I.

Two HRP injections involve the dorsocaudal part of the median raphe nucleus. In one of these cases (*RHI-45*; Fig. 14B) the HRP deposit is quite asymmetric, and in this experiment retrogradely labeled cells are found in the IP-R and IP-RL, predominantly ipsilateral to the greatest width of the injection site. Furthermore, cells are labeled in the IP-DM on both sides but in greater number ipsilaterally (Fig. 14B). Farther caudally scattered labeled cells appear in the IP-I and IP-L.

*Conclusions* with respect to descending projections: The results of these anterograde- and retrograde-tracing experiments suggest at least a few conclusions concerning the topographic arrangements of descending interpeduncular projections. The HRP evidence indicates that each half of the unpaired parvicellular subnucleus IP-R projects mainly, though not exclusively, to the ipsilateral half of the caudal central gray substance (cf. *RHI-82*; Fig. 14A). These IP-R efferents most probably terminate in the autoradiographically demonstrated zone surrounding and caudally enclosing the dorsal tegmental nucleus (Fig. 12B,C). A contralateral IP projection to the region of the dorsal tegmental nucleus, as evident in cases *RRIP-4, -11,* and *-14,* stems from the IP-DL and IP-DM (cf. *RHI-73, -82*). The fibers from the contralateral IP-DM presumably terminate medially in the central gray, in the vicinity of the dorsal tegmental nucleus (cf. *RRIP-4* and *RHI-73*). The projection

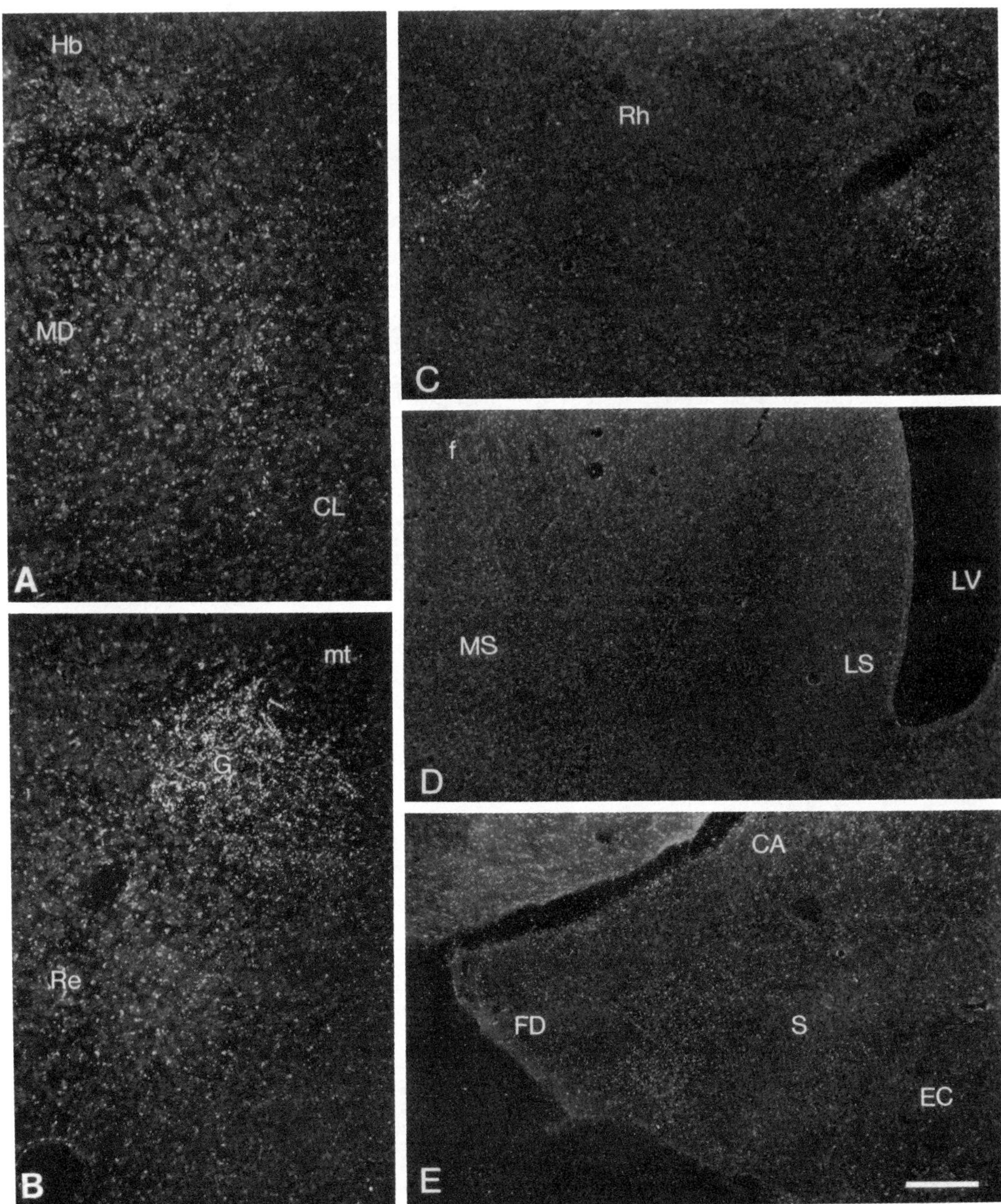


Fig. 13. Darkfield photomicrographs of ascending IP projections. A. Narrow band of fiber labeling in the lateral part of the mediodorsal thalamic nucleus. Midline is to the left. Case *RRIP-14*. B. Relatively strong fiber labeling in the nucleus gelatinosus of the thalamus. Third ventricle appears in left lower corner. Case *RRIP-14*. C. Bilateral fiber labeling in the rostral thalamus at the level of the nucleus rhomboidalis. Case *RIP-14*. D. Fiber labeling in the medial and lateral septal nuclei. Case *RIP-11*. E. Fiber labeling in the temporal tip of the hippocampus. Case *RRIP-11*. Scale bar represents 200 μm.

to the contralateral dorsolateral tegmental nucleus is likely to originate in the IP-L and IP-DL (*RRIP-11* and *-14*). The existence of additional efferents from the IP-L, IP-DL, and IP-DM to regions other than those here mentioned cannot, however, be excluded.

It cannot be determined from the present experiments whether the median raphe nucleus receives projections from all IP subnuclei except the IP-A, as suggested by *RHI-45* (Fig. 14B), or merely provides a passage way for some or all such projections to the caudal central gray substance. The retrograde tracing experiments demonstrate that the projection from the IP to the median raphe nucleus is predominantly ipsilateral. The projection to the dorsal raphe nucleus appears to originate mainly ipsilaterally, from all subnuclei of the IP except IP-A (cf. *RHI-31*).

### Ascending projections

*Hippocampus.* Almost all cases in which the injection of tritiated leucine is located in the caudal half of the IP show some fiber labeling in the hippocampal formation. This labeling is most conspicuous in cases *RRIP-1* and *-11*, in both of which the injection site is centered in the IP-A (Fig. 10 B,C,). As in case *RRIP-15* described above, labeling in the hippocampus mainly involves the ventral limb (Fig. 13E, case *RRIP-11*), but in the dorsal limb of the hippocampus (nearer the septum) labeling appears to be above background as well. The interpeduncular efferents seem to be distributed primarily to the hilus region and the molecular layer of the fascia dentata, and in lesser number to the stratum radiatum of CA3 (Fig. 13E).

In retrograde tracing experiments, massive injections of HRP or WGA-HRP placed in the hippocampus were invariably found to label a large proportion of the neurons in the IP-A, as well as a few of the large neurons in the caudal part of the IP-L (Figs. 15C, 16C). Many additional labeled cells appeared in such cases immediately caudal to the IP in the median raphe nucleus, in the caudal part of the ventral tegmental area, and in the pontine reticular formation (Fig. 16D). These labeled neurons exhibit morphological characteristics remarkably similar to those of the labeled cells in the IP-A (Fig. 16C,D).

The fact that hippocampal HRP injections consistently labeled a number of cells in the caudal part of the IP-L is compatible with the presence of some anterograde labeling in the hippocampus in autoradiographic experiments *RRIP-13* and *-14*. In these cases the injection site was centered in the IP-C and IP-I, to some extent involved the IP-L, but did not spread into the IP-A (Fig. 10B,C).

*Septal region.* The present autoradiographic findings suggest that efferents from the IP to the ventral part of the lateral septal nucleus, the medial septal nucleus, and the nucleus of the diagonal band of Broca likewise originate for the most part in the IP-A. The largest number of such fibers were labeled in cases *RRIP-1* and *-11*, in which the center of the injection site includes the IP-A (Fig. 13D; *RRIP-11*), while injections centered in other parts of the IP and not, or only minimally, involving the IP-A, labeled few if any fibers in the septum.

The pattern of labeling in the basal forebrain in cases *RRIP-2* and *-9*, in which the isotope deposit was confined to the ventral tegmental area immediately adjacent to the IP (Fig. 10A), agrees with an earlier report by Beckstead et al. ('79). The distribution of labeled fibers in the septum

in both these cases differs markedly from that of the interpeduncular efferents described above. Moreover, such fibers are distributed over a much larger area of the basal forebrain that includes the nucleus accumbens and the olfactory tubercle.

These conclusions from autoradiographic evidence are confirmed by the results of injections of HRP or WGA-HRP in the septal area. Two HRP injections, *RHI-26* and *-28*, are located in the rostral half of the septum and involve also the medial part of the nucleus accumbens. In both cases, retrogradely labeled cells in the IP are confined to the IP-A except for a few scattered labeled cells in the caudal part of the IP-L (Fig. 15B). Continuous with the cell labeling in the IP-A retrogradely filled neurons appear in the median raphe nucleus, as also noted in cases of intrahippocampal injection. In accord with the anterograde findings, the rostral, unlabeled pole of the IP is dorsally and rostrally spanned by cell labeling in the ventral tegmental area and interfascicular nucleus. In case *RHI-38* a large HRP injection in the caudal half of the septum, involving the fornix, labeled numerous cells not only in the IP-A but also in the caudal part of IP-L.

*Thalamus.* Anterograde fiber labeling in the thalamus, as charted in *RRIP-15* (Fig. 11), is also observed in several other cases of tritiated leucine injection in the interpeduncular nucleus, most clearly in case *RRIP-14*. In this case the injection site includes the IP-R, IP-C, IP-I, IP-DL, and IP-L (Fig. 10B). The difference between the injection site of *RRIP-14* and that of several cases in which fiber labeling in the thalamus is sparse or absent is the involvement of the IP-L in *RRIP-14* (Fig. 10B). Labeling in the rostral thalamus in case *RRIP-14* is strongest in the nucleus gelatinosus (Fig. 13B). At this rostral level some labeling is seen in the midline nuclei, in particular in the nucleus rhomboideus (Fig. 13C). A few labeled fibers accompany the mammillothalamic tract farther rostrally and dorsally and appear to terminate in the narrow zone between the nuclei anteroventralis and ventralis anterior of the thalamus.

Another thalamic target for labeled interpeduncular efferents in case *RRIP-14* is the mediodorsal nucleus (Fig. 13A). Labeling in this nucleus is restricted to the lateral one-third but does not extend entirely to the lateral border of the nucleus.

The medial half of the lateral habenular nucleus exhibits sparse fiber labeling, but it is not possible to determine whether any of it actually marks axon terminals. It seems certain that an interpedunculohabenular connection, if it exists at all, is of very small volume.

Findings in retrograde experiments indicate that interpedunculothalamic projections originate predominantly or even exclusively in the lateral columns of the IP. In case *RHI-24* with a small iontophoretic injection of HRP centered in the nucleus gelatinosus of the thalamus, immediately medial to the mammillothalamic tract, labeled IP cells are exclusively found in the IP-DL of both sides, but in largest number contralaterally (Fig. 15E).

The results of HRP or WGA-HRP injections in the mediodorsal thalamic nucleus indicate that, in agreement with the anterograde data, only injections involving the lateral part of the nucleus result in retrograde filling of cells in the IP. By far the greatest number of labeled cells are found in case *RHI-59*, in which an iontophoretic injection of WGA-HRP fills almost the entire mediodorsal nucleus.

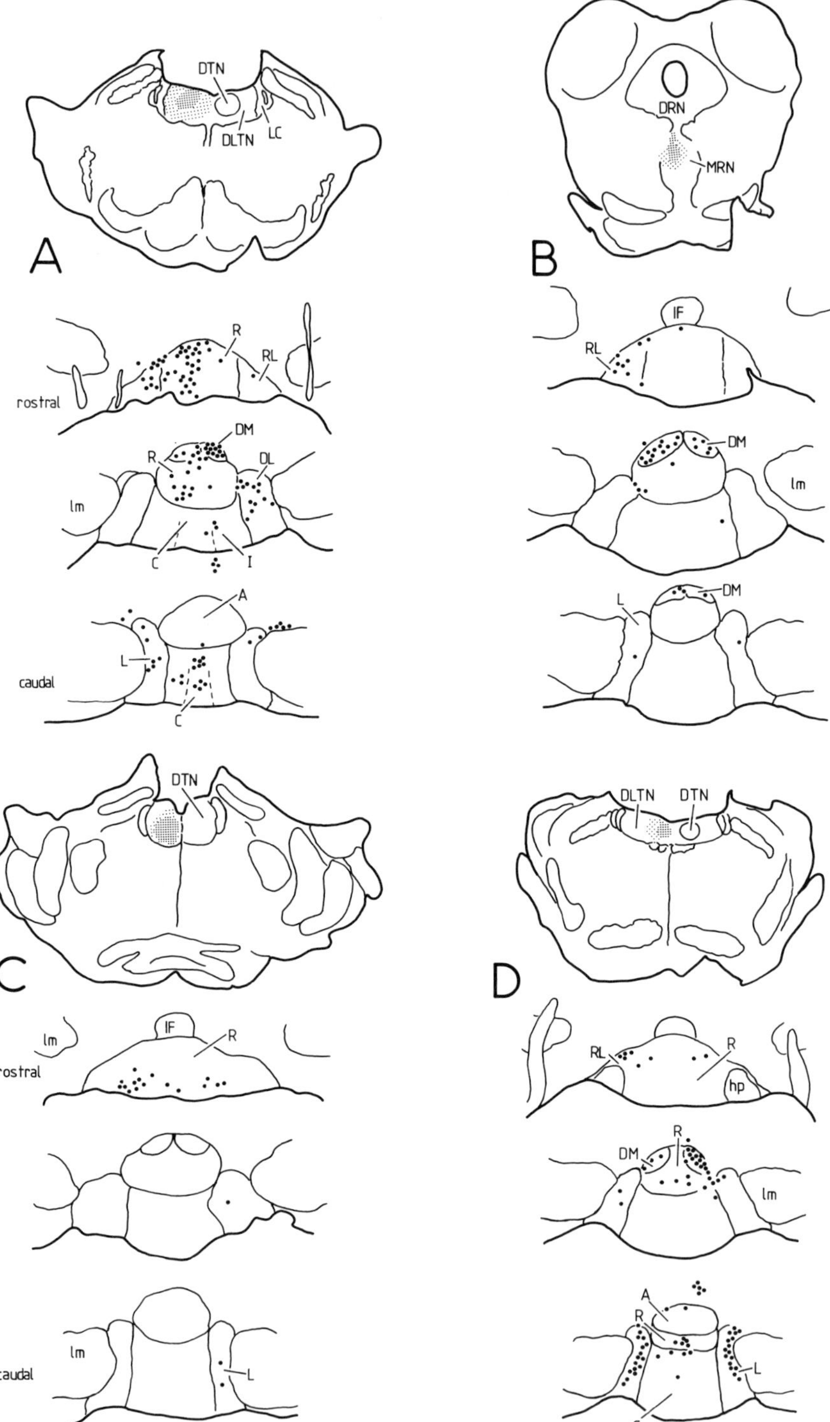

Fig. 14. Chartings of IP neurons retrogradely filled with HRP or WGA-HRP injected in the caudal central gray substance (A, C and D) and the median raphe nucleus (B; case *RHI-45*). All levels shown in A–D can, from rostral to caudal, be compared with the three rostrocaudal levels represented in Figures 1A–C and 2A–C. A. Large injection of WGA-HRP in the caudal central gray (*RHI-82*). C. Small HRP injection in the caudal part of the dorsal tegmental nucleus (*RHI-49*). D. Small WGA-HRP injection in the rostral part of the dorsal tegmental nucleus and adjacent regions (*RHI-73*).

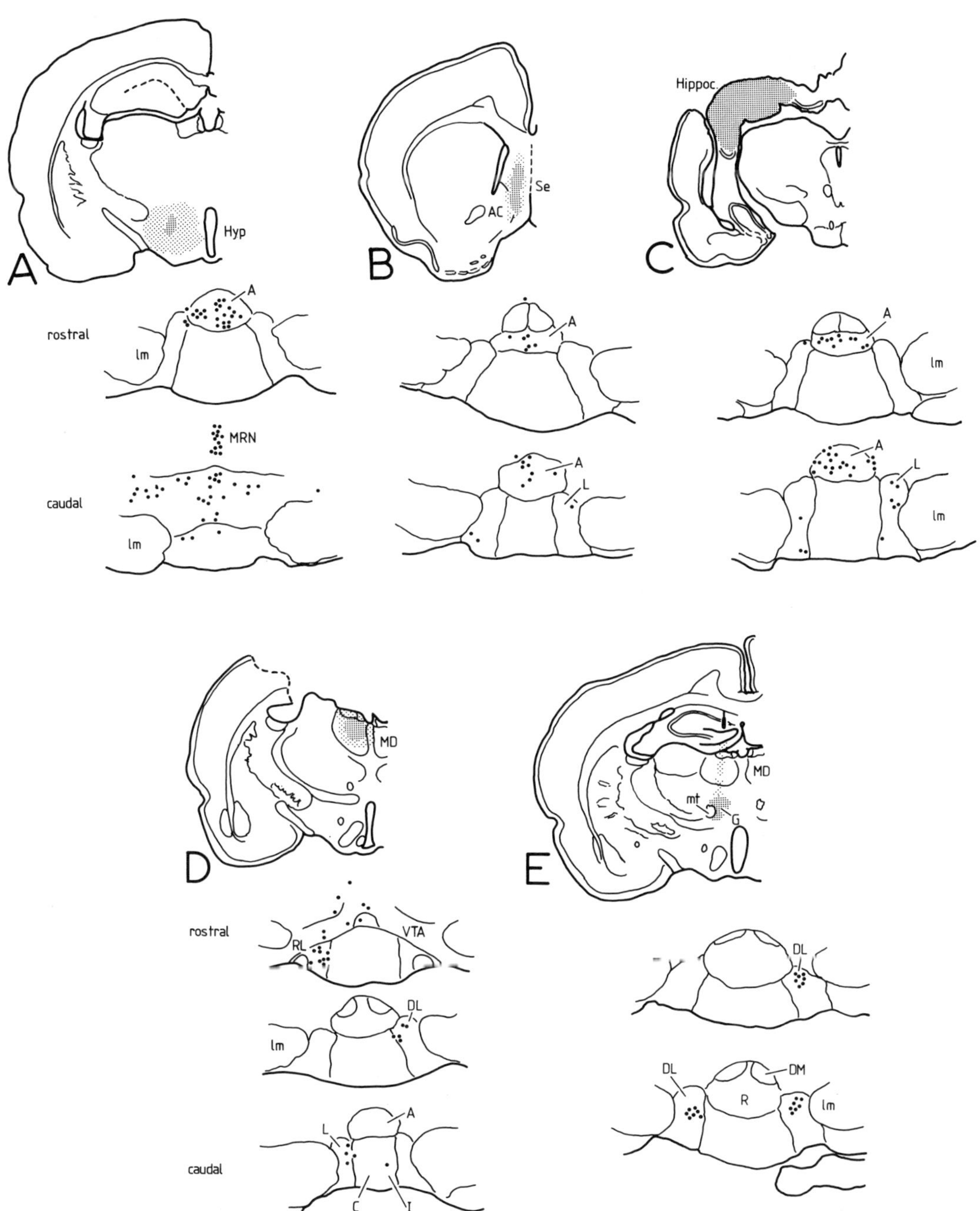

Fig. 15. Chartings of IP cells labeled by HRP injected in different telencephalic and diencephalic IP projection areas: A. Hypothalamic HRP injection. Prominent retrograde cell labeling in IP-A. The rostral level depicted here is at approximately the same level as Figures 1C and 2C. A level just caudal to IP is also shown to demonstrate the continuity of IP-A labeling with labeling in the caudally adjacent raphe region (MRN) (*RHI-23*). B. Injection in the septal nuclei (*RHI-26*). The levels shown here are taken from the caudal IP at approximately the level in Figures 1C and 2C. C. Large mechanical injection in the hippocampal formation (*RHI-42*). About the same rostrocaudal levels as shown in Figure 15B. D. WGA-HRP injection in the mediodorsal thalamic nucleus (*RHI-59*). The levels shown here approximately correspond to Figures 1A–C and 2A–C. E. Injection in the nucleus gelatinosus of the thalamus (*RHI-24*). These two levels are at approximately the same longitude as in Figures 1B and 2B.

371

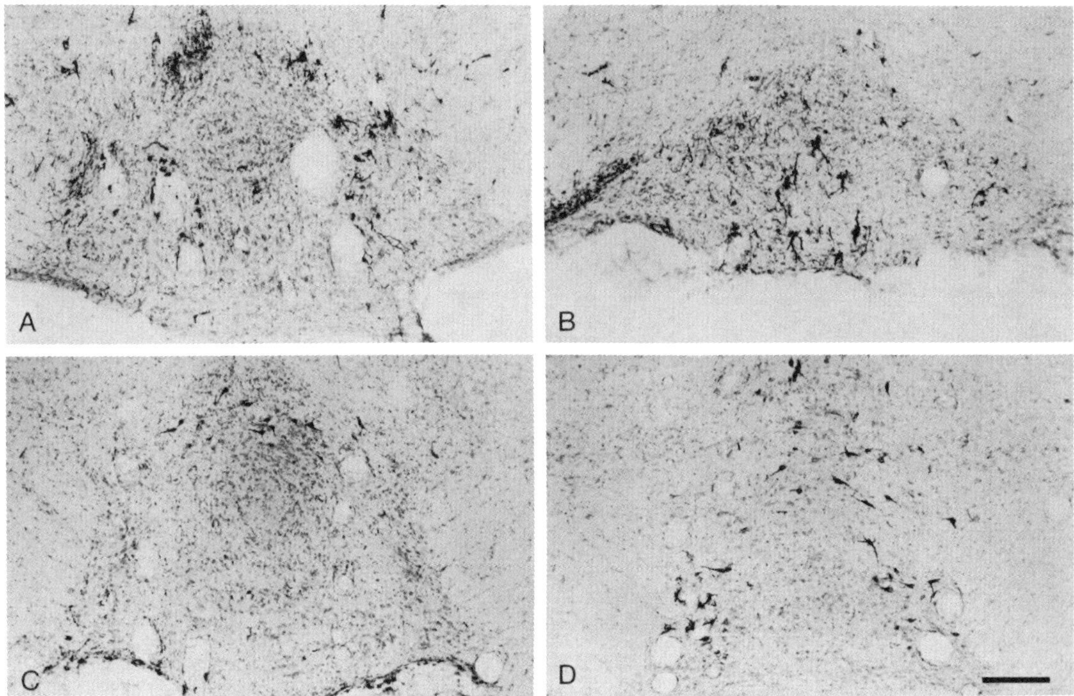

Fig. 16. Photomicrographs of retrogradely labeled IP cells. A,B. HRP injection in the caudal central gray substance (*RHI-40*). Micrograph A is taken at about the same level as Figures 1B and 2B; micrograph B at the level of Figures 1A and 2A. C,D. HRP injection in the hippocampal formation (*RHI-42*). Micrograph C is taken at the same level as Figures 1C and 2C; D is slightly more caudal. C. Labeled cells in IP-A. D. Labeled neurons in IP-L, the median raphe nucleus and adjacent VTA. Note the similar appearance of the labeled cells in C and D (case *RHI-42*). Scale bar represents 200 μm.

Retrogradely labeled cells are most numerous in the IP-RL, ipsilateral to the injection site. Further labeled neurons appear in the contralateral IP-DL, and more caudally also in the ipsilateral IP-L (Fig. 15D). When all available cases are considered together, it appears that the rostral lateral subnucleus IP-RL is most consistently labeled by HRP deposited in the mediodorsal nucleus.

*Hypothalamus.* Although in almost all autoradiographic experiments the hypothalamus contained anterogradely labeled fibers, and injections of HRP in the lateral hypothalamus resulted in retrogradely labeled cells in the IP, the present experiments provide no unequivocal evidence for an interpedunculohypothalamic connection. Anterograde labeling in the hypothalamus is sparse along the entire trajectory of the medial forebrain bundle. Evidence of very weak offsets from the main fiber path to the medial hypothalamus was found in several cases, but no clear termination areas appeared in any of these. Injections of HRP in the lateral hypothalamus (cases *RHI-23* and *-36*) were found to label numerous cells in the IP-A and sporadic cells in the caudal part of the IP-L (Fig. 15A). It is possible that this labeling resulted from interruption of interpeduncular fibers ascending through the medial forebrain bundle to the septum and hippocampus. An actual termination of interpeduncular efferents in the caudal hypothalamus, just dorsal to the mammillary bodies, however, seems likely. In this supramammillary region labeling in most autoradiographic cases is denser than it is at more caudal and rostral levels.

## Distribution of afferent fibers within IP

The present material includes several cases in which small deposits of HRP or WGA-HRP were placed in the IP. The results of these experiments are in general agreement with recent accounts of IP afferents based on similar experiments (rat: Marchand et al., '80; Contestabile and Flumerfelt, '81; cat: Hayakawa and Zyo, '82). In summary, following injections of retrograde tracers in the IP, retrogradely filled neurons appeared in greatest number in the medial habenular nucleus, the median and dorsal raphe nuclei, and the caudal part of the central gray substance, i.e., dorsomedial to the dorsal tegmental nucleus of Gudden. Fewer labeled cells were found in the lateral hypothalamus, the nucleus of the diagonal band of Broca, the triangular nucleus of the septum, the lateral pontine tegmentum, and the region of the parabigeminal nucleus. Since the evidence obtained by HRP injections in the IP has been adequately dealt with in the aforementioned papers, a detailed account of the present material will be omitted. Instead, several additional experiments will be described in which anterograde tracers were injected in regions known to project to the IP. These anterograde tracing experiments in general reveal that those subnuclei of the IP that were shown in the foregoing experiments to have unique efferent connections are also projected upon in a more or less selective fashion. The next section will deal with the distribution within the IP of fibers labeled by either WGA-HRP or by tritiated amino acids injected into, respectively, the caudal central gray substance, median raphe nucleus, parabigeminal nucleus, habenula, hypothalamus, and the nucleus of the diagonal band. A short additional account will be given of intranuclear connections suggested by the findings in the anterograde tracing material.

*Caudal central gray substance.* This term is here used to denote that part of the central gray substance which composes the dorsal tegmental nucleus of Gudden and the nucleus tegmenti dorsalis lateralis, as well as the generally less well-defined cell configurations in which these nuclei lie embedded.

As a consequence of the intricate composition and small size of the region, none of the isotope injections placed in the caudal central gray substance in the present study can confidently be localized to any one particular cell group. Conclusions concerning the projections from each individual cell group can therefore be drawn only with much reservation. Nonetheless, some prominent contrasts between individual cases were noted in the fiber labeling within the IP, and these differences appear to convey at least a general impression of how the projection from the caudal central gray substance to the IP is organized. To illustrate that apparent organization, three of the 16 available cases will be described in some detail.

In case *RRLM-18* the uptake site of the injected tritiated leucine (as determined under brightfield illumination) is very largely confined to the caudal half of the dorsal tegmental nucleus, spreading slightly beyond the latter's caudal, medial, and dorsal borders, but remaining entirely unilaterally (Fig. 17A). As illustrated by Figure 17A, most of the fibers labeled in this case pass rostrally and ventrally over two separate pathways. One of these is a fairly compact, mainly *contralateral* fiber system taking a straight paramedian course through the median raphe nucleus, and thus approaching the IP from the dorsocaudal side. The other is a largely *ipsilateral* system composed of a multitude of slender, loosely grouped arcuate fascicles that trace a wide lateral curve through the caudal midbrain tegmentum and reach the base of the brain to form the mammillary peduncle after passing in large part through the medial one-third of the medial lemniscus and the pars compacta of the substantia nigra. The former of these two principal efferent systems traces—in the opposite direction—the path followed by the interpedunculotegmental tract of Ganser. The other, arciform, system is the main tributary of the mammillary peduncle, but in passing rostralward alongside the IP issues numerous fibers that enter the nucleus from the lateral side. Each of these principal fiber systems has its sparser counterpart on the opposite side (Fig. 17A). The partial decussations allowing for such bilaterality are diffusely scattered from the ventral region of the central gray substance down to the base of the median raphe nucleus.

Combined with further fiber crossings within the IP itself, these diffuse tegmental decussations result in a pre-dominantly contralateral distribution of IP afferents from the region of the dorsal tegmental nucleus. This is particularly evident in the rostral half of the IP where the contralateral half of the IP-R is marked by extremely dense fiber labeling (Fig. 17A) that decreases abruptly both on the medial and lateral side along an almost straight vertical plane; the half of the IP-R ipsilateral to the isotope injection exhibits only moderate fiber labeling. A conspicuous left-right difference favoring the contralateral side is shown also by the IP-A and IP-DM (Fig. 17A): both these cell groups display a wide lateral zone of dense labeling on the contralateral side but are only moderately labeled ipsilateral to the injection. A left-right contrast is less pronounced in the lateral IP columns: both the IP-L and IP-DL exhibit moderately dense labeling on the contralateral side and light-to-moderate labeling ipsilaterally; the IP-RL is only lightly labeled on both sides (Fig. 17A). The IP-C and IP-I in this case display only sparse labeling that appears confined largely to the cell-poor zones between their cell columns and is of about equal density on both sides.

*Similar cases.* Findings nearly identical to the foregoing were made in case *RRLM-24* (Fig. 18A,B). As in *RRLM-18*, the injection in this case was strictly unilateral, but it was similar and involved little more than the caudal one-third of the dorsal tegmental nucleus, spreading somewhat beyond the latter's caudal pole.

*Contrasting cases.* Findings different from those described above were recorded in case *RRLM-6* in which the isotope injection involves the rostral two-thirds but not the caudal third of the dorsal tegmental nucleus. The injection is larger than in the foregoing cases (Fig. 17B) and has spread into the adjacent nucleus tegmenti dorsalis lateralis, as well as across the midplane into the contralateral dorsal tegmental nucleus (not illustrated). The resulting fiber labeling is more nearly equal in volume bilaterally than was the case in *RRLM-18* and -24, but is otherwise similar, at least in the caudal half of the IP(Fig. 18D). In the rostral half, however, a marked difference appears: whereas in the two foregoing cases the contralateral half of the IP-R at this level exhibited profuse labeling, in this case only a narrow strip marking the IP-R's lateral border is densely labeled. At the same rostral level the IP-RL of both sides is much more massively labeled than in the foregoing cases (Figs. 17B, 18C).

*Case RRLM-22* provides a further significant topographic detail. In this case the isotope injection is centered in the nucleus tegmenti dorsalis lateralis but also involves adjoining parts of the locus coeruleus and of the middle one-third of the dorsal tegmental nucleus. In contrast to the foregoing cases, only sparse fiber labeling marks the median cell groups IP-A, IP-DM, and IP-R. The only dense labeling in this case appears in the contralateral lateral column (IP-L, IP-DL, and IP-RL) and is particularly massive in the IP-DL. The fiber labeling in the IP-L of the ipsilateral side is of only moderate density.

*In summary,* the following conclusions can be drawn concerning the projection from the caudal central gray substance to the IP. First, the projection as a whole is predominantly crossed. Second, although it involves all subnuclei of the IP, it is not distributed to all in equal volume: the most massive components of the connection pass to the contralateral half of the IP-R, most likely from the caudal one-third of the dorsal tegmental nucleus, and to the contralateral IP-L and IP-DL, probably in part from the dorsal tegmental nucleus but in larger volume from the

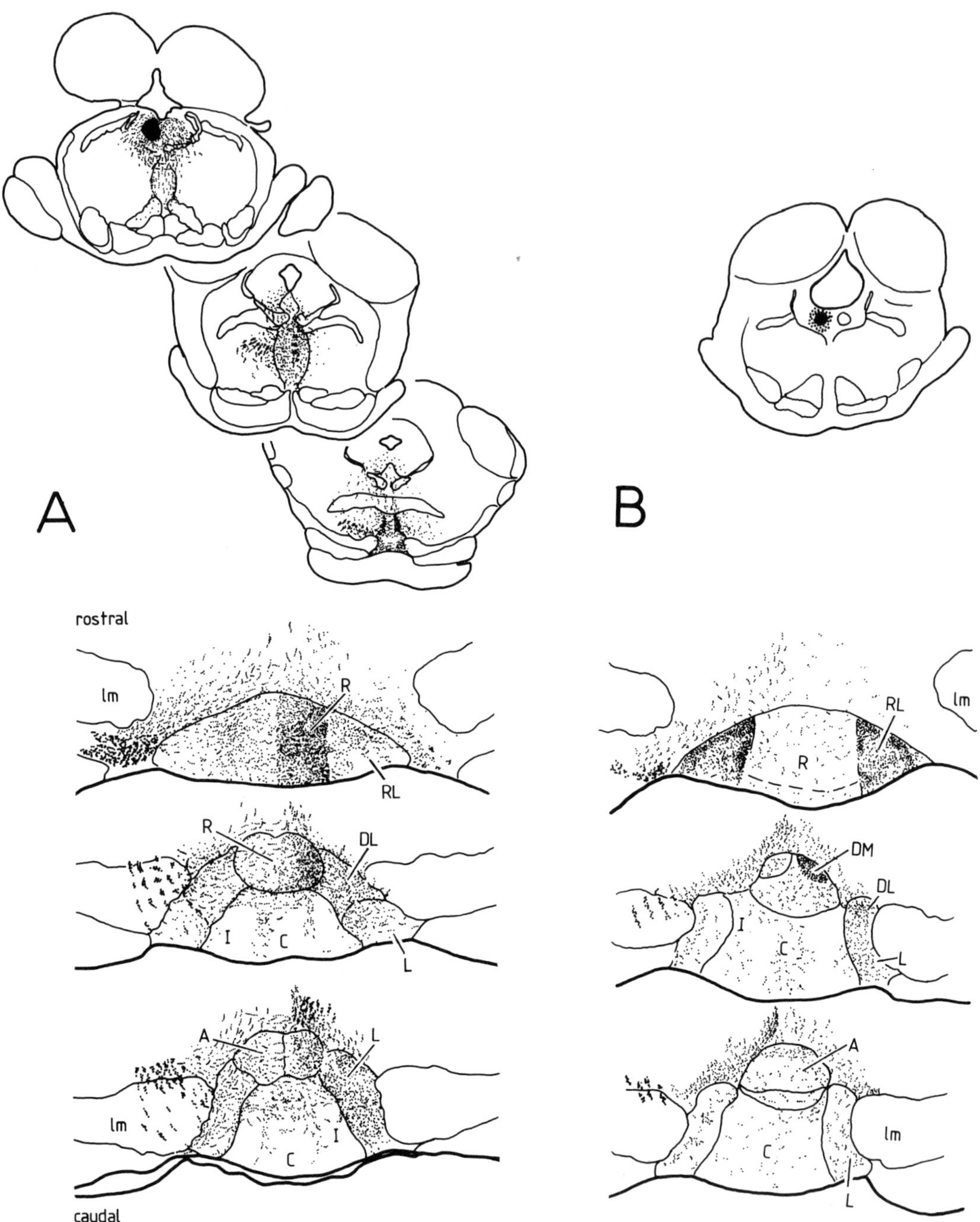



Fig. 17. Chartings of the autoradiograms of anterograde fiber labeling in IP in a case of: (A) injection of tritiated leucine in the caudal part of the dorsal tegmental nucleus (case *RRLM-18*). Note the predominantly contralateral labeling of IP-R; (B) tritiated leucine injection in a more rostral part of the caudal central gray substance, including the dorsal and dorsolateral tegmental nuclei (DTN and DLTN; Case *RRLM-6*). The rostrocaudal levels represented in both A and B are approximately the same as those in Figures 1A–C and 2A–C.

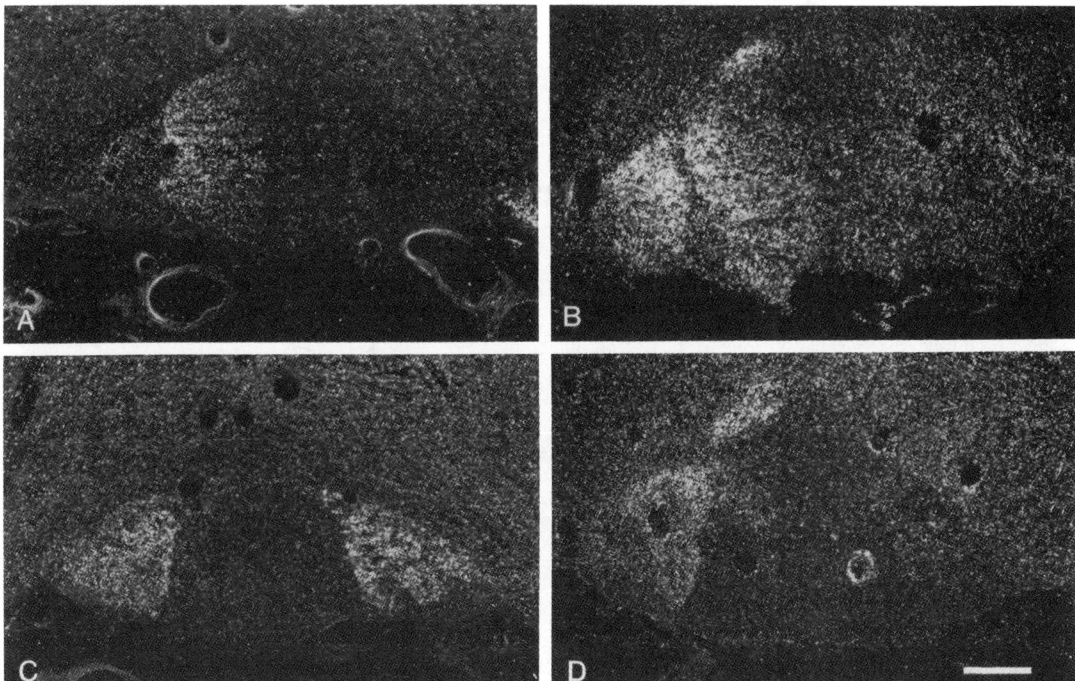

Fig. 18.  Darkfield photomicrographs of the anterograde fiber labeling in IP following injections in the caudal gray substance. A. Injection at caudal levels of the dorsal tegmental nucleus. Rostral pole of IP. Case *RRLM-24* (compare Fig. 17A). Note that the injection is contralateral to the heaviest labeling in IP. B. Large injection in the dorsal tegmental nucleus and adjacent structures. Case *RRLM-7*. Note dense labeling in both IP-R and IP-RL, and also that injection site is contralateral to the most heavy label-ing in IP. Approximately the same level as Figures 1B and 2B. C. Injection at the rostral level of the dorsal tegmental nucleus of Gudden. Anterograde labeling in the rostral pole of IP in IP-RL. Case *RRLM-6* (compare Fig. 1A). D. Same case as C, midstrocaudal level of IP (compare Figs. 1B and 2B). Case *RRLM-6* (compare Fig. 17B). Note that the strong labeling in IP-DL and IP-DM is contralateral to the injection site. Scale bar 200 $\mu$m.

nucleus tegmenti dorsalis lateralis. Further substantial projections, most likely arising from the caudal half of the dorsal tegmental nucleus, involve the contralateral halves of the IP-A and IP-DM. The subnuclei IP-C and IP-I appear to be only sparsely innervated by the caudal central gray.

*Median raphe nucleus.*  An injection of tritiated amino acids was placed in the median raphe nucleus in three cases, all of which showed comparable patterns of antero-grade labeling in the IP. In case *RRIP-16* the center of the injection site is located asymmetrically in the dorsocaudal part of the median raphe nucleus, immediately ventral to the level of the ventral tegmental nucleus of Gudden (Fig. 19). Anterograde labeling in the IP for the greater part is located in the IP-C and IP-I bilaterally with a slightly higher density on the side of main involvement of the median raphe nucleus. Labeling is sparse over the IP-A, IP-R, and IP-DM, and only little denser over the IP-L. Especially in the dorsally located subnuclei IP-R, IP-DM, and IP-A it is difficult to determine whether the sparse labeling represents any fiber terminals or exclusively fi-bers passing to the IP-C and IP-I.

In experiment *RRIP-17* the injection site is located more ventrally in the median raphe nucleus, only a short dis-tance dorsocaudal to the IP. In this case likewise, most of the label is concentrated over the IP-C and IP-I, but the IP-L is somewhat more densely labeled than it is in *RRIP-16*.

*Habenular complex.*  In the following paragraphs a col-lective account will be given of the distribution of haben-ulo-interpeduncular fibers labeled anterogradely by an injection of WGA-HRP (case *RHb-9*) in the habenular com-plex. The injection involves nearly the entire habenula together with adjacent parts of the lateral and mediodorsal thalamic nuclei (Fig. 20A). Labeled fibers follow the ipsi-lateral fasciculus retroflexus and enter the rostral lateral subnucleus IP-RL from the rostral and lateral side. From the densely labeled lateral subnuclei most—perhaps all—labeled fibers sweep across the cell groups IP-R, IP-I, and IP-C into the medial half of the contralateral IP-L where many, at least, describe a sharp reverse turn that leads them across the IP for a second time.

The labeling picture indicates that afferents from the habenula are distributed to all parts of the IP, but the apparent density of their termination varies among the subdivisions (Fig. 20D–F). Very dense labeling is found bilaterally throughout the IP-R and ipsilaterally in the lateral subnuclei; in the latter subdivisions it is concentrated in three or four regions delineated by less densely labeled zones. The contralateral lateral subnuclei show only moderate to dense fiber labeling in their lateral half; the dense label concentration in the medial half is difficult to interpret because of the great crowding of recurving fibers in this part. Fiber labeling in the IP-I and IP-C, the main traverse of crossing and recrossing fasciculus retroflexus fibers, is of moderate density, equal on the left and right. Of the two caudodorsal subnuclei of the IP, the parvicellular IP-DM shows very dense labeling bilaterally. The magnocellular IP-A, on the other hand, is much less massively labeled (Fig. 20F).

*In summary*, it appears certain that the habenulo-interpeduncular tract is distributed to all subdivisions of the IP. The present findings suggest, moreover, that it terminates in greater volume in some subdivisions (IP-R, IP-DM, and ipsilateral lateral column) than in others. However, a *caveat* must here be added. It is obvious, even on cursory inspection, that the habenulo-interpeduncular tract in addition to its majority of thin fibers (estimated caliber about 0.5 $\mu$m) contains a considerable number of much thicker axons (estimated caliber 2–2.5 $\mu$m). These two fiber categories are not evenly distributed within the IP: the thick axons seem distributed preferentially to the IP-R, IP-DM, and the ipsilateral lateral subnuclei, i.e., the subdivisions in which the label concentration is highest. In light-microscopic experiments by anterograde HRP labeling the relative volume of a connection is judged by the apparent packing density of labeled axon arborizations and synaptic endstructures in the target area. Since this density is determined not only by the number but also by the caliber of the labeled structues, the statements made in the foregoing account have at best descriptive value but only limited quantitative significance. It is possible, for example, that the comparatively light-appearing IP-I, IP-C, and IP-A in actuality receive a proportionally greater number of fibers from the habenula than do the heavily labeled IP-R, IP-DM, and IP-L. Further studies will be needed to determine the comparative volume of each individual component of the habenulo-interpeduncular connection.

*Lateral hypothalamic region.* The present material includes several cases in which tritiated amino acids were placed in the lateral hypothalamus. In case *RRSI-87* the center of the isotope injection lies in a frontal plane approximately corresponding to the rostral pole of the ventromedial hypothalamic nucleus (Fig. 21A). Massive fiber labeling follows the medial forebrain bundle to the midbrain where labeling is most concentrated ipsilaterally in the ventral tegmental area and in and alongside the median raphe nucleus. The labeling of these fields is augmented by fibers that have reached the midbrain by way of the stria medullaris and fasciculus retroflexus.

The labeling pattern suggests that the lateral hypothalamus is not a main source of afferents to the interpeduncular nucleus. In the rostral half of the nucleus labeling is sparse, but from about its midlevel caudalward increasing numbers of labeled fibers enter the nucleus from the dorsolaterally adjacent ventral tegmental area (Fig. 21A).

---

Fig. 19. Chartings of anterograde fiber labeling in the interpeduncular nucleus in a case of tritiated leucine injection in the median raphe nuclei. Case *RRIP-16*.

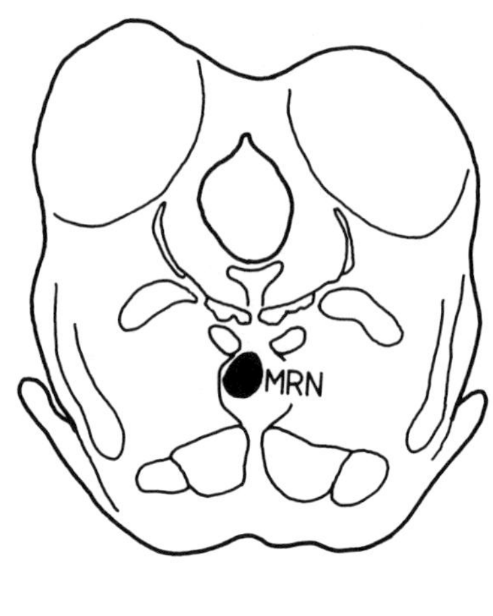


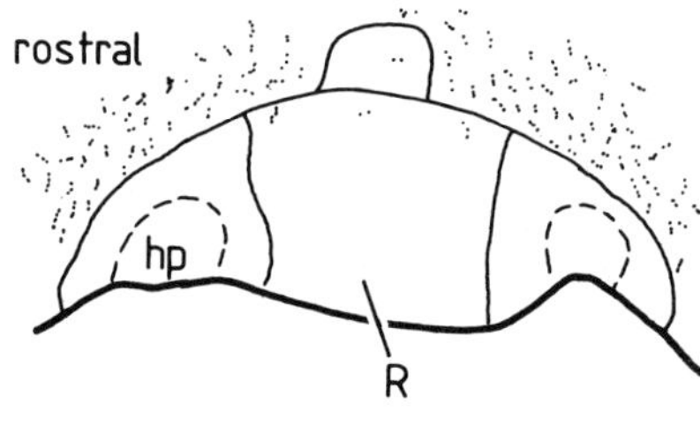


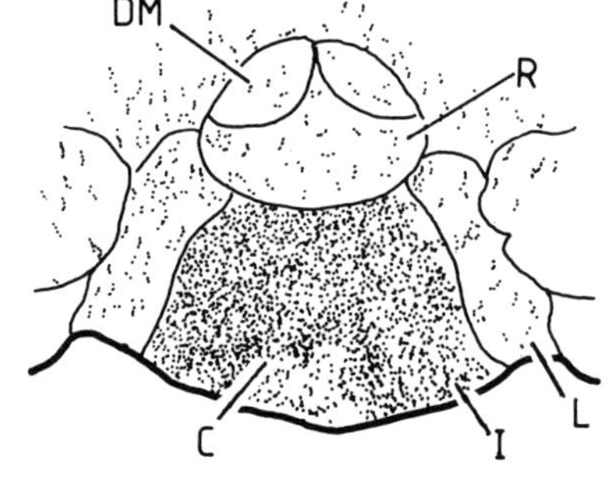


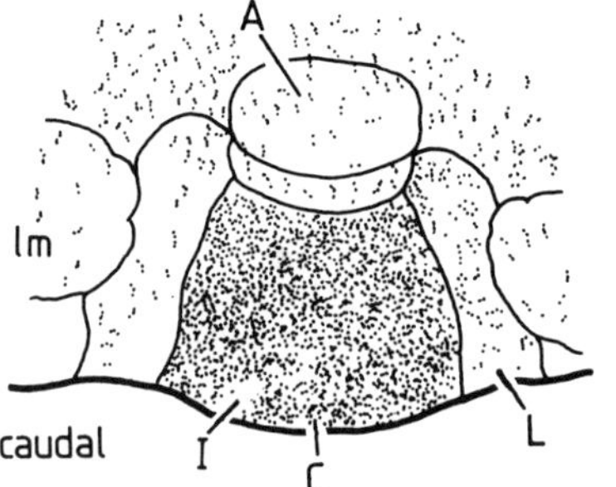

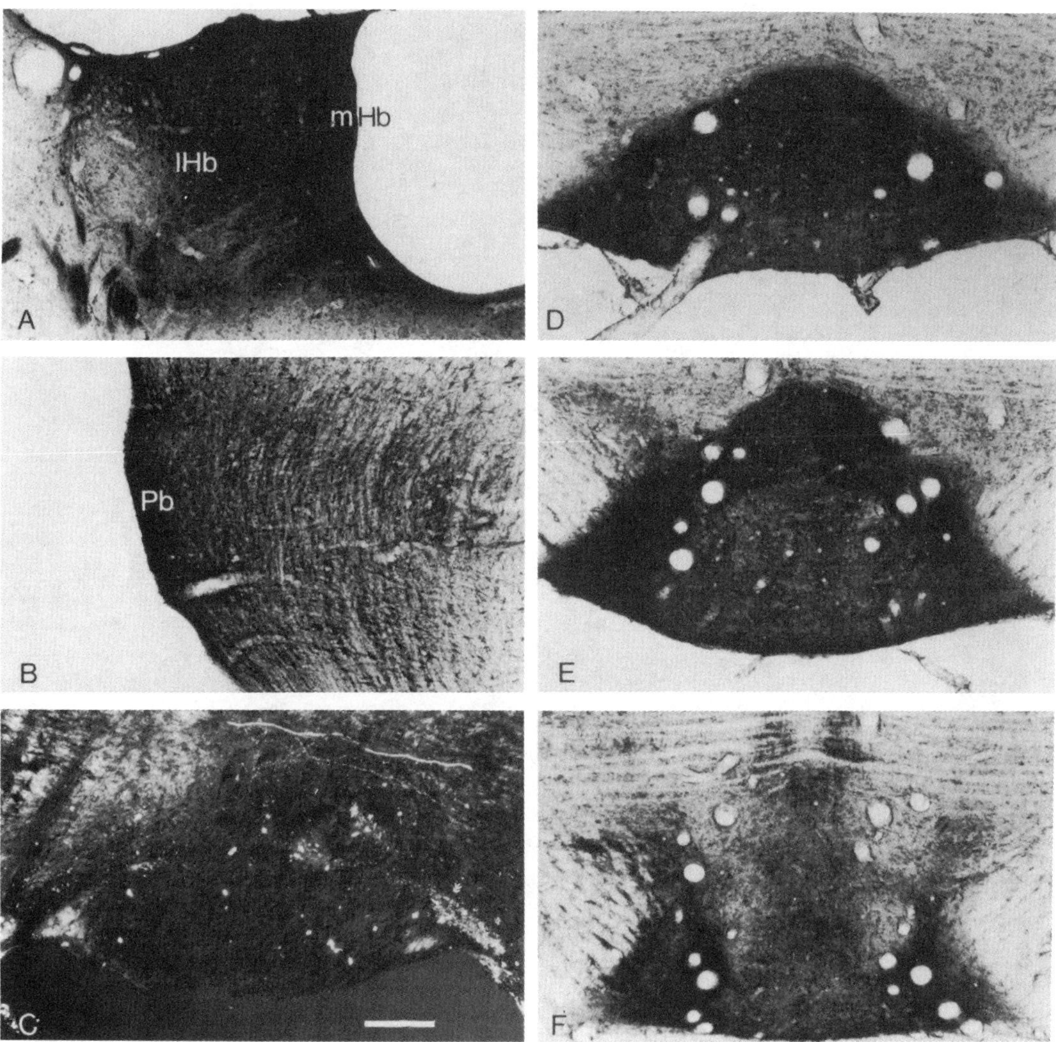



Fig. 20. Photomicrographs of the injection site and the resulting anterograde labeling in the interpeduncular nucleus in a case of WGA-HRP injection of, respectively, the habenula (A,D–F) and the parabigeminal nucleus (B,C). A. Injection of the habenular complex; case *RHb 9*. D–F. Anterograde labeling in IP in brightfield illumination. Rostrocaudal levels in D–F are at the same longitudes as Figures 1A–C and 2A–C. B. Injection site in the parabigeminal region; case *RHLM-11*. C. Dense anterograde labeling in the rostral pole of IP in IP-RL indicated by white double-headed arrows. Darkfield-polarized light. Scale bar (200 $\mu$m) in 20C applies to all micrographs.

Upon entering, these fibers for the most part follow the border between the dorsal and lateral subnuclei, which show some sparse labeling bilaterally. Most labeled fibers enter the IP-I, where their main distribution bilaterally involves a narrow zone along the dorsal and lateral borders of this subnucleus. In the caudal third of the IP, bilateral labeling largely coincides with the three vertical cell columns that occupy the intermediate subnuclei IP-I and the median subnucleus IP-C.

*Nucleus of the diagonal band (NDB).* The following description is based mainly upon case *RRSI-75*, one of four cases of tritiated amino acid injection in the nucleus of the diagonal band. In *RRSI-75* the injection is located at the base of the septum and involves primarily the transition between the vertical and the horizontal limb of the nucleus (Fig. 21B).

Labeled fibers reach the midbrain exclusively by way of the medial forebrain bundle and, much like lateral hypo-

                                              

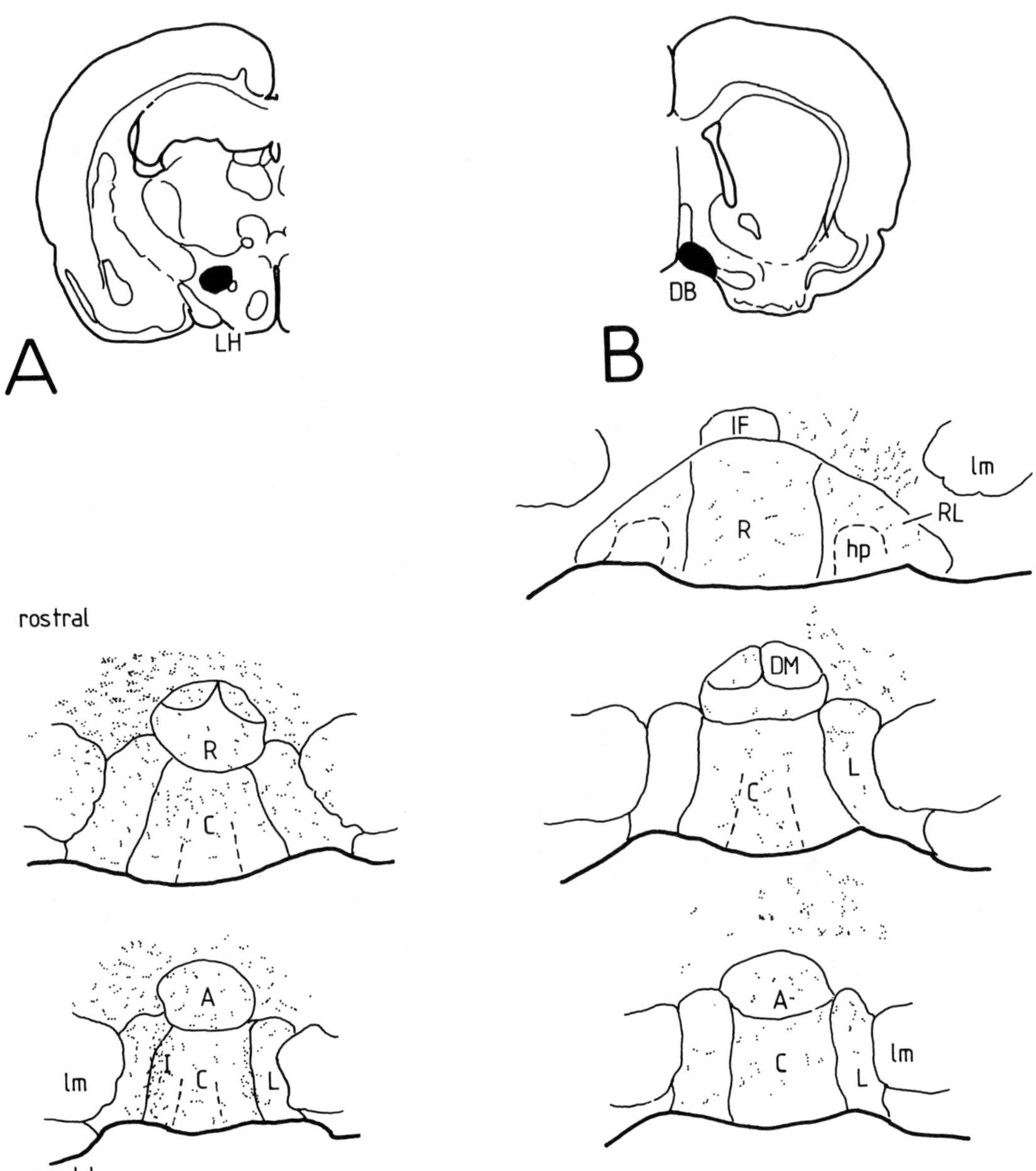


Fig. 21. Chartings of anterograde labeling in IP following an injection of tritiated leucine in the lateral hypothalamic region (A) and in the nucleus of the diagonal band of Broca (B). Rostrocaudal levels represented here approximately correspond to Figures 1A–C and 2A–C.

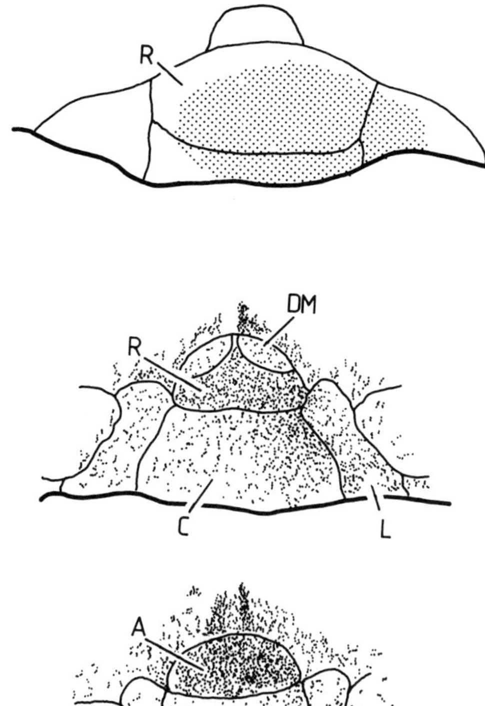



Fig. 22. Chartings of anterograde labeling in the caudal part of IP following a small tritiated leucine injection in the rostral pole of IP (top of figure) (IP-R). Case *RRIP-6*. The two caudal levels can be compared with Figures 1B,C and 2B,C.

thalamic efferents, are distributed over the ventral tegmental area, from which a modest number enter the interpeduncular nucleus (Fig. 21B). In contrast to hypothalamic fibers, labeled NDB efferents enter the IP throughout its length. Their main distribution involves the dorsal lateral subnucleus IP-DL, almost exclusively on the ipsilateral side, and the IP-I, IP-C, and IP-R bilaterally. Even in these IP divisions, labeling must be rated sparse, and it thus appears that the NDB, like the lateral hypothalamic region, is only a minor source of afferents to the IP.

*Parabigeminal region.* The parabigemino-interpeduncular projection of the rat is shown most clearly in case *RHLM-11* in which a small injection of WGA-HRP fills the entire parabigeminal nucleus and spreads into adjacent parts of the nuclei peripeduncularis, parabrachialis dorsalis, and lemnisci lateralis (Fig. 20B). In this case a dense plexus of labeled axon terminals selectively fills a narrow strip of the interpeduncular nucleus immediately within the dorsolateral border of the rostral lateral subnucleus (IP-RL). This circumscript field of labeling is present on both sides but is denser contralateral to the injection (Fig. 20C). A slight distance farther caudally much sparser labeled fibers are found in the IP-DL, more numerous contra- than ipsilaterally. Most or all of the terminal labeling appears to be associated with thin fibers approaching the interpeduncular nucleus from the lateral side and reaching the contralateral IP-DL and IP-RL by skirting the dorsal border of the IP.

The findings in both of the two available cases of successful injection clearly indicate that the parabigeminal region is the source of a modest projection to the IP that is remarkable chiefly for its limited, sharply defined distribution within the nucleus.

*Intranuclear connections.* Owing to the diffusion of tracer around the injection site it is difficult to identify short-axon connections by the autoradiographic method. Nevertheless, some of the iontophoretic injections of tritiated leucine in the IP are small enough to permit some cautious remarks about connections between different parts of the IP. In experiment *RRIP-6* the injection site includes the rostral part of the IP-R (Figs. 10A, 22). The arrangement of silver grains over the caudal half of the IP suggests labeling that is not due to mere diffusion. The caudal half of the IP-L, especially on the side where most of the label is deposited, is more densely labeled than the adjacent medial lemniscus (Fig. 22). Likewise, the IP-A appears to be more heavily labeled than would be expected from diffusion or labeling of passing fibers, whereas the IP-DM by contrast contains only sparse label. Silver grains are predominantly distributed over the neuropil of the IP-A, and the labeling becomes less intense at the boundaries of this subnucleus (Fig. 22). This observation clearly suggests a projection from the IP-R to the IP-A. Other experiments not here reported provide similar evidence of a projection from the IP-A to the IP-C and IP-L. Further connections intrinsic to the IP cannot be recognized in the present material.

## DISCUSSION

As could be expected from preexisting data, the present study has revealed a complexity of cytoarchitecture, connectivity, and histochemical parcellation that is remarkable for a structure as small and circumscript as the interpeduncular nucleus. Not surprising in view of such intricacy, the numerous studies done thus far—including the present one—have left many questions concerning each of these three major aspects of interpeduncular anatomy unanswered. As will become evident from the discussion to follow, such questions are particularly likely to be encountered when attempts are made to correlate cytoarchitectural with connectional data, and the latter with the histochemical aspects of the interpeduncular nucleus. In the first part of this discussion, the cytoarchitectural subdivision and the fiber connections of the IP will be considered together. A later section will deal with the problem of correlating these more strictly morphological data with the distribution of acetylcholinesterase activity, and of substance P–like and enkephalin-like immunoreactivity within the IP. Finally, some aspects of the functional role of the interpeduncular nucleus will be considered in the light of the present connectional findings.

To facilitate a comparison of the various hodological and histochemical aspects of the different subnuclei of the inter-

TABLE 1. Fiber Connections and Histochemical Characteristics of IP Subnuclei[1]

| | Rostral IP-R | Dorsal-medial IP-DM | Apical IP-A | Central IP-C | Intermediate IP-I | Rostral-lateral IP-RL | Dorsal-lateral IP-DL | Lateral IP-L |
|---|---|---|---|---|---|---|---|---|
| Acetylcholinesterase activity | + + | + + + | + | ± | + | ±/+ + | + + | +/+ + |
| Substance P–like immunoreactivity | ☆ + | + + | + | ± | ± | +/+ + + | + | +/+ + |
| Enkephalinlike immunoreactivity | ☆ + + | − | ☆ + | ☆ + | − | ± | + + | ± |
| Serotoninlike immunoreactivity | + | + + + | ☆ + + | − | − | ± | ± | ☆ + |
| Efferents | DTR + + +<br>DLTN + +<br>MRN + | DTR● + +<br><br>MRN + +<br>DRN + + | LH +<br><br><br><br>Se + +<br>HIP + + | DTR +<br><br>MRN + | DTR ±<br><br>MRN ± | DTR + +<br><br>MRN +<br>DRN + +<br>MD(thal) + | DTR + +<br><br>MRN +<br>DRN + +<br>MD(thal) +<br>G(thal)● + + | DTR + +<br>DLTN + +<br>MRN ±<br>DRN +<br><br>Se +<br>HIP + + |
| Afferents | DTR● + + +<br><br>DRN ±<br>Hab + + +<br><br>DB ± | DTR● + + +<br><br>DRN● + + +<br>Hab + + +<br>LH + | DTR +/+ +<br>MRN ±<br><br>Hab +<br><br>DB +<br>IP-R + | DTR +<br>MRN + +<br><br>Hab + +<br>LH +<br>DB +<br>IP-R ± | DTR +<br>MRN + +<br><br>Hab + +<br>LH +<br>DB +<br>IP-R ± | DTR +<br><br>DRN + +<br>Hab + +/+ + +<br>LH +<br>DB +<br><br>Pb + | DTR + +<br><br>DRN + + +<br>Hab + +<br>LH +<br>DB ±<br><br>Pb + | DTR + +<br>MRN +<br><br>Hab + +/+ + +<br>LH ±<br>DB ±<br>IP-R ± |

[1]The relative density of anterograde, respectively retrograde labeling is indicated for each subnucleus by plus-signs (between 3–0). Cell groups containing immunoreactive cell bodies are marked by a white-cored star and projections showing a contralateral predominance are indicated by a solid black circle.
[2]−, no; ±, very light; +, light/sparse; + +, moderate; + + +, heavy; ☆, cell bodies; ●, contralateral predominance.

peduncular nucleus, the data available at present are compiled in Table 1 in a semiquantitative way. Figure 23 summarizes the main features of the afferent and efferent connections of the IP in a more graphic manner.

## Cytoarchitecture and fiber connections

Although architectonically the interpeduncular nucleus (IP) appears as a circumscript morphological entity into the ventral part of the mesencephalon, parts of what now seems certain to belong to the IP have been included in the ventral tegmental area or the raphe nuclei (e.g., Palkovits and Jacobowitz, '74; Cuello et al., '78; Steinbusch, '81). The distinctness and unity of the IP with respect to its surrounding structures are expressed most explicitly, perhaps, by its afferent connections, in particular by the massive habenular input that distributes over all parts of the IP (see Fig. 20). Nevertheless, when further architectonic and hodological criteria are applied, it is clear that within the IP a number of more or less independent units can be recognized, each having its own unique cytoarchitecture and combination of afferent and efferent connections. The parcellation proposed by Ives ('71) for the IP of rodents provides a basic scheme, in which further subdivisions can be made. Ives's pattern of organization of the IP implies a central region (pars medialis) flanked laterally on either side by a lateral cell column (pars lateralis) and covered dorsally by two unpaired nuclei that form a bridge between the two lateral columns. The more rostral of these dorsal subnuclei is composed of the smallest cells in the IP (pars dorsalis parvicellularis of Ives; see also Hamill and Lenn, '84), whereas the caudal group is formed by neurons that are notably larger and differ in shape from those encountered farther rostrally (pars dorsalis magnocellularis). Although within these four basic cell groups delineated by Ives further subdivisions have been proposed and confirmed (Hamill and Lenn, '84; Hemmendinger and Moore, '84; Lenn and Hamill, '84; the present study), in several of the observa-

tions here described Ives's fundamental pattern is apparent (e.g., acetylcholinesterase staining, substance P immunoreactivity, connectivity of the IP with the habenula, the raphe nuclei, and the caudal central gray substance). Therefore, although in the present study the recently proposed compromise nomenclature of Lenn and Hamill ('84) has been adopted, the cytoarchitecture and fiber connections of the various subnuclei will be dealt with in a sequential order that follows the scheme proposed by Ives ('71).

*The rostral and dorsal medial subnuclei (IP-R and IP-DM).* The only important discrepancy between the parcellation of the IP here followed and the recently proposed scheme of Lenn and Hamill ('84), which is mainly based on a number of recent detailed studies (Hamill and Lenn, '84; Hamill et al., '84; Hemmendinger and Moore, '84), concerns the rostral subnucleus. In the present study an area is distinguished in the caudal half of the rostral subnucleus (as defined by Lenn and Hamill, '84) that differs cytoarchitectonically from the rest of the subnucleus. In Nissl-stained material this area at middle longitudes of the IP takes the shape of a paired ovoid cell cluster in the dorsal part of the IP that consists of neurons slightly smaller and more closely spaced than those found more rostrally. Owing to their position these cell clusters are here called the dorsal medial subnuclei (IP-DM) to distinguish them from the rest of the unpaired rostral subnucleus (IP-R). IP-DM is bordered caudally by the more magnocellular apical subnucleus (IP-A); depending on the plane of sectioning, the IP-DM may appear as the more dorsally placed of the two last-mentioned nuclei. The present study has provided several reasons besides cytoarchitecture to delineate the IP-DM from the IP-R (see Figs. 1–7, Table 1). The two differ with respect to myelin content, acetylcholinesterase activity, and substance P–like, enkephalin-like, and serotonin-like immunoreactivity. Furthermore, although the IP-R and IP-DM are both reciprocally connected with the median and dorsal raphe nuclei and the caudal central gray substance, and

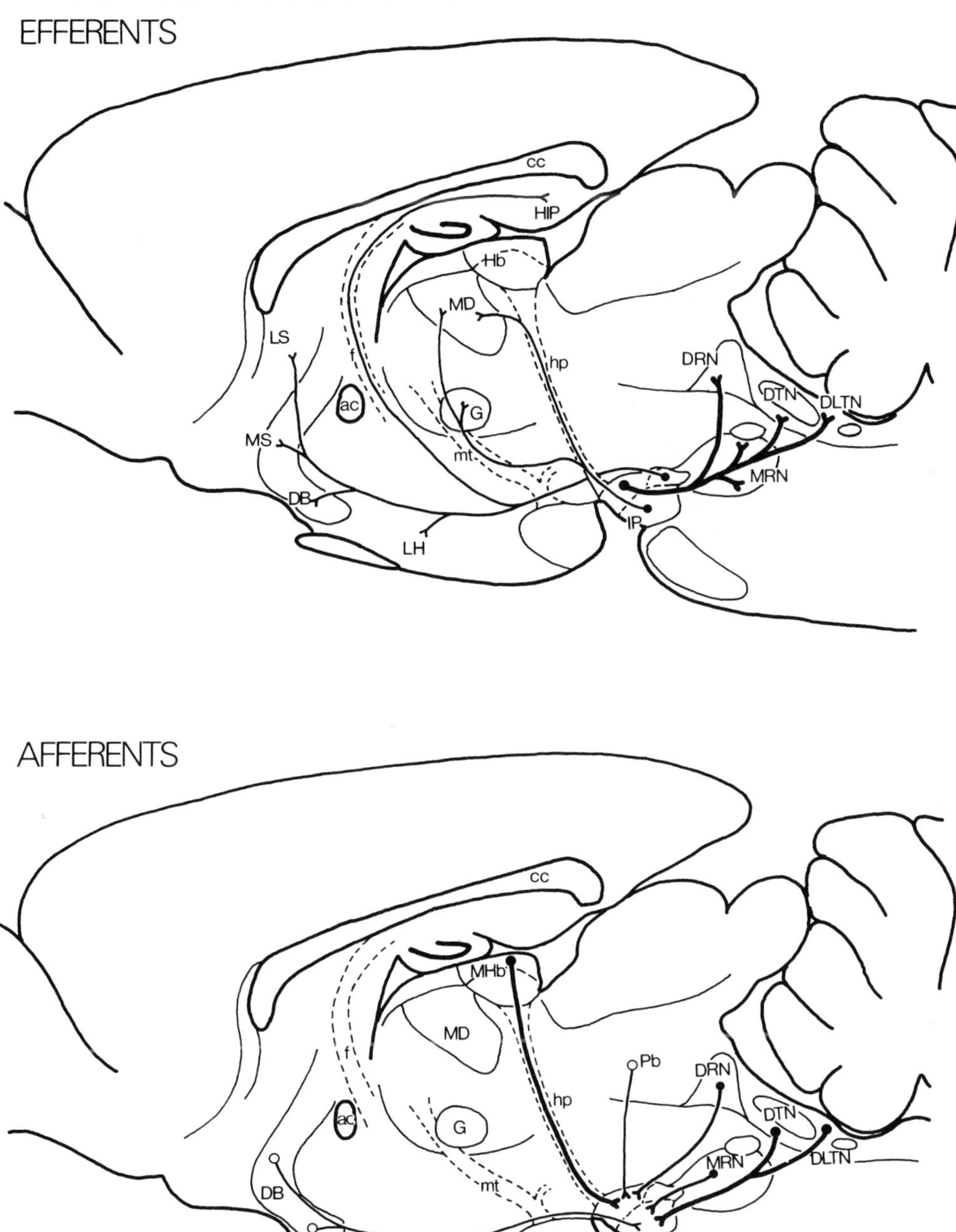


Fig. 23. Summary diagrams of the efferent and afferent connections of the interpeduncular nucleus, represented in a sagittal view of the rat brain.

both receive fibers from the medial habenular nucleus, these connections differ with respect to volume and topography (see Table 1). Though Hamill et al. ('84) and Hemmendinger and Moore ('84) did not distinguish between the IP-R and IP-DM as separate entities, they found differences in the distribution of substance P, enkephalin and serotonin immunoreactivity between the two regions similar to those here described. Moreover, as indicated in the schemes of Hamill et al. ('84), differences in cholecystokinin (CCK) and vasoactive intestinal polypeptide (VIP) immunoreactivity might also distinguish the IP-R from the IP-DM.

The present findings indicate that the main efferent projections of the IP-R and IP-DM follow the interpedunculotegmental tract to the median and dorsal raphe nuclei and the caudal central gray substance. These projections appear to terminate in greatest density around the dorsal and ventral tegmental nuclei of Gudden, but only sparsely within the borders of these cell groups. This pattern is very similar to that followed by the direct projections to these regions from the lateral habenular nucleus (Herkenham and Nauta, '79). Similarly, the reciprocating projection from the caudal central gray substance to the interpeduncular nucleus for the greater part does not originate in the circumscript dorsal tegmental nucleus itself but immediately dorsomedial to it in the nucleus incertus and in the dorsolateral tegmental nucleus (Marchand et al., '80; Hamill and Jacobowitz, '84; the present observations).

It is interesting to note that, according to the present retrograde-labeling evidence, the projections from the IP-R and IP-DM to the raphe nuclei and the caudal central gray substance are for the most part ipsilateral, with the exception of the fibers from the IP-DM to the dorsolateral tegmental nucleus, which are mainly crossed (Table 1). This contrasts with the predominantly crossed disposition of the reciprocating projections from the caudal central gray to the IP-R and IP-DM. Isotope injections involving the caudal half of the central gray substance or, more rostrally, its medial part, result in a very characteristic labeling pattern: in the midline a surprisingly sharp border appears between the heavy contralateral and much sparser ipsilateral labeling in the IP-R, a subnucleus that by all other histological or hodological criteria is an unpaired structure. Except for recent autoradiographic evidence (Hemmendinger and Moore, '84), this remarkable laterality has not been noted in earlier reports on these projections (Nauta and Kuypers, '58; Morest, '61; Ban and Zyo, '63; Briggs and Kaelber, '71), which is most likely due to the inherent restrictions of the lesion techniques used in these earlier studies. It is apparent from the present study that the efferent fibers from one-half of the caudal central gray follow four routes to the IP and more rostral structures: the ipsi- and contralateral mammillary peduncles and the ipsi- and contralateral interpedunculotegmental tracts. As the decussating fibers in part travel over some distance through the contralateral half of the caudal central gray substance, lesions of the central gray may well interrupt these fibers, thus obscuring the laterality of the projection. The functional significance of the topographic arrangement of the reciprocal connections between the caudal central gray and the interpeduncular nucleus remains unclear. However, it is apparent that one-half of the caudal central gray may affect its contralateral counterpart both directly via crossing homologous projections (Groenewegen et al., '83c; Groenewegen and Van Dijk, '84; Liu et al., '84) and, more massively, by way of the contralateral half of the IP-R.

The results of the present anterograde-tracing experiments suggest a further point-to-point relationship between the caudal central gray and the IP. Evidence was found that the most medial part of the dorsal tegmental region projects (largely contralaterally) to the medial part of the IP-R, while more lateral parts of the region project to a more lateral zone of the IP-R. This suggestion of an orderly medial-to-lateral topography has been further substantiated in recent experiments in our laboratory by the method of immunohistochemical localization of anterogradely transported *Phaseolus vulgaris* lectin (Gerfen and Sawchenko, '84). These experiments (Groenewegen and Van Dijk, '84) have shown that the dorsal tegmental afferents of the IP are in large part sagittally oriented in the IP-R and establish an orderly mediolateral relationship between the dorsal tegmental region of one side and the contralateral half of the IP-R. As a consequence of this sagittal disposition, the dorsal-tegmental afferents are oriented perpendicular to the fibers composing the other main input to the IP-R: the habenulo-interpeduncular tract, which originates in the medial habenular nucleus and distributes to the IP-R in transverse-horizontal planes (Cajal, '11; Herkenham and Nauta, '79). The projection of the medial habenula to the IP-R, likewise, is topographically organized: a ventral-to-dorsal gradient in the IP-R corresponds to a medial-to-lateral gradient in the medial habenular nucleus (Herkenham and Nauta, '79; Contestabile and Flumerfelt, '81; Marchand et al., '80). This nonrandom fiber arrangement suggests the possibility of a systematic integration of the inputs conveyed by the two most massive afferent connections of the interpeduncular nucleus.

***The apical subnucleus (IP-A).*** The apical subnucleus (IP-A) forms a distinct subnucleus in the caudodorsal IP, embedded partially in the more rostrally located subnuclei IP-R and IP-DM. Furthermore, the IP-A is caudally continuous with the median raphe nucleus. Although a number of earlier reports likewise consider the IP-A part of the interpeduncular nucleus (cat: Berman, '68; Kapadia and DeLanerolle, '84; rat: Baisden et al., '79; Riley and Moore, '81; Hamill and Lenn, '84; Hemmendinger and Moore, '84), several other authors include the IP-A in the raphe nuclei (Palkovits and Jacobowitz, '74; their "nucleus linearis oralis," a term also adopted by Ljungdahl et al., '78; and Steinbusch, '81, '84). The reason to regard the IP-A as part of the interpeduncular complex in the present study is its topographic relationship with the other subnuclei of the IP and the fact that its constituent neurons, although morphologically very similar to the cells of the median raphe nucleus, show a preferential, horizontal orientation and are closely spaced, in contrast to the caudally adjacent raphe cells, which are more loosely packed and lack any preferential orientation. However, a clear caudal border of the IP-A is not evident. Moreover, in a number of other respects, to be dealt with next, the IP-A resembles the raphe nuclei.

In agreement with earlier reports (Baisden et al. '79, Wyss et al., '79; Riley and Moore, '80; Shibata and Suzuki, '84) the present findings indicate that projections from the IP-A ascend to the septum and hippocampus, and possibly also to the hypothalamus. It should be emphasized, however, that similar projections originate in the caudal part of the lateral subnucleus of the IP (IP-L) and in the median raphe nucleus, a matter also discussed in the above-mentioned studies. Furthermore, in accord with several immunohistochemical studies (rat: Steinbusch, '81; Hamill et al., '84; Hemmendinger and Moore, '84; cat: Kapadia and De-

Lanerolle, '84a), the present study showed that both the IP-A and the mesencephalic raphe nuclei contain cell bodies immunoreactive for serotonin. By retrograde tracing of fluorescent dyes, combined with immunohistochemical staining for serotonin, Köhler et al. ('82) have demonstrated that part of the fibers to the septum are serotoninergic and originate in the dorsal and median raphe nuclei and in the nucleus linearis oralis (the present IP-A). Accordingly, injections of retrograde fluorescent tracers in the ventral part of the hippocampus were found to result in retrograde labeling of some IP-A cells exhibiting serotonin-like immunoreactivity (Groenewegen and Steinbusch, '84), indicating that the projection from the IP to the ventral hippocampus is, at least in part, serotoninergic.

According to the present findings the IP-A receives fibers predominantly from the caudal central gray substance (Table 1) while an additional but relatively sparse input comes from the habenula. Moreover, although the present autoradiographic findings cannot be taken as conclusive, there is some evidence of an intrinsic projection from the IP-R to IP-A, a connection that the latter subnucleus shares with the median raphe nucleus.

***The central and intermediate subnuclei (IP-C and IP-I).*** The central nucleus (IP-C) and the two paramedian intermediate subnuclei (IP-I) together constitute the core region, or pars medialis (Ives, '71), of the IP. It is noteworthy that these three centrally placed subnuclei differ in several respects from the surrounding lateral, dorsal, and rostral IP subnuclei. For example, they are collectively distinguished from adjacent subnuclei by a low content in acetylcholinesterase and a weak or even negative substance P–like and serotonin-like immunoreactivity (Hamill et al., '84; Hemmendinger and Moore, '84; Rotter and Jacobowitz, '84; Artymyshyn and Murray, '85; the present study, Table 1). This is not to suggest that the IP-C and IP-I are indistinguishable from each other also by other criteria. The two subnuclei have different cyto- and myeloarchitectonic features, and a further reason to distinguish between them is the unique presence of crest-synapses in the IP-I (Lenn, '76; Lenn et al., '83; Murray et al., '79; Hamill and Lenn, '84). Additional differences between the IP-C and IP-I found in the present study mainly concern the distribution of enkephalin immunoreactivity and some slight contrasts in their afferent and efferent connectivity. Enkephalin is absent in the IP-I, whereas sparse positive fibers are present throughout the IP-C, with a few enkephalin-stained neurons in its caudodorsal part, adjacent to the IP-A. According to the present evidence, the IP-C and IP-I receive their main input from the habenula and from the median raphe nucleus, while sparser afferents come from cells in the dorsal tegmental region, the nucleus of the diagonal band, and the hypothalamus (Table 1). Habenular fibers traverse the rostral IP-C and IP-I in transverse-horizontal planes as described previously (Cajal, '11; Herkenham and Nauta, '79). In their caudal part the distribution of habenular afferents appears to be more irregularly clustered, with a particularly dense termination in the IP-I. In agreement with a recent retrograde-tracing study (Shibata and Suzuki, '84), the present study has shown that efferent IP fibers originating in the IP-C and to a lesser degree in the IP-I are directed mainly to the caudal central gray substance and the median raphe nucleus. It must, however, be noted that, compared with other subnuclei of the IP, only a relatively small number of IP-C and IP-I cells are labeled by tracer injections in the above-mentioned areas, and that these cells are largely confined to caudal regions of these

subnuclei. This suggests the possibility that most of the neurons of the IP-C and IP-I subserve intrinsic connections within the IP.

***The lateral subnuclei (IP-RL, IP-DL, and IP-L).*** Although Ives ('71) considered the pars lateralis (PL) of the interpeduncular nucleus a cytoarchitectonic entity, reasons can be found to subdivide this lateral part of the IP into several subunits. In agreement with recent studies of the IP (cat: Berman, '68; rat: Hamill and Lenn, '84; Hemmendinger and Moore, '84), the lateral IP in the present account has been subdivided into a rostral (IP-RL), a dorsal (IP-DL), and a main caudal (IP-L) subnucleus on the basis of differences in cytoarchitecture, histochemistry, and connectivity (Table 1). The borders between these subnuclei are not sharply defined. In the cell-poor medial zone that forms the border between the lateral subnuclei and the intermediate subnucleus IP-I a multitude of habenulo-interpeduncular fibers collect, recurving to traverse the medially located subnuclei IP-R, IP-C, and IP-I in their peculiar "figure 8" course through the interpeduncular nucleus (Herkenham and Nauta, '79; the present study). Caudally, subnucleus IP-L cannot be sharply delimited from the pontine gray and ventral tegmental area, a vagueness comparable to that of the caudal border of the IP-A.

The present evidence indicates that the lateral part of the IP contributes to both the ascending and descending projections of the IP. All three of its subnuclei project ipsilaterally to the caudal central gray substance, including the dorsal raphe nucleus. Although minor differences may exist between the projections from different parts of the lateral IP to the central gray, no evidence of a prominent topographic relationship could be found. Projections to the median raphe nucleus appear to arise mainly from the rostral IP-RL (Table 1). These descending projections of the lateral subnuclei are reciprocated by fibers from the caudal central gray substance and dorsal raphe nucleus, and in lesser number also from the median raphe nucleus. It must be noted that these return projections to the lateral subnuclei are mainly crossed and that they originate more rostrally and laterally in the caudal central gray than do the projections to the IP-R, in keeping with the orderly mediolateral organization of the central-gray projections to the IP that was emphasized earlier in this discussion. In addition to the afferents from the caudal central gray and the raphe nuclei, the lateral subnuclei receive a massive input from the habenula, in particular from the dorsal part of the medial habenular nucleus (Herkenham and Nauta, '79; the present study).

An interesting result of the present study is the evidence that all ascending IP efferents originate either in the IP-A, as discussed above, or in one or more of the different lateral subnuclei. Thus, the present retrograde-labeling experiments showed that fibers to the septum and hippocampus arise not only from the IP-A but also from the caudal part of the IP-L. These findings agree with the earlier accounts of Wyss et al. ('79), Baisden et al. ('79), and Shibata and Suzuki ('84). Köhler et al. ('82) have demonstrated that the projection to the septum from the IP-L—a cell group that these authors include among the raphe nuclei—is serotoninergic at least in part. Recent experiments combining immunohistochemistry and retrograde tracing of fluorescent dyes have shown that this is also true for the projection from the IP-L to the hippocampus (Groenewegen and Steinbusch, '84).

The origin of the IP projections to the thalamus appears to be restricted to the more rostral of the lateral subnuclei, the IP-DL and IP-RL. A widely distributed interpedunculothalamic pathway was reported from studies that used degeneration techniques in the rat and cat (Massopust and Thompson, '62; Smaha and Kaelber, '73). However, the results of the present autoradiographic tracing experiments, in agreement with the study of Stofer and Edwards ('78) in the cat, indicate that interpeduncular fibers to the thalamus reach only the nuclei mediodorsalis and gelatinosus; no such fibers could with certainty be traced to the habenula. As regards the projection from the IP to the mediodorsal nucleus, our findings in the rat differ somewhat from those of Velayos and Reinoso-Suarez ('82) in the cat. In the latter species the projection for the most part appears to innervate the medial portion of the mediodorsal nucleus. In addition, in the cat its origin within the IP is not restricted to the lateral subnuclei, as appears to be the case in the rat. According to our findings, IP efferents in the rat completely avoid the medial, limbic-system-related part of the mediodorsal nucleus and remain confined to the latter's lateral subdivision. This lateral zone receives additional afferents from the caudal central gray substance, the pars reticulata of the substantia nigra, and the deep layers of the superior colliculus (Groenewegen and Nauta, '82, '86).

Projections from the interpeduncular nucleus to the nucleus gelatinosus thalami, which forms part of the nucleus submedius of other authors (e.g., Craig and Burton, '81; Craig et al., '82; Price and Slotnick, '83), appear not to have been reported before. It is remarkable that this projection originates almost exclusively in the IP-DL and that it exhibits a contralateral predominance (Table 1). It must be noted that its termination is restricted to the most rostral part of the nucleus gelatinosus, a region probably corresponding to the anterior part of the nucleus submedius (SMa of Craig et al., '82). The projection area thus lies immediately ventral to, but completely avoids, the dorsal area (SMd) which according to Craig and Burton ('81) receives a substantial projection from the pars caudalis of the spinal trigeminal nucleus and the dorsal horn of the spinal cord, and is likely to be involved in nociception. According to Craig et al. ('82) the nucleus gelatinosus is reciprocally connected with the ventrolateral orbital cortex. The recent report of Price and Slotnick ('83) indicates that the anterior part of the nucleus submedius (SMa; our nucleus gelatinosus) may differ from its more caudal parts by being less clearly associated with the olfactory system than are the latter.

## Interpeduncular connections and cholinergic transmission

In accord with a number of earlier reports, beginning with the publication by Lewis and Shute ('67), the present study has demonstrated a high activity of acetylcholinesterase (AChE), differentially distributed over different parts of the IP (Table 1; Hamill and Lenn, '84; Flumerfelt and Contestabile, '82; Hemmendinger and Moore, '84; Rotter and Jacobowitz, '84). It must be emphasized, however, that any relation of this AChE activity pattern to the cholinergic innervation pattern of the IP remains to be established. First, the presence of a high activity of AChE may be a prerequisite for a cholinergic system but cannot by itself reliably identify a cell group or its efferents as cholinergic (Lehman and Fibiger '79; Fibiger, '82; Wainer et al., '84; Butcher and Woolf, '84). Second, recent studies have indicated that AChE may also play a role in the degradation of

several peptides that are abundantly present in the IP (Chubb et al., '80, '84). According to the lesion experiments of Flumerfelt and Contestabile ('82), most of the AChE activity of the interpeduncular nucleus appears to be intrinsic, i.e., associated with cells and their processes within the IP proper. Even total interruption of the fasciculus retroflexus, very likely to include the cholinergic pathways to the IP from the habenula and, potentially, from the basal forebrain, has only little effect on the AChE activity level in the IP. Although there is still much debate on the issue of whether the habenulo-interpeduncular pathway is cholinergic or not (reviews: Fibiger, '82; Butcher and Woolf, '84; Wainer et al., '84), the findings of Flumerfelt and Contestabile ('82) indicate that it does contain AChE, even though not enough to account for more than a small proportion of the AChE activity in the IP.

The fact that the IP is the site of a massive cholinergic mechanism is evident from earlier studies in which the nucleus was found to be the site of the highest activity of the acetylcholine-synthetizing enzyme choline acetyltransferase (ChAT) and of the highest choline uptake in the entire brain (Palkovits et al., '74; Sorimachi et al., '74; Leranth et al., '75; Kuhar et al., '75; Cheney et al., '75; Hattori et al., '77; Sastry et al., '79). The actual presence of ChAT in fibers and terminals within the IP was recently confirmed with immunohistochemical techniques (cat: Kimura et al., '80; rat: Sofroniew et al., '82; Houser et al., '83). According to the immunohistochemical study of Houser et al. ('83) in the rat, there appears to be a rather specific distribution of intense ChAT immunoreactivity in the IP with respect to its subnuclei. In their material, heavy ChAT staining marks the rostral subnucleus IP-R and the laterally adjacent IP-RL (their Fig. 2); the latter subnucleus contains the habenulo-interpeduncular tract, which is also stained. Caudally in the IP only the central region, encompassing the subnuclei IP-C and IP-I, is heavily stained (their Fig. 5A). In addition, it was established by Houser et al. ('83) that the ventral part of the medial habenular nucleus contains ChAT-positive neurons. This location of ChAT-positive neurons in the habenula, as well as the distribution pattern of ChAT-positive fibers and terminals over the IP subnuclei, corresponds rather well with the topographical organization in the habenulo-interpeduncular pathway: the ventral part of the medial habenula, for example, is known to project to the medial IP and the medial part of the medial habenula to ventral portions of the IP (Herkenham and Nauta, '79; Marchand et al., '80; Contestabile and Flumerfelt, '81; Hayakawa and Zyo, '82). This similarity strongly suggests a cholinergic nature of the habenulo-interpeduncular pathway, a notion supported by the results of a study using transmitter-specific retrograde labeling (Villani et al., '83). In that study injections of tritiated choline in the IP resulted in heavy labeling of neurons in the medial habenula.

Unfortunately, no data are available on the relation between the distribution patterns of ChAT immunoreactivity and cholinergic receptor sites. In a recent study Rotter and Jacobowitz ('84) have shown rather specific and extensively overlapping distributions in the IP for both muscarinic and nicotinic receptor sites. When compared with the immunohistochemical study of Houser et al. ('83), there appears to be only partial overlap between ChAT immunoreactivity and the distribution of cholinergic receptors as determined by Rotter and Jacobowitz ('84; see also Clarke et al., '84).

Besides emphasizing the habenula as a likely source of cholinergic input to the IP, possible alternative origins of cholinergic fibers may be considered. Structures that are known to contain cholinergic cells (Kimura et al., '80, '84; Sofroniew et al., '82, '83; Houser et al., '83; Mesulam et al., '83; Armstrong et al., '83) and have been shown to distribute fibers to the IP include the nucleus of the diagonal band and the dorsolateral tegmental nucleus (Contestabile and Flumerfelt, '81; Hayakawa and Zyo, '82; Hamill and Jacobowitz, '84; the present study). Both anterograde and retrograde studies have shown that neurons in the diagonal band project to the IP (Domesick, '76; Contestabile and Flumerfelt, '81; Hayakawa and Zyo, '82; Hamill and Fass, '84; Hamill and Jacobowitz, '84; the present study). The present evidence suggests that this projection is distributed chiefly to the ipsilateral IP-DL and to the IP-R, IP-C, and IP-I bilaterally. It also suggests, however, that fibers from the nucleus of the diagonal band are among the sparsest IP afferents, and that they travel mainly, if not entirely, with the medial forebrain bundle rather than with the stria medullaris and fasciculus retroflexus. These findings cannot satisfactorily explain reports of a marked decrease of ChAT activity in the IP following lesions of the diagonal-band region (Gottesfeld and Jacobowitz, '78) or interruption of the stria medullaris (Kataoka et al., '77; Gottesfeld and Jacobowitz, '78).

Cholinergic neurons in the dorsal tegmental region have been identified with several methods (Fibiger, '82; Mesulam et al., '83; Houser et al., '83). The suggestion that the dorsolateral tegmental nucleus may be a source of cholinergic input to the IP is compatible with the notion, dealt with earlier in this discussion, that the former nucleus, as a lateral component of the dorsal tegmental region, projects to lateral parts of the interpeduncular nucleus, the region of the IP that is rich in cholinergic receptors (Rotter and Jacobowitz, '84). The fibers establishing this connection travel in part by way of the interpedunculotegmental tract (Groenewegen et al., '83c; Groenewegen and Van Dijk, '84; the present data), but whether they are AChE-positive is at present controversial. Wilson and Watson ('80) were the first to report that the interpedunculotegmental tract is strongly AChE-positive, an observation confirmed in the present study. Since the interpeduncular nucleus appears to contain no cholinergic cells (Fibiger, '82; Kimura et al., '80; Houser et al., '83), it is unlikely that the AChE activity in the tract reflects a cholinergic projection from the IP to the dorsal tegmental region. Neither, however, does it appear to signify a cholinergic projection of opposite polarity: lesions of the dorsal tegmental region, even when large and bilateral, do not perceptibly affect the AChE positivity of the interpedunculotegmental tracts (unpublished observations). At present it therefore seems most likely that the AChE activity is bound either to the fibers of the interpedunculotegmental tract or to some hitherto unidentified element or elements interstitial to the tract.

### Substance P–like immunoreactivity

According to the present evidence all subnuclei of the interpeduncular nucleus, with the exception of the IP-C and IP-I, contain substance P–positive fibers and terminals in densities ranging from light to heavy (Table 1). In the IP-RL most of the immunoreactivity appears to be bound to the habenulo-interpeduncular tract, but reactive terminals may also be present. In addition to fibers, substance P–positive cell bodies were found in the IP-R both in rats pretreated with colchicine and in untreated animals. The present findings largely agree with the results of previous immunohistochemical studies (rat: Ljungdahl et al., '78; Cuello and Kanazawa, '78; Hamill et al., '84; Hemmendinger and Moore, '84; Artymyshyn and Murray, '85; cat: Kapadia and DeLanerolle, '84a,b). In the present study a clear correlation could be made between the cytoarchitecture of the IP and the pattern of substance P–like immunoreactivity. In combination with the results of our tracing experiments, such correlation provides some basis for speculation upon the possible sources of substance P immunoreactivity in the IP.

The study of Sakanaka et al. ('82) has provided circumstantial evidence that most of the substance P–positive fibers in the IP are of extrinsic origin: in the ontogeny of the IP such fibers appear much earlier than do substance P–positive cell bodies. This is consistent with the finding of substance P–like immunoreactivity in the habenulo-interpeduncular tract (Mroz et al., '76; Hong et al., '76). It must be noted that only the dorsal part of the medial habenular nucleus contains substance P–positive neurons (Cuello and Kanazawa, '78; Ljungdahl et al., '78; Shinoda et al., '84). There is autoradiographic evidence (Herkenham and Nauta, '79) that this part of the medial habenular nucleus projects most heavily to lateral parts of the IP and thus could account for the strong substance P–like immunoreactivity in the IP-L. Lesions of the habenula or the fasciculus retroflexus result in a considerable reduction of the substance P positivity in the lateral subnuclei of the IP (Emson et al., '77; Cuello et al., '78; Artymyshyn and Murray, '85). In a study in which retrograde fluorescent tracing was combined with staining for substance P immunoreactivity, the habenular source of a substance P projection to the IP could recently be substantiated (Hamill and Jacobowitz, '84).

The finding that the loss of immunoreactivity in the IP following habenular lesions is less than complete suggests, however, the existence of an additional source of substance P–positive fibers to the IP. The present autoradiographic evidence indicates that the IP-L receives a massive innervation not only from the habenula but also from the caudal region of the central gray substance (Table 1). In this region, the dorsolateral tegmental nucleus and the caudal extension of the dorsal raphe nucleus contain numerous substance P–positive neurons (rat: Ljungdahl et al., '78; Sakanaka et al., '81, '82; cat: Moss and Basbaum, '83). In ontogeny, according to Sakanaka et al. ('82), such neurons appear much earlier in the dorsolateral tegmental nucleus than in the IP. It may thus be speculated that the dorsolateral tegmental nucleus contributes to the substance P innervation of the IP-L and possibly also to that of the IP-R and IP-DM (cf. Table 1).

In the context of these findings it is of interest to note the presence of substance P–like immunoreactivity in the interpedunculotegmental tract. It is apparent from the present autoradiographic experiments that efferent fibers of the IP-R—the subdivision of the IP containing the substance P–positive neurons—travel in the interpedunculotegmental tract to reach the caudal central gray substance. The localization of substance P–positive fibers around the dorsal tegmental nucleus and in the dorsolateral tegmental nucleus (rat: Ljungdahl et al., '78; Cuello and Kanazawa, '78; cat: Moss and Basbaum, '83) corresponds closely to the distribution of interpedunculotegmental fibers labeled by tritiated leucine deposited in the interpeduncular nucleus. Accordingly, retrograde fluorescent tracers deposited in the dorsal tegmental region have been found to label a number of substance P–immunoreactive neurons in the IP-R (Huitinga et al., '85; Groenewegen, and Huitinga, '85).

A further interesting aspect of the pattern of substance P–like immunoreactivity in the IP may be noted here. In agreement with the recent report of Rotter and Jacobowitz ('84) the present study has demonstrated a remarkable overlap in the respective distributions of substance P–like immunoreactivity and acetylcholinesterase, not only in the various subnuclei of the IP but also in the habenulo-interpeduncular and interpedunculotegmental tracts (compare Figs. 4, 5, 8; Table 1). Chubb et al. ('80) have reported evidence that acetylcholinesterase hydrolyzes substance P in vitro, that these two substances comigrate in polyacrylamide gels, and that they are topographically closely associated with each other in the dorsal horn of the chick spinal cord. On the basis of these findings Chubb et al. ('80) suggested that acetylcholinesterase may play a role in the biological degradation of substance P. As a consequence of that suggestion, the possibility must be considered that the close colocalization of AChE with substance P–like immunoreactivity in the IP and some of its efferent fiber systems reflects a functional coupling of these two biological agents.

### Enkephalin-like immunoreactivity

Numerous previous studies have demonstrated cell bodies, fibers, and terminals in the interpeduncular nucleus that exhibit an enkephalin-like immunoreactivity (rat: Watson et al., '77; Sar et al., '78; Uhl et al., '79; Wamsley et al., '80; Finley et al., '81; Khachaturian et al., '83; Hemmendinger and Moore, '84; cat: Kapadia and DeLanerolle, '84a). In concurrence with these findings, a high concentration of opiate receptors in the IP has been reported (Pert et al., '76; Atweh and Kuhar, '77; Herkenham and Pert, '80). According to Goodman et al. ('80) the interpeduncular nucleus is especially rich in mu-receptors, associated with the analgesic action of opiates, while delta-receptors are much sparser in the IP.

In most of the reports mentioned above the distribution of enkephalin-like immunoreactivity or opiate receptors within the IP is not specified. According to the present findings, enkephalin-positive cell bodies are almost entirely confined to the IP-R; but some additional cells were found in the IP-A and the caudodorsal part of the IP-C. This localization of enkephalin-positive cell bodies, as well as the findings of an enkephalin-positive fiber plexus in the IP-R, IP-A and IP-DL, is in general agreement with the recent reports of Hemmendinger and Moore ('84) and Hamill et al. ('84). Other subnuclei of the IP show little if any immunoreactivity for leu- or met-enkephalin. The distribution of enkephalin-positive fibers largely coincides with that of opiate receptors as determined by the specific binding of diprenorphine (Pert et al., '76; Atweh and Kuhar, '77) or naloxone (Herkenham and Pert, '80). The last-mentioned authors have emphasized that the opiate-receptor distribution in the habenula, interpeduncular nucleus, and raphe nuclei closely corresponds to the topographic organization of the fiber connections sequentially linking these structures. In view of that correspondence, it would seem conceivable that the enkephalin-positive fibers and terminals in the IP are, at least in part, of extrinsic origin, and most likely that they come from the caudal central gray substance. Most immunohistochemical studies have indicated the presence of enkephalin-positive neurons in the caudal extension of the dorsal raphe nucleus, as well as around the dorsal tegmental nucleus (e.g., Uhl et al., '79). The present findings and other observations (Groenewegen et al., '83c; Groenewegen and VanDijk, '84) indicate that fibers from these regions innervate the IP-R and IP-DL, the subnuclei of the IP containing most of the enkephalin-positive fibers. A combined immunohistochemical-retrograde-tracing study by Hamill and Jacobowitz ('84) supports the notion of an enkephalinergic projection from the dorsal tegmental region to the IP. Conversely, the enkephalin-positive neurons in the IP-R are most likely to project to the caudal central gray substance, where both enkephalin-positive fibers (rat: Uhl et al., '79; cat: Moss et al., '83) and opiate receptors (Pert et al., '76; Simantov et al., '77) are located around the dorsal tegmental nucleus. In confirmation, fluorescent retrograde tracers injected in the dorsal tegmental region have recently have been found to label a population of enkephalin-immunoreactive neurons in the IP-R (Huitinga et al., '85; Groenewegen and Huitinga, '85).

### Functional considerations

On the basis of current data the interpeduncular nucleus (IP) can be viewed first and foremost as a highly differentiated processing station placed in the path of the fasciculus retroflexus, the route from the habenular complex to the median raphe nucleus (MRN) and to the dorsal tegmental region (DTR, encompassing the dorsal raphe nucleus, the tegmental nuclei of Gudden, and the nucleus tegmenti dorsalis lateralis). From a functional point of view it may be significant that the interpeduncular nucleus almost selectively interrupts, apparently with all its subdivisions, that component of the fasciculus retroflexus that originates from the medial habenular nucleus. Since the medial habenular nucleus in turn is innervated very largely by cell groups of the supra- and retrocommissural septum, and appears to receive few if any afferents from its immediate neighbor, the lateral habenular nucleus (cf. Herkenham and Nauta, '77, '79), it seems likely that the habenulo-interpeduncular connection conveys information of predominantly hippocampal antecedents. The lateral habenular nucleus is innervated by a greater diversity of forebrain sites, but its substantial contribution to the fasciculus retroflexus largely (if perhaps not entirely) bypasses the interpeduncular nucleus and is distributed mainly to the median raphe nucleus and dorsal tegmental region (Herkenham and Nauta, '79). It is nonetheless likely that the lateral habenular nucleus can modulate the mechanisms of the interpeduncular nucleus at least indirectly by way of the voluminous (and reciprocated) projection from the MRN and DTR that provides the IP with what appears to be its second-most massive input. Like the habenulo-interpeduncular tract, this projection appears to involve all of the currently recognized subdivisions of the IP.

To what extent (and to what effect) these two main afferent systems interact with each other at the level of the IP appears to be an unexplored question. Thus far, only the integral effect of habenula stimulation on the activity of single IP units appears to have been examined, apparently without systematic attempts to differentiate between afferent conduction directly from the habenula by the habenulo-interpeduncular tract, or indirectly by way of the dorsal tegmental region. In such studies both Lake ('73) and Sastry ('78) found the effect of habenula stimulation upon single IP units to be predominantly excitatory. Both these authors reported, moreover, that most (more than 80% according to Lake) of the IP units recorded from, whether facilitated or inhibited by habenula stimulation, were excited by acetylcholine delivered by microiontophoresis; Sastry found this is true also of substance P, the second neural messenger identified in the habenulo-interpeduncular tract. No similar studies examining the effects of electrical stimulation of the dorsal tegmental region upon IP units have come to our attention; such studies presumably would be complicated by the fact that the projection to the IP from the DTR, in contrast to that from the habenula, is heavily reciprocated.

The largely excitatory effect of habenula stimulation upon IP neurons is remarkable in view of Wang and Aghajanian's ('77) finding that such stimulation has a strongly inhibitory effect on serotonin neurons, at least in the dorsal raphe region. Wang and Aghajanian found this inhibition to be blocked by iontophoretically delivered picrotoxin, and therefore suggested that it is transmitted by GABAergic fibers. Whether the effect in the raphe nuclei is mediated directly by the lateral habenular nucleus as Wang and Aghajanian suggested or by intrinsic GABA neurons in the raphe nuclei proper (McGeer et al., '79) must for the present remain an open question. From the observations of Maciewicz et al. ('81) it appears that in the cat the entire population of raphe neurons is not inhibited by habenula stimulation but rather that a separate group receives an excitatory input from the IP. Whether the cells of origin of this IP projection in turn are influenced by the excitatory input from the habenula is, as yet, unknown. In any case, an active involvement of the interpeduncular nucleus in the mechanisms of inhibition, and facilitation of the serotonin innervation of the forebrain seems a distinct possibility.

It is difficult to correlate these findings at the single-unit level with the organismic aspects of interpeduncular-nucleus function. In view of its placement in the brain's circuitry it seems safe to assume that the IP is involved in some, at least, of the functions attributed to the "dorsal diencephalic conduction system" (Sutherland, '82) descending from the forebrain to the mesencephalon by way of the habenular complex. Not surprisingly, a wide-ranging and in part quite controversial literature recently reviewed by Sutherland ('82) indicated that these functions affect endocrine (in particular thyroid) and autonomic (respiratory, cardivascular, and intestinal) as well as general behavioral mechanisms such as mating behavior, avoidance response, and intracranial self-stimulation. Judged by the great longitudinal spread of its efferents, the IP would seem capable of exerting an effect on these phenomena both at the level of the hippocampus and throughout the extent of the septo-hypothalamo-mesencephalic continuum (Nauta and Domesick, '81) implicated in the subcortical circuitry of the limbic system.

## ACKNOWLEDGMENTS

This study extended over 7 years and was supported over different parts of that period by one or more of the following, gratefully acknowledged grants: a stipend from the Netherlands Organization for the Advancement of Pure Research (ZWO) to H.J.G.; PHS grant 5R23NS20467 to S.N.H.; Public Health Service grants PO1MH31154 and RO1NS19945 to W.J.H.N.; and National Science Foundation grants BNS7681227, BNS8007905, and BNS8306284 to W.J.H.N. It is a pleasure to acknowledge the excellent assistance received from Miss Diane Major, Mrs. J. Voerman, Mrs. R.E. Clark, and Mrs. Rita Stoevelaar-de Zeeuw, as well as from Mr. S.H. Speelman and Mr. Dirk de Jong, who made the photographs illustrating this paper. Drs. Anthony Lohman and Fokje Russchen are thanked for critically reading the manuscript.

## LITERATURE CITED

Ahlenius, S., and W.J.H. Nauta (1980) Afferent and efferent connections of the interpeduncular nucleus in the rat as revealed by neuroanatomical tracer techniques. Neurosci. Lett. Suppl. 5:188 (Abstract).

Armstrong, D.M., C.B. Saper, A.I. Levey, B.H. Wainer, and R.D. Ferry (1983) Distribution of cholinergic neurons in rat brain: Demonstrated by the immunohistological localization of choline acetyltransferase. J. Comp. Neurol. 216:53–68.

Artymyshyn, R., and M. Murray (1985) Substance P in the interpeduncular nucleus of the rat: Normal distribution and the effects of deafferentiation. J. Comp. Neurol. 231:78–90.

Atweh, S.F., and M.J. Kuhar (1977) Autoradiographic localization of opiate receptors in rat brain. II. The brain stem. Brain Res. 129:1–12.

Azmitia, E.C., and M. Segal (1978) An autoradiographic analysis of the differential ascending projections of the dorsal and median raphe nuclei in the rat. J. Comp. Neurol. 179:641–667.

Baisden, R.H., D.B. Hoover, and R.J. Cowie (1979) Retrograde demonstration of hippocampal afferents from the interpeduncular and reuniens nuclei. Neurosci. Lett. 13:105–109.

Ban, T., and K. Zyo (1963) Experimental studies on the mammillary peduncle and mammillotegmental tracts in the rabbit. Med. J. Osaka Univ. 13:241–270.

Beckstead, R.M., V.B. Domesick, and W.J.H. Nauta (1979) Efferent connections of the substantia nigra and ventral tegmental area in the rat. Brain Res. 175:191–217.

Berman, A.L. (1968) The Brain Stem of the Cat. Madison: University of Wisconsin Press.

Bobillier, P., F. Petitjean, D. Salvert, M. Ligier, and S. Seguin (1975) Differential projections of the nucleus raphe dorsalis and nucleus raphe centralis as revealed by autoradiography. Brain Res. 85:205–210.

Briggs, T.L., and W.W. Kaelber (1971) Efferent fiber connections of the dorsal and deep tegmental nuclei of Gudden. An experimental study in the cat. Brain Res. 29:17–29.

Butcher, L.L., and N.J. Woolf (1984) Histochemical distribution of acetylcholinesterase in the central nervous system: Clues to the localization of cholinergic neurons. In: A. Björklund, F. Hökfelt, and M.J. Kuhar (eds): Handbook of Chemical Neuroanatomy. Amsterdam: Elsevier, Vol. 3, Part II, pp. 1–50.

Cajal, S. Ramón y (1911) Histologie du Système Nerveux de l'Homme et des Vertébrés. Maloine, Paris, Vol. II.

Cheney, D.L., H.F. Lefevre, and G. Racagni (1975) Choline acetyltransferase activity and mass fragmentography measurement of acetylcholine in specific nuclei and tracts of rat brain. Neuropharmacology 14:801–809.

Chubb, I.D., A.J. Hodgson, and G.H. White (1980) Acetylcholinesterase hydrolyzes substance P. Neuroscience 5:2065–2072.

Chubb, I.W., E. Ranieri, G.H. White, and A.J. Hodgson (1983) The enkephalins are amongst the peptides hydrolyzed by purified acetylcholinesterase. Neuroscience 10:1369–1377.

Clarke, P.B.S., C.B. Pert, and A. Pert (1984) Autoradiographic distribution of nicotine receptors in rat brain. Brain Res. 323:390–395.

Conrad, L.C.A., C.M. Leonard, and D. Pfaff (1974) Connections of the median and dorsal raphe nuclei in the rat: An autoradiographic and degeneration study. J. Comp. Neurol. 156:179–205.

Contestabile, A., and B. Flumerfelt (1981) Afferent connections of the interpeduncular nucleus and the topographic organization of the habenulo-interpeduncular pathway: An HRP study in the rat. J. Comp. Neurol. 196:253–270.

Craig, A.D., and H. Burton (1981) Spinal and medullary lamina projection to nucleus submedius in medial thalamus: A possible pain centre. J. Neurophysiol. 45:443–466.

Craig, A.D., S.J. Wiegand, and J.L. Price (1982) The thalamo-cortical projection of the nucleus submedius in the cat. J. Comp. Neurol. 206:28–48.

Cuello, A.C., P.C. Emson, G. Paxinos, and T. Jessell (1978) Substance P containing and cholinergic projections from the habenula. Brain Res. 149:413–429.

Cuello, A.C., and I. Kanazawa (1978) The distribution of substance P immunoreactive fibers in the rat central nervous system. J. Comp. Neurol. *178:*129–156.

Dahlström, A., and K. Fuxe (1964) Evidence for the existence of monoamine-containing neurons in the central nervous system—I Demonstration of monoamines in the cell bodies of brain stem neurons. Acta Physiol. Scand. *62,* Suppl. *232:*1–55.

Domesick, V.B. (1976) Projections of the nucleus of the diagonal band of Broca in the rat. Anat. Rec. *184:*391–392 (Abstract).

Emson, P.C., A.C. Cuello, G. Paxinos, T. Jessel, and L.L. Iversen (1977) The origin of substance P and acetylcholine projection to the ventral tegmental area and interpeduncular nucleus in the rat. Acta Physiol. Scand. Suppl. *452:*43–46.

Fibiger, H.C. (1982) The organization and some projections of cholinergic neurons of the mammalian brain. Brain Res. Rev. *4:*327–388.

Finley, J.C.W., J.L. Maderdrut, and P. Petrusz (1981) The immunocytochemical localization of enkephalin in the central nervous system of the rat. J. Comp. Neurol. *198:*541–565.

Flumerfelt, B.A., and A. Contestabile (1982) Acetylcholinesterase histochemistry of the habenulo-interpeduncular pathway in the rat and the effects of electrolytic and kainic acid lesions. Anat. Embryol. *163:*435–446.

Ganser, S. (1882) Vergleichend-anatomische Studien über das Gehirn des Maulwurfs. Morphol. Jahrb. *7:*591–725.

Geneser-Jensen, F.A., and T.W. Blackstad (1971) Distribution of acetylcholinesterase in the hippocampal region of the guinea pig. I Enthorhinal area, parasubiculum and presubiculum. Z. Zellforsch. Mikrosk. Anat. *114:*460–481.

Gerfen, C.R., and P.E. Sawchenko (1982) Immunohistochemical localization of axonally transported PHA-L lectin to demonstrate the fine morphological details of efferent connections in the CNS. Soc. Neurosci. Abstr. *8:*786.

Gerfen, C.R., and P.E. Sawchenko (1984) An anterograde neuroanatomical tracing method that shows the detailed morphology of neurons, their axons and terminals: Immunohistochemical localization of an axonally transported plant lectin, Phaseolus vulgaris-leucoagglutinin. Brain Res. *290:*219–238.

Goodman, R.R., S.H. Snyder, M.J. Kuhar, and W.S. Young (1980) Differentiation of delta and mu opiate receptor localizations by light microscopic autoradiography. Proc. Natl. Acad. Sci. U.S.A. *77:*6239–6243.

Gottesfeld, Z., and D.M. Jacobowitz (1978) Cholinergic projection of the diagonal band to the interpeduncular nucleus of the rat brain. Brain Res. *156:*329–332.

Graham, R.C., and M.J. Karnovsky (1966) The early stages of absorption of injected horseradish peroxidase in the proximal tubulus of mouse kidney: Ultrastructural cytochemistry by a new technique. J. Histochem. Cytochem. *14:*291–302.

Groenewegen, H.J., and W.J.H. Nauta (1982) Afferent and efferent connections of the mediodorsal thalamic nucleus in the rat. Neuosci. Lett. Suppl. *10:*217–218 (Abstract).

Groenewegen, H.J., and H.W.M. Steinbusch (1984) Serotonergic and nonserotonergic projections from the interpeduncular nucleus to the ventral hippocampus in the rat. Neurosci. Lett. *51:*19–24.

Groenewegen, H.J., and C.A. Van Dijk (1984) Efferent connections of the dorsal tegmental region in the rat, studied by means of anterograde transport of the lectin Phaseolus vulgaris-leucoagglutinin (PHA-L). Brain Res. *304:*367–371.

Groenewegen, H.J., and I. Huitinga (1985) Peptidergic projections from the interpeduncular nucleus to the dorsal tegmental region in the rat. Neurosci. Lett. Suppl. *22:*111 (Abstract).

Groenewegen, H.J., and W.J.H. Nauta (1986) Organization of the afferent connections of the mediodorsal thalamic nucleus in the rat, related to the mediodorsal-prefrontal topography. Neuroscience (Submitted).

Groenewegen, H.J., S.N. Haber, and W.J.H. Nauta (1983a) Structure and efferent connections of the interpeduncular nucleus in the rat. An immunohistochemical and neuroanatomical tracer study. Neurosci. Lett. Suppl. *14:*145 (Abstract).

Groenewegen, H.J., S.N. Haber, and W.J.H. Nauta (1983b) Organization and efferent connections of the interpeduncular nucleus in the rat. J. Anat. *137:*398–399 (Abstract).

Groenewegen, H.J., N.W. Kowall, and W.J.H. Nauta (1983c) Efferent connections of the dorsal tegmental region in the rat. Soc. f. Neurosci. Abstr. *9:*516.

Haber, S., and R. Elde (1982) The distribution of enkephalin immunoreac-

tive fibers and terminals in the monkey central nervous system: An immunohistochemical study. Neuroscience *7:*1049–1095.

Hamill, G.S., and B. Fass (1984) Differential distribution of diagonal band afferents to subnuclei of the interpeduncular nucleus in rats. Neurosci. Lett. *48:*43–48.

Hamill, G.S., and D.M. Jacobowitz (1984) A study of afferent projections to the rat interpeduncular nucleus. Brain Res. Bull. *13:*527–539.

Hamill, G.S., and N.J. Lenn (1984) The subnuclear organization of the rat interpeduncular nucleus: A light and electron microscopic study. J. Comp. Neurol. *222:*396–408.

Hamill, G.S., and N.J. Lenn (1981) A morphological and morphometric analysis of the subnuclear structure of the interpeduncular nucleus in the rat. Soc. f. Neurosci. Abstr. *7:*83.

Hamill, G.S., and N.J. Lenn (1984) The subnuclear organization of the rat interpeduncular nucleus: A light and electron microscopic study. J. Comp. Neurol. *222:*396–408.

Hamill, G.S., J.A. Olschowska, N.J. Lenn, and D.M. Jacobowitz (1984) The subnuclear distribution of substance P, cholecytokinin, vasoactive intestinal peptide, somatostatin, leu-enkephalin, dopamine-B-hydroxylase and serotonin in the rat interpeduncular nucleus. J. Comp. Neurol. *226:*580–596.

Hattori, T., E.G. McGeer, V.K. Singh, and P.L. McGeer (1977) Cholinergic synapse of the interpeduncular nucleus. Exp. Neurol. *55:*666–679.

Hayakawa, T., and K. Zyo (1982) Organization of the habenulo-interpeduncular connections in cats: A horseradish peroxidase study. Brain Res. *240:*3–11.

Hemmendinger, L.M., and R.Y. Moore (1984) Interpeduncular nucleus organization in the rat: Cytoarchitecture and histochemical analysis. Brain Res. Bull. *13:*163–179.

Hemmendinger, L.M., and R.Y. Moore (1982) The interpeduncular nucleus in the rat: Cytoarchitecture and cytochemistry. Soc. f. Neurosci. Abstr. *8:*665.

Herkenham, M., and W.J.H. Nauta (1977) Afferent connections of the habenular nuclei in the rat. A horseradish peroxidase study, with a note on the fiber-of-passage problem. J. Comp. Neurol. *173:*123–146.

Herkenham, M., and W.J.H. Nauta (1979) Efferent connections of the habenular nuclei in the rat. J. Comp. Neurol. *187:*19–48.

Herkenham, M., and C.B. Pert (1980) In vitro autoradiography of opiate receptors in rat brain suggests loci of "opiatergic" pathways. Proc. Natl. Acad. Sci. U.S.A. *77:*5532–5536.

Hess, D.T., and G. Schneider (1981) Advantages of polarization microscopy in horseradish peroxidase neurochemistry. J. Histochem. Cytochem. *29:*1448–1450.

Ho, R.H., and L.R. De Palatis (1980) Substance P immunoreactivity in the median eminence of the North American opossum and domestic fowl. Brain Res. *189:*565–569.

Hong, J.S., E. Costa, and H.-Y.T. Yang (1976) Effects of habenular lesions on the substance P content of various brain regions. Brain Res. *118:*523–525.

Houser, C.R., G.D. Crawford, R.P. Barber, P.M. Salvaterra, and J.E. Vaughn (1983) Organization and morphological characteristics of cholinergic neurons: An immunohistochemical study with a monoclonal antibody to choline acetyltransferase. Brain Res. *226:*97–119.

Huitinga, I., C.A. van Dijk, and H.J. Groenewegen (1985) Substance P- and enkephalin-containing projections from the interpeduncular nucleus to the dorsal tegmental region in the rat. Neurosci. Lett. *162:*311–316.

Illing, R.B., and H. Wässle (1979) Visualization of HRP reaction product using the polarization microscope. Neurosci. Lett. *13:*7–11.

Ives, W.R. (1971) The interpeduncular complex of selected rodents. J. Comp. Neurol. *141:*77–94.

Kapadia, S.E., and N.C. De Lanerolle (1984a) Population of substance P, met-enkephalin and serotonin-immunoreactive neurons in the interpeduncular nucleus of cat. Brain Res. *302:*33–43.

Kapadia, S.E., and N.C. De Lanerolle (1984b) Substance P neuronal organization in the median region of the interpeduncular nucleus of the cat: An electron microscopic analysis. Neuroscience *12:*1229–1242.

Kataoka, K., Y. Nakamura, and R. Hassler (1973) Habenulo-interpeduncular tract: A possible cholinergic neuron in the rat brain. Brain Res. *62:*264–267.

Khachaturian, H., M.E. Lewis, and S.J. Watson (1983) Enkephalin systems in diencephalon and brain stem of the rat. J. Comp. Neurol. *220:*310–320.

Kimura, H., P.L. McGeer, J.H. Peng, and E.G. McGeer (1980) The central cholinergic system studied by choline acetyltransferase immunohistochemistry in the cat. J. Comp. Neurol. 200:151–201.

Kimura, H., P.L. McGeer, and J.H. Peng (1984) Choline acetyltransferase containing neurons in the rat brain. In: A. Björklund, T. Hökfelt, and M.J. Kuhar (eds): Handbook of Chemical Neuroanatomy. Amsterdam: Elsevier, Vol. 3, Part II, pp. 51–67.

Köhler, C., V. Chan-Palay, and H. Steinbusch (1982) The distribution and origin of serotonin-containing fibers in the septal area: A combined immunohistochemical and fluorescent tracing study in the rat. J. Comp. Neurol. 209:91–111.

Kuhar, M., R.N. DeHaven, H.I. Yamamura, H. Rommelspacher, and V.R. Simon (1975) Further evidence for cholinergic habenulo-interpeduncular neurons: Pharmacologic and functional characteristics. Brain Res. 97:265–275.

Lake, N. (1973) Studies of the habenulo-interpeduncular pathway in cats. Exp. Neurol. 41:113–132.

Lehman, J., and H.C. Fibiger (1979) Acetylcholinesterase and the cholinergic neuron. Life Sci. 25:1939–1947.

Lenn, N.J. (1976) Synapses in the interpeduncular nucleus: Electron microscopy of normal and habenula lesioned rats. J. Comp. Neurol. 166:73–100.

Lenn, N.J., and G.S. Hamill (1984) Subdivisions of the interpeduncular nucleus: A proposed nomenclature. Brain Res. Bull. 13:203–204.

Lenn, N.J., V. Wong, and G.S. Hamill (1983) Left-right pairing at the crest synapses of rat interpeduncular nucleus. Neuroscience 9:383–389.

Leranth, C.S., M. Browstein, L. Zaborsky, Z.S. Jaranyi, and M. Palkovits (1975) Morphological and biochemical changes in the rat interpeduncular nucleus following the transection of the habenulo-interpeduncular tract. Brain Res. 99:124–128.

Lewis, P.R., and C.C.D. Shute (1967) The cholinergic limbic system: Projections to the hippocampal formation, medial cortex, nuclei of the ascending cholinergic reticular system, and the subfornical organ and supraoptic crest. Brain 90:521–542.

Liu, R., L. Chang and G. Wickern (1984) The dorsal tegmental nucleus: An axoplasmic transport study. Brain Res. 310:123–132.

Ljungdahl, Å., T. Hökfelt, and G. Nilsson (1978) Distribution of substance P-like immunoreactivity in the central nervous system of the rat. I. Cell bodies and nerve terminals. Neuroscience 3:861–943.

Maciewicz, R., W.E. Foote, and J. Bry (1981) Excitatory projection from the interpeduncular nucleus to central superior raphe neurons. Brain Res. 225:179–183.

Maley, B., and R. Elde (1982) Immunohistochemical localization of putative neurotransmitters within the feline nucleus tractus solitarius. Neuroscience 7:2469–2490.

Marchand, E.R., J.N. Riley, and R.Y. Moore (1980) Interpeduncular nucleus afferents in the rat. Brain Res. 193:339–352.

Massopust, L.C., and R. Thompson (1962) A new interpedunculo-diencephalic pathway in rats and cats. J. Comp. Neurol. 118:97–105.

McGeer, E.G., U. Scherer-Singer, and E.A. Singh (1979) Confirmatory data on habenular projections. Brain Res. 168:375–376.

Merker, B.H. (1983) Silver staining of cell bodies by means of physical development. J. Neurosci. Methods 9:235–241.

Mesulam, M.M. (1976) The blue reaction product in horseradish peroxidase neurochemistry: Incubation parameters and visibility. J. Histochem. Cytochem. 24:1273–1280.

Mesulam, M.M. (1978) Tetramethyl benzidine for horseradish peroxidase neurochemistry: A non-carcinogenic blue reaction product with superior sensitivity for visualizing neural afferents and efferents. J. Histochem. Cytochem. 26:106–117.

Mesulam, M.M., E.J. Mufson, B.H. Wainer, and A.I. Levey (1983) Central cholinergic pathways in the rat: An overview based on an alternative nomenclature (Ch1–Ch6). Neuroscience 10:1185–1201.

Miller, R.J., K.J. Chang, B. Cooper, and P. Cuatrecasas (1978) Radioimmunoassay and characterization of enkephalins in rat tissues. J. Biol. Chem. 253:531–538.

Miller, R.J., and V.M. Pickel (1980) The distribution and functions of the enkephalins. J. Histochem. Cytochem. 28:903–917.

Mitchell, R. (1963) Connections of the habenula and the interpeduncular nucleus in the cat. J. Comp. Neurol. 121:441–457.

Morest, D.K. (1961) Connexions of the dorsal tegmental nucleus in rat and rabbit. J. Anat. 95:229–246.

Moss, M.S., and A.I. Basbaum (1983) The peptidergic organization of the cat periaqueductal grey. II. The distribution of immunoreactive substance P and vasoactive intestinal polypeptide. J. Neurosci. 3:1437–1449.

Moss, M.S., E.J. Glazer, and A.I. Basbaum (1983) The peptidergic organization of the cat periaqueductal grey. I. The distribution of immunoreactive enkephalin-containing neurons and terminals. J. Neurosci. 3:603–616.

Mroz, E.A., M.J. Brownstein, and S.E. Leeman (1976) Evidence for substance P in the habenulo-interpeduncular tract. Brain Res. 113:597–599.

Murray, M., J. Zimmer, and G. Raisman (1979) Quantitative electron microscopic evidence for reinnervation in the adult rat interpeduncular nucleus after lesions of the fasciculus retroflexus. J. Comp. Neurol. 187:447–468.

Nauta, W.J.H. (1958) Hippocampal projections and related neural pathways to the midbrain in the cat. Brain 81:319–341.

Nauta, W.J.H., and H.G.J.M. Kuypers (1958) Some ascending pathways in the brain stem reticular formation. In H.H. Jasper, L.D. Proctor, R.S. Knighton, W.C. Noshay, and R.T. Costello (eds): Reticular Formation of the Brain. London: Churchill Ltd. pp. 3–30.

Nauta, W.J.H., and V.B. Domesick (1981) Ramifications of the limbic system. In S. Matthysse (ed): Psychiatry and the Biology of the Human Brain: A Symposium Dedicated to Seymour S. Kety. Amsterdam, Elsevier, pp. 165–188.

Palkovits, M., and D.M. Jacobowitz (1974) Topographic atlas of catecholamine and acetylcholinesterase-containing neurons in the rat brain. II. Hindbrain, mesencephalon, rhombencephalon. J. Comp. Neurol. 157:29–41.

Palkovits, M., J.M. Saavedra, R.M. Kobayashi, and M. Brownstein (1974) Choline acetyltransferase content of limbic nuclei of the rat. Brain Res. 79:443–450.

Pert, C.B., M.J. Kuhar, and S.H. Snyder (1976) Opiate receptor: Autoradiographic localization in rat brain. Proc. Natl. Acad. Sci. U.S.A. 73:3729–3733.

Price, J.L., and B.M. Slotnick (1983) Dual olfactory representation in the rat thalamus: An anatomical and electrophysiological study. J. Comp. Neurol. 215:63–77.

Riley, J.N., and R.Y. Moore (1981) Diencephalic and brainstem afferents to the hippocampal formation of the rat. Brain Res. Bull. 6:437–444.

Rotter, A., and D.M. Jacobowitz (1984) Localization of substance P, acetylcholinesterase and muscarinic receptors and alpha-bungarotoxin binding sites in the rat interpeduncular nucleus. Brian Res. Bull. 12:83–94.

Sakanaka, M., S. Shiosaka, K. Takatsuki, S. Inagaki, H. Takagi, E. Senba, Y. Kawai, Y. Hara, H. Iida, M. Minagawa, T. Matsuzaki, and M. Tohyama (1981) Evidence for the existence of a substance P-containing pathway from the nucleus laterodorsalis tegmenti (Castaldi) to the lateral septal area in the rat. Brain Res. 230:351–355.

Sakanaka, M., S. Inagaki, E. Senba, H. Takagi, K. Takatsuki, Y. Kawai, H. Iida, Y. Hara, and M. Tohyama (1982) Ontogeny of substance P-containing neuron system of the rat: Immunohistochemical analysis. II. Lower brain stem. Neuroscience 7:1097–1126.

Sar, M., W.E. Stumpf, R.J. Miller, K.J. Chang, and P. Cuatracasas (1978) Immunohistochemical localization of enkephalin in rat brain and spinal cord. J. Comp. Neurol. 182:17–38.

Sastry, B.R. (1978) Effects of substance P, acetylcholine and stimulation of habenula on rat interpeduncular neuronal activity. Brain Res. 144:404–410.

Sastry, B.R., S.E. Zialkowski, L.M. Hansen, J.P. Kavanagh, and E.M. Envoy (1979) Acetylcholine release in interpeduncular nucleus following stimulation of habenula. Brain Res. 164:334–337.

Shibata, H., and T. Suzuki (1984) Efferent projections of the interpeduncular complex in the rat, with special reference to its subnuclei: A retrograde horseradish peroxidase study. Brain Res. 296:345–349.

Shinoda, K., S. Inagaki, S. Shiosaka, J. Kohuo, and M. Tohyama (1984) Experimental immunohistochemical studies on the substance P neuron system in the lateral habenular nucleus of the rat: Distribution and origins. J. Comp. Neurol. 222:578–588.

Simantov, R., M.J. Kuhar, G.R. Uhl, and S.H. Snyder (1977) Opioid peptide enkephalin: Immunohistochemical mapping in rat central nervous system. Proc. Natl. Acad. Sci. U.S.A. 74:2167–2171.

Smaha, L.A., and W.W. Kaelber (1973) Efferent fiber projections of the habenula and the interpeduncular nucleus. An experimental study in the opossum and cat. Exp. Brain Res. 16:291–308.

Sofroniew, M.V., F. Eckenstein, and A.C. Cuello (1983) Immunohistochemistry of cholinergic neurons in the rat brain and spinal cord. Neurosci. Lett. Suppl. 14:348 (Abstract).

Sofroniew, M.V., F. Eckenstein, H. Thoenen, and A.C. Cuello (1982) Topography of choline acetyltransferase-containing neurons in the forebrain of the rat. Neurosci. Lett. 33:7–12.

Sorimachi, M., and K. Kataoka (1974) Choline uptake by nerve terminals: A sensitive and a specific marker of cholinergic innervation. Brain Res. 72:350–353.

Steinbusch, H.W.M. (1981) Distribution of serotonin-immunoreactivity in the central nervous system of the rat—Cell bodies and terminals. Neuroscience 6:557–618.

Steinbusch, H.W.M. (1984) Serotonin-immunoreactive neurons and their projections in the CNS. In A. Björklund, T. Hökfelt, and M.J. Kuhar (eds): Handbook of Chemical Neuroanatomy. Amsterdam: Elsevier, Vol. 3, Part II, pp. 68–125.

Sternberger, L.A. (1979) Immunocytochemistry. New York: Wiley.

Stofer, D.W., and S.B. Edwards (1978) Organization and efferent projection of the interpeduncular complex in the cat. Soc. f. Neurosci. Abstr. 4:228.

Sutherland, R.J. (1982) The dorsal diencephalic conduction system: A review of the anatomy and functions of the habenular complex. Neurosci. Biobehav. Rev. 6:1–13.

Uhl, G.R., R.R. Goodman, M.J. Kuhar, S.R. Childers, and S.H. Snyder (1979) Immunohistochemical mapping of enkephalin containing cell bodies, fibers and nerve terminals in the brain stem of the rat. Brain Res. 166:75–94.

Villani, L., A. Contestabile, and F. Fonnum (1983) Autoradiographic labeling of the cholinergic habenulo-interpeduncular projection. Neurosci.

Lett. 42:261–267.

Velayos, J.L., and F. Reinoso-Suarez (1982) Topographic organization of the brain stem afferents to the mediodorsal thalamic nucleus. J. Comp. Neurol. 206:17–27.

Wainer, B.H., A.I. Levey, E.J. Mufson, and M.-M. Mesulam (1984) Cholinergic systems in mammalian brain identified with antibodies against choline acetyltransferase. Rev. Neurochem. Int. 6:163–182.

Wamsley, J.K., W.S. Yong, and M.J. Kuhar (1980) Immunohistochemical localization of enkephalin in rat forebrain. Brain Res. 190:153–174.

Wang, R.Y., and G.K. Aghajanian (1977) Physiological evidence for habenula as major link between forebrain and midbrain. Science 197:89–91.

Watson, S.J., H. Akil, S. Sullivan, and J.D. Barchas (1977) Immunohistochemical localization of methionine enkephalin: Preliminary observations. Life Sci. 21:733–738.

Wilson, P.M., and C. Watson (1980) Acetylcholinesterase staining in an interpedunculotegmental pathway in four species: Procavia capensis (dassie), Cavia procellus (guinea pig), Trichosurus vulpecula (brushtail possum) and Rattus rattus (hooded rat). Brain Res. 201:418–422.

Wyss, J.M., L.W. Swanson, and W.M. Cowan (1979) A study of subcortical afferents to the hippocampal formation in the rat. Neuroscience 4:463–476.

# IV

# BASAL GANGLIA

Reprinted from
*Brain Research*
Elsevier Publishing Company, Amsterdam
Printed in the Netherlands

## Review Article

# PROJECTIONS OF THE LENTIFORM NUCLEUS IN THE MONKEY

WALLE J. H. NAUTA* AND WILLIAM R. MEHLER**

*Department of Neurophysiology, Walter Reed Army Institute of Research, Washington, D.C. (U.S.A.)*

(Received August 19th, 1965)

The fiber projections of the corpus striatum have been the subject of intensive study and controversy for almost a century. This is particularly true of the major efferent connections represented by the ansa lenticularis. To early neurological explorers, this striking fiber system, sweeping across main subdivisions of the diencephalon in tortuous and recurrent trajectories, presented anatomical complexities that confounded even such distinguished investigators as Meynert, von Monakow, and von Bechterew. Although these early explorers set the stage for all subsequent studies by providing a definition and nomenclature of some of the major fiber systems related to the corpus striatum, their observations in normal human brains could furnish no reliable information concerning the origin of the ansa lenticularis, and only gross and conflicting evidence concerning its distribution. To this extent it is possible to agree with the Ransons' indictment[62] of the literature on the corpus striatum as '. . . . voluminous and full of contradictions and misinformation'.

A more experimentally oriented period in the analysis of striatal and pallidal connections was ushered in by Wilson's[81] classic Marchi experiments in the monkey and by the neuropathological studies of Cécile and Oscar Vogt[73,74]. Wilson's study forms the prototype of numerous later experimental investigations, including the present one, and set an example of careful surgical planning by the use of a lateral stereotaxic approach to the lentiform nucleus, designed to avoid involvement of the internal capsule. If it were fair in hindsight to criticize Wilson's study, it could be said that precisely this fear of being misled by corticofugal fiber degeneration led Wilson to reject his significant evidence of the massive pallidothalamic connection identified later by other workers, and to consider the red nucleus to be the principal terminus of

---

*Present address: Department of Psychology, Massachusetts Institute of Technology, Cambridge, Mass. (U.S.A.).
**Present address: Neurobiology Branch, Ames Research Center, Mountain View, Calif. (U.S.A.).

the pallidofugal projection. Nonetheless, Wilson's work furnished the first experimental evidence against a striatal, and in favor of a pallidal origin of the ansa lenticularis. It also verified the existence of a substantial ansal component distributed to the subthalamic nucleus.

The three decades following Wilson's paper produced numerous further reports on the fiber connections of the basal ganglia. From this period date the main attempts to determine the descending pathways linking the corpus striatum with mesencephalic and rhombencephalic levels of the brain stem tegmentum. The numerous reports on this subject, to be reviewed in more detail below, have led to conflicting conclusions janging between, on the one hand, a complete absence of direct lenticulofugal prorections below rostral mesencephalic levels, and, on the other hand, the existence of long pathways extending well into the hindbrain and including direct connections with cranial motor nuclei. Late in the same period, increasingly convincing evidence of a massive projection of the corpus striatum to the rostral part of the ventral thalamic nucleus accumulated. In view of the fact that Cécile Vogt already in 1909 had identified the continuity between the lenticular and thalamic fasciculi, it is remarkable that the connection was not widely recognized until the late nineteen-thirties.

Of the later studies fashioned after Wilson's experimental model, those of the Ransons[62,63] stand out by a judicious interpretation of Marchi and retrograde degeneration experiments. Most recent authors, despite Wilson's observations, had persisted in von Monakow's[39] notion of a mixed striatal and pallidal origin of the ansa lenticularis. The Ransons, however, confirmed Wilson's suggestion that the ansa contains no striatal efferents and introduced evidence indicating the large cells composing the internal pallidal segment as the source of pallidothalamic fibers, and the smaller-celled[12] external pallidal segment as the origin of the ansal component to the subthalamic nucleus.

The observations made in the course of the present study confirm the conclusions of the Ransons concerning the origin of the ansa lenticularis. Several data hitherto not available, however, can now be added to the earlier reports. By the use of experimental silver techniques it has been possible to determine the caudal extent and distribution of the pallidofugal projection in somewhat greater detail than appears to have been attainable by the Marchi method. Furthermore, the present study has yielded evidence suggesting that the pallidothalamic projection involves not only the ventral thalamic nucleus, but also the centre médian. These and other findings to be reported below were considered significant enough to justify publication.

MATERIAL AND METHODS

This report is based on a series of 13 monkeys (*Macaca mulatta*) in which unilateral stereotaxic lesions were placed in different subdivisions of the lentiform nucleus. In an attempt to avoid damage to the internal capsule, and following the example of Wilson[81] and Glees[15], all lesions were placed by the aid of an electrode introduced from the lateral side and, if possible, through the Sylvian fissure. This precaution naturally does not lessen the risk of ischaemic necrosis due to the occlusion of blood vessels near the electrode tip. The observations in the two cases in which such incidental lesions were found to have involved the internal capsule (cases MGB 10 and 11, see Table I) have been accepted only in so far as they corroborated findings made in comparable cases showing no capsular damage.

The animals were sacrificed by an overdose of Nembutal following a survival period of 9–15 days. The perfused and formalin-fixed brains were sectioned on the freezing microtome, two in the sagittal plane, one in the horizontal, and the remaining 11 in the frontal plane. A minimum of 4 sections per millimeter were impregnated for degenerating nerve fibers following either the Nauta–Gygax[48] or the Albrecht–Fernstrom[1] technique, and immediately neighboring sections were stained for cell bodies with cresylechtviolet.

The observed patterns of fiber degeneration in each case were recorded in projected drawings of selected silver sections, using recognizable blood vessels and other incidental landmarks as points of orientation. Relevant cytoarchitectonic boundaries were ascertained by projecting matching Nissl sections over the drawings thus obtained.

In the following account the degeneration patterns observed in 4 cases are illustrated in some detail. The findings in these cases are summarized in Table I, which also lists the observations in all remaining cases. As an extensive verbal description of the illustrated material seemed superfluous, the main part of this paper is presented in the form of a discussion of the present findings in the context of a historical survey.

### OBSERVATIONS AND DISCUSSION

*Efferent connections of the striatum.* There appears to be unanimous agreement with Grünstein's[16] experimental verification in the cat and rabbit of the fine fibered putaminal projection to the globus pallidus described in normal material by von Bechterew (ref. 3, p. 550). A similar projection arising in the caudate nucleus had been demonstrated earlier in the cat by Probst[61]; the existence of this connection likewise has been confirmed repeatedly by later workers.

Evidence of striatal efferents traversing the globus pallidus and cerebral peduncle and terminating in the substantia nigra appears to have been reported first by Edinger[8]. Striatonigral connections were later traced by Riese[64,65] in normal brains of various cetaceans, as well as in human brains with large porencephalic defects. These findings were confirmed by Rundles and Papez's[67] observations in a human brain. More recently, Szábo's[69] experimental studies demonstrated topographically organized caudatonigral and putaminonigral connections in the monkey. In the cat, a projection from the caudate nucleus to the rostromedial part of the substantia nigra was demonstrated by Voneida[76]. In the same species, however, Johnson and Clemente[26] had found no evidence of a comparable projection originating from the putamen. Contrary to their report, one of us (W.R.M., unpublished) has traced fiber degeneration, in the cat, from putaminal lesions to the caudolateral part of the substantia nigra. The latter observation, considered together with Voneida's report, offers strong indication that topographically organized striatonigral projections exist in the cat as well as in the monkey.

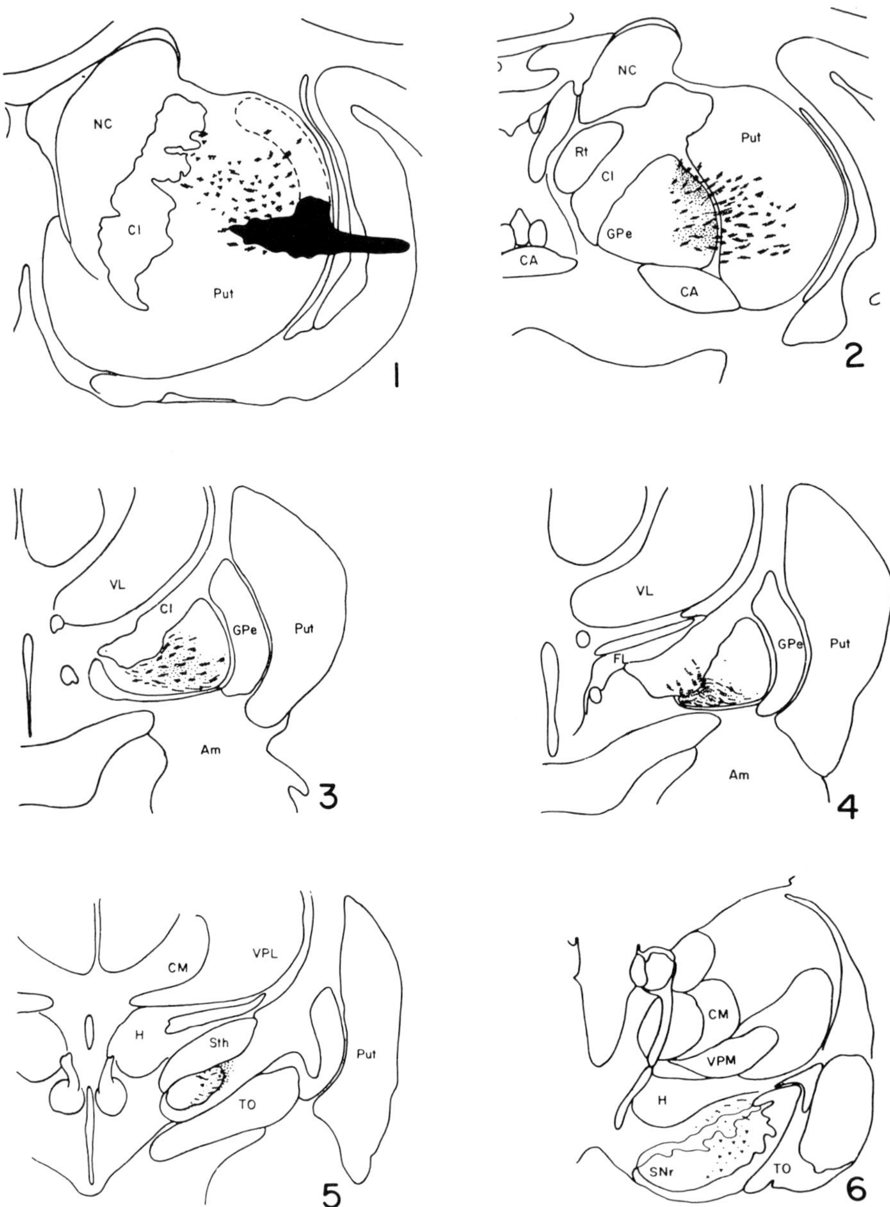

Figs. 1–6. Fiber degeneration observed following lesion of the putamen (case MGB 2). The electrolytic defect is shown in jet-black in Fig. 1; indicated in broken lines is an ischemic extension which nowhere exceeded the limits of the putamen. The *coarser dots* indicate degenerating *fibers of passage*; evidence of degenerating *terminal* axon ramifications is recorded in *finer stipple*. Note the radial disposition of the degenerating putaminofugal fascicles (Wilson's "pencils") in the globus pallidus (Figs. 2 and 3), the continuation of some of these fascicles through the cerebral peduncle (Fig. 4) and their distribution in the substantia nigra (Figs. 5 and 6). Abbreviations: see page 38.

No convincing evidence of striatofugal fibers to other brain stem structures appears to exist. Wallenberg[77], from a re-examination of Edinger and Fischer's[9] brain of an acerebral human infant, reported evidence of a putaminofugal fiber system descending in the central tegmental tract as far caudally as the inferior olive. Muskens[44] on the basis of apparently inadequately controlled experimental observations in the cat, concurred with Wallenberg's conclusions, but neither Riese's[65] studies nor Weisschedel's[78] painstaking analysis of the central tegmental tract, involving the study of several further porencephalic human brains, produced evidence to support the notion of a putaminal contribution to this heterogenous fiber system.

A further putamino-mesencephalic connection was suggested by Verhaart's[70] interpretation of Winkler's[82] (see below) bundle 'X' as a putaminofugal projection to the lateral pontomesencephalic reticular formation. Furthermore, the suggestion of putaminal projections to a variety of structures in the fore- and midbrain is implicit in those concepts according to which the putamen contributes fibers to the ansa lenticularis[12,29,31,39,83].

*The present findings.* The results of our own experiments (see Figs. 1–6) have confirmed only the existence of fine fibered putaminopallidal (Figs. 1–4) and putaminonigral (Figs. 5–6) connections. In agreement with the descriptions of von Bechterew[3], Wilson[81], C. and O. Vogt[74], Ranson and Ranson[62], and Mettler[37], no putaminal efferents could be identified in the ansa lenticularis; nor could any such fibers be traced caudally beyond the confines of the substantia nigra. The present experiments have thus provided no indication of any projection from the putamen to the thalamus, subthalamic region, or pontomesencephalic tegmentum.

In accord with Szábo's[69] previous findings, the differential distribution of fiber degeneration following variously located lesions of the putamen indicates a fairly well defined topographical organization in the putaminal projections to both the globus pallidus and substantia nigra. The putaminopallidal projection terminates in both pallidal segments; according to Szábo[69] it does not involve the dorsomedial quarter of the globus pallidus. As each individual fiber of the connection appears to terminate in close vicinity to the putaminofugal fiber bundle of which it forms part, the typical mosaic of the connection must show a roughly radial design closely corresponding to the generally converging disposition of these bundles (Wilson's 'pencils'). The putaminonigral connection almost exclusively involves the pars reticulata of the substantia nigra, as observed previously by Rundles and Papez[67] and Szábo[69], but there is some evidence of sporadic putaminal efferents terminating in the pars compacta, and a few fibers of this connection appear to be distributed along the dorsal margin of the pars compacta (Fig. 6). It is possible that part of the fibers distributed ventral and dorsal to the pars compacta terminate in connection with dendrites of compacta cells. The present findings suggest that the distribution area of the putaminonigral fibers extends throughout the anteroposterior extent of the substantia nigra, but does not involve a rostromedial segment presumably corresponding to the distribution of the caudatonigral projection outlined by Szábo[69].

*Ansa lenticularis.* In one of his classical papers, von Monakow[39] defined the ansa lenticularis (Linsenkernschlinge) as '.... the sum of the fiber masses which come from the region of the lentiform nucleus, penetrate the cerebral peduncle, and gain the subthalamic region and the medial divisions of the thalamus' (p. 29). In this massive fiber complex von Monakow distinguished three components: (1) a dorsal division (often described separately as fasciculus lenticularis) formed by numerous thick fibers which traverse the internal capsule and contribute to the horizontal fiber stratum forming the dorsal capsule of the subthalamic nucleus commonly labelled fasciculus lenticularis or field $H_2$ of Forel; (2) a middle division (sometimes called the subthalamic fasciculus) perforating the cerebral peduncle caudal and ventral to the dorsal division to enter the subthalamic nucleus, and (3) a ventral division, the

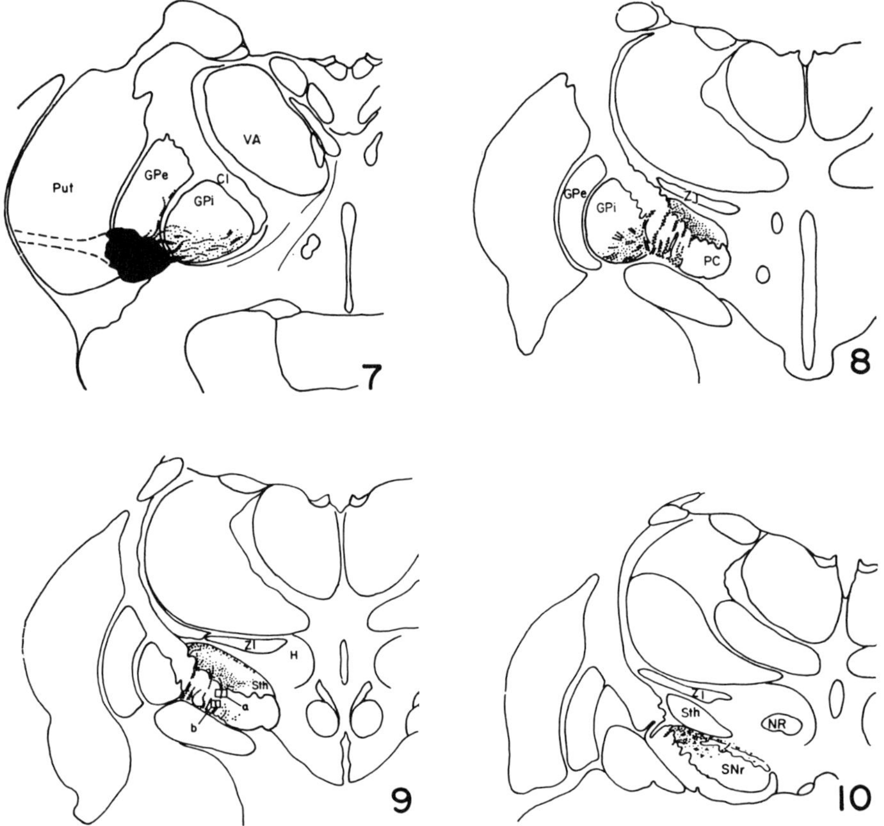


Figs. 7–10. Fiber degeneration following lesion of the external segment of the globus pallidus. Case MGB 12. Symbols as in Figs. 1–6. Abbreviations: see page 38. As any other surgical lesion of the globus pallidus, defects of this localization unavoidably interrupt numerous striatofugal fibers. The terminal degeneration in the internal pallidal segment (Fig. 8) and substantia nigra (Fig. 10) may be due entirely to this circumstance. The only degeneration clearly attributable to the external pallidal lesion affects fibers of von Monakow's middle division (see text) of the ansa lenticularis, distributed to the subthalamic nucleus (Figs. 8 and 9). These pallidosubthalamic fibers, after traversing the internal pallidal segment, perforate the cerebral peduncle as dorsal components of Edinger's 'Kammsystem' (a in Fig. 9, photograph in Fig. 31) and can be distinguished by their heavier calibre from the more ventrally located fine-fibered perforating bundles of the putaminonigral connection (b in Fig. 9, photograph in Fig. 33). Note the absence of fiber degeneration in the tegmental fields of Forel.

ansa lenticularis *sensu strictiori*[38] of most textbook descriptions which forms the well-known stratum of thick fibers ventral to the internal segment of the globus pallidus, turns dorsally around the medial edge of the internal capsule, and in its further course mingles its fibers with those of the dorsal division, '. . . . so dass sie von diesem schwer zu trennen ist'[39].

It must be noted that von Monakow's description did not include the fine fibered caudatonigral and putaminonigral pathways which were identified only several decades later. Had von Monakow been aware of the projection, he might well have included it in his definition of the ansa lenticularis, for the striatonigral fibers follow the path of his 'middle division of the ansa' over a considerable distance, and can hardly be distinguished from the latter proximal to the region where both begin to traverse the cerebral peduncle. However, in order to avoid confusion, and at the risk of perpetuating a rather arbitrary restriction, von Monakow's definition of the ansa lenticularis will be adhered to, and hence, striatonigral fibers are excluded from this account of the ansa lenticularis. It must be emphasized that different denotations given the term 'ansa lenticularis' can easily lead to misunderstanding. Not all previous authors have stated whether their use of the term corresponded to Meynert's more restricted, or von Monakow's broader definition of the ansa lenticularis.

As regards the origin of the ansa, von Monakow believed that the dorsal and middle divisions received fibers mostly from the putamen, whereas the ventral division would arise in all parts of the lentiform nucleus. This concept of a mixed striatal and pallidal origin of the ansa was accepted by Dejerine[7], Foix and Nicolesco[12], Kodama[29], Lhermitte and Trelles[31], and more recently by Woodburne *et al.*[83]. None of these later reports, however, appears to have been based on original experimental evidence in support of von Monakow's notion.

Contrary to von Monakow, von Bechterew (ref. 3, p. 552) summarily stated that it appeared to him that the ansa lenticularis originated exclusively from the globus pallidus. Von Bechterew's view received support from the results of experimental studies in the monkey[62,81] and cat [26], as well as from studies of normal and patholog ical simian and human brains[19,37,67,74].

A further specification regarding the origin of the ansa was proposed by Ranson *et al.*[63] who, from experimental studies in the monkey, concluded that the ansa arose only from the internal segment of the globus pallidus, with the exception of the ansal component distributed to the subthalamic nucleus (von Monakow's middle division) which they found to originate from the external pallidal segment. Papez[54] concurred with this concept*. C. and O. Vogt had earlier reported heavy cell

---

* Laursen[30], on the basis of observations in his case M7, objected to the Ransons' conclusion[63] that the ansa originated exclusively from the internal pallidal segment. However, as Laursen himself pointed out, his argument is open to the criticism that M7, in addition to the lesion of the putamen and external pallidal segment showed considerable damage to various cortical areas and to the internal capsule. In the same paper, Whitlock and Nauta's[80] study is erroneously referred to for evidence of temporal cortical contributions to the 'ansa'. Actually, the study in question produced evidence of temporal cortical efferents in the ansa *peduncularis* instead of the ansa lenticularis (see the present discussion of the ansa peduncularis).

loss in the subthalamic nucleus in cases of isolated destruction of the external pallidal segment, an observation which led them to suggest that the two structures were at least preferentially interconnected.

Our present findings support the conclusions of the Ransons. In agreement with von Bechterew's original interpretation, lesions in the putamen (see Figs. 1–6) were found to cause no fiber degeneration in any of the ansal components described by von Monakow. Lesions extending medially far enough to involve the external pallidal segment (see Figs. 7–10) failed to produce degeneration in von Monakow's dorsal and ventral ansal divisions, but did cause profuse degeneration of pallidal fibers distributed via the middle division to the subthalamic nucleus. Only those lesions that involved the internal pallidal segment were found to elicit fiber degeneration in the dorsal and ventral ansal division and in the subthalamic fiber fields of Forel, extending into the thalamus and midbrain tegmentum (see Figs. 11–20). It thus appears evident: (a) that the ansa lenticularis as defined by von Monakow originates exclusively from the globus pallidus; (b) that the origin of ansal components distributed by way of Forel's tegmental fields to the thalamus and mesencephalic tegmentum is restricted to the internal pallidal segment, and (c) that the external pallidal segment contributes only to the middle division of the ansa, *i.e.* to the pallidal projection upon the subthalamic nucleus (whether the internal segment contributes any fibers to this projection cannot be stated with certainty).

*The 'comb system'.* This term (Kammsystem des Fusses) was introduced by Edinger[8] to denote the characteristic pallisades of slender fiber bundles traversing the cerebral peduncle at frontal levels involving the subthalamic nucleus and the rostral part of the substantia nigra (see, for example, Figs. 8–10). Edinger correctly considered the bundles to represent a lenticulonigral ('striopeduncular') connection. Fibers of opposite polarity, especially such ascending from the substantia nigra to the lentiform nucleus have also been suggested to contribute to the comb system[66]. It should be pointed out that the comb system as described above is no more than the caudal part of a wider system of perforant fibers related to the caudate and lentiform nuclei and traversing the internal capsule and the cerebral peduncle in the region described by Flechsig (ref. 11, p. 37) as 'das Durchflechtungsgebiet der inneren Kapsel'. It is rostrally contiguous with the system of considerably thicker perforating fibers of von Monakow's dorsal division of the ansa lenticularis, which in their continuation contribute to the formation of field $H_2$.

As regards descending components, the present findings indicate that the comb system is made up largely of two fiber categories, *viz.* the middle division of the ansa lenticularis (*i.e.* the pallidofugal pathway to the subthalamic nucleus), and the striatonigral projections. The relatively thick pallidosubthalamic fibers traverse the peduncle in the dorsomedial direction at a steep angle, whereas the thinner striatonigral fibers follow a flatter, more caudally directed transpeduncular trajectory and thus tend to accumulate in fascicles ventral to the pallidosubthalamic bundles (see the present case MGB 12: Figs. 9, 31, 33).

*Ansa peduncularis and substantia innominata.* It is appropriate at this point to emphasize the importance of distinguishing the ansa *lenticularis* from the closely neighboring but apparently unrelated system of the so-called ansa *peduncularis* of Meynert[38]. The latter is a massive fiber group extending from the amygdalopiriform complex medialward through the sublenticular grey matter forming the substantia innominata of Reichert. In this region the system divides into two components, one of which continues medialward to the preoptico-hypothalamic region, whereas the other turns dorsally around the medial edge of the internal capsule, immediately medial to the similarly oriented curve of the ventral division of the ansa lenticularis, and enters the thalamus from the ventral side as the so-called inferior thalamic peduncle. These statements have no more than general descriptive value, for the composition of the ansa peduncularis is extremely complex. An excellent account of its topography, based on a refined technique of gross dissection, was given recently by Klingler and Gloor[27]. Experimental studies have identified the following components in the system (refs. 22, 27, 45–47, 80): (a) fibers from the amygdalopiriform complex to the substantia innominata and lateral preoptico-hypothalamic region; (b) amygdalothalamic fibers mostly to the medial component of the mediodorsal thalamic nucleus; (c) fibers from the orbitofrontal cortex to the thalamus and hypothalamus; (d) fibers from the mediodorsal thalamic nucleus to the lateral preoptico-hypothalamic region and amygdala; (e) hypothalamo-amygdalar fibers, and (f) fibers from the temporal cortex to the mediodorsal thalamic nucleus. Closely related to the ansa peduncularis is the so-called tractus corticohabenularis lateralis (Gurdjian[17]; identical with Cajal's (ref. 4, Fig. 274, Vol. II) stria medullaris) which appears to originate in the lateral preoptico-hypothalamic region, swings over the rostral pole of the thalamus, and follows the stria medullaris to the lateral habenular nucleus.

Despite their close mutual proximity, the ansae lenticularis and peduncularis appear to represent two entirely different neural mechanisms. The ansa lenticularis forms the main and apparently the only pathway connecting the globus pallidus with the thalamus, subthalamic region, and midbrain tegmentum. The ansa peduncularis, by contrast, appears to be a composite system of largely reciprocal connections between the components of a continuum of grey matter that extends from the amygdalopiriform complex and the adjoining substantia innominata to the preoptico-hypothalamic region and a medial thalamic zone encompassing the medial component of the mediodorsal nucleus.

The substantia innominata* has occasionally been considered a contributor to the ansa lenticularis[83]. This suggestion is difficult to verify experimentally, for the topographical position of the substantia innominata virtually precludes the placement of lesions that do not involve parts of the globus pallidus. In the present case, MGB 6, such concomitant damage was confined to the most ventral part of the external pallidal segment, while the major lesion occupied the region of the nucleus basalis of Ganser, without involving the adjacent ventral division of the ansa lenticularis. As illustrated by Figs. 21–25, no fiber degeneration was found in this case in either the ventral or dorsal ansal divisions, and the circumscript partial degeneration of the

---

* The nuclear configuration of the substantia innominata is discussed in the accounts of Foix and Nicolesco[12] and Papez and Aronson[55].

middle division leading to the subthalamic nucleus seems adequately explained by the electrode tract piercing the external pallidal segment. The most massive fiber degeneration observed in this case clearly outlines the extent of the ansa peduncularis and suggests that the substantia innominata is closely related to that fiber system

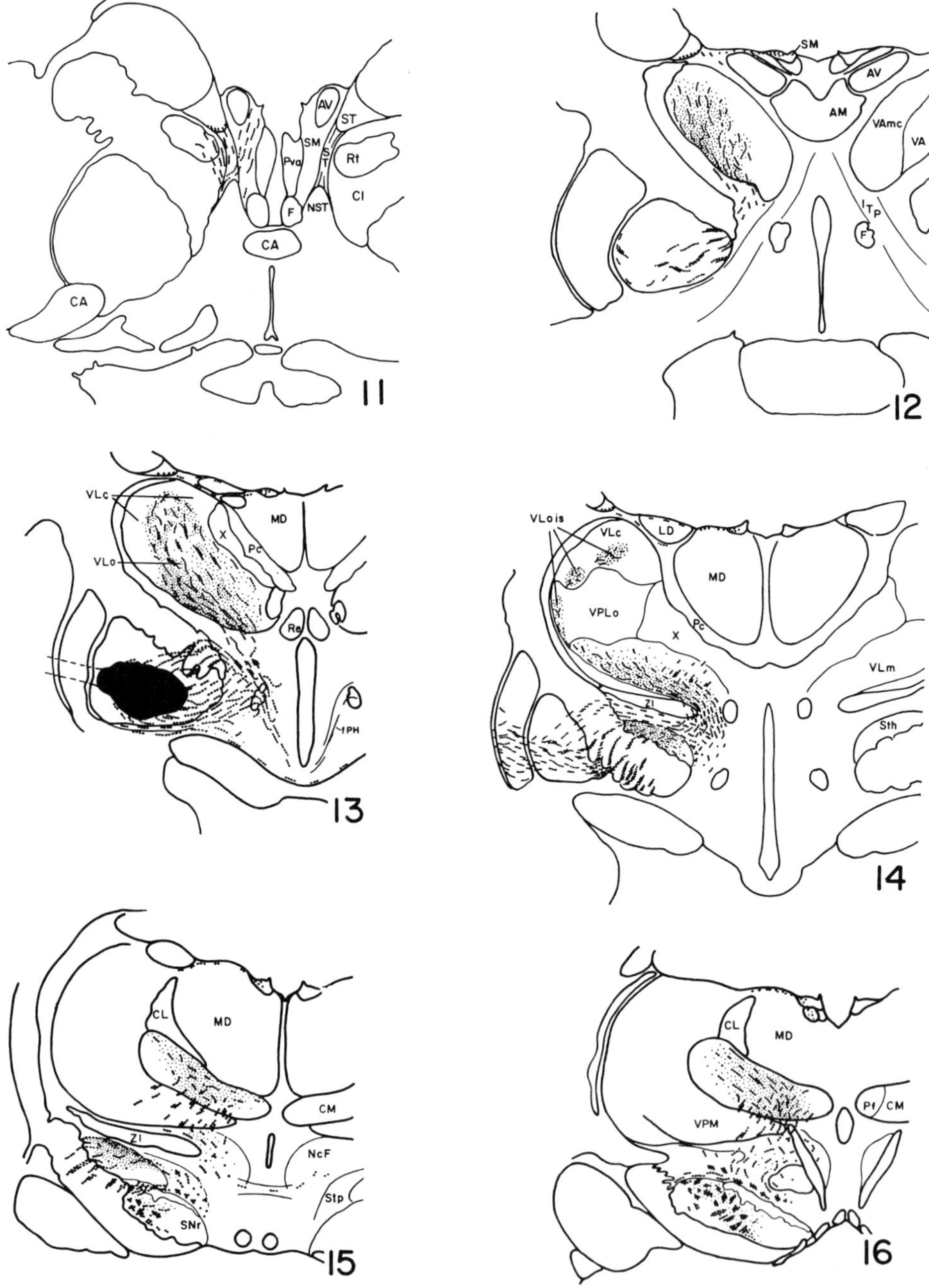

Figs. 11–16. For legend see next page.

rather than to the ansa lenticularis. However, the same case, MGB 6, demonstrates an additional degenerating fiber group of considerable volume, *viz.* one that extends caudalward along the medial edge of the internal capsule to an area of dense distribution lateral to the mammillary body (Fig. 24). The mesencephalic extent of part

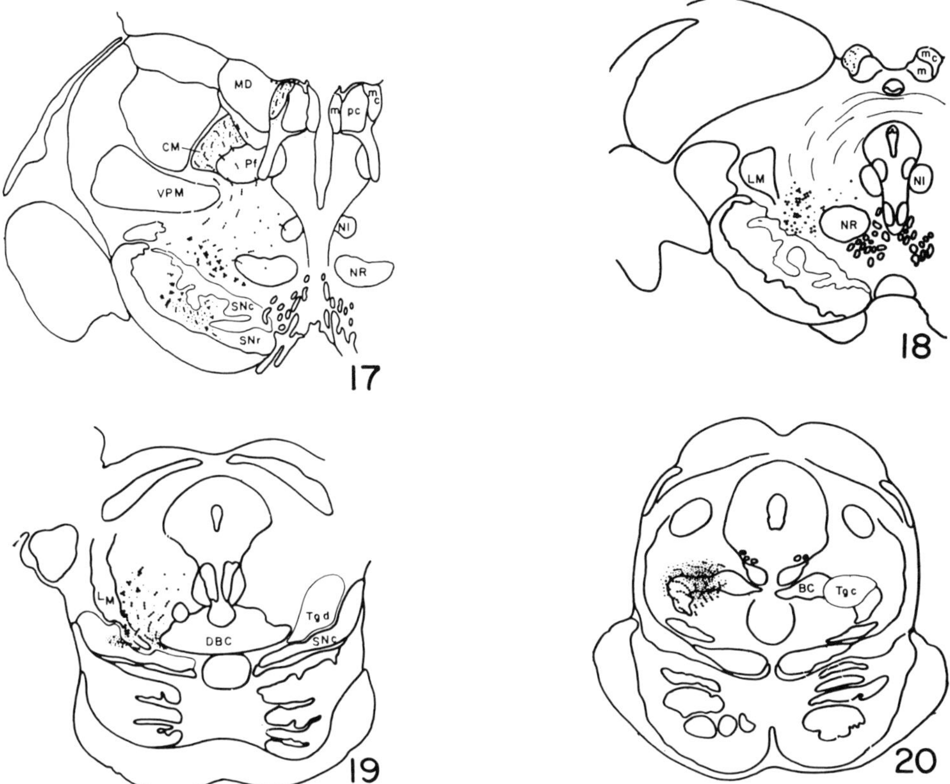

Figs. 11–20. Fiber degeneration following lesion of the internal pallidal segment. Case MGB 8 (see also Fig. 33). Symbols as in Figs. 1–6. Abbreviations see page 38. Lesions in this location involve components of all major lenticulofugal pathways; hence the fiber degenerations observed in the subthalamic nucleus (Figs. 14 and 15) and substantia nigra (Figs. 15–17) could be due entirely to the interruption of efferents from, respectively, the putamen and external pallidal segment (compare the two preceding cases illustrated). Characteristic of lesions involving the internal pallidal segment is the degeneration of thick fibers of both ventral and dorsal divisions of the ansa lenticularis (Fig. 13) extending into Forel's tegmental fields. From here, the degeneration spreads in several directions:

(1) The largest number of degenerating fibers loop dorsally in the thalamic fasciculus (Fig. 14) to end in the nuclei ventralis lateralis (VLm and VLo; Figs. 13 and 14) and ventralis anterior thalami (Fig. 12); in Fig. 14 note the slender fascicles separating at right angles from the main body of the thalamic fasciculus, and extending in nearly straight lines dorsally, caudally, and medially to their terminal ramifications in the 'centre médian' (Figs. 15 and 16).

(2) A discrete bundle of ansal fibers curves ventrally over and through the fornix bundle into the hypothalamus (Fig. 13); no evidence of a hypothalamic termination of these fibers is apparent.

(3) Degenerating fascicles extending caudally along the lateral side of the red nucleus into the mesencephalic tegmentum are distributed mainly to the nucleus tegmenti pedunculopontinus and terminate with particular density in the pars compacta of this nucleus (Fig. 20).

(4) Scattered degenerating fibers sweep over the rostral pole of the thalamus (Fig. 11) and turn caudalward in the stratum zonale thalami to terminate in the lateral habenular nucleus (Figs. 16–18).

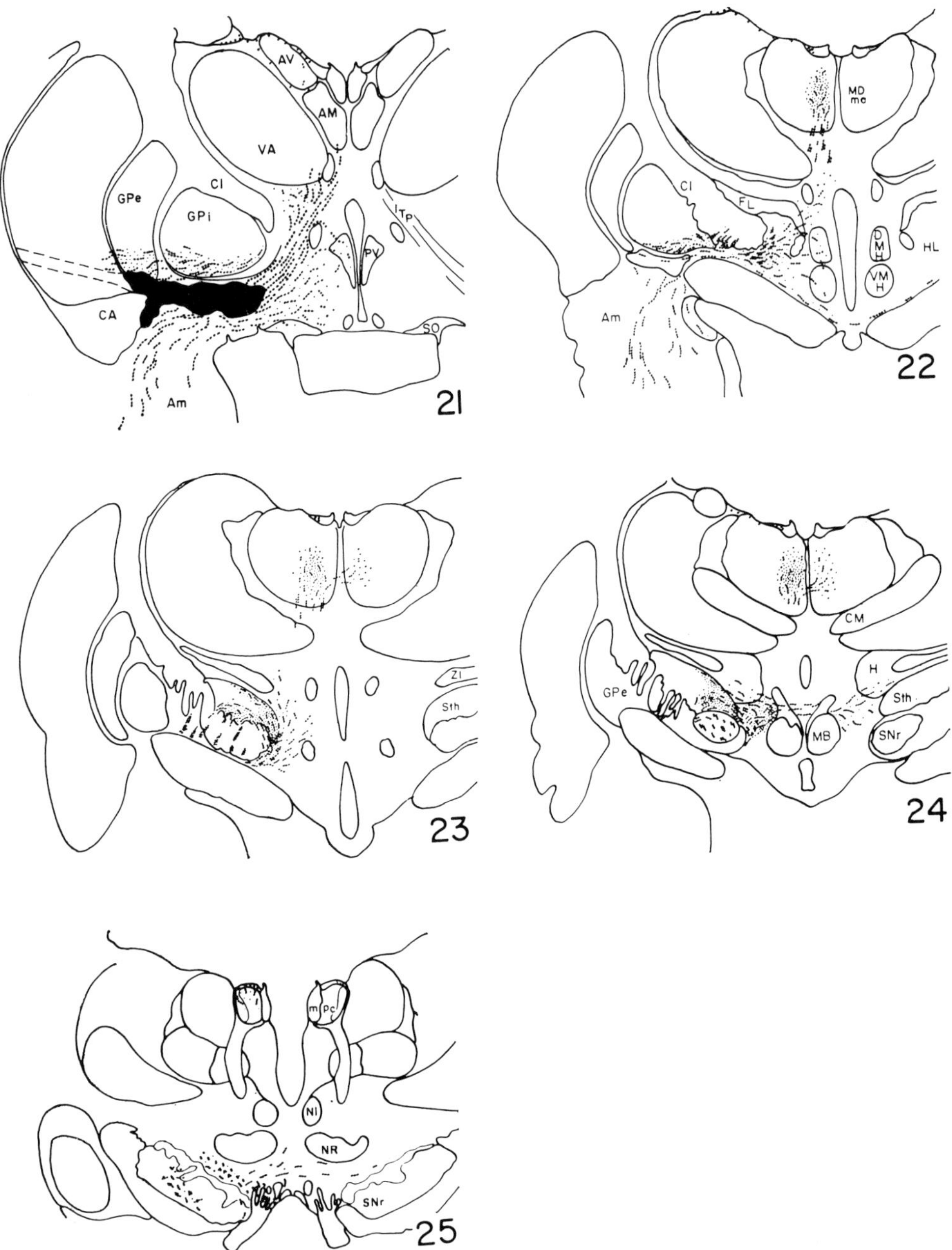


Figs. 21–25. Fiber degeneration following lesion of the substantia innominata (case MGB 6). Note the electrode tract traversing the putamen, and the extension of the electrolytic defect involving the ventral corner of the external pallidal segment (Fig. 21). These incidental lesions are the probable cause of the fiber degeneration extending via the comb system (Figs. 22 and 23) to the subthalamic nucleus (Figs. 23 and 24) and the substantia nigra (Fig. 25). No degenerating fibers have appeared in the ventral or dorsal divisions of the ansa lenticularis. The most massive fiber degeneration observed in this case involves the ansa peduncularis with its intricate and largely reciprocal fiber connections extending between the amygdalopiriform complex and the mediodorsal thalamic nucleus and hypothalamus (see Nauta[47]). Apparently related to the ansa peduncularis is a fasciculated fiber system which extends caudally along the medial edge of the cerebral peduncle; it loses numerous fibers in the lateral hypothalamus, and continues into the mesencephalic tegmentum. At the level of Figs. 23 and 24, several fascicles of this fiber system can be seen entering the subthalamic nucleus through its medial aspect. Symbols as in Figs. 1–6. Abbreviations see page 38.

404

of this fiber system will not be discussed here, but it is important to note that the fiber group issues fascicles which curve dorsally around the medial margin of the cerebral peduncle and terminate densely in the most medial part of the subthalamic nucleus (Figs. 23–24). Whether the fiber system in question actually originates in the substantia innominata cannot be ascertained from the present material. The findings in MGB 6 do, however, suggest the existence of a non-pallidal projection to the subthalamic nucleus that could conceivably originate in the substantia innominata. Apart from this suggestion, the present findings contain no evidence that the substantia innominata contributes to the ansa lenticularis; moreover, they specifically disprove such contributions to the dorsal and ventral divisions of the ansa.

*Pallidothalamic connections.* Flechsig[11], apparently under the impression that the ansa lenticularis was, in part at least, continuous with the inferior thalamic peduncle, appears to have been the first to suggest the existence of a thalamic connection in the ansa lenticularis. Von Monakow[39] strongly emphasized the lenticulothalamic connection, and actually considered both the ventral and dorsal division of the ansa to be distributed chiefly to ventral and anterior regions of the thalamus. Both Flechsig and von Monakow failed to identify the thalamic fasciculus (field $H_1$ of Forel) as a major trajectory in the path between the lentiform nucleus and the thalamus; their accounts vividly illustrate the intensely confusing complexity of the fiber picture with which the early explorers of the subthalamic region found themselves confronted. It was not until 1909 that Cécile Vogt[72], in a careful study of a horizontally sectioned infant brain, first demonstrated the now generally recognized loop-like continuity between the fasciculi lenticularis and thalamicus. C. Vogt identified the anterior part of the ventral nucleus ('vtl') as the thalamic region involved in the connection, but neither she nor C. and O. Vogt a decade later[74] were able to state the polarity of the pathway. It is remarkable that Wilson[81] found degeneration in the thalamic fasciculus in cases of pallidal lesion, but discounted it on the suspicion that it represented a corticothalamic pathway. Wilson's account of pallidothalamic degeneration is limited to fibers crossing the ventral one-third of the internal capsule and extending medialward across the thalamus, 'diffusing out in the neighborhood of the internal nucleus'. Contrary to later authors, Wilson considered the pallidothalamic pathway to constitute only a minor portion of the total pallidofugal projection.

*(a) Pallidal projections to the ventral thalamic nucleus.* The predominantly thalamopetal orientation of C. Vogt's ansiform fiber system appears to have been demonstrated first by Inui's[21] Marchi experiments in the rabbit. Inui not only traced massive fiber degeneration from pallidal lesions via the ansa lenticularis to the rostral part of the ventral thalamic nucleus, but also reported that lesions in that nucleus caused cell loss in the globus pallidus*. Papez and Stotler[56], in a study of normal

---

* Inui's report presents several puzzling aspects. His term 'globus pallidus' apparently refers to the external pallidal segment. Nowhere in his publication does he mention the entopeduncular nucleus (the probable subprimate homologue of the internal pallidal segment), a very small structure in the rabbit. The pallidal cell degeneration following his lesions was observed bilaterally. It seems certain that Inui's findings in Marchi material were heavily contaminated by corticofugal fiber degeneration. Nevertheless, Inui's study historically appears as the first report of experimentally induced degeneration of pallidothalamic fibers in Forel's fields.

TABLE I

A TABULAR SUMMARY OF FINDINGS FOLLOWING VARIOUSLY LOCALIZED LESIONS OF THE LENTIFORM NUCLEUS

| Case MGB- | 1 | 2 (14) | 3 | 4 | 6 | 7 | 8 | 10 * | 11 * | 12 | 13 | Cerebello-thalamic[35] |
|---|---|---|---|---|---|---|---|---|---|---|---|---|
| VA | 0 | 0 | 0 | 0 | 0 | ++ | ++++ | 0 | ++++ | 0 | 0 | + |
| VAmc | 0 | 0 | 0 | 0 | 0 | 0 | 0 | 0 | 0 | 0 | 0 | + |
| VLo | 0 | 0 | 0 | 0 | (+) | ++ | ++++ | 0 | ++++ | 0 | 0 | + |
| VLm | 0 | 0 | 0 | 0 | (—) | ++ | ++++ | 0 | ++++ | 0 | 0 | + |
| VLc | 0 | 0 | 0 | 0 | 0 | 0 | 0 | 0 | 0 | 0 | 0 | + |
| VLps | 0 | 0 | 0 | 0 | 0 | 0 | 0 | 0 | 0 | 0 | 0 | + |
| MDmc | 0 | 0 | 0 | 0 | +++ | +++ | 0 | 0 | 0 | 0 | (—) | 0 |
| CL | 0 | 0 | 0 | 0 | 0 | 0 | 0 | 0 | 0 | 0 | 0 | + |
| CM dm | 0 | 0 | 0 | 0 | 0 | + | +++ | 0 | ++ | 0 | 0 | 0 |
| vl | 0 | 0 | 0 | 0 | 0 | (—) | ++ | 0 | +++ | 0 | 0 | 0 |
| NcF | 0 | 0 | 0 | 0 | (—) | + | + | 0 | + | 0 | 0 | + |
| ZI | 0 | 0 | 0 | 0 | 0 | 0 | 0 | 0 | 0 | 0 | 0 | + |
| Sth | + | 0 | (+) | (+) | + | + | +++ | ++ | +++ | +++ | + | 0 |
| (top.) | 1 | | or-1 | 1 | v-m | v-m | 1 | or-ce | or-ce | 1 | or-m | |
| SNr | ++ | ++ | + | ++ | + | ++ | ++ | + | ++ | ++ | ++ | 0 |
| SNc | (+) | (+) | (+) | (+) | (—) | ? | (+) | (+) | (+) | ? | (—) | 0 |
| NR | 0 | 0 | 0 | 0 | 0 | 0 | 0 | 0 | 0 | 0 | 0 | + |
| Tgd | 0 | 0 | 0 | 0 | 0 | (—) | (+) | 0 | (+) | 0 | 0 | 0 |
| Tgc | 0 | 0 | 0 | 0 | 0 | ++ | ++++ | 0 | ++++ | 0 | 0 | 0 |
| Hab. lat. | | | | | | | | | | | | |
| mc | (—) | 0 | 0 | 0 | (+) | ++ | +++ | 0 | +++ | 0 | 0 | 0 |
| pc | 0 | 0 | 0 | 0 | + | + | 0 | 0 | (—) | 0 | 0 | 0 |

Symbols are as follows: 0: no evidence of terminal degeneration; (—): not entirely negative; (+) slight but significant terminal degeneration; + to ++++ denote increasing densities of terminal degeneration; ce = central; dm = dorsal medial; vl = ventrolateral; or = oral. Other abbreviations see page 38. CL refers to the nucleus centralis lateralis as defined by Mehler *et al.*[36] and Mehler[34]. The extreme right-hand column, included for the purpose of comparison, lists the terminal degeneration observed following deep cerebellar lesions; the symbols in this column merely state the presence or absence of degeneration and have no comparative-quantitative significance. Asterisks (cases MGB 10 and 11) indicate involvement of the internal capsule in the electrolytic   esion.

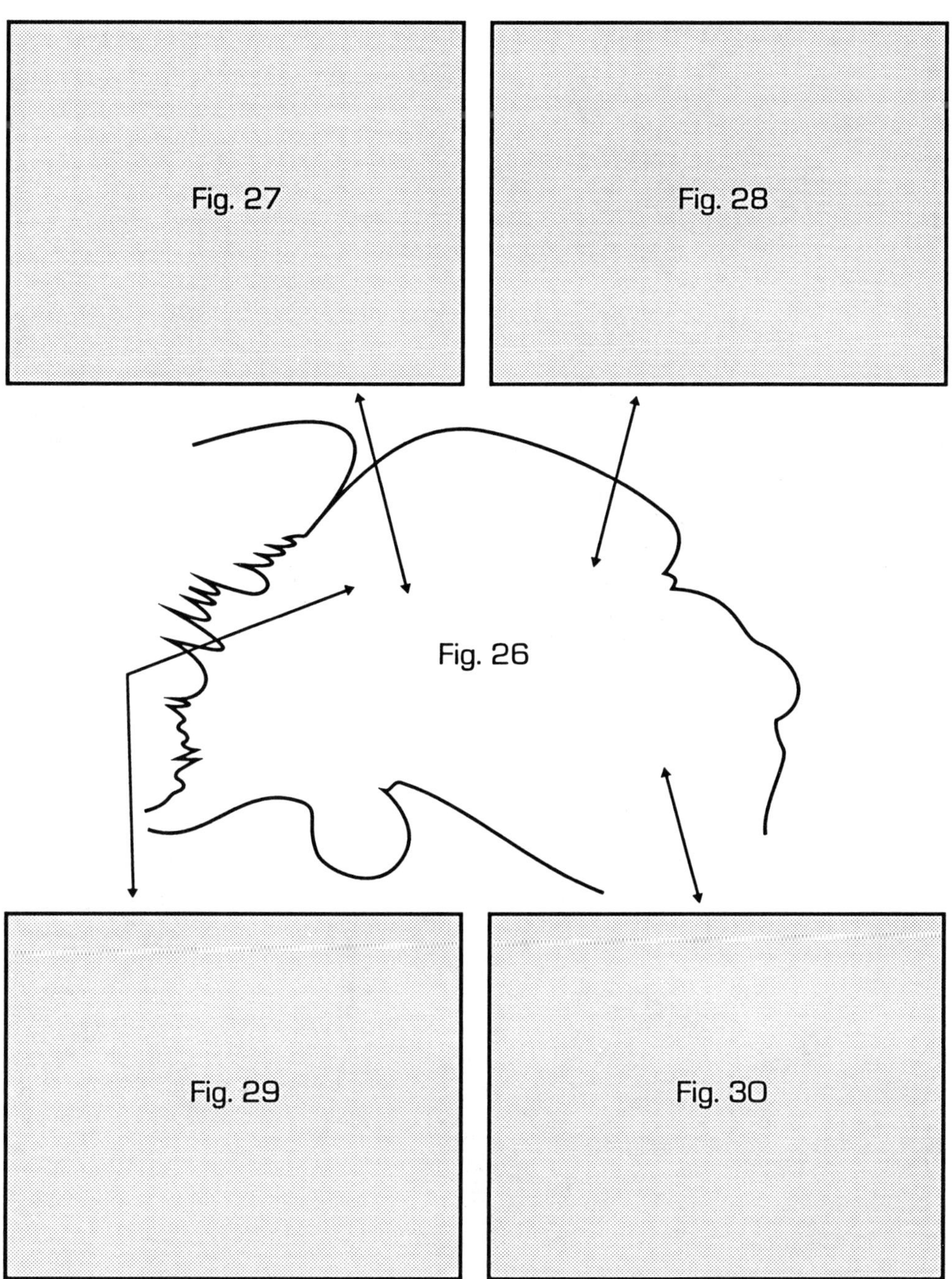

Figures 26–30 are on the next two pages.

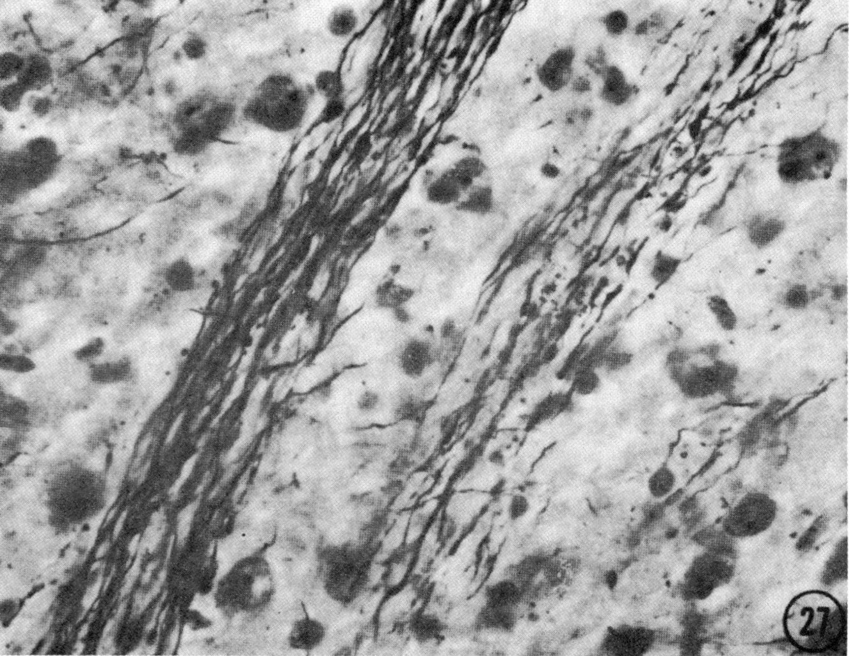
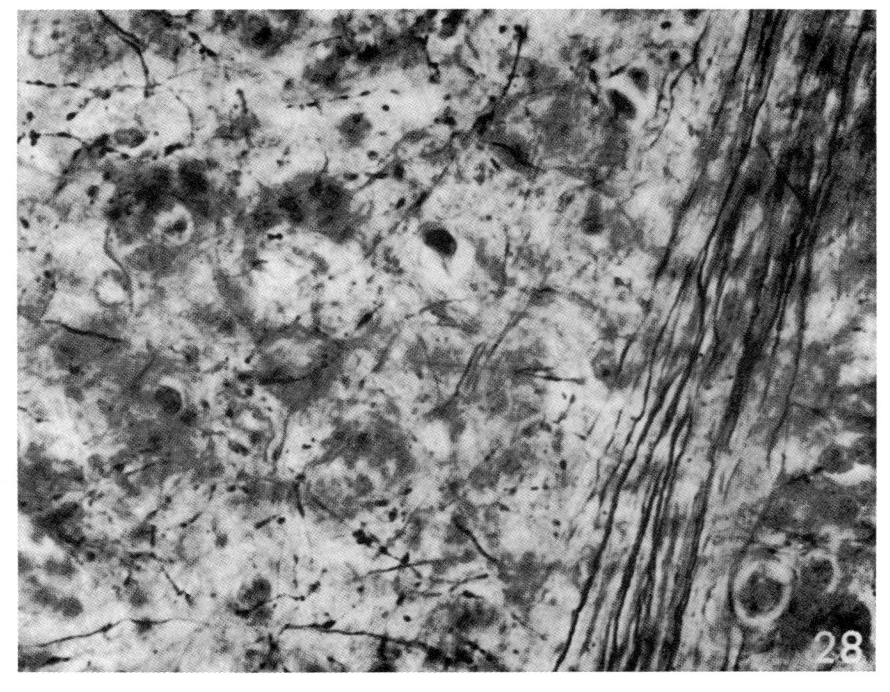
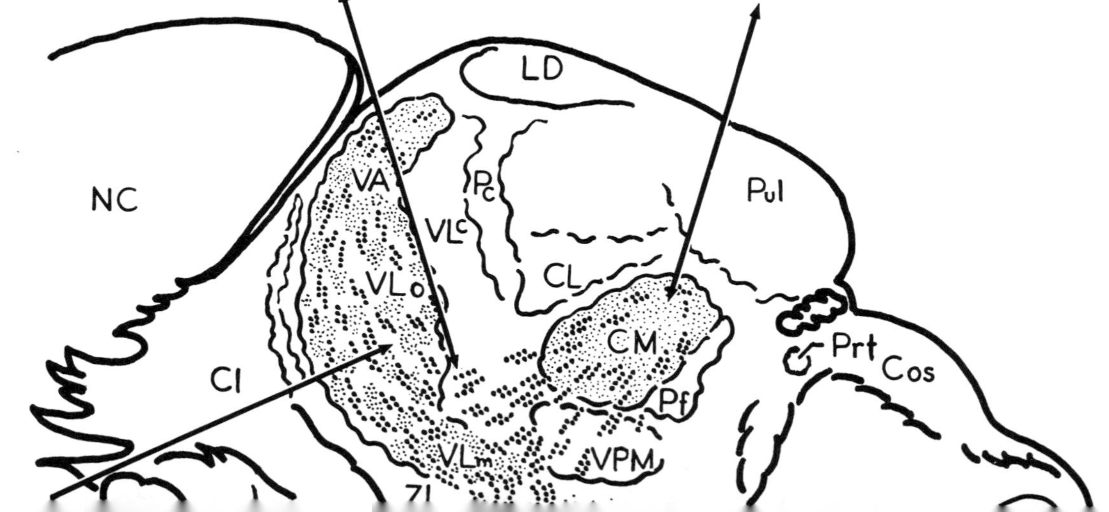

NC
VA
Pc
VLc
VLo
CL
LD
Pul
Cl
CM
Pf
Prt
Cos
VLm
VPM
ZI

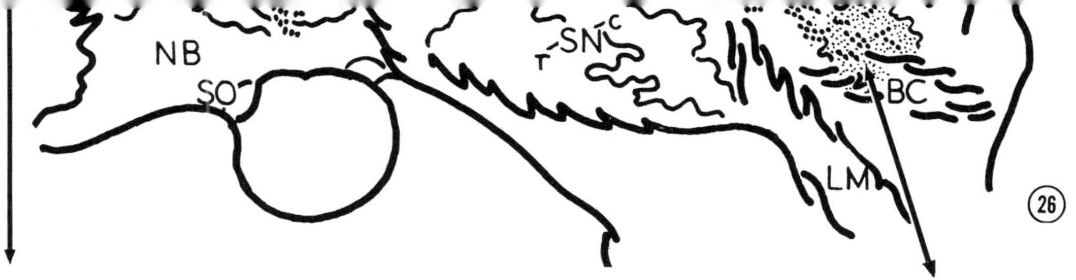

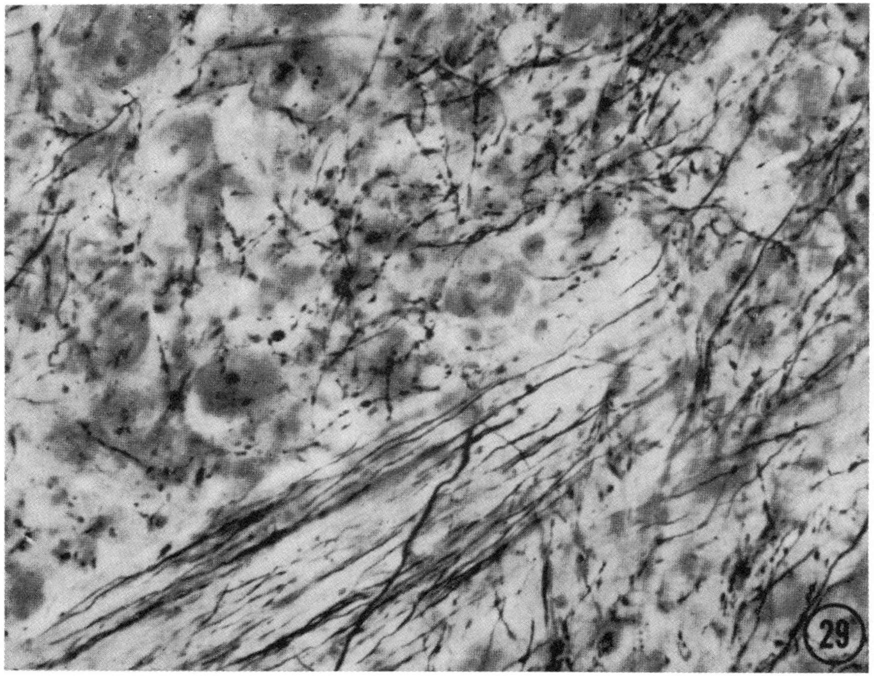

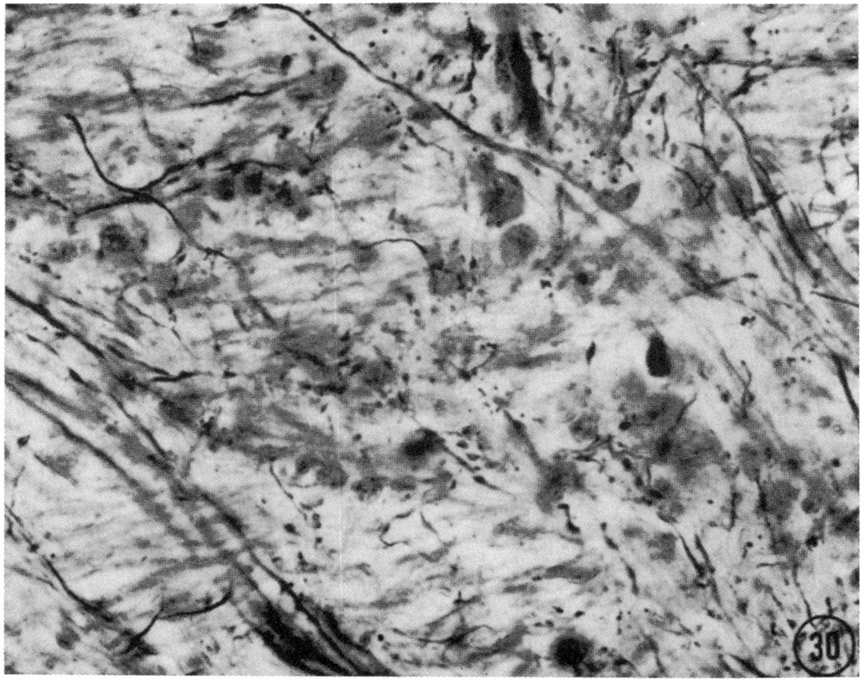

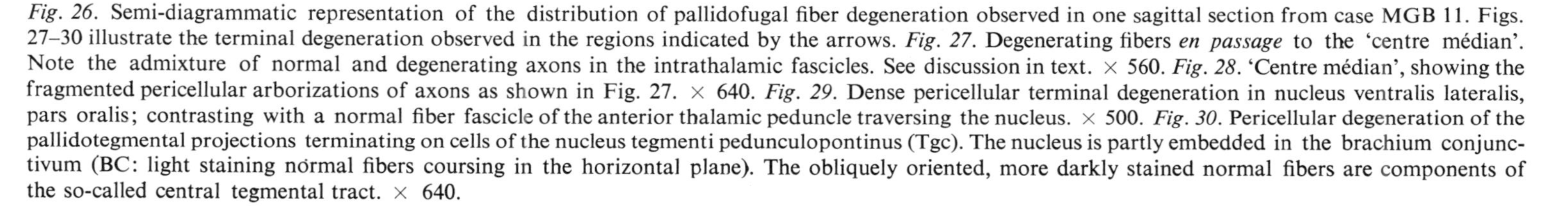



*Fig. 26.* Semi-diagrammatic representation of the distribution of pallidofugal fiber degeneration observed in one sagittal section from case MGB 11. Figs. 27–30 illustrate the terminal degeneration observed in the regions indicated by the arrows. *Fig. 27.* Degenerating fibers *en passage* to the 'centre médian'. Note the admixture of normal and degenerating axons in the intrathalamic fascicles. See discussion in text. × 560. *Fig. 28.* 'Centre médian', showing the fragmented pericellular arborizations of axons as shown in Fig. 27. × 640. *Fig. 29.* Dense pericellular terminal degeneration in nucleus ventralis lateralis, pars oralis; contrasting with a normal fiber fascicle of the anterior thalamic peduncle traversing the nucleus. × 500. *Fig. 30.* Pericellular degeneration of the pallidotegmental projections terminating on cells of the nucleus tegmenti pedunculopontinus (Tgc). The nucleus is partly embedded in the brachium conjunctivum (BC: light staining normal fibers coursing in the horizontal plane). The obliquely oriented, more darkly stained normal fibers are components of the so-called central tegmental tract. × 640.

human and monkey material, likewise interpreted the thalamic fasciculus as a pallidothalamic fiber system. Experimental evidence confirming Inui's report was obtained in the monkey by Verhaart[70], Ranson *et al.*[63], and Glees[15], all of whom traced Marchi degeneration from pallidal lesions through fields $H_2$ and $H_1$ to the nucleus ventralis anterior. Glees[15] pointed out that the projection also involved the ventrolateral thalamic nucleus. Our present findings are in accord with this statement, but also permit a more detailed definition of the various subdivisions of the ventral nucleus involved in the projection. As illustrated by Figs. 12–14, pallidofugal fiber degeneration could be traced only to the pars principalis (VA in Olszewski's[51] terminology) of the nucleus ventralis anterior, and avoided the more medially situated magnocellular subdivision VAmc (Fig. 12). In the ventrolateral nucleus, fiber degeneration appeared only among the characteristically clustered cells of the pars oralis (VLo, Fig. 13), and in the relatively small-celled ventromedial region (VLm, Fig. 14), no degenerating fibers could be traced to Olszewski's subdivisions X (Fig. 13) and VLc (Figs. 13–14). As will be documented in detail elsewhere, these findings also indicate that the area of convergence of pallidal and cerebellar projections to the ventral thalamic nucleus is somewhat wider than was suggested by Hassler[18].

Johnson and Clemente[26] reported experimental evidence of pallidothalamic connections in the cat, but their statement that the projection in that animal is restricted to the ventrolateral nucleus is not in full agreement with our own observations in several control experiments in the cat. In two cases in which stereotaxic lesions were placed in the entopeduncular nucleus (the apparent homologue of the primate internal pallidal segment) by lateral approach, pallidothalamic degeneration was found to be distributed, as in the monkey, to both the nuclei ventralis anterior and ventrolateralis.

Glees' report[15] of pallidothalamic projections originating in the external pallidal segment is incompatible with the findings of the Ransons[63] and the present observations. The reasons for Glees' conclusions are not clearly stated and appear to have been derived from a case (M.P.L. 6) in which the additional finding of dense Marchi degeneration in the caudal hypothalamus suggests involvement of structures besides the putamen and external pallidal segment. Hassler[19], although acknowledging the internal pallidal segment as the main source of pallidothalamic fibers, nevertheless suggested from his findings in human pathological material that the rostral one-third of the external pallidal segment contributes some fibers to the pallidothalamic projection. The present study involves 5 cases with lesions confined to the putamen and external pallidal segment (cases MGB 1, 3, 10, 12, and 13, see Table I) one of which (MGB 13) involved rostral parts of the external pallidal segment. In all these cases, only putaminonigral fibers and pallidal fibers to the subthalamic nucleus were found degenerated. In agreement with the Ransons' findings we can only conclude that the pallidothalamic connection in the monkey appears to originate exclusively from the internal pallidal segment.

*(b) 'Centre médian'.* We have been unable to find explicit earlier mention of the massive pallidofugal projection to the 'centre médian' demonstrated by our present experiments (see Figs. 14–17). This may be due in part to the common difficulty of

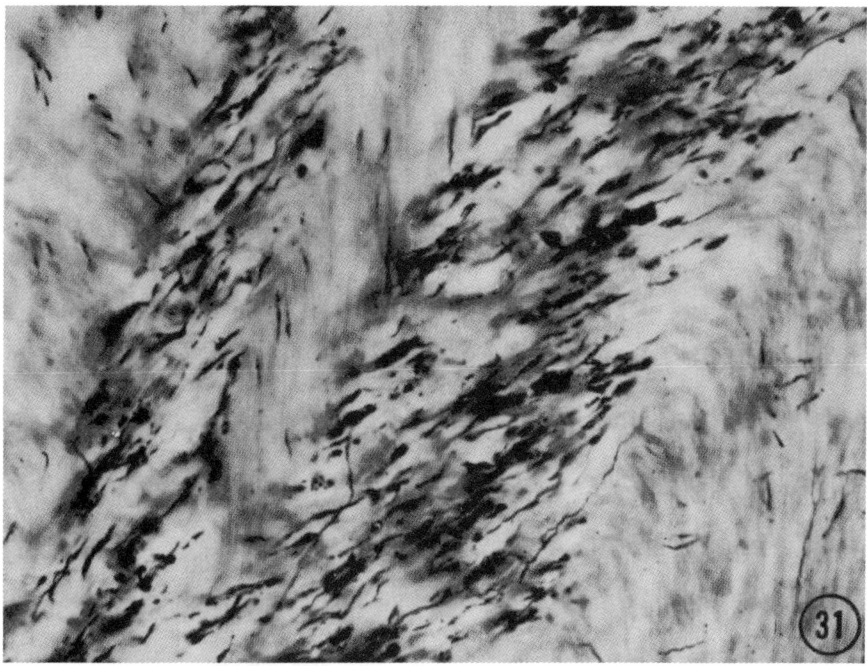

Fig. 31. Degenerating coarse-fibered bundles of the middle division of the ansa lenticularis (pallido-subthalamic fibers) traversing the cerebral peduncle as dorsal components of the 'comb system'. The field shown corresponds to block a in text-figure 9. Compare with Fig. 33 (case MGB 12). × 640.

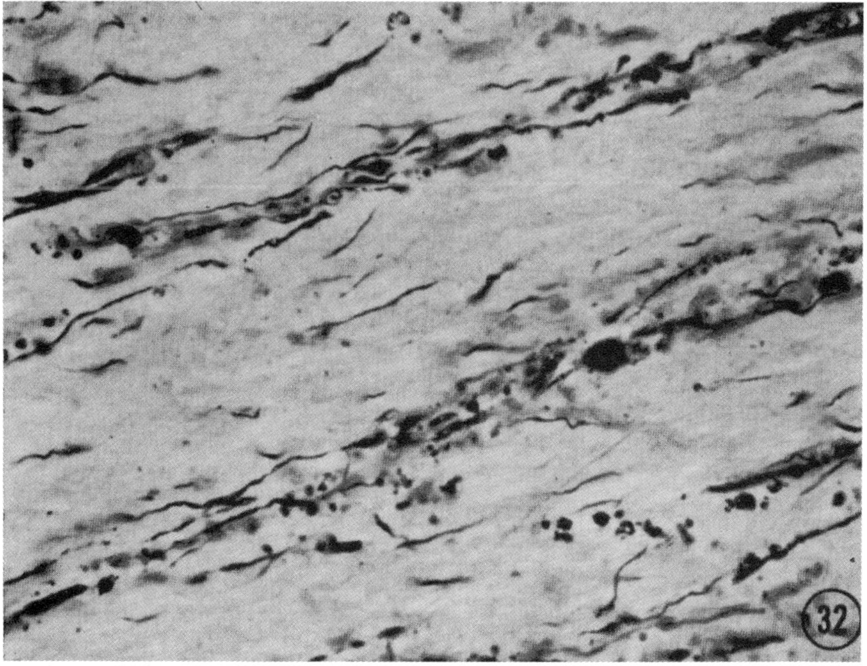

Fig. 32. Degenerating thick axons of the ventral division of the ansa lenticularis (case MGB 8). × 640

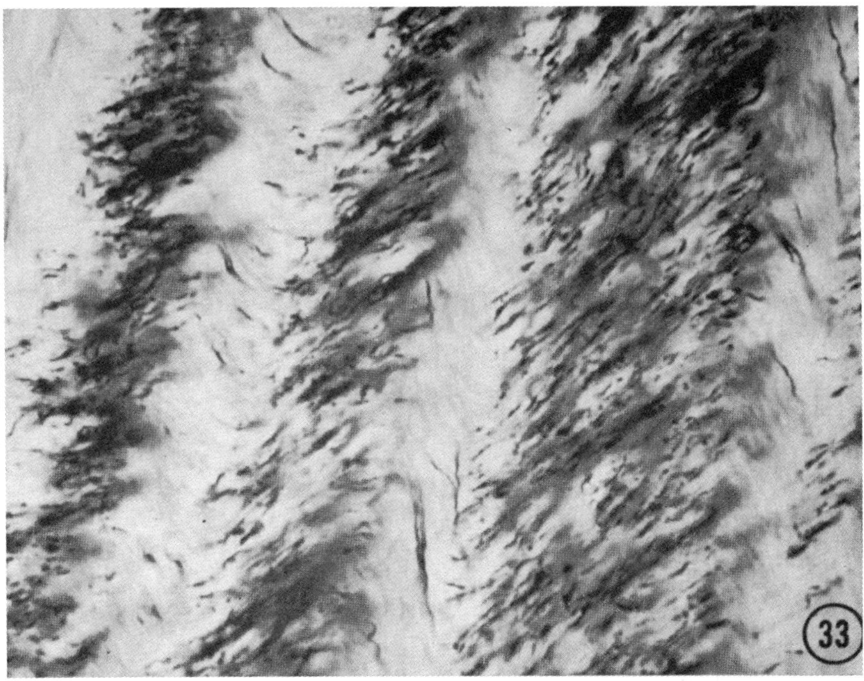

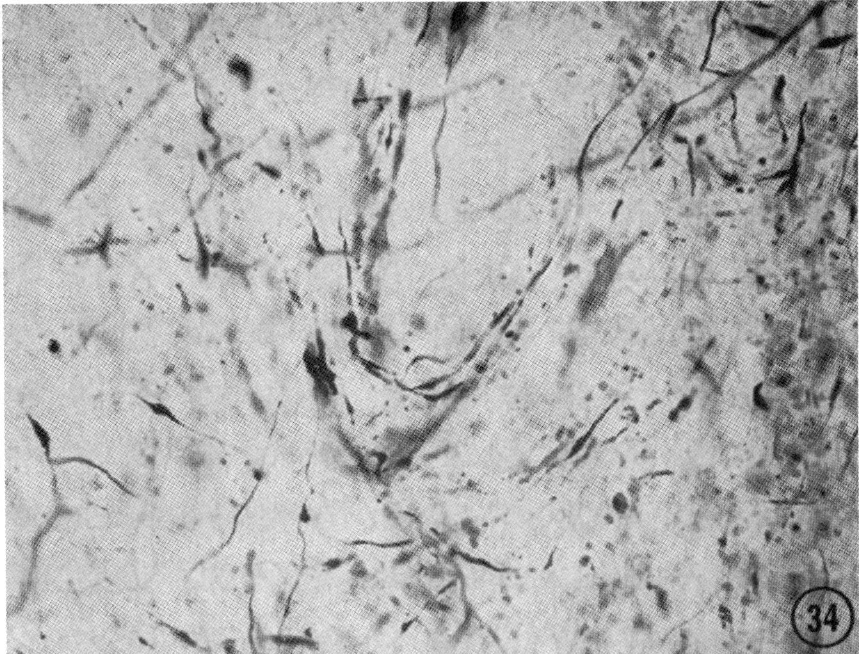

Fig. 33. Degenerating bundles of fine putaminonigral fibers composing the ventral part of the comb system (block b in text-figure 9). Same slide as Fig. 31 (case MGB 12). × 640. Fig. 34. Degenerating fibers of the 'pallidohypothalamic tract'. Sagittal section, × 500 (case MGB 11). Note the recurrent curve in some of the fibers.

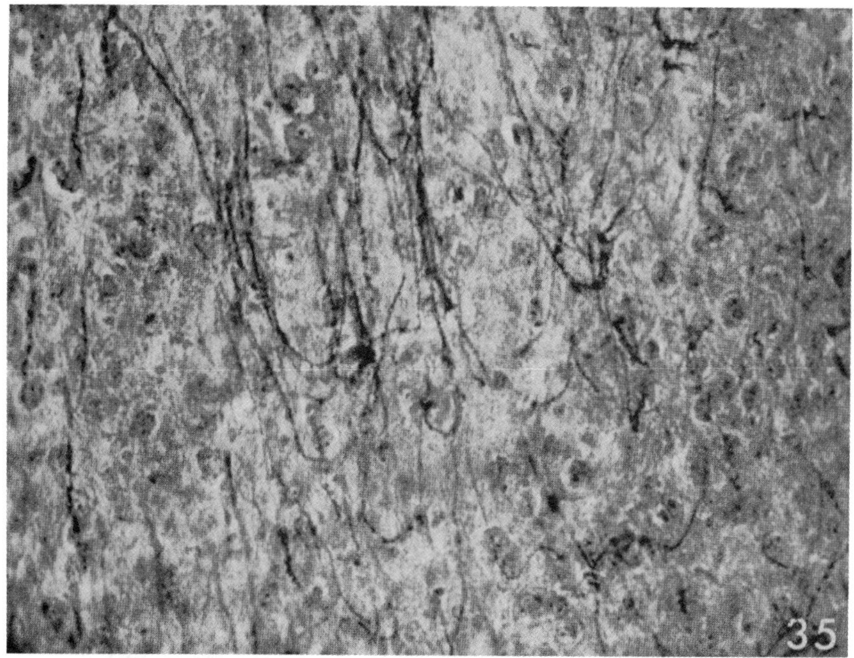

Fig. 35. Pallidohypothalamic 'loops' as appearing in a sagittal section of a chimpanzee brain. Weil stain, × 350.

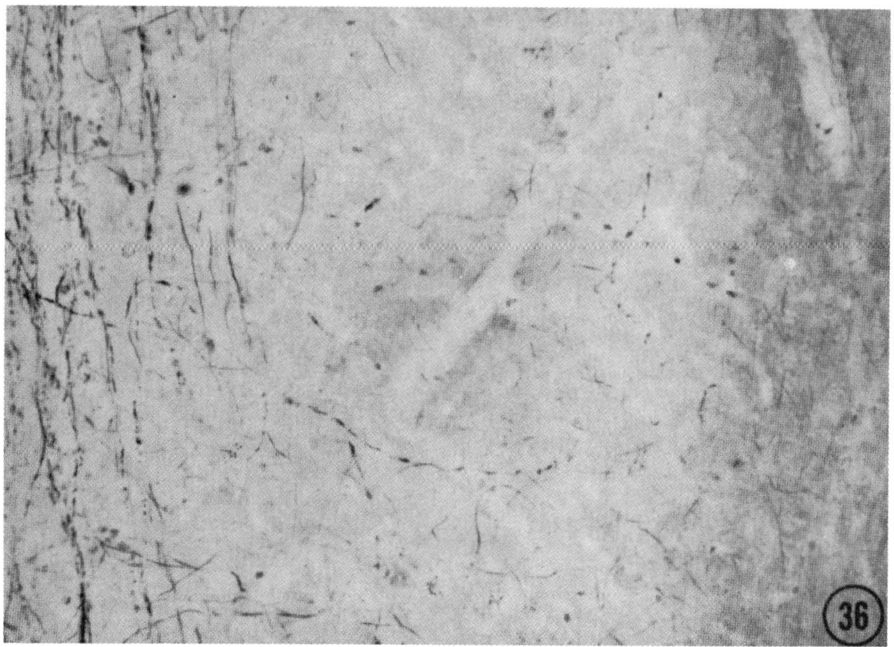

Fig. 36. Low-power view of fiber degeneration in the pallidohypothalamic tract. In frontal sections only a few 'loops' are evident. Sporadic fibers can be seen entering the medial hypothalamic zone (right half of the field shown), but evidence of terminal fiber degeneration appears neither here nor elsewhere in the hypothalamus (case MGB 8). × 200.

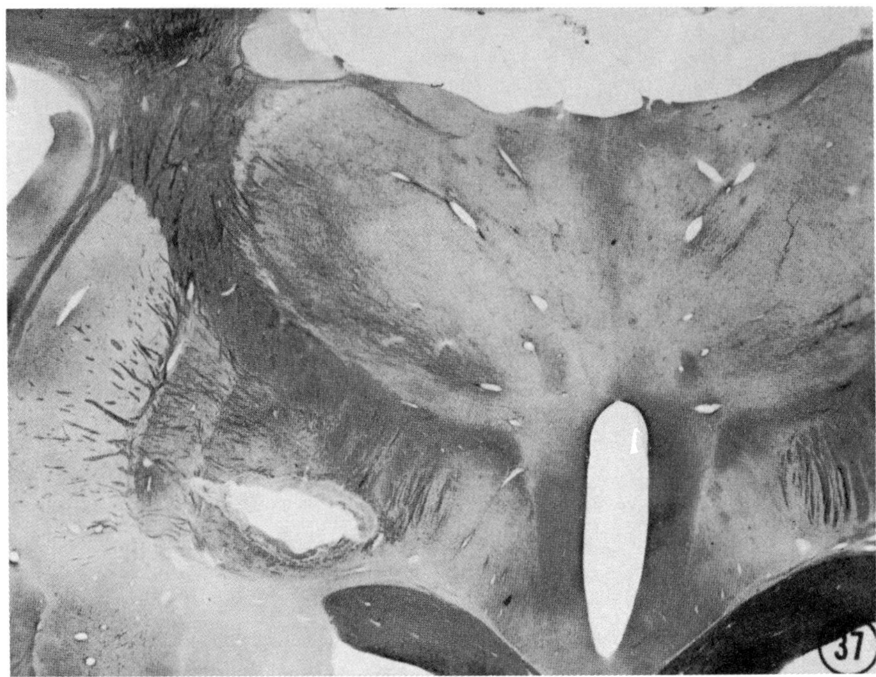

Fig. 37. Low-power photomicrograph of a section through the ,maximum extent of the lesion in case MGB 8. × 5.

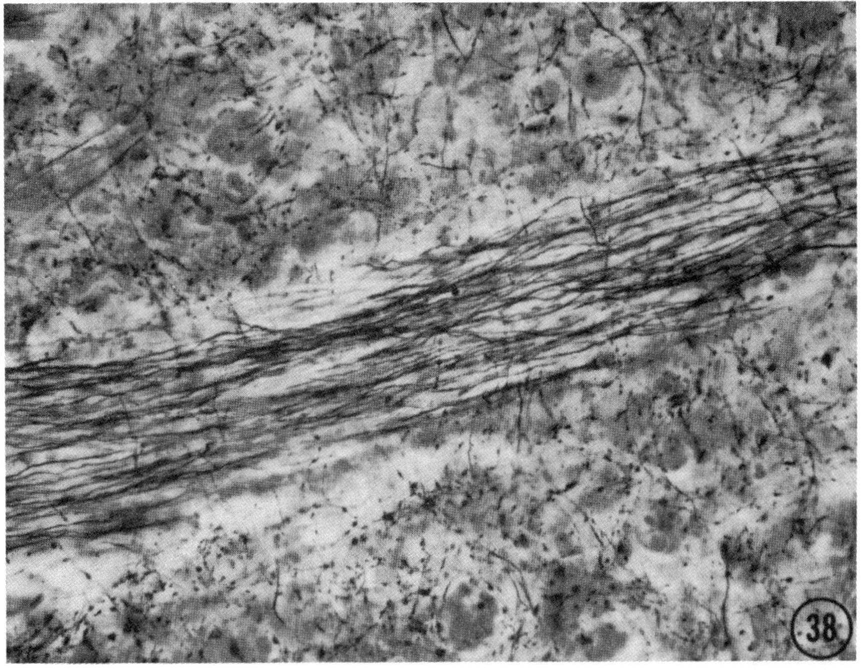

Fig. 38. Dense pericellular degeneration in nucleus ventralis lateralis, pars oralis, traversed by a fascicle of normal fibers. Sagittal section (case MGB 11). × 350.

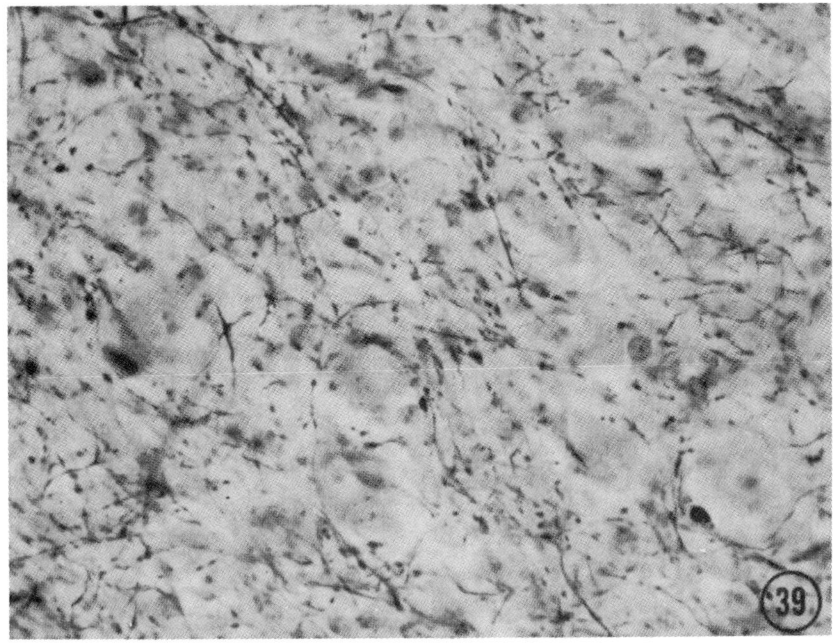

Fig. 39. Pericellular degeneration of fibers from the internal pallidal segment in the nucleus ventralis anterior (case MGB 8). × 450.

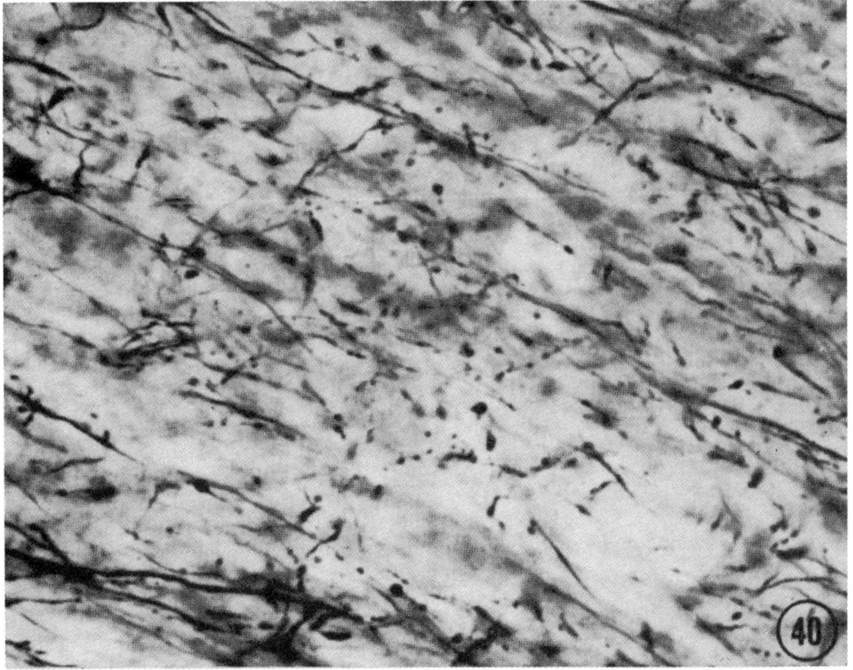

Fig. 40. Degeneration of fine terminal fibers in the 'centre médian' (case MGB 8). × 660.

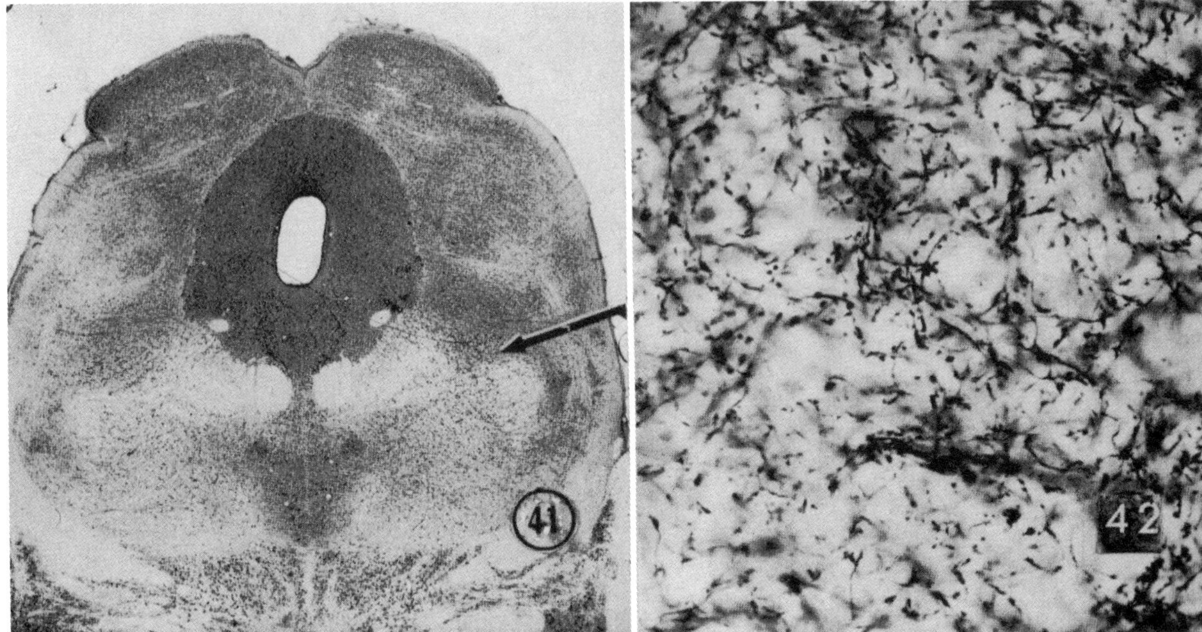

Fig. 41. Large-celled nucleus tegmenti pedunculopontinus, pars compacta (arrow) in a cresylecht-violet-stained section adjacent to the silver-impregnated section shown in Fig. 43 (case MGB 8). × 5.
Fig. 42. High-power view of terminal degeneration indicated by arrow in Fig. 43. × 350.

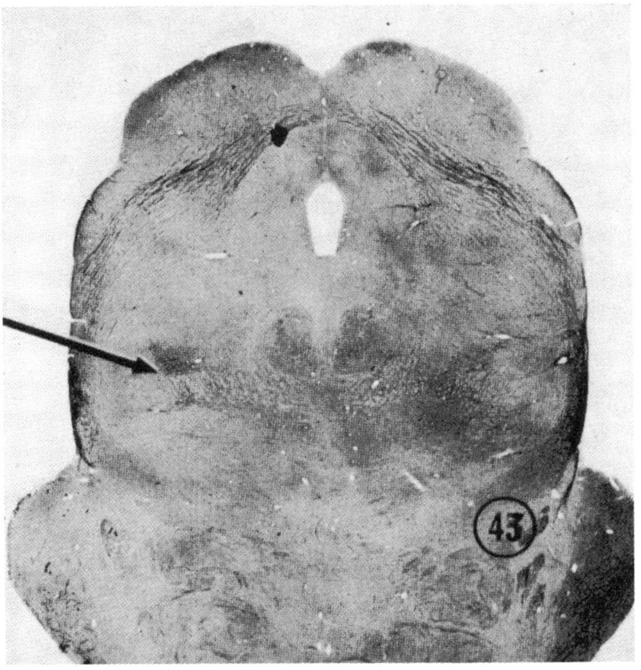

translating older anatomical descriptions into terms of current thalamic terminology. For example, Inui[21] mentions pallidofugal fiber degeneration not only to the rostras part of the ventral nucleus but also to more caudal and medial thalamic regions. In one of his figures he indicates, without comment, fiber degeneration in a cell group labeled 'CM', apparently adjoining the lateral border of the fasciculus retroflexus. Wilson's[81] description raises a similar question. The absence of a relevant illustration in his report leaves doubt as to whether his term 'neighborhood of the internal nucleus' can have referred to the 'centre médian'.

As shown clearly in our cases MGB 8 and 11, the pallidal projection to the 'centre médian' is established mainly by a large number of fascicles separating from the thalamic fasciculus in a straight caudodorsomedial course (see Figs. 14–16). The nearly rectangular mode of offset suggests the possibility that at least part of the fibers in question are collaterals of the fibers composing the massive pallidofugal pathway to the nuclei ventralis anterior and ventrolateralis. Fig. 27 shows the degenerating fibers in the fascicles to be accompanied by numerous normal fibers. Part of these may be pallidofugal fibers not affected by the lesion, but it appears likely that some of the normal fibers are components of a recently identified projection from area 4 of the precentral gyrus to the 'centre médian'[57].

The controversial literature concerning the afferent connections of the 'centre médian' cannot be discussed here in adequate detail and is therefore reviewed elsewhere (see Mehler[34]). On the basis of the present findings it can be stated, however, that the pallidal projection to the 'centre médian' is quite massive and must contribute substantially to the neuropil filling the nucleus. Most of the pallidofugal fibers in question enter the 'centre médian' in fascicles from the rostroventral side. It is interesting in this connection that, according to Olszewski's[51] findings, the number of discrete myelinated fiber bundles in the 'centre médian' is considerable in rostral parts of the nucleus, but gradually diminishes in the caudal direction, the most caudal part of the 'centre médian' containing no such organized bundles at all.

Evidence reported by C. and O. Vogt[75], Nauta and Whitlock[49], and Powell and Cowan[60] indicates that the 'centre médian' projects predominantly to the putamen. The present finding that it also receives a massive projection from the globus pallidus suggests the existence of a reciprocal lenticulothalamic connection: putamen–globus pallidus–'centre médian'–putamen. This circuit appears to represent a neural mechanism either linked collaterally or placed in parallel to the previously recognized pallidothalamo-cortical pathway involving the ventrolateral thalamic nucleus and the precentral gyrus[54].

*(c) Pallidohabenular connections.* Fibers from the globus pallidus to the habenular complex have been recognized, by virtue of differential staining characteristics, by Marburg[33] in normal and pathological human material. Experimental evidence of this connection appears to have been reported in only two publications[63,81]. Wilson, in some of his cases of pallidal lesion in the monkey, noted degeneration ol fibers on the dorsal surface of the thalamus. He identified these fibers as componentf of the 'stratum subcaudatum of Sachs', but did not state their distribution. The Ransons[63] confirmed Wilson's[81] findings but gave a much more detailed account in

which the degeneration is described as scattered over the width of the stratum zonale thalami, converging caudally into the stria medullaris, and terminating in the lateral nucleus of the habenula.

The present findings, illustrated in case MGB 8 (see Figs. 11–18), confirm the description of the Ransons and afford some further statements concerning the trajectory and distribution of the apparent pallidohabenular connection. Firstly, the fibers in question are of the same calibre as those of the dorsal and ventral divisions of the ansa lenticularis. They appear to become separated from the main body of the ansa near the apex of the globus pallidus. It is difficult to trace this fiber group in continuity, presumably because of irregular and divergent rostral trajectories through and around the medial part of the internal capsule. Curving dorsally over the rostral curvature of the thalamus, the fibers mingle with other fiber systems covering the rostral thalamic pole (Fig. 11) and divide into two main contingents which join the stria medullaris and stria terminalis respectively, whereas scattered fibers become incorporated into the stratum zonale between the two striae. Secondly, the terminal arborization of the system appears to be limited largely to the lateral, magnocellular component[52] of the lateral habenular nucleus (Figs. 17–18).

The Ransons, although inclined to consider the connection in question to be of pallidal origin, cautiously pointed out that a pallidohabenular pathway could have been mirrored in their experiments by involvement of fibers from the amygdaloid complex passing through the globus pallidus. Although later studies of the amygdalofugal projection in the monkey[46] have failed to disclose a direct amygdalohabenular connection, the possibility that pallidal lesions may involve fibers of non-pallidal origin in passage to the habenula nevertheless deserves further study, especially because the habenula is known to receive afferents from several basal forebrain areas in close proximity to the lentiform nucleus.

To be considered among the possible sources of such fibers is the large-celled nucleus basalis of Ganser ('nucleus of the ansa peduncularis' of Meynert, see ref. 55), for this cell group lies in close contact with the ventral division of the ansa lenticularis and is, moreover, known to extend with scattered cells into the external medullary lamina of the lentiform nucleus[12]. In cases MGB 6 and MGB 7 with extensive involvement of the nucleus basalis, habenulopetal fiber degeneration similar to that observed in case MGB 8 was indeed observed (see Figs. 21–25). This degeneration, however, was no greater in volume than that following lesions of the internal pallidal segment; moreover, in contrast to the latter its distribution was found to involve largely the medial, relatively parvicellular subdivision[52] of the lateral habenular nucleus. These observations suggest that the habenulopetal fibers degenerating following lesions of the internal pallidal segment do not originate in the amygdalopiriform complex, and are not identical with those affected by lesions in the sublenticular region. However, other possible sources of habenulopetal projection need to be investigated before the present suggestive evidence of a direct pallidohabenular connection can be considered conclusive.

*Projections from the corpus striatum to the hypothalamus*

*(a) Striatohypothalamic connections.* No claims of direct projections from the striatum to the hypothalamus appear to exist. In the normal opossum brain, however, Loo[32] described a contribution from the caudate nucleus to the medial forebrain bundle, and it appears likely that such caudatofugal fibers would terminate, at least in part, in the lateral preoptic region and hypothalamus. Experimental studies of the caudatofugal projection[69,76] have failed to verify this suggestion, but the possibility remains that caudatohypothalamic fibers arise from the most ventral part of the head of the caudate nucleus, a region hitherto neglected in experimental work.

*(b) Pallidohypothalamic connections.* Meynert[38] and Forel[13] appear to have been the first to describe a slender but conspicuous fiber bundle which projects medially from field H, loops over and through the fornix bundle, and curves steeply ventral-ward into the medial zone of the infundibular region (see case MGB 8; Fig. 13). After these earliest descriptions, the bundle has been mentioned under various names (see Kodama[28] and Vidal[71]). Following Bard and Rioch's[2] description in the cat it became widely known as the pallidohypothalamic tract[22,26,62,71,83]. Bard and Rioch did not identify any particular hypothalamic cell group as the recipient of the pallido-hypothalamic bundle, and stated only that the fibers appear to lose their myelin sheath near the ventromedial nucleus. Vidal[71], using the Marchi technique, experi-mentally confirmed the pallidal origin of the bundle, but neither he nor Ingram[22] claimed to have found histological evidence of its termination in either the ventro-medial nucleus or elsewhere in the hypothalamus. More recently, however, Johnson and Clemente[26] from an experimental study by the Nauta–Gygax method in the cat, reported evidence of degenerating fibers of the pallidohypothalamic tract among the cells of the ventromedial nucleus. Unfortunately, Johnson and Clemente did not illustrate this evidence.

Although the present experiments have confirmed the pallidal origin of the bundle, they have provided no evidence that the pallidohypothalamic tract in the monkey actually terminates in the hypothalamus. Fiber degeneration appeared in the bundle following lesions of either the medial segment of the globus pallidus or the ventral division of the ansa lenticularis (Figs. 13 and 36) but no histological signs of preterminal degeneration were found in such cases in the ventromedial nucleus or elsewhere in the hypothalamus. It is unlikely that this negative finding is attributable to a technical failure, for the terminal arborizations of fibers as coarse as those of the pallidohypothalamic tract are generally easy to identify with the technique used in this study. An unexpected explanation was suggested by a careful study of serial sagittal sections of normal monkey and chimpanzee brains. In such material numerous fibers of the pallidohypothalamic tract can be seen to describe sharp, fish-hook-like curves at various dorsoventral levels in the hypothalamus, apparently initiating a recurrent course toward the subthalamic region (Fig. 34–35). Although an intrahypo-thalamic termination of some individual fibers of the bundle cannot be ruled out completely, the present findings suggest the possibility that the so-called pallido-hypothalamic tract is composed largely of aberrant fibers of the ansa lenticularis rejoining the remainder of the system following a longer or shorter descent into the hypothalamus. A few isolated fibers of the bundle, however, continue ventralward, cross the midline in Meynert's commissure, and extend laterally underneath the internal capsule (Fig. 13). The destination of these few decussating fibers could not be ascertained; it still remains conceivable that they terminate in the contralateral globus pallidus as originally suggested by Darkschewitsch and Pribytkow[6].

Evidence of a further pallidohypothalamic connection was reported by Glees[15]. In two of his experimental monkeys Glees observed dense Marchi degeneration in the region between the mammillary body and the subthalamic nucleus, extending caudally as far as the interpeduncular nucleus. In the present study, comparable findings were made in only two cases (MGB 6 and MGB 7, see Figs. 21–25), both of which showed extensive involvement of the substantia innominata. In both these cases the preterminal degeneration lateral to the mammillary body clearly represented a pathway extending caudally from the lesion via the most lateral part of the medial forebrain bundle instead of following the ansa or fasciculus lenticularis. From our findings the conclusion appears inescapable that the fiber degeneration in question resulted from the involvement of a sublenticular pathway, not directly related to the globus pallidus and possibly originating in the substantia innominata. The extent of the lesion in at least one of Glees' cases in question (M.S.L. 1) supports this conclusion. It should also be noted that lesions involving the sublenticular grey matter also unavoidably involve the ansa peduncularis which contains numerous further hypothalamopetal fibers of non-pallidal provenance (see the present discussion of the ansa peduncularis).

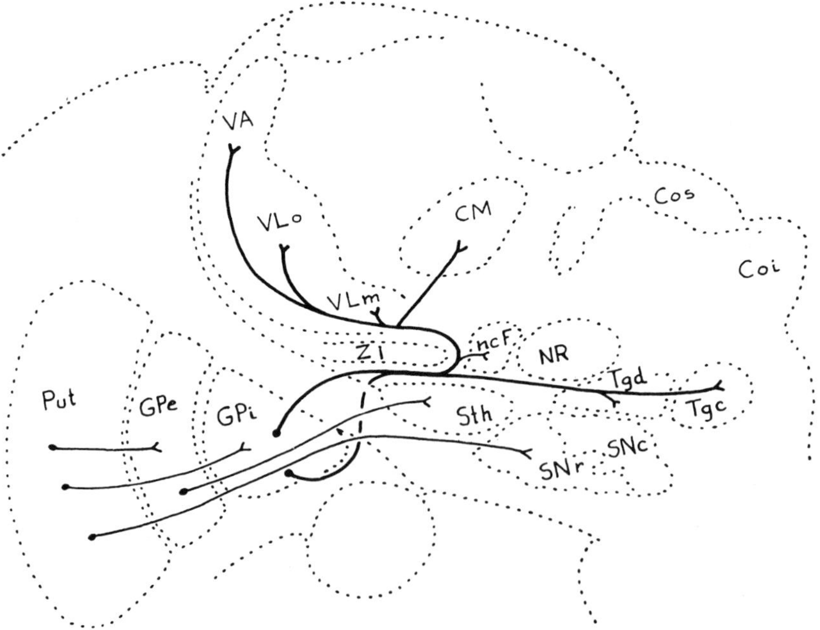


Fig. 44. A diagrammatic representation of various lenticulofugal fiber systems discussed in the text. The brain stem is oriented sagittally, the lentiform nucleus frontally. Not shown is the apparent pallidohabenular connection. Abbreviations see page 38.

*Pallidal projections to the subthalamic region*

*(a) Subthalamic nucleus.* Connections of the globus pallidus with the subthalamic nucleus of Luys appear to have been reported first by Wernicke[79] and
Flechsig[11]. Von Monakow[39] included the connection in his broad definition of the
ansa lenticularis and labelled it the middle division of the ansa. Wilson's[81] Marchi
studies in the monkey furnished the first experimental verification of pallidosubthalamic fibers. Wilson described the connection as the 'second division of the ansa
lenticularis' and stated that it arose from the globus pallidus with fibers finer than
those forming the remainder of the ansa. C. and O. Vogt[74] noted heavy cell loss in
the subthalamic nucleus in cases of pathological defects confined to the external
pallidal segment, and concluded that the two structures must be interconnected at
least preferentially. Ranson and Ranson[62] found indirect support for the Vogts' view
in their observation that subthalamic lesions restricted to the ansa and fasciculus
lenticularis" (*i.e.* von Monakow's ventral and dorsal ansal divisions), and not involving the subthalamic nucleus, caused cell loss only in the internal pallidal segment.
In a later study by the Marchi technique the Ransons[63] furnished more direct evidence
that pallidosubthalamic fibers arise from the external segment of the globus pallidus.

In agreement with the earlier conclusions of the Vogts and the Ransons, the
present findings (see, for example, case MGB 12 illustrated by Figs. 7–10) clearly
show that the subthalamic nucleus receives a massive projection from the external
pallidal segment. This projection possibly represents the only extrinsic efferent
connection of the external segment; in any case, in accord with the Ransons' statements[63], the present observations indicate that the external segment does not contribute to the dorsal or ventral division of the ansa and hence, appears to take no part
in the pallidal projections to the thalamus and mesodiencephalic tegmentum. The
existence of additional pallidosubthalamic fibers arising in the internal pallidal
segment cannot be ruled out by experiments such as the present, for lesions in that
part of the globus pallidus unavoidably interrupt fibers passing to the subthalamic
nucleus from the external segment.

A comparison of the present cases of pallidal lesion confined to the external
segment indicates the existence of an orderly rostrocaudal arrangement in the projection to the subthalamic nucleus. The material available was, however, too restricted
to afford an adequate picture of further details of the topographical organization,
particularly because no lesions were placed in the far rostral and caudal quarters of
the globus pallidus. It was nevertheless apparent that, in all cases of pallidal lesions,
preterminal degeneration was densest in the lateral two-thirds, and slight or absent in
the area approximately corresponding to the medial quarter of the subthalamic
nucleus. Further experiments will be needed to determine whether this medial subdivision receives afferents from other pallidal regions; however, as described in the
discussion of the ansa peduncularis, there is evidence that it receives a non-pallidal
projection, possibly originating in the substantia innominata. The evidence for this
connection is apparent from a comparison of case MGB 12, showing a lesion in the
most ventral part of the external pallidal segment (Figs. 8–9), with MGB 6 in which

the same pallidal region was involved in a lesion extending into the substantia innominata (Figs. 23–24).

*(b) Zona incerta.* The zona incerta has been mentioned only infrequently in relation to the lenticulofugal projection. Wilson[81], on the strength of negative Marchi evidence denied the existence of lenticulo-incertal connections. Kodama[28] and Woodburne *et al.*[83], on the other hand, described fibers of the fasciculus lenticularis terminating in the zona incerta. The present findings, in accord with Wilson's observations, have furnished no evidence of putamino- or pallido-incertal fiber connections. The medial part of the zona incerta is traversed by some thick fibers of the fasciculus lenticularis in their dorsal curve to the thalamus, but no signs of termination of such fibers in the structure were observed. Findings in other cases studied in this laboratory suggest that the zona incerta receives its main projections from the cerebellum through the brachium conjunctivum, and from a large variety of cortical regions[34,57].

*(c) Nuclei of the tegmental fields of Forel.* A termination of the ansa lenticularis fibers among the cells of Forel's field H (nucleus campi Foreli) appears to have been accepted first by C. and O. Vogt[73]. Wilson[81] mentioned the prerubral cell group in the path of the ansa lenticularis, but did not explicitly identify it as a locus of termination of lenticulofugal fibers. Kodama[28] expressed skepticism regarding pallido-fugal fibers to the nucleus campi Foreli, but Papez and Stotler[56] emphasized the importance of the prerubral field as a way-station in the pallidorubral connection. Evidence of pallidal projections to the prerubral field was likewise reported by Woodburne *et al.*[83] in normal monkey brains, and by Johnson and Clemente[26] in an experimental study in the cat.

The present experiments have supplied further evidence of ansa lenticularis fibers distributed to Forel's tegmental fields. Most of these fibers seem to terminate caudal to the level at which the fasciculus thalamicus begins to separate itself from the remainder of the ansa (Figs. 15–16). The dorsomedial tegmental field[37] and its caudal extension immediately dorsal to the rostral pole of the red nucleus appear to be the main recipients of this pallidofugal projection. The volume of the connection is difficult to estimate, especially at more rostral levels where field H is densely packed with degenerating thick and tortuous fibers of passage.

*Pallidomesencephalic connections.* The various viewpoints in regard to this controversial subject will be briefly reviewed, with comments based on present findings.

*(a) Pallidorubral connections.* In the earlier reports of Wilson[81] and C. and O. Vogt[73,74] particular emphasis was placed upon the pallidal projections via field H to the red nucleus. The existence of a direct pallidorubral connection was thereafter widely accepted[12,24,31,65,82], but the notion did not long remain uncontested. Von Monakow[39] had earlier stated his inability to ascertain a continuity between the ansa lenticularis and the capsule of the red nucleus, and again in his celebrated monograph (1909–1910) had not mentioned[40] the globus pallidus among the sources of afferent fibers to the red nucleus. Muskens[43], Morgan[41], and Kodama[29] expressed skepticism in regard to Wilson's pallidorubral projection, and the existence of the connection

was later categorically denied by Verhaart[70] and Papez and Stotler[56]. Papez and Stotler nevertheless regarded the red nucleus as an important way-station in the descending discharge of the globus pallidus, assuming that the nucleus receives pallidal impulses via short neuronal links arising in the prerubral field. The Ransons[63] likewise failed to find significant evidence of direct pallidorubral connections in their Marchi experiments in the monkey, and Mettler[37] omitted the connection in his summary diagram of striatal and pallidal projections. Recently, however, Johnson and Clemente[26] reported experimental evidence of a direct pallidorubral connection in the cat.

Our present findings, in accord with most of the more recent previous studies, suggest that direct pallidorubral fibers are extremely sparse in the monkey (see, for example, case MGB 8: Figs. 16–20). The few that could be identified appeared to be isolated thick fibers spreading medially from the fasciculus lenticularis (field $H_2$).

Although the present experiments have provided no indication of a significant direct pallidorubral connection, it is clear that the possibility of a prominent involvement of the red nucleus and its descending efferent pathways in the transmission of pallidofugal impulses cannot be dismissed on the basis of the cumulative strength of the currently available negative evidence alone. Limitations in the technique and the scope of all experiments reported thus far, including those in the present study, have precluded any experimental evidence of intermediate links between the ansa lenticularis and the red nucleus, such as the prerubrorubral tract proposed by Papez and Stotler[56]. The possible existence of such indirect connections would seem to merit further investigation.

*(b) Pallidonigral connections.* Wilson[81] reported experimental evidence of a small number of pallidonigral fibers in the monkey. Kodama[28], on the basis of observations in human pathological material, accepted the existence of pallidal projections to the pars reticulata of the substantia nigra. Poppi[58] and Winkler[82] included pallidonigral fibers in their respective pallidotegmental systems (see below). Recently, the connection was suggested again by Johnson and Clemente's experimental findings in the cat[26].

The evidence obtained in the present study is rather limited, and the question must be left open in how far it presents a true picture of the pallidonigral relationship. No reliable evidence of a pallidal projection to the pars reticulata of the substantia nigra was found: the fiber degeneration observed in this structure resembles that caused by lesions of the putamen and could conceivably be due entirely to interference with the putaminonigral fibers traversing the globus pallidus. However, in contrast to the findings in cases of putaminal lesions, a small and circumscript area of dense preterminal degeneration was found in the caudal one-third of the pars compacta of the substantia nigra in both the case of MGB 8, with a lesion of the internal pallidal segment (see Fig. 19) and MGB 7 with a sublenticular defect involving the ventral division of the ansa lenticularis. The fibers in question appear to be components of the ventrolateral pallidotegmental fiber system which passes caudalward through the dorsally adjoining nucleus tegmenti pedunculopontinus (see below). The volume of this apparent pallidonigral connection cannot be estimated, for no case of complete destruction of the globus pallidus was studied. However, the limited and insular distribution of its fibers, confined to a small part of the pars compacta, suggests that the projection is discrete and precisely localized, rather than massive and diffuse.

*(c) Other pallidomesencephalic connections.* Wilson[81] did not mention pallidofugal fibers to any mesencephalic structure other than the red nucleus and the substantia nigra. Ranson *et al.*[63] concluded from their Marchi experiments in the monkey that the vast majority of the fibers of the ansa and fasciculus lenticularis did not extend into the mesencephalon. However, evidence of further pallidotegmental connections was reported from several other studies. In general, each of these reports indicated the existence of either one or both of two pallidofugal pathways extending caudally from the subthalamic tegmental fields: (1) a fiber group entering the midbrain tegmentum dorsal to the red nucleus and, according to some authors, continuing caudally through the dorsomedial tegmental area occupied by the so-called central tegmental tract and the medial longitudinal fasciculus, and (2) a fiber system passing ventral and lateral to the red nucleus to the ventrolateral regions of the pontomesencephalic tegmentum.

*Pallidofugal pathways associated with the dorsomedial tegmentum.* In their well-known diagram, C. and O. Vogt[73,74] incorporated a projection from the pallidum into the ipsilateral nucleus of Darkschewitsch and the contralateral nucleus interstitialis. A similar connection had been postulated several years earlier by Muskens[42] as a component of the reciprocal connections which this author believed to exist between the corpus striatum and the vestibular nuclei. Muskens' and the Vogts' conclusions were supported by Jakob[24], Riese[64], and Morgan[41], and more recently by Woodburne *et al.*[83] and Johnson and Clemente[26]. Riese[64] and Morgan[41] also reported evidence of pallidofugal fibers descending in the medial longitudinal fasciculus.

It appears likely that the afore-mentioned descriptions can be considered together with those of pallidal efferents descending in the central tegmental tract, for both must refer to pallidofugal fibers entering the midbrain tegmentum dorsal to the red nucleus. Foix and Nicolesco[12] mentioned a small bundle of such fibers, coming from field $H_2$ and passing dorsal to the red nucleus toward the midbrain tegmentum. Pallidofugal components of the central tegmental tract descending as far caudally as the inferior olive were described by Winkler[82]. Weisschedel[78] accepted the existence of a pallidal contribution to the central tegmental tract, but stated that its volume seemed to be smaller than had been suggested by Winkler. Woodburne *et al.*[83] in a study of normal monkey material, confirmed the existence of fibers belonging to the general system of the ansa lenticularis which follow a central tegmental pathway to the mesencephalic and bulbar reticular formation and inferior olive, but they expressed caution in interpreting the system as a direct pallidal projection, suggesting that the pathway might be interrupted, at least in part, in the zona incerta. The pathway as described by the above-mentioned authors appears to correspond, in part at least, to the rubro-reticulo-olivary connections indicated by Papez and Stotler[56] as potential (*i.e.* transsynaptic) conveyors of pallidofugal impulses. Papez and Stotler, however, made no mention of direct pallidofugal projections to the mesencephalic and bulbar tegmentum or inferior olive.

*Ventrolateral pallidotegmental pathways.* In his re-examination of Edinger and

Fischer's[9] unique case of a human infant with complete cystic atrophy of both cortical hemispheres, Wallenberg[77] found evidence suggesting the existence of a pallidofugal fiber system traversing the lateral one-third of the substantia nigra and joining the medial lemniscus in a descending course to the lateral pontine tegmentum. Prior to Wallenberg's study, the fiber group in question, Schlesinger's 'lateral pontine bundles', had been thought to represent a cortical projection. Wallenberg's interpretation of the lateral pontine bundles as a pallidofugal fiber tract was promptly supported by Jakob[25] and later by Poppi[58] and Poppi and Imber[59] who labelled the system 'pallidopeduncular bundle'. Wallenberg's pallidofugal fiber system appears to be identical to a large extent with Winkler's[82] 'bundle X', described as a group of fine-fibered fascicles of lenticular, and probably pallidal origin traversing the subthalamic nucleus and substantia nigra and continuing caudally in close association with the medial lemniscus. It is also comparable to the 'lateral striobulbar fasciculus' observed by Morgan[41] in an experimental study in the cat. Morgan, however, appears to have remained alone in his identification of direct pallidonuclear connections—*viz.* to the oculomotor, hypoglossal, masticatory, facial and ambiguus nuclei—in the system. More recently, the existence of pallidofugal fibers to the ventrolateral regions of the pontomesencephalic tegmentum has been confirmed by two further studies, neither of which however, suggested that the fibers in question followed the transnigral course indicated by Wallenberg's and Winkler's descriptions: in normal monkey brains Woodburne *et al.*[83] traced a component of the ansa lenticularis caudalward along the lateral aspect of the red nucleus to ventral and lateral parts of the cell region described by Huber *et al.*[20] as nucleus mesencephalicus profundus. A similar distribution of pallidofugal fibers was observed in Johnson and Clemente's[26] experimental study in the cat. It appears that Papez and Stotler[56] and Papez[54] described the same general fiber system in normal human brains as arising in the substantia nigra and subthalamic nucleus instead of the pallidum. Previously, Papez[53] had labelled the fibers in question as the 'subthalamicotegmental tract', under the impression that they originated from the subthalamic nucleus. It is of interest to note Papez's additional observation that this fiber system is particularly well-developed in the horse (see below: Gillilan[14]).

*The present findings.* The present experiments have furnished only scant evidence of a pallidofugal projection to the central tegmental region of the mesencephalon. In case MGB 8 (Figs. 15–20) scattered fibers could be traced caudalward through the region dorsal to the red nucleus into the general system of the central tegmental tract, but, as already suggested by Weisschedel[78], their number is very small, and it is augmented only slightly by contributions from the ventrolateral pallidotegmental fiber system. In correspondence to the paucity of pallidofugal fibers passing dorsal to the red nucleus, and contrary to the accounts of the Vogts and others, no evidence of a significant direct pallidal projection to the nuclei of Cajal (nucleus interstitialis) and Darkschewitsch was found. Neither could the presence of pallidofugal fibers in the medial longitudinal fasciculus be confirmed.

According to the present findings, the ventrolateral pallidotegmental fiber system represents the main direct pallidomesencephalic connection (Figs. 15–20, 26).

The observations regarding this fiber system in case MGB 8 appear to correspond to the anatomical description of Woodburne *et al.*[83]. The fascicles of the tract first become identifiable where they separate from field $H_2$ and collect dorsal and medial to the subthalamic nucleus. The fiber group passes caudally lateral and ventral to the red nucleus as a band of slightly scattered fascicles, partly mingling with the medial lemniscus. Only a few fibers appear to end at this rostral mesencephalic level. At frontal levels involving the caudal pole of the red nucleus, the fascicles occupy a ventrolateral position in the tegmentum, immediately central to the stratum lemnisci. With the exception of a few fibers issued dorsomedially into more central tegmental regions, and a component distributed to the pars compacta of the substantia nigra, the ventrolateral pallidotegmental tract appears to terminate almost exclusively in the cell region termed nucleus tegmenti pedunculopontinus by Olszewski and Baxter[52] in the human brain. A particularly dense termination of pallidofugal fibers is evident in the caudal subnucleus compactus, a relatively magnocellular component partly embedded in the brachium conjunctivum a short distance rostral to the isthmus (see Figs. 20, 41–43 of the present paper). The large multipolar cells of this region were already noted in the monkey by Papez[54], who believed them to be the recipients of nigrotegmental fibers. Crosby and Woodburne[5], in a study of the midbrain tegmentum of monkey and man, included the large cells in the pars lateralis of their nucleus mesencephalicus profundus (*cf.* their Figs. 11–12). Gillilan[14] in a concurrent study found this large-celled region particularly well-developed in the horse brain. Crosby and Woodburne[5] suggested that the phylogenetic increase in the development of the large cells might be compensatory to the concomitant regression of the magnocellular component of the red nucleus. Noback[50] identified the cell group in the gorilla and in accordance with Olszewski and Baxter's terminology labelled it nucleus tegmenti pedunculopontinus, subnucleus compactus. In a recent cytoarchitectonic study of the human midbrain, Feremutsch and Simma[10] describe comparable cells which they include in the caudolateral region of their nucleus centralis reticularis pontis. Except for differences in nomenclature, the description given by these authors is comparable to the earlier accounts of Jacobsohn[23], Stern[68], and Olszewski and Baxter[52].

No pallidofugal fibers could be followed caudally beyond the region of dense termination in the nucleus tegmenti pedunculopontinus. Thus, the present findings have furnished no evidence of significant direct pallidal projections to brain stem levels below the mesencephalon.

ACKNOWLEDGEMENTS

The authors are indebted to the late Mrs. Mildred H. Albrecht and to Messrs. Gordon Fletcher and Curtis King for their consistently excellent technical support. The photomicrographs were made by Miss Sally L. Craig.

Grateful appreciation is expressed to Dr. David McKenzie Rioch for the years of advice, support and encouragement that made this and other studies possible.

*Abbreviations in Figures*

| | | | | |
|---|---|---|---|---|
| AM | = n. anterior medialis thalami | | | cellularis |
| Am | = amygdaloid complex | | Pf | = n. parafascicularis |
| AV | = n. anterior ventralis thalami | | Prt | = pretectal region |
| BC | = brachium conjunctivum | | Pul | = pulvinar |
| CA | = commissura anterior | | Put | = putamen |
| CI | = capsula interna | | PV | = n. paraventricularis hypothalami |
| CL | = n. centralis lateralis thalami | | Pva | = n. paraventricularis anterior |
| CM | = 'centre médian' | | Re | = n. reuniens |
| Coi | = colliculus inferior | | Rt | = n. reticularis thalami |
| Cos | = colliculus superior | | SM | = stria medullaris |
| DBC | = decussation of brachium con-junctivum | | SNc | = substantia nigra, pars compacta |
| | | | SNr | = substantia nigra, p. reticulata |
| DMH | = n. dorsomedialis hypothalami | | SO | = n. supraopticus |
| F | = columna fornicis | | ST | = stria terminalis |
| FL | = fasciculus lenticularis | | Sth | = subthalamic nucleus |
| fPH | = pallido-hypothalamic tract | | Stp | = subthalamo-prerubral field |
| GPe | = globus pallidus, external segment | | Tgc | = n. tegmenti pedunculopontinus, p. compacta |
| GPi | = globus pallidus, internal segment | | | |
| H | = field H of Forel | | Tgd | = n. tegmenti pedunculopontinus, p dissipata |
| HL | = lateral hypothalamic region | | | |
| ITP | = inferior thalamic peduncle | | TO | = optic tract |
| LD | = nucleus lateralis dorsalis thalami | | VA | = n. ventralis anterior thalami |
| LM | = medial lemniscus | | VAmc | = n. ventralis anterior, p. magno-cellularis |
| m | = n. medialis habenulae | | | |
| MB | = mammillary body | | VL | = n. ventralis lateralis thalami |
| mc | = n. lateralis habenulae, p. magno-cellularis | | VLc | = n. ventralis lateralis, p. caudalis |
| | | | VLm | = n. ventralis lateralis, p. medialis |
| MD | = n. mediodorsalis thalami | | VLo | = n. ventralis lateralis, p. oralis |
| MDmc | = n. mediodorsalis, p. magno-cellularis | | VLois | = VLo islets in Vlc |
| | | | VLps | = caudal, small-celled area of VLc |
| NB | = n. basalis (n. ansae peduncularis) | | VMH | = n. ventromedialis hypothalami |
| NC | = n. caudatus | | VPL | = n. ventralis posterior lateralis thalami |
| NcF | = n. campi Foreli (prerubral field) | | | |
| NI | = n. interstitialis (Cajal) | | VPLo | = n. ventralis posterior lateralis, p. oralis |
| NR | = n. ruber | | | |
| PC | = cerebral peduncle | | VPM | = n. ventralis posterior medialis |
| NST | = bednucleus of stria terminalis | | X | — Zono X of n. ventralis laternalis |
| Pc | = n. paracentralis thalami | | ZI | = zona incerta |
| pc | = n. lateralis habenulae, p. parvi- | | | |

SUMMARY

The fiber degenerations resulting from variously located lesions of the lentiform nucleus were studied in the rhesus monkey by the aid of the Nauta–Gygax and Albrecht–Fernstrom techniques. The following observations were made.

*(1) Putaminofugal connections.* Thin fibers originating in the putamen and composing Wilson's 'pencil' bundles traverse the globus pallidus, converging toward the medial point of the lentiform nucleus. The majority of these fibers terminate in both segments of the globus pallidus, but a considerable number continue caudalward, perforating the cerebral peduncle as ventral components of Edinger's comb system, and terminate in lateral parts of the substantia nigra, pars reticulata.

*(2) Pallidofugal connections.* The ansa lenticularis as defined by von Monakow

originates exclusively from the globus pallidus. Its middle division, composed of fibers of medium calibre, arises in the external pallidal segment and traverses the cerebral peduncle as the dorsal component of the comb system to end in the subthalamic nucleus. The thick-fibered dorsal and ventral ansal divisions arise in the internal pallidal segment and combine to form the fasciculus lenticularis which represents the only apparent direct connection of the globus pallidus with the thalamus and the mesencephalic tegmentum.

(a) Pallidothalamic fibers follow successively the lenticular and thalamic fasciculi and are distributed to the nuclei ventralis lateralis (subnuclei medialis and oralis of Olszewski and Baxter; none to Zone X and subnucleus caudalis) and ventralis anterior (except subnucleus VAmc). A considerable number of thinner fibers, possibly collaterals of those to VL and VA, terminate in the 'centre médian'; this connection appears to close a potential transthalamic circuit: putamen–globus pallidus–'centre médian'–putamen.

(b) There is suggestive evidence of pallidofugal fibers following the stratum zonale thalami to the habenula.

(c) Pallidohypothalamic connections could not be identified. Most, and possibly all, of the ansal fibers composing the so-called pallidohypothalamic tract loop back into Forel's fields after a shorter or longer descent into the hypothalamus.

(d) Fibers of the fasciculus lenticularis by-passing the thalamus are distributed to the nucleus of Forel's field H (prerubral field). Longer fibers of the same category pass caudalward lateral and ventral to the red nucleus and terminate in the nucleus tegmenti pedunculopontinus, particularly in the latter's caudal subnucleus compactus (terminology of Olszewski and Baxter). A few such pallidomesencephalic fibers appear to end in a small circumscript caudal area of the substantia nigra, pars compacta. No evidence was obtained of pallidotegmental fibers extending caudally beyond the mesencephalon.

(e) Pallidal efferents to the zona incerta could not be identified. Only sporadic pallidofugal fibers could be followed to the red nucleus, nucleus interstitialis, and nucleus of Darkschewitsch.

## REFERENCES


1 ALBRECHT, M. H., AND FERNSTROM, R. C., A modified Nauta–Gygax method for human brain and spinal cord, *Stain Technol.*, 34 (1959) 91–94.
2 BARD, P., AND RIOCH, D. McK., A study of four cats deprived of neocortex and additional portions of the forebrain, *Bull. Johns Hopk. Hosp.*, 60 (1937) 73–106.
3 BECHTEREW, W. VON, *Die Leitungsbahnen in Gehirn und Rückenmark*, Georgi, Leipzig, 1899.
4 CAJAL, S. RAMÓN Y, *Histologie du Système Nerveux de l'Homme et des Vertébrés*, Vol. II, Maloine, Paris, 1911.
5 CROSBY, E. C., AND WOODBURNE, R. T., The nuclear pattern of non-tectal portions of the midbrain and isthmus in primates, *J. comp. Neurol.*, 78 (1943) 441–520.
6 DARKSCHEWITSCH, L., AND PRIBYTKOW, G., Ueber die Fasersysteme am Boden des dritten Hirnventrikels, *Neurol. Zbl.*, 10 (1891) 417–429.
7 DEJERINE, J., *Anatomie des Centres Nerveux*, Tome 2, Rueff, Paris, 1901.
8 EDINGER, L., *Vorlesungen über den Bau der nervösen Zentralorgane*, 8th Ed., Vol. I, Vogel, Leipzig, 1911.

9 EDINGER, L., AND FISCHER, B., Ein Mensch ohne Grosshirn, *Pflügers Arch. ges. Physiol.*, 152 (1913) 535–561.
10 FEREMUTSCH, E., AND SIMMA, K., Die Formatio reticularis mesencephali und die Regio praetectalis des Menschen, *Z. Anat. Entwickl.-Gesch.*, 122 (1961) 289–313.
11 FLECHSIG, P., Zur Anatomie und Entwicklungsgeschichte der Leitungsbahnen im Grosshirn des Menschen, *Arch. Anat. Physiol. Anat. Abth.*, (1881) 12–75.
12 FOIX, C., AND NICOLESCO, J., *Anatomie Cérébrale. Les Noyaux Gris Centraux et la Région Mésencéphalo-Sous-Optique*, Masson, Paris, 1925.
13 FOREL, A., Untersuchungen über die Haubenregion und ihre oberen Verknüpfungen im Gehirne des Menschen und einiger Säugethiere, mit Beiträgen zu den Methoden der Gehirnuntersuchung, *Arch. Psychiat. Nervenkr.*, 7 (1877) 393–495.
14 GILLILAN, L. A., The nuclear pattern of non-tectal portions of the isthmus and midbrain in ungulates, *J. comp. Neurol.*, 78 (1943) 289–364.
15 GLEES, P., The interrelation of the strio-pallidum and the thalamus in the macaque monkey *Brain*, 68 (1945) 331–346.
16 GRÜNSTEIN, A. M., Zur Frage von den Leitungsbahnen des Corpus Striatum, *Neurol. Zbl.*, 30 (1911) 659–665.
17 GURDJIAN, E. S., The diencephalon of the albino rat, *J. comp. Neurol.*, 43 (1927) 1–114.
18 HASSLER, R., Ueber die afferenten Bahnen und Thalamuskerne des motorischen Systems des Grosshirns. I. Mitteilung. Bindearm und Fasciculus thalamicus. *Arch. Psychiat. Neurol.*, 182 (1949) 759–785.
19 HASSLER, R., II. Mitteilung. Weitere Bahnen aus Pallidum, Ruber, vestibulärem System zum Thalamus; Uebersicht und Besprechung der Ergebnisse. *Arch. Psychiat. Neurol.*, 182 (1949) 786–818.
20 HUBER, G. C., CROSBY, E. C., WOODBURNE, R. T., GILLILAN, L. A., BROWN, J. O., AND TAMTHAI, B., The mammalian midbrain and isthmus regions. I. The nuclear pattern, *J. comp. Neurol.*, 78 (1943) 129–534.
21 INUI, S., Ueber die strio- und pallido-thalamischen Fasern beim Kaninchen. *Okayama Igakkai Zasshi*, 39 (1927) 2064–2074.
22 INGRAM, W. R., Nuclear organization and chief connections of the primate hypothalamus. *Res. Publ. Ass. nerv. ment. Dis.*, 20 (1940) 195–244.
23 JACOBSOHN, L., Ueber die Kerne des menschlichen Hirnstammes. *Anhang z. d. Abhandl. kön. preuss. Akad. Wiss., Physik.-Math. Kl.*, 1909.
24 JAKOB, A., Die Extrapyramidalen Erkrankungen. *Monogr. a. d. Gesamtgeb. Neur. Psychiat.*, 37, Springer, Berlin, 1923.
25 JAKOB, A., The anatomy, clinical syndromes and physiology of the extrapyramidal system. *A.M. A. Arch. Neurol. Psychiat.*, 13 (1925) 596–620.
26 JOHNSON, T. N., AND CLEMENTE, C. D., An experimental study of the fiber connections between the putamen, globus pallidus, ventral thalamus, and midbrain tegmentum in the cat, *J. comp. Neurol.*, 113 (1959) 83–101.
27 KLINGLER, J., AND GLOOR, P., The connections of the amygdala and of the anterior temporal cortex in the human brain, *J. comp. Neurol.*, 115 (1960) 333–369.
28 KODAMA, S., Ueber die sogenannten Basalganglien (morphogenetische und pathologisch-anatomische Untersuchungen), *Schweiz. Arch. Neurol. Psychiat.*, 23 (1928) 38–100.
29 KODAMA, S., Ueber die sogenannten Basalganglien, *Schweiz. Arch. Neurol. Psychiat.*, 23 (1929) 179–258.
30 LAURSEN, A. M., An experimental study of the pathways from the basal ganglia, *J. comp. Neurol.*, 102 (1955) 1–25.
31 LHERMITTE, J., AND TRELLES, O., Physiologie et physio-pathologie du corps strié et des formations sous-thalamiques, *Encéphale*, 27 (1932) 235–256.
32 LOO, Y. T., The forebrain of the opossum, *Didelphys virginiana*. Part II. Histology, *J. comp. Neurol.*, 52 (1931) 1–148.
33 MARBURG, O., The structure and fiber connections of the human habenula, *J. comp. Neurol.*, 80 (1944) 211–233.
34 MEHLER, W. R., Further notes on the centre median nucleus of Luys. In D. P. PURPURA, M. B. CARPENTER AND M. D. YAHR (Eds.), *Thalamic Integration of Sensory and Motor Activities*, Columbia University Press, New York, In Press.
35 MEHLER, W. R., VERNIER, V. G., AND NAUTA, W. J. H., Efferent projections from dentate and interpositus nuclei in primates, *Anat. Rec.*, 130 (1958) 430 (Abstract).

36 MEHLER, W. R., FEFERMAN, M. E., AND NAUTA, W. J. H., Ascending axon degeneration following anterolateral cordotomy. An experimental study in the monkey, *Brain*, 83 (1960) 718–750.
37 METTLER, F. A., Fiber connections of the corpus striatum of the monkey and baboon, *J. comp. Neurol.*, 82 (1945) 169–204.
38 MEYNERT, TH., Vom Gehirne der Säugethiere. In S. STRICKER (Ed.), *Handbuch der Lehre von den Geweben*, Engelmann, Leipzig, 1872.
39 MONAKOW, C. VON, Experimentelle und pathologische-anatomische Untersuchungen über die Haubenregion, den Sehhügel und die Regio subthalamica, nebst Beiträgen zur Kenntnis früherworbener Gross- und Kleinhirndefekte, *Arch. Psychiat. Nervenkr.*, 27 (1895) 1–128.
40 MONAKOW, C. VON, Der rote Kern des Menschen und der Tiere. *Arb. a.d. Hirnanat. Inst. Zurich*, J. F. Bergmann, Wiesbaden, 1909, pp. 64–116.
41 MORGAN, L. O., The corpus striatum. A study of secondary degenerations following lesions in man and symptoms and acute degenerations following experimental lesions in cats. *A.M.A. Arch. Neurol. Psychiat.*, 18 (1927) 495–549.
42 MUSKENS, L. J. J., An anatomico-physiological study of the posterior longitudinal bundle in its relation to forced movements, *Brain*, 36 (1914) 352–426.
43 MUSKENS, L. J. J., The central connections of the vestibular nuclei with the corpus striatum and their significance for ocular movements and for locomotion, *Brain*, 45 (1922) 454–478.
44 MUSKENS, L. J. J., Experimentelle und klinische Untersuchungen über die Verbindungen der unteren Olive und ihre Bedeutung für die Fallrichtung, *Arch. Psychiat. Nervenkr.*, 102 (1934) 558–613.
45 NAUTA, W. J. H., Hippocampal projections and related neural pathways to the midbrain in the cat, *Brain*, 81 (1958) 319–340.
46 NAUTA, W. J. H., Fiber degeneration following lesion of the amygdaloid complex in the monkey, *J. Anat. (Lond.)*, 95 (1961) 515–531.
47 NAUTA, W. J. H., Neural associations of the amygdaloid complex in the monkey, *Brain*, 85 (1962) 505–520.
48 NAUTA, W. J. H., AND GYGAX, P. A., Silver impregnation of degenerating axons in the central nervous system: a modified technic, *Stain Technol.*, 29 (1954) 91–93.
49 NAUTA, W. J. H., AND WHITLOCK, D. G., Anatomical analysis of the non-specific thalamic projection system. In J. DELAFRESNAYE (Ed.), *Brain Mechanisms and Consciousness*, Blackwell, Oxford, 1954.
50 NOBACK, C., Brain of a gorilla. II. Brain stem nuclei, *J. comp. Neurol.*, 111 (1959) 345–385.
51 OLSZEWSKI, J., *The Thalamus of Macaca mulatta*, Karger, Basel, 1952.
52 OLSZEWSKI, J., AND BAXTER, D., *Cytoarchitecture of the Human Brain Stem*, Lippincott, Philadelphia, 1954.
53 PAPEZ, J. W., Reciprocal connections of the striatum and pallidum in the brain of *Pithecus (Macacus) rhesus. J. comp. Neurol.*, 69 (1938) 329–349.
54 PAPEZ, J. W., A summary of fiber connections of the basal ganglia with each other and with other portions of the brain. *Res. Publ. Ass. nerv. ment. Dis.*, 21 (1941) 21–68.
55 PAPEZ, J. W., AND ARONSON, L. R., Thalamic nuclei of *Pithecus (Macacus) rhesus*. I. Ventral thalamus, *Arch. Neurol. Psychiat.*, 32 (1934) 1–26.
56 PAPEZ, J. W., AND STOTLER, W., Connections of the red nucleus, *A.M.A. Neurol. Psychiat.*, 44 (1940) 776–791.
57 PETRAS, J. M., Some fiber connections of the precentral cortex (areas 4 and 6) with the diencephalon in monkey (*Macaca mulatta*), *Anat. Rec.*, 148 (2) (1964) 322.
58 POPPI, U., Ueber die Fasersysteme der Substantia nigra, *Arb. a.d. neur. Inst. a.d. Wien. Univ.*, 29 (1927) 8–49.
59 POPPI, U., AND IMBER, I., Studium der pallidalen und cortico-pontinen Haubenbahnen in einem Fall von Porencephalie, *Z. ges. Neurol. Psychiat.*, 138 (1932) 503–514.
60 POWELL, T. P. S., AND COWAN, W. M., A study of thalamo-striate relations in the monkey, *Brain*, 79 (1932) 364–390.
61 PROBST, M., Ueber die Rinden-Sehhügelfasern des Riechfeldes, über das Gewölbe, die Zwinge, die Randbogenfasern, über die Schweifenkernfaserung, und über die Vertheilung der Pyramidenfasern im Pyramidenareal, *Arch. Anat. Physiol., Anat. Abth.*, H II–V (1903) 138–152.
62 RANSON, S. W., AND RANSON, M., Pallidofugal fibers in the monkey, *A.M.A. Arch. Neurol. Psychiat.*, 42 (1939) 1059–1067.
63 RANSON, S. W., RANSON, JR., S. W., AND RANSON, M., Fiber connections of the corpus striatum as seen in Marchi preparations, *A.M.A. Arch. Neurol. Psychiat.*, 46 (1941) 230–249.

64 RIESE, W., Zur vergleichenden Anatomie der striofugalen Faserung, *Anat. Anz.*, 57 (1924) 487–494.

65 RIESE, W., Beiträge zur Faseranatomie der Stammganglien, *J. Psychol. Neurol.*, 31 (1924) 81–122.

66 ROSEGAY, H., An experimental investigation of the connections between the corpus striatum and substantia nigra in the cat, *J. comp. Neurol.*, 80 (1944) 293–321.

67 RUNDLES, R. W., AND PAPEZ, J. W., Connections between the striatum and substantia nigra in a human brain, *Arch. Neurol. Psychiat.*, 38 (1937) 550–563.

68 STERN, K., Der Zellaufbau des menschlichen Mittelhirns, *Z. ges. Neurol. Psychiat.*, 154 (1936) 521–598.

69 SZÁBO, J., Topical distribution of the striatal efferents in the monkey, *Exp. Neurol.*, 5 (1962) 21–36.

70 VERHAART, W. J. C., A comparison between the corpus striatum and the red nucleus as subcortical centra of the cerebral motor system, *Psychiat. neurol. Bl. (Amst.)*, 32 (1938) 676–737.

71 VIDAL, F., Pallidohypothalamic tract, or X bundle of Meynert, in the rhesus monkey, *A.M.A. Arch. Neurol. Psychiat.*, 44 (1940) 1219–1223.

72 VOGT, C., La myéloarchitecture du thalamus du cercopithèque, *J. Psychol. Neurol.*, 12 (Ergänzungsheft) (1909) 285–324.

73 VOGT, C., AND VOGT, O., Zur Kenntnis der pathologischen Veränderungen des Striatum und des Pallidum und zur Pathophysiologie der dabei auftretenden Krankheitserscheinungen, *S.-B. Heidelb. Akad. Wiss.*, 10, B14 (1919) 1–56.

74 VOGT, C., AND VOGT, O., Zur Lehre der Erkrankungen des striären Systems, *J. Psychol. Neurol. (Lpz.)*, 25 (1920) 627–846.

75 VOGT, C., AND VOGT, O., Thalamusstudien I–III. I. Zur Einführung. II. Homogenität und Grenzgestaltung der Grisea des Thalamus. III. Das Griseum centrale (centrum medianum Luys), *J. Psychol. Neurol. (Lpz.)*, 50 (1941) 31–154.

76 VONEIDA, T., An experimental study of the course and destination of fibers arising in the head of the caudate nucleus in the cat and monkey. *J. comp. Neurol.*, 115 (1960) 75–87.

77 WALLENBERG, A., Beiträge zur Kenntnis der zentrifugalen Bahnen des Striatum und Pallidum beim Menschen, *Dtsch. Z. Nervenheilk.*, 77 (1923) 201–202.

78 WEISSCHEDEL, E., Die zentrale Haubenbahn und ihre Bedeutung für das extrapyramidal-motorische System, *Arch. Psychiat. Nervenkr.*, 107 (1937) 443–579.

79 WERNICKE, C., *Lehrbuch der Gehirnkrankheiten für Aerzte und Studirende*, Vol. 2, Fischer, Kassel and Berlin, 1881.

80 WHITLOCK, D. G., AND NAUTA, W. J. H., Subcortical projections from the temporal neocortex in *Macaca mulatta*, *J. comp. Neurol.*, 106 (1956) 183–212.

81 WILSON, S. A. K., An experimental research into the anatomy and physiology of the corpus striatum, *Brain*, 36 (1914) 427–492.

82 WINKLER, C., *Manuel de Neurologie*, Tome 1, Part. 4, Bohn, Haarlem, 1929.

83 WOODBURNE, R. T., CROSBY, E. C., AND MCCOTTER, R. E., The mammalian midbrain and isthmus regions. Part II. The fiber connections. A. The relations of the tegmentum of the midbrain with the basal ganglia in *Macaca mulatta*, *J. comp. Neurol.*, 85 (1946) 67–92.

*Neuroscience* Vol. 3, pp. 385–401. Pergamon Press Ltd. 1978. Printed in Great Britain.
© IBRO

0306-4522/78/0501-0385$02.00/0

# EFFERENT CONNECTIONS AND NIGRAL AFFERENTS OF THE NUCLEUS ACCUMBENS SEPTI IN THE RAT

W. J. H. Nauta, G. P. Smith,[1] R. L. M. Faull[2] and
Valerie B. Domesick

From the Department of Psychology, Massachusetts Institute of Technology,
Cambridge, MA 02139, and the Mailman Research Laboratories,
McLean Hospital, Belmont, MA 02178, U.S.A.

**Abstract**—The results of this study by the methods of autoradiographic fiber-tracing and retrograde cell-labeling confirm earlier reports of accumbens projections to the globus pallidus and to dorsal strata of the medial half of the substantia nigra. Also in accord with previous autoradiographic evidence, sparser projections could be traced to a variety of subcortical structures implicated in the circuitry of the limbic system: bed nucleus of the stria terminalis, septum, preoptic region, hypothalamus, ventral tegmental area, nuclei paratenialis and mediodorsalis thalami, and lateral habenular nucleus. Contrary to earlier reports, striatopallidal fibers from the accumbens were found to be distributed largely to the subcommissural part of the external pallidal segment (ventral pallidum of Heimer & Wilson, 1975) and to avoid almost entirely the internal pallidal segment (entopeduncular nucleus). Mesencephalic projections from the accumbens largely coincide with those from the preoptic region and hypothalamus; like the latter they prominently involve the region of the out-lying nigral cell groups A10 and A8 and extend caudally beyond the nigral complex to the cuneiform and parabrachial regions of the tegmentum as well as to caudoventral parts of the central grey substance.

Horseradish peroxidase injected into the nucleus accumbens labels numerous neurons in the region of cell group A10 and in the supralemniscal 'retrorubral nucleus', but only sporadic cells in the pars compacta of the substantia nigra proper. It thus appears that the accumbens projects to a region of the nigral complex considerably larger than that from which it receives nigrostriatal fibers, and hence, that the nigro–striato–nigral circuit associated with the accumbens is not organized in a mode of simple point-for-point reciprocity.

The problem of delimiting and accumbens from the rest of the striatum is examined by comparing cases of tracer injection into various discrete loci within the ventral zone of the striatum.

The term, nucleus accumbens, introduced by Ziehen more than 70 yr ago, refers to an anterior, ventromedial part of the striatum that surrounds the bottom of the anterior horn of the lateral ventricle and extends over some distance dorsally into the lateral part of the septum. Initially distinguished from the rest of the striatum only because of its remarkable topographic relationship with the septum, the nucleus accumbens in later years proved to be a distinctive striatal region also in view of its afferent and efferent connections. First, over the past 30 yr or so evidence has accumulated that this striatal region, in contrast to the larger remainder of the striatum, receives telencephalic afferents from the hippocampus and amygdala rather than from the neocortex. Second, biochemical and histofluorescence studies have shown that the dopamine-containing fibers innervating the accumbens region of the striatum largely originate from a dorsally out-lying nigral cell group called A10

by Dahlström & Fuxe (1964) rather than from the substantia nigra *sensu strictiori*. Third, and most recently, autoradiographic findings have indicated that the nucleus accumbens, in contrast to other parts of the striatum, projects not only to the globus pallidus and substantia nigra but also to a variety of subcortical structures implicated in the circuitry of the limbic system.

The study here reported was prompted by several questions that arose from an examination of our own autoradiographic and retrograde-labeling material. Our observations indicated that the distribution of accumbens efferents that extend caudally beyond the substantia nigra had not been charted adequately. Moreover, it appeared that no previous study of accumbens connections had included an identification, by retrograde labeling, of the region or regions of the nigral complex that project to the accumbens; hence, no information concerning the degree of regional reciprocity in the nigro–striato–nigral relationships of the accumbens appeared to be available. Finally, previous reports have included no information as to whether the reported striatofugal connections are indeed uniquely characteristic of the anterior, ventromedial region of the striatum referred to as nucleus accumbens.

<hr>

[1] Permanent address: Department of Psychiatry, New York Hospital-Cornell Medical Center, White Plains, NY 10605, U.S.A.

[2] Permanent address: Department of Anatomy, University of Auckland, New Zealand.

*Abbreviation*: HRP, horseradish peroxide.

## EXPERIMENTAL PROCEDURES

The observations reported here were made in a material consisting of some 80 rat brains in which small deposits of either an equal-parts mixture of tritiated leucine and tritiated proline (New England Nuclear), or horseradish peroxidase (HRP, Sigma type VI) had been placed in one of a wide variety of loci within the striatum by microelectrophoresis. In many of these animals (RH series) the radioactive amino-acid mixture had been injected on the right side, the HRP in the symmetrical locus on the left. Additional cases were selected from a series (RR) in which only unilateral isotope deposits had been placed in one of a variety of basal forebrain structures.

The experimental protocol was as follows. All operations were done under stereotaxic guidance, after the animals had been deeply anesthetized with Equithesin (Salisbury-Jensen). In the autoradiographic experiments, the isotopes were ejected from a glass micropipette (tip diameter 10–20 µm) in which they were contained in a 0.01 N acetic-acid medium at a concentration of 20 µCi/µl. The driving force, delivered by a constant-current source (Midgard Electronics, Watertown, MA), consisted in a 0.5–0.8 µA positive direct current, pulsed at a rate of 7 s on–7 s off, and applied for a total duration of 2–5 min. A similar procedure was followed for the HRP injections, with the exception that the pipette-fluid was a 13% solution of HRP in Tris-HCl buffer of pH 8.6, the expulsion force a 1–1.5 µA positive current applied continuously for 10–20 min.

In the exclusively autoradiographic experiments of the RR series, the animals were given an overdose of barbiturate 3–15 days after surgery, and were perfused with 10% formalin-saline. Following post-fixation for 7 days, the brain in each case was embedded in albumin-gelatin, and sectioned on a freezing microtome at 25 µm. The sections were mounted in four alternating series, coated with Kodak NTB-2 emulsion, stored in lightproof boxes in a freezer at −20°C for, respectively, 4, 6, 8 and 12 weeks, developed in Kodak D-19 developer at 16°C, fixed, and counterstrained with cresyl violet. In the combined autoradiographic–HRP experiments, the animals were perfused 24–48 h after surgery with a solution of 4% paraformaldehyde and 5% sucrose in 0.1 M phosphate buffer at pH 7.4; the brains were post-fixed for 8 h, washed in 10% sucrose at pH 7.4 for 8 h, and sectioned on a freezing microtome at 50 µm. Every second section was processed as required for the diaminobenzidine reaction, while the remaining sections were mounted for autoradiography as described above.

## RESULTS

### I. *Efferent connections of the nucleus accumbens septi*

Since, with exceptions to be noted below, the results of the autoradiographic experiments were quite comparable, they can be illustrated by a single case, RR 86. In this and several other instances the isotope deposit was purposely placed in a rather rostral part of the nucleus accumbens, so as to avoid involvement of the bed nucleus of the stria terminalis with which the nucleus accumbens is caudally contiguous. The primary injection site, characterized by tissue damage and complete label-saturation, is indicated in Fig. 1A in jet black. The shaded area surrounding it in the drawing indicates the region in which neuronal perikarya were clearly distinguishable by a label density higher than that of the intervening neuropil; presumably, this entire region represents the tissue volume in which label had been sequestered by neurons.

Sections immediately posterior to the injection site show several heavily labeled, caudo-ventrally oriented fiber bundles most of which, about 1.5 mm farther caudally, pass below the anterior commissure into the dorsal part of the substantia innominata (Fig. 1B). At this level, the fascicular labeling pattern changes abruptly into one characteristic of massive terminal arborization. The region marked by this more diffuse but extremely dense grain distribution corresponds to an anterior, subcommissural protrusion of the globus pallidus identified, and named *ventral pallidum* (VP in Fig. 1B) by HEIMER & WILSON (1975).

At the same rostro-caudal level labeled fibers are beginning to spread medially and dorsally from the main path into the adjacent lateral preoptic region and bed nucleus of the stria terminalis, both of which are marked by generally diffuse and light labeling. The longest of these diverging fibers extend medially into the medial and periventricular zones of the preoptic region, and dorsally into the lateral septal nucleus.

At levels slightly farther caudally (Fig. 1C–D) moderately dense labeling appears in a ventromedial

---

FIG. 1. A charting of radioactively labeled fibers appearing at various levels of the forebrain in case RR 86. The injection site is indicated in jet black in A; together with its shaded surround this area represents the presumable uptake site of the injected tritiated amino acids leucine and proline. Abbreviations: AC, anterior commissure; AL, nucleus lateralis amygdalae; AON, medial accessory optic nucleus; CF, columna fornicis; Cl, claustrum; CP, pedunculus cerebri; DBC, decussation of brachium conjunctivum; Ep, entopeduncular nucleus; F, fornix; FR, fasciculus retroflexus; GP, globus pallidus; HL, lateral habenular nucleus; HtL, lateral hypothalamic region; IP, interpeduncular nucleus; MB, mammillary body; MD, mediodorsal nucleus; ML, medial lemniscus; MP, mammillary peduncle; MT, mammillothalamic tract; NB, nucleus basalis (preopticus magnocellularis); NCS, nucleus centralis superior (medial raphe nucleus); ND, nucleus of the diagonal band; NDk, nucleus of Darkschewitsch; NI, interstitial nucleus of Cajal; NR, nucleus ruber; NST, bed nucleus of stria terminalis; PL, lateral preoptic nucleus; PM, medial preoptic area; Pt, paratenial nucleus; SM, stria medullaris; SNc, substantia nigra, pars compacta; SNr, substantia nigra, pars reticulata; St, subthalamic nucleus; VMH, nucleus ventromedialis hypothalami; VP, subcommissural pallidum (ventral pallidum of Heimer and Wilson); VTA: ventral tegmental area; IV, trochlear nucleus.

A
D
GP
SM
Pt
F
NST
CI
HtL
B
E
AC
NST
GP
MD
PL
VP
Ep
PM
ND
NB
AL
HtL
F
VMH
C
F
CF
HL
GP
AC
NST
MD
CI
St
MT
HtL
NB
ND
F

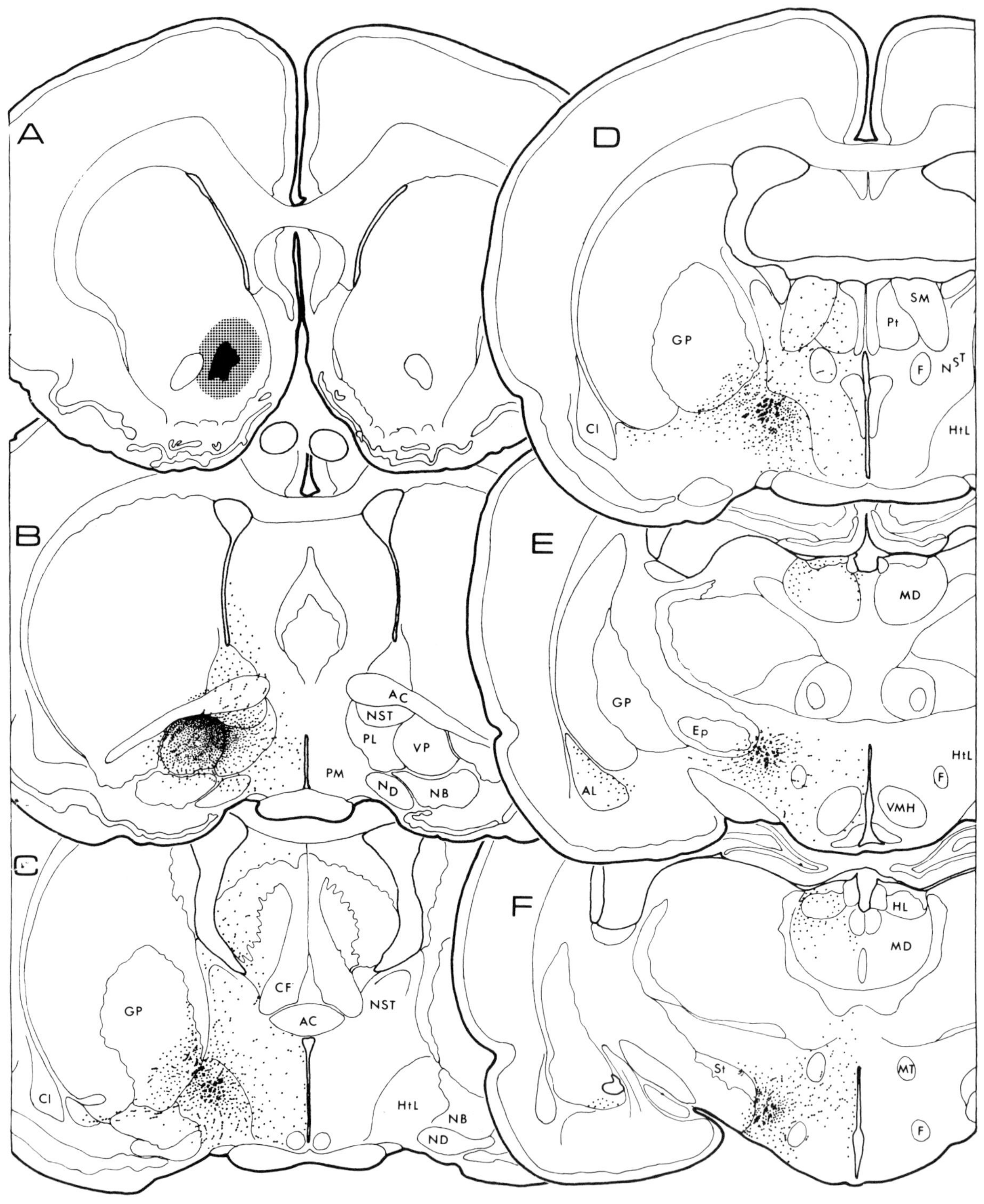

FIG. 1.

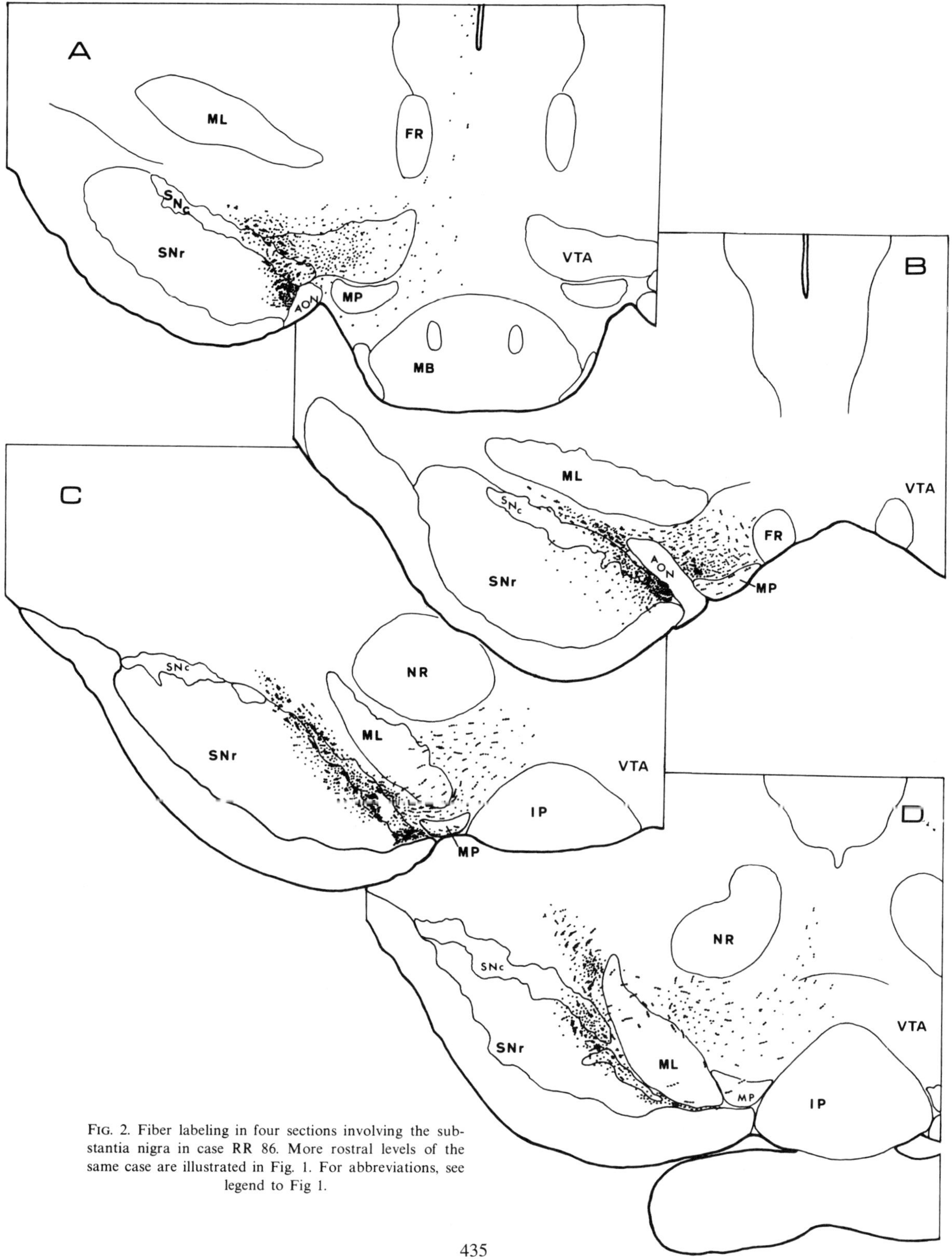


FIG. 2. Fiber labeling in four sections involving the substantia nigra in case RR 86. More rostral levels of the same case are illustrated in Fig. 1. For abbreviations, see legend to Fig 1.

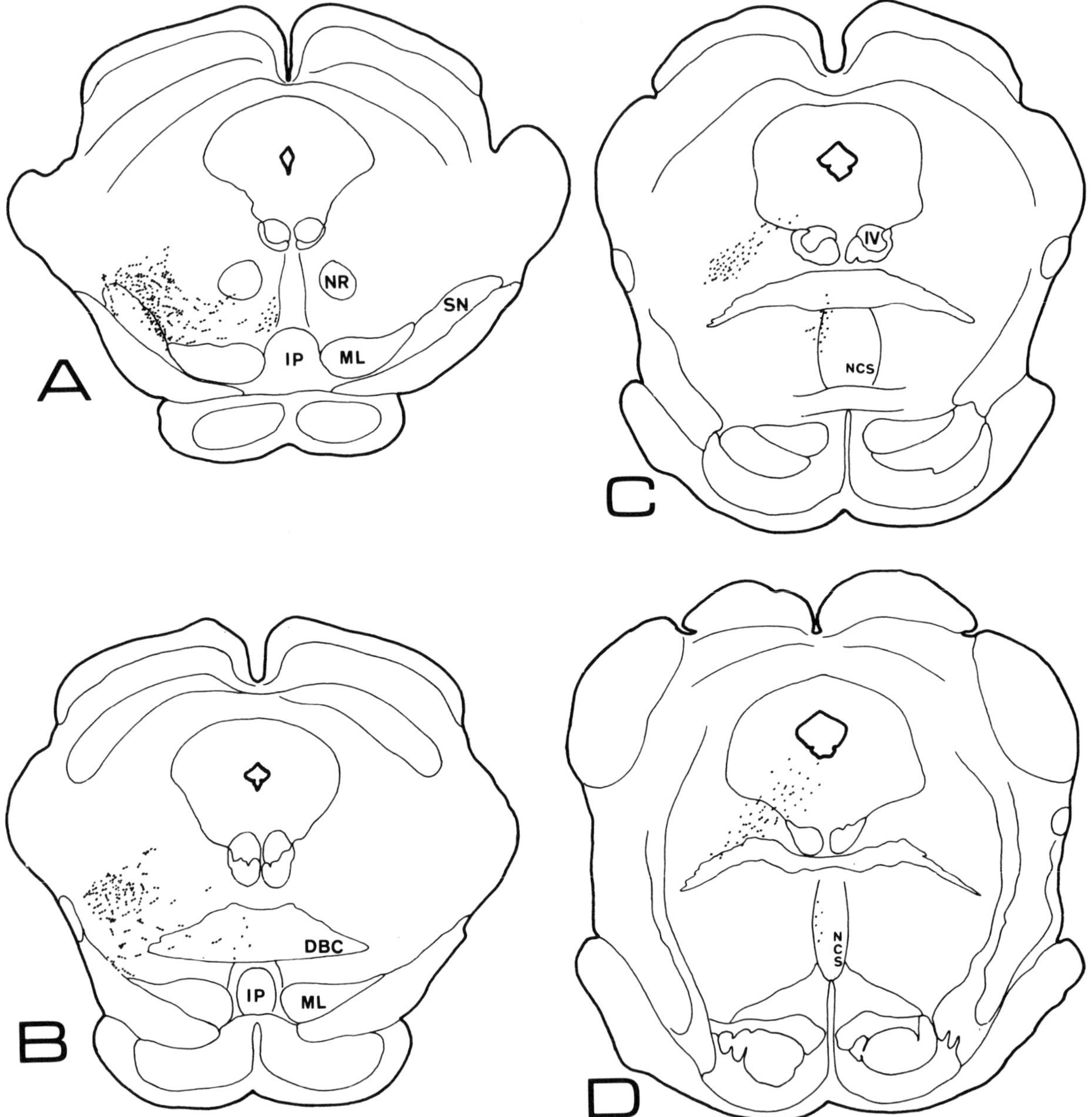



Fig. 3. Fiber labeling in the caudal half of the midbrain in case RR 86. More rostral levels of the same case are illustrated in Figs 1–2. For abbreviations, see legend to Fig. 1.

sector of the main body of the globus pallidus. This labeling appears to correspond to the few labeled fascicles that passed from the injection site caudally over, rather than under, the anterior commissure. The same sections show a substantial group of labeled fascicles that have emerged from the densely labeled sub-commissural pallidum and have joined the dorso-lateral part of the medial forebrain bundle. Enclosed in the latter they extend through the length of the lateral hypothalamic region and beyond it as far as the most caudal levels of the midbrain. Throughout its hypothalamic trajectory this group of longitudinal fascicles issues fibers in various directions. With few exceptions such off-sets are diffusely disposed rather than organized in discrete fiber groups. For example, scattered labeled fibers oriented in the dorsomedial direction join the stria medullaris at rostral hypothalamic levels (Fig. 1D) and can be followed from here to the paratenial nucleus, to a dorsolateral part of the mediodorsal nucleus, and to a medial part of the lateral habenular nucleus (Fig. 1D–F). At about the same levels another, somewhat better defined group of labeled fibers extends laterally through the sublenticular region to the lateral amygdaloid nucleus (Fig. 1C–E). Finally, throughout the length of the hypothalamus fibers spread out medially from the labeled fascicles over large regions of the medial and periventricular hypothalamic zones. A short distance rostral to the mammillary body (Fig. 1F) some of these medially directed fibers turn dorsalward into the dorsal hypothalamic area and the nucleus reuniens thalami; farther caudally such dorsal off-sets from the labeled fascicles in the medial forebrain bundle extend into the posterior hypothalamic nucleus and to a contiguous rostroventral part of the central grey substance (Fig. 2A).

It is remarkable that the descending accumbens fibers labeled in this case appear to avoid the entopeduncular nucleus. The main group of labeled fascicles skirts the medial margin of the cerebral peduncle, and only sporadic labeled fibers appear at or within the medial border of the entopeduncular nucleus (Fig. 1E).

The mesencephalic extension of the accumbens projection is illustrated in Figs 2 and 3. Figure 2 has been drawn on a larger scale so as to provide a better resolution of the complex pattern of fiber labelling in and near the substantia nigra. In charting the sections represented in this figure the utmost attention was paid to topographic accuracy; as a consequence, several of the larger bloodvessels that served as charting landmarks are truthfully represented as voids in the labeling pattern (see for instance Fig. 2C).

Figure 2A illustrates the labeling pattern at the level of the meso-diencephalic transition. The accumbens efferents at this level have maintained the arrangement in which they passed caudally through the lateral hypothalamic region: their lateral members form a group of closely spaced fascicles, while the remainder composes a more diffusely disposed medial fiber group that occupies the lateral part of the ventral tegmental area. The ventral half of the lateral, fasciculated fiber group at this level has entered the medial corner of the substantia nigra; in successively more caudal sections these intranigral fascicles are seen to spread laterally, distributing their fibers to the medial half of the substantia nigra's pars compacta and to a narrow, immediately subjacent zone of the pars reticulata. More dorsally situated fascicles likewise spread lateralward, but their major passage leads immediately dorsal to the substantia nigra rather than through it (Fig. 2B–C). Even though it

seems likely that some fibers of this supranigral group, also, ultimately terminate in the substantia nigra, a majority of its fascicles clearly by-pass the substantia nigra. At the level of Fig. 2D such fascicles have begun to accumulate at and beyond the lateral margin of the medial lemniscus, well dorsal to the substantia nigra. A slight distance farther caudally (Fig. 3A), these fascicles appear as a concentrated group in the ventrolateral tegmental region that corresponds to the location of the dopamine cell group A8 identified by DAHLSTRÖM & FUXE (1964), UNGERSTEDT (1971) and PALKOVITS & JACOBOWITZ (1974) (compare Fig. 3A with Fig. 6 which illustrates cell labeling in the region of A8). Caudal to this level the fiber group begins to curve in the mediodorsal direction and in its further course caudalward it successively traverses the cuneiform and parabrachial regions of the midbrain tegmentum (Fig. 3B–D); the longest of its fibers invade the ventral half of the central grey substance at levels behind the trochlear nucleus (Fig. 3D).

As to the more diffusely arranged fibers that compose a medial subdivision of the system of descending accumbens-efferents, some at least appear to follow a sagittal course through the ventral tegmental area. Since this medial group of fibers is only vaguely delimited from the more lateral, fasciculated fiber group, it is difficult to determine whether some of its fibers may perhaps join the latter group in a more laterally oriented course through and over the substantia nigra. This possibility is suggested especially by the fiber arrangements at the levels of Fig. 2C where fibers pass transversely through and underneath the medial lemniscus, and 3A where similar transverse fibers pass dorsal to the lemniscus. However, since the polarity of these bridging fibers cannot be determined it seems equally possible that they are medially directed off-sets from the lateral group, comparable to the fibers issuing medialward from the lateral-hypothalamic trajectory of the main pathway.

At levels caudal to the substantia nigra (Fig. 3B–D) the relatively few remaining fibers of the medial group are confined to a narrow paramedian zone of the tegmentum corresponding in part to the lateral half of the nucleus centralis tegmenti superior (median raphe nucleus).

*Other cases.* Case RR 86 described and charted above was compared with six further cases of nucleus-accumbens injection, as well as with several cases in which the injection had been placed in more lateral and caudal parts of the deepest striatal region. Four of these comparison cases are illustrated in Fig. 4. The following observations seem noteworthy.

1. The projection from the antero-ventral region of the striatum to the thalamic nuclei paratenialis and mediodorsalis appears to originate largely in medial parts of the nucleus accumbens: fewer fibers to these cell groups were labeled in RH 4 (Fig. 4) than in RR 86, and none at all in the remaining cases illustrated in Fig. 4.

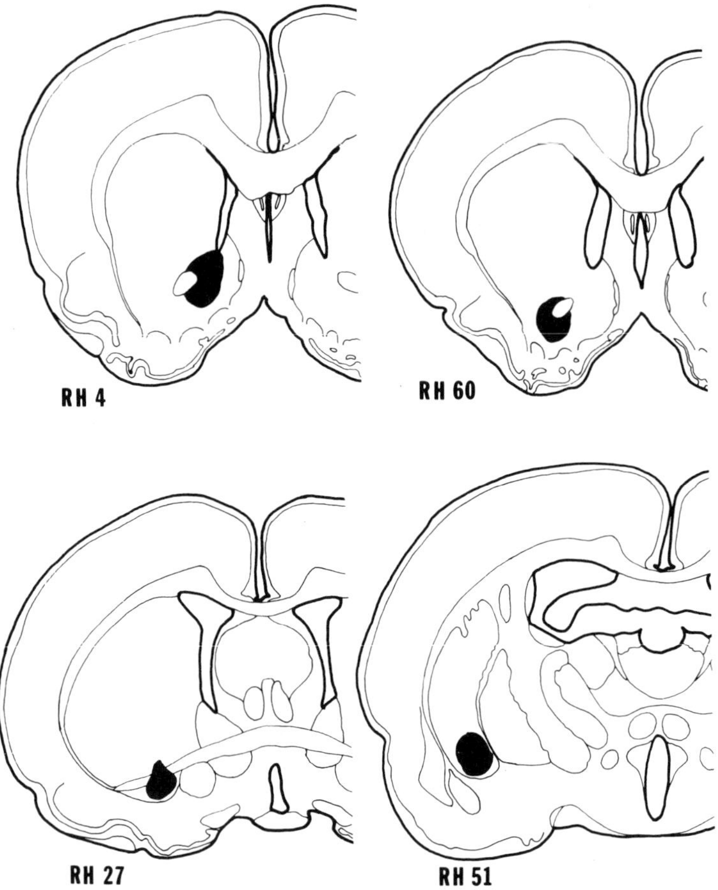



FIG. 4. The presumable uptake site of tritiated leucine and proline in four of the comparison cases discussed in the text.

2. Accumbens efferents to the lateral habenular nucleus, sparse in any case, were labeled only in cases in which the injection site involved the most medial part of the nucleus; none were labeled in any of the cases shown in Fig. 4.

3. With the exception of RH 51, all cases illustrated in Fig. 4 showed some fiber labeling in the bed nucleus of the stria terminalis; the lateral septal nucleus, however, was found labeled only in case RH 4.

4. In all cases except RH 51 the subcommissural 'ventral pallidum' was densely labeled. In case RH 51 the striatopallidal fiber labeling involves a caudal and ventral region of the pallidum adjoining the injection site.

5. In all cases except RH 51 labeled striatofugal fibers extending beyond the globus pallidus almost exclusively followed the medial forebrain bundle. Accordingly, only sporadic fiber labeling appeared in the entopeduncular nucleus. In RH 51, likewise, labeled fibers very largely avoided this internal seg-

ment of the globus pallidus, in this case by passing lateral and ventral to it in the cerebral peduncle.

6. Fibers spreading from the medial forebrain bundle to more medial parts of the hypothalamus were labeled in largest number by the most medially placed accumbens injections. Such fibers were identifiable in RH 4 and RH 60, but none were found labeled in RH 27 and RH 51. However, in RH 27 scattered labeled fibers did appear in the medial preoptic region.

7. In all cases shown in Fig. 4, including RH 51, labeled striatonigral fibers were distributed to the most dorsal zone of the pars reticulata and to the overlying pars compacta, as well as to the ventrolateral tegmental area corresponding to the location of cell group A8. Labeled fibers in the region of cell group A10 (ventral tegmental area) appeared in all cases except RH 51, but were most numerous in the cases of medially placed accumbens injection.

8. Labeled fibers continuing beyond A8 (the most

caudal component of the nigral complex) into central tegmental regions and the central grey substance were found in all cases shown in Fig. 4 except RH 51.

9. Small isotope-injections of the anterior striatum centered dorsal to a horizontal plane passing through the floor of the ventricle label no fibers to either the limbic diencephalon or any mesencephalic region outside the substantia nigra proper.

## II. *Nigral afferents of the nucleus accumbens septi*

Of the seven cases in which HRP had been injected into the nucleus accumbens, only one needs to be presented here. In this case, RH 29, the enzyme deposit (insert, Fig. 5) corresponded in its location almost exactly to the injection site in the preceding autoradiographic experiment RR 86 described above, and the two cases thus can be considered, in a sense, each other's counterpart. In order further to facilitate comparison of cell- with fiber-labeling, the rostral mesencephalic levels of RH 29 represented by Fig. 5 were chosen so as to match as closely as possible those illustrated in Fig. 2. The great majority of cells labeled in case RH 29 were labeled vividly enough to be plainly visible in low-power darkfield photomicrographs; a smaller number, although readily recognized under low power as labeled cells, required retouching for adequate recording on the photographic prints. Cells containing granules too small in number to provide an outline of the cell body were discounted.

As shown by Fig. 5, most of the HRP-positive cells are located in the ventral tegmental area ipsilateral to the injection. Together, the labeled cells shown in Fig. 5A–D occupy approximately the same region in which histochemical methods have revealed the presence of a distinct group of out-lying nigral dopamine cells; DAHLSTRÖM & FUXE (1964) labeled this cell group A10, and the present findings are compatible with the subsequent experimental evidence of ANDÉN, DAHLSTRÖM, FUXE, LARSSON, OLSON & UNGERSTEDT (1966) and UNGERSTEDT (1971) that A10 projects preferentially at least to the nucleus accumbens. Figure 5A clearly shows the broad continuity of the labeled cell group with the substantia nigra's pars compacta (cell group A9 in the listing by DAHLSTRÖM & FUXE (1964) of catecholamine cell groups); in fact, the rostral part of the labeled group can be described as a somewhat wedge-shaped dorsomedial extrusion from the medial half of the pars compacta. Farther caudally (Figs 5B–C), however, this continuity becomes less marked, and at the level of Fig. 5D where the labeled cell group extends lateralward over the dorsal borders of the interpeduncular nucleus and medial lemniscus, its confluence with the pars compacts is limited to a narrow passage around the lateral margin of the lemniscus. The irregularly structured group of labeled cells overlying the medial lemniscus appears to correspond to the cell group labeled 'retrorubral nucleus' by BERMAN (1968) in the cat. In the same supralemniscal region UNGERSTEDT (1971)

and PALKOVITS & JACOBOWITZ (1974) indicate catecholamine cells which they consider to form part of a second group of out-lying nigral cells, cell group A8 of DAHLSTRÖM & FUXE (1964). The lateral part of this supralemniscal group extends caudalward slightly beyond the caudal pole of the substantia nigra proper; at such caudal transverse levels it appears as an independent nucleus, but serial examination shows it to be continuous both with the pars compacta and, via the supralemniscal cell bridges, with cell group A10.

In no case of HRP injection confined to the nucleus accumbens as conventionally defined were labeled cells found in the region corresponding to the more caudal part of A8, but in several others in which the HRP was deposited in ventral parts of the striatum caudal and lateral to the nucleus accumbens, the region is clearly marked by cell labeling. In order to complete the illustration of the extranigral tegmental regions in which neurons can be labeled by intrastriatal HRP injection, a photograph (Fig. 6) is included of the cell labeling at a level just caudal to the substantia nigra in one such case (RH 27) in which the HRP deposit was confined to the small ventrolateral striatal pocket protruding beneath the anterior commissure (insert to Fig. 6). In this case, the cell labeling at more rostral levels of the midbrain was remarkably similar to that observed in case RH 29 (Fig. 5), but additional labeled cells, distributed in a more or less triangular array, appeared at the level immediately caudal to the substantia nigra proper (Fig. 6). The round, fairly compact group of labeled cells shown by Fig. 6 in the ventrolateral tegmentum corresponds in its location to dopamine cell group A8 as charted by DAHLSTRÖM & FUXE (1964), UNGERSTEDT (1971) and PALKOVITS & JACOBOWITZ (1974). From it, trails of more loosely arranged labeled tegmental cells extend in both the dorsomedial and ventromedial direction; it is interesting that PALKOVITS & JACOBOWITZ (1974) in their chartings indicate a scattering of catecholamine cells in the same central tegmental regions. The labeled cells in the tegmental mid-plane may represent the caudal extreme of dopamine cell group A10, but those in the central grey substance form part of the dorsal raphe nucleus.

It is important to note that in case RH 29, as in all other cases of HRP injection limited to the nucleus accumbens, only few HRP-positive cells appear in the substantia nigra proper. Those that are present are confined to the medial half of the pars compacta. Besides being few in number, most of these cells are only weakly labeled and would have been invisible in the photographs shown in Fig. 5 without retouching. The apparent sparseness of compacta cells that project to the nucleus accumbens contrasts markedly with the considerable volume of the projection from the nucleus accumbens to the corresponding part of the pars compacta (Fig. 2). It seems unlikely that this contrast is due to a significant size difference between

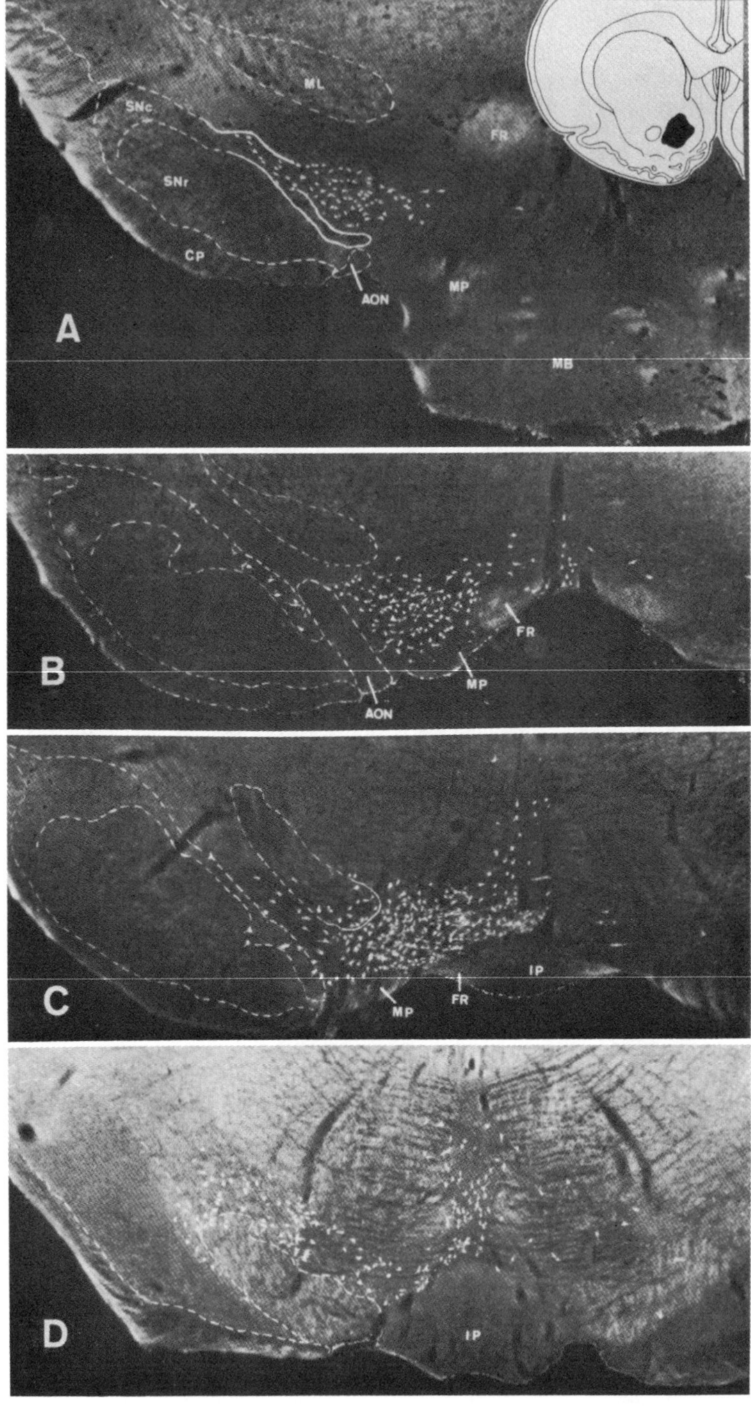



FIG. 5. Retrograde cell-labeling in the nigral complex in a case (RH 29) of horseradish peroxide injection confined to the nucleus accumbens. The injection site is indicated in the insert to frame A. Darkfield photographs, retouched as stated in the text. Note cell labeling in the supralemniscal 'retrorubral nucleus' in frame D, and the paucity of labeled cells in the pars compacta at all levels of the substantia nigra. For abbreviations, see legend to Fig. 1.

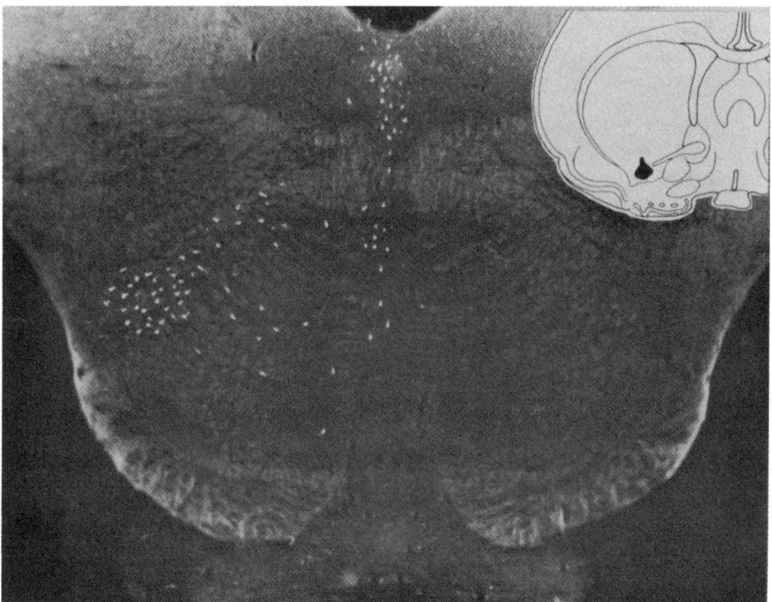

FIG. 6. Retrograde cell-labeling in the midbrain tegmentum, closely behind the caudal pole of the substantia nigra proper, in case RH 27 (insert). Darkfield photograph, retouched as explained in the text. In this case, the cell-labeling at more rostral levels was very similar to that in case RH 29 shown in Fig. 5. The round group of labeled cells in the ventrolateral tegmentum corresponds in its location to the caudal part of cell group A8.

the respective uptake sites of HRP and radioactive amino acids, for the number of labeled compacta cells is not much larger in cases of more massive HRP injection of the nucleus accumbens, and increases markedly only when the injection site is centered in more dorsal striatal regions.

*Control experiments.* Dopamine fibers ascending from the nigral complex in the rat are known to be distributed not only to the striatum, central nucleus of the amygdala, and bed nucleus of the stria terminalis (UNGERSTEDT, 1971), but also to some cortical regions, in particular the anteromedial cortex and entorhinal area (LINDVALL, BJÖRKLUND, MOORE & STENEVI, 1974; BECKSTEAD, 1976). Since it seemed possible that some of the nigrocortical fibers destined for the anteromedial cortex course through the striatum, and thus could have taken up HRP at its injection site in the nucleus accumbens, a few control experiments were done to determine whether, and to what extent, the cell labeling following intrastriatal HRP injection might reflect a nigrocortical rather than a nigrostriatal projection. In these experiments, the entire expanse of medial cortex from the frontal pole caudalward to the anterior limit of the retrosplenial granular cortex was injected in two rats with HRP delivered in the longitudinal series of six injections placed at intervals of about one millimeter.

In accord with the report of BECKSTEAD (1976), the very large HRP deposit in both cases was found to have labeled cells in the ventral tegmental area and in the medial half of the substantia nigra, pars compacta. These cells were much fewer in number than those labeled by much smaller accumbens injections, and their distribution nowhere exceeded the limits of the cell groups labeled in case RH 29 (Fig. 5). The conclusion drawn from these control experiments is, that even if some of the tegmental cells labeled by HRP deposited in the nucleus accumbens were nigrocortical neurons, they could not have affected significantly either the topography or the quantity of the cell labeling resulting from HRP injection in the nucleus accumbens. Their number in all likelihood would be very small in any case, since, according to the description of LINDVALL & BJÖRKLUND (1974), nigrocortical dopamine fibers circumvent rather than traverse the striatum in their passage to the cortex.

## DISCUSSION

The results of the present study are in general accord with previous observations concerning the neural circuitry associated with the nucleus accumbens septi. They do, however, add to the existing descriptions certain new details that may be of some relevance to the still evolving concepts of corpus striatum, limbic system, and the relationship between these two major forebrain components.

*Efferent connections*

The nucleus accumbens corresponds to, or forms part of, the anterior subdivision of a fairly narrow ventral zone of the striatum from which the longest and apparently also the most diverse striatofugal projections originate. This general statement is based upon comparisons of the cases of accumbens injection here described with a large additional autoradiographic material which, although it cannot be reviewed here in detail, needs to be referred to from time to time in this discussion in order to forestall the erroneous impression that all of the connections reported in the descriptive part of this paper are unique to that rostral, ventromedial striatal region which is somewhat vaguely and arbitrarily delineated from the remainder of the striatum as nucleus accumbens septi.

The 'striatal' nature of the nucleus accumbens, indicated by the cytoarchitecture of the nucleus but nonetheless a subject of some controversy in the past, has been emphasized by SWANSON & COWAN (1975) on the basis of their original autoradiographic evidence in the rat that the nucleus, like the remainder of the striatum, projects to the globus pallidus and substantia nigra. Similar findings were reported by CONRAD & PFAFF (1976a), likewise in the rat, and by POWELL & LEMAN (1976) in the squirrel monkey, but both these studies brought evidence that the nucleus accumbens has additional efferent connections with several structures implicated in the circuitry of the limbic system, in particular with the septum, preoptic region and hypothalamus. The present findings are in general accord with these observations.

1. *Projections to the globus pallidus.* In both the report of SWANSON & COWAN (1975) and CONRAD & PFAFF (1976a), the accumbens projection to the globus pallidus in the rat is described and illustrated as being distributed to an anterior and ventral part of the globus pallidus, at levels closely behind the anterior commissure. The present findings confirm this description (Fig. 1C–D), but it is remarkable that neither of the two reports contains mention of the very much denser terminal distribution of accumbens efferents in the rostral, subcommissural extension of the globus pallidus (Fig. 1B). CONRAD & PFAFF (1976a) interpreted the labeled fibers in this subcommissural region as fibers in passage to more caudal parts of the forebrain. It appears that only WILSON (1972) and WILLIAMS, CROSSMAN & SLATER (1977), working with the Fink–Heimer method, have recognized a profuse terminal distribution of accumbens efferents in this part of the forebrain. The region in question is widely considered part of the lateral preoptic region (see, for example, WILLIAMS *et al.* (1977); also Fig. 18 of KÖNIG & KLIPPEL (1963)); SWANSON (1976) and SWANSON & COWAN (1976) labeled it 'substantia innominata'. On the basis of the electron-microscopic observations of HEIMER & WILSON (1976) it must be identified as a rostrally directed, subcommissural spur of the globus pallidus. In accordance with the nomenclature of HEIMER & WILSON (1975), it is here labeled *ventral pallidum* (VP in

Fig. 1B). The present autoradiographic findings suggest that it is the brain region in which accumbens efferents terminate in greatest density.[1]

*Entopeduncular nucleus.* Upon emerging from the ventral pallidum the descending accumbens efferents become incorporated into the medial forebrain bundle. Enclosed in this fiber system they skirt the medial margin of the cerebral peduncle, and thus almost entirely by-pass the entopeduncular nucleus. On this point the present findings conflict with those of CONRAD & PFAFF (1976a). This negative conclusion appears to apply not only to the nucleus accumbens but more in general to the most ventral zone at least of the entire striatum: in none of the cases of ventral striatum injection illustrated in Fig. 4 was there more than sporadic labeling of the entopeduncular nucleus. Our autoradiographic material includes cases of more dorsally placed intrastriatal isotope injection in which, by contrast, the entopeduncular nucleus is densely labeled. The present findings naturally do not rule out the possibility that some accumbens efferents synapse with medially out-lying dendrites or even out-lying cell bodies of the entopeduncular nucleus. They do, however, suggest that the deepest zone of the striatum, including the nucleus accumbens, is not a major source of direct input to this medial pallidal segment, and thus, in contrast to other parts of the striatum, seems likely to have only little direct influence upon those pallidal neurons from which originate the ansa lenticularis fibers to the ventral nucleus of the thalamus, parafascicular nucleus, lateral habenular nucleus and midbrain tegmentum.

2. *Basal forebrain; thalamus.* As emphasized by CONRAD & PFAFF (1976a) and POWELL & LEMAN (1976), an unusual aspect of the striatofugal projection arising from the nucleus accumbens is that it includes, in addition to striatopallidal and striatonigral connections common to all parts of the striatum, fibers distributed to regions implicated in the circuitry of the limbic system, in particular the continuum formed by the septum, preoptic region and hypothalamus. The findings in the present autoradiographic case RR 86 confirm these reports. This same case also demonstrates the accumbens projection to the nuclei paratenialis and mediodorsalis thalami reported earlier by CONRAD & PFAFF (1976a), as well as additional accumbens efferents to the lateral habenular nucleus and the lateral nucleus of the amygdala. It must be emphasized, however, that all these projections from

the accumbens to the thalamus and basal forebrain seem rather sparse and diffuse when compared with those to the globus pallidus and nigral complex. Moreover, a comparison of RR 86 with other cases (Fig. 4) suggests that they do not originate evenly from all parts of the accumbens region. Thus, all of the injections illustrated in Fig. 4 except the most caudally placed one (case RH 51) labeled fibers to the bed nucleus of the stria terminalis, but labeling of the lateral septal region was noted only in case RH 4. The origin of accumbens efferents to the thalamus seems largely if not entirely confined to medial parts of the nucleus: in the material shown in Fig. 4 such fibers were labeled only in case RH 4. The sparse projection to the lateral habenular nucleus appears to originate exclusively in a narrow zone of the nucleus accumbens near its septal margin, i.e. in the same region in which the few accumbens neurons labeled by HRP deposited in the lateral habenular nucleus were found (HERKENHAM & NAUTA, 1977).

Striatal projections to the preoptic region and hypothalamus appear to originate throughout the most ventral zone of the striatum from the level of the anterior commissure forward. According to the present findings, efferents from this wide ventral striatal region almost exclusively follow the medial forebrain bundle in their descent to the midbrain. As stated previously by SWANSON & COWAN (1975), it is difficult to determine whether any of these fibers establish collateral or *de passage* contacts in their course through the lateral hypothalamus. However, it seems certain that the striatofugal fiber system descending from the accumbens region issues fibers medialward that spread rather diffusely over more medial zones of the preoptic region and hypothalamus, apparently (Fig. 1E) avoiding the ventromedial and arcuate nuclei as noted previously by CONRAD & PFAFF (1976a). A comparison of various cases of ventral-striatal isotope injection suggests that these medially distributed fibers originate in largest number in medial parts of the ventral striatum; the small subcommissural pocket of the striatum (case RH 27, Fig. 4) appears to project such fibers only to the medial preoptic region.

3. *Accumbens projections to the midbrain.* It appears that no previous report has drawn attention to the similarity between the mesencephalic projection from the nucleus accumbens and that from the preoptic region and hypothalamus. In all of the studies published thus far the accumbens projection to the substantia nigra seems to have been the major focus of interest, and only fragments of the remaining mesencephalic distribution of accumbens efferents are illustrated in these papers. A more systematic charting such as that shown by Figs 2–3 reveals that

---

[1] A subcommissural part of the external pallidal segment is by no means unique to the rat, and occurs also in primates, including man (see, for example, p. 212 of the atlas of RILEY 1943). In the rat, in contrast to primates, however, the larger remainder of the globus pallidus lies almost entirely behind the level of the anterior commissure.

---

FIG. 7. Juxtaposition of fiber-labeling patterns at nigral levels in case RR 86 (left column; same chartings as shown in Fig. 2) and in a case (RR 89, insert) of injection of tritiated leucine and proline placed in the lateral hypothalamic region. For abbreviations, see legend to Fig. 1.

RR 86
RR 89
ML
FR
SNc
SNr
AON
MP
MB
V
ML
SNc
AON
SNr
FR
MP
V
ND
NI
RN
ML
FR
VTA
AON
SNc
MP
SNr
Cp
SNc
SNr
NR
ML
IP
MP
RN
VTA
ML
IP
MP
SNr
CP
N III
V
SNc
SNr
ML
MP
IP
RN
VTA
ML
IP
MP
SN
CP

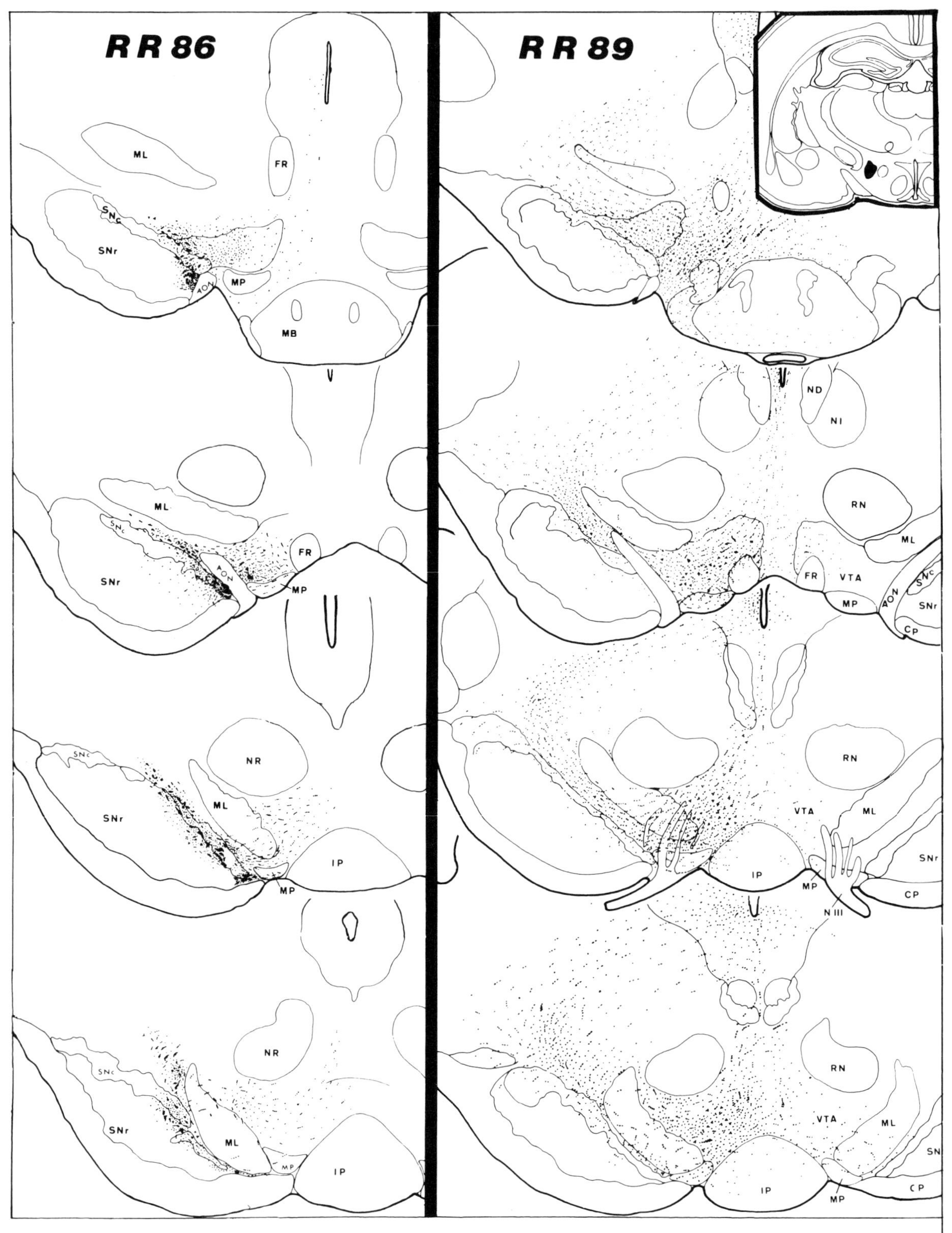

Fig. 7.

accumbens fibers follow the medial forebrain bundle not only in their hypothalamic trajectory, but also in their further descent through nearly the full length of the midbrain.

Fibers descending in the medial forebrain bundle to the mesencephalic tegmentum were noted first by BISCHOFF (1900) in the hedgehog and by WALLENBERG (1902) in the rabbit. The fundamental distribution pattern of these descending fibers was determined only much later by the aid of axon-degeneration techniques (GUILLERY, 1957; NAUTA, 1958). More recently, those originating in the preoptic (SWANSON, 1976; CONRAD & PFAFF, 1976b) and hypothalamic regions (CONRAD & PFAFF, 1976c; NAUTA & DOMESICK, 1977) have been charted by the autoradiographic method. Both of these experimental methods allow a distinction to be made between a medial and a lateral fiber group: the medial group maintains the sagittal orientation of its hypothalamic trajectory and is distributed to the ventral tegmental area and the paramedian zone of the midbrain; the more lateral fiber contingent spreads over the dorsal border of the substantia nigra in a caudolateral direction, and at the caudal pole of the substantia nigra initiates a mediodorsal curve which eventually leads these fibers to the cuneiform and parabrachial regions of the midbrain tegmentum and to the central grey substance at and behind the level of the trochlear nucleus. This summary description omits many important details, one of which is of particular significance to the subject here discussed: the autoradiographic method has revealed that a few preoptic (SWANSON, 1976) and hypothalamic (NAUTA & DOMESICK, 1977) components of the more lateral, supranigral fiber contingent actually invade and apparently terminate in the pars compacta and in a narrow, immediately subjacent dorsal zone of the pars reticulata of the substantia nigra (Fig. 7, right-hand column).

The findings in our autoradiographic material indicate that the mesencephalic distribution of striatofugal fibers originating in the most ventral zone of the anterior half of the striatum corresponds closely to that of the descending preoptic and hypothalamic components of the medial forebrain bundle. In case RR 86 here illustrated, for example, labeled accumbens efferents appear only in midbrain regions in which other experiments have demonstrated the presence of similarly disposed preoptico-mesencephalic (SWANSON, 1976; CONRAD & PFAFF, 1976b) and hypothalamo-mesencephalic (NAUTA, 1958; CONRAD & PFAFF, 1976c; NAUTA & DOMESICK, 1977) fibers. Figure 7 summarizes this coincident distribution at a sampling of frontal levels, but also illustrates an important quantitative contrast between the two fiber categories: whereas preoptic and hypothalamic fibers form a stratum immediately dorsal to the substantia nigra, and invade the latter's dorsal parts with only relatively sparse off-sets from their main group, the corresponding stratum of accumbens fibers takes a slightly more ventral route that leads a large number

of its more ventral fibers tangentially through, rather than over, the dorsal zone of the substantia nigra. The present findings, in agreement with those of SWANSON & COWAN (1975); CONRAD & PFAFF (1976a) and POWELL & LEMAN (1976), indicate that many of these transnigral fibers establish either terminal or collateral contacts in the medial half of the pars compacta as well as in a subjacent zone of the pars reticulata, i.e. in the same nigral strata that also receive much sparser preoptic and hypothalamic afferents.

Accumbens efferents are comparable to preoptic and hypothalamic efferents also in their topographic relationship with the tegmento-striatal cell groups lying outside the substantia nigra proper, in locations corresponding to dopamine cell groups A10 and A8 (Figs 5–6). The entire range of these medially and dorsally out-lying nigral cell aggregates lies embedded among hypothalamic efferents (NAUTA & DOMESICK, 1977), and the present findings indicate that they have a similar relationship with the fibers descending from the nucleus accumbens. It must be noted, however, that in the region of the out-lying nigral cell groups fibers of preoptic and hypothalamic origin appear to outnumber accumbens efferents. By contrast, as noted above, in the dorsal strata of the substantia nigra proper a converse ratio seems to prevail.

*Nigral afferents of the accumbens.* According to biochemical and histochemical observations (ANDÉN *et al.*, 1967; UNGERSTEDT, 1971; LINDVALL & BJÖRKLUND, 1974) the dopamine afferents of the nucleus accumbens originate from cell group A10 in the ventral tegmental area. The present retrograde-labeling evidence supports this notion and indicates that only few such fibers originate from the pars compacta of the substantia nigra proper (Fig. 5). However, it furthermore suggests that many additional fibers to the accumbens originate from the retrorubral nucleus (cf. BERMAN, 1968), a supralemniscal cell aggregate (Fig. 5D) corresponding in location to a dopamine cell group considered the rostral part of A8 by PALKOVITS & JACOBOWITZ (1974). The more caudal parts of this cell region, however, appear to project preferentially at least to ventral striatal regions lateral and caudal to the nucleus accumbens as conventionally defined (compare Figs 5 and 6).

*Nigrostriatal vs striatonigral connections of the accumbens.* It is remarkable that the accumbens appears to project to a nigral region larger than that from which it receives its nigral afferents. According to the present evidence accumbens efferents are distributed not only to the region of cell group A10 and the anterior part of A8 but also, and apparently in greatest density, to the medial half of the pars compacta in which only few cells are labeled by HRP deposited in the accumbens, and to the region of posterior A8 in which such injections fail to label any cells. This suggests that the nigro–striato–nigral circuit associated with the accumbens is not organized in a mode of point-for-point reciprocity. More specifically, it appears that the accumbens, itself receiving nigral afferents from the region of cell groups A10 and the rostral part of A8, in its return projection not only reciprocates this afferent connection, but in addition projects in major volume to the medial half of the pars compacta and to the region of the more caudal parts of A8. Since observations in other autoradiographic cases in our collection indicate that the medial half of the pars compacta projects to nearly all of the medial half of the striatum (DOMESICK, BECKSTEAD & NAUTA, 1976; R. M. BECKSTEAD, V. B. DOMESICK & W. J. H. NAUTA, unpublished observations), while cell group A8 projects preferentially to a more ventrolateral striatal region, the arrangement of the circuit suggests that impulses from the nucleus accumbens can be transmitted directly to a region of the nigral complex large enough to affect, presumably by dopamine transmission, the activity state of all but a dorso-lateral segment of the striatum.

It is important to note that the notion of reciprocity in the relationship between substantia nigra and striatum at present cannot be extended beyond the regional to the cellular level. HATTORI, FIBIGER & McGEER (1974) have reported experimental electron microscopic evidence suggesting that striatonigral fibers synapse very largely with non-dopamine neurons, and thus arguing against any interpretation of the striatonigral connection as an uncomplicated reciprocation of the dopaminergic nigrostriatal projection. However, since the experiments of HATTORI et al. (1975) appear not to have involved the most ventral striatal region, it seems uncertain whether their conclusion is valid also for that part of the striatonigral projection that originates from the nucleus accumbens. That it might not be is suggested by the observation that the striatonigral connection is organized so as to reverse the dorso–ventral coordinate: the most dorsal striatal regions project to the most ventral zone of the substantia nigra's pars reticulata, while successively more ventral zones of the striatum project to successively more dorsal nigral strata; only the most ventral striatal zone—including, but by no means limited to, the nucleus accumbens septi as conventionally defined—projects predominantly into the pars compacta and, in addition, to the region of the medially and dorsally out-lying nigral cell groups A10 and A8 (DOMESICK, 1977; FAULL, V. B. DOMESICK & W. J. H. NAUTA, unpublished observations). In the light of these light-microscopic findings the likelihood of direct synaptic termination on dopamine neurons would seem considerably greater for striatonigral fibers originating in the deepest striatal zone than for those fibers having more dorsal striatal origins. It seems clear that further data on this aspect of the striatonigral connection are needed before the reciprocity of the nigro–striato–nigral relationship can be specified in adequate detail. It therefore should be emphasized that the currently available autoradiographic evidence clarifies no more than the regional distribution of accumbens fibers in the nigral complex, and must leave questions as to

the ultimate closure of the nigro–striato–nigral circuit unanswered.

It nonetheless seems likely that the remarkably selective distribution of the various components of the striatonigral connection has a functional significance. In the case of the nucleus accumbens particularly interesting functional corollaries are suggested by the fact that this part of the striatum receives its dopamine afferents almost exclusively from cell groups A10 and A8. Both of these cell groups lie in the main path of the medial forebrain bundle (NAUTA & DOMESICK, 1977) and therefore are likely to receive major inputs from diencephalic and mesencephalic structures implicated in the circuitry of the limbic system. This would suggest that the nucleus accumbens is associated with the limbic system not only directly by its known afferents from the hippocampal formation (FOX, 1943; SPRAGUE & MEYER, 1950; WEBSTER, 1961; CARMAN, COWAN & POWELL, 1963; RAISMAN, COWAN & POWELL, 1966) and amygdala (DE OLMOS & INGRAM, 1972) but also indirectly by way of its dopaminergic innervation by cell groups A10 and A8. It must be noted, however, that the innervation of the striatum by the region of cell group A10 far exceeds the conventionally accepted bounds of the nucleus accumbens (DOMESICK et al., 1976; R. M. BECKSTEAD, V. B. DOMESICK & W. J. H. NAUTA, unpublished observations). Observations in our autoradiographic and retrograde-labeling material, to be published elsewhere in detail, indicate that the projections from cell groups A10 and A8 together involve to some degree the entire expanse of the striatum except, apparently, for a limited dorsolateral striatal region. Both of these out-lying nigral cell groups, however, project most heavily by far to the most ventral zone of the entire striatum, and it would thus seem tempting to consider this deepest striatal zone as a 'limbic striatum' in the sense that much of its dopamine input is likely to stand under quite major control of subcortical limbic circuits. It is interesting that this is the same zone from which the striatofugal projection to the most dorsal strata of the substantia nigra, and to the region of cell groups A10 and A8, originates.

*Delimitation of the nucleus accumbens.* The foregoing considerations naturally raise questions concerning the identity of the nucleus accumbens. As first introduced by Ziehen (cited by ARIENS KAPPERS & THEUNISSEN, 1907) the term nucleus accumbens broadly referred to that part of the striatum that surrounds the bottom of the frontal horn of the lateral ventricle. Neither Ziehen nor later workers appear to have been able to define the region in question more precisely, and some (e.g. SMITH, 1930) have emphasized the difficulty of demarcating it from the remainder of the striatum by cytoarchitectural criteria. As to other characteristics, in myelin-stained preparations the accumbens region of the striatum in rodent species stands out by a striking sparseness of perforating fascicles of the internal-capsule radiation, but since this is true of nearly the entire longitudinal extent of the ventral striatal region it cannot serve as an aid in delimiting the nucleus accumbens. In the context of this study it therefore should be asked if the striatal region in question could perhaps be defined more accurately on the basis of distinctive afferent or efferent relationships.

The nucleus accumbens clearly cannot be characterized simply as 'that part of the striatum that receives its nigral afferents from the region of cell group A10'. The A10 region apparently projects to a much larger part of the striatum, and in greatest density by far to a ventral zone extending throughout the length of the striatum. However, there are reasons to draw a distinction between the part of the ventral striatal zone that extends rostral to the anterior commissure, and the part caudal to the commissure. According to the present data (see, for example, Figs 5–6) the more anterior part receives partially overlapping nigral afferents *almost exclusively* from the region of cell groups A10 and A8. Findings in other cases in our autoradiographic and retrograde-labeling material indicate that in more caudal parts of the ventral striatal zone, by contrast (for example, in that part injected in RH 51, Fig. 4), fibers from the region of cell groups A10 and A8 overlap with equally or more numerous fibers from the lateral half of the substantia nigra proper.

On the basis of this difference, the nucleus accumbens could be defined as that part of the ventral striatum that receives nigral afferents most nearly exclusively from the region of the out-lying nigral cell groups A10 and A8. This definition, however, would hardly seem satisfactory, based as it is on a difference of degree rather than on an exclusive criterion. It is possible that the nucleus accumbens could be more accurately defined by its afferents from the hippocampus and amygdala, but the total distribution of these direct limbic projections to the striatum appears as yet not to have been determined. The same remarks apply to the thalamic afferents of the nucleus accumbens which, as demonstrated by Cowan &

Powell (1955) and Swanson & Cowan (1975), originate from the nucleus paratenialis. A third possibility to be considered is, that the nucleus may be most sharply identified as that part of the striatum all points of which have efferent connections with any (or any combination) of the following subcortical structures: bed nucleus of the stria terminalis, lateral septal nucleus, preoptic region, hypothalamus, amygdala, ventral tegmental area, and levels of the midbrain tegmentum and central grey substance caudal to the substantia nigra. However, the striatal district giving rise to such projections cannot be outlined on the basis of current data alone.

In conclusion, the results of the present study provide no sharp identity of a striatal region that could be delineated as nucleus accumbens from the rest of the striatum. They do suggest that the striatal region in question forms an anterior part of a ventral striatal zone which, extending throughout the length of the striatum, projects to the most dorsal strata of the substantia nigra as well as to the region of cell groups A8 and A10. This antero–ventral striatal region has additional projections to the subcommissural part of the globus pallidus (ventral pallidum of Heimer & Wilson, 1975) and to a variety of forebrain and midbrain structures implicated in the circuitry of the limbic system. Its nigral afferents originate very largely from the region of cell groups A10 and A8, and apparently only in minor number from the pars compacta. More detailed studies will be required to determine whether this striatal region, or any part of it, can be delimited from the rest of the striatum by any structural or connectional criterion, or alternatively, whether it blends into neighboring parts of the striatum without any clear boundary.

*Acknowledgements*—This study was supported by USPHS Grants NB 06542 and MH 25515, by NSF Grant BNS 76-81227, and by a Harkness Fellowship from the Commonwealth Fund of New York and an Overseas Research Fellowship from the Medical Research Council of New Zealand to RLMF.

## REFERENCES

Andén N.-E., Dahlström A., Fuxe K., Larsson K., Olson L. & Ungerstedt U. (1966) Ascending monoamine neurons to the telencephalon and diencephalon. *Acta physiol. scand.* **67,** 313–326.

Ariens Kappers C. U. & Theunissen W. F. (1907) Die Phylogenese des Rhinencephalons, des Corpus striatum und der Vorderhirnkommissuren. *Folia neurobiol.* **1,** 173–288.

Beckstead R. M. (1976) Convergent thalamic and mesencephalic projections to the anterior medial cortex in the rat. *J. comp. Neurol.* **166,** 403–416.

Berman A. L. (1968) *The Brain Stem of the Cat. A Cytoarchitectonic Atlas with Stereotaxic Coordinates.* The University of Wisconsin Press, Madison.

Bischoff E. (1900) Beitrag zur Anatomie des Igelgehirnes. *Anat. Anz.* **18,** 348–358.

Carman J. B., Cowan W. M. & Powell T. P. S. (1963) The organization of the corticostriate connections in the rabbit. *Brain* **86,** 525–562.

Conrad L. C. A. & Pfaff D. W. (1976*a*) Autoradiographic tracing of nucleus accumbens efferents in the rat. *Brain Res.* **113,** 589–596.

Conrad L. C. A. & Pfaff D. W. (1976*b*) Efferents from medial basal forebrain and hypothalamus in the rat—I. An autoradiographic study of the medial preoptic area. *J. comp. Neurol.* **169,** 185–220.

CONRAD L. C. A. & PFAFF D. W. (1976c) Efferents from medial basal forebrain and hypothalamus in the rat—II. An autoradiographic study of the anterior hypothalamus. *J. comp. Neurol.* **169**, 221–262.

COWAN W. M. & POWELL T. P. S. (1955) The projections of the midline and intralaminar nuclei of the thalamus of the rabbit. *J. Neurol. Neurosurg. Psychiat* **18**, 266–279.

DAHLSTRÖM A. & FUXE K. (1964) Evidence for the existence of monoamine-containing neurons in the central nervous system—I. Demonstration of monoamines in the cell bodies of brainstem neurons. *Acta physiol. scand.* **62**, Suppl. **232**, 1–55.

DE OLMOS J. S. & INGRAM W. R. (1972) The projection field of the stria terminalis in the rat brain: an experimental study. *J. comp. Neurol.* **146**, 303–334.

DOMESICK V. B. (1977) The topographic organization of the striatonigral connection in the rat. (Abstr.) *Anat. Rec.* **187**, 567.

DOMESICK V. B., BECKSTEAD R. M. & NAUTA W. J. H. (1976) Some ascending and descending projections of the substantia nigra and ventral tegmental area in the rat. *Neuroscience Absts* **II**, part 1, p. 61.

FOX C. A. (1943) The stria terminalis, longitudinal association bundle and precommissural fornix fibers in the cat. *J. comp. Neurol.* **79**, 277–295.

GUILLERY, R. W. (1957) Degeneration in the hypothalamic connexions of the albino rat. *J. anat.* **91**, 91–115.

HATTORI T., FIBIGER H. C. & MCGEER P. L. (1975) Demonstration of a pallido-nigral projection innervating dopaminergic neurons. *J. comp. Neurol.* **162**, 487–504.

HEIMER L. & WILSON R. D. (1975) The subcortical projections of the allocortex: Similarities in the neural associations of the hippocampus, the piriform cortex, and the neocortex. In *Golgi Centennial Symposium* (ed. SANTINI M.), pp. 177–193. Raven, New York.

HERKENHAM M. & NAUTA W. J. H. (1977) Afferent connections of the habenular nuclei in the rat. A horseradish peroxidase study, with a note on the fiber-of-passage problem. *J. comp. Neurol.* **173**, 123–146.

KÖNIG J. F. R. & KLIPPEL R. A. (1963) *The Rat Brain, a Stereotaxic Atlas of the Forebrain and Lower Parts of the Brain Stem.* R. E. Krieger, Huntington, N.Y.

LINDVALL O. & BJÖRKLUND A. (1974) The organization of the ascending catecholamine neuron systems in the rat brain as revealed by the glyoxylic acid fluorescence method. *Acta physiol. scand., Suppl.* **412**, 1–48.

LINDVALL O., BJÖRKLUND A., MOORE R. Y. & STENEVI U. (1974) Mesencephalic dopamine neurons projecting to neocortex. *Brain Res.* **81**, 325–331.

NAUTA W. J. H. (1958) Hippocampal projections and related neural pathways to the midbrain in the cat. *Brain* **81**, 319–340.

NAUTA W. J. H. & DOMESICK V. B. (1977) Cross-roads of limbic and striatal circuitry: hypothalamo-nigral connections. In *Limbic Mechanisms: The Continuing Evolution of the Limbic System Concept* (eds LIVINGSTON K. E. & HORNYKIEWICZ O.). Plenum Press, New York.

PALKOVITS M. & JACOBOWITZ D. M. (1974) Topographic atlas of catecholamine and acetylcholinesterase-containing neurons in the rat brain—II. Hindbrain (mesencephalon, rhombencephalon). *J. comp. Neurol.* **157**, 29–42.

POWELL E. W. & LEMAN R. B. (1976) Connections of the nucleus accumbens. *Brain Res.* **105**, 389–403.

RAISMAN G., COWAN W. M. & POWELL T. P. S. (1966) An experimental analysis of the efferent projections of the hippocampus. *Brain* **89**, 83–108.

RILEY H. A. (1943) *An Atlas of the Basal Ganglia, Brain Stem and Spinal Cord.* Williams & Wilkins, Baltimore.

SMITH O. C. (1930) The corpus striatum, amygdala, and stria terminalis of *Tamandua tetradactyla. J. comp. Neurol.* **51**, 65–127.

SPRAGUE J. M. & MEYER M. (1950) An experimental study of the fornix in the rabbit. *J. Anat.* **84**, 354–368.

SWANSON L. W. (1976) An autoradiographic study of the efferent connections of the preoptic region in the rat. *J. comp. Neurol.* **167**, 227–256.

SWANSON L. W. & COWAN W. M. (1975) A note on the connections and development of the nucleus accumbens. *Brain Res.* **92**, 324–330.

SWANSON L. W. & COWAN W. M. (1976) Autoradiographic studies on the development and connections of the septal area in the rat. In *The Septal Nuclei* (ed. DE FRANCE J. F.), pp. 37–64. Plenum Press, New York.

UNGERSTEDT U. (1971) Stereotaxic mapping of the monoamine pathways in the rat brain. *Acta physiol. scand.* **197**, Suppl. **367**, 1–48.

WALLENBERG A. (1902) Das basale Riechbündel des Kaninchens. *Anat. Anz.* **20**, 175–187.

WEBSTER K. E. (1961) Cortico-striate interrelations in the albino rat. *J. Anat.* **95**, 532–545.

WILLIAMS D. J., CROSSMAN A. R. & SLATER P. (1977) The efferent projections of the nucleus accumbens in the rat. *Brain Res.* **130**, 217–227.

WILSON R. D. (1972) Efferent connections of the nucleus accumbens in the rat. Master's thesis, Massachusetts Institute of Technology.


(*Accepted 27 December* 1977)

*Brain Research*, 175 (1979) 191–217
© Elsevier/North-Holland Biomedical Press



# Research Reports

EFFERENT CONNECTIONS OF THE SUBSTANTIA NIGRA AND VENTRAL TEGMENTAL AREA IN THE RAT

ROBERT M. BECKSTEAD, VALERIE B. DOMESICK and WALLE J. H. NAUTA*

*Department of Psychology, Massachusetts Institute of Technology, Cambridge, Mass. 02139 and (V.B.D.) Mailman Research Center, McLean Hospital, Belmont, Mass. (U.S.A.)*

(Accepted December 28th, 1978)

SUMMARY

Small injections of tritiated leucine and proline confined to the ventral tegmental area (AVT) were found to label fibers ascending: (a) to the entire ventromedial half of the striatum, but most massively to the ventral striatal zone that includes the nucleus accumbens; (b) to the thalamus: lateral habenular nucleus, nuclei reuniens and centralis medius, and the most medial zone of the mediodorsal nucleus; (c) to the posterior hypothalamic nucleus and possibly the lateral hypothalamic and preoptic region; (d) to the nuclei amygdalae centralis, lateralis and medialis; (e) to the bed nucleus of the stria terminalis, the nucleus of the diagonal band, and the medial half of the lateral septal nucleus; (f) to the anteromedial (frontocingulate) cortex; and (g) to the entorhinal area. Further AVT efferents descend to the medial half of the midbrain tegmentum including an anterior region of the median raphe nucleus, to the ventral half of the central grey substance including the dorsal raphe nucleus, to the parabrachial nuclei, and to the locus coeruleus.

Similar injections centered in the pars compacta of the substantia nigra (SNC) label fibers that are distributed in the striatum in an orderly medial-to-lateral arrangement, and almost entirely avoid the nucleus accumbens and olfactory tubercle. With the exception of the lateral quarter of the substantia nigra, which apparently does not project to the extreme rostral pole of the striatum, each small SNC locus, regardless of its anteroposterior localization, distributes nigrostriatal fibers throughout the length of the striatum. Descending SNC efferents are distributed to the same general regions that receive descending AVT projections, except that no SNC fibers appear to enter the locus coeruleus.

Isotope injections confined to the pars reticulata (SNR) label sparse nigrostriatal

* To whom correspondence should be addressed.

fibers, and numerous nigrothalamic fibers ascending mainly to the nucleus ventro-medialis and in lesser number to the parafascicular nucleus and the paralamellar zone of the nucleus mediodorsalis. Descending SNR fibers leave the nigra as a voluminous fiber bundle that bifurcates into a large nigrotectal and a smaller nigrotegmental component, the latter terminating largely in the pedunculopontine nucleus of the pontomesencephalic tegmentum.

---

## INTRODUCTION

As late as 1964 it could be argued that no efferent connection of the substantia nigra had been conclusively demonstrated. In that year, Andén et al.[3] by the newly developed monoamine–histofluorescence method supplied the first direct anatomical evidence of a massive nigrostriatal projection originating from nigral dopamine neurons. This accomplishment prompted numerous further studies by the histofluorescence method, in the course of which it was found that nigral dopamine fibers innervate not only the striatum but also other forebrain structures, in particular the amygdala, the septal region, the anterior limbic cortex, and the entorhinal area.

Curiously, the discovery of other, apparently largely non-dopaminergic, nigral efferents also began in 1964 when Cole et al.[13] identified a substantial nigral projection to the ventromedial nucleus of the thalamus. Subsequent studies produced the first evidence of projections from the nigral complex to structures below the forebrain, namely, to the mesencephalic and pontine tegmentum[32,40,61], the superior colliculus[30, 31,41], and the locus coeruleus[63].

Except for the nigrotectal connection, nigromesencephalic projections thus far appear to have been charted only in fiber-degeneration material. The autoradiographic study here reported adds several details to the existing descriptions of these descending nigral efferents. In addition, this paper reports some hitherto undescribed nigro-thalamic connections, and documents findings concerning the topographic organization of the nigrostriatal projection that were reported previously in abstract form[18].

## MATERIALS AND METHODS

The projections of the ventral tegmental area and substantia nigra were traced autoradiographically in 32 adult albino rats (Charles River Laboratories) of both sexes. Each rat received a single small injection of [³H]proline and [³H]leucine (New England Nuclear) in either the ventral tegmental area or various medial-to-lateral divisions of the pars compacta. In additional cases injections were placed in bordering cell areas and in the substantia nigra's pars reticulata. The isotopes were delivered by microelectrophoresis from a stereotaxically inserted glass micropipette filled with a 20 $\mu$Ci/$\mu$l solution of the isotopes in 0.1 M acetic acid; the driving force was supplied by passing a cathodal, direct current of 0.8 $\mu$A, delivered in 7.5 Hz square-wave pulses, between the pipette solution and a ground wire attached to exposed periosteum. In order to obtain injections small enough for the purpose of this study, pipettes of small internal tip diameter (10–15 $\mu$m) were used, as were low current intensities, and injection times of no more than 3 min.

After survival times ranging from 5 to 10 days the rats were once more deeply anesthetized, and perfused transcardially with 10% formol–saline. The brains were immediately dissected out and stored first in the same fixative for 10 days, then in 30% sucrose–formalin for an additional 4 days before being embedded in albumin–gelatin and sectioned in the frontal plane at 25 $\mu$m on a freezing microtome. Every third section was mounted on a gelatin-coated slide, coated with Kodak NTB-2 emulsion and stored in a freezer at $-20$ °C for 4–16 weeks before being developed, fixed, and counterstained with cresylecht violet.

Each section was systematically scanned under both bright- and dark-field illumination for radioactively labeled fibers. In all illustrated cases the animal had been sacrificed 8–10 days after surgery, and the photographic emulsion had been exposed to the underlying radioactive sections for 12 weeks. The injection sites are collectively indicated in Fig. 1A–C.

RESULTS

*I. Definition of the substantia nigra*

It is necessary to deal briefly with the terminology used here. The term substantia nigra refers to a complex structure that lies immediately dorsal to the cerebral peduncle and is composed largely of medium-sized cells of a fairly uniform type which, however, in Nissl material appear more darkly stained and much more closely spaced in the dorsal stratum, the pars compacta (SNC), than in the larger subjacent pars reticulata (SNR). In the rat, the pars compacta is demarcated from the pars reticulata along a fairly even plane, except for the lateral quarter of the nigra (often separately labeled pars lateralis) where the border between the two subdivisions becomes quite indistinct.

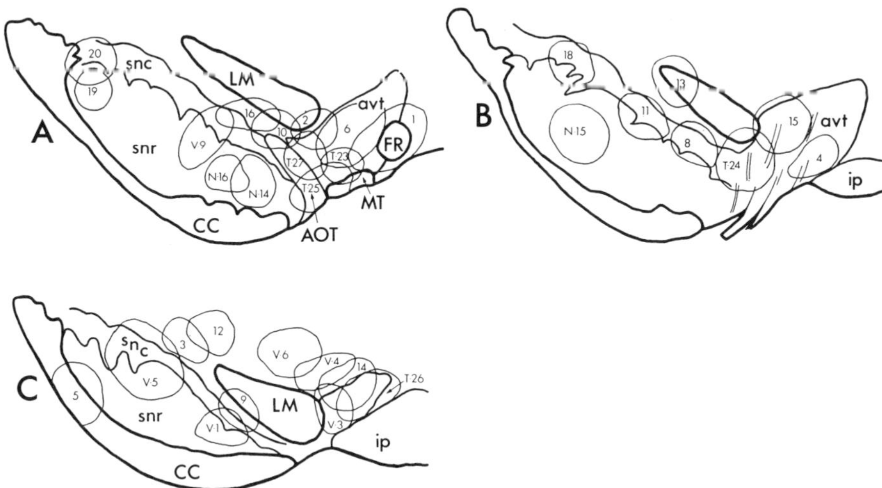


Fig. 1. Line drawings illustrating the approximate position and extent of the injection site in each of the 32 experiments done in this study. Four cases selected from this material are illustrated photographically in Fig. 2.

As demonstrated with the histofluorescence method, the pars compacta is composed largely if not entirely of dopamine neurons, whereas most, but not all, of the cells of the pars reticulata are non-dopaminergic. The same technique revealed that, in the rat at least, the pars compacta (dopamine-cell group A9 of Dahlström and Fuxe[15]) does not have the flat dorsal border traditionally ascribed to it, and instead emits two large dorsal excrescences which, however, do not have the high cell-packing density of the pars compacta and consequently in Nissl-stained material had only rarely been recognized as components of the nigral complex. The larger medial one of these dorsally protruding cell masses, dopamine-cell group A10, extends dorso-medially into the ventral tegmental area of Tsai, whereas a smaller, more caudal and lateral extrusion from the pars compacta, dopamine-cell group A8, invades a ventrolateral tegmental region near the caudal pole of the substantia nigra. At caudal levels of the nigra, cell groups A10 and A8 are interconnected by an irregular array of cells that extends transversely over the dorsal border of the medial lemniscus; long before histofluorescence studies showed that at least many of its cells synthesize dopamine, this supralemniscal cell group had been identified in Nissl material under various names, most recently as *retrorubral nucleus* (see Berman[8]).

Since all these dopamine cell groups, A9, A10, A8 and the retrorubral nucleus, project predominantly to the striatum, many of their cells can be labeled retrogradely by intrastriatal injections of horseradish peroxidase (HRP). In a recent series of experiments[53,55] HRP deposited in the most ventral zone of the striatum was found to label numerous cells in the region of A10, A8, and the retrorubral nucleus; as the outlines of the labeled cell groups practically coincided with those of the dopamine cell groups identified in histofluorescence studies[53] it was concluded that non-dopaminergic tegmentostriatal neurons, if any exist and can be labeled by HRP, must be located largely within the boundaries of the nigral dopamine cell groups.

Since both histofluorescence and retrograde-labeling methods clearly show cell groups A10 and A8 to be continuous with the pars compacta (A9) as well as with each other, it seems appropriate to view the labels A10, A9, A8 and retrorubral nucleus as terms of primarily topographic value, each specifying a particular subdivision of a coherent nigral cell complex. This view is not contradicted by the fact that there are some notable differences between the respective afferent and efferent relationships of these subdivisions, for they all project predominantly to the striatum.

The present report deals only with those nigrofugal projections that originate in the pars compacta, pars reticulata, and ventral tegmental area. Projections arising from more caudal parts of the nigral complex, in particular the regions of cell group A8 and the retrorubral nucleus, will be described separately elsewhere (Lénard and Nauta, in preparation).

*II. Projections of the Ventral Tegmental Area (AVT)*

In Fig. 3 the fibers labeled by an injection in the ventral tegmental area (AVT) are charted in frontal sections. The injection site in this case (RVT6, photographically illustrated in Fig. 2A) was centered in an anterior AVT region rostral to the interpeduncular nucleus and separated from the pars compacta by the medial terminal

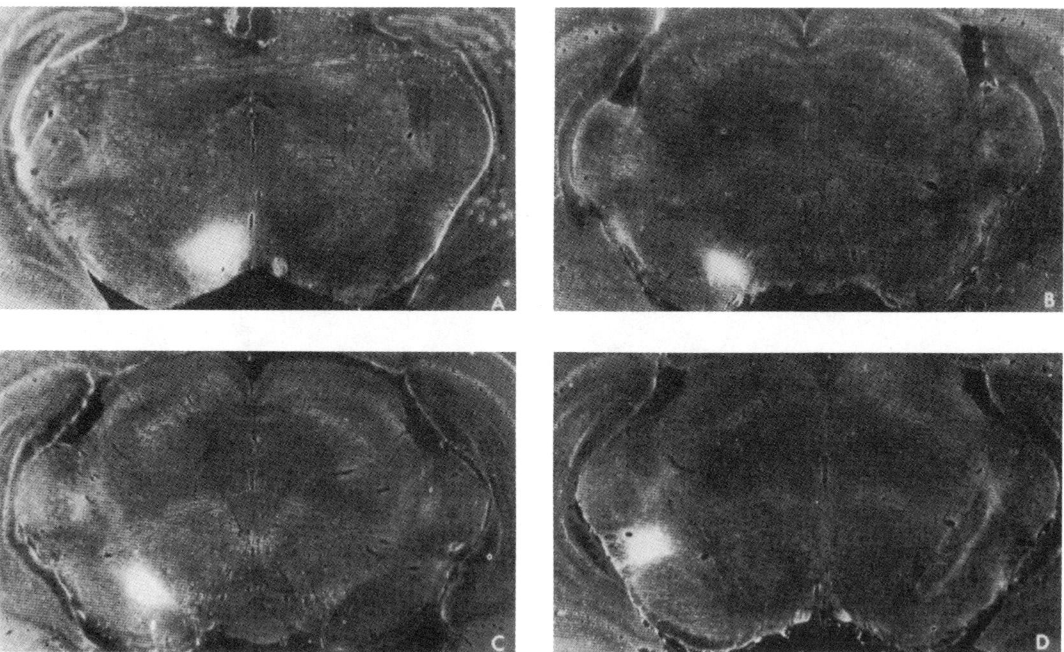

Fig. 2. Dark-field photographs of the injection sites in cases RVT6(A), RVT8(B), RVT11(C) and RVT-20(D) described in the text. The injection sites, shown at their maximal extent, appear somewhat amplified by the glare from the dense silver grain marking their center.

nucleus of the accessory optic tract. Because neurons exhibiting signs of label-uptake were seen only in the AVT, it seems likely that the fibers labeled in this case originated in AVT and not in the pars compacta. However, injections placed in more caudal parts of AVT less distinctly separated from the pars compacta labeled fibers of essentially the same distribution. The lateral supramammillary area rostrally contiguous with AVT was not involved in the injection: control injections involving that region invariably labeled fibers in the fascia dentata, and no such labeling was found in case RVT6.

From the injection site numerous labeled axons extended rostrally into the medial forebrain bundle, whereas others spread lateralwards over the substantia nigra, and a third contingent passed dorsocaudally into median and paramedian parts of the midbrain tegmentum. A small group of labeled fibers was seen to cross the midline to descend in a pattern similar to that of the ipsilateral descending fibers.

*Lateral and ascending projections of AVT.* Numerous labeled fibers spread laterally from the injection site in AVT to the ipsilateral substantia nigra in which they were distributed over the entire rostrocaudal extent of the pars compacta, and in lesser number to the most dorsal zone of the pars reticulata (Fig. 3J–L). In its spread lateralwards the labeling of the nigra became progressively more confined to the pars compacta (Fig. 3K). Its rather homogeneous distribution suggests a termination of AVT efferents in contact with either somata of compacta neurons or compacta dendrites oriented parallel to the dorsal border of the nigra.

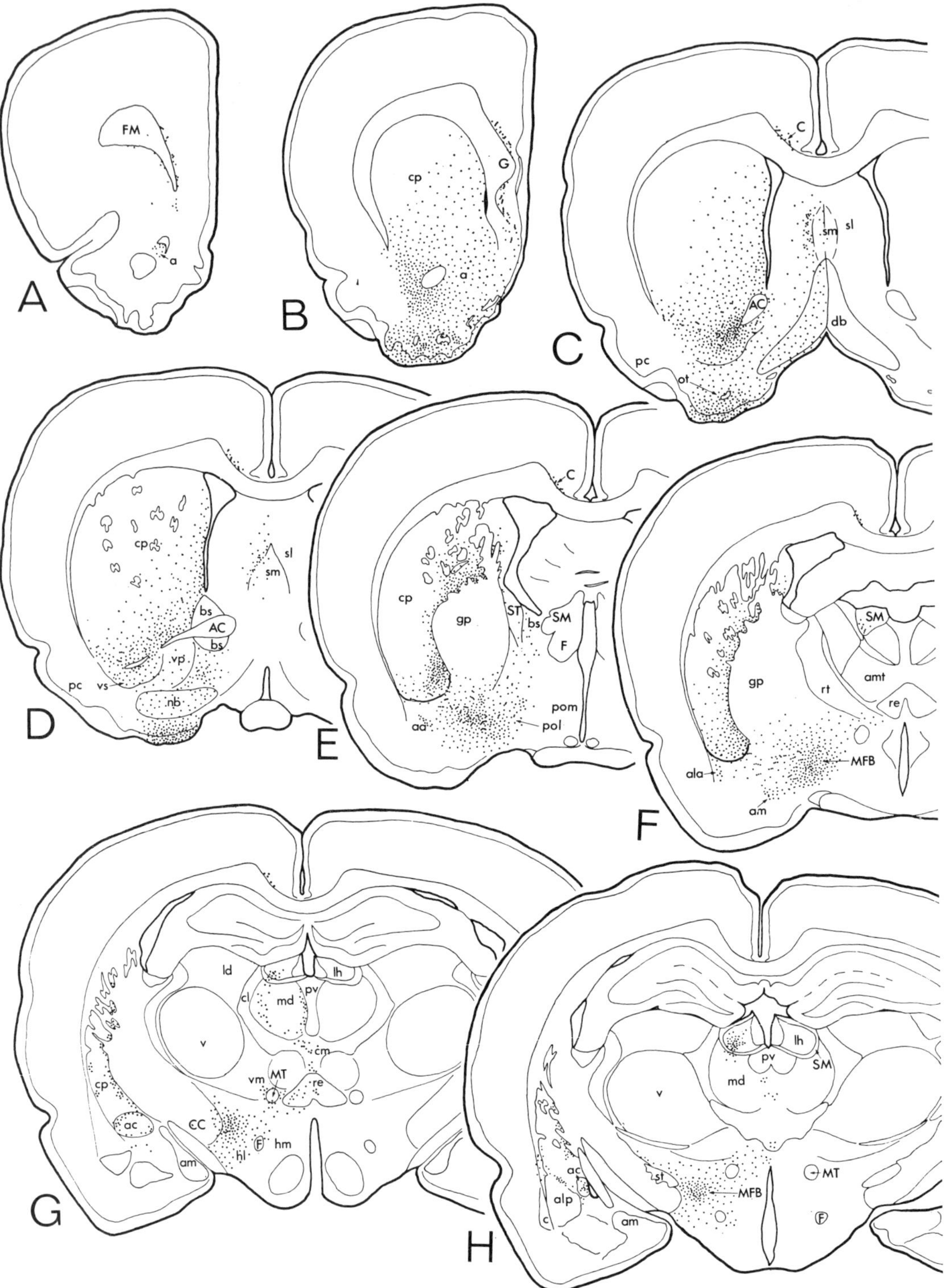

Fig. 3. Serial chartings of the fiber labeling resulting from an isotope injection of the ventral tegmental area (case RVT6). The injection site is indicated in K, and photographically illustrated in Fig. 2A.

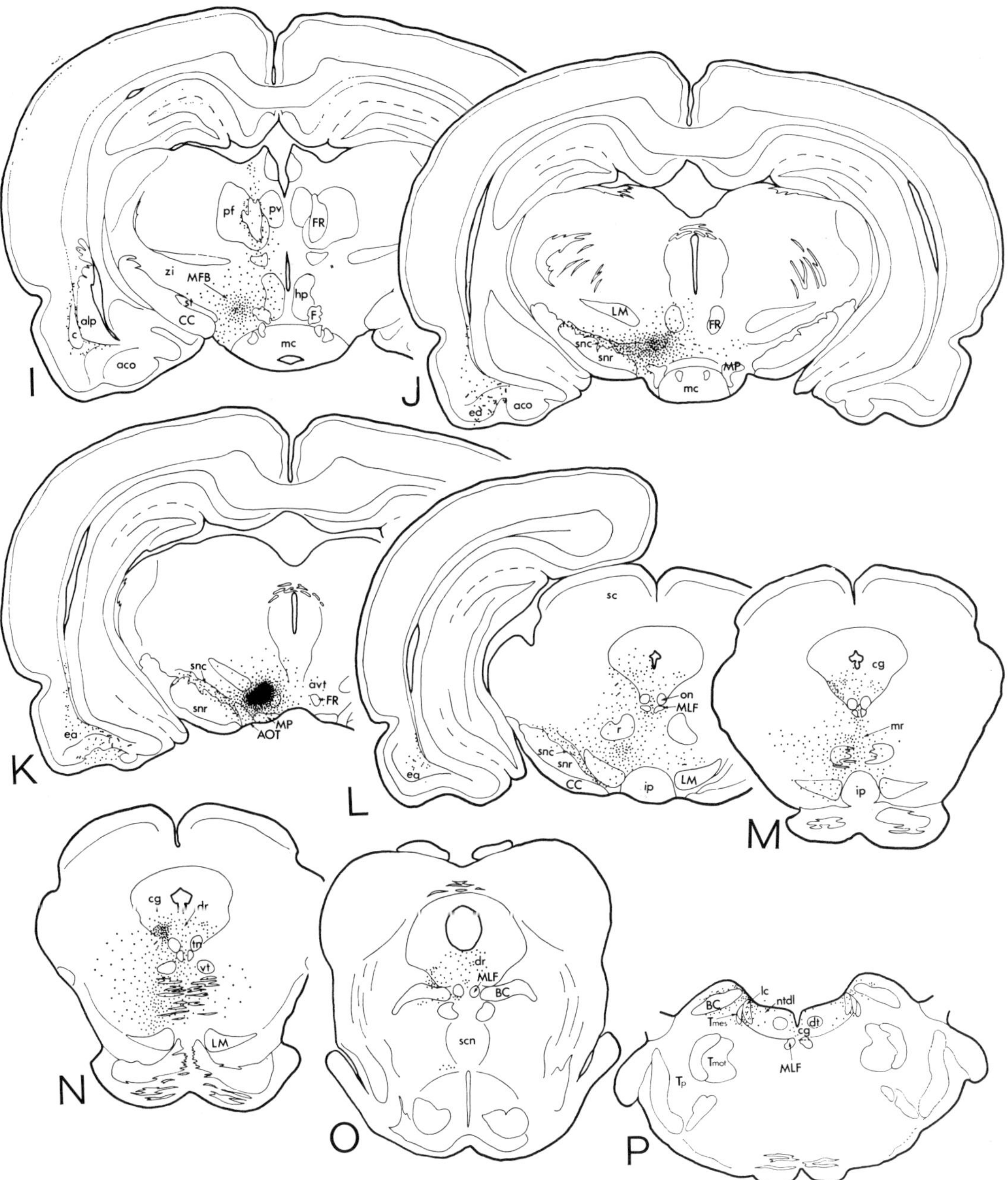

The subthalamic nucleus, which abuts the rostral pole of the nigra, was free of label (Fig. 3H, I).

*Diencephalic projections of AVT.* Labeled axons ascending from the injection site largely followed the ipsilateral medial forebrain bundle in which they were rather evenly distributed, with only a slight tendency to cluster near the core of the bundle. At the level where the fasciculus retroflexus passes medial to the medial forebrain bundle a few labeled fibers deviated dorsally from the medial forebrain bundle. These fibers could be traced alongside the fasciculus retroflexus (Fig. 3I, J) to the lateral half of the lateral habenular nucleus (Fig. 3H; 9B).

Because of the diffuse distribution of label throughout the length of the medial forebrain bundle it is uncertain whether any AVT fibers actually terminate in the lateral hypothalamic region. However, some labeled fibers could be followed medial-wards from the bundle into the posterior hypothalamic nucleus (Fig. 3I). Farther rostrally, further off-sets from the main ascending path passed dorsalwards through the subthalamic region to the extreme ventromedial part of the ventromedial thalamic nucleus, the nucleus reuniens, the nucleus centralis medius, and the marginal zone of the mediodorsal nucleus. In the last-mentioned nucleus, labeling was densest medially, i.e. in the region bordering the periventricular nucleus (Fig. 3G; 9C).

Near the level of the rostral pole of the thalamus, a third contingent of dorsal off-sets from the main group of labeled fibers spread into the preoptic part of the bed nucleus of the stria terminalis (Fig. 3E); more anterior parts of the bed nucleus were free of label. Finally, at levels rostral to the transverse limb of the anterior commissure diffusely arranged labeled fibers from the medial forebrain bundle entered the nucleus of the diagonal band of Broca and the medial half of the lateral septal nucleus (Fig. 3C, D).

*AVT projections to the striatum.* At middle hypothalamic levels (Fig. 3F) labeled fibers began to spread laterally from the medial forebrain bundle to enter the striatum from the ventral side. Farther rostrally, where the globus pallidus gradually dwindles and ultimately disappears, the striatum progressively expands in the medial direction and, from the level of the anterior commissure forward, labeled fibers enter the base of the striatum without deflecting laterally from the sagittal path of the medial forebrain bundle.

Within the striatum labeled AVT fibers were widely distributed. In addition to the nucleus accumbens and olfactory tubercle, the entire fundus region of the striatum was labeled, and at all rostrocaudal levels labeling extended from this most basal region dorsalwards in gradually diminishing density through nearly the entire dorsoventral dimension of the striatum, avoiding only the dorsolateral half of the striatum. Thus, AVT fibers appear to innervate nearly all of the ventromedial half of the striatum (Figs. 3B–F; 9A), even though their largest number terminate in the most ventral striatal region including, at rostral levels, the nucleus accumbens.

At rostral levels the labeling of the nucleus accumbens continued ventrally over the entire extent of the olfactory tubercle (Fig. 3B), marking all layers of this structure but avoiding the islands of Calleja. Farther caudally, a progressively widening zone of sparsely labeled substantia innominata intervened between the two structures (Fig. 3C, D); it could not be determined whether the nucleus basalis (magnocellular preoptic nucleus) and ventral pallidum (Heimer and Wilson[34]) lying within this zone of sparser labeling, receive or are only traversed by AVT efferents.

*AVT projections to the cortex.* At levels rostral to the callosal genu some of the most medial labeled fibers in the medial forebrain bundle turned dorsally, traversed the most anterior part of the septum, and deep to the hippocampal rudiment gathered into slender fascicles which curved caudalwards, joining the fasciculus cinguli in which they could be followed to a level as far caudally as the rostral border of the granular retrosplenial cortex (Fig. 3B–G). Some fibers of the same group described a wider curve around the genu through more anterior regions of the medial cortex (Fig. 3A). Throughout its precallosal and supracallosal course, sparse label spread from the fasciculus cinguli over the deep layers of the medial cortex. At far rostral levels (Fig. 3A) a very small number of labeled fibers extended into the cortex forming the fundus of the rhinal sulcus; these likewise appeared to be distributed to the deepest cortical layer.

Apart from the deep strata of the anteromedial and sulcal cortex, all layers of the rostral half of the entorhinal area were labeled (Fig. 3J, K). Labeled AVT efferents reached this area by two routes: one fiber group by curving caudalwards in the external and extreme capsules from the base of the striatum, a more medial second contingent by passing caudally through the amygdaloid complex.

*AVT projections to the amygdala.* Since the anterior half of the amygdaloid complex is traversed by fibers passing to the striatum it is difficult to determine in which amygdaloid nuclei AVT fibers actually terminate. At more caudal levels it seems certain that the central nucleus receives such fibers (Fig. 3G, H), and farther rostrally some termination of AVT fibers in the lateral and medial amygdaloid nuclei seems likely (Fig. 3F). It could not be determined whether further AVT efferents are distributed to the anterior amygdaloid area.

*Descending projections of AVT.* Labeled fibers descending from AVT spread caudally and dorsally over a wide medial zone of the midbrain tegmentum extending caudally to the level of the rostral pole of the inferior colliculus (Fig. 3L–N). Labeling throughout this tegmental region was densest near the midline, but it did not spread caudally far enough to involve more than the rostral margin (Fig. 3L–N) of the median raphe nucleus. Although largely ipsilateral, it sparsely marked also the corresponding contralateral region.

Apparently, some of the labeled fibers in this anterior tegmental field continued dorsally into the ventral region of the central gray substance, including the dorsal raphe nucleus (Fig. 3L–O). A single dense, wedge-shaped cluster of labeled fibers appeared alongside the trochlear nucleus (Fig. 3N). At more caudal levels, the dorsal and ventral peribrachial areas showed sparse labeling (Fig. 3O–P) that continued medially and caudally over the locus coeruleus and an adjacent lateral part of the nucleus tegmenti dorsalis lateralis (Fig. 3P). The dorsal and ventral tegmental nuclei of Gudden were free from label, and neither did label above background appear in the brain stem caudal to the locus coeruleus.

### *III. Projections of the Substantia Nigra, Pars Compacta*

Owing to the thinness of the pars compacta (SNC) it proved impossible even with the microelectrophoretic method to obtain injections that involved exclusively SNC cells. Nonetheless, in several cases the small injection of the pars compacta appeared only minimally to involve the underlying pars reticulata or dorsally adjacent parts of the tegmentum. Three such cases will be described below. In one of these,

RVT8 (Figs. 1B and 2B), the injection was located in the medial one-quarter of SNC and also involved the lateral fringe of AVT. In a second case, RVT11 (Figs. 1B and 2C), it involved approximately the middle third, and in the third case, RVT20 (Figs. 1A and 2D), the lateral quarter ('pars lateralis') of SNC.

*Case RVT8.* From the injection site in the medial part of SNC, labeled axons passed rostrally in the medial forebrain bundle and the medial margin of the cerebral peduncle (Fig. 4E, F). It could not be determined whether any of those that follow the medial forebrain bundle actually terminate in the lateral hypothalamic region. At caudal hypothalamic levels, a group of labeled fibers separated from the medial forebrain bundle and passed rostrodorsally to the medial part of the ventromedial thalamic nucleus and, bilaterally, to the ventromedial part of the mediodorsal nucleus (Fig. 4E). Farther forward, at the level of the anterior hypothalamus and preoptic region, most of the remaining labeled fibers curved away dorsolaterally from the medial forebrain bundle; after traversing the globus pallidus, these fibers entered the medial and dorsal regions of the caudatoputamen (Fig. 4D). A few labeled fibers, presumably originating from the lateral part of AVT involved in the injection, continued forward in the medial forebrain bundle to the nucleus accumbens and olfactory tubercle (Fig. 4A, B). Most of the labeled nigrostriatal fibers were distributed within the medial half of the striatum, most densely in the striatal zone adjoining the lateral ventricle (Fig. 4A–D). No labeling marked the septal nuclei, nucleus of the diagonal band, or nucleus basalis, even though the lateral part of AVT was involved in the injection. This negative finding suggests that only the more medial parts of AVT project to these structures. Neither did any labeling above background mark any cortical area or the amygdala.

*Case RVT11.* From the injection site, centered in the pars compacta about the middle of the latter's width (Fig. 5G; see also Fig. 2C), labeled fibers ascended in the medial forebrain bundle and cerebral peduncle. A larger proportion of these nigral efferents followed the peduncle in this case than in the foregoing case of more medially placed injection (Fig. 5E); moreover, all labeled nigrostriatal fibers curved laterally at a more caudal level, so that no labeled fibers continued in the medial forebrain bundle beyond mid-hypothalamic levels (Fig. 5E). Although labeled nigrostriatal fibers were distributed over nearly the entire caudatoputamen, they were most highly concentrated in a zone extending vertically almost throughout the dorsoventral diameter of the caudatoputamen, about midway between the latter's medial and lateral borders (Fig. 5A–D). It is important to note that in this case no labeled fibers appeared in the nucleus accumbens (Fig. 5A). In the thalamus, labeled fibers spread over the medial two-thirds of the ventromedial nucleus, and bilaterally over the basal part of the mediodorsal nucleus (Fig. 5E).

*Case RVT20.* The isotope injection in this case was centered in the most lateral part of SNC (Fig. 6E; see also Fig. 2D). From it, labeled fibers ascended in the lateral part of the medial forebrain bundle and the medial part of the cerebral peduncle to be distributed most densely to the lateral part of the caudatoputamen (Fig. 6A–C). It is interesting to note that the labeling did not involve the rostral pole of the striatum. This case and the comparable case RVT18 (Fig. 1B) of isotope injection confined to

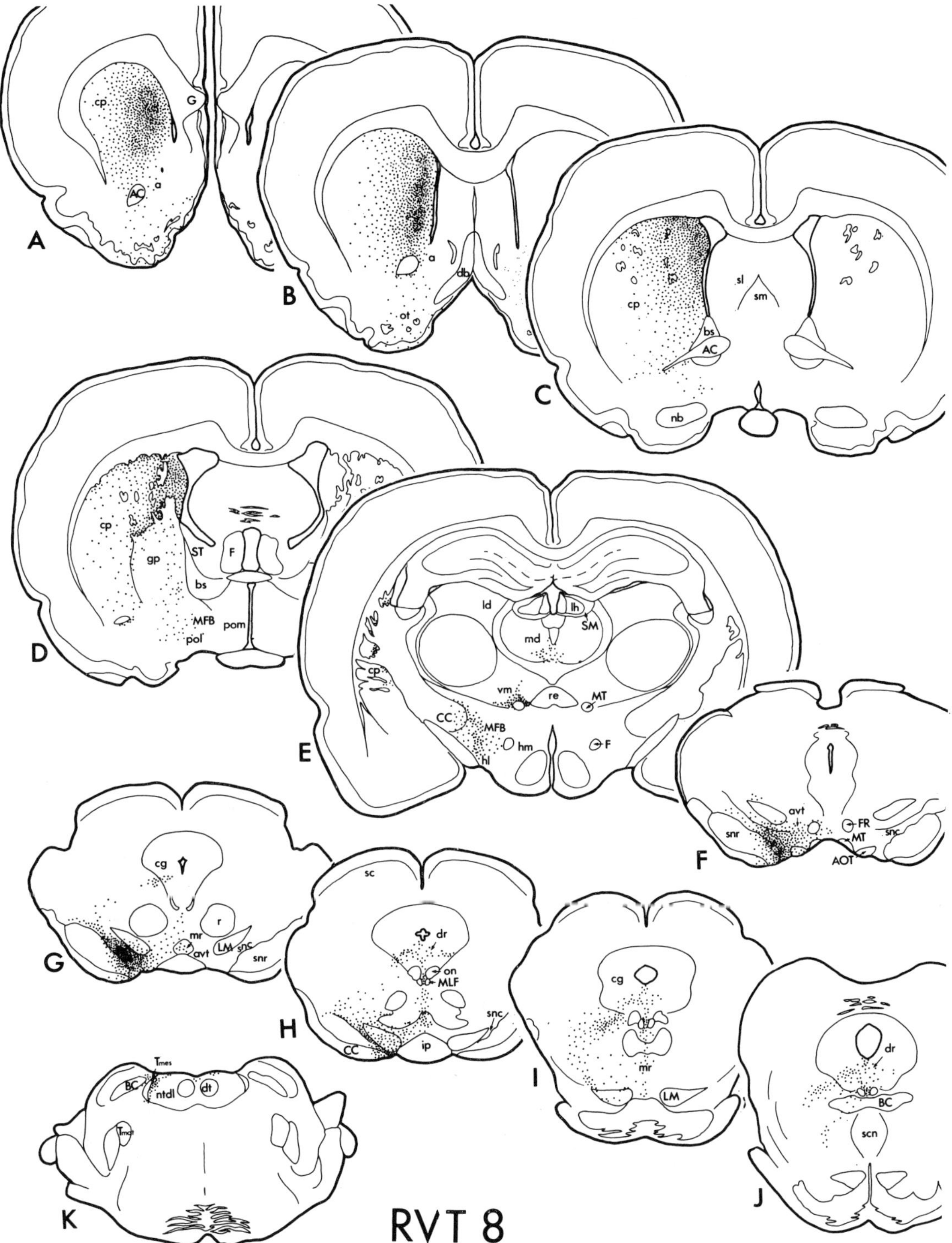


Fig. 4. Fiber labeling resulting from an isotope injection centered in the medial part of the pars compacta but also involving a lateral part of the ventral tegmental area (case RVT8). The injection site is indicated in G, and photographically illustrated in Fig. 2B.

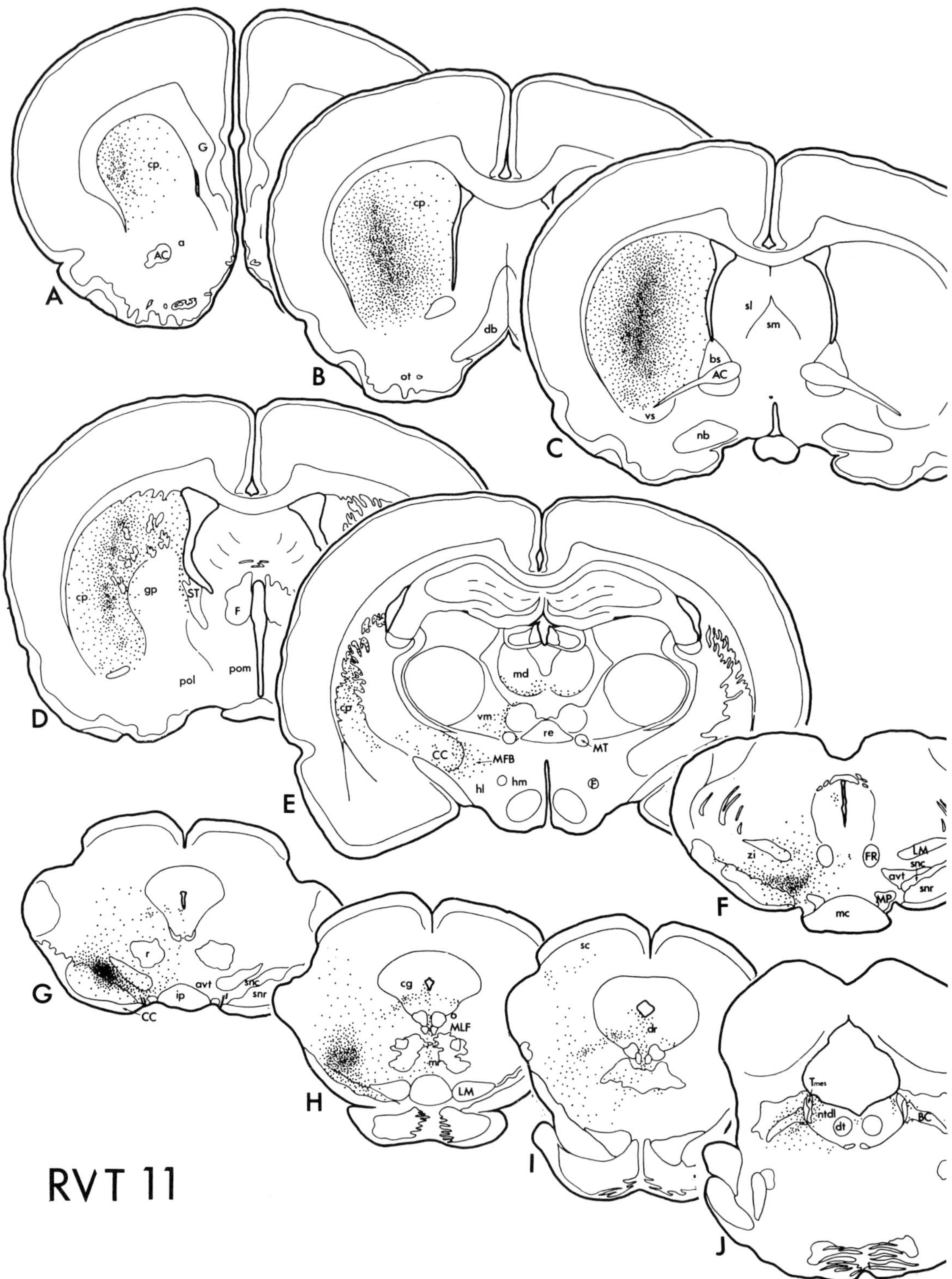


Fig. 5. Fiber labeling caused by an isotope injection centered at a middle latitude of the pars compacta (case RVT11). Injection site indicated in G, and photographically illustrated in Fig. 2C.

the lateral one-quarter of the SNC form the only present exceptions to the apparent rule that injections of the pars compacta, regardless of their rostrocaudal location, label fibers distributed throughout the length of the striatum.

Labeled fibers passing to the thalamus through Forel's fields in this case were distributed mainly to the lateral two-thirds of the ventromedial nucleus and the extreme lateral part of the mediodorsal nucleus (Fig. 6D). Thus, in both striatum and thalamus the label was distributed farther laterally than it was in the two preceding cases of more medially located nigra injection.

It must be noted in passing that in this case of far-lateral pars compacta injection considerable fiber labeling marked the hypothalamus and amygdala (Fig. 6D). Since

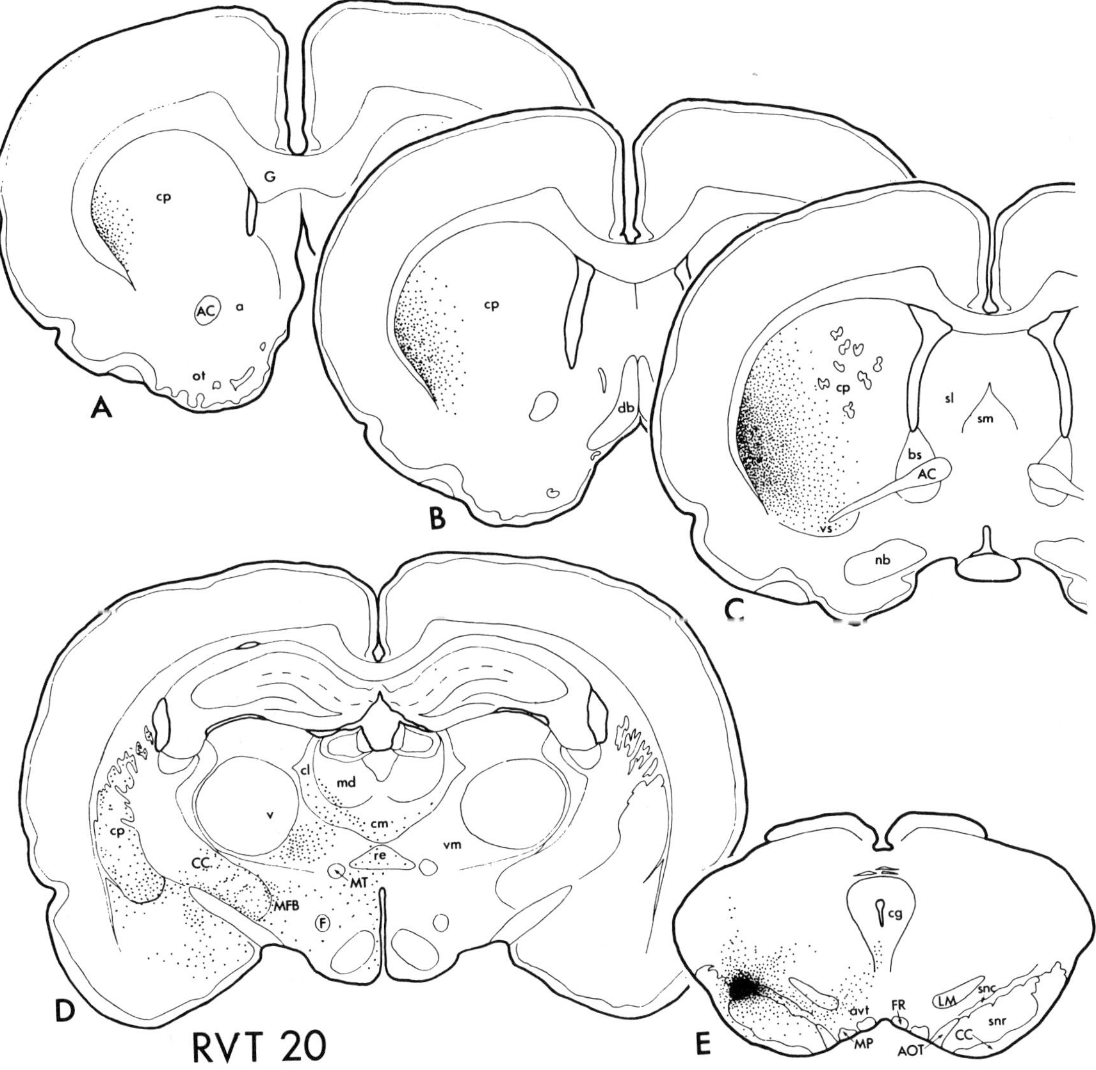


Fig. 6. Fiber labeling caused by an isotope injection centered near the lateral margin of the substantia nigra (case RVT20). Injection site indicated in E, and photographically illustrated in Fig. 2D.

there is reason to suspect that these fibers arise from tegmental cells outside the nigral complex they will here be disregarded, to be dealt with in a later report.

*Descending projections of the substantia nigra, pars compacta.* Labeled fibers descending from all 3 illustrated SNC-injection sites (cases RVT8, 11 and 20) swept dorsally and caudally around both the medial and the lateral margin of the medial lemniscus (Figs. 4G, H; 5G). Those following the medial route passed through the ventral tegmental area and along the mid-plane of the tegmentum to be distributed mainly to the rostral part of the median raphe nucleus and to the dorsal raphe nucleus whose entire fountain-shaped cross-section was marked by grain (Figs. 4H–J; 5H, I). Labeled fibers following this medial course were more numerous in cases of medial than in cases of more lateral SNC injection. By contrast, injections in lateral parts of SNC labeled a larger number of fibers that curve around the lateral margin of the medial lemniscus. Such fibers in their further caudodorsal course successively traversed the central, cuneiform and parabrachial regions of the tegmentum. In passage to the last of these regions the fibers perforate the brachium conjunctivum, and it seems possible that some at least establish synaptic contacts with neurons of the nucleus tegmenti pedunculopontinus, pars compacta, that lie embedded amongst the fibers of the brachium (see below). The longest of these labeled descending fibers could be followed to the ipsilateral, and in small number also to the contralateral, ventral quadrant of the central grey substance in which their distribution was restricted to the lateral half of the nucleus tegmenti dorsalis lateralis (ntdl in Figs. 4K and 5J). In contrast to injections of the ventral tegmental area, none of the injections centered in the pars compacta had labeled fibers to the locus coeruleus.

## IV. Injections of the Pars Reticulata (SNR)

The ascending pathways labeled by isotope injections confined to the pars reticulata (cases 5, N14–16) differed in several respects from those traced from the pars compacta or ventral tegmental area. Some labeled SNR efferents entered the cerebral peduncle directly ventral to the injection site, and continued rostralwards in that same segment of the peduncle. All of these peduncular fibers separated from the peduncle at various frontal levels, traversed the globus pallidus and entered the caudatoputamen from the medial side. It must be emphasized that their number was much smaller than that of the nigrostriatal fibers labeled by injections centered in the pars compacta, even in cases N14 and N15 in which the injection site was quite large. The low density of labeling made it difficult to determine the intrastriatal distribution of these nigral efferents accurately, but it nonetheless seemed that their mediolateral topography corresponds to that of the nigrostriatal fibers originating in the pars compacta. It is of interest to note that none of the 4 pars reticulata injections had labeled fibers to the nucleus accumbens or the olfactory tubercle. Neither did any fiber labeling in these cases mark the amygdala, septum, anteromedial cortex, or entorhinal area.

A larger number of SNR efferents, instead of entering the cerebral peduncle, passed forward and dorsally from the substantia nigra through the subthalamic region. Most of these nigral efferents compose a fairly circumscript fiber group that enters the thalamus along the lateral side of the mammillothalamic tract to be distributed mainly to the ventromedial nucleus and in much smaller number to the paralamellar zone of the mediodorsal nucleus; the latter of these distributions is to some extent bilateral. A considerably smaller number of labeled nigrothalamic fibers, not grouped into a distinct bundle, rose more steeply dorsalwards from the substantia nigra to the parafascicular nucleus. Some of these fibers appeared merely to pass through this caudal thalamic region to enter the rostrally adjacent paralamellar zone of the mediodorsal nucleus, but the labeling pattern left little doubt that the largest number actually terminate in the parafascicular nucleus.

*Descending projections of the pars reticulata.* Isotope injections confined to the pars reticulata labeled fibers in the mesencephalon in a pattern quite distinct from that described above. Even when such injections were placed in the medial half of the nigra as here exemplified by case RSN16 (Fig. 7), only few labeled fibers traced a medial route into the midbrain tegmentum; medial and central regions of the tegmentum consequently exhibited only sparse and diffuse labeling, except for a small and circumscript region of high label density in the rostral midbrain ventral and medial to the medial longitudinal fasciculus (Fig. 7A, B). Most fibers labeled by SNR injections, after descending over some distance within the nigra proper, entered the overlying tegmentum loosely grouped into a substantial and fairly circumscript fiber tract that traverses the lateral tegmental region in a dorsomedial direction (Fig. 7C). Near the lateral corner of the central grey substance this fiber group divides into: (a) a dorsolaterally oriented branch that enters the lateral quarter of the superior colliculus and distributes itself to the deeper collicular layers in the laminar cluster-pattern first noted by Graybiel and Sciascia[31]; and (b) a smaller descending branch that extends to the level immediately behind the trochlear nucleus where it terminates in a dorsal tegmental region straddling the brachium conjunctivum and corresponding in location to the nucleus tegmenti pedunculopontinus, pars compacta, of the monkey (tpc in Fig. 7D; cf. ref. 53). In contrast with the cases of SNC injection, only sparse and diffuse labeling marked the central grey substance in all cases of SNR injection.

DISCUSSION

*The Nigrostriatal Projection*

Although observations of rapid cell-atrophy in the substantia nigra in cases of extensive striatal destruction[22,26,39,49,62] had strongly suggested the existence of a prominent nigrostriatal projection, attempts to demonstrate the connection directly in fiber-degeneration experiments long went unrewarded and succeeded only after the introduction of the Fink–Heimer method in 1966[33,51,64]. Already several years earlier, however, Andén et al.[3] by the monoamine–histofluorescence method had provided the first direct evidence of a nigrostriatal fiber system originating from the dopamine neurons of the substantia nigra. Subsequent observations in histofluorescence and biochemical studies[15,28,38], as well as electrophysiological evidence[14,27,41,48,58,70] confirmed the existence of a prominent nigrostriatal projection, and in 1974 Lindvall and Björklund[45] charted the anatomical trajectory of the system in detail by the aid of the glyoxylic acid–histofluorescence technique.

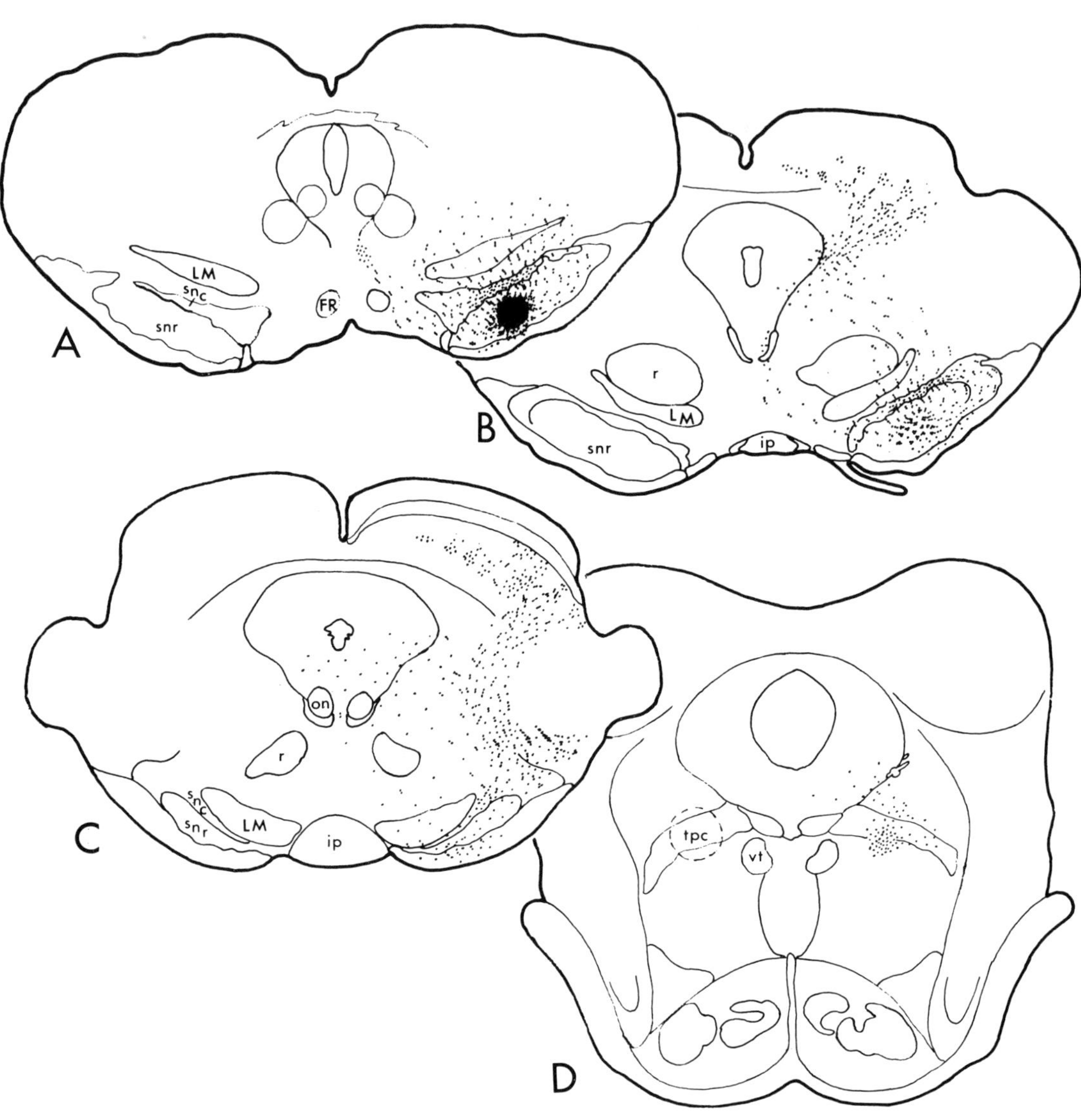



Fig. 7. Nigral efferents to other mesencephalic structures, labeled by an isotope injection confined to the pars reticulata. Case RSN16.

Despite the excellent visualization of the nigrostriatal projection afforded by it, the histofluorescence method has been of only limited value in attempts to determine the topographic organization of the projection. Nonetheless, in combination with biochemical analysis it allowed Andén et al.[4] and Ungerstedt[68] to identify a fundamental dualism in the nigrostriatal system. According to these authors, the nigrostriatal dopamine system can be subdivided into: (1) a nigrostriatal component properly speaking originating from the dopamine neurons of the substantia nigra's pars compacta (dopamine cell group A9 of Dahlström and Fuxe[15]) and distributed to the larger part of the striatum; and (2) a mesolimbic system that arises from dopamine cell group A10. The term, mesolimbic system, appears to have been based on Ungerstedt's[68] observation — confirming earlier findings by Andén et al.[4] — that cell group A10 projects in greatest volume to the olfactory tubercle and nucleus accumbens septi, components of the striatum[34] that receive their major telencephalic input from the hippocampal formation[24,59,65] and amygdala[16,44]. Ungerstedt noted additional projections from A10 to the central nucleus of the amygdala and the bed nucleus of the stria terminalis. His view of A10 as a dopamine cell group preferentially associated with the limbic system received further support from later evidence that A10 also projects to two regions of the limbic cortex: the entorhinal area and the anteromedial cortex[6,29,46,67].

*The projection from AVT to the striatum.* Although the present findings with respect to the striatopetal projection from the ventral tegmental area are largely compatible with earlier descriptions of the mesolimbic system, they include some observations that have been reported from only few earlier studies[18,19]. Most important, perhaps, among these is the evidence that the distribution of the AVT projection to the striatum, instead of being limited to the olfactory tubercle and nucleus accumbens as conventionally defined[4,68], involves the full length of the striatum. Moreover, although the projection is distributed in greatest volume to the fundus region of the striatum (of which the nucleus accumbens as conventionally defined is an anteromedial component), it expands dorsally from this most ventral striatal zone so as to involve, in lesser volume, most of the ventromedial half of the striatum. In comparing this finding with the histochemical and biochemical observations of Andén et al.[4] and Ungerstedt[68], it must be borne in mind that the autoradiographic method cannot characterize fibers with respect to their histochemical properties, and that it therefore must be considered possible that the AVT fibers extending into striatal regions dorsal to the nucleus accumbens and olfactory tubercle originate from AVT cells other than those composing dopamine cell group A10. It none the less seems likely that the dorsal expansion of the projection in question is at least in part dopaminergic, for Tassin et al.[66] reported that destruction of the AVT is followed by significant dopamine loss not only in the nucleus accumbens and olfactory tubercle but also in the larger overlying part of the caudatoputamen.

*Projections to the striatum from the substantia nigra proper.* A comparison of the present cases plainly shows that the nigrostriatal connection, as reported previously[18, 19], is organized in a regular manner so that medial parts of the nigra project to medial, and more lateral parts to more lateral regions of the striatum. The reciprocating striatonigral connection recently has been shown likewise to maintain an orderly medial-to-medial and lateral-to-lateral organization[10,17].

A further point to be emphasized is the apparent absence of a similar gradient along the sagittal axis; instead, the present findings suggest that the nigrostriatal fibers originating in each small locus of the nigra are distributed over most or all of the length of the striatum. The only exception to this rule noted in the present study is that the nigrostriatal projection from the most lateral quarter of the nigra appears to avoid the anterior pole of the striatum.

Owing to the thinness (3–5 cells) of the pars compacta sensu strictiori, no dorsoventral gradient could be recognized in the striatal projection originating in that part of the nigral complex. However, if the substantia nigra is considered to comprise also the dorsally protruding dopamine cell groups A10 and A8, it is evident that the nigrostriatal projection is organized so as to *invert* the dorsoventral axis: the regions of both A10 and A8 project in greatest volume to the most ventral parts of the striatum[19,53,55], whereas the more ventrally placed pars compacta projects to more dorsal striatal regions. It is interesting to note that a similar inversion of the dorsoventral axis has been found in the reciprocating striatonigral projection[17].

Notably, fewer nigrostriatal fibers were labeled by isotope injections confined to the substantia nigra's pars reticulata (SNR) than by injections centered in the SNC. It could not be determined to what extent those labeled by SNR injections originated from cells within the SNR or from SNC cells that had incorporated the isotopes through one or more dendrites protruding into SNR. However, histofluorescence studies have shown that, even though the majority by far of the dopamine cells of the substantia nigra proper lie within the pars compacta, the pars reticulata is by no means without such cells (e.g. Palkovits and Jacobowitz[56]). Moreover, depending on their placement, intrastriatal deposits of horseradish peroxidase may label a considerable number of nigrostriatal cells in the pars reticulata. These findings leave the possibility open that some or even all of the nigrostriatal fibers originating in the pars reticulata arise from dopamine neurons, but they give no clue whether the pars reticulata may be a source also of non-dopaminergic innervation of the striatum.

Finally, it must be emphasized that the nucleus accumbens and olfactory tubercle appear to receive only very few fibers from the substantia nigra proper. This is documented by the negative findings in the nucleus accumbens and olfactory tubercle in RVT11 and RVT20 (Figs. 5 and 6), the only cases here illustrated in which the injection was centered in SNC without involving the region of cell group A10. A similar conclusion was drawn from a recent study by the HRP method[55]. It thus appears that the respective striatopetal projections of the SN and AVT do not, or only minimally, overlap each other in the nucleus accumbens and olfactory tubercle. These two projections are, however, in overlap throughout most of the remainder of the medial half of the striatum (compare RVT6 and RVT11) including the entire striatal fundus caudal to the anterior commissure.

*The Nigrothalamic Projection*

*Thalamic projections from the substantia nigra proper.* A projection from the substantia nigra to the ventromedial nucleus of the thalamus was first demonstrated by Cole et al.[13] in the cat and monkey, and later confirmed in the rat[20,23,32], in the cat[60], and in the monkey[11,12,50]. Carpenter et al.[11] in an autoradiographic study in the monkey found that the distribution of nigrothalamic fibers also involves the paralamellar zone of the mediodorsal nucleus. The present findings show that this is the case also in the rat, and that this nigrothalamic projection is distributed in an orderly medial-to-lateral arrangement. Moreover, the present study revealed a relatively sparse projection from the substantia nigra proper to the parafascicular nucleus, confirming a conclusion drawn from findings in a recent HRP study[2].

It seems certain from various data that these thalamopetal projections from the substantia nigra proper originate largely at least from non-dopaminergic neurons in the pars reticulata: (a) in the present experiments many more nigrothalamic fibers were labeled by isotope injections confined to SNR than by injections centered in SNC; (b) the projections appear not to have been observed in studies by histofluorescence methods; (c) intrathecal injection of 6-hydroxydopamine (6-OHDA) has been reported to result in degeneration of nigrostriatal but not nigrothalamic fibers[32]; and (d) horseradish peroxidase injected into either the ventromedial[21] or parafascicular[2] nucleus has been reported to label cells almost exclusively in the pars reticulata.

According to Faull and Mehler[21], the nigral neurons projecting to the ventromedial thalamic nucleus are located largely in dorsal parts of the pars reticulata. As a consequence, some nigrothalamic neurons at least are likely to be involved by isotope injections centered in the immediately overlying pars compacta. The fiber labeling in various parts of the ventromedial and of the mediodorsal thalamic nucleus in cases RVT8 and RVT20 probably resulted from such unintentional involvement of the pars reticulata.

*Thalamic projections from the ventral tegmental area.* Examining the catecholamine innervation of the rat's thalamus, Lindvall et al.[47] found dopamine fibers only in caudal parts of the thalamus, and suggested that they arose from dopamine cell groups A11 and A13 of Dahlström and Fuxe[15]. In the present study, labeled fibers were traced from the ventral tegmental area to a variety of medial thalamic nuclei: the lateral habenular nucleus, nucleus reuniens, nucleus centralis medius, and mediodorsal nucleus. In the last-mentioned nucleus, labeling was densest along the medial border; it seems possible that AVT fiber distribution to this paramedian zone may account for the dopamine fibers noted by Fuxe[28] in a region identified by him as the periventricular thalamic nucleus. The AVT projection that follows the fasciculus retroflexus to the lateral habenular nucleus explains earlier findings of cell labeling in the AVT by horseradish peroxidase deposited in that nucleus[37]. This projection likewise is likely to include dopamine fibers, for destruction of dopamine cell groups A8, A9 and A10 has been reported to cause a 75% reduction in the dopamine content of the habenular complex[43]. The present autoradiographic suggestion that AVT also projects to the nucleus reuniens supplements the recent finding that horseradish peroxidase injected into that thalamic nucleus labels AVT cells[36]. An AVT projection to the nucleus centralis medius, however, appears not to have been suggested previously.

*Projections to Limbic Structures*

Almost all ascending AVT efferents follow the medial forebrain bundle, but the evenness of the labeling throughout the length of this fiber system makes it difficult to determine whether any AVT fibers actually terminate along the bundle's path through the lateral hypothalamus and preoptic region. The evidence that, farther rostrally, some of the more medial AVT efferents are distributed to the nucleus of the diagonal band and to a medial part of the lateral septal nucleus agrees well with earlier anatomical descriptions[19,45], and also with reports that these regions contain moderate amounts of dopamine[9,69]. Fuxe[28] already earlier had suggested that the catechol-

210

amine fibers found by him in the nucleus of the diagonal band and in the dorsolateral part of the interstitial nucleus of the stria terminalis were dopamine fibers. According to our present evidence, AVT efferents to the septal region and the interstitial nucleus of the stria terminalis originate exclusively in medial parts of the region: an isotope injection involving more lateral parts of AVT (case RVT8, Fig. 4) labeled no such fibers. The AVT projection to the nucleus accumbens and olfactory tubercle, by contrast, appears to originate in both medial and lateral regions of the ventral tegmental area.

Fuxe[28] also was the first to note dopamine fibers in the central nucleus of the amygdala. Ungerstedt[68] later showed that these fibers originate from dopamine cell group A10 in the ventral tegmental area. Later biochemical assay studies suggested that the dopamine innervation of the amygdala also involves the lateral nucleus[43] and anterior amygdaloid area[9]. The present autoradiographic evidence agrees well with the histochemical and biochemical data: labeled fibers could be followed from an injection site in the anteromedial part of AVT (case RVT6, Fig. 3) not only to the central amygdaloid nucleus but also to the anterior amygdaloid area, to the anterolateral nucleus, and to the medial rim of the posterolateral nucleus. It must be emphasized that the small injection of case RVT6 seems unlikely to have labeled more than a small part of the AVT projection to the amygdala. However, the negative findings in case RVT8 suggest that this projection, like that from AVT to the septal region, originates exclusively in more medial parts of the ventral tegmental area.

The autoradiographic evidence of AVT efferents in the anteromedial cortex agrees with the biochemical findings of Thierry et al.[67] as well as with earlier anatomical evidence reported by Lindvall et al.[46] and Berger et al.[7] from histofluorescence studies, and by Beckstead[5] from HRP experiments. The present study and that of Fallon and Moore[19] have raised to 3 the number of independent methods by which this mesencephalocortical projection has been shown to be distributed to a continuous cortical expanse anterior and dorsal to the genu corporis callosi. According to the present evidence, AVT efferents are distributed to only the deeper layers of this cortical region, but the label density in our material may have been too low to demonstrate the intracortical distribution of these fibers accurately. To a slightly lesser degree, the same uncertainty applies to the fibers traced from the AVT to the entorhinal area, in good accord with findings in histofluorescence[45] and HRP[6] studies. The present findings suggest that this projection is distributed throughout the thickness of the entorhinal cortex without apparent preference for any particular cell layer.

*Descending Projections of the Nigral Complex*

*The nigrotectal connection.* Graybiel[30] and Graybiel and Sciascia[31] showed that nigrotectal fibers in the cat originate in the pars reticulata and are distributed in a distinct band pattern to layers of the superior colliculus below the stratum opticum. A similarly distributed nigrotectal projection was demonstrated in the monkey[42] and by the present experiments in the rat. From an HRP study in the rat Faull and Mehler[21] concluded that most nigrotectal neurons lie in the ventral half of the pars reticulata, segregated from the nigrothalamic neurons whose cell bodies preferentially occupy the pars reticulata's dorsal half. The present autoradiographic findings cannot define the origin of the nigrotectal projection in such detail, but it is noteworthy that in case RVT11 (Fig. 5) a small injection centered in SNC and involving only a narrow dorsal zone of SNR had labeled nigrotectal fibers; in this case some nigrotectal neurons may have incorporated label through dorsally oriented dendrites.

*Nigral projections to the tegmentum and central grey substance.* Evidence of such projections was reported from HRP studies in the cat[40,61] and monkey[40]. Some years earlier Hedreen[32] in a fiber degeneration study in the rat had identified a nigral projection to the nucleus tegmenti pedunculopontinus (TPC) of the dorsal ponto-mesencephalic tegmentum. The present findings confirm Hedreen's report, and also indicate that the nigrofugal pathway to TPC is closely associated with the nigrotectal projection. Its separation from the latter at a nearly right angle suggests that it may be composed to some degree of collaterals from nigrotectal fibers. From Hedreen's finding that it resists 6-OHDA it seems certain that the nigral projection to TPC, like the nigrotectal connection, is at least largely non-dopaminergic, and this is indicated also by the present evidence that it originates in the pars reticulata. It must be noted, however, that the TPC is traversed by some of the fibers passing from the pars compacta and ventral tegmental area to the parabrachial nuclei and central grey substance, and this may explain that HRP injected into the TPC may label cells not only in SNR but also in SNC and AVT (unpublished control experiments).

In the present study no nigral efferents could be traced caudally beyond the level of the locus coeruleus. In the cat, however, Rinvik et al.[61] found some cells in the medial half of the substantia nigra labeled by an HRP injection as far caudal as the medulla oblongata.

Following HRP injections in either the mesencephalic tegmentum or central grey substance Rinvik et al.[61] and Hopkins and Niessen[40] found nigral cell labeling restricted to the pars reticulata. This is difficult to reconcile with the present evidence that the central grey substance receives far more fibers from the pars compacta than from the pars reticulata, and that numerous efferents from the pars compacta pass caudalwards through the midbrain tegmentum. To summarize the present observations: all isotope injections restricted to AVT or centered in SNC, labeled fibers spreading caudally through most of the medial two-thirds of the tegmental cross-section in a somewhat orderly medial-to-lateral arrangement. The gradual decrease in their number as they extend caudalwards suggests that many are distributed to the tegmental reticular formation. The tegmental raphe nuclei seem only marginally involved by this nigrotegmental projection, in contrast to the dorsal raphe nucleus which receives fibers from all AVT and SNC sites involved in these experiments. A projection from SNC to the dorsal raphe nucleus had been suggested earlier by findings in an HRP study[57]. Farther caudally, the nucleus tegmenti dorsalis lateralis is a convergence site of such fibers, but fibers to the locus coeruleus were labeled only by injections involving the medial two-thirds of the ventral tegmental area. This AVT projection to the locus coeruleus was recognized earlier in a fiber degeneration study by Simon and Le Moal[63].

It must be emphasized that the chemical nature of nigrofugal fibers descending from the pars compacta and ventral tegmental area is uncertain. Biochemical assays[28,43] have indicated the presence of dopamine-containing structures in various extranigral regions of the brain stem, but it appears that thus far these findings have not been correlated with any particular fiber connection.

*General Comment*

It is evident from the foregoing account that the nigral complex has a sphere of direct influence that extends from the cerebral cortex caudalwards as far as the pontomesencephalic border region. Within this wide range, dopaminergic nigral projections most massively affect the striatum and various parts of the limbic forebrain, but the present evidence raises the possibility that the nigral dopamine system has lesser ramifications that involve the thalamus — in particular the lateral half of the lateral habenular nucleus, certain midline nuclei, and the most medial zone of the mediodorsal nucleus — as well as certain mesencephalic and upper-pontine structures including the dorsal raphe nucleus and the locus coeruleus. These and other nigral relationships are summarized in Fig. 8.

Of no less interest, even though apparently less widespread, are the non-dopaminergic nigrofugal projections originating in the pars reticulata and distributed largely to the ventromedial thalamic nucleus, superior colliculus, and pedunculo-

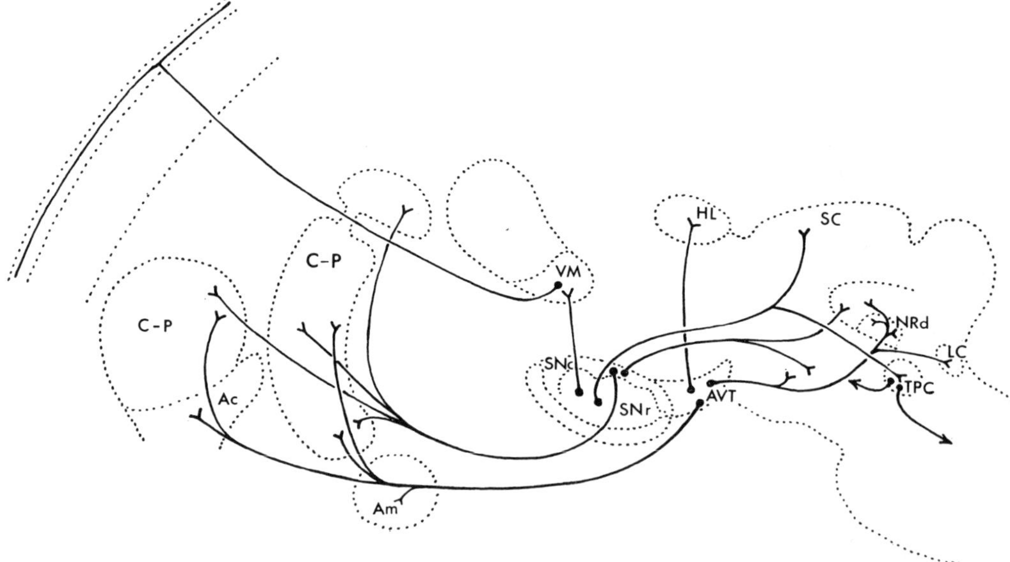


Fig. 8. Diagrammatic representation of the efferent relationships of the pars compacta, pars reticulata, and ventral tegmental area. Not included in the diagram are the dopaminergic AVT projections to the anteromedial and entorhinal cortex, but the presumable trans-thalamic nigrofugal pathway to the plexiform layer of the neocortical convexity[36] (left upper corner) has been indicated. Abbreviations: Ac, nucleus accumbens; Am, amygdala; AVT, ventral tegmental area; C-P, caudatoputamen; HL, lateral habenular nucleus; LC, locus coeruleus; NRd, dorsal raphe nucleus; SC, superior colliculus; SNc and SNr, substantia nigra, pars compacta and pars reticulata; TPC, pedunculopontine nucleus.

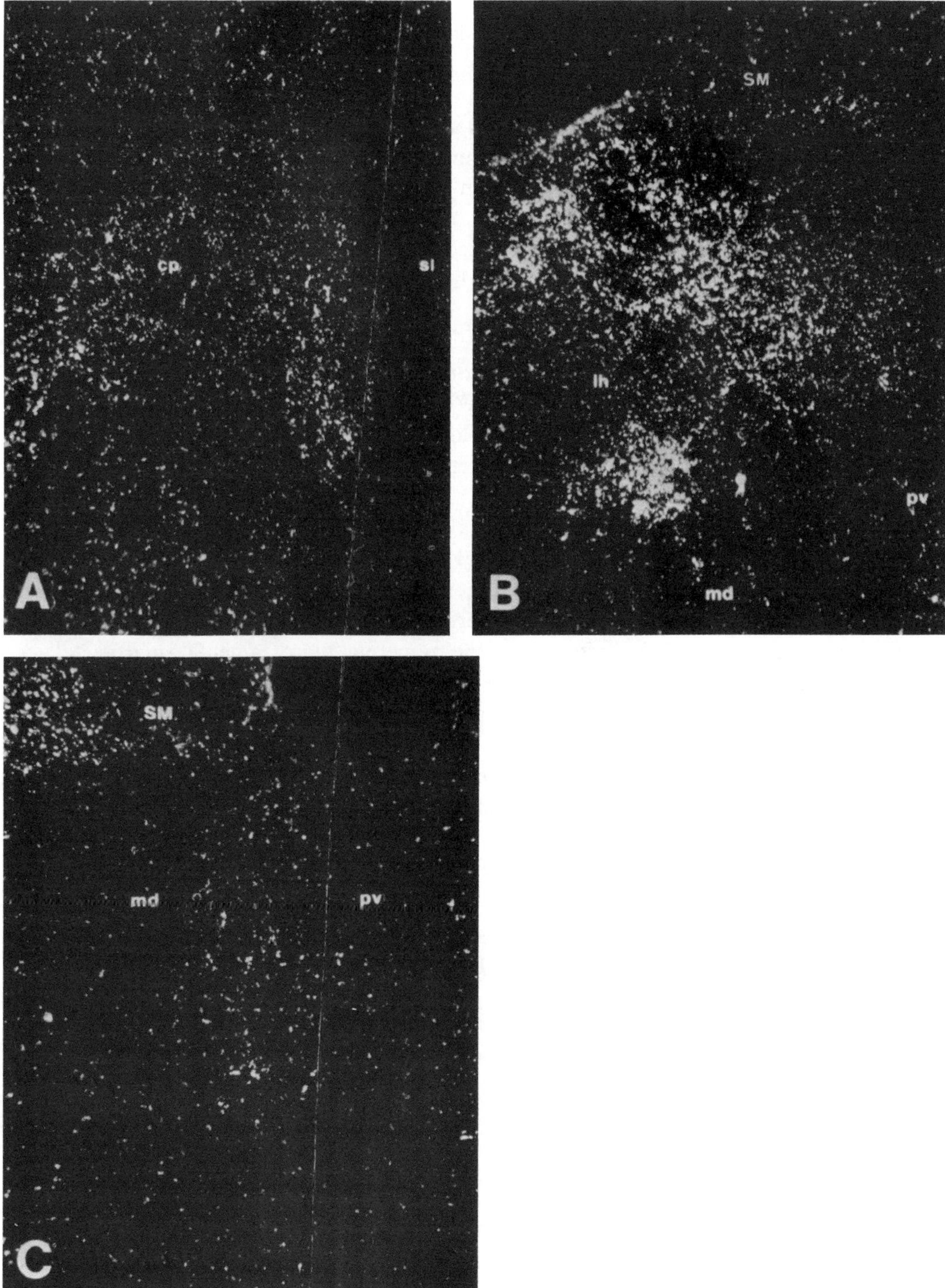

Fig. 9. Dark-field photographs showing fiber labeling in: A: the medial zone of the caudatoputamen dorsal to the nucleus accumbens; B: the lateral habenular nucleus; and C: the most medial zone of the mediodorsal thalamic nucleus following an isotope injection in the ventral tegmental area (case RVT6, described in the text).

214

pontine nucleus. Since the ventromedial thalamic nucleus in the rat projects to the plexiform layer (layer I) of all but the caudal third or so of the neocortex[35], it seems possible that the substantia nigra by way of its nigrothalamic projection can affect not only the somatic motor mechanisms but also sensory and associative functions of the cortex. Moreover, by its projections to the superior colliculus and pedunculopontine nucleus the pars reticulata could serve as a link in a pathway from the striatum to the bulbospinal motor apparatus: the deeper strata of the superior colliculus give rise to tectobulbar and tectospinal tracts, and the pedunculopontine nucleus seems likely to project caudalwards at least as far as the rhombencephalic tegmentum, much like other regions of the pontomesencephalic tegmentum. It is noteworthy in this connection that the pedunculopontine nucleus receives descending projections not only from the substantia nigra but also from the internal segment of the globus pallidus[54] and from the subthalamic nucleus[52].

The foregoing considerations caution against an interpretation of the striato-nigral connection as composing primarily a 'return-loop' modulating the reciprocating dopaminergic nigrostriatal mechanism. With the apparent exception of that component which orginates from the most ventral regions of the striatum (including the nucleus accumbens), the massive striatonigral pathway terminates in the pars reticulata[17]. It is difficult to imagine that within that nigral subdivision it synapses only with outlying pars compacta dendrites and/or interneurons closing the nigro-striato-nigral circuit, and more likely that it affects also the numerous nigrotegmental, nigrotectal and nigrothalamic neurons of the pars reticulata. These considerations suggest a role of the striatonigral projection as a neural link extending the influence of the dopaminergic nigrostriatal system to the thalamus and midbrain via the non-dopaminergic nigral projections arising from the pars reticulata. From this point of view the striatonigral system appears comparable to the striatopallidal projection which extends the effects of the dopamine system to the thalamus and midbrain via the non-dopaminergic projections (ansa lenticularis) of the globus pallidus. It is interesting in this context that Fox et al.[25] have reported electron microscopic evidence strongly suggesting that striatopallidal and striatonigral fibers are identical in the sense that the striatonigral connection is established largely by the attenuated end-stretches of the same striatofugal fibers that compose the striatopallidal projection. However, comparable though the striatopallidal and striatonigral systems would thus appear to be, an important contrast remains: whereas the striatonigral connection is recipro-cated by the dopaminergic nigrostriatal projection, thus far no similar massive 'return-loop' from the globus pallidus to the striatum appears to have been demon-strated.


ACKNOWLEDGEMENTS

This study was made possible by National Science Foundation Grant BNS 76-81227, and United States Public Health Service Grant MH25515. R.M.B. was supported by USPHS Training Grant ST01 GM01064-13 BHS.

ABBREVIATIONS

a, nucleus accumbens; aa, nucleus anterior amygdalae; ac, nucleus centralis amygdalae; aco, nucleus corticalis amygdalae; ala, nucleus lateralis amygdalae; alp, nucleus lateralis posterior amygdalae; am, nucleus medialis amygdalae; avt, ventral tegmental area; bs, bed nucleus of stria terminalis; c, claustrum; cg, substantia grisea centralis; cl, nucleus centralis lateralis; cm, nucleus centralis medius; cp, caudatoputamen; db, nucleus of the diagonal band; dr, dorsal raphe nucleus; dt, dorsal tegmental nucleus; ea, entorhinal area; gp, globus pallidus; hl, lateral hypothalamic region; hm, medial hypothalamic region; hp, posterior hypothalamic nucleus; ip, interpeduncular nucleus; lc, locus coeruleus; lh, lateral habenular nucleus; mc, mammillary body; md, mediodorsal nucleus; mr, medial raphe nucleus; nb, nucleus basalis; ntdl, nucleus tegmenti dorsalis lateralis; on, oculomotor nucleus; ot, olfactory tubercle; pf, parafascicular nucleus; pol, lateral preoptic region; pom, medial preoptic region; pv, periventricular nucleus; r, red nucleus; re, nucleus reuniens; snc, nucleus tegmenti, centralis superior; sl, lateral septal nucleus; sm, medial septal nucleus; snc, substantia nigra, pars compacta; snr, substantia nigra, pars reticulata; st, subthalamic nucleus; tn, trochlear nucleus; tpc, nucleus tegmenti pedunculopontinus, pars compacta; vm, ventromedial nucleus; vp, ventral (subcommissural) pallidum; vs, ventral (subcommissural) striatum; vt, ventral tegmental nucleus; zi, zona incerta; AC, anterior commissure; AOT, medial accessory optic tract and nucleus; BC, brachium conjunctivum; C, fasciculus cinguli; CC, crus cerebri; F, fornix; FR, fasciculus retroflexus; G, genu corporis callosi; LM, medial lemniscus; MFB, medial forebrain bundle; MP, mammillary peduncle; MT, mammillothalamic tract; ST, stria terminalis; Tmes, mesencephalic tract of trigeminus; Tmot, motor nucleus of trigeminus.

REFERENCES


1 Afifi, A. K. and Kaelber, W. W., Efferent connections of the substantia nigra in the cat, *Exp. Neurol.*, 11 (1965) 474–482.
2 Ahlenius, S., Potentiation by haloperidol of the catalepsy produced by lesions in the parafascicular nucleus of the rat, *Brain Research*, 150 (1978) 648–652.
3 Andén, N.-E., Carlsson, A., Dahlström, A., Fuxe, K., Hillarp, N. A. and Larsson, K., Demonstration and mapping out of nigro-neostriatal dopamine neurons, *Life Sci.*, 3 (1964) 523–530.
4 Andén, N.-E., Dahlström, A., Fuxe, K., Larsson, K., Olson, L. and Ungerstedt, U., Ascending monoamine neurons to the telencephalon and diencephalon, *Acta physiol. scand.*, 67 (1966) 313–326.
5 Beckstead, R. M., Convergent thalamic and mesencephalic projections to the anterior medial cortex in the rat, *J. comp. Neurol.*, 166 (1976) 403–416.
6 Beckstead, R. M., Afferent connections of the entorhinal area in the rat as demonstrated by retrograde cell-labeling with horseradish peroxidase, *Brain Research*, 152 (1978) 249–264.
7 Berger, B., Thierry, A. M., Tassin, J. P. and Myre, M. A., Dopaminergic innervation of the rat prefrontal cortex: a fluorescence histochemical study, *Brain Research*, 106 (1976) 133–145.
8 Berman, A. L., *The Brain Stem of the Cat. A Cytoarchitectonic Atlas with Stereotaxic Coordinates*, University of Wisconsin Press, Madison, 1968.
9 Brownstein, M., Saavedra, J. M. and Palkovits, M., Norepinephrine and dopamine in the limbic system of the rat, *Brain Research*, 79 (1974) 431–436.
10 Bunney, B. S. and Aghajanian, G. K., The precise localization of nigral afferents in the rat as determined by a retrograde tracing technique, *Brain Research*, 117 (1976) 423–436.
11 Carpenter, M. B., Nakano, K. and Kim, R., Nigrothalamic projections in the monkey demonstrated by autoradiographic technics, *J. comp. Neurol.*, 165 (1976) 401–416.
12 Carpenter, M. B. and Peter, P., Nigrostriatal and nigrothalamic fibers in the rhesus monkey, *J. comp. Neurol.*, 144 (1972) 93–116.
13 Cole, M., Nauta, W. J. H. and Mehler, W. R., The ascending efferent projections of the substantia nigra, *Trans. Amer. neurol. Ass.*, 89 (1964) 74–78.
14 Connor, J. D., Caudate unit responses to nigral stimuli: evidence for a possible nigro-neostriatal pathway, *Science*, 130 (1968) 899–900.
15 Dahlström, A. and Fuxe, K., Evidence for the existence of monoamine-containing neurons in the central nervous system, *Acta physiol. scand.*, Suppl. 232 (1964) 1–55.
16 De Olmos, J. S. and Ingram, W. R., The projection field of the stria terminalis in the rat brain, *J. comp. Neurol.*, 146 (1972) 303–334.
17 Domesick, V. B., The topographic organization of the striatonigral connection in the rat (Abstr.), *Anat. Rec.*, 187 (1977) 567.

18 Domesick, V. B., Beckstead, R. M. and Nauta, W. J. H., Some ascending and descending projections of the substantia nigra and ventral tegmental area in the rat, *Neurosci. Abstr.*, II (1976) 61.
19 Fallon, J. H. and Moore, R. Y., Catecholamine innervation of the basal forebrain. IV. Topography of the dopamine projection to the basal forebrain and neostriatum, *J. comp. Neurol.*, 180 (1978) 545–580.
20 Faull, R. L. M. and Carman, J. B., Ascending projections of the substantia nigra in the rat, *J. comp. Neurol.*, 132 (1968) 73–92.
21 Faull, R. L. M. and Mehler, W. R., Studies of the fiber connections of the substantia nigra in the rat using the method of retrograde transport of horseradish peroxidase, *Neurosci. Abstr.*, II (1976) 62.
22 Ferraro, A., The connections of the pars suboculomotorica of the substantia nigra, *Arch. Neurol. Psychiat. (Chic.)*, 19 (1928) 177–180.
23 Fibiger, H. C., Pudritz, R. E. McGeer, P. L. and McGeer, E. G., Axonal transport in nigro-striatal and nigro-thalamic neurons: effects of medial forebrain bundle lesions and 6-hydroxydopamine, *J. Neurochem.*, 19 (1972) 1697–1708.
24 Fox, C. A., The stria terminalis, longitudinal association bundle and precommissural fornix fibers in the cat, *J. comp. Neurol.*, 79 (1943) 277–295.
25 Fox, C. A., Rafols, J. A. and Cowan, W. M., Computer measurements of axis cylinder of radial fibers and 'comb' bundle fibers, *J. comp. Neurol.*, 159 (1975) 201–224.
26 Fox, C. A. and Schmitz, J. T., The substantia nigra and the entopeduncular nucleus in the cat, *J. comp. Neurol.*, 80 (1944) 323–334.
27 Frigyesi, T. L. and Purpura, D. F., Electrophysiological analysis of reciprocal caudatonigral relations, *Brain Research*, 6 (1967) 440–456.
28 Fuxe, K., Evidence for the existence of monoamine nerve terminals in the central nervous system. IV. Distribution of dopamine nerve terminals in the central nervous system, *Acta physiol. scand.*, 64, Suppl. 247 (1965) 39–85.
29 Fuxe, K., Hökfelt, T., Johansson, O., Jonsson, G., Lidbrink, P. and Ljungdahl, A., The origin of the dopamine nerve terminals in limbic and frontal cortex. Evidence for mesocortical dopamine neurons, *Brain Research*, 82 (1974) 349–355.
30 Graybiel, A. M., Organization of the nigrotectal connection: an experimental tracer study in the cat, *Brain Research*, 143 (1978) 339–348.
31 Graybiel, A. M. and Sciascia, T. R., Origin and distribution of nigrotectal fibers in the cat, *Neurosci. Abstr.*, 1 (1975) 174.
32 Hedreen, J. C., Separate demonstration of dopaminergic and non-dopaminergic projections of substantia nigra in the rat (Abstr.), *Anat. Rec.*, 169, Vol. 2 (1971) 338.
33 Hedreen, J. C., and Chalmers, J. P., Neuronal degeneration in the rat brain induced by 6-hydroxydopamine: a histological and biochemical study, *Brain Research*, 47 (1972) 1–36.
34 Heimer, L. and Wilson, R. D., The subcortical projections of the allocortex: similarities in the neural associations of the hippocampus, the piriform cortex, and the neocortex. In M. Santini (Ed.), *Golgi Centennial Symposium*, Raven Press, New York, 1975, pp. 177–193.
35 Herkenham, M., The nigro-thalamo-cortical connection mediated by the nucleus ventralis medialis thalami: evidence for a wide cortical distribution in the rat, *Anat. Rec.*, 184 no. 3 (1976) 426.
36 Herkenham, M., The connections of the nucleus reuniens thalami: evidence for a direct thalamo-hippocampal pathway in the rat, *J. comp. Neurol.*, 177 (1977) 589–610.
37 Herkenham, M. and Nauta, W. J. H., Afferent connections of the habenular nuclei in the rat. A horseradish peroxidase study, with a note on the fiber-of-passage problem, *J. comp. Neurol.*, 173 (1977) 123–146.
38 Hökfelt, T. and Ungerstedt, U., Electron and fluorescence microscopical studies on the nucleus caudatus putamen of the rat after unilateral lesions of ascending nigro-neostriatal dopamine neurons, *Acta physiol. scand.*, 76 (1969) 415–426.
39 Holmes, J. M., The nervous system of a dog without a forebrain, *J. Physiol. (Lond.)*, 27 (1901) 1–25.
40 Hopkins, D. A. and Niessen, L. W., Substantia nigra projections to the reticular formation, superior colliculus and central gray in the rat, cat and monkey, *Neurosci. Lett.*, 2 (1976) 253–259.
41 Hull, C. D., Barnardi, G. and Buchwald, N. A., Intracellular responses of caudate neurons to brain stem stimulation, *Brain Research*, 22 (1970) 163–179.
42 Jayaraman, A., Batton, B. R. IIIrd and Carpenter, M. B., Nigrotectal projections in the monkey: an autoradiographic study, *Brain Research*, 135 (1977) 147–152.

43 Kizer, J. S., Palkovits, M. and Brownstein, M. J., The projections of the A8, A9 and A10 dopaminergic cell bodies: evidence for a nigral-hypothalamic-median eminence pathway, *Brain Research*, 108 (1976) 363–370.
44 Krettek, J. E. and Price, J. L., Projections from the amygdala to the perirhinal and entorhinal cortices and the subiculum, *Brain Research*, 71 (1974) 150–154.
45 Lindvall, O. and Björklund, A., The organization of the ascending catecholamine neuron systems in the rat brain, *Acta physiol. scand.*, Suppl. 412 (1974) 1–48.
46 Lindvall, O., Björklund, A., Moore, R. Y. and Stenevi, U., Mesencephalic dopamine neurons projecting to neocortex, *Brain Research*, 81 (1974) 325–331.
47 Lindvall, O., Björklund, A., Nobin, A. and Stenevi, U., The adrenergic innervation of the rat thalamus as revealed by the glyoxylic acid method, *J. comp. Neurol.*, 154 (1974) 317–348.
48 McLennon, H., The release of dopamine from the putamen, *Experientia (Basel)*, 21 (1965) 725.
49 Mettler, F. A., Extensive unilateral cerebral removals in the primate: physiologic effects and resultant degeneration, *J. comp. Neurol.*, 79 (1943) 185–245.
50 Mettler, F. A., Nigrofugal connections in the primate brain, *J. comp. Neurol.*, 138 (1970) 291–320.
51 Moore, R. Y., Bhatnagar, R. K. and Heller, A., Anatomical and chemical studies of a nigro-necstriatal projection in the cat, *Brain Research*, 30 (1971) 119–135.
52 Nauta, H. J. W. and Cole, M., Efferent projections of the subthalamic nucleus: an autoradiographic study in monkey and cat, *J. comp. Neurol.*, 180 (1978) 1–16.
53 Nauta, W. J. H. and Domesick, V. B., Crossroads of limbic and striatal circuitry: hypothalamo-nigral connections. In K. E. Livingston and O. Hornykiewicz (Eds.), *Limbic Mechanisms*, Plenum Press, London, 1978.
54 Nauta, W. J. H. and Mehler, W. R., Projections of the lentiform nucleus in the monkey, *Brain Research*, 1 (1966) 3–42.
55 Nauta, W. J. H., Smith, G. P., Faull, R. L. M. and Domesick, V. B., Efferent connections and nigral afferents of the nucleus accumbens septi in the rat, *Neuroscience*, 3 (1978) 385–401.
56 Palkovits, M. and Jacobowitz, D. M., Topographic atlas of catecholamine and acetylcholinesterase-containing neurons in the rat brain. II. Hindbrain (mesencephalon, rhombencephalon), *J. comp. Neurol.*, 157 (1974) 29–52.
57 Pasquier, D. A., Kemper, T. L., Forbes, W. B. and Morgane, P. J., Dorsal raphe, substantia nigra and locus coeruleus: interconnections with each other and the neostriatum, *Brain Res. Bull.*, 2 (1977) 323–339.
58 Portig, P. J. and Vogt, M., Activation of a dopaminergic nigro-striatal pathway, *J. Physiol. (Lond.)*, 197 (1968) 20–21.
59 Raisman, J., Cowan, W. M. and Powell, T. P. S., An experimental analysis of the efferent projection of the hippocampus, *Brain*, 89 (1966) 83–108.
60 Rinvik, E., Demonstration of nigro-thalamic connections in the cat by retrograde transport of horseradish peroxidase, *Brain Research*, 90 (1975) 313–318.
61 Rinvik, E., Grovofá, I. and Otterson, O. P., Demonstration of nigrotectal and nigroreticular projections in the cat by axonal transport of proteins, *Brain Research*, 112 (1976) 388–394.
62 Rosegay, H., An experimental investigation of the connections between the corpus striatum and substantia nigra in the cat, *J. comp. Neurol.*, 80 (1944) 293–321.
63 Simon, H. and Le Moal, M., Demonstration by the Fink–Heimer impregnation method of a ventral mesencephalic-locus coeruleus projection in the rat, *Experientia, (Basel)*, 33 (1977) 614–615.
64 Simon, H., Le Moal, M., Galey, D. and Cardo, B., Silver impregnation of dopaminergic systems after radiofrequency and 6-OHDA lesions of the rat ventral tegmentum, *Brain Research*, 115 (1976) 215–231.
65 Sprague, J. M. and Meyer, M., An experimental study of the fornix in the rabbit *J. Anat. (Lond.)*, 84 (1950) 354–368.
66 Tassin, J. P., Cheramy, A., Blanc, G., Thierry, A. M. and Glowinski, J., Topographical distribution of dopaminergic innervation and of dopaminergic receptors in the rat striatum. I. Microestimation of [3H]dopamine uptake and dopamine content in microdiscs, *Brain Research*, 107 (1976) 291–301.
67 Thierry, A. M., Blanc, G., Sobel, A., Stinus, L. and Glowinski, J., Dopaminergic terminals in the rat cortex, *Science*, 180 (1973) 499–501.
68 Ungerstedt, U., Stereotaxic mapping of the monoamine pathways in the rat brain, *Acta physiol. scand.*, 197, Suppl. 367 (1971) 1–48.
69 Versteeg, D. H. G., van der Gugten, J., de Jong, W. and Palkovits, M., Regional concentrations of noradrenaline and dopamine in rat brain, *Brain Research*, 113 (1976) 563–574.
70 York, D. H., Possible dopaminergic pathway from substantia nigra to putamen, *Brain Research*, 20 (1969) 233–249.

*Neuroscience* Vol. 9, No. 2, pp. 245–260, 1983
Printed in Great Britain

0306-4522/83 $3.00 + 0.00
Pergamon Press Ltd
© 1983 IBRO

# RAMIFICATIONS OF THE GLOBUS PALLIDUS IN THE RAT AS INDICATED BY PATTERNS OF IMMUNOHISTOCHEMISTRY

S. N. Haber and W. J. H. Nauta

Department of Psychology and Brain Science, Massachusetts Institute of Technology, Cambridge,
MA 02139 and Mailman Research Center, McLean Hospital, Belmont, MA 02178, U.S.A.

**Abstract**—An attempt has been made to redefine the borders of the globus pallidus by the aid of the unique pattern of enkephalin-like and substance P-like immunoreactivity characterizing the pallidum of both monkey[6] and rat. In preparations immunoreacted for these two peptides by the peroxidase-antiperoxidase histochemical method of Sternberger[22] this pattern appears in the form of ribbon-like fibers (here called "woolly fibers") that have been interpreted by Haber and Elde[6] as unstained pallidal elements (dendrites and cell bodies) each enmeshed by a plexus of thin, enkephalin- or substance P-positive striatopallidal fibers. A dense enkephalin-positive woolly-fiber plexus fills the entire external pallidal segment as conventionally defined (here called "dorsal pallidum") and extends from there in various, generally ventral, directions. The most massive, rostral extension defines the subcommissural or "ventral pallidum" of Heimer and Wilson[10] and expands from there ventralward into the olfactory tubercle, supporting Heimer's[9] suggestion that many of the large cells of the tubercle are pallidal neurons. Further extensions from the enkephalin-positive dorsal pallidum plexus invade the ventral striatal region (including the nucleus accumbens), a dorsal region of the amygdala, and the bed nucleus of the stria terminalis. Substance P-positive woolly fibers, like their enkephalin-positive counterparts, fill the ventral pallidum and invade the olfactory tubercle, but avoid all except a small rostroventral part of the dorsal pallidum, and do not invade the striatum, the amygdala, or the bed nucleus of the stria terminalis. On the other hand, the dense substance P-positive woolly-fiber plexus filling the internal pallidal segment (entopeduncular nucleus) expands medialward into the lateral hypothalamic region. The entopeduncular nucleus invades the hypothalamus also with a loose plexus of enkephalin-positive woolly fibers.

It is suggested that woolly fibers extending outward beyond the conventionally recognized borders of the pallidum represent pallidal elements innervated by enkephalin or substance P-positive fibers arising from ventromedial striatal regions in turn innervated by limbic structures.

The caudal part of the globus pallidus* appears in frontal sections of the rat brain as a somewhat lens-shaped nucleus wedged between the striatum (caudatoputamen) on the lateral side, the internal capsule medially. As it extends rostralward the nucleus widens, reaching its greatest width between 1 and 0.5 mm caudal to the anterior commissure, then rapidly tapering off. Until recently, the globus pallidus of the rat was considered to have its rostral limit at the caudal border of the transverse limb of the anterior commissure. In 1975, however, Heimer and Wilson[10] reported that a region ventral and rostral to the anterior commissure is composed of large to medium-sized cell bodies emitting stout, long and sparsely ramified, smooth dendrites each enveloped by an almost uninterrupted sheet of axon terminals. Since the combination of these features previously had been found by Fox *et al.*[5] to characterize the cells of the globus pallidus, Heimer and Wilson[10] suggested that this subcommissural region—earlier interpreted as part of the preoptic region or, alternatively, of the substantia innominata—was in effect a rostral extension of the globus pallidus. For this subcommissural pallidal subdivision they introduced the term "ventral pallidum". Consistent with their interpretation was the fact that Wilson[27] in an earlier fiber-degeneration study had identified the same subcommissural region as a main terminus for striatofugal fibers, namely, efferents from the nucleus accumbens, a finding later confirmed in fiber-degeneration experiments[26] and in an autoradiographic study.[19]

In a later review, Heimer[9] suggested that the subcommissural extension of the globus pallidus is larger than initially realized, and specifically, that it might extend ventrally far enough to include some, at least, of the large cells of the olfactory tubercle. The notion that many of these large cells are pallidal neurons rather than outlying cells of the preoptic-hypothalamic region or nucleus of the diagonal band, as had been suggested previously[17] has received strong histochemical and histological support. Switzer and Hill[23] reported that the region of the olfactory tubercle containing such cells shares with the globus pallidus the characteristics of a high iron content, and Ribak and Fallon[21] found that the electron-microscopic appearance of these large cells corresponds to that of pallidal neurons. Further evidence supporting the notion comes from the im-

---

*Unless otherwise specified, the term "globus pallidus" here refers exclusively to the external segment of the pallidal complex (the "globus pallidus" of atlas descriptions), the internal segment being represented in the rat by the entopeduncular nucleus.

*Abbreviations*: PAP, peroxidase-antiperoxidase; PBS/T, phosphate-buffered saline containing Triton X-100; SP, substance P.

munohistochemical observations to be reported here. Briefly stated, the present study has demonstrated in the olfactory tubercle a pattern of enkephalin and substance-P immunoreactivity that appears to be uniquely pallidal.

In a recent study by the immunohistofluorescence method in the monkey brain[6,7] enkephalin-like immunoreactivity was found to appear in three distinct patterns of staining, one of which takes the form of networks of long ribbon-like fibers (here to be called "woolly fibers"). Each of these peculiar fibrous structures appears to consist of thin, enkephalin-positive fibers wrapped around a long, unstained, thick core element (Figs 1D–E). This pattern, first noted in the external segment of the globus pallidus of the monkey, seemed remarkably comparable to the Golgi pictures by which Fox *et al.*[5] documented their observation that the neurons of the globus pallidus and of the pars reticulata of the substantia nigra are ensheathed in a meshwork of thin, beaded fibers of striatal origin (Figs 1A–C). This similarity led Haber and Elde[6] to interpret the ribbon-like "woolly fibers" as pallidal dendrites entwined by enkephalin-positive axons.

In the same study,[6] woolly fibers were found to characterize not only the immunoreactivity pattern of the globus pallidus to enkephalin, but also that to substance-P antisera. In a following section of this paper an account will be given of the distribution of the apparently uniquely pallidal pattern of enkephalin and substance-P immunoreactivity in the rat, together with some observations on the manner in which it is affected by striatal lesions. Part of the findings to be reported below were previously reported in abstract form elsewhere.[8]

## EXPERIMENTAL PROCEDURES

Deeply anesthetized adult rats were perfused through the heart with a solution containing 4% paraformaldehyde and 1.5% sucrose in 0.1 M phosphate buffer at pH 7.2 and 4°C. The brains were removed, postfixed for 90 min, transferred to increasing concentrations of sucrose (5, 10, 20%) and finally stored in 30% sucrose in phosphate buffer for 1–3 days. The brains were then frozen and cut on a sliding microtome at 50 $\mu$m.

*Immunohistochemistry.* Antibodies to enkephalin and substance P were generously provided by Dr R. P. Elde, University of Minnesota, and Dr R. Ho, University of Ohio, respectively. Antisera for met-enkephalin[7] and substance P[12] were generated in response to immunogens produced by linking the peptides to bovine thyroglobulin and keyhole limpet hemocyanin, respectively, by means of glutaraldehyde coupling. The antiserum to met-enkephalin was shown by radioimmunoassay to cross-react less than 0.1% with leu-enkephalin or beta-endorphin. This antiserum was also tested at a 1/100 dilution for its ability to bind to a series of peptides coupled to cyanogen bromide-activated Sepharose 4B (Pharmacia). The antiserum is specific in that it did not cross-react with any of the other peptides (in particular, endogenous opiates or substance P) coupled to Sepharose beads. Absorption controls for both met-enkephalin and substance P were prepared by incubating each antiserum overnight with synthetic met-enkephalin or substance P (10 $\mu$g/ml diluted antiserum). The peroxidase-

antiperoxidase (PAP) technique of Sternberger[22] was used throughout the present study. The sections were incubated overnight at 4°C with a primary antiserum diluted 1/500 with 0.3% Triton X-100 in phosphate-buffered saline (hereafter called PBS/T). After rinsing, the tissue was incubated in goat anti-rabbit IgG (1/100 in PBS/T) for 1 h at room temperature, then rinsed and incubated in rabbit PAP (1/500 in PBS/T) for 1 h at 22°C. After rinsing with Tris (0.05 M, pH 7.6) the tissue was incubated in 3,3'-diaminobenzidine tetrahydrochloride (0.5 mg/ml in Tris buffer containing 0.01% $H_2O_2$) for approximately 7 min at 22°C. The sections were rinsed, dehydrated and coverslipped with Permount. Alternate sections were stained with Cresyl Violet, or processed for enkephalin-like and substance P-like immunoreactivity, respectively.

*Lesions.* Two types of lesions were employed. Suction lesions were used to destroy the rostral tip of the n. accumbens anterior to the caudate nucleus, and to destroy the caudate nucleus dorsal to the n. accumbens, lateral to the stria terminalis, and rostral to the anterior tip of the globus pallidus. Lesions localized to more caudal parts of the nucleus accumbens were produced electrolytically so as to avoid massive damage to the overlying caudatoputamen; a Grass LM4 lesion maker, and currents of about 0.15 mA were used for this purpose.

## RESULTS

### Enkephalin-like immunoreactivity

*Intrastriatal woolly fibers.* Almost throughout its longitudinal extent the striatum is invaded by slender, long, enkephalin-positive strands emanating from the dense intrapallidal plexus of woolly fibers. Each of these strands is composed of one or at most a few woolly fibers. In most parts of the striatum such strands occur only sporadically, and they are particularly rare in dorsal striatal regions. The most ventral part of the striatum, however, is invaded by numerous woolly fibers. At levels corresponding to the middle length of the amygdala, some of these woolly fibers continue ventralward to contribute to the fairly dense intra-amygdaloid plexus to be described below. Rostral to the anterior commissure, the striatal region invaded by enkephalin-positive strands comprises the nucleus accumbens and laterally adjacent parts of the ventral striatal region, and extends dorsally from there over a short distance along the lateral wall of the lateral ventricle. Many intrastriatal woolly fibers at these levels rostral to the main body of the globus pallidus issue from a plexus filling the subcommissural or ventral pallidum (Fig. 2A). It is of interest that some of these strands can be traced over distances that seem too great to be spanned by a single pallidal neuron. It has not been possible in enkephalin material to identify pallidal cell bodies outlying along the intrastriatal strands, but it must be kept in mind that pallidal neurons themselves are unstained in such material.

*Woolly fibers in the amygdala.* At antero-posterior levels corresponding to middle longitudes of the amygdala, an especially prominent bundle of woolly fibers issues from the ventrolateral corner of the globus pallidus into the laterally adjacent ventral pocket of the striatum. Some of the fibers composing this bundle remain within the striatum, but others, upon reaching the border between striatum and lateral amygdaloid nucleus, form intricate tangles from which fibers emerge that pass ventromedially into the zone separating the central from the basolateral amygdaloid nucleus (Fig. 2C). In this zone woolly fibers accumulate to form a dense plexus from which fibers loop medially around the ventral border of the central amygdaloid nucleus where they join with a second contingent of more loosely arranged intra-amygdaloid woolly fibers that emerge from a more medial, infrapeduncular site along the base of the globus pallidus (arrowhead in Fig. 2C). Thus, at some transverse levels at least, the central amygdaloid nucleus appears completely surrounded by woolly fibers. It must be noted that some of these fibers actually traverse, rather than circumvent, the central nucleus. Also noteworthy is that none appear to extend into more ventral regions of the amygdala.

*Woolly fibers in the infrapeduncular region and hypothalamus.* At a level with the intra-amygdaloid woolly-fiber plexus, the base of the globus pallidus extends far medially as a plate of scattered cells underneath the cerebral peduncle. In sections stained for enkephalin, this extension corresponds to a dense stratum of ENK-positive woolly fibers (Fig. 2C), gradually tapering as it expands medially along the undersurface of the peduncle and the rostral half of the entopeduncular nucleus embedded in the latter. A few woolly fibers deviate from this stratum dorsally into the otherwise nearly enkephalin-negative entopeduncular nucleus. A larger number curve dorsomedially around the medial margin of the peduncle, traverse the lateral hypothalamic region, and form a rather loose plexus lateral and dorsal to the fornix bundle (Fig. 2D). It is remarkable that this plexus remains confined to antero-posterior levels corresponding to the entopeduncular nucleus.

Although at first glance the intrahypothalamic ENK-positive woolly fibers here described would seem to emerge from the plexus filling the infrapeduncular extension of the globus pallidus or external pallidal segment, it becomes likely upon closer examination that they issue from a medial part of the entopeduncular nucleus or internal pallidal segment. These woolly fibers differ in appearance from those composing the bulk of the infrapeduncular stratum of ENK-positive woolly fibers: their plexus of entwining ENK-positive fibers seems sparser and more sharply defined. It must be noted, however, that only very few of these ENK-positive woolly fibers can be traced back into the entopeduncular nucleus across its ventral border, and that the ENK content of the nucleus in the rat is very low throughout.

*Woolly fibers in the bed nucleus of the stria terminalis.* A short distance caudal to the crossing of the anterior commissure, ENK-positive woolly fibers from the medial wing of the ventral pallidum (see below) extend steeply dorsalward to form a thin stratum covering the medial surface of the internal capsule (Fig. 3A). Slender strands of woolly fibers issue from this ENK-positive stratum medialward into the bed nucleus of the stria terminalis in which they compose an irregular plexus that is densest in the lateral half of the nucleus (Fig. 2B).

*Woolly fibers in the rostrobasal forebrain.* The widest and most massive expansion of the pallidal woolly-fiber pattern outside the conventionally recognized borders of the globus pallidus occurs at rostral levels of the basal forebrain, more especially in the region stretching from behind the transverse limb of the anterior commissure ventrally and rostrally into the olfactory tubercle. For a proper understanding of the woolly-fiber distribution in this region it is necessary first to consider briefly the cellular organization of the olfactory tubercle.

*Cytoarchitecture of the olfactory tubercle.* As shown by Fig. 5B, in the olfactory tubercle of the rat the greatest number of large cells lie embedded in the prominent bundles composing the deep, fenestrated fiber stratum (Dejerine's "deep olfactory radiations") that incompletely demarcates the olfactory tubercle from the base of the overlying striatum. Commingling of fibers and large cells has led to the designation "deep white and magnocellular grey stratum" for this deepest, fibrocellular layer of the olfactory tubercle.[17] The discontinuities in this fibrocellular stratum correspond to the "striatal cell bridges" of Heimer and Wilson,[10] each of which is a narrow column of striatal cells establishing a continuity of the striatum with the next, intermediate cell layer of the olfactory tubercle.

This highly polymorphous middle stratum is made up of: (a) a core population of generally medium-sized cells of striatal appearance, unevenly distributed and partly arranged in fairly dense, irregularly shaped, often conical columns that appear to be the widening ventral continuations of the "striatal cell bridges"; (b) the conspicuous islands of Calleja, dense aggregates of very small, darkly staining cells generally situated immediately below the deep fibrocellular layer in the intervals between the columns of medium-sized striatal cells; and (c) scattered large cells of the same type also found in the deepest, fibrocellular stratum. Such more superficial large cells tend to be closely associated with the islands of Calleja, lying alongside, or even enclosed within the cell-poor center of, these dense and often cup-shaped granule-cell aggregates. As some islands of Calleja extend ventrally far enough to protrude through the most superficial, pyramidal, cell layer into the molecular layer, large cells associated with them may be encountered even in this most superficial position within the tubercle. It must be noted that some, and perhaps all of the most superficial islands of Calleja in reality represent protrusions of the middle (subpyramidal) layer of the tubercle through the pyramidal into the molecular layer; only a peripheral zone of such protrusions is composed of the characteristic granule cells.

The appearance of the region ventral to the anterior commissure in frontal sections stained for enkephalin is shown by Fig. 3. Dorsal and caudalmost in the region (Fig. 3B) the subcommissural area first identified by Heimer and Wilson[10] as a ventral extension of the globus pallidus ("ventral pallidum") appears. As outlined by its content of woolly fibers, its cross-section has the shape of a wide-brimmed hat, the dome of which lies immediately beneath the anterior commissure, flanked medially by the bed nucleus of the stria terminalis and laterally by the subcommissural striatum, while the upswept brim extends medially and laterally around the base of these two neighboring structures. Stretching ventrally from the lateral, substriatal part of the brim, Fig. 3B shows a number of long, columnar arrays of woolly fibers, separated from each other by what appear to be slender columns of striatal tissue. At their base, these woolly-fiber columns fuse into the horizontally-placed, thick but fenestrated plate of woolly fibers that corresponds in position to the deepest stratum of the olfactory tubercle: the fibrocellular layer of deep olfactory radiations.

Slightly farther rostrally, at the level of Fig. 3C, the bed nucleus of the stria terminalis has been replaced by the nucleus accumbens. At this level, the ventral pallidum, as outlined by its high content of enkephalin-positive woolly fibers, has become reduced to a thick, flat plate that extends obliquely along the ventromedial border of the nucleus accumbens, and is directly continuous ventrally with the horizontally disposed fenestrated stratum of woolly fibers in the depth of the olfactory tubercle. Comparison of a frontal enkephalin-stained section with a neighboring Nissl-stained section (Figs 5A and B) shows that the discontinuous "woolly-fiber" layer is identical with the fibrocellular layer or layer of deep olfactory radiations of the olfactory tubercle. Figure 5A, in addition, shows a number of single woolly fibers extending from the main plexus dorsalward into the overlying ventral striatum, as well as ventrally into the striatal cell bridges and the intermediate, subpyramidal cell layer of the olfactory tubercle. Some woolly fibers can be followed ventrally as far as the pyramidal layer of the tubercle (Fig. 5D).

Finally, it is of interest to note the strikingly low ENK-like reactivity of the islands of Calleja. Whereas all of the remaining parts of the striatum and olfactory tubercle exhibit a notable—even though regionally variable—background of diffuse ENK-like positivity, the islands of Calleja stand out as circumscript, almost entirely blank areas in tissue sections stained for enkephalin (Figs 3C, 5A). Equally noteworthy is that they are conspicuously avoided by the woolly fibers protruding into the middle layer of the tubercle from the main plexus in the deepest layer.

*Substance P-like immunoreactivity*

Staining of the globus pallidus for substance P-like immunoreactivity (SP) is lighter than that for enkephalin, but similar in that it appears in the form of woolly fibers. However, whereas enkephalin-positive woolly fibers fill the entire expanse of the globus pallidus, SP-positive woolly fibers are largely confined to the ventral, subcommissural pallidal region and—with an exception to be mentioned below—are very sparse in the main body of the globus pallidus.

*Rostrobasal forebrain regions.* As shown by Figs 3F–H the distribution of SP-positive woolly fibers in the region stretching rostroventrally from behind the anterior commissure into the olfactory tubercle largely corresponds to that of ENK-positive woolly fibers. Beneath the transverse limb of the commissure SP-positive woolly fibers, like their ENK-positive counterparts, form a dense and widespread, bell- or hat-shaped plexus that is continuous with a fenestrated plate of woolly fibers corresponding in position to the deepest, fibrocellular stratum of the olfactory tubercle; from this stratum, strands of woolly fibers extend ventrally into more superficial layers of the tubercle.

The region of transition of the ventral pallidum with the main body of the globus pallidus (here to be referred to as "dorsal pallidum") is shown in frontal section in Figs 3A and E. The difference in distribution between ENK-like and SP-like immunoreactivity is obvious from a comparison of these two adjacent sections: whereas strong ENK-like reactivity continues undiminished from the ventral throughout the dorsal pallidum, strong SP-like reactivity extends dorsally over no more than a small anteroventromedial sector of the dorsal pallidum, then rapidly declines; in the large remainder of the dorsal pallidum SP-positive woolly fibers are rare.

Figure 4B illustrates the appearance of the SP-positive woolly-fiber pattern in sagittal section, and permits comparison with the ENK pattern shown in an adjacent section from the same brain (Fig. 4A). The difference in dorsal extent of the two woolly-fiber patterns behind the transverse limb of the anterior commissure (i.e. into the dorsal pallidum) is particularly striking in this plane of sectioning. The same Figure, incidentally, also illustrates the contrasting immunoreactivity of the substantia nigra's pars reticulata to ENK and SP, respectively. The strongly positive SP-like reaction of this nigral subdivision appears, as it does in the entopeduncular nucleus and ventral pallidum, in the form of SP-positive woolly fibers.

Although very similar to ENK-like immunoreactivity in its distribution in the rostrobasal forebrain region, SP-like reactivity differs considerably from the former in other parts of the forebrain. The contrasting ENK-like and SP-like immunoreactivities of the dorsal pallidum were already noted above, but several further differences in the respective distributions of ENK-positive and SP-positive woolly fibers remain to be mentioned.

*Striatum and amygdala.* Except for the large number of SP-positive woolly-fiber strands in the largely striatal intermediate layer of the olfactory tubercle, such strands are much sparser than their ENK-positive counterparts in all other striatal regions. Only few strands extend into the nucleus accumbens and adjoining ventral striatal parts from the SP-positive woolly-fiber plexus in the deep, fibrocellular layer of the olfactory tubercle. At more caudal, retrocommissural levels intrastriatal SP-positive woolly fibers are completely lacking, in accord with the pronounced sparseness of such fibers in the adjoining globus pallidus. Also in contrast with ENK-immunoreacted material, no plexus of woolly fibers appears around the central amygdaloid nucleus in sections stained for SP.

*Bed nucleus of the stria terminalis.* No SP-positive woolly fibers invade the bed nucleus of the stria terminalis as ENK-positive woolly fibers do from the medial wing of the ventral pallidum.

*Entopeduncular nucleus and hypothalamus.* The entopeduncular nucleus is almost entirely ENK-negative, but it is pervaded by a dense plexus of SP-positive woolly fibers that extends medially beyond the medial margin of the cerebral peduncle across most of the lateral half of the lateral hypothalamic region. This intrahypothalamic plexus of SP-positive woolly fibers must represent dendrites and/or cell bodies of the internal (entopeduncular) pallidal segment. As mentioned earlier, this seems true also of the intrahypothalamic ENK-positive woolly fibers which invade the hypothalamus by looping dorsally from beneath the medial margin of the cerebral peduncle.

*The effect of striatal lesions on pallidal enkephalin-like and substance P-like immunoreactivity*

In 6 rats, lesions were placed in various parts of the striatum, either electrolytically or by aspiration through the dorsal or ventral brain surface. These lesions were deliberately limited to rostral striatal levels in order to avoid primary involvement of the globus pallidus.

In a case of 12-day survival from an extensive ablation (Fig. 6A) localized to the rostro-dorsal striatum, a large, circumscript defect appeared in the ENK woolly-fiber content of the dorsal pallidum, the main body of the globus pallidus (Fig. 6B). Located dorsolaterally in rostral sections of the globus pallidus, this void gradually shifted medialward in successively more caudal sections, in accordance with the generally radial orientation of striatopallidal fibers. Except for a slight extension of the defect into the most caudal and dorsal part of the ventral pallidum (probably attributable to involvement of a ventrolateral striatal region in the lesion), no depletion of

ENK or SP-positive woolly fibers was found in subcommissural parts of the pallidum.

By contrast, a well-defined area of depletion of both ENK and SP woolly-fiber staining was found in the ventral pallidum in a case in which the rostral half of the nucleus accumbens had been removed 14 days earlier by aspiration through the base of the brain. In this case the staining defect extended about 600 $\mu$m rostral to the commissure and 150 $\mu$m caudal to it. Electrolytic lesions placed farther caudally in the nucleus accumbens eliminated ENK and SP-immunoreactivity over the same longitudinal range, but these voids in the staining pattern were located farther dorsally in the ventral pallidum, suggesting a dorsoventral topography in the striatal afferentation of the ventral pallidum. An example of such more dorsally located staining defects is shown in Figs 6D–E, together with the causative electrolytic accumbens lesion (Fig. 6C). The region of woolly-fiber depletion in this case coincides closely with that region of the ventral pallidum which was shown in autoradiographic experiments[19] to receive the densest fiber projection from the accumbens area here destroyed.

It must be emphasized that none of the staining defects found in the present lesion experiments involved the fenestrated sheet of woolly fibers pervading the deepest, fibrocellular stratum of the olfactory tubercle. It seems most likely that both the ENK and SP-positive fibers entwining the dendrites of the large cells in this stratum originate largely from striatal elements in the more superficial layers of the olfactory tubercle.

### DISCUSSION

A fundamental assumption made in the present study is that the woolly fibers characterizing both enkephalin-like (ENK) and substance P-like (SP) immunoreactivity of the globus pallidus are indeed, as suggested by Haber and Elde,[6] pallidal dendrites ensleeved in a plexus of thin striatal efferents containing the immunoreactive substance. Arguments for this premise are: (a) the striking histological resemblance of woolly fibers to the pictures illustrating the classical Golgi study of the globus pallidus and substantia nigra by Fox *et al.*;[5] (b) the present evidence that woolly fibers disappear from pallidal regions deafferented by striatal lesions—a finding compatible with histochemical evidence associating enkephalin with the striatopallidal projection;[2,3,13] and (c) the fact that wherever in immunoreacted sections woolly fibers are found in large numbers outside the conventionally recognized borders of the globus pallidus, adjacent Nissl sections reveal the presence of large cells which, in some locations at least (e.g. the ventral pallidum;[10] olfactory tubercle[21]) have been reported to be ultrastructurally similar to pallidal neurons.

The last-mentioned of these three arguments is, perhaps, the most vulnerable. The woolly fiber almost

Fig. 1. Light-microscopic appearance of striatal efferents and their plexus formations enmeshing dendrites of pallidal and nigral neurons. A–C: these axo-dendritic complexes as originally described and illustrated by Fox *et al.*;[5] D–E: as appearing in material processed for enkephalin-like immunoreactivity. Note difference between single ENK-positive striatal axons (sf in D and E) and "woolly fibers" (wf). A–C are reproduced from Ref. 5 by kind permission of Akademie Verlag, Berlin.

Abbreviations used in Figures

ac: anterior commissure
amb: nucleus basalis lateralis amygdalae
amc: nucleus centralis amygdalae
cp: cerebral peduncle
dp: dorsal pallidum (body of globus pallidus)
fx: fornix
iC: island of Calleja
lp: lamina pyramidalis of olfactory tubercle
nac: nucleus accumbens septi
ndb: nucleus of diagonal band
nst: bed nucleus of stria terminalis
ot: olfactory tubercle
put: putamen
sf: single enkephalin-positive fiber
vp: ventral pallidum
vs: ventral region of striatum
wf: enkephalin-positive woolly fiber

Fig. 2. Enkephalin-positive woolly-fiber plexuses: A in ventral pallidum (vp), ramifying into ventral striatum with thin strands or solitary woolly fibers; B in bed nucleus of stria terminalis; C in dorsal pallidum (dp), ventral putamen (put), and dorsal region of amygdala surrounding nucleus centralis amygdalae (amc); arrowhead indicates ENK-positive woolly-fiber group issuing from infrapeduncular extension of dorsal pallidum; D in hypothalamic region lateral and dorsal to fornix (fx) (arrowheads).
Other abbreviations: amb: nucleus amygdalae basalis lateralis; cp: cerebral peduncle.

Fig. 3. Composite photograph juxtaposing ENK-like (left-hand column: A–D) and substance P-like immunoreactivity (right-hand column: E–H) at 4 levels of the rat's forebrain. Each horizontal pair of photographs represents two closely neighboring sections. In the 3 lower pairs (B–F, C–G and D–H) note the strikingly similar distributions of, respectively, the ENK-positive and the SP-positive woolly-fiber plexus outlining the ventral pallidum (vp) in the region ventral to the anterior commissure. In the uppermost pair (A–E), however, note a conspicuous difference: the dense ENK-positive woolly-fiber plexus of the ventral pallidum continues undiminished throughout the dorsal pallidum (A: dp), whereas the SP-positive plexus (E) extends over no more than a short distance into the dorsal pallidum. In the same frames note also that the SP-positive woolly-fiber plexus does not extend along the medial side of the internal capsule (E) as the ENK-positive plexus does (A); accordingly, no SP-positive woolly fibers appear in the bed nucleus of the stria terminalis.

Fig. 4. A: sagittal section showing continuity of enkephalin-positive woolly-fiber plexus filling dorsal (main) pallidum (dp) with that filling ventral pallidum (vp) and beyond it with that marking fibrocellular stratum of olfactory tubercle. Note conspicuously low ENK-like immunoreactivity of island of Calleja (cf. Fig. 5A). B: sagittal section closely neighboring that shown in A, but processed for SP-like immunoreactivity. Note similarity of SP-positive woolly-fiber plexus marking ventral pallidum and olfactory tubercle with its ENK-positive counterpart shown in A. Also note low SP-like immunoreactivity of dorsal pallidum.

Fig. 5. Enkephalin-positive (A) and substance P-positive (C) woolly-fiber plexus of the olfactory tubercle. Note that both plexuses lie in the deepest, fibrocellular layer of the tubercle identified by large, widely scattered cells in the closely adjacent Nissl section shown in B. Also note: (1) woolly fibers extend from both plexuses ventrally alongside the islands of Calleja (iC in A); some ENK-positive woolly fibers may extend as far as the pyramidal layer (lp) of the tubercle (D); (2) other ENK-positive (but few if any SP-positive) woolly fibers extend from the main plexus dorsally into the overlying ventral striatal region (cf. Fig. 2A); (3) markedly low ENK-like immunoreactivity of the islands of Calleja. Asterisks mark two blood vessels appearing in A–C.

Fig. 6. Effect of striatal lesions. The extensive aspiration lesion of the caudatoputamen shown in A caused a large, circumscript void in the ENK-positive woolly-fiber plexus normally pervading the entire dorsal pallidum (B); this void extends over a short distance into the ventral pallidum, reflecting involvement of ventrolateral striatal parts in the lesion. The smaller electrolytic lesion of the nucleus accumbens shown in C results in a small, equally circumscript defect in both the ENK-positive (D) and the SP-positive plexus (E) filling central parts of the ventral pallidum (arrowheads in D and E); no similar defect appeared in this case in the dorsal pallidum.

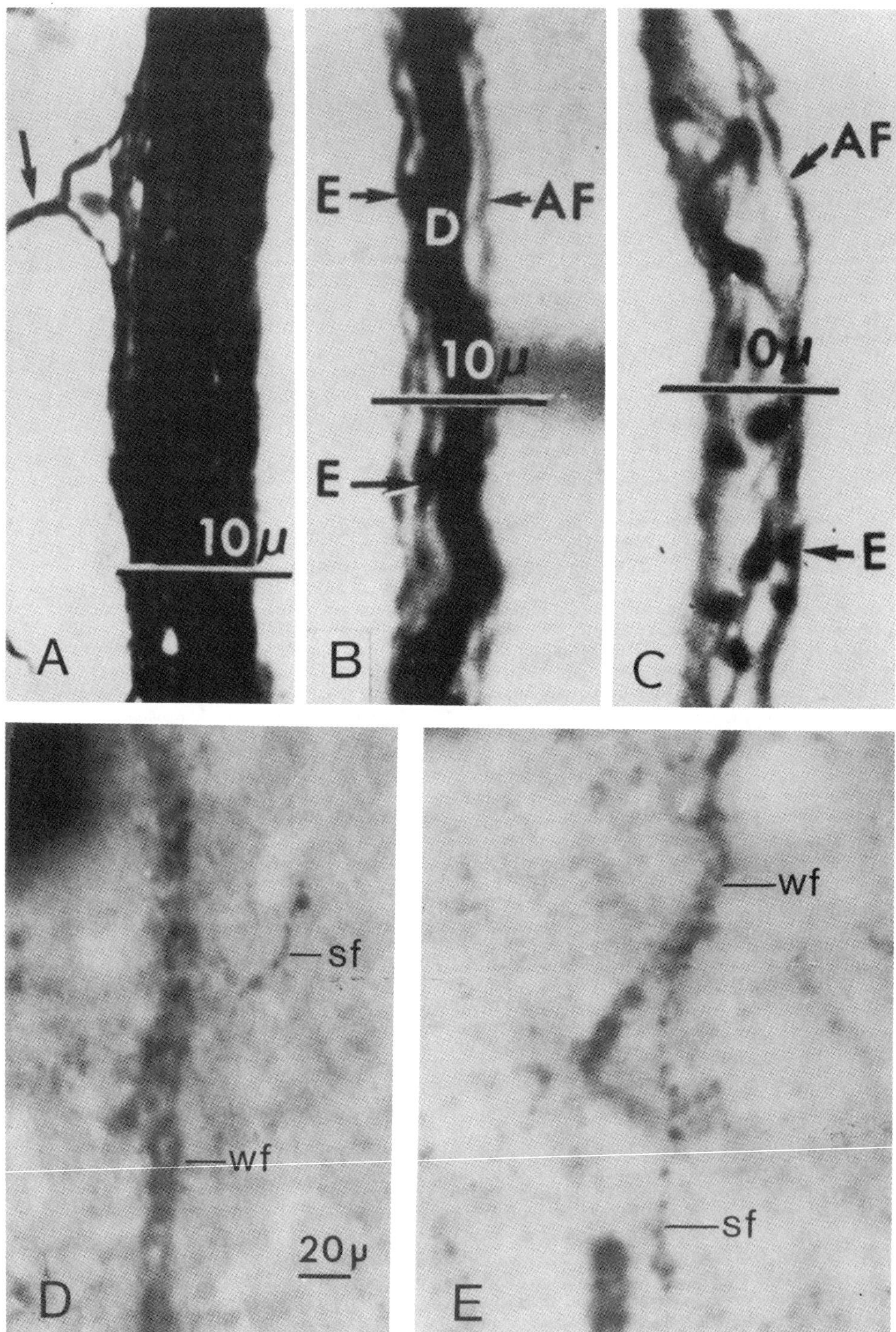



Fig. 1.

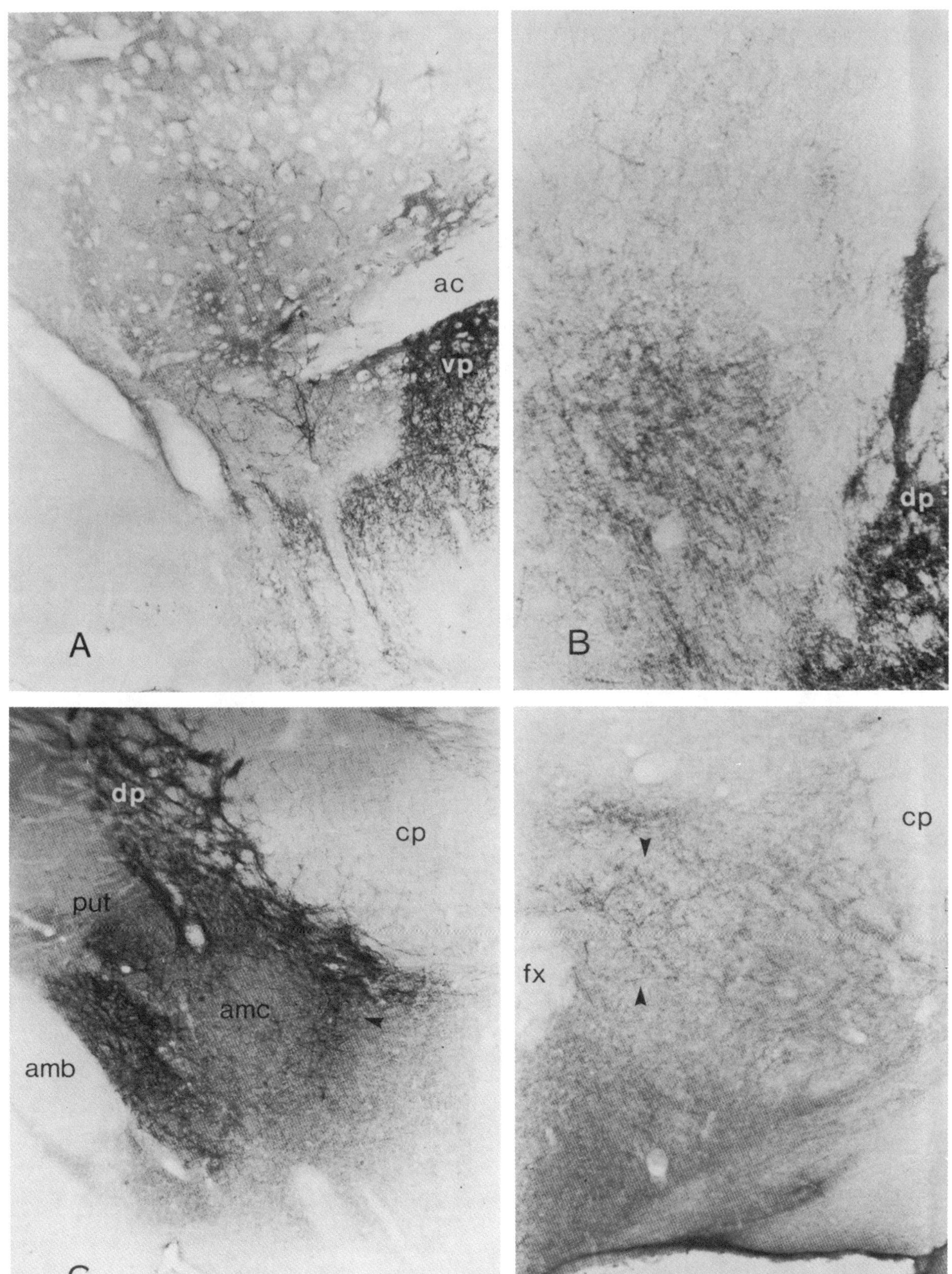

Fig. 2.

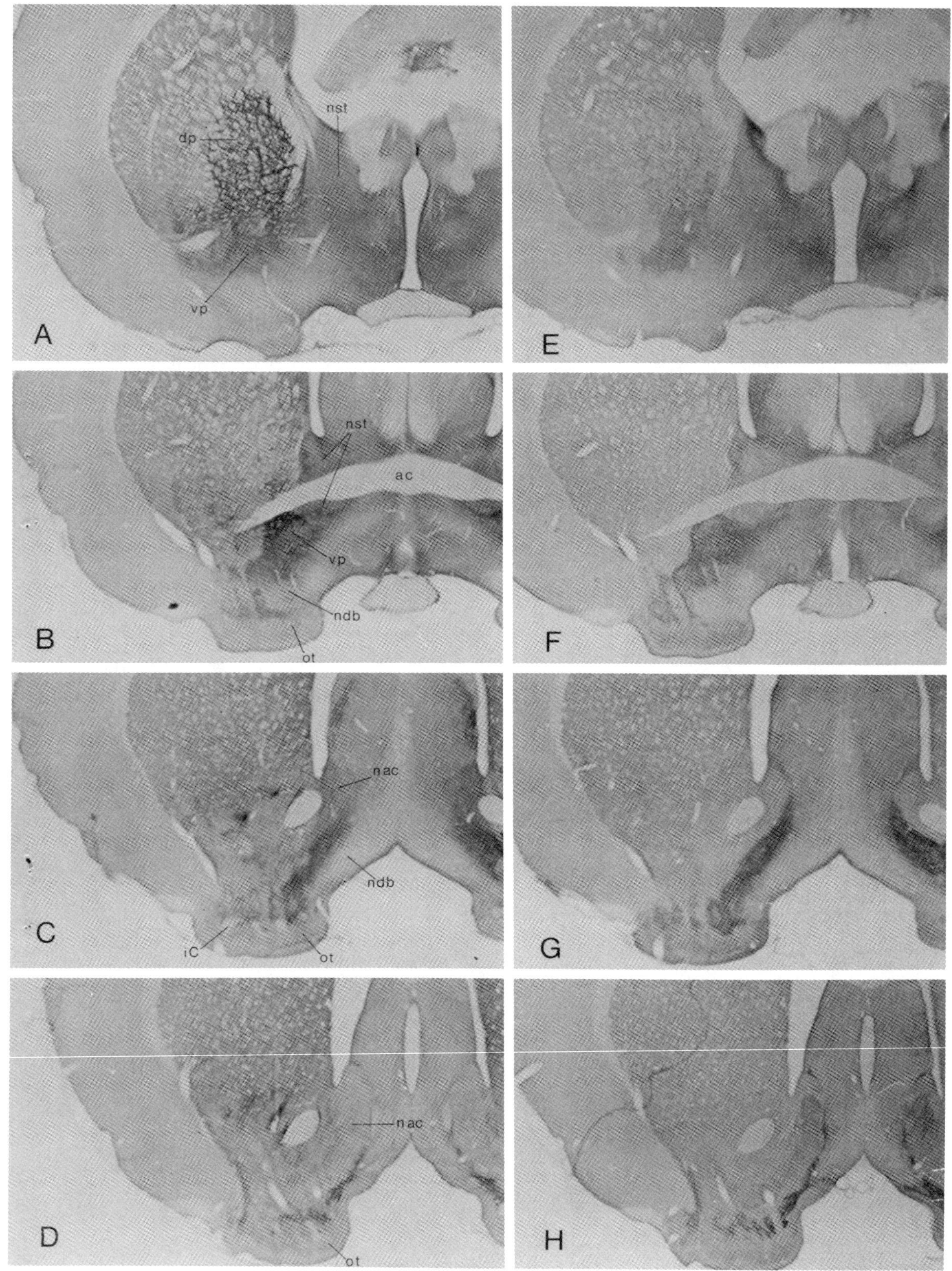



Fig. 3.

484

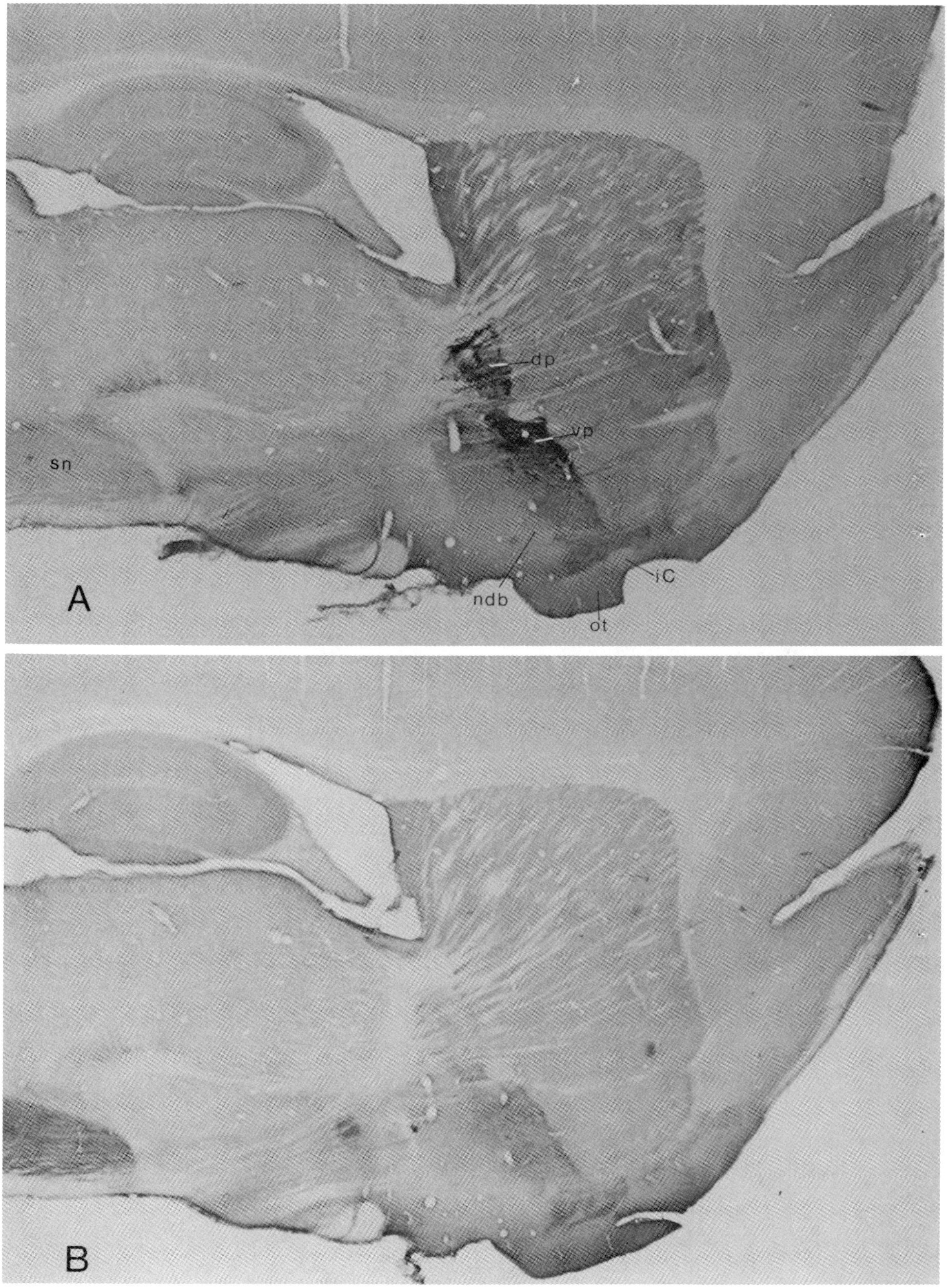

Fig. 4.

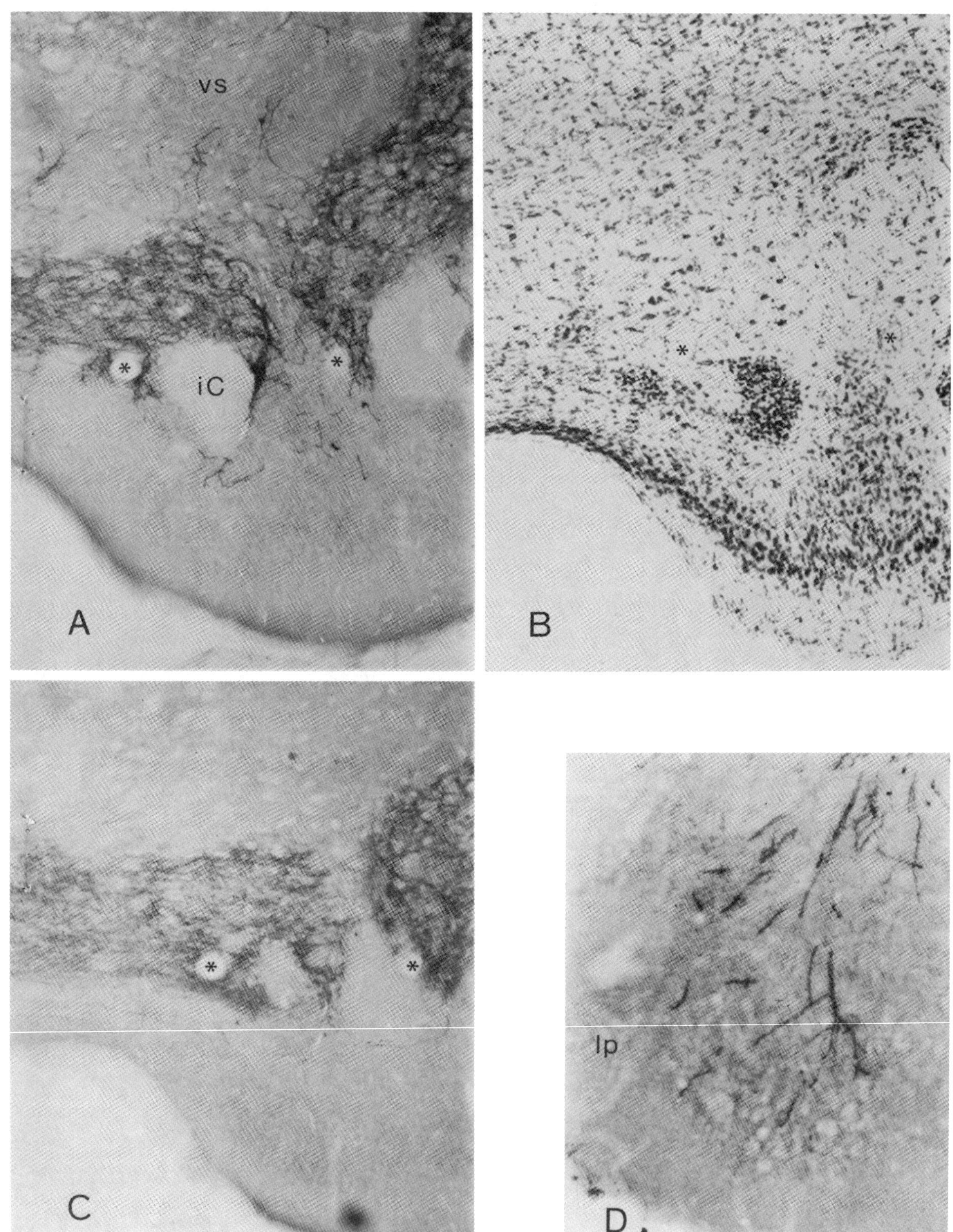



Fig. 5.

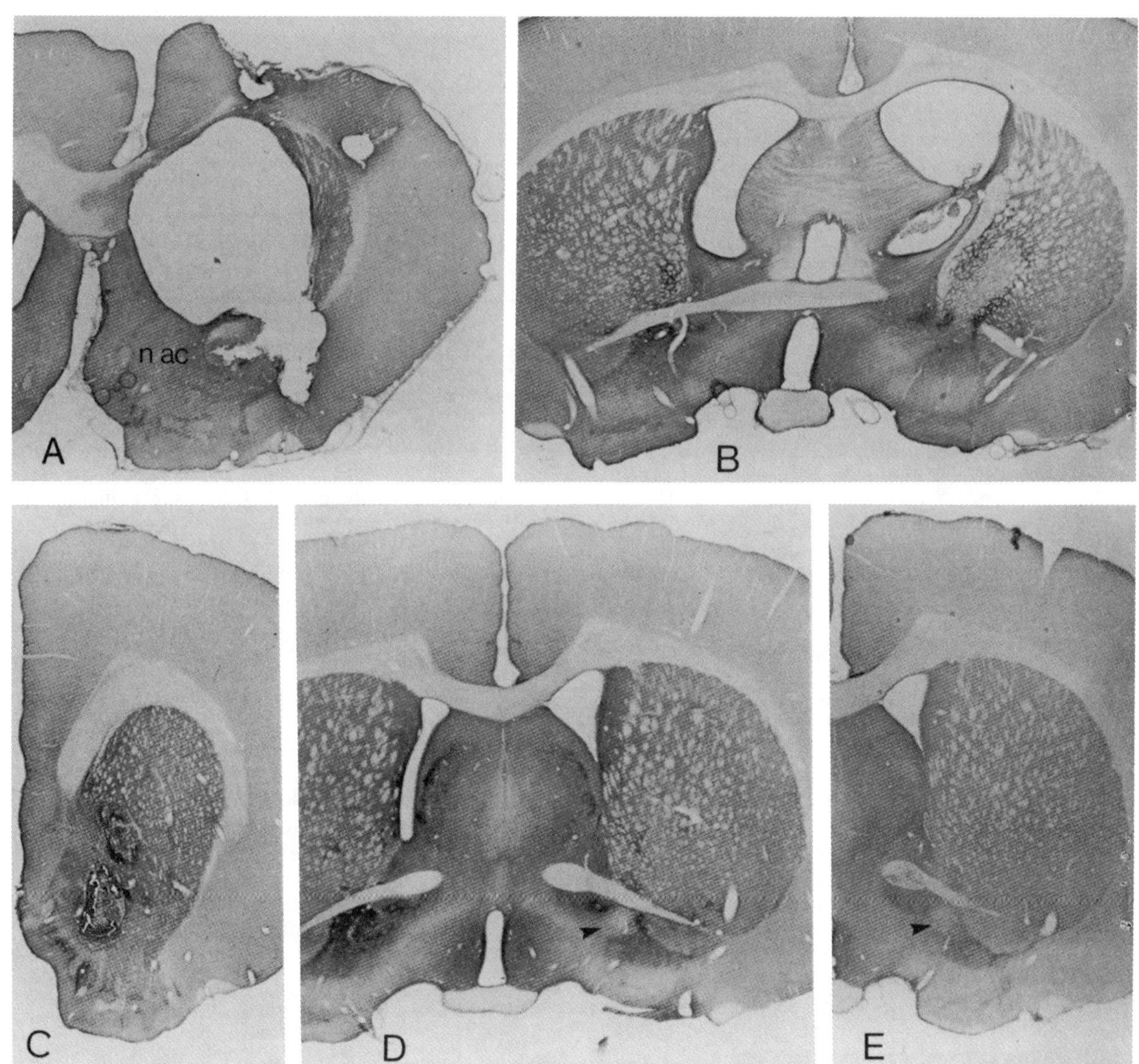

Fig. 6.

certainly represents a particular form of axo-dendritic junction, but it is only its presynaptic side, the striatal projection fiber, that is ENK- or SP-positive. Thus far, the dendritic process forming the unstained core of the woolly fiber has not been positively identified as pallidal, and as long as this is the case, some uncertainty must remain whether striatal ENK and SP efferents can enmesh only pallidal dendrites, or dendrites of non-pallidal neurons as well. In the present study the former assumption has been made, for the reason that all woolly fibers appearing outside the conventionally recognized borders of the globus pallidus formed part of woolly-fiber plexuses directly continuous with the massive plexus pervading the body of the globus pallidus. That is to say: all woolly fibers seen in this study appeared to be part of a continuous plexus ramifying away from the globus pallidus as conventionally outlined.

As reported above, such ramifications of the ENK-positive pallidal woolly-fiber pattern extend into (a) ventral parts of the striatum; (b) a limited dorsal region of the amygdaloid complex; (c) a hypo-thalamic region lateral and dorsal to the fornix; (d) a lateral part of the bed nucleus of the stria terminalis, and, (e) a large rostrobasal forebrain region stretching from behind the anterior commis-sure rostrally and ventralward into the olfactory tubercle. To this enumeration the following com-ments are added.

Intrastriatal enkephalin positive woolly fibers are largely confined to ventral regions of the striatum. At retrocommissural levels, many become deflected ven-tralward and combine to form a fairly dense plexus around the rostral half of the central amygdaloid nucleus. Since no woolly fibers extend farther ven-trally into the amygdala, the question arises whether perhaps this plexus in reality marks the most ventral part of the striatum, a part in which the central amygdaloid nucleus lies embedded. This suggestion, however, is only partially supported by observations in sections stained for acetylcholinesterase. In such material, the characteristic dark shade of the striatum continues ventralward over the rostral pole and lat-eral side of the central amygdaloid nucleus, but not along its medial border and base. It thus appears that the ENK-positive woolly-fiber plexus only partially reflects a striatal surround of the central amygdaloid nucleus, and that the medial part of the plexus, at least, lies in amygdaloid rather than striatal territory. It must, moreover, be noted that the central nucleus cannot on the basis of its known efferent connections be considered part of the striatum (cf. for example ref. 16). It may, therefore, be significant that a number of ENK-positive woolly fibers traverse, rather than encircle, the central nucleus, and that the nucleus contains a plexus of fine ENK-positive fibers that is at least as dense as that pervading the striatum. This suggests the possibility that some pallidal neurons may derive their ENK-fiber innervation in whole or part from a non-striatal source, in this

instance the central amygdaloid nucleus. However, as long as this remains unproven it would seem more likely that intra-amygdaloid woolly fibers are out-lying pallidal elements innervated by ENK-positive efferents from the most ventral striatal region.

*Woolly fibers in the hypothalamus and bed nucleus of the stria terminalis.* Similar questions may arise in regard to two other non-striatal regions invaded by woolly fibers: the hypothalamus and the bed nucleus of the stria terminalis. Both these regions contain numerous fine, beaded, ENK-positive fibers that could supply or contribute to the fiber plexus en-meshing invading pallidal dendrites.

As described in foregoing sections of this paper, both ENK-positive and SP-positive woolly fibers invade the hypothalamus from the lateral side. The two plexuses formed by these fibers overlap only little: whereas ENK-positive woolly fibers form a rather loose plexus dorsal and lateral to the fornix bundle, SP-positive intrahypothalamic woolly fibers seem confined to a dense plexus in the lateral hypo-thalamic region, adjacent to the entopeduncular nu-cleus. Both the SP-positive and the ENK-positive intrahypothalamic plexus appear to represent pallidal elements (dendrites and/or cell bodies) protruding into the hypothalamus from the entopeduncular nu-cleus (internal pallidal segment). While this is readily evident in the case of the SP-positive plexus, which in fact is a medial expansion of the dense plexus of SP-positive woolly fibers filling the entopeduncular nucleus, it could seem paradoxical for the ENK-positive plexus since the rat's entopeduncular nucleus contains only sporadic ENK-positive woolly fibers. In the monkey, however, a fairly dense plexus of ENK-positive woolly fibers has been noted in the medial portion of the otherwise ENK-poor internal pallidal segment.[6,7] It is conceivable that in the rat the neurons corresponding to this part of the monkey's internal pallidal segment are extruded from the ento-peduncular nucleus into hypothalamic locations.

The significance of pallidal incursions into the hypothalamus is difficult to judge. In primates, axons arising from the internal pallidal segment are known to enter the hypothalamus—largely, it seems, in passage to other destinations[18]—but no pallidal den-drites or cell bodies appear to have been positively identified in the hypothalamus of any species. None-theless, the existence of intrahypothalamic pallidal neurons as here indicated by hypothalamic woolly-fiber plexuses has been suspected before on the basis of findings in retrograde-labelling experiments.[11,25] Such pallidal elements invading the lateral hypo-thalamic region must lie in the path of the medial forebrain bundle. As this complex fiber system is known to include numerous fibers descending to the midbrain from the nucleus accumbens[1,19,20,24] and adjoining parts of the basomedial striatum, it seems possible that despite their outlying position intra-hypothalamic pallidal dendrites are innervated by striatal projection fibers. The alternate or additional possibility of non-striatal afferents to these outlying pallidal elements cannot, however, be ruled out at present.

Immediately behind the crossing of the anterior commissure the bed nucleus of the stria terminalis is invaded by ENK-positive—but no SP-positive—woolly fibers from the medial wing of the ventral pallidum. As described earlier in this paper, these fibers form a loose plexus that is densest in the lateral half of the nucleus. Before it is concluded that these outlying pallidal dendrites and/or neurons must be innervated by amygdalofugal ENK-positive fibers travelling in the stria terminalis, it should be considered that their location corresponds closely to that of the anterograde fiber labelling found in the bed nucleus after an injection of a rostromedial part of the nucleus accumbens with tritiated amino acids (cf. Fig. 1C of ref. 19). Therefore, these woolly fibers, like those in the lateral hypothalamus, could represent outlying pallidal elements innervated by the nucleus accumbens.

*Woolly fibers in the rostrobasal forebrain.* The present evidence of a massive plexus of both ENK-positive and SP-positive woolly fibers in the basal forebrain region stretching from behind the transverse limb of the anterior commissure rostral- and ventralward is entirely compatible with Heimer's[9] notion of a "ventral pallidum" extending from the main body of the globus pallidus basalward into the subcommissural region. In its most caudal and dorsal extent, at the level where the anterior commissure crosses the midplane (Figs 3B and F), the woolly-fiber plexus appears in transverse sections immunoreacted for either ENK or SP in the shape of a bell or wide-brimmed hat. Only the dome of this shape was originally identified as a subcommissural continuation of the globus pallidus;[10] without the benefit of the present immunohistochemical methods its wide expansions laterally underneath the subcommissural pocket of the striatum and medially along the base of the bed nucleus of the stria terminalis would have been difficult or impossible to delineate. Neither would its ventral extensions passing along the lateral margin of the almost entirely ENK and SP-negative nucleus of the diagonal band into the fibrocellular layer of the olfactory tubercle (Figs 3B and F) have been easy to distinguish in Nissl material from the likewise magnocellular nucleus of the diagonal band.

The continuation of the subcommissural woolly-fiber plexus into the deepest, fibrocellular layer of the olfactory tubercle strongly supports Heimer's[9] suggestion that many, at least, of the large cells of the olfactory tubercle are pallidal neurons. The present histochemical evidence for this view is the more compelling as numerous woolly fibers extend from the main plexus in the deepest layer into more superficial strata of the tubercle which contain lesser numbers of similar large cells. Electron-microscopic studies[21] have indeed shown that at least those large cells that lie within or adjacent to islands of Calleja

are structurally indistinguishable from pallidal neurons as described by Fox *et al.*[5] The same studies[21] showed that the granule cells composing the islands of Calleja share several unique structural characteristics with the medium-sized cells of the striatum, including a close (40–60 Å) spacing of cells suggestive of ephaptic functional linkage between neighboring striatal cell bodies. Judged by these structural characteristics, as well as by their high dopamine content,[14] the islands of Calleja appear to be true striatal structures, even though sharply distinguished from the rest of the striatum by the unique combination of an almost exclusively parvicellular cytoarchitecture with an almost entirely negative immunoreactivity for enkephalin (Figs 3C, 5A).

No similar studies appear to have focussed on the irregularly columnar aggregates of medium-sized neurons that separate the islands of Calleja from each other. However, the striking resemblance of their component cells to medium-sized striatal neurons, as well as their continuity with the striatum by way of the "striatal cell bridges" and the fact that they are invaded by ENK-positive and SP-positive woolly fibers (Fig. 5A) all suggest that these aggregates of medium-sized neurons likewise are components of the striatum, and likewise sources of striatal efferents innervating large neurons of the olfactory tubercle.

*The border between ventral and dorsal pallidum.* In view of the direct continuity of the ventral pallidum with the main body of the globus pallidus or dorsal pallidum it is remarkable that the two pallidal regions are histochemically dissimilar. Both share a dense plexus of ENK-positive woolly fibers, but the ventral pallidum alone is pervaded by an additional, almost equally dense plexus of SP-positive woolly fibers that extends dorsally behind the anterior commissure over no more than a small rostral, ventromedial part of the dorsal pallidum. The large remainder of the dorsal pallidum contains only sparse SP-positive woolly fibers.

Earlier studies by fiber-degeneration and autoradiographic techniques have shown that the central region of the ventral pallidum receives its massive striatopallidal projection from the nucleus accumbens, and the results of the present lesion experiments bear out this relationship. Additional findings in an autoradiographic study in progress in this laboratory suggest that the remaining parts of the ventral pallidum, its wing-like medial and lateral expansions and its limited extension into the rostroventral part of the dorsal pallidum, likewise receive their striatal afferents from ventromedial regions of the rostral striatum. Since these regions compose a striatal district converged upon by a variety of afferents originating from within the circuitry of the limbic system—in particular from the hippocampal formation, amygdala, ventral tegmental area, dorsal raphe nucleus, and frontocingulate cortex[15,15a]—it appears that the substance P-positive striatal projection to (the external segment of) the globus pallidus originates, for the most part at least, from a limbic system-afferented striatal subdivision. The distribution of this projection, as here documented by Figs 3E–H and 4B, is largely confined to the ventral pallidum, and it is therefore tempting to regard the latter as a limbic system-associated component of the external pallidal segment.

Two further comments must here be made. First, it would plainly be incorrect to consider the entire substance P-positive striatopallidal projection as the exclusive conveyor of limbic outputs to the pallidum. Substance P-positive striatofugal projections originate not only from limbic-afferented but also from all other parts of the striatum, and are distributed not only to the ventral pallidum but also to the entire extent of the internal pallidal segment and substantia nigra, pars reticulata. Second, in contrast to the substance P-positive striatopallidal projection, its enkephalin-positive counterpart is distributed not only to the ventral but also to the larger dorsal or main subdivision of the external pallidal segment; the findings in our lesion experiments (cf. Figs 6C–E)

show that only in the ventral pallidum do these two projections converge. It is thus clear that the massive connection from the limbic-afferented striatum to the ventral palliou...n is established by substance P-positive and enkephalin-positive striatofugal fibers almost equally. In remarkable contrast, the remaining out-lying pallidal ramifications identified in this study—those extending into the striatum, amygdaloid region and bed nucleus of the stria terminalis—receive few if any substance P-positive fibers and thus could be identified only by virtue of their enkephalin-positive afferents. As pointed our earlier in this discussion, circumstantial evidence suggests that these lesser pallidal ramifications, much like the ventral pallidum, receive such afferents mainly or even exclusively from ventral, limbic system-afferented parts of the striatum.

*Acknowledgements*—We thank Mr P. Paskevich for his valuable assistance. This study was made possible by NSF Grant BNS80-07905, and NIH Grants 5-P01-MH-31154 and F 32-DA05179.

## REFERENCES

1. Conrad L. C. A. and Pfaff D. W. (1976) Autoradiographic tracing of nucleus accumbens efferents in the rat. *Brain Res.* **113**, 589–596.
2. Cuello A. C. and Paxinos G. (1978) Evidence for a long leu-enkephalin striopallidal pathway in the rat brain. *Nature, Lond.* **271**, 178–180.
3. Del Fiacco M., Paxinos G. and Cuello A. C. (1982) Neostriatal enkephalin-immunoreactive neurons project to the globus pallidus. *Brain Res.* **231**, 1–17.
4. Fallon J. H. and Ribak C. E. (1980) Multiple neurotransmitter studies in the island of Calleja complex of the basal forebrain—III. Connections, correlations and reservations. *Soc. Neurosci. Abs* **6**, 114.
5. Fox C. A., Andrade H. N., LuQui I. J. and Rafols J. A. (1974) The primate globus pallidus. A Golgi and electron microscope study. *J. Hirnforsch.* **15**, 75–93.
6. Haber S. and Elde R. (1981) Correlation between met-enkephalin and substance P immunoreactivity in the primate globus pallidus. *Neuroscience* **6**, 1291–1298.
7. Haber S. and Elde R. (1982) The distribution of enkephalin immunoreactive fibers and terminals in the monkey central nervous system: an immunohistochemical study. *Neuroscience* **7**, 1049–1095.
8. Haber S. and Nauta W. J. H. (1981) Substance P, but not enkephalin, immunoreactivity distinguishes ventral from dorsal pallidum. *Soc. Neurosci. Abs* **7**.
9. Heimer L. (1978) The olfactory cortex and the ventral striatum. In *Limbic Mechanisms* (eds Livingston K. E. and Hornykiewicz O.) pp. 95–187. Plenum Press, New York.
10. Heimer L. and Wilson R. D. (1975) The subcortical projections of the allocortex: similarities in the neural associations of the hippocampus, the piriform cortex, and the neocortex. In *Golgi Centennial Symposium* (ed. Santini M.) pp. 177–193. Raven Press, New York.
11. Herkenham M. and Nauta W. J. H. (1977) Afferent connections of the habenular nuclei in the rat. A horseradish peroxidase study, with a note on the fiber-of-passage problem. *J. comp. Neurol.* **173**, 123–146.
12. Ho R. H. and DePalatis L. R. (1980) Substance P immunoreactivity in the median eminence of the North American opossum and domestic fowl. *Brain Res.* **189**, 565–569.
13. Hong T. S., Yang H. Y. T. and Costa E. (1977) On location of methionine enkephalin neurons in rat striatum. *Neuropharmacology* **16**, 450–453.
14. Isaacs S., Fallon J. H. and Ribak C. E. (1980) Multiple neurotransmitter studies in the island of Calleja complex of the basal forebrain—I. Light-microscopy. *Soc. Neurosci. Abs* **6**, 114.
15. Kelley A. E., Domesick V. B. and Nauta W. J. H. (1982) The amygdalostriatal projection of the rat—an anatomical study by anterograde and retrograde tracing methods. *Neuroscience* **7**, 615–630.
15a. Kelley A. E. and Domesick V. B. (1982) The distribution of the projection from the hippocampal formation to the nucleus accumbens in the rat: an anterograde- and retrograde-horseradish peroxidase study. *Neuroscience* **7**, 2321–2335.
16. Krettek J. E. and Price J. C. (1978) Amygdaloid projections to subcortical structures within the basal forebrain and brainstem of the rat and cat. *J. comp. Neurol.* **178**, 225–254.
17. Nauta W. J. H. and Haymaker W. (1969) Hypothalamic nuclei and fiber connections. In *The Hypothalamus* (eds Haymaker W., Anderson E. and Nauta W. J. H.) Chap. 4, pp. 136–209. C. C. Thomas, Springfield, IL.
18. Nauta W. J. H. and Mehler W. R. (1966) Projections of the lentiform nucleus in the monkey. *Brain Res.* **1**, 3–42.
19. Nauta W. J. H., Smith G. P., Faull R. L. M. and Domesick V. B. (1978) Efferent connections and nigral afferents of the nucleus accumbens in the monkey. *Neuroscience* **3**, 385–401.
20. Powell E. W. and Leman R. B. (1976) Connections of the nucleus accumbens. *Brain Res.* **105**, 389–403.

21. Ribak C. E. and Fallon J. H. The island of Calleja complex of rat basal forebrain—I. Light- and electron-microscopic observations. *J. comp. Neurol.* in press.
22. Sternberger L. A. (1979) *Immunocytochemistry.* Wiley, New York.
23. Switzer R. C. and Hill J. M. (1979) Globus pallidus component in olfactory tubercle: evidence based on iron distribution. *Soc. Neurosci. Abs* **5,** 79.
24. Swanson L. W. and Cowan W. M. (1975) A note on the connections and development of the nucleus accumbens. *Brain Res.* **92,** 324–330.
25. Van der Kooy D. and Carter D. A. (1981) The organization of the efferent projections and striatal afferents of the entopeduncular nucleus and adjacent areas in the rat. *Brain Res.* **211,** 15–36.
26. Williams D. J., Crossman A. R. and Slater P. (1977) The efferent projections of the nucleus accumbens in the rat. *Brain Res.* **130,** 217–227.
27. Wilson R. D. (1972) Efferent connections of the nucleus accumbens in the rat. Master's thesis, Massachusetts Institute of Technology.

(*Accepted 6 November* 1982)

*Note added in proof:* Upon completion of this paper for publication, an article appeared by S. Witzer *et al.* (*Neuroscience* **7,** 1891–1904, 1982), describing the ventral pallidum of the rat and including an illustration of the distribution of ENK-positive staining in that brain region.

# V

# CROSSROADS OF LIMBIC AND STRIATAL CIRCUITRY

*Neuroscience* Vol. 7, No. 3, pp. 615 to 630, 1982
Printed in Great Britain

0306-4522/82/030615-15$03.00/0
Pergamon Press Ltd
© 1982 IBRO

# THE AMYGDALOSTRIATAL PROJECTION IN THE RAT—AN ANATOMICAL STUDY BY ANTEROGRADE AND RETROGRADE TRACING METHODS

A. E. KELLEY,* V. B. DOMESICK and W. J. H. NAUTA

Laboratories for Psychiatric Research, McLean Hospital, Belmont, MA 02178 and
Department of Psychology and Brain Science, Massachusetts Institute of Technology,
Cambridge, MA 02139, U.S.A.

**Abstract**—Tritiated leucine and proline injected into the amygdaloid complex was found to label a voluminous amygdalostriatal fiber system which is distributed to all parts of the striatum except an antero–dorsolateral striatal sector. The connection is established by way of the longitudinal association bundle as well as the stria terminalis, and includes a modest (10–15%), symmetrically distributed contralateral component conveyed by the anterior commissure. Both autoradiographic findings and subsequent observations in retrograde cell-labelling (horseradish peroxidase) material indicate that the amygdalostriatal projection originates mainly from the nucleus basalis lateralis amygdalae, in much lesser volume from the nucleus basalis medialis, and minimally from the nucleus lateralis amygdalae; no other contributing amygdaloid cell group could be identified.

A comparison of the present findings with earlier reports indicates that the amygdalostriatal projection widely overlaps the striatal projections from the ventral tegmental area, the mesencephalic raphe nuclei and the prefrontal cortex. Like the amygdalostriatal projection, these striatal afferents largely or entirely avoid the antero-dorsolateral striatal quadrant, which thus appears to be the striatal region most sparsely innervated by afferents originating from structures within the circuitry of the limbic system. Findings in additional autoradiographic material identify this relatively non-limbic striatal quadrant as the main region of distribution of the corticostriatal projection from the sensorimotor cortex.

This report grew out of a study addressing the question whether it might be possible to delineate the nucleus accumbens from the remainder of the striatum on the basis of coextensive distribution of two or more of its afferent fiber systems. An autoradiographic examination of the amygdalostriatal projection was included in the study for the reason that afferents to the nucleus accumbens from the amygdala had been reported earlier.[5,6,19] In this phase of the study, evidence appeared of an unexpectedly massive and widespread system of amygdala efferents distributed not only to the nucleus accumbens and olfactory tubercle but also to most of the larger remainder of the striatum. This finding prompted the present, somewhat more systematic and detailed investigation of the origin and distribution of the amygdalostriatal projection. Part of the data to be reported below were published earlier in abstract form elsewhere.[18]

## EXPERIMENTAL PROCEDURES

The present report is based on experiments in 37 adult female albino rats of the Charles River CD strain. In 19 of these, a smaller or larger deposit of tritiated L-leucine and L-proline was placed in one (or, in 4 cases, several) of

various locations within the amygdala. Prior to surgery, each animal was deeply anesthetized by intraperitoneal injection of 0.4 ml/100 g body weight of a commercial mixture of chloral hydrate and barbiturate (Chloropent, Fort Dodge Laboratories). Two different stereotaxic approaches to the amygdala were followed: in the first 4 cases, a glass micropipette measuring 10–20 μm internal diameter at the tip, and containing equal parts of tritiated leucine and tritiated proline (New England Nuclear Corporation) in a medium of 0.01 N acetic acid (20 μCi/μl), was lowered vertically through the dorsal brain surface to the intended depth. The labelled amino acids were then delivered by passing a positive current of 0.5–1 μA, pulsed at a rate of 5 s on/5 s off, through the solution for 8–15 min. In the remaining 15 rats, a similar procedure was followed except that a lateral approach was followed, avoiding passage of the pipette through the striatum. These rats placed in a head-holder designed so as to permit the head to be rotated 45° sideways. After exposure of the rhinal sulcus by resection of the squama temporalis and zygomatic arch, the pipette was inserted through or immediately below the sulcus at an angle of 15–25°, supplementing the angle of approach to 60–70° from vertical and thus allowing the pipette to enter the amygdala without traversing any adjacent structures other than the perirhinal cortex and the claustrum.

In the 18 remaining experiments, small injections of horseradish peroxidase (HRP) were placed in various parts of the striatum by electrophoresis of the enzyme from a micropipette containing a 13% solution of HRP (Sigma VI) Tris-buffered at pH 8.6.[14] All injections were placed in the right half of the brain.

The autoradiographic experiments proceeded as follows.

---

* Present address: Dr A. E. Kelley, Laboratoire de Neurobiologie des Comportements, Université de Bordeaux II, 146 Rue Leo Saignat, 33076 Bordeaux Cedex, France.

*Abbreviations:* DAB, diaminobenzidine; HRP, horseradish peroxidase; TMB, tetramethylbenzidine.

The animals were given an overdose of Chloropent 5–8 days postoperatively and were perfused through the heart with saline, followed by 10% formol-saline. The brains were post-fixed for 5–7 days, and then transferred to 30% sucrose-formalin for 3 days. They were next encased in albumin-gelatin and sectioned in the frontal plane on a freezing microtome at 25 μm. The sections were mounted, defatted, coated with Kodak NTB-2 emulsion, stored in light- and moisture-proof boxes for 12–16 weeks at −20°C, and, after being developed in Kodak D-19 at 16°C, counterstained with Cresyl violet.

Animals which received HRP injections were deeply anesthetized 24–48 h later, and perfused with a solution of 5% sucrose, 1.25% glutaraldehyde and 0.5% paraformaldehyde in 0.1 M phosphate buffer at pH 7.4. The brains were stored overnight in a sucrose-phosphate buffer solution and then sectioned at 50 μm. Alternating series of sections were processed for the diaminobenzidine (DAB) and tetramethylbenzidine (TMB) reaction.[23] The DAB-reacted sections were counterstained with Cresyl violet, the TMB material with neutral red. Subsequently all slides were examined microscopically in both bright- and dark-field illumination.

### RESULTS

*Autoradiographic experiments*

The first indication that the amydalostriatal projection might be more widely distributed than was believed thus far was found in a series of four cases (RR 1–4) in which multiple injections of tritiated amino acids had been placed in the amygdaloid complex. The most massive fiber labelling seen in this study appeared in one such case (RR 3) in which an aggregate of four large injections, located along two vertical pipette tracks, involved to various extents all of the amygdaloid cell groups (Fig. 1). Although less suitable for detailed charting, the case is here illustrated by some photographs intended to document the density of intrastriatal fiber labelling (Figs 1–2). This labelling was quite substantial throughout the caudal half of the striatum (Fig. 2A,C), and especially massive in the dorsal and ventral one-quarter of the striatal cross-section. From the level of the anterior commissure rostrally, its density progressively declined from the dorsolateral border of the striatum inward (Fig. 2D), so that at levels rostral to the genu corporis callosi little more than the nucleus accumbens, the ventral striatal region lateral to it and the olfactory tubercle remained densely labelled. Further noteworthy in this case was a substantial number of labelled fibers crossing in the anterior commissure to be distributed to the contralateral striatum in a largely symmetrical but considerably less dense pattern. Finally, it appeared from this case that labelled fibers entered the striatum not only from the ventral side but also in large numbers by way of the stria terminalis (Fig. 2C).

In case RR 3, the pipette tracks passed vertically through a lateral part of the thalamus and a caudal part of the striatum, respectively. To exclude the possibility that the intrastriatal fiber labelling had been caused by radioactive label contaminating one or both tracks, the material was supplemented with cases in which a lateral approach to the amygdala was followed, as described in the preceding section of this paper. From an examination of the 15 cases prepared by this procedure, it became evident that the intrastriatal fiber labelling seen in case RR 3 was not attributable to spillage of label along pipette tracks, and hence must have represented an amygdalostriatal projection. A case documenting this conclusion is described next.

*Case RR 18.* The tritiated leucine and proline injection in this case involved the rostral part of the basolateral amygdaloid nucleus, and spared a larger caudal part of the nucleus. As shown in Fig. 3A in jet black, the region within which neurons showed evidence of having sequestered radioactive label was limited to a lateral part of the basolateral nucleus and largely excluded the medial half and basal zone of the nucleus.

From the injection site labelled fibers extend in several directions, but for the present purpose only those passing to the striatum and some adjoining subcortical structures will be considered in the following description. For a more impartial account of amygdalofugal connections in the rat reference is made to the earlier, detailed autoradiographic study by Krettek & Price.[19]

The amygdalostriatal fibers labelled in this and other cases enter the striatum for the most part from the ventral side. Many such fibers combine with other labelled amygdala efferents to form a substantial, sagittally oriented fiber group, apparently a dorsal subdivision of the longitudinal association bundle, in which they pass rostralward along the ventral border of the striatum (Fig. 3B–D). From this communal substriatal bundle they enter the striatum either directly through its ventral border, or after passing dorsally over variable distances in the white matter laminae bordering the caudatoputamen medially and laterally (Fig. 3A–C). Some fibers may follow a more medial route leading through the globus pallidus (see below).

A second, in this case much smaller, contingent of labelled fibers appears to enter the most dorsal reaches of the caudatoputamen as a component of the stria terminalis (Fig. 3A–B). This fiber group forms part of a ventral stratum of the stria apparently composed of efferents from the basal amygdaloid nuclei. The modest size of this more dorsal group of labelled fibers in case RR 18 may be a factor in the considerably lesser labelling density of the dorsal half compared with the ventral half of the striatum, a difference that persists throughout the length of the striatum. The labelling in the ventral half of the striatum can be rated moderate-to-dense; its pattern is uneven, with denser and sparser patches in a configuration that seems to correspond only partly to the brush-like disposition of internal-capsule bundles within the caudatoputamen.

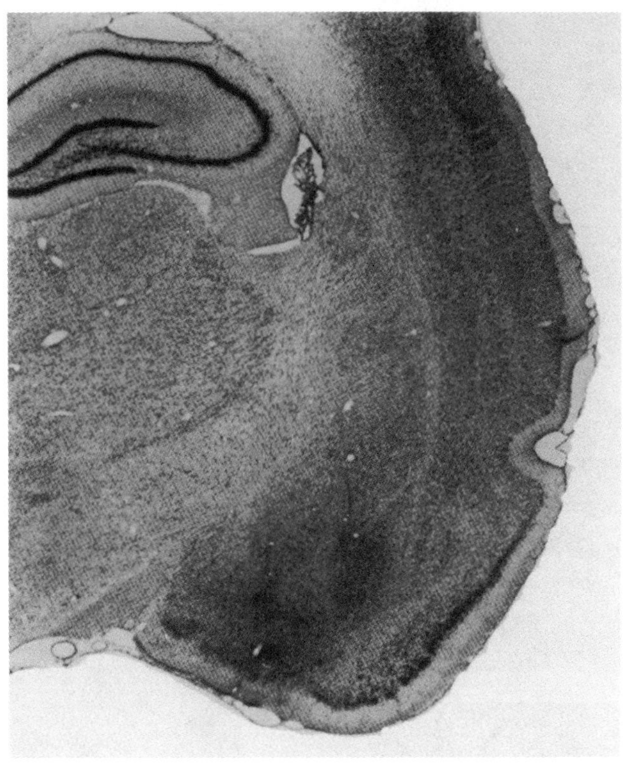

Fig. 1. Bright-field photograph of the multiple intra-amyglaloid injection of tritiated leucine and tritiated proline in case RR 3. The resulting fiber labelling is represented photographically in Fig. 2.

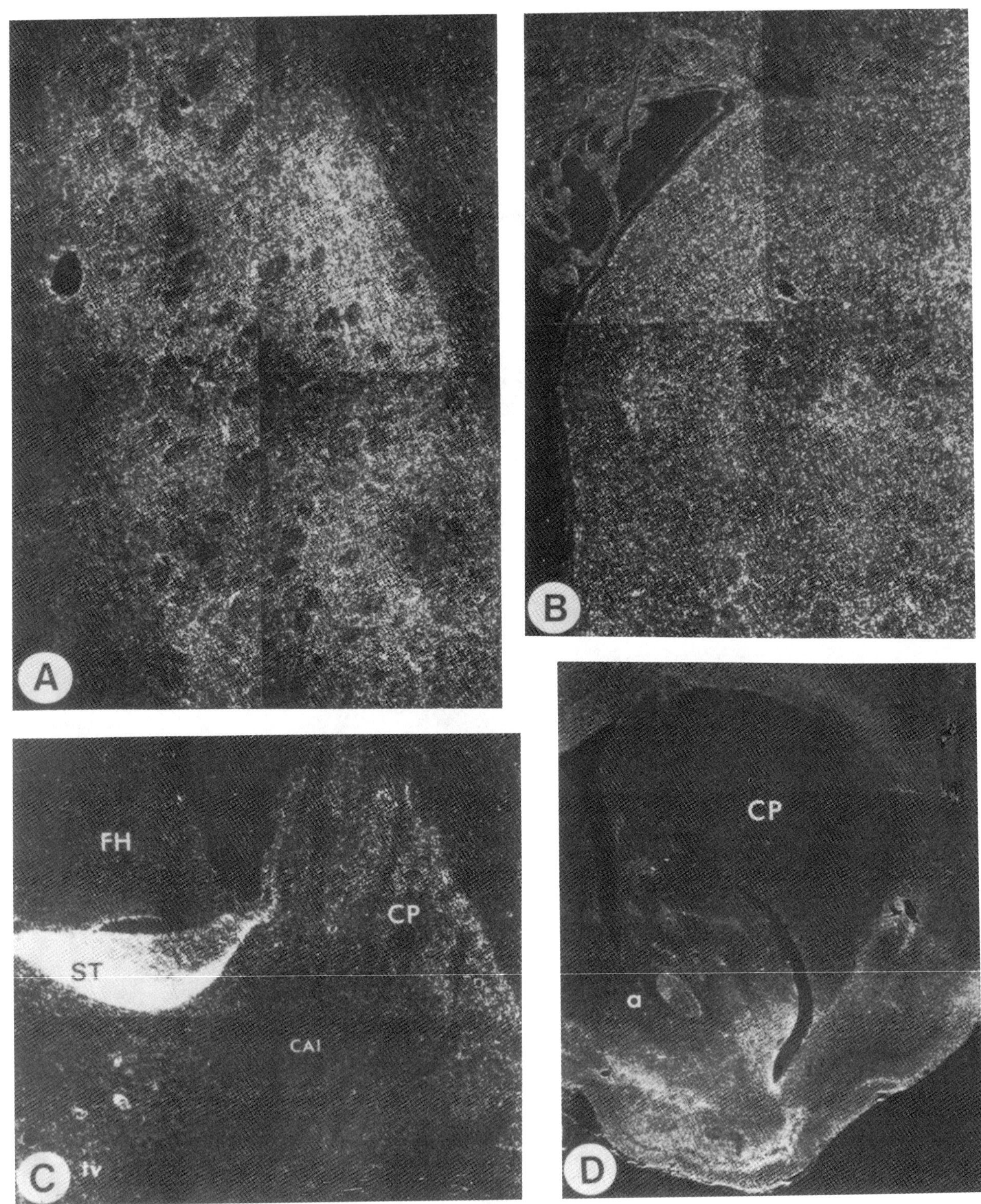


Fig. 2.

From the level of Fig. 3D forward, the density of the intrastriatal fiber labelling markedly diminishes from the laterodorsal border of the striatum inward. As a result, the region of dense labelling becomes progressively more confined to the ventromedial half of the striatum (e.g. Fig. 3E) and is finally reduced largely to the region of the nucleus accumbens (Fig. 3E–F). It is noteworthy that the labelling of the accumbens region is patchy despite the sparseness of internal-capsule bundles in this striatal subdivision, and furthermore, that it almost entirely avoids the most medial portion of the accumbens that extends dorsally into the septum (Fig. 3E).

Further noteworthy is the considerable number of labelled fibers crossing in the anterior commissure to the contralateral striatum (Fig. 3C–D). Fibers of both the longitudinal association bundle and the stria terminalis appear to contribute to this decussation, but it seems likely in this case that the former account for most of the fiber labelling in the contralateral striatum.

Finally, it must be noted that the ipsilateral globus pallidus exhibits some fiber labelling. Especially at caudal levels (Figs 3A–B) labelled fibers can be seen to curve dorsalward into the globus pallidus from the same substriatal fiber system that also conveys the larger part of the amygdalostriatal projection. From its rather diffuse pattern it is difficult to determine whether the intrapallidal labelling represents fibers merely traversing the globus pallidus in passage to the striatum, or an actual amygdalopallidal projection. However, the presence of a small number of labelled fibers also in the contralateral pallidum, in particular its ventral (subcommissural) portion (Fig. 3D), seems to suggest that some amygdalofugal fibers actually terminate in the globus pallidus.

*Case RR 2.* This case is here included for the supplementary evidence it provides concerning the amygdalofugal projection to the dorsal quarter of the caudatoputamen. In the foregoing case RR 18 with an isotope injection involving the lateral part of the basolateral nucleus this striatal region contained only relatively sparse labelled fibers, raising the question from which part of the amygdala its massive amygdaloid input indicated by the findings in case RR 3 might originate. Case RR 2 is one of three instances in which, in contrast to RR 18, the isotope injection involved a medial rather than lateral part of the basolateral amygdaloid nucleus. It must be emphasized that this case was complicated by a second injection massively involving the medial amygdaloid nucleus (Fig. 4). Although this additional injection had a marked effect upon the fiber labelling in several forebrain structures, in particular the stria terminalis, its bed nucleus and the hypothalamus, findings to be reported below suggest that it did not significantly contribute to the intrastriatal fiber labelling.

As illustrated by Fig. 4, the intrastriatal fiber labelling in RR 2 is to some extent the converse of that found in RR 18 (Fig. 3), in the sense that it is here the dorsal rather than the ventral third of the striatum which is most densely labelled at, and caudal to, the level of the anterior commissure (Fig. 4, A–B). At the level shown in Fig. 4B, labelling below the dorsal third is fairly dense only in the most ventral striatal region, including the subcommissural striatal pocket, and sparse in the middle third. Farther rostrally the contrast between the two charted cases becomes less pronounced. As in RR 18 (Fig. 3) most of the nucleus-accumbens region shows substantial but uneven labelling, while its dorsomedial part nearest the septum is only very sparsely labelled (Fig. 4C–D). Further similarities between the two charted cases are found in (a) the sparseness of fiber labelling in a large dorsal and lateral striatal region at levels rostral to the anterior commissure (Fig. 4C–D), (b) the considerable and largely symmetrical labelling of the contralateral striatum, and (c) the invasion of the striatum by labelled fibers of the longitudinal association bundle. Some, at least, of the latter, ventrally entering amyg-

---

Fig. 2. Fiber labelling resulting from the multiple isotope injection shown in Fig. 1. Frames *A* and *B* show the labelling density in the dorsal striatal region at two levels (A caudal to B) behind the plane of the anterior commissure; *C* a fascicle of labelled fibers entering the dorsal striatal region from the stria terminalis; *D* a low-power view of the fiber labelling in the rostral striatum (note (1) paucity of labelling in the dorsolateral half of the striatum, and (2) pronounced clustering of label especially in the medial striatal half).

Abbreviations used in figures:

| | | | | |
|---|---|---|---|---|
| a | nucleus accumbens; | | cp | caudatoputamen (striatum); |
| abl | nucleus basalis lateralis; | | FH | fimbria fornicis; |
| abm | nucleus basalis medialis amygdalae; | | gp | globus pallidus; |
| ac | nucleus centralis amygdalae; | | nb | nucleus basalis (preopticus magno- |
| AC | commissura anterior; | | | cellularis); |
| aco | nucleus corticalis amygdalae; | | ot | olfactory tubercle; |
| al | nucleus lateralis amygdalae; | | pom | medial preoptic area; |
| am | nucleus medialis amygdalae; | | SM | stria medullaris; |
| bs | bed nucleus of stria terminalis; | | TOL | lateral olfactory tract; |
| CAI | capsula interna; | | VL | lateral ventricle; |
| cl | claustrum; | | vp | ventral pallidum. |

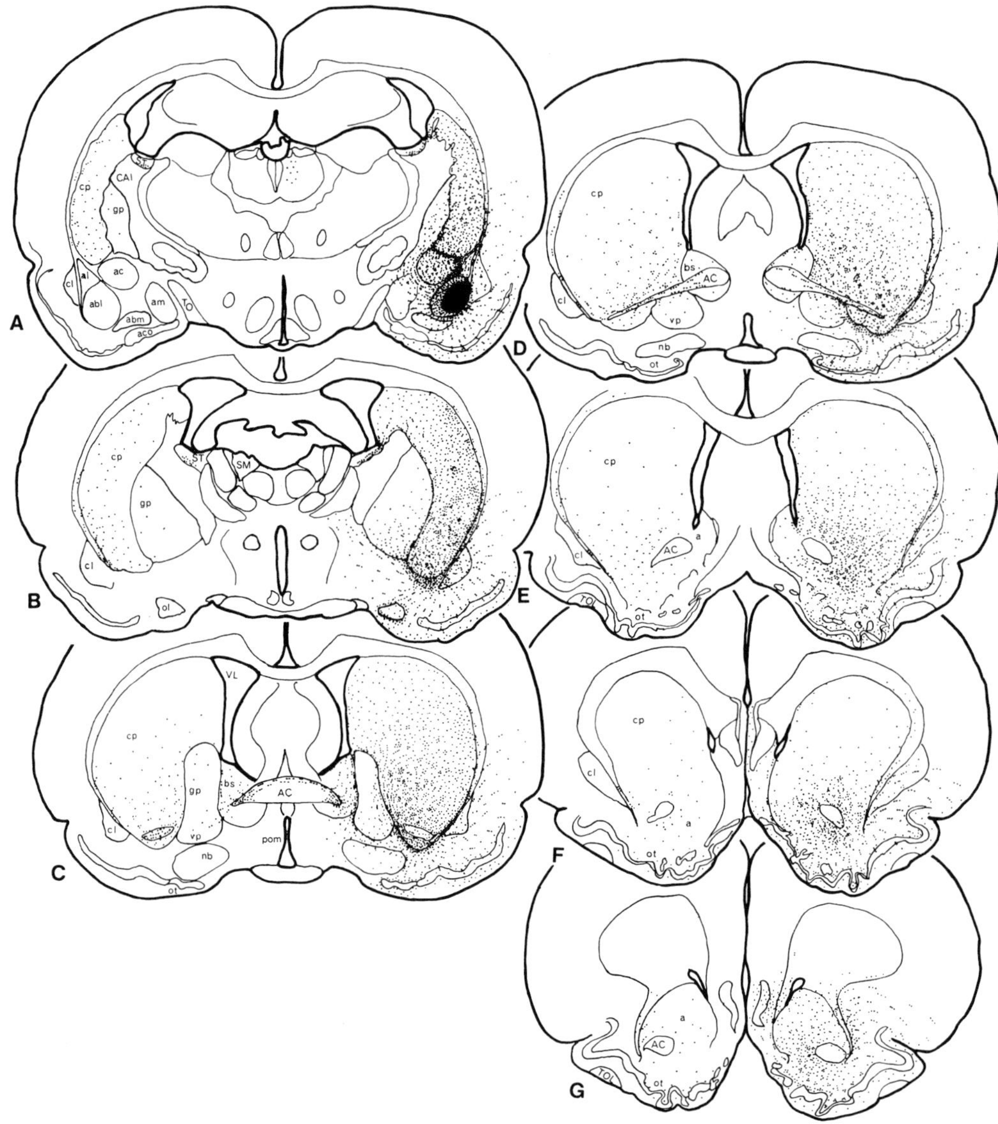

Fig. 3. Fiber labelling in case RR 18, as described in the text. Indicated in jet black in A is the region within which cell bodies showed evidence of having sequestered the labelled amino acid. For abbreviations see Fig. 2.

dalofugal fibers appear to extend dorsalward far enough to contribute to the dense fiber labelling in the dorsal third of the striatum. However, it must be emphasized that the number of labelled fibers that appear to enter the striatum from the stria terminalis is considerably greater in this case than in RR 18. It seems likely that the denser labelling of the dorsal third of the striatum is due largely to these labelled stria fibers. As in case RR 18 they appear to enter the striatum from a labelled ventrolateral stratum of the

stria terminalis that in this case is paralleled by—but largely separated from—a large and compact group of labelled fibers occupying medial and dorsal positions on the stria's cross section (Fig. 4A) and almost certainly consisting of efferents from more central and medial amygdaloid cell groups.

Patterns of intrastriatal fiber labelling much like that described in the foregoing case RR 2 were found in RR 11 and RR 19. Both these cases had in common with RR 2 that the injection site involved the medial

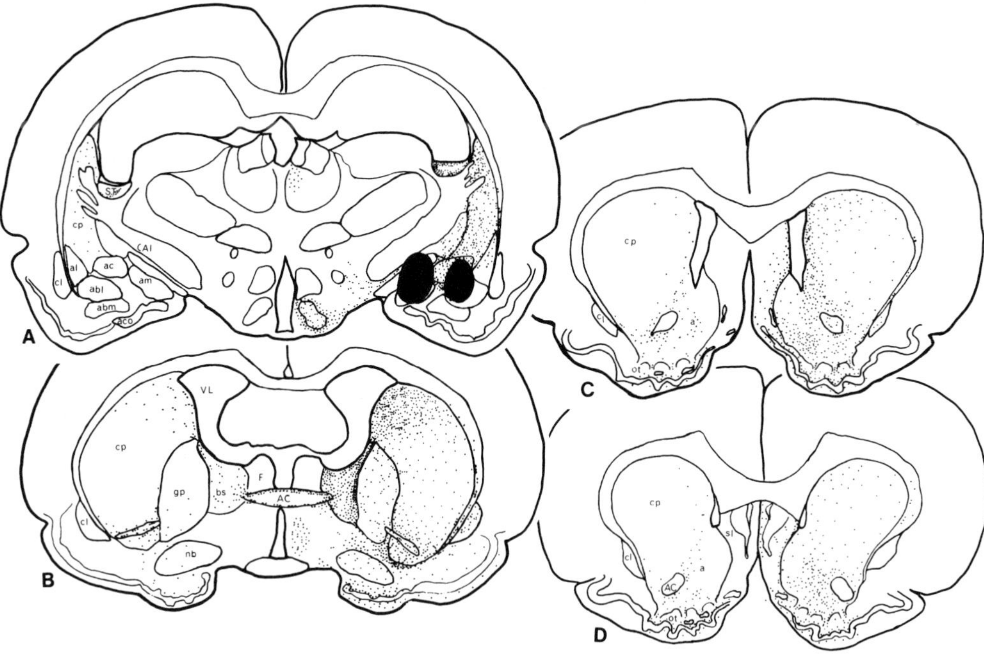

Fig. 4. Fiber labelling in case RR 2, as described in the text. Injection sites (i.e. regions within which cell bodies showed evidence of active label uptake) have been indicated in jet black in A. For abbreviations see Fig. 2.

half, but largely avoided the lateral half of the basolateral nucleus. The injection in both cases was comparable to that in RR 2 also by involving other amygdaloid cell groups: in RR 11 the medial nucleus, in RR 19 the basomedial and cortical nuclei.

All factors considered, the two cases here charted appeared to suggest the basolateral amygdaloid nucleus as the main origin of the amygdalostriatal projection. They also seemed to suggest that medial parts of the basolateral nucleus project more strongly to dorsal regions, lateral parts in particular to ventral regions of the striatum, and both parts of the nucleus about equally to the most ventral striatal zone including, at rostral levels, most of the accumbens region. However, it could not be determined from the available autoradiographic material whether other more medially situated cell groups, such as the medial and basomedial amygdaloid nuclei and the intra-amygdaloid portion of the bed nucleus of the stria terminalis, contribute to the projection. The following retrograde-labelling experiments were done in an attempt to clear up this issue.

*Experiments with horseradish peroxidase*

HRP-positive cells were found in the amygdaloid complex in nearly all cases of intrastriatal HRP injection. The number of these cells varied markedly from case to case, and the variation generally was in good accord with the density gradients in the amygdalostriatal projection demonstrated by the autoradiographic experiments. For example, labelled amygdala cells consistently were numerous in cases of injection involving the caudal striatum, or ventral regions of the rostral striatum, and sparse or even absent altogether in cases of injection localized to the dorsolateral half of the rostral striatum, the region consistently exhibiting the sparsest fiber labelling in the autoradiographic material.

By far the largest number of amygdala cells labelled by intrastriatal HRP injection were found in the basolateral nucleus. In several cases, this nucleus was the only amygdaloid cell group to contain HRP-positive cells, but in other cases, in particular such of caudodorsal (e.g. case RH 14, Fig. 5) or rostromedial injection, a significant number of labelled cells appeared also in the basomedial nucleus. Moreover, in several cases (e.g. RH 14 and 18, Fig. 5), a few scattered labelled cells were found in the lateral nucleus. In no instance were any detected in the medial, central, or cortical nuclei, the intercalated cell mass, the bed nucleus of the stria terminalis, or the anterior amygdaloid area. With some reservation, imposed by the lingering possibility of false-negative results, these findings can be interpreted as evidence that the amygalostriatal projection originates largely, though not entirely, from the basolateral nucleus.

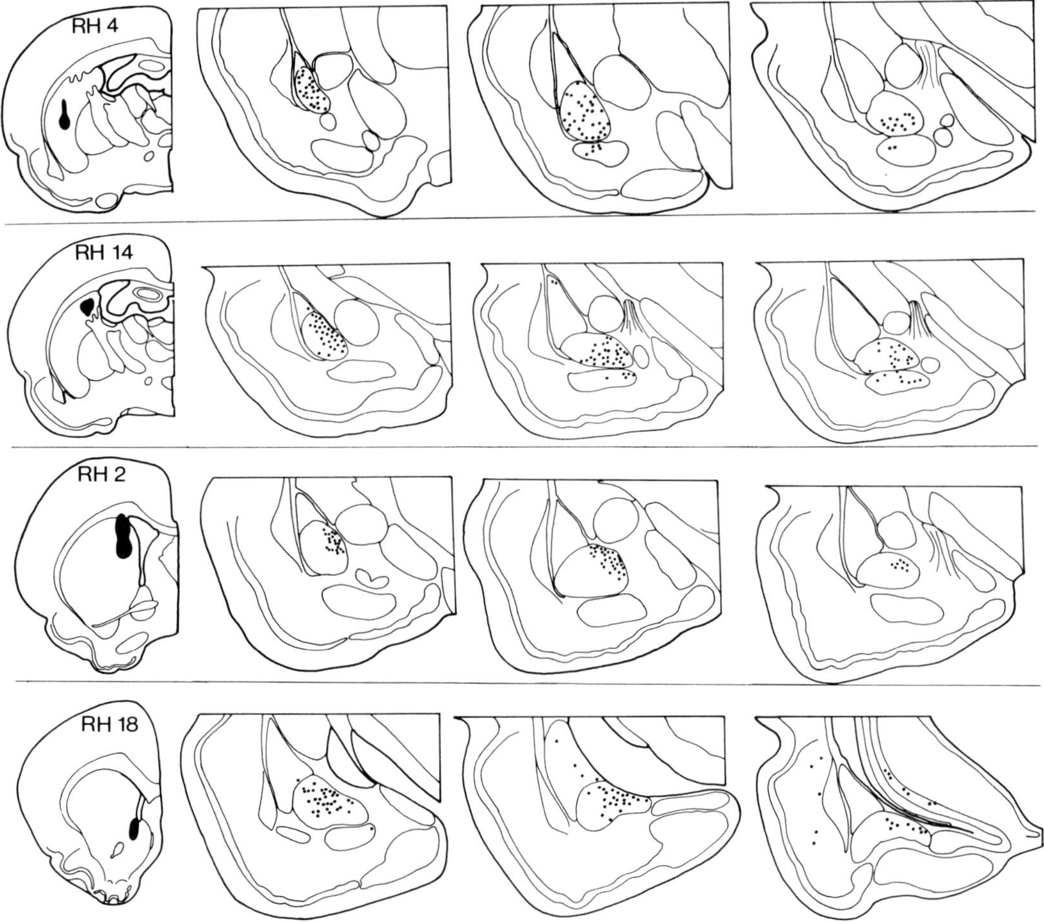

Fig. 5. Four cases of intrastriatal horseradish peroxidase injection. In each case the three amygdala sections shown have been arranged sequentially from rostral (left) to caudal. Each dot represents a single labelled cell. Note: (1) cell labelling mainly in the nucleus basalis lateralis in all cases, with additional labelled cells in the nucleus basalis medialis in RH 4 and RH 14 and in the nucleus lateralis in RH 14 and RH 18; (2) labelled cells located predominantly in medial parts of nucleus basalis lateralis in RH 14 and RH 2 with a dorsal striatal injection; (3) cell labelling in the nucleus basalis lateralis pars caudalis only in RH 18 with an injection of the nucleus accumbens.[27] In the same drawing, note additional cell labelling in the entorhinal area and subiculum hippocampi.[15,27]

Compatible with such an interpretation are the autoradiographic findings which, as emphasized earlier in this paper, suggested that the amygdalo-striatal projection is topographically organized so that the dorsal-to-ventral coordinate of the caudal half of the striatum, at least, corresponds to a medial-to-lateral coordinate of the basolateral amygdaloid nucleus. In remarkably close agreement with this auto-radiographic suggestion, the basolateral-nucleus cells found labelled by HRP injected into the dorsal quarter of the striatum were confined almost entirely to the medial half of the nucleus (RH 2 and 14, Fig. 5), whereas by contrast an injection placed at middle depth of the caudal striatum labelled about equal numbers of cells in the lateral and medial half of the nucleus (RH 4, Fig. 5). In view of the autoradio-graphic evidence that many amygdalostriatal fibers enter the striatum from the ventral side as components of a substriatal fiber system, it would seem likely that an HRP injection at middle depth of the striatum would affect not only those amygdalofugal fibers terminating at and about the injection site but also those passing beyond it to more dorsal striatal regions. An injection localized to the dorsal quarter of the striatum (RH 14, Fig. 5), by contrast, would pre-sumably label only fibers distributed to that region;

the autoradiographic evidence suggests that such fibers are conveyed not only by the substriatal bundle, but also, and especially, by the stria terminalis, and according to the present HRP findings they originate mainly in the medial part of the basolateral nucleus.

Several further observations concerning the origin and topography of the amygdalostriatal projection are noteworthy. First, it is remarkable that in all but two cases cell labelling in the basolateral amygdaloid nucleus was confined to the rostral two-thirds or so of the nucleus. As a rule, few if any labelled cells were found in the nucleus caudal to the frontal plane in which the stria terminalis first appears as a compact fiber bundle at the medial margin of the amygdala (a level approximately corresponding to Krettek & Price's[20] Fig. 9). The two exceptions were the two cases of accumbens injection, in both of which cell labelling extended caudalward in the basolateral nucleus well beyond the level of the temporal tip of the lateral ventricle (RH 18, Fig. 5). In both these cases, the rostral pole of the basolateral nucleus was free of labelled cells, as it also was in other cases of HRP injection in the most rostral parts of the striatum. These findings suggest a weak reversed-sagittal gradient in the projection of the basolateral nucleus to the striatum, in the sense that the cell population projecting to the rostral one-third or so of the striatum extends farther caudalward in the nucleus than does the population projecting to more caudal striatal regions. The latter population, by contrast, extends farther rostralward so as to include the rostral pole of the basolateral nucleus. The present material includes no case in which cell labelling exclusively involved either the rostral pole or the caudal one-third of the nucleus.

Second, the present material yielded no clear topographic picture of the apparently much smaller part of the amygdalostriatal projection that arises from the basomedial and lateral amygdaloid nuclei. The most numerous basomedial-nucleus neurons were found labelled in a case of accumbens injection, but HRP deposits in dorsal parts of the caudal striatum likewise labelled a considerable number of such cells (RH 14, Fig. 5). In most of the remaining cases only sporadic HRP-positive cells, if any, appeared in the basomedial nucleus.

As to the striatal afferents from the lateral amygdaloid nucleus, the present material failed to suggest any particular topographic pattern in this apparently quite minor amygdalofugal projection.

In most cases in which the intrastriatal HRP deposit had labelled cells in the ipsilateral nucleus basalis lateralis, a smaller number of cells were found labelled in the same nucleus on the contralateral side (Fig. 6). This finding accords well with the autoradiographic evidence that the amygdalostriatal projection has a substantial contralateral component. The distribution of the cell labelling within the contralateral basolateral nucleus generally corresponded closely to that of the ipsilateral labelling. Very rarely were labelled cells found in the contralateral nuclei lateralis and basomedialis; since cell labelling in these nuclei in most cases was sparse even on the ipsilateral side, not much significance can be attached to this finding.

## DISCUSSION

Experimental evidence of any amygdalostriatal projection appears to have accumulated only during the past 20 years. In an early fiber-degeneration study in the monkey amygdalofugal fibers could be traced to a ventrolateral region of the putamen.[25] Later studies in the rat by improved silver-degeneration methods[5,6] demonstrated an amygdalofugal projection to the ventromedial striatal region designated nucleus accumbens. In a subsequent autoradiographic study in the rat and cat Krettek & Price[19] likewise reported a substantial amygdalofugal projection to the nucleus accumbens and olfactory tubercle, but they noted that the projection also involves the most ventral part of the striatum caudal to the nucleus accumbens.

Considered together, these anatomical findings seemed to indicate that the distribution of amygdalostriatal fibers is confined to the most ventral region of the striatum. The first observation suggesting a wider spread of the projection appears to have been made in an electrophysiological study in the rat by Dafny, Dauth & Gilman[4] in the form of short-latency responses of caudate-nucleus units to electrical stimulation of the amygdala. A few years later the notion of an amygdalofugal projection involving more dorsal regions of the striatum received anatomical support from Royce's[31] observation of retrograde cell labelling in the basolateral amygdaloid nucleus of the cat following injection of HRP in the caudate nucleus. From a similar but more extensive study of striatal afferents in the rat, Veening, Cornelissen & Lieven[34] likewise reported evidence of a projection from the basolateral amygdaloid nucleus to a striatal region well outside the nucleus accumbens. Veening et al.,[34] however, suggested that this projection is of small size and involves only a centrodorsal part of the caudato-putamen.

The findings in the present autoradiographic and retrograde-labelling material suggest that the amygdalostriatal projection is considerably more widespread than could have been inferred from previous observations. Its distribution involves all of the caudal striatum, i.e. that part of the striatum extending caudal to the level of the anterior commissure, alongside the globus pallidus. From the level of the anterior commissure forward, the amygdalofugal projection to the dorsolateral half of the striatum rapidly thins out, while that to the ventromedial half remains dense. The conclusion that the projection avoids (or, at least, only very sparsely involves) the antero-dorsolateral quarter of the striatum is based on the sparseness of autoradiographic fiber labelling in this striatal region even in the case of massive, multiple-focus isotope injection of the amygdala (RR 3, Figs 1–2), as well as on the observation that HRP deposited in the region labels few if any cells in the amygdala even though vividly labelling numerous cells of the substantia nigra.

*Comparison with distribution of other striatal afferents*

It is of interest to compare the distribution of the amygdalostriatal projection with that of two other known afferent striatal connections, namely, those originating from the ventral tegmental area and prefrontal cortex, respectively. As to the striatal afferents from the ventral tegmental area, it is remarkable how widely they overlap the amygdalostriatal projection, especially in the rostral half of the striatum where both projections are largely restricted to the ventromedial half of the striatum.[3,11] Caudal to the level of the anterior commissure the overlap appears to be less extensive: whereas at these levels, the amygdalostriatal projection involves the entire striatal cross-section the projection from the ventral tegmental area appears largely restricted to the most ventral part of the striatum and a contiguous strip along the medial border of the striatum.[3]

A second system of striatal afferents strongly resembling the amygdalofugal projection in its distribution to the rostral striatum arises from the anteromedial or frontocingulate cortex (areas 32, 25 and 24) as well as from the cortex forming the upper bank of the rostral one-third or so of the rhinal sulcus.[2] Both of these cortical regions are projected upon by the mediodorsal nucleus of the thalamus, and thus, despite their agranular nature, can be interpreted as homologues of the prefrontal or granular frontal cortex of primates.[22] According to the chartings of Beckstead,[2] the sulcal cortex as well as the pregenual areas 32 and 25 project to the entire ventromedial half of the anterior striatum. This also seems true of the pregenual portion of area 24, but the projection from the caudal, supracallosal part of area 24 appears to be restricted to a medial zone of the striatum alongside the lateral ventricle, and to avoid the nucleus accumbens and olfactory tubercle.

Beckstead[2] emphasized the nearly completely overlapping distribution of prefrontal-cortex and ventral-tegmental efferents in the antero-ventromedial quadrant of the striatum, and the present study indicates the amygdala as the source of a third major afferent system selectively distributed within the same striatal quadrant.

It is noteworthy that the striatal afferents from both the amygdala and ventral tegmental area originate from within the circuitry of the limbic system, and that a third limbic projection, originating from the hippocampal formation and conveyed by the fornix, likewise is distributed in the antero-ventromedial striatal quadrant, although its distribution appears to be limited more strictly to a medial part of the accumbens region.[8,16] All of these afferents almost entirely avoid a large dorsolateral sector of the rostral striatum. It must be emphasized that no region similarly avoided by all known limbic afferents appears to exist in the caudal striatum. Even though in this large striatal expanse alongside the globus pallidus the ventral tegmental projection remains confined to a ventromedial distribution,[3] and the prefrontal projection involves no more than small dorsal and ventral regions,[2] the amygdalostriatal projection extends over the entire striatal cross-section. The topographic relationships discussed in the foregoing account are illustrated in Fig. 7.

For the sake of completeness, the striatal afferents from the mesencephalic raphe nuclei should be mentioned in this discussion. Since both the median and the dorsal raphe nucleus are heavily afferented by the medial forebrain bundle and fasciculus retroflexus, the raphe projection to the striatum could be conceived as originating, much like the ventral-tegmental projection, from within the circuitry of the limbic system. Unfortunately, there appears to be little consensus regarding its intrastriatal distribution. In a microassay study, Terneaux, Héry, Bourgoin, Adrien, Glowinski & Hamon[33] found serotonin levels highest in ventrocaudal parts of the striatum and lowest in the dorsorostral area. However, from an autoradiographic study Moore, Halaris & Jones[24] reported a quite uniform distribution of mesencephalic-raphe efferents throughout the striatum. The autoradiographic chartings of Azmitia & Segal,[1] by contrast, appear to suggest a distribution of raphe efferents exclusively to the caudal striatum, the nucleus accumbens and the lateral half of the rostral striatum. In turn, both of the latter two reports conflict, in part at least, with the results of a study by Parent, Descarries & Beaudet[28] involving intraventricular injection of tritiated serotonin, according to which serotonin fibers are distributed to all of the caudal striatum, the nucleus accumbens and the medial half of the rostral striatum, but avoid the latter's lateral half in a distribution somewhat comparable to that of the amygdalostriatal projection. In an extensive study by the aid of serotonin-immunohistochemistry Steinbusch[32] found serotonin fibers in the rat's rostral striatum most numerous in the nucleus accumbens-olfactory tubercle complex and in adjoining ventrolateral regions of the striatum; his charts indicate a distinctly low concentration of serotonin fibers in the anterodorsolateral striatal quadrant. All considered, Steinbusch's[32] report appears to suggest that intrastriatal serotonin fibers are distributed mainly within the region also receiving projections from the ventral tegmental area, prefrontal cortex, and amygdala.

*Subdivision of the striatum into 'limbic' and 'non-limbic' compartments*

The present findings clearly indicate that the part of the striatum receiving afferents of limbic origin is considerably larger than was assumed thus far. Up to the present, only the region composed of the nucleus accumbens and olfactory tubercle has been generally regarded as a limbic compartment of the striatum,

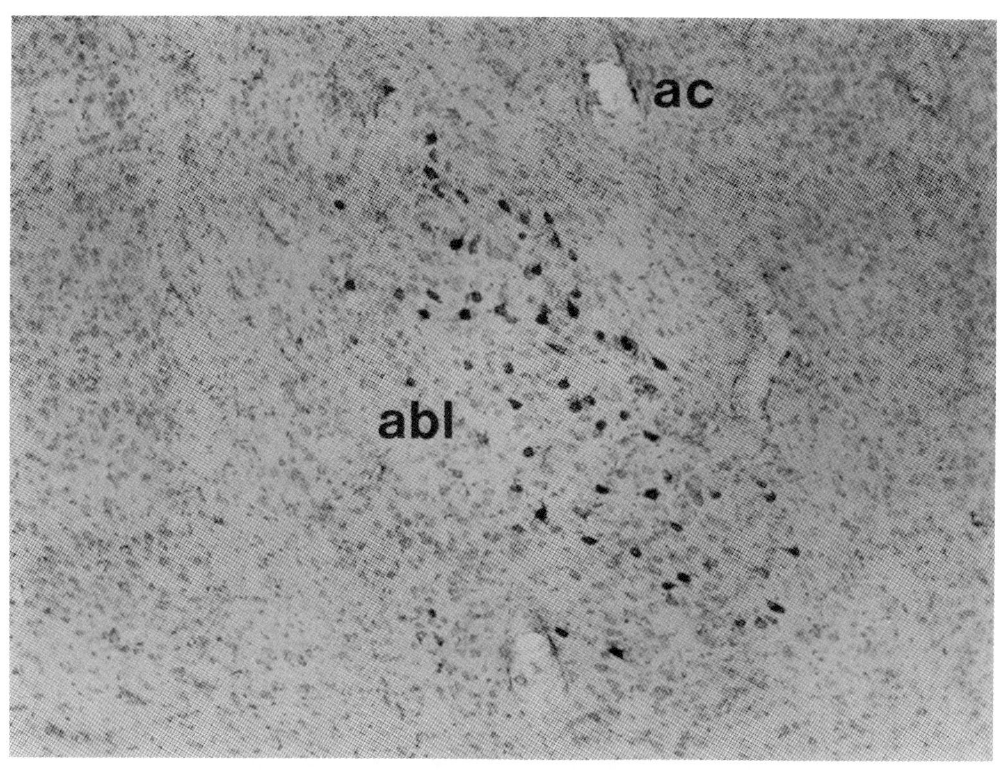

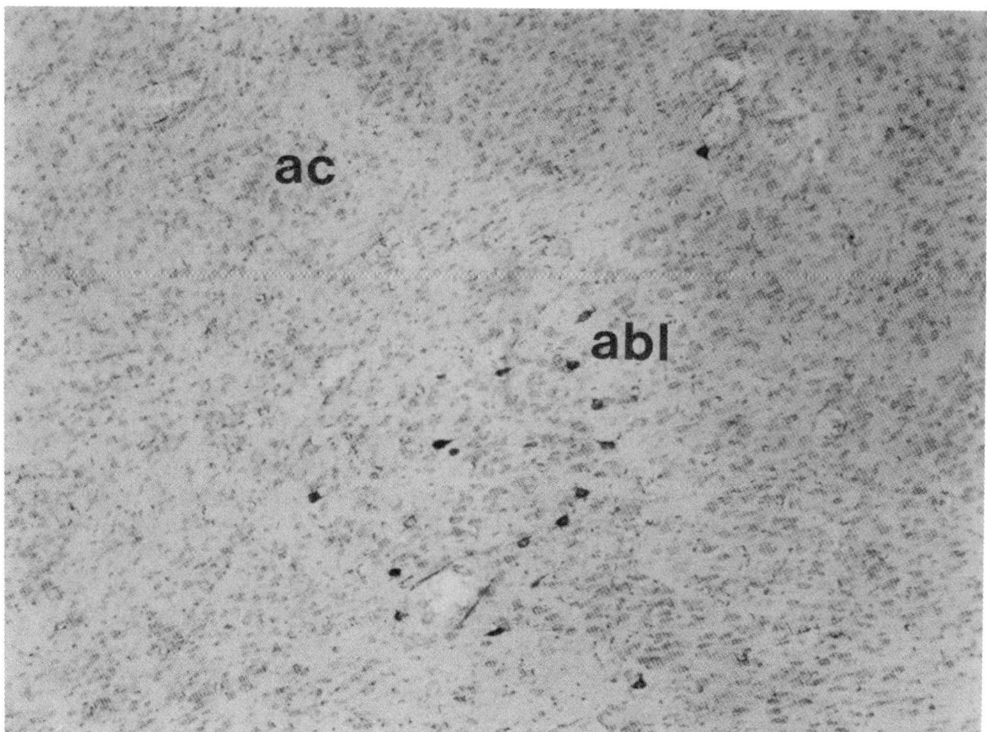

Fig. 6. Retrograde cell labelling in the ipsilateral (above) and contralateral nucleus basolateralis amygdalae in case RH 14 (cf. Fig. 5) tetramethylbenzidine procedure; counterstaining with neutral red.

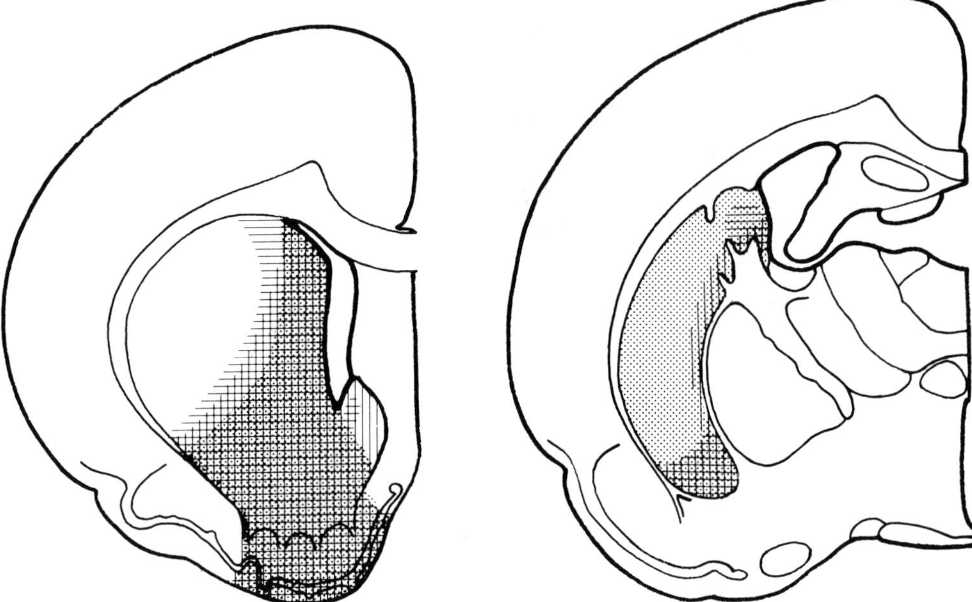

Fig. 7. Semidiagrammatic representation of three projections to the striatum: from the ventral tegmental area (vertical lines), the prefrontal cortex (horizontal lines), and the amygdala (stipple pattern). The projection from the ventral tegmental area has been adapted from Beckstead et al.,[3] that from the prefrontal cortex from Beckstead.[2] No attempt has been made in this Figure to indicate irregularities ('clustering') in the distribution patterns, prominent especially in the prefrontal and amygdalar projections.

receiving dopamine fibers from the ventral tegmental area rather than substantia nigra in the strict sense, and furthermore, projected upon directly by the hippocampal formation and amygdala. Already some years ago, however, this notion was drawn in question when it was found that the ventral tegmental projection is distributed well beyond the accumbens region, and in fact involves the entire ventromedial half of the rostral striatum as well as a somewhat smaller proportion of the caudal striatum.[3] The present data indicate an even wider distribution of limbic afferents, and appear to warrant the statement that the basolateral complex of the amygdala, in partial overlap with other limbic afferents, voluminously innervates all of the striatum except the antero-dorsolateral quadrant, which receives only very few amygdalar afferents.

The apparent subdivision of the striatum into a large limbic system-afferented and a smaller non-limbic compartment might well become more meaningful from a functional point of view if it could be correlated with the corticostriatal projection pattern. Unfortunately, current knowledge of corticostriatal topography in the rat is too fragmentary for such correlation. Nonetheless, some statements can here be made with respect to the corticostriatal projection originating from the sensorimotor cortex.

Figure 8 shows the intrastriatal fiber labelling resulting from an injection of tritiated leucine in the rostral half of the sensorimotor cortex. At the frontal level of Fig. 8B, the injection involves most of the medial, agranular strip of the sensorimotor field. The case shown is one of a series of 6 in each of which a single such injection was placed in this agranular region at a site varying between 2–6 mm behind the frontal pole. Despite the considerable longitudinal variation, a remarkably similar pattern of corticostriatal fiber labelling appeared in all 6 cases. Fiber labelling invariably was densest and most widespread in the dorsolateral region of the rostral striatum (A–B) and extended from there caudalward in tail-like fashion almost throughout the length of the caudal striatum (C–E). It is interesting to note that the projection was strongly bilateral in all cases, as previously reported for the striatal projection from the motor cortex in the monkey.[21]

It appears from these observations that the corticostriatal projection from the medial, agranular portion of the sensorimotor cortex, at least, involves mainly the 'non-limbic' antero-dorsolateral sector of the striatum, and in lesser volume the caudal striatum in which it must overlap at least the amygdalostriatal projection. It is not possible at present to state which other corticostriatal projections, if any, share this bimodal distribution pattern. Nor can much be said

                    A. E. Kelley, V. B. Domesick and W. J. H. Nauta

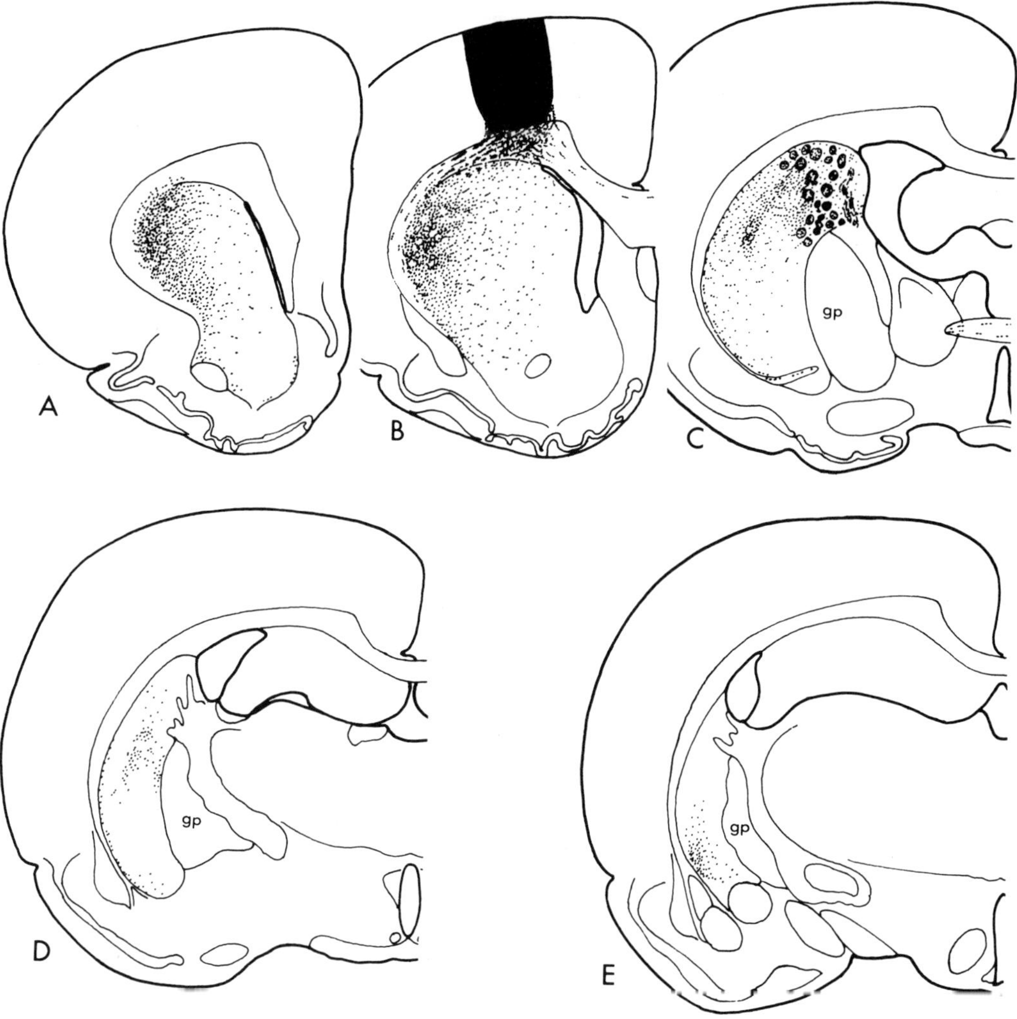

Fig. 8. Ipsilateral intrastriatal fiber labelling following a tritiated-leucine and proline injection localized to the medial, agranular portion of the sensorimotor cortex at approximately 4 mm behind the frontal pole. Fiber labelling in internal capsule and thalamus has been deleted from these chartings. Compare with Fig. 7.

about the manner in which limbic and neocortical projections overlap in the striatum. Do both classes of afferents synapse with the same, or with different populations of striatal neurons? These questions obviously require much further study, but even in the absence of answers it seems certain that the presence or absence of limbic afferents in a particular region of the striatum affects the repertoire of mechanisms modulating its function.

A subdivision of the striatum into a limbic and a non-limbic compartment as here suggested is of course extremely elementary and leaves ample room for further parcellation. For example, even when con-sidering exclusively its limbic afferents the large limbic compartment can hardly be considered homoge-neously innervated (cf. Figs 2D; 7), and the mosaic of its afferentation is undoubtedly further complicated by cortical and thalamic projections overlapping and interdigitating with the limbic inputs. The non-limbic compartment receives no or only sparse fibers from the amygdala, hippocampal formation, cingulate cor-tex, ventral tegmental area and mesencephalic sero-tonin cell groups, but it may nonetheless prove to be intricately subdivided by its cortical, thalamic and nigral afferents. A parcellation of the striatum into anatomically defined subregions may well be the basis of the regional differentiation of striatal function suggested by behavioral findings.[7,9,10,26,30] It must be emphasized, however, that any attempt to designate and delineate such subdivisions would be premature in the absence of a more complete chart of corticostriatal and thalamostriatal projections.

*Possible functional implications*

Until now, the functions of striatum and amygdala have been treated separately in the literature. The present evidence permits some speculation about a functional linkage of the two structures. That the amygdala plays a role in the neural mechanisms underlying motivation and adaptive behavior seems well documented (see, for example, Goddard;[13] Gloor;[12] Jones & Mishkin;[17] Richardson[29]), but the link between the amygdala and a motor mechanism necessary for behavioral expression has remained obscure. Since the striatum appears to be involved in the initiation and patterning of somatic motor behavior, it would seem possible that the amygdalostriatal projection provides an access route of the amygdala to the somatic motor system. In view of the large volume and wide intrastriatal distribution of the connection it is tempting to speculate that the functional mode of a large part of the striatum may be decisively influenced by the animal's affects and motivational sets.

*Acknowledgements*—This study was made possible by NIMH Grant 5 P01 MH 31154 and NSF Grant BNS 80-07905. The technical assistance of Mr P. Paskevich is gratefully acknowledged.


## REFERENCES

1. Azmitia E. C. & Segal M. (1978) An autoradiographic analysis of the differential ascending projections of the dorsal and median raphe nuclei in the rat. *J. comp. Neurol.* **179**, 641–668.
2. Beckstead R. M. (1979) An autoradiographic examination of corticocortical and subcortical projections of the mediodorsal-projection (prefrontal) cortex in the rat. *J. comp. Neurol.* **184**, 43–62.
3. Beckstead R. M., Domesick V. B. & Nauta W. J. H. (1979) Efferent connections of the substantia nigra and ventral tegmental area in the rat. *Brain Res.* **175**, 191–217.
4. Dafny N., Dauth G. & Gilman S. (1975) A direct input from amygdaloid complex to caudate nucleus of the rat. *Expl Brain Res.* **23**, 203–210.
5. De Olmos J. (1972) The amygdaloid projection field in the rat as studied with the cupric-silver method. In *The Neurobiology of the Amygdala* (ed. Eleftheriou B. E.) pp. 145–204. Plenum Press, New York.
6. De Olmos J. & Ingram W. R. (1972) The projection field of the stria terminalis in the rat brain. *J. comp. Neurol.* **146**, 303–334.
7. Divac I. & Oberg R. G. E. (1979) Current conceptions of neostriatal functions. In *The Neostriatum* (eds Divac I. & Oberg R. G. E.) pp. 215–230. Pergamon Press, Oxford.
8. Domesick V. B. (1981) Similarities and contrasts in the connections of the nucleus accumbens and caudatoputamen. In *The Nucleus Accumbens* (eds de France J. & Chronister R.). HAER Institute Press, Rockland, Maine.
9. Dunnett S. B. & Iversen S. D. (1980) Regulatory impairments following selective kainic acid lesions of the neostriatum. *Behav. Brain Res.* **1**, 497–506.
10. Dunnett S. B. & Iversen S. D. (1981) Learning impairments following selective kainic acid-induced lesions within the neostriatum of rats. *Behav. Brain Res.* **2**, 189–209.
11. Fallon J. H. & Moore R. Y. (1978) Catecholamine innervation of the basal forebrain—IV. Topography of the dopamine projection to the basal forebrain and neostriatum. *J. comp. Neurol.* **180**, 545–580.
12. Gloor P. (1972) Inputs and outputs of the amygdala: what the amygdala is trying to tell the rest of the brain. In *The Neurobiology of the Amygdala* (ed. Eleftheriou B. E.) pp. 189–209. Plenum Press, New York.
13. Goddard G. V. (1964) Functions of the amygdala. *Psychol. Bull.* **62**, 89–109.
14. Graybiel A. M. & Devor M. (1974) A microelectrophoretic delivery technique for use with horseradish peroxidase. *Brain Res.* **68**, 167–173.
15. Groenewegen H. J., Arnolds D. E. A. T. & Lopes da Silva F. H. (1981) Afferent connections of the nucleus accumbens in the cat, with special emphasis on the projections from the hippocampal region—an anatomical and electrophysiological study. In *The Nucleus Accumbens* (eds de France J. & Chronister R.). HAER Institute Press, Rockland, Maine.
16. Heimer L. & Wilson R. (1975) The subcortical projections of the allocortex: similarities in the neural associations of the hippocampus, the piriform cortex, and the neocortex. In *Golgi Centennial Symposium* (ed. Santini M.) pp. 177–193. Raven Press, New York.
17. Jones B. & Mishkin M. (1972) Limbic lesions and the problem of stimulus-reinforcement associations. *Expl Neurol.* **36**, 362–377.
18. Kelley A. E., Domesick V. B. & Nauta W. J. H. (1981) The amygdalostriatal projection in the rat. *Anat. Rec.* **199, No. 3**, 134–135.
19. Krettek J. E. & Price J. L. (1978a) Amygdaloid projections to subcortical structures within the basal forebrain and brainstem in the rat and cat. *J. comp. Neurol.* **178**, 225–254.
20. Krettek J. E. & Price J. L. (1978b) A description of the amygdaloid complex in the rat and cat with observations on intra-amygdaloid axonal connections. *J. comp. Neurol.* **178**, 255–280.
21. Künzle H. (1975) Bilateral projections from precentral motor cortex to the putamen and other parts of the basal ganglia. An autoradiographic study in *Macaca fascicularis*. *Brain Res.* **88**, 195–209.

22. Leonard C. M. (1969) The prefrontal cortex of the rat—I. Cortical projection of the mediodorsal nucleus—II. Efferent connections. *Brain Res.* **12,** 321–346.
23. Mesulam M. M. (1978) Tetramethylbenzidine for horseradish peroxidase neurohistochemistry: a non carcinogenic blue reaction product with superior sensitivity for visualizing neural afferents and efferents. *J. Histochem. Cytochem.* **26,** 106–117.
24. Moore R. Y., Halaris A. E. & Jones B. E. (1978) Serotonin neurons of the midbrain raphe: ascending projections. *J. comp. Neurol.* **180,** 417–438.
25. Nauta W. J. H. (1961) Fibre degeneration following lesions of the amygdaloid complex in the monkey. *J. Anat., Lond.* **95,** 515–531.
26. Neill D. B. & Herndon J. G. (1978) Anatomical specificity within rat striatum for the dopaminergic modulation of DRL responding and activity. *Brain Res.* **153,** 529–538.
27. Newman R. & Winans S. S. (1980) An experimental study of the ventral striatum in the golden hamster—I. Neuronal connections of the nucleus accumbens. *J. comp. Neurol.* **191,** 167–198.
28. Parent A., Descarries L. & Beaudet A. (1981) Organization of ascending serotonin systems in the adult rat brain. A radioautographic study after intraventricular administration of [$^3$H]5-hydroxytryptamine. *Neuroscience* **6,** 115–138.
29. Richardson J. S. (1973) The amygdala: historical and functional analysis. *Acta Neurobiol. Exp.* **33,** 623–648.
30. Rosvold H. E. (1972) The frontal lobe system: cortical-subcortical interrelationships. *Acta Neurobiol. Exp.* **32,** 439–460.
31. Royce G. J. (1978) Cells of origin of subcortical afferents to the caudate nucleus: A horseradish peroxidase study in the cat. *Brain Res.* **153,** 465–475.
32. Steinbusch H. W. M. (1981) Distribution of serotonin-immunoreactivity in the central nervous system of the rat—cell bodies and terminals. *Neuroscience* **6,** 557–618.
33. Terneaux J. P., Héry F., Bourgoin S., Adrien J., Glowinski J. & Hamon M. (1977) The topographical distribution of serotoninergic terminals in the neostriatum of the rat and the caudate nucleus of the cat. *Brain Res.* **121,** 311–326.
34. Veening J. G., Cornelissen F. M. & Lieven P. A. J. M. (1980) The topical organization of the afferents to the caudatoputamen of the rat. A horseradish peroxidase study. *Neuroscience* **5,** 1253–1268.

(*Accepted* 10 *October* 1981)

# VI

# CEREBRAL CORTEX

Reprinted from
*Brain Research*
Elsevier Publishing Company, Amsterdam
Printed in the Netherlands

# Research Reports

---

# A NOTE ON THE TERMINATION OF COMMISSURAL FIBERS IN THE NEOCORTEX

LENNART HEIMER, FORD F. EBNER* AND WALLE J. H. NAUTA

*Department of Psychology, Massachusetts Institute of Technology, Cambridge, Mass., and (F.F.E.)
Division of Biological and Medical Sciences, Brown University, Providence, R.I. (U.S.A.)*

(Received December 27th, 1966)

The general arrangement of fibers in the corpus callosum shows a strongly symmetrical (homotopic) pattern characteristic of a true commissure. Connections between non-symmetrical (heterotopic) areas also have been demonstrated for many regions, both by anatomical studies using the Marchi technique[13,14,16,24,26,29] and by electrophysiological methods[6,18]. The contribution of recently developed silver techniques, notably the suppressive Nauta-Gygax method[23], has been to emphasize the extent of non-homotopic connections[3,8,9,19,21]. Some clearly delineated areas of neocortex appear not to receive any commissural fibers[1,2,6,8,9,17,18,21,28]; outstanding examples of such areas are 'visual cortex' or area 17, and the hand and foot portions of somatosensory cortex. It is generally presumed that each cortical area contributes fibers to the corpus callosum, and the sparse histological evidence available indicates that the cells of origin reside in cortical layers III through VI[5,12,25].

Although from these studies a fairly clear picture of the general topography of the commissural fibers has emerged, details of their mode of termination have been more difficult to establish. Golgi studies have indicated that the commissural fibers terminate in all layers[5,15,27], and some electrophysiological studies also seem to indicate both a superficial and a deep distribution of the callosal fibers[7,11]. Villaverde[30–32] who used a reduced silver method in addition to the Golgi method, made the interesting observation that commissural fibers have a laminar distribution rather than being distributed diffusely throughout the layers. He furthermore noted local differences in the laminar pattern; in the area precentralis granularis, for instance, the commissural fibers appeared to end particularly in layers V and I, whereas in the area precentralis agranularis callosal fibers appeared to terminate largely in layer I. Villaverde repeatedly emphasized, however, the difficulty of obtaining adequate impregnation of the thin commissural fibers; some fibers described as commissural may also have been confounded with other fiber types.

Experimental-anatomical studies so far have been remarkably unsuccessful in

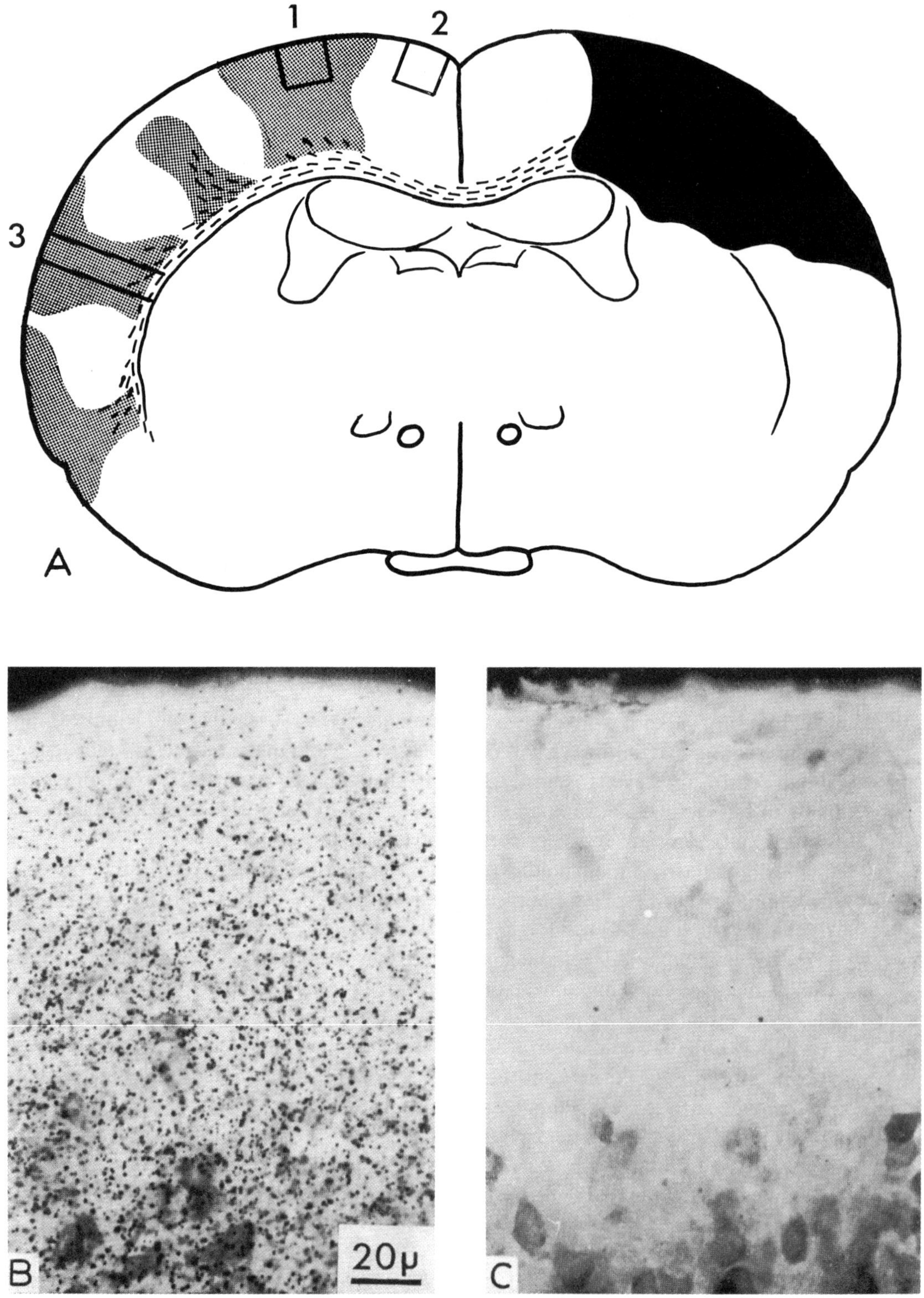

   

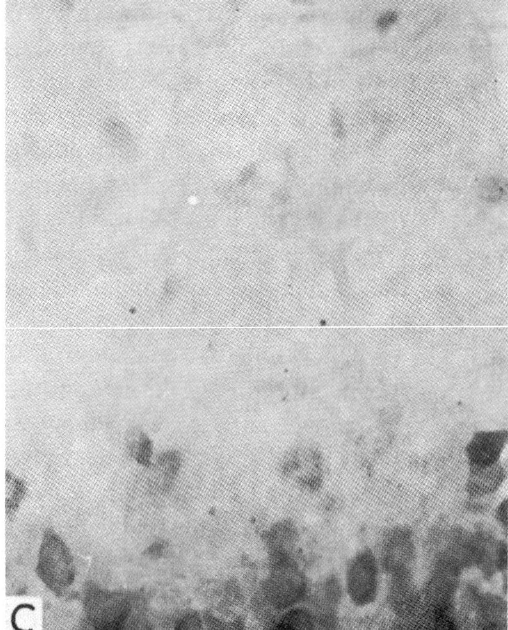

1
2
3
A
B
20µ
C

corroborating Villaverde's laminar distribution pattern of callosal fibers. Blackstad[4], by using the original Nauta[20,22] method, was able to demonstrate a well-defined laminar pattern in the termination of the hippocampal commissure, but unfortunately the same technique proved less suitable for the identification of terminal degeneration in the neocortex. Numerous studies employing the later 'suppressive' Nauta–Gygax[23] method consistently have suggested that callosal fibers are distributed predominantly in deep cortical layers VI–IV, involve the more superficial layers III and II to a much lesser extent and do not reach the first or plexiform layer[3,8,9,12,19,21]. Besides being at variance with the interpretations of Golgi material, this evidence conflicted markedly with electrophysiological observations which indicated a considerable (according to some, predominant[5,7]) termination of callosal fibers in the superficial cortical layers.

The present short communication reports some observations resulting from a preliminary re-investigation of the controversy by the aid of a recently developed modification of the original Nauta method[10]. As this new technique has been found to yield better pictures of the terminal distribution, including the synaptic endstructures, of degenerating axons than are obtained by the Nauta–Gygax method, it was expected to furnish a more complete picture of the callosal fiber distribution than has been attainable by degeneration methods used heretofore.

Our findings in three rats and three opossums, hemidecorticated 3–11 days prior to sacrifice, strongly suggest that the distribution of callosal afferents may involve all cortical layers. Whereas in certain areas of the intact cortical hemisphere a dense terminal degeneration was found fairly evenly distributed throughout the thickness of the cortex, other regions showed evidence of a more stratified distribution pattern recalling Villaverde's observations.

Fig. 1A shows the distribution of callosal fibers in a frontal plane through the caudal part of the optic chiasma. The lesion (jet-black), which was made 5 days before sacrifice, involved corpus callosum in addition to neocortex. The resulting terminal degeneration in the contralateral intact hemisphere was almost entirely confined to the four shaded areas. The quantity of the degeneration, however, varied between the different columns as well as between different layers within the same column. An extremely dense terminal degeneration is often found in the superficial layers I and II (Fig. 1B). The distribution pattern within one of the columns (3) is shown in Fig. 2. A dense terminal degeneration is found in layers I and II, a considerably sparser terminal degeneration in layers III and VI, while the remaining layers IV and V, although traversed by degenerating fibers of passage, contain only sporadic structures identifiable as degenerating axon terminals. At present it is an open question in how far these regional differences may reflect an organization of the callosal projection in the form

---

Fig. 1. A, Schematic representation of the commissural degeneration following a large unilateral cortex lesion (jet-black, right side of drawing) which was made 5 days before sacrifice. Terminal degeneration in the intact cortical hemisphere (left side of drawing) was almost completely confined to the four shaded areas. B, Microphotograph corresponding to the insert square 1 shown in A, and illustrating a dense terminal degeneration in layers I and II. The heavily impregnated pia mater is seen at the top of the figure. C, Microphotograph showing absence of degenerating axonal endstructures in layers I and II of the region outlined by square 2 in A. Compare with B. Fink–Heimer method (procedure II), counterstained with cresylechtviolet. × 550.

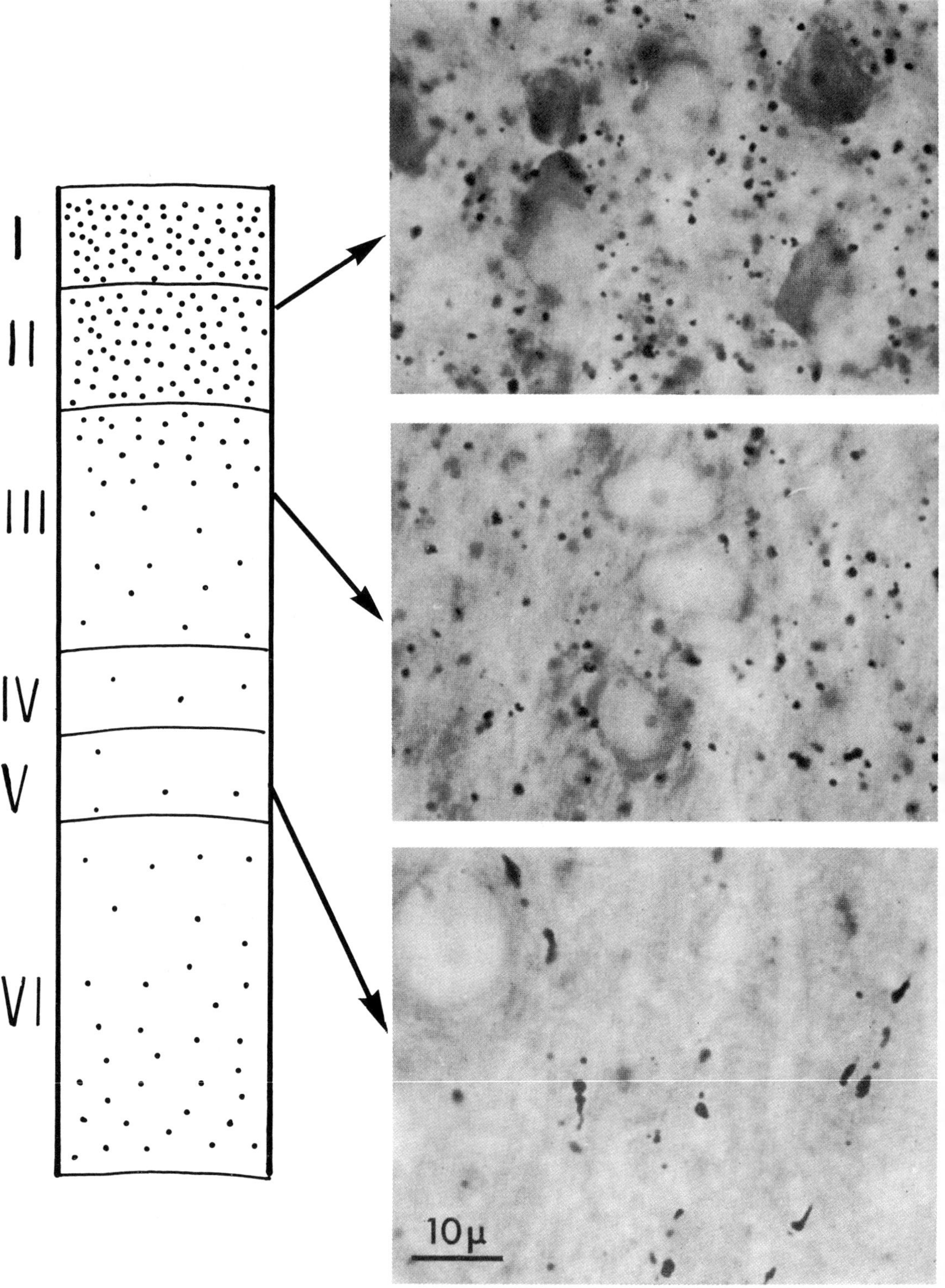

Fig. 2. The degeneration pattern within different layers of column 3 shown in Fig. 1A. The relative density of terminal degeneration in the various cortical layers is marked with dots in the schematic figure to the left. Fink–Heimer method (procedure II), counterstained with cresylechtviolet. × 1400.

of columns or other stereometric configurations oriented at variable angles to the plane of sectioning. Obviously, a complete account of the three-dimensional callosal distribution pattern will have to await much more detailed analysis.

Besides regional differences in the relative quantity of callosal fiber termination in different cortical layers, conspicuous and locally steep gradients in the apparent total number of callosal afferents are encountered. A striking contrast, for example, exists in the rat between cortical area 17 in which extremely few degenerating axons can be identified, and a laterally adjoining narrow 'strip' (presumably corresponding to Krieg's area 18a) which shows dense terminal degeneration in all layers.

The findings here reported, however preliminary, tend to resolve the existing controversy between the conclusions derived from studies by the Golgi and neurophysiological methods on the one hand, and the suppressive Nauta–Gygax technique on the other. It now appears likely that the latter category of evidence was incomplete as a result of failure of the Nauta–Gygax method to provide adequate pictures of terminal degeneration in the superficial layers of the cerebral cortex. The present findings, obtained by a suitably modified silver impregnation process, substantiate Villaverde's description, published more than 30 years ago, of callosal fibers distributed in regionally variable laminar patterns. They are compatible with the published neurophysiological evidence of callosal fiber termination in superficial cortical layers. Further neurophysiological studies may, it is hoped, elucidate the functional significance of the pronounced regional and laminar variations which appear to characterize the distribution of this massive interhemispheric pathway.

## SUMMARY

A study of callosal fiber degeneration in the rat and opossum by the aid of the Fink–Heimer silver technique has led to the conclusion that the intracortical distribution of the corpus callosum involves all cortical layers. Evidence was found of regional differences in the intracortical distribution pattern; in some regions a stratified mode of callosal fiber termination was apparent.

## ACKNOWLEDGEMENTS

This work was done while one of us (L.H.) was supported by a Training Fellowship from the Foundations Fund for Research in Psychiatry, New Haven, Conn.

These studies were supported in large part by N.I.H. Grants NB 06551 and NB 06542.

The authors gratefully acknowledge the technical assistance of Miss E. W. Swick and Mr. Robert P. Fink.

## REFERENCES

1 BAILEY, P., BONIN, G. V., GAROL, H. W., AND McCULLOCH, W. S., Functional organization of temporal lobe of monkey (*Macaca mulatta*) and chimpanzee (*Pan Satyrus*), *J. Neurophysiol.*, 6 (1943) 121–128.

2 BAILEY, P., BONIN, G. V., DAVIS, E. W., GAROL, H. W., McCULLOCH, W. S., ROSEMAN, E., AND
   SILVEIRA, A., Functional organization of the medial aspect of the primate cortex, *J. Neurophysiol.*,
   7 (1944) 51–55.
3 BLACK, P., AND MEYERS, R. E., Connections of occipital lobe in the monkey, *Anat. Rec.*, 142 (1964)
   216.
4 BLACKSTAD, T. W., Commissural connections of the hippocampal region in the rat, with special
   reference to their mode of termination, *J. comp. Neurol.*, 105 (1956) 417–537.
5 CHANG, H.-T., Cortical response to activity of callosal neurons, *J. Neurophysiol.*, 16 (1953) 117–
   131.
6 CURTIS, H. J., Intercortical connections of corpus callosum as indicated by evoked potentials,
   *J. Neurophysiol.*, 3 (1940) 407–413.
7 CURTIS, H. J., An analvsis of cortical potentials mediated by the corpus callosum, *J. Neurophysiol.*,
   3 (1940) 414–422.
8 EBNER, F. F., AND MYERS, R. E., Commissural connections in the neocortex of monkey, *Anat.
   Rec.*, 142 (1962) 229.
9 EBNER, F. F., AND MYERS, R. E., Distribution of corpus callosum and anterior commissure in
   cat and raccoon, *J. comp. Neurol.*, 124 (1965) 353–365.
10 FINK, R. P., AND HEIMER, L., Two methods for selective silver impregnation of degenerating axons
   and their synaptic endings in the central nervous system, *Brain Research*, 4 (1967) 369–374.
11 GRAFSTEIN, B., Organization of callosal connections in suprasylvian gyrus of cat, *J. Neurophysiol.*,
   22 (1959) 504–515.
12 JACOBSON, S., Intralaminar, interlaminar, callosal, and thalamocortical connections in frontal and
   parietal areas of the albino rat cerebral cortex, *J. comp. Neurol.*, 124 (1965) 131–145.
13 JANSEN, J., Experimental investigations of the associational connections of the cerebral cortex,
   with special reference to the conditions in the frontal lobe, *Norske Vid.-Akad., Oslo, Arh. 1. Mat.-
   Nat. Kl.*, 1 (1937) 1–21.
14 KRIEG, W. J. S., Connections of the cerebral cortex 1. The albino rat, (c) extrinsic connections,
   *J. comp. Neurol.*, 86 (1947) 267–394.
15 LORENTE DE NÓ, R., La corteza cerebral del raton, *Trab. Lab. Invest. biol. Univ. Madrid*, 20 (1922)
   41–78.
16 METTLER, F. A., Corticofugal connections of the cerebral cortex, *Arch. Neurol. Psychiat. (Chic.)*,
   35 (1936) 1338–1344.
17 McCULLOCH, W. S., Cortico-cortical connections. In C. P. BUCY (Ed.), *The Precentral Motor
   Cortex*, Univ. of Illinois Press, Urbana, 1944, pp. 212–242.
18 McCULLOCH, W. S., AND GAROL, H. W., Cortical origin and distribution of corpus callosum and
   anterior commissure in the monkey (*Macaca mulatta*), *J. Neurophysiol.*, 4 (1941) 555–563.
19 NAUTA, W. J. H., Terminal distribution of some afferent fiber systems in the cerebral cortex, *Anat.
   Rec.*, 118 (1954) 333.
20 NAUTA, W. J. H., Über die sogenannte terminale Degeneration im Zentralnervensystem und ihre
   Darstellung durch Silberimprägnation, *Schweiz. Arch. Neurol. Psychiat.*, 66 (1950) 353–376.
21 NAUTA, W. J. H., AND BUCHER, V. M., Efferent connections of the striate cortex in the albino rat,
   *J. comp. Neurol.*, 100 (1954) 257–286.
22 NAUTA, W. J. H., AND GYGAX, P. A., Silver impregnation of degenerating axon terminals in the
   central nervous system. (1) Technic; (2) Chemical notes, *Stain Technol.*, 26 (1951) 5–11.
23 NAUTA, W. J. H., AND GYGAX, P. A., Silver impregnation of degenerating axons in the central
   nervous system: a modified technique, *Stain Technol.*, 29 (1954) 91–93.
24 PEELE, T. L., Cytoarchitecture of individual parietal areas in the monkey (*Macaca mulatta*) and the
   distribution of the efferent fibers, *J. comp. Neurol.*, 77 (1942) 693–737.
25 PINES, L. J., AND MAIMAN, R. M., Cells of origin of fibers of the corpus callosum, *Arch. Neurol.
   Psychiat. (Chic.)*, 42 (1939) 1076–1081.
26 POLIAK, S., An experimental study of association, callosal, and projection fibers of the cerebral
   cortex of the cat, *J. comp. Neurol.*, 44 (1927) 197–258.
27 RAMÓN Y CAJAL, S., Sur la structure de l'écorce cérébrale de quelques mammifères, *Cellule*, 7
   (1891) 125-176.
28 VAN VALKENBURG, C. T., Experimental and pathological anatomical researches on the corpus
   callosum, *Brain*, 36 (1913) 119–165.
29 VILLAVERDE, J. M. DE, Les connexions commissurales des régions postérieures du cerveau du
   lapin, *Trab. Lab. Invest. biol. Univ. Madrid*, 22 (1924) 99–141.

30 VILLAVERDE, J. M. DE, Sur la terminaison des fibres calleuses dans l'écorce cérébrale, *Trab. Lab. Invest. biol. Univ. Madrid*, 27 (1932) 275–297.
31 VILLAVERDE, J. M. DE, Quelques détails sur la manière dont les fibres calleuses se distribuent dans l'écorce cérébrale, *Trab. Lab. Invest. biol. Univ. Madrid*, 27 (1932) 345–375.
32 VILLAVERDE, J. M. DE, Contribution à la connaissance du système commissural de l'écorce motrice de la chauve-souris, *Trab. Lab. Invest. biol. Univ. Madrid*, 28 (1932) 75–101.

# 2 A General Profile of the Vertebrate Brain, with Sidelights on the Ancestry of Cerebral Cortex

WALLE J. H. NAUTA and HARVEY J. KARTEN

THE MOST elementary tenet of the theory of evolution is that animal specification followed a temporal sequence such that one order of species developed from another, and in time gave rise to one or more further orders. The reconstruction of the "tree of evolution," one of the most constantly pursued goals of biology, is attended by numerous difficulties, foremost among which is the circumstance that existing forms of life represent little more than "leaves on the ends of branches" of a tree, the trunk and limbs of which have long been extinct. Virtually all extant animals appear to be specialized forms that have diverged in greater or lesser degree from any of the identified or presumed mainlines of evolution. The identification of such "mainlines," furthermore, is often highly uncertain, the more so because several vertebrate classes appear to have evolved not from one, but from several ancestors. Modern amphibians, for example, are suspected of representing several developmental lines originating from various piscine forms. Similarly, monotremes, marsupials, and placental animals may represent parallel phyletic lines among mammals, each evolving from a different reptilian ancestor.

These and other constraints of evolutionary biology, reviewed more systematically elsewhere in this volume, apply to comparative neurology no less strictly than to other phylogenetic disciplines. Current knowledge of brain evolution is, however, limited by two additional circumstances. In the first place, neural tissues rapidly disintegrate after death, and there is thus no hope that any but indirect and relatively superficial information about the brain's architecture in extinct forms will ever be obtained. Second, and for the moment no less restricting, intensive systematic studies of brain organization so far have been limited very largely to mammalian species—a consequence, perhaps, of the fact that neuroanatomy received a major early impetus from the neuropsychiatric clinic. It would be unfair to say that nonmammalian brains have been neglected, but the enormous amount of work devoted to such brains during the past century has resulted almost exclusively in normal anatomical descriptions and, until quite recently, has included only sporadic studies by the more rigorous experimental methods that have been used so profitably in the study of mammalian brain organization.

As a result of this incongruity, our understanding of nonmammalian brains is really quite limited, too much so to permit far-reaching conclusions as to fundamental organizational differences among the brains of different classes. For example, the brain of a frog clearly differs from that of a lizard with respect to external features and, at a slightly less macroscopic level, with respect to the relative size and topography of nerve-cell groups and "fiber systems" (axon bundles). In the absence of more detailed knowledge of interneuronal relationships, however, it is impossible to determine whether such differences represent dissimilarities in

---

WALLE J. H. NAUTA and HARVEY J. KARTEN Department of Psychology, Massachusetts Institute of Technology, Cambridge, Massachusetts

fundamental structural organization, or signify no more than variations in the relative degree of development of various central nervous subsystems common to both classes. Clearly, an enormous amount of experimental work—anatomical, physiological, biochemical, and behavioral—remains to be done before the fundamental steps of brain evolution and their functional corollaries can be fully recognized. It seems likely that future studies will lead to the identification of several crucial determinants of brain evolution. Such determinants may consist in the emergence somewhere in phylogeny of a novel neuronal element, or in a re-grouping of neurons or of synaptic junctions, possibly in response to the development of a new afferent relationship with one or another part of the brain, or in the first appearance of a distinctive chemical specification of a particular neuronal system, or even in a restructuring of glia-neuron relationships, leading to a significantly different partitioning of brain tissue.

Far from dealing with all these presumable aspects of brain evolution, the present chapter is focused on one of the central issues of comparative neurology, namely, the problem of *homology of neuronal systems*. This term needs the following explanation. The evolution of the vertebrate brain is characterized primarily not by a linear increase in the size of the central nervous system or in the number of its constituent neurons, but, instead, by more or less progressive modification of neuronal subsystems. Such tendencies are especially noticeable in the forebrain, where structural rearrangements can be profound enough to cause severe problems in determining which, if any, of a diversity of neuron groups found in a more recently evolved class of vertebrates corresponds to a particular cell group or system of interrelated cell groups in the ancestral class. A particularly precipitous recomposition of the vertebrate forebrain takes place at the transition between reptiles and mammals. Both birds and mammals have modified the ancestral reptilian forebrain structure, but, whereas in birds the resulting organization still appears relatively comparable to that of modern reptiles, the forebrain of modern mammals has deviated so much from the presumed precursory pattern that it can no longer be compared readily with the latter by routine cytoarchitectonic and fibroarchitectonic comparison. Instead, such identifications require detailed experimental studies of afferent and efferent relationships, as well as a search for common embryological derivations. The more refined, recent, histochemical techniques may provide an additional approach to the homology problem in brain evolution.

In the last part of this chapter, the search for interphyletic homologies in brain evolution is illustrated by some recent studies conducted by experimental-anatomical and histochemical methods. To place the results of these studies in proper perspective, however, it is necessary first to consider some general structural features of the vertebrate central nervous system. Needless to say, the following brief account can provide no more than a generalized survey of this highly complex subject. Its main purpose is to identify some of the major organizational features that distinguish the forebrains of mammals from those of nonmammalian forms. The scheme of the nonmammalian brain given below is based largely on recent experimental findings in birds and reptiles; in the absence of comparable data for other nonmammalian classes, it is impossible to say to what extent its fundamental aspects are valid for a generalization of "the nonmammalian vertebrate brain."

## General overview of the vertebrate brain

GROSS MORPHOLOGICAL FEATURES *The nonmammalian brain.* Some major general features of the central nervous system of nonmammalian vertebrates are represented schematically in Figure 1A. The central nervous system of all vertebrates can be subdivided into an unpaired, bilaterally symmetrical *neuraxis* and a paired subdivision, the *cerebral hemisphere,* or telencephalon. The neuraxial subdivision consists of (1) the spinal cord, (2) the rhombencephalon with the associated cerebellum, (3) the mesencephalon or midbrain, and (4) the diencephalon. In the diencephalon, three subdivisions are distinguished: *thalamus, subthalamus* (not shown in Figure 1), and *hypothalamus*. It represents the unpaired middle part of the prosencephalon, lateral parts of which become extruded in early embryonic development to form the cerebral hemisphere. The term *forebrain,* used repeatedly in the following account, is a literal translation of prosencephalon, and invariably refers to the complex formed by the diencephalon and cerebral hemisphere.

In the cerebral hemisphere, three major neural territories can be recognized:

1. The *olfactory system,* largely composed of primitive cortical formations *—the olfactory bulb, which receives the axons of the olfactory receptor cells, and a variety of other cortical regions that receive the projections from the olfac-

---

* The term *cortex* denotes variously complex neural organizations at the surface of the brain that are distinguished from subcortical (noncortical) cell organizations by the following combination of characteristics: (1) neuronal cell bodies are arranged in more or less clearly distinguishable layers, varying in number from one or two in the allocortex (see below) to five or six in the neocortex; (2) there is an additional, most superficial layer (plexiform or molecular layer), which essentially is a fiber stratum containing numerous axons and dendritic processes but few neuronal cell bodies; and (3) numerous neurons in at least the most superficial cell layer (in the neocortex, all cell layers) emit at least one major "apical" dendrite perpendicularly toward the brain surface; many such dendrites reach into the first (plexiform) layer even from the deepest cell layer of the neocortex.

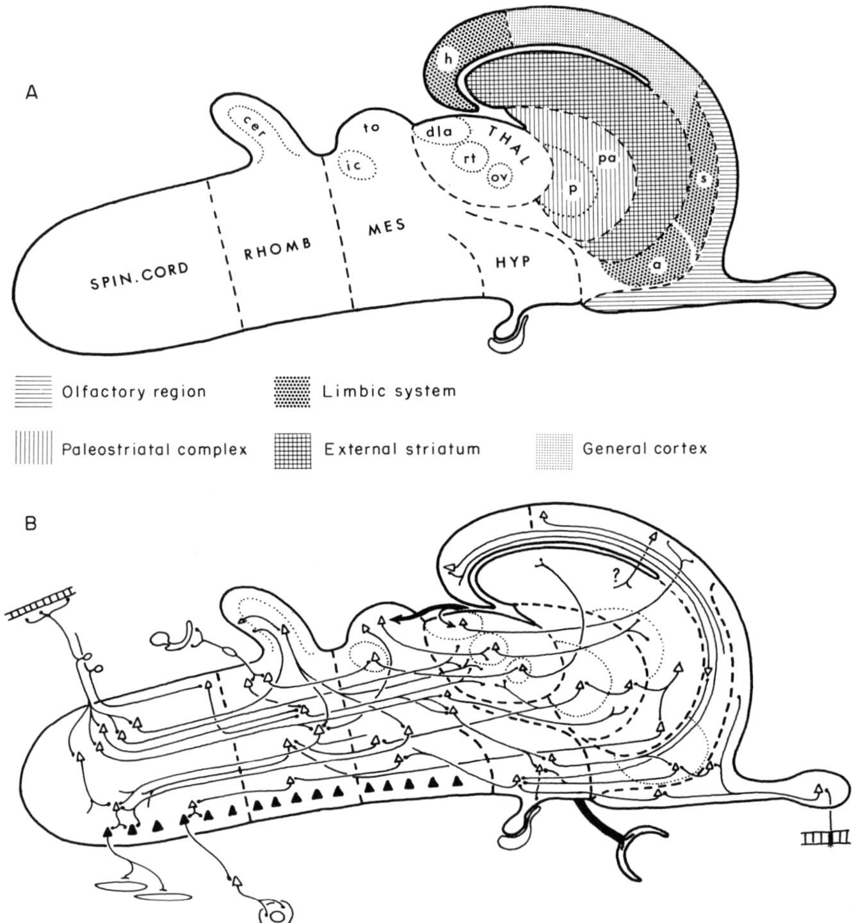



FIGURE 1 Diagramatic representation of the central nervous system of a nonmammalian vertebrate in longitudinal view. The diagrams are based on recent findings (explained in the text) with respect to connectional patterns in avian and reptilian brains in particular; it is uncertain to what extent they are valid for other nonmammalian vertebrate orders. A. General overview, showing the principal divisions of the neuraxis and, in various line and stipple patterns, the major territories of the cerebral hemisphere as itemized in the text. B. Schematic illustration of some major neural conduction pathways. For reasons of convenience, ascending fiber systems have been indicated in the dorsal half of the spinal cord and brainstem, descending pathways in the ventral half (in the forebrain this separation could not be maintained). Motor neurons are schematically indicated as a row of solid black triangles near the ventral periphery of spinal cord and brainstem. Only three primary sensory neurons are included in the diagram: one enters the spinal cord (left upper corner of diagram), a second one represents the auditory nerve (immediately behind the cerebellum), and a third one is an olfactory receptor cell (right lower extreme of drawing). All other neurons appearing in the diagram are indicated as open triangles and, strictly speaking, are intermediate between sensory and motor neurons. It should be kept in mind that the diagram represents an attempt to express some apparent principles rather than actual complexities of neuronal alignment. *Abbreviations:* a, amygdala; cer, cerebellum; dla, dorsolateralis anterior of thalamus; h, hippocampus; HYP, hypothalamus; ic, inferior colliculus; MES, mesencephalon; ov, nucleus ovoidalis thalami; p, paleostriatum primitivum; pa, paleostriatum augmentatum; RHOMB, rhombencephalon; rt, nucleus rotundus thalami; s, septum; SPIN. CORD, spinal cord; THAL, thalamus; to, optic tectum.

THE VERTEBRATE BRAIN    9

tory bulb and form a major basal part of the cerebral hemisphere, the piriform lobe.

2. The *corpus striatum,* which forms the central core of the cerebral hemisphere and (in reptiles and birds, at least) can be subdivided readily into two major territories, here to be named *internal striatum* (paleostriatum); and *external striatum,* each, in turn, composed of more or less clearly demarcated neuronal territories. In the avian brain, for example, the internal striatum consists of a magnocellular medial segment (the paleostriatum primitivum) and a larger lateral aggregate of smaller neurons (the paleostriatum augmentatum), and the external striatum is sharply subdivided into at least four segments—the ectostriatum, archistriatum, neostriatum, and hyperstriatum. Evidence mentioned in the final section of this paper suggests that the external striatum is a neuronal territory typical of nonmammalian phyla, and is absent as such from mammals.

3. *The limbic system.* This term, derived from mammalian neuroanatomy, is preferable to the more traditional name *rhinencephalon,* which too exclusively came to imply a specific relationship with the olfactory system. The term denotes a heterogeneous array of structures, some of which (the hippocampus and adjacent cortical regions) compose a considerable dorsomedial part of the pallium (cortical mantle of the hemisphere) separated from the more ventrally situated olfactory cortex by a smaller or larger expanse of corticoid tissue (dorsolateral or general cortex). Together with the olfactory cortex and the general cortex, the limbic cortexes are often classified as *allocortex,* a term used to distinguish these primordial forms of cortex, with only a small number of cell layers, from the multilayered *isocortex* or neocortex, which appears as an additional form of cortex only in the pallium of mammals. The limbic system, however, also includes noncortical formations such as the amygdaloid complex and septum.

Much like the olfactory apparatus, the limbic system is characterized by a fairly stable evolutionary history. Despite considerable phylogenetic modification of several of its components, the amygdala in particular, its character of a major forebrain complex connected with the hypothalamus remains preserved throughout vertebrate evolution.

*The mammalian brain.* Figure 2A epitomizes the gross morphological features of the mammalian brain. The diagram emphasizes only two of the major characteristics that distinguish the brain of the mammal from that of its phylogenetic prototypes, as well as from the avian brian, which represents a contemporary but far more conservative evolution from the ancestral reptilian model.

The most outstanding feature of the mammalian brain is the replacement of the "general cortex" by a large expanse of novel type, the multilayered *isocortex* (often called *neocortex*), a development that crowds the more primordial allocortical formations of the limbic system (hippocampus

and associated regions) relatively farther toward the edge (*limbus*) of the mantle. At the same time, the corpus striatum, although still massive, shows a marked reduction in relative size. This reduction, as is documented below, is attributable to an apparent loss of the external striatum, the structure accounting for the larger part of the striatal mass in reptiles and birds.

Even the most cursory survey of the external features of the vertebrate brain reveals a striking phylogenetic increase in the relative size of the forebrain (Figure 3). A small rostral prominence in the fish brain, the forebrain becomes progressively larger in the amphibian-reptilian sequence, and in birds and mammals (believed to have branched off from reptilian ancestors) forms the largest subdivision of the central nervous system. Progressive phylogenetic modification of the forebrain is, however, a matter not only of relative size but also of internal structure. Interphyletic differences in the configuration of the cerebral hemisphere, especially that of the corpus striatum and the pallium, pose the most challenging problems in attempts to trace the course of brain evolution. Such differences are paralleled by equally profound changes in the architecture of the diencephalon, especially that of the thalamus.

By comparison, more caudal parts of the brain, particularly the rhombencephalon, exhibit a much greater constancy of organization. Even though considerable interphyletic and species differences are encountered in comparative studies of these more caudal brain regions, such variations are only rarely so profound that homologies cannot be recognized readily. A functional corollary of this contrast in the structural evolution of forebrain and hindbrain suggests itself. Whereas the *intrinsic* mechanisms of the rhombencephalon and spinal cord appear to be involved largely with the maintenance of primordial stabilities, such as those of the internal milieu (internal homeostasis) and somatic posture (stability in space), the intrinsic function that most typifies the more rostral parts of the brain, and especially the forebrain, appears to lie in the perception of goals and goal priorities, as well as in the patterning of behavioral strategies serving the pursuit of these goals (stability in time, or stability of goal-directed behavior).

Regardless of class or species, it is likely that, in order to discharge this function, the forebrain has access to, and possesses mechanisms for, the integration of all available forms of sensory information concerning the environments, both internal and external, of the organism. It is unfortunate that comparative data on the subject of sensory afflux to the forebrain are at a premium; unfortunate because any concept of interphyletic homology in central nervous organization, if it is to be better than speculative, should be based first and foremost on reliable identification of similarities in the afferent and efferent connections of forebrain components.

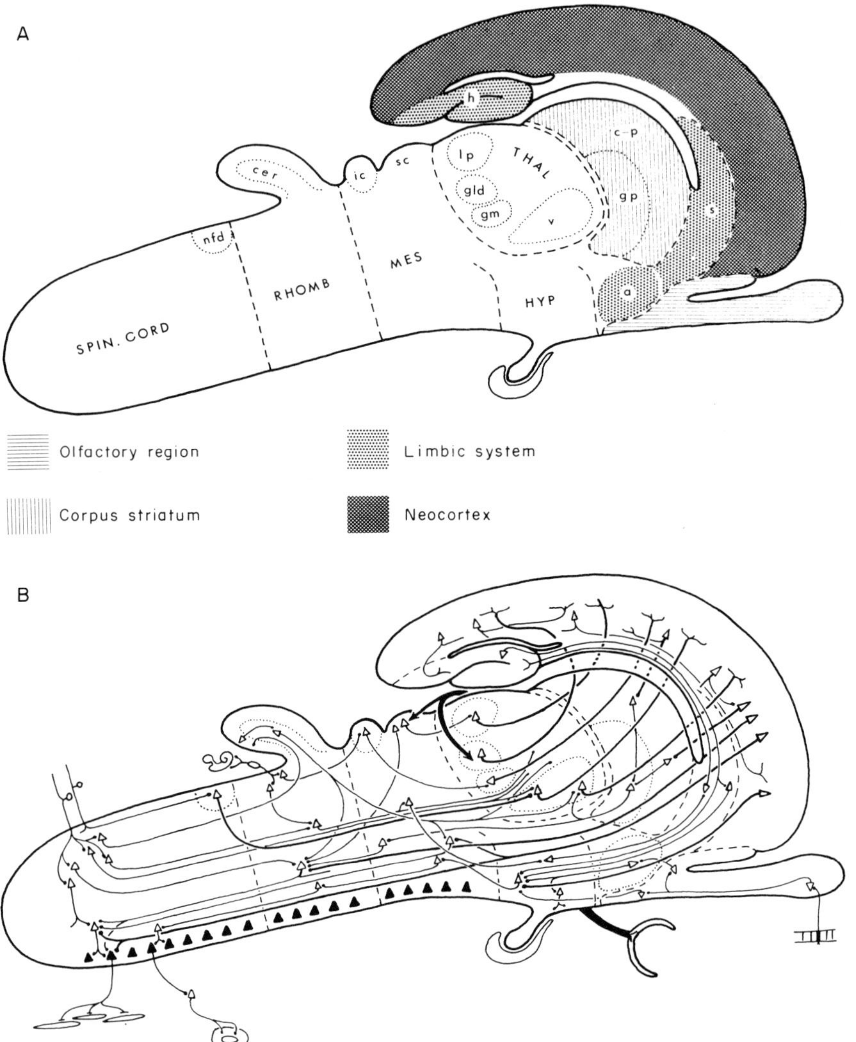



FIGURE 2　Schematic drawings representing the mammalian brain. In comparisons with the diagrams of the nonmammalian brain shown in Figure 1, the most pronounced differences appear in the composition of the pallial mantle in which the general cortex has become replaced by neocortex; the apparent absence of the nonmammalian external striatum; and the appearance of a circumscript somatic sensory nucleus (v) in the thalamus receiving, among other somatic sensory lemnisci, part of the spinothalamic tract and most of the medial lemniscus originating in the nuclei of the dorsal funiculus (nfd). Major conduction pathways afferent and efferent to the neocortex have been indicated in slightly bolder line. *Abbreviations:* a, amygdala; cer, cerebellum; c-p, caudoputamen; gld, lateral geniculate body; gm, medial geniculate body; gp, globus pallidus; h, hippocampus; HYP, hypothalamus; ic, inferior colliculus; lp, nucleus lateralis posterior of thalamus; MES, mesencephalon; nfd, nuclei of the dorsal funiculus; RHOMB, rhombencephalon; s, septum; sc, superior colliculus; SPIN. CORD, spinal cord; THAL, thalamus.

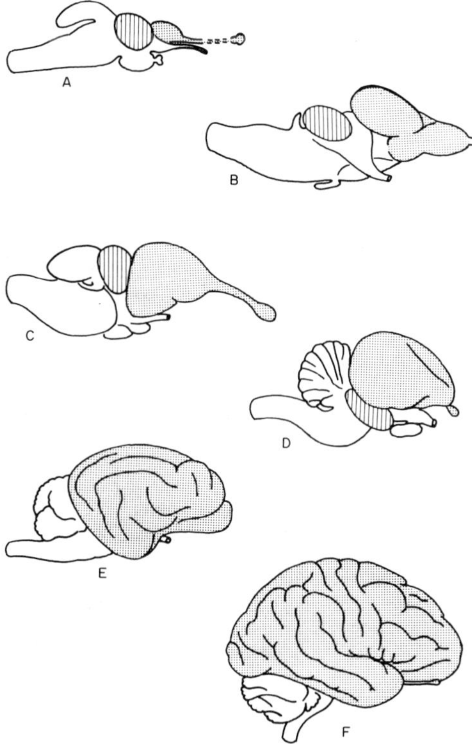

FIGURE 3 A series of side views of brains of various vertebrate orders, to illustrate the progressive increase in relative size of the cerebral hemisphere (stippled). A, codfish; B, frog; C, alligator: D, pigeon; E, cat; F, man. The optic tectum, indicated by vertical shading, is obscured from view in the two mammalian forms by the overlying cerebral hemisphere. The drawings, made on vastly different scales, are based in part on an illustration appearing in Strong and Elwyn, 1964.

INTERNAL STRUCTURE OF THE VERTEBRATE BRAIN  *General considerations.*   The central nervous system of all vertebrates, at its most elementary level, can be characterized as a *three-neuron nervous system*. Presumably evolved from one- and two-neuron stages of primitive Metazoa, the three-neuron organization is characterized by having a net of synaptically interconnected *intermediate* neurons interposed between the primary sensory nerve cells and the motor neurons. The appearance of this "great intermediate net" (Herrick, 1922) undoubtedly marks a fundamental step in the evolution of the nervous system. It became virtually the sole recipient of sensory messages from the environments of the organism (it is by-passed only by the direct, i.e., monosynaptic, reflex connections between primary sensory neurons and motor neurons, possibly a relatively late development of particular adaptive value for terrestrial vertebrates). By virtue of its intermediate position, it must have permitted a phenomenal amplification of the limited possibilities for sensory processing and integration afforded by the more primitive one- and two-neuron organizations.

It must be pointed out that the neuronal population of the vertebrate *central nervous system* (brain and spinal cord) consists exclusively of intermediate neurons and motor neurons: the cell bodies of the primary sensory neurons are outside the central organ, even though naturally their centrally directed axons must enter (and often extend far into) the brain and spinal cord to establish synaptic contacts.

Of modest size at the outset, the net of intermediate neurons rapidly expands in phylogeny so that in all vertebrates it forms the largest part by far of the central neuron population. In higher primates, the total number of brain cells is commonly estimated to be of the order of 10 billion ($10^{10}$), of which number no more than a few millions are motor neurons. These figures suggest that many mammalian brains contain at least as many as 2000 intermediate neurons to each motor neuron, a surprising ratio when it is realized that it is only by way of motor neurons that the activity of the central nervous system can be expressed in movement, whether in the form of a simple reflex or a complex, goal-directed behavior. A numerical relationship of this order suggests a very high degree of convergence of central neuronal conduction pathways toward the motor neurons, and emphasizes the appropriateness of Sherrington's characterization of the motor neuron as the "final common pathway" of the nervous system.

It is clear from these numerical data alone that the complexity of the brain is, in essence, the complexity of the intermediate net of neurons. Moreover, because the neurons composing the intermediate net account for something like 99.95 per cent of the neuron population of the mammalian brain and spinal cord, the term *intermediate net* becomes nearly synonymous with the central nervous system as a whole and, hence, loses much of its practical usefulness in discussions of neural organization. From a conceptual point of view, however, the term is nonetheless important, for it recalls the essential fact that an overwhelming majority of the neurons that make up the brain and spinal cord (in fact, all but the motor neurons) cannot be classified as either sensory or motor in the strict sense. This is true even though some intermediate neurons obviously are more directly involved in the receipt and processing of sensory information conveyed into the organization by true (i.e., primary) sensory neurons than are others that may, by contrast, have relatively direct (i.e., oligosynaptic) efferent connections with motor neurons. Such differences in relative position vis-à-vis the afferent and efferent portals of the central nervous system have made it customary to refer to certain groups of intermediate neurons as "sensory" and to others as "motor," but it should be emphasized that the terms in this application indicate relative rather than absolute characteristics of such cell groups.

Some basic organizational features of the nonmammalian central nervous system are illustrated in Figure 1B. For reasons of convenience, the motor neurons (solid black triangles) are shown aligned near the ventral surface of the spinal cord and brainstem. As shown in the diagram, motor neurons (strictly, neurons the axons of which leave the central nervous system and innervate peripheral effector tissues either directly, as in the case of skeletal musculature, or by the mediation of a second, peripheral motor-neuronal element, as in the case of smooth musculature and gland cells of the viscera) are limited to the spinal cord, rhombencephalon, and mesencephalon. Not shown in the diagram is a variant form of motor neuron that is limited to the diencephalon and is typical of the hypothalamic connection with the anterior lobe of the pituitary complex. The axon of this unique type of effector neuron does not leave the central nervous system, but terminates in close relationship to a superficial plexus of blood capillaries, the effluent venous vessels of which ("hypothalamo-pituitary portal system") enter the anterior lobe of the pituitary complex, in which they break up into a second capillary plexus. The effect of these hypothalamo-pituitary effector neurons on the gland is thought to be conveyed by means of neural transmitter substances ("releasing factors") released into the capillaries and reaching the anterior pituitary by way of the portal blood vessels.

Of the far more numerous intermediate neurons, only a small sample has been indicated, all in the form of open triangles. The sensory neurons are characterized by having cell bodies that are outside the central nervous system, usually in a peripheral sensory ganglion, but, in the unique case of the olfactory system, in the surface epithelium proper. The optic system represents a special case: the retina is a forwarded outpost of the brain itself, and impulses elicited by photic stimulation of the retinal receptor cells have passed at least two synapses in the retina before arriving at the midbrain and diencephalon by way of the optic tract (heavy arrow in Figure 1B immediately behind the cerebral hemisphere).

*Sensory Systems* SENSORY MECHANISMS OF SPINAL CORD AND RHOMBENCEPHALON Figure 1B diagramatically illustrates a common model of the possible neuronal sequences followed by impulses transmitted to the central nervous system over primary sensory neurons entering the spinal cord and rhombencephalon. At the far left of the diagram the central axonal process of one primary sensory neuron forming part of a spinal segmental nerve is shown distributed to a group of intermediate neurons in the dorsal part of the gray matter of the spinal cord. Such cell groups of the intermediate net, in direct contact with primary sorysen neurons and hence "first in line" in terms of sensory processing mechanisms, are often termed *secondary sensory nuclei*.

The pool of secondary sensory neurons that come into contact with each sensory nerve shows considerable local differentiation, exemplified in the mammalian spinal cord by the different cellular characteristics of the substantia gelatinosa, nucleus proprius, and columna dorsalis Clarkei, respectively, and in the rhombencephalon by marked architectonic differences between various subdivisions of the cochlear nucleus. There are only scattered data with respect to the question whether such differentiations correspond to a selective processing of one particular component (submodality) of the incoming information by each subdivision of the secondary sensory complex or, alternatively, to a parallel processing of each submodality by various specialized secondary sensory cell groups. In the cochlear-nucleus complex, the latter of these alternatives appears to prevail.

The axons arising from second sensory nuclei form the so-called secondary sensory pathways, conduction channels which extend the spread of sensory information far beyond the first processing station. These secondary sensory channels can extend in any or all of the following directions: (1) the *local reflex channel* to motoneurons, either directly or by the intermediary of one or more internuncial neurons; (2) the *cerebellar channel* leading to the cortex of the cerebellum; (3) the *lemniscal channel*.

The term *lemniscus* refers somewhat loosely to fiber systems originating from secondary sensory cell groups and ascending toward the forebrain, in particular to the thalamus. The lemniscal pathway arising from the spinal cord in the nonmammalian animals largely corresponds to the so-called *spinothalamic tract* of mammalian forms. This pathway is sometimes referred to as the spinal paleolemniscus to distinguish it from a parallel spinal neolemniscus or *medial lemniscus,* originating from the so-called nuclei of the dorsal funiculus (nfd, in Figure 2A), a massive conduction system in mammals that appears to be either absent from or only weakly developed in nonmammalian forms. The spinal paleolemniscus is characterized by having a widespread distribution in the brainstem. Most of its fibers terminate in the reticular formation of the rhombencephalon and mesencephalon, and only a minority extend beyond these levels to the thalamus. The relatively small thalamic component of the paleolemniscus is probably augmented by tertiary fibers arising from neurons of the reticular formation, but, as these neurons are believed to receive neural afflux also from several sources other than the spinal cord, such tertiary links cannot be regarded as specific carriers of impulse pat-

THE VERTEBRATE BRAIN   13

terns originating in the spinal cord or in any other source of secondary sensory pathways.

It is noteworthy that no well-defined thalamic cell group for the receipt and processing of spinal (i.e., somesthetic and visceroceptive) impulses has been identified in nonmammalian forms. Findings in experimental studies in birds and reptiles suggest that the relatively sparse spinothalamic fibers are distributed among a diffuse cell population in the caudal thalamus rather than in a characteristic, circumscript thalamic nucleus. This apparent absence of a well-defined homologue of the mammalian ventrobasal thalamic nucleus is the more remarkable as circumscript thalamic cell groups associated with the visual and auditory systems have recently been identified in birds and reptiles (see below).

With considerable individual variation, the central conduction pathways associated with sensory *cranial nerves* are patterned after the spinal cord model. With the exception of the optic and olfactory tracts—which, strictly speaking, neither are true cranial nerves nor are composed of primary sensory neurons—all the sensory cranial nerves have their central distribution in secondary sensory cell groups in the rhombencephalon. These first-order sensory processing stations include the complex of the *trigeminal nucleus* receiving somatic sensory afflux from the face region and the linings of the oral and nasal cavities, the *nucleus of the solitary tract* receiving impulses originating in taste receptors and in visceroreceptors (mechano-receptors and chemoreceptors in the walls of internal organs, particularly the cardiovascular system, the digestive, and the respiratory tracts), and the *vestibular and cochlear nuclei,* the impulse afflux of which originates in highly specialized receptor epithelia of the membranous labyrinth.

Each of these secondary sensory nuclei originates a local reflex channel, sometimes of great complexity, as in the cases of the vestibular nuclei and the nucleus of the solitary tract. The vestibular nuclei give rise to a prominent cerebellar channel, but cerebellar projections arise also from other secondary sensory nuclei of the rhombencephalon, with the possible exception of the nucleus of the solitary tract. As to lemniscal conduction channels, a variety of long, ascending fiber systems arising from rhombencephalic sensory nuclei has been identified, but in nonmammalian forms none of these is known to extend directly to the thalamus. In mammals, several fiber groups ascend from the trigeminal nucleus to the thalamus, at least one of which forms a component of the medial lemniscus, but similar pathways so far have not been demonstrated in nonmammalian forms.

It is not clear to what extent the avian *quinto-frontal tract* (Wallenberg) is comparable to any component of the trigeminal lemniscus of mammals. Originating from the principal nucleus of the secondary sensory trigeminal complex, this peculiar fiber system terminates in the so-called *nucleus basalis* near the ventral surface of the forebrain, a cell group whose relationship to the thalamus, if any, is unknown.

THE AUDITORY SYSTEM    Of particular relevance, in view of recent findings mentioned below, is the thalamic representation of the auditory system. Neither in mammals nor in any nonmammalian form thus far examined have lemniscal fibers from the cochlear nuclei (*lateral lemniscus*) been traced directly to the thalamus; none appears to extend rostrally beyond the *inferior colliculus* of the mammalian midbrain tectum (ic, in Figure 2A), or, in nonmammalian forms, beyond the nucleus mesencephali lateralis dorsalis (ic, in Figure 1A), the apparent homologue of the mammalian inferior colliculus. From this mesencephalic structure, however, a massive fiber system, the brachium of the inferior colliculus, leads to the thalamus. In mammals, this last link in the auditory path ascending to the forebrain terminates in the *medial geniculate body* (gm, in Figure 2A), a major complex of thalamic cell groups from which originates a massive fiber radiation to the auditory region of the neocortex. Only recently has the circumscript *nucleus ovoidalis* (ov, in Figure 1A) in the thalamus of reptiles and birds been identified as the nonmammalian counterpart of the medial geniculate body, in the sense that it appears to be, like the mammalian nucleus, the major thalamic receiving station of the ascending auditory conduction system. It must be stressed, however, that the nucleus ovoidalis, unlike the mammalian medial geniculate body, does not appear to project to any part of the pallial mantle. Instead, in birds it has recently been found to have its major efferent relationship with the external striatum, more specifically the so-called "field L" (Rose, 1914) of the neostriatum (Figure 4B).

CENTRAL MECHANISMS OF VISION    Composed of the efferent fibers of the retina, the optic tract is the only major afferent system that is distributed from the sensory receptor organ immediately to the mesencephalon and diencephalon. In all vertebrate classes, a large number of retinal fibers (in nonmammalians the majority by far) terminate in the dorsal part of the midbrain, the tectum mesencephali. In nonmammalian vertebrates, the recipient region forms the *optic tectum* (to, in Figure 1A), a large, multilaminate structure in most classes, and especially highly differentiated in birds. In mammals, the optic tectum appears in modified and generally reduced form as the *superior colliculus* (sc, in Figure 2A).

The mammalian superior colliculus, as well as the nonmammalian optic tectum, emits a massive descending fiber system to the brainstem reticular formation (tecto-bulbar tract) and beyond it to at least the upper segments of the spinal cord (tecto-spinal tract). On the basis of physiological observations, it appears likely that, by these descending connections, the tectum is prominently involved in the conjugated eye- and body-axis (neck or whole body) movements subserving the tracking of moving objects.

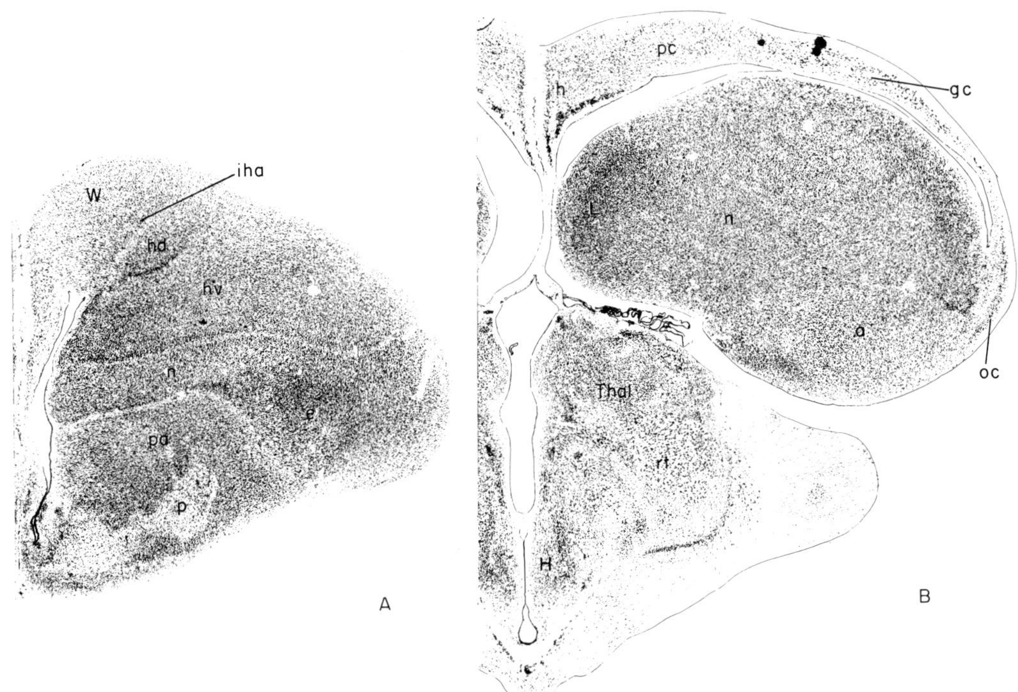


FIGURE 4  Two transverse sections of the pigeon brain, stained for cell bodies by the Nissl method. A, a rostral level of the cerebral hemisphere; B, a more caudal region. Also shown in B is the diencephalon (lower half of the figure). *Abbreviations*: a, archistriatum; e, ectostriatum; gc, general cortex; h, hippocampus; H, hypothalamus; hd, hyper-striatum dorsale; hv, hyperstriatum ventrale; iha, nucleus intercalatus of hyperstriatum accessorium; L, "field L" of neostriatum; n, neostriatum; oc, olfactory cortex; p, pale-ostriatum primitivum; pa, paleostriatum augmentatum; pc, parahippocampal cortex; rt, nucleus rotundus thalami; THAL, thalamus; W, Wulst.

Only recently has an important *ascending* connection of the optic tectum been identified. In mammals, the pathway in question extends from the superior colliculus to the thalamus, where it terminates largely in the so-called *nucleus lateralis posterior* (lp, in Figure 2A), a cell group projecting to widespread (and thus far inadequately delineated) regions of the neocortex, probably including at least part of the "visual cortex," defined below. In birds and reptiles, the thalamic cell group receiving the ascending projection from the optic tectum is the *nucleus rotundus* (rt, in Figures 1A and 4B). The largest and most conspicuous thalamic nucleus in these non-mammalian forms, the nucleus rotundus has recently been found in birds to project to a well-defined subdivision of the external striatum, the *ectostriatum*.

Judged by its structural characteristics and efferent con-nections, the optic tectum appears certain to convey highly coded information from the retina, not only to effector mechanisms of the brainstem and spinal cord, but also to the forebrain. The path over the mesencephalic tectum, how-ever, is not the only route by which retinal impulses can be transmitted to the forebrain. In most (and possibly all) vertebrates, at least part of the fibers of the optic tract are distributed directly to the thalamus. Of modest volume in most nonmammalian forms, this direct retinothalamic con-nection in mammals becomes a major conduction route for retinal impulses. In most primates, for example, it has con-siderably outgrown the retinotectal pathway. The thalamic region involved in this connection in mammals is the *dorsal nucleus of the lateral geniculate body* (gld, in Figure 2A), a well-defined—and in primates and other advanced forms strikingly laminated—cell group from which originates the massive *optic radiation* distributed to a large occipital region of the neocortex, the so-called *visual cortex*.

In birds and reptiles, the thalamic cell region receiving direct retinal afflux has recently been identified as the *nucleus dorsolateralis anterior* (dla, in Figure 1A), a relatively indistinct territory in the pigeon, but a prominent, circumscript, and highly differentiated complex in the owl. There is convincing evidence, both anatomical and physiological, that this cell region of the avian thalamus projects selectively to a remarkable territory of the cerebral hemisphere, situated in the region of transition between the most dorsal zone of the external striatum (hyperstriatum) and the pallial mantle. The region in question occupies a substantial rostral part of the external prominence of the avian cerebral hemisphere known as the "Wulst." It is characterized by a sharply defined layer of small neurons (granule cells) quite reminiscent of the granular fourth layer of the mammalian visual cortex (Figures 4A and 7).

An important conclusion to be drawn from the foregoing accounting of central sensory systems in nonmammalian forms is that, contrary to earlier views, *the cerebral hemisphere of birds and reptiles receives well-defined projections from specific visual and auditory nuclei in the thalamus. Unlike their mammalian counterparts, however, these thalamo-cerebral afferents, with the exception of the retino-thalamo-cerebral pathway to the Wulst, do not (or at least not immediately) involve the pallial mantle; instead, they are each distributed to a circumscript region of the external striatum.* The question of the extent to which these systems of sensory conduction to the hemisphere are comparable to those of mammals is discussed further in a final section of this chapter.

Neuronal Systems of Nonspecific Afferentation
In the foregoing account of sensory conduction mechanisms believed to be relatively highly modality-specific, only occasional mention has been made of conduction systems of lesser specificity—in other words, systems composed of neurons receiving afferents from more than one single source. It seems likely that, in the neuraxis of all vertebrates, such multi-afferented neurons outnumber the more nearly unimodal nerve cells. Apparently all vertebrates, besides depending heavily for their orientation in the external environment on "labeled lines" conveying information of high spatial and temporal precision, also need for their survival other, more diffusely afferented, neuronal systems.

In vertebrates, neurons corresponding to this general description are found throughout the brainstem and spinal cord. In the rhombencephalon and mesencephalon they occupy major parts of the cross-section; the circumstance that a large part of their territory is traversed at these levels by numerous disseminated fascicles of myelinated fibers has led to the term "formatio reticularis alba et grisea," later

modified to *brainstem reticular formation*. The systemic connotation of the term reticular formation, however, cannot but arbitrarily be limited to regions that have such a "reticulated" gross appearance. On the basis of continuity of fiber connections and similarity of cytological characteristics, it would seem arbitrary not to extend the "brainstem reticular formation" forward beyond the midbrain into certain regions of the thalamus, and through the hypothalamus into the septal region, or down beyond the rhombencephalon into the spinal cord.

Such continuity, needless to say, does not imply functional equivalence throughout. The septal region may be comparable to the rhombencephalic tegmentum in terms of general cellular characteristics; it may be reciprocally connected with the latter by polysynaptic chains of neurons, but the fact remains that neurons of the rhombencephalic reticular formation receive direct afferents from the spinal cord, from the cerebellum, from such secondary sensory cell groups as the vestibular nuclei and the nucleus of the solitary tract, and from the sensorimotor cortex, but those of the septum, by contrast, receive a major afferent fiber system from the hippocampus. Clearly, the reticular formation, however continuous throughout the neuraxis, is by no means an undifferentiated neural continuum; witness also the considerable number of its cytoarchitectural subdivisions that have been recognized in studies by the Nissl method.

The reticular formation is quite generally characterized (1) by being composed of neurons with long, poorly ramified dendrites (a pecularity that has led Ramón-Moliner and Nauta [1966] to suggest the term "isodendritic core" as a substitute for "reticular formation") and (2) by a tendency of its conduction lines to be organized in the form of polysynaptic chains among which, however, some longer axonal lines are not completely lacking. Conduction systems of such predominantly polysynaptic composition can be described as "open systems" in the sense that each line at each synaptic interruption is open to at least several further afferent inputs. Convergence of various different afferent systems on a single neuron indeed appears to be among the most prominent characteristics of the brainstem reticular formation (Figures 1B and 2B include several examples of such heterogeneously afferented neurons).

Understandably, this mode of organization is often described as "nonspecific" or "diffuse." It deserves emphasis, however, that several major *effector manifestations* of the reticular formation cannot be considered diffuse. Among such functions are the maintenance of the respiratory rhythm and other homeostatic processes that require exquisitely specific rather than diffuse neural guidance. Furthermore, a majority of the "motor" systems descending from the forebrain converge on the pool of motor neurons by way of neurons of the brainstem reticular formation. Such reticular intermediaries can be interpreted as *interneurons* of the motor system in the same sense that adheres to this term as applied to groups of neurons near, and channeling impulses to, motoneuronal cell groups in the spinal cord and brainstem (see below). In this context, likewise, the reticular formation, despite its apparently diffuse and "open" organization, can hardly be considered diffuse functionally.

The same considerations apply to *ascending* conduction pathways involving the reticular formation. Although several sensory systems, notably the somatic (in mammals at least), visual, and auditory, have well-developed lemniscal pathways to the forebrain, others appear to lack such relatively "closed" ascending conduction lines, and it therefore must be assumed that information transmitted by such systems can reach the forebrain only by way of intermediary neurons in the rhombencephalic and mesencephalic reticular formation. Such extralemniscal pathways exist in all sensory systems, but they appear to be the only route to the forebrain available to, for example, the visceroceptive system represented by the nucleus of the solitary tract. The information conveyed by this afferent system must be of fundamental significance to the hypothalamic and limbic organizations governing visceral and endocrine effector mechanisms, and it would seem strange, at first glance, that its transmission to the forebrain should be less than direct.

In view of the foregoing considerations, a general interpretation of the reticular formation would seem to be based more appropriately on the effector characteristics of the organization than on the complexity and apparent diffuseness of its afferent relationships. Clearly, several major functional manifestations of the reticular formation must be classified as highly specific, and the "openness" of the structure would in these functions appear to subserve convergent integration, rather than diffusion, of multiple forms of information. Nonetheless, other functional aspects of the reticular formation, notably its well-documented central role in mechanisms determining the general activity state of the organism, could be considered "diffuse," at least from a phenomenological point of view. In this latter class of functioning, the reticular formation appears to have the significance of a general adjustment mechanism of the central nervous system itself, capable of modulating the processing of information in both afferent and efferent systems, and thereby the responsiveness of the organism to its external environment.

The great complexity and wide ramification of its neuronal concatenations have naturally interfered with a detailed anatomical analysis of the reticular formation. Polysynaptic, "short-neuron" systems are much more difficult to trace out by either anatomical or physiological methods than are the more nearly "closed" long-axon conduction pathways. Only relatively long links in the reticular organization have therefore been identified anatomically, and what is known about them has been derived almost

exclusively from studies of mammalian species, the cat and rat in particular. Some of the longer conduction lines involving the reticular formation have been indicated schematically in Figures 1B and 2B, but it must be stressed that these diagrams emphasize the apparent principle rather than the details of the reticular organization. Interphyletic differences in the organization of this widespread neural apparatus are almost certain to exist, but comparative data on the subject are too scarce to supply a basis for phylogenetic conclusions.

THE MOTOR SYSTEM   This term is here used in its customary loose sense to indicate not only the motor neurons but also the neural systems descending upon these "final common pathway" elements in a generally convergent manner.

From quantitative anatomical impressions, it appears likely that most of the neural instructions issued to motor neurons in the spinal cord and brainstem come from a large population of nerve cells, the so-called *interneurons*. Neurons of this category are numerous in both the ventral horn and zona intermedia of the spinal cord, as well as in the dorsolateral (parvicellular) region of the rhombencephalic tegmentum. Together with the motor neurons, these generally smaller "pre-motoneuronal" elements compose a functional system that could be termed the *lower motor system,* an organization consisting in turn of local functional units here to be named *local motor apparatus,* for the reason that each appears to be organized in correspondence with the patterns of movement peculiar to the skeletomuscular mechanism that it controls (arm different from leg, facial musculature different from external eye muscles, and so forth).

The notion of a lower motor system composed of motor neurons and the associated interneurons is based, among other considerations, on the observation that the so-called motor pathways (with considerable variation) appear to terminate for the most part among the pools of interneurons, and to a lesser extent in direct synaptic contact with motor neurons. Such fiber systems converge on the lower motor apparatus from virtually all levels of the brain. In mammalian forms, the better-known of these systems arise from the rhombencephalon (reticular formation and vestibular nuclei), the mesencephalon (superior colliculus and red nucleus), and the sensorimotor region of the neocortex.

Several of these fiber systems, in particular the reticulospinal and reticulobulbar tracts, originate in cell territories converged on by afferents from diverse brain regions, and could thus be viewed as common last links of confluent encephalomotor systems. For example, the magnocellular medial region of the rhombencephalic reticular formation, which gives rise to a substantial reticulospinal pathway, is converged on by fibers from area 6, i.e., the trunk region of the sensorimotor cortex, and from medial, apparently also trunk-related components of the cerebellum. The red nucleus, originating in the rubrospinal and rubrobulbar tracts, likewise receives major fiber afflux from the sensorimotor cortex and the cerebellum, but the cortical and cerebellar regions from which these afferents arise appear to be involved in the motor mechanisms of face and extremities rather than trunk.

Other major effector systems appear to affect the lower motor apparatus only by way of more polysynaptic pathways of transreticular conduction. For example, the principal efferent path descending from the corpus striatum, the *ansa lenticularis* arising from the globus pallidus, does not extend directly beyond the mesencephalon, and it appears likely that impulses traveling over this pathway to the lower motor apparatus must pass at least one additional synapse in the rhombencephalic reticular formation. A similar polysynaptic organization appears to exist in the pathways leading from the forebrain to the lower motor apparatus of the *visceral motor system*. Such pathways arise largely from the hypothalamus, a region of the diencephalon receiving most of its afferent supply from the limbic and olfactory systems (in mammals, from the frontal neocortex as well) of the cerebral hemisphere, and from the mesencephalic reticular formation. No fibers of these descending hypothalamic tracts have been traced beyond the mesencephalon, and it therefore seems likely that the conduction of hypothalamic impulses to the lower visceromotor levels involves synaptic transmissions in the mesencephalic and rhombencephalic reticular formation. Judged by physiological observations (and contrary to the diagramatic representation in Figures 1B and 2B) the course followed by these transreticular visceromotor pathways is quite diffuse in the sense that they are spread over a large part of the cross-section of the brainstem.

Much less is known of the organization of the central motor system in nonmammalians. From normal anatomical descriptions as well as from sparse experimental studies, it appears likely that, at least in birds and reptiles, the descending fiber systems are largely comparable to those of mammals. Comprehensive comparisons, however, with the mammalian motor system must await more detailed experimental studies in a large sample of nonmammalian forms. If the highly variable modes of propulsion characterizing individual species be considered, fundamental variations in the organization of the central motor apparatus seem almost certain to exist, but it cannot be predicted whether such differences will be found to have their major anatomical expression at the level of the lower motor system, at higher levels, or in the disposition of descending conduction systems.

Despite this dearth of comparative data, there is one observation with respect to descending conduction pathways that might hold a clue applicable to the problem of forebrain homology. It is a remarkable fact that only one of

the major fiber systems descending from the mammalian forebrain, namely the neocortical projection, has been found to extend directly (i.e., without synaptic interruption) beyond the mesencephalon. Neither the ansa lenticularis originating from the corpus striatum, nor the conduction pathways descending from the limbic system and hypothalamus, appear to exceed the caudal limits of the mesencephalic reticular formation (Figure 2B). By contrast, a considerable proportion of the corticofugal projection bypasses the mesencephalic level and distributes itself directly to the pontine nuclei—which, in turn, project to the cerebellum (this cortico-ponto-cerebellar system is not included in Figure 2B)—and to the rhombencephalic tegmentum and spinal cord. In most mammalian species, the neocortical efferents to the rhombencephalon and spinal cord form a compact, more or less prominent fiber bundle on the ventral surface of the medulla oblongata, the so-called *pyramidal tract*.

There is some recent evidence that in birds and reptiles, likewise, the fiber systems descending from the limbic system and hypothalamus, as well as the conduction pathways arising from those striatal subdivisions (internal striatum) that are comparable to the mammalian corpus striatum, do not extend directly below the mesencephalon. Naturally, as birds and reptiles do not have a neocortex, it is debatable whether any descending fiber system in such nonmammalian forms could be considered homologous with the mammalian pyramidal tract. Nonetheless, experimental studies in birds have produced evidence of a long fiber system descending from the forebrain and extending directly beyond the midbrain to the rhombencephalic tegmentum and to at least the upper segments of the spinal cord. It is remarkable that this avian cerebrobulbar and cerebrospinal projection system, in contrast to the mammalian pyramidal tract, does not originate in the pallial mantle but, instead, in a dorsal region of the so-called *archistriatum* (the archistriatum, labeled a in Figure 4B, is a ventral component of the "external striatum"; its ventrocaudal part appears to be homologous with the mammalian amygdala). In its course and distribution, this avian fiber system bears a marked resemblance to the corticobulbar fiber bundle described in ungulates by Bagley (1922) and Haartsen and Verhaart (1967).

## Structural and functional homology in the vertebrate forebrain

In the foregoing account, reference was made to a striking contrast between the evolution of the forebrain (diencephalon and cerebral hemisphere) and that of the mesencephalon, rhombencephalon, and spinal cord. The latter, more caudal, subdivisions of the central nervous system have evolved in a fairly stable manner that allows relatively easy interphyletic comparisons to be made for most of the motor and secondary sensory nuclei, and even for major subdivisions of the brainstem reticular formation. The evolution of the forebrain, by contrast, is characterized by considerable architectural modification, which culminates in the abrupt emergence of a neocortex in the pallial mantle of mammals, an event accompanied by seemingly profound changes in the composition of the thalamus and corpus striatum.

The following discussion is addressed largely to the radical departure in organization that distinguishes the mammalian brain from its probable ancestral prototypes.

*Histochemical observations*  The principal features distinguishing the forebrain of mammals from that of nonmammalian forms are vividly illustrated by a comparison with the avian forebrain. The comparison is the more interesting, as birds (which share with mammals a reptilian ancestry) appear to have further elaborated, rather than modified, the basic plan of the reptilian forebrain.

Figure 4 shows two transverse sections of the forebrain of the pigeon. It will be noted that by far the largest part of the section is occupied by structures bearing designations ending with the suffix "striatum," whereas only a small amount of pallium appears. The large number of striatal subdivisions is striking when a bird's forebrain is compared with that of a rat (Figure 5): in this, as in all other mammalian forms, only two major striatal territories are evident, viz., the *caudoputamen* and the *globus pallidus*. It has been widely assumed in the past that the corpus striatum of birds is, in its totality, homologous with, even though far more differentiated than, that of mammals. This assumption was based exclusively on rather gross morphological criteria, such as the striated appearance of the gray matter, comparable positions of the structures in question with respect to the ventricle of the forebrain, and comparably subcortical (rather than cortical) arrangement of the component neurons. Furthermore, the absence of a structure in the pallium of reptiles and birds that could be compared with the mammalian neocortex was taken to imply a virtually *de novo* genesis in mammals of huge populations of specific sensory neurons in the telencephalon, without precedence in ancestral forms.

Reasonable as these assumptions appeared to be, recent

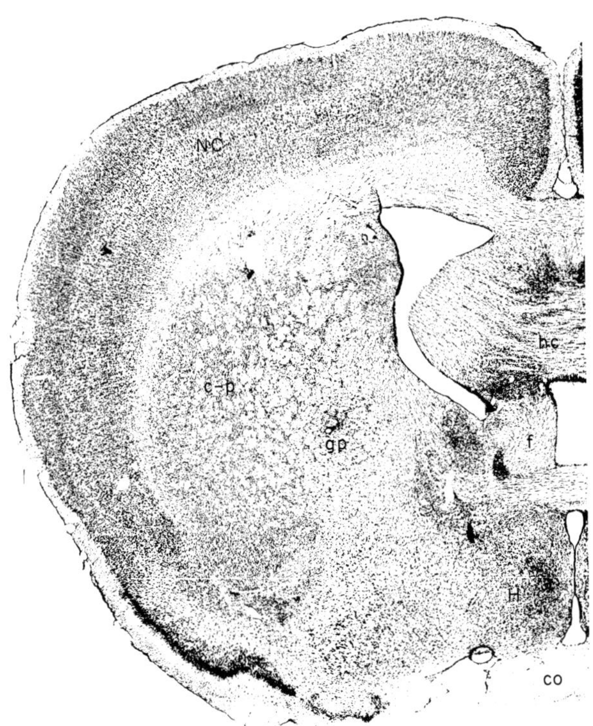


FIGURE 5  Transverse section through the rostral part of the forebrain of a small mammal (rat). Note that the mammalian corpus striatum consists of the caudoputamen (c-p) and globus pallidus (gp), and that it is adjoined laterally and dorsally by the neocortex (NC). The olfactory cortex, characterized by a dark band of pyramidal cells, can be seen at the left lower margin of the photograph. The hippocampus does not appear at this rostral level, but two of its major fiber systems, the hippocampal commissure (hc) and fornix (f) are visible in the section. *Other abbreviations:* co, optic chiasm; H, hypothalamus.

histochemical observations have made it questionable whether, indeed, the entire corpus striatum of birds is comparable to the mammalian corpus striatum. It is well-established that the mammalian caudoputamen contains exceedingly large amounts of both dopamine and acetylcholinesterase. Dopamine in the telencephalon is confined almost exclusively to the caudoputamen (c-p, in Figures 2A and 5), but acetylcholinesterase, although present throughout the telencephalon, is particularly highly concentrated in the same region. Recent studies of the pigeon (Juorio and Vogt, 1967; Karten and Iversen, unpublished) have yielded the unexpected observation that, in the avian telencephalon (Figure 6), such high concentrations of dopamine and acetylcholinesterase are characteristic of the *paleostriatum augmentatum* (pa, in Figure 1B). Furthermore, subsequent experimental studies (Karten, 1969) produced evidence that this particular subdivision of the avian striatum, much like the mammalian caudoputamen, has massive efferent connections with the large-celled, most internal striatal component, which is known in birds as the *paleostriatum primitivum* (p, in Figure 1B), in mammals as the globus pallidus (gp, in Figures 2A and 5). A homology of the avian paleostriatum primitivum with the mammalian globus pallidus is suggested by the observation that each in its respective class is characterized by having a selectively high content of iron compounds, and both project by way of comparable, well-defined fiber systems (the ansa lenticularis) to the mesencephalic tegmentum.

Considered together, these independent forms of evidence strongly indicate that the mammalian caudoputamen corresponds to the avian paleostriatum augmentatum, and the globus pallidus of the mammal to the paleostriatum primitivum of birds. Implied in this notion is the conclusion that the mammalian corpus striatum (caudoputamen and globus pallidus) corresponds to no more than the internal striatum (paleostriatal complex) of birds, and the question arises: What, then, is the nature of the huge and differentiated "external striatum" that overlies the avian paleostriatal complex?

That the external striatum of reptiles and birds appears to have no readily identifiable counterpart in the mammalian brain—identifiable in the gross-morphological sense of similarity in relative position with respect to the ventricle, for example—should not lead to the *a priori* conclusion that this vast neuronal population somehow became "lost" in the transition from the ancestral reptiles to their descendent mammalian forms. Before such a conclusion is drawn, an alternative possibility should be considered, namely, that the neurons composing the external striatum of birds and reptiles have come to occupy a radically different position in the mammalian brain. There is no direct evidence of such translocation, but circumstantial support for the notion would be provided if cell groups in the mammalian brain could be identified, the neural associations and physiological properties of which were comparable to those of the external striatum of nonmammalian forms. Such indirect evidence is indeed available.

All the recent anatomical findings point to the neocortex as the mammalian brain structure most comparable to the nonmammalian external striatum with respect to afferent and efferent relationships. Some of the principal findings supporting this comparison are mentioned in the foregoing overview of the vertebrate brain. In the following account, several further details of the neural connections in question are briefly reviewed.

*The representation of the visual system in the avian forebrain*
As mentioned earlier in this chapter, two separate circumscript pathways appear to be available in birds for the transmission of visual information to the forebrain. One of these leads from the retina to the optic tectum, and from there, via the nucleus rotundus of the thalamus (Karten and Revzin, 1966), to a circumscript subdivision of the external striatum, the ectostriatum (Revzin and Karten, 1966/67). In more detailed studies (Karten and Hodos, in press), the distribution of this tecto-thalamo-ectostriatal conduction pathway was found to be restricted to a cytologically

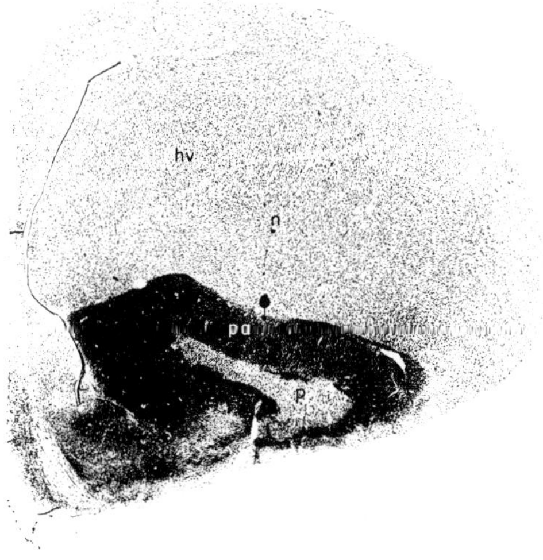


FIGURE 6 Transverse section of the pigeon forebrain stained for acetylcholinesterase by Koelle's method, and counterstained for cells by the Nissl technique. The histochemical reaction identifies the paleostriatum augmentatum (pa) as the forebrain structure containing by far the highest content of acetylcholinesterase. *Other abbreviations:* hv, hyperstriatum ventrale; n, neostriatum; p, paleostriatum primitivum.

distinct central core of the ectostriatum; a more peripheral "periectostriatal" zone, however, receives projections from this central core, as also do the adjoining neostriatum and a laminated population of cells on the dorsolateral surface of the hemisphere. That this tectofugal conduction system indeed plays a vital role in a bird's visual function has been demonstrated by the finding that destruction of the nucleus rotundus in the pigeon causes gross disturbances of visual discrimination, the exact nature of which is currently being investigated (Hodos and Karten, 1966; in press). Curiously, in a recent study Revzin (1967) found the neurons of the nucleus rotundus responsive to photic stimuli delivered at any point within an extremely large part of the visual field, a characteristic contrasting markedly with the generally more restricted receptive-field properties of neurons in other parts of the forebrain (see below).

The tecto-thalamo-ectostriatal conduction pathway of birds bears a resemblance to a conduction route in the mammalian brain leading from the superior colliculus to a relatively well-defined cell group in the caudal part of the thalamus, the nucleus lateralis posterior (Altman and Carpenter, 1961; Morest, 1965; Abplanalp, 1968; Schneider, 1969). This nucleus of the mammalian thalamus appears to project to widespread regions of the neocortex. These regions have not yet been adequately defined, but Snyder and Diamond (1968) have recently reported evidence that the nucleus projects in part to a peripheral region of the visual cortex, the so-called circumstriate belt (the term "striate," in this context, refers not to the corpus striatum but to the central region of the visual cortex, called striate cortex because of a particularly well-developed fiber plexus in its fourth layer, the stripe of Gennari).

The comparison between the respective representations of the visual system in the forebrains of birds and mammals cannot end at this point, for in both classes another important route is available for the conduction of visual impulses to the forebrain. As pointed out in the foregoing overview, in mammals this pathway leads from the retina over the lateral geniculate body of the thalamus to the visual region of the neocortex (striate cortex together with the surrounding "circumstriate belt"). Long considered a unique feature of the mammalian visual system, this massive pathway recently has been found to have an analogue in the avian brain, in the form of a conduction route from the retina to the thalamic nucleus dorsolateralis anterior (Cowan et al., 1961; Karten and Nauta, 1968), a cell group that, in turn, projects to a well-defined region of the "Wulst" of the cerebral hemisphere (Karten and Nauta, 1968). In reptiles, as well, a pathway from the retina to the dorsal thalamus has been identified (Knapp and Kang, 1968; Hall and Ebner, 1969), but in this class the corresponding projection on the cerebral hemisphere has not yet been determined.

The region of the avian Wulst that receives the retino-thalamo-cerebral projection is difficult to classify. Extremely well developed in the owl, this composite region is characterized by the presence of a distinct layer of closely spaced small neurons (granule cells), termed the *nucleus intercalatus of the hyperstriatum accessorium* (iha, in Figure 4A and Figure 7) and quite comparable in appearance to the conspicuous granular fourth layer of the mammalian striate cortex. The termination of the thalamic projection, however, is not entirely confined to this small-celled layer, for it also involves a deeper stratum of larger cells, the *hyperstriatum dorsale* (hd, in Figures 4 and 7). The thalamic projection to this remarkable cell region appears to have a precise ("point-for-point") topographical organization, much like the mammalian geniculo-cortical projection. The parallel between the visual region of the avian Wulst and the mammalian visual cortex is further emphasized by the recent observation (Revzin, 1969) that the neurons of the Wulst of pigeons have several of the physiological characteristics (small receptive fields, columnar organization of neurons with comparable stimulus requirements) that were earlier identified in the neuronal population of the mammalian visual cortex (Hubel and Wiesel, 1962).

The two routes for the conduction of visual impulses to

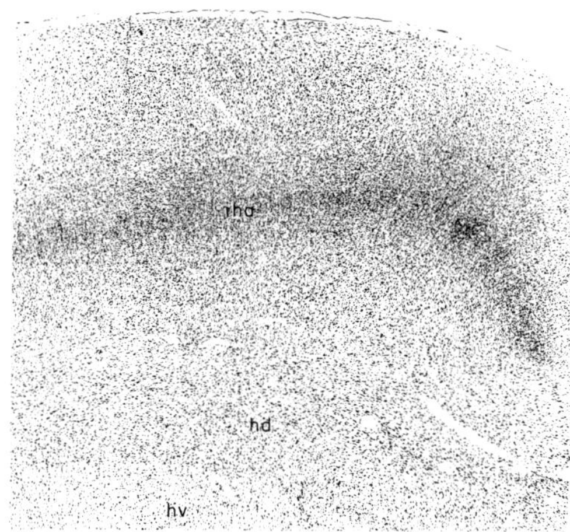

FIGURE 7  Transverse section through the region of the "Wulst" in the owl. Compare with the corresponding region of the pigeon, shown at lower magnification in Figure 4A, and note the great development of the granular layer iha ("nucleus intercalatus of the accessory hyperstriatum") in the owl. Also appearing in the photograph are the hyperstriatum dorsale (hd) and part of the hyperstriatum ventrale (hv).

the avian forebrain are illustrated in diagramatic form by Figure 8.

*Auditory structures in the avian forebrain* The auditory conduction pathway to the forebrain in birds is described briefly in an earlier section of this chapter and is illustrated in semidiagramatic form by Figure 9. As in mammals, a massive lateral lemniscus originating from the cochlear nuclei of the rhombencephalon (Am, Al, Av and Mm, Ml and Mvl) ascends to the mesencephalon, where it terminates in the nucleus mesencephali lateralis dorsalis (MLD), the apparent avian homologue of the mammalian inferior colliculus (Karten, 1967). From this mesencephalic cell group, a well-defined tractus ovoidalis, apparently homologous with the mammalian brachium of the inferior colliculus (BCI), extends forward to terminate in the nucleus ovoidalis (Ov) (Karten, 1967), a circumscript thalamic cell group, which, in turn, projects to field L, a medial region of the neostriatum characterized by a high density of its neuronal population (Karten, 1968; see also Figure 4B).

The problem of forebrain homology appears in sharp

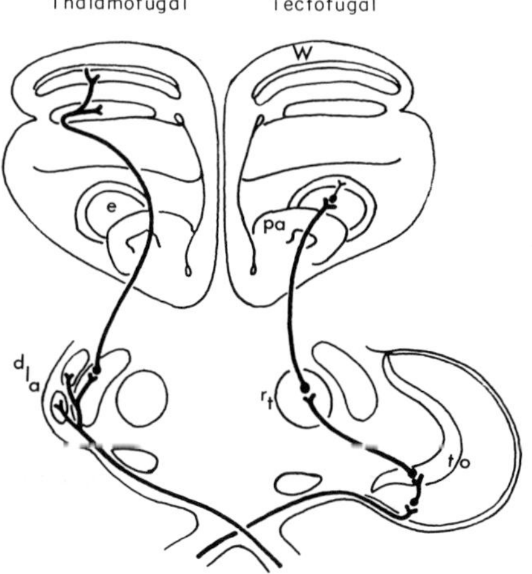


FIGURE 8 Schematic drawing of the two principal pathways leading from the retina to the cerebral hemisphere in the owl. All retinal efferents, forming the optic tract, cross in the optic chiasm to the opposite side. The right half of the diagram shows the path over the optic tectum (to) and nucleus rotundus thalami (rt) to the ectostriatum (e), a ventral component of the external striatum. On the left, the conduction route over the nucleus dorsolateralis anterior thalami (dla) is shown; this pathway is distributed to various cell layers of the "Wulst" (W), a complex part of the pallial mantle. The drawing is oriented in the frontal plane. pa, paleostriatum augmentatum.

outline when the avian auditory and visual systems are compared with those of mammals. It appears likely enough that the avian nucleus ovoidalis thalami corresponds to the mammalian medial geniculate body, for each of these nuclei in its vertebrate class is the major thalamic recipient of the ascending auditory lemniscus. Major problems are, however, encountered in extending the suggested homology to the respective efferent relationships of the two thalamic cell groups: whereas the medial geniculate body projects primarily to the auditory area of the neocortex, the nucleus ovoidalis has its major efferent connection, not with any region of the avian pallial mantle, but with field L of the distinctly noncortical neostriatum. Many of the same problems arise in attempts to identify homologous structures in avian and mammalian visual systems. As in the auditory realm, homologies appear fairly self-evident up to the thalamic level: there is good reason to compare the avian nucleus rotundus to the mammalian nucleus lateralis posterior on the basis of common afferent relationships with the mesencephalic tectum, and the avian nucleus dorsolateralis anterior to the lateral geniculate body of mammals by the criterion that each in its class appears to be virtually the only thalamic cell group to receive afferents directly from the retina. Nonetheless, beyond the thalamus, comparison becomes questionable, for the mammalian nuclei project mainly to various components of the visual neocortex, whereas the avian nucleus rotundus projects to the noncortical ectostriatum, and the nucleus dorsolateralis anterior to a region that forms a transition between the hyperstriatum and the pallial mantle (the Wulst).

### Discussion and conclusions

The major conclusions to be drawn from the foregoing account can be stated as follows:

1. The "external striatum" of reptiles and birds is absent as such from mammals, in which, conversely, a "neocortex" appears in the pallial mantle.

2. The nonmammalian external striatum is comparable to the mammalian neocortex in the sense that, as is the latter, it is the major recipient of specific visual and auditory projections from the thalamus, and appears to be the only forebrain component to project directly to the rhombencephalon and the spinal cord.

Naturally, in view of these conclusions, it is tempting to suggest that the mutual exclusiveness of the two cerebral structures may be more than coincidental. In fact, it would seem reasonable to postulate that *the same neurons that in reptiles and birds compose the large external striatum, in mammals have come to occupy the pallial mantle and form a major proportion of the cell population of the neocortex.*

It is interesting that the same hypothesis arrived at here by histochemical evidence and comparisons of the respective

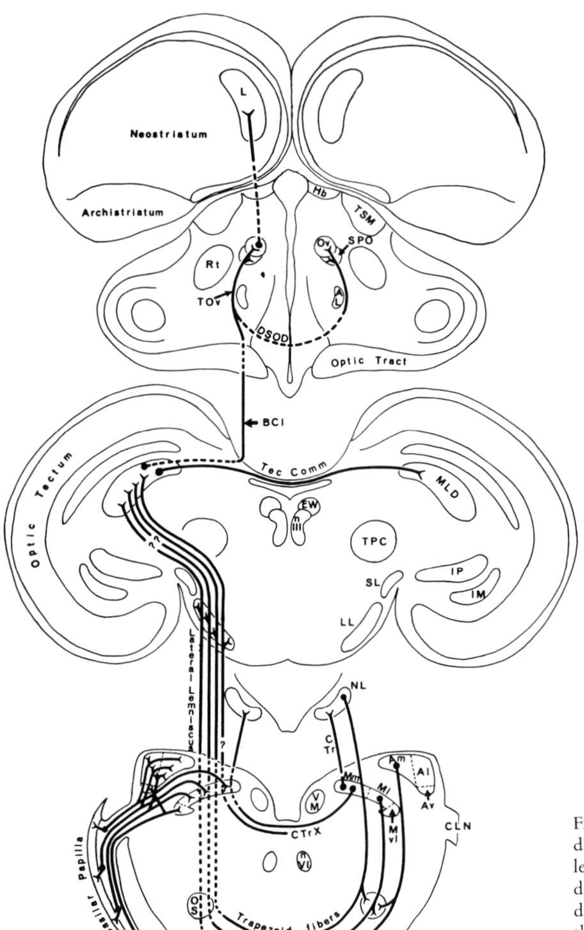


Figure 9 Diagramatic representation of the principal auditory conduction pathways of the pigeon. Shown on the left side of the lower drawing is the auditory nerve and its distribution to the cochlear nuclei. The remainder of the diagram illustrates the conduction pathway ascending from the cochlear nuclei to the nucleus ovoidalis thalami, and thence to field L of the neostriatum. (From Boord, 1969).

neural connections was postulated earlier by Källén (1962) on the basis of comparative-embryological findings. The external striatum is known to arise in embryonic development by cell proliferation in the so-called dorsal ventricular ridge. In nonmammalian forms, its neuroblasts mature *in situ,* that is to say, without radical migration away from their matrix. A dorsal ventricular ridge also appears as a zone of intensive cell proliferation in early mammalian embryos, but in later stages of embryonic development the structure gradually dwindles. Is it possible that, as sche-

matically indicated in Figure 10, the neuroblasts generated in this region in the mammalian embryo migrate around the lateral corner of telencephalic ventricle and invade the pallial mantle? Direct proof of such migration is not available and would be difficult to obtain, for it would require *selective* radioactive labeling of the dorsal ventricular ridge, followed at later developmental stages by the identification of labeled neuroblasts in the pallial mantle.

In the absence of such direct proof, the present hypothesis can be supported only by indirect evidence, in particular by

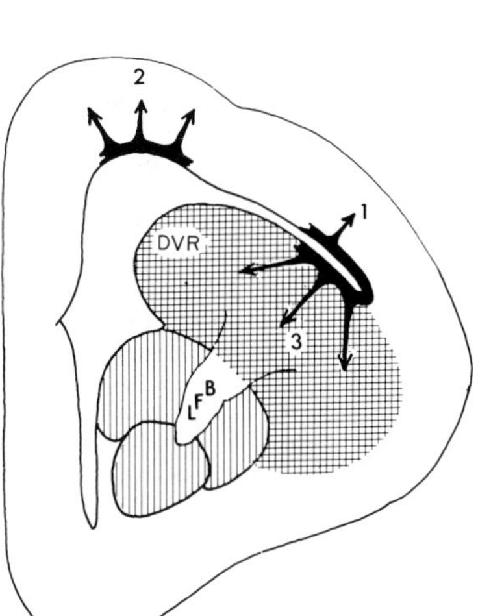
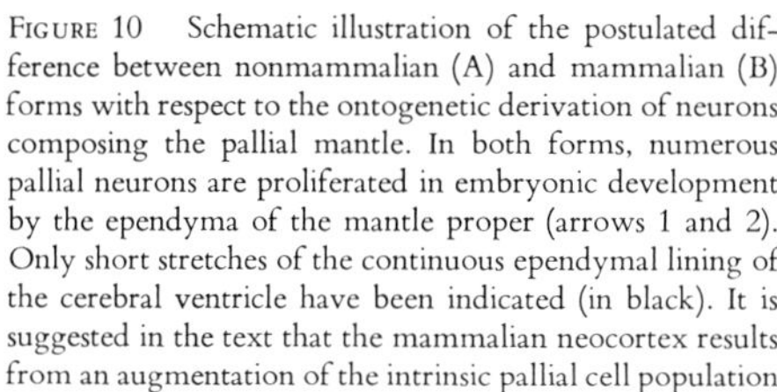
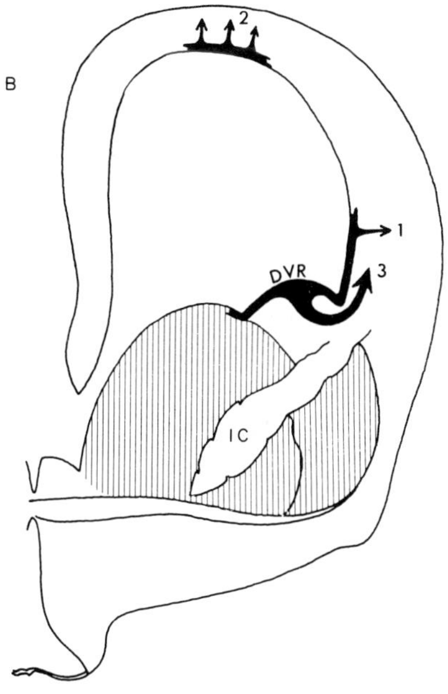


FIGURE 10   Schematic illustration of the postulated difference between nonmammalian (A) and mammalian (B) forms with respect to the ontogenetic derivation of neurons composing the pallial mantle. In both forms, numerous pallial neurons are proliferated in embryonic development by the ependyma of the mantle proper (arrows 1 and 2). Only short stretches of the continuous ependymal lining of the cerebral ventricle have been indicated (in black). It is suggested in the text that the mammalian neocortex results from an augmentation of the intrinsic pallial cell population by neuroblasts proliferated by the ependyma of the dorsal ventricular ridge (DVR, arrow 3 in B); the corresponding proliferation in nonmammalian forms produces the external striatum (arrows 3, in A). As in Figures 1 and 2, the paleostriatal complex of the nonmammalian brain and the caudoputamen-globus pallidus of the mammal are indicated by vertical shading; the nonmammalian external striatum is shown by cross-hatching. *Abbreviations:* DVR, dorsal ventricular ridge; IC, internal capsule; LFB, lateral forebrain bundle.

the observation that major neural connections of the nonmammalian external striatum are strikingly comparable to those of the mammalian neocortex. Obviously, the hypothesis is based on two assumptions: first, that the neurons originating in a particular matrix, or even in particular territories thereof, are somehow so specified as to be allowed synaptic connections only with particular neuronal systems elsewhere in the brain, and second, that such specifications remain, in principle, unchanged across class boundaries, in this case across the transition from reptiles to mammals. Some support for the first of these assumptions is supplied by a recent observation in the golden hamster suggesting that the fibers of the developing optic tract are subject to considerable territorial constraints in establishing their synaptic contacts (Schneider and Nauta, 1969). The second

of the aforementioned premises is at present only inferential and, although supported by the present data, cannot yet be fully evaluated.

It should be emphasized that the present notion does not imply that the whole of the mammalian neocortex is homologous with the nonmammalian external striatum. It is possible, and even likely, that the neocortex contains large numbers of neurons that were proliferated by the ependyma of the pallial mantle proper. It would therefore seem more appropriate to suggest that the neuronal population of the external striatum has become incorporated into the mammalian cortex, and thus to formulate the notion of *homologous neurons* rather than homologous composite structures. For example, it would seem conceivable that field L of the nonmammalian external striatum corresponds only to those neurons of the mammalian auditory cortex that stand in direct synaptic contact with the fibers of the auditory thalamocortical radiation.

A further point to be emphasized is the uncertainty as to whether all the sensory areas of the neocortex contain neuronal aggregates that have homologous counterparts in the nonmammalian external striatum. As stated earlier in this chapter, no circumscript cell territories specifically representing the somatic sensory system have been identified so far either in the thalamus or in the external striatum of nonmammalian forms. Future studies may yet show such cell groupings to exist, but until such time it must be considered possible that the somatic sensory cortex of the mammal has no functional correspondent in the nonmammalian external striatum.

For somewhat different reasons, the same doubt could be entertained with respect to the area striata of the mammalian visual cortex. As pointed out earlier, this principal recipient of the retino-geniculo-cortical conduction system has a parallel of sorts in the avian Wulst, a complex structure that straddles the transition between the external striatum and the pallium. Observations in the chick embryo (Kuhlenbeck, 1938; Northcutt, 1969) indicate that the Wulst develops from the ependyma of the pallium rather than that of the external striatum, a conclusion supported by the recent finding that in turtles (which, as do all reptiles, lack a structure readily comparable to the avian Wulst) the corresponding thalamic nucleus (dorsolateralis anterior) projects to the pallial mantle (personal communication from Dr. Ford F. Ebner). Taken together, these data suggest that the mammalian striate cortex has its ancestral prototype in the pallial mantle, rather than in the external striatum of nonmammalian forms.

The evidence reported and discussed in the foregoing parts of this paper suggests that the mammalian neocortex should not be regarded as a huge mass of neurons that had no precedence in ancestral forms. The comparison of mammalian ascending and descending systems with their nonmammalian counterparts leads to a quite different supposition, namely, that the neocortex resulted from a translocation of large masses of neurons, which in ancestral forms occupied subcortical stations, in particular the region of the external striatum. The indisputable uniqueness of the neocortex, according to this view, would reside not in any novel origin, but rather in the *alignment* of its neural components.

The question naturally arises as to what benefits might accrue to organisms in which subcortical cell groups are realigned so as to become integrated into a cortical organization. It could be argued that the characteristic "stacking" of neurons in layers, combined with the parallel, palisade-like orientation of apical dendrites, entails greater economy in the expenditure of axonal material, and thereby permits a greater number of neurons per unit volume to be connected with one another in cortical tissue than in subcortical gray matter. This point may have some relevance in a consideration of the respective evolutionary potentialities of cortical and noncortical neural tissues, but it remains to be seen whether the neocortical neurons of such mammals as the hedgehog or opossum are more numerous or more intricately interconnected than are the neurons of the formidable and differentiated external striatum of such avian forms as the owl. Moreover, it appears that whatever functional advantages may result from a cortical reorganization of subcortical gray matter are not necessarily manifested (as pointed out by Hodos elsewhere in this volume) either in a greater capacity for sensory discrimination or in greater refinements of movement. The essential functional characteristics of neocortical tissue as opposed to subcortical neuronal groupings have escaped definition thus far, but there is reason to hope that some pertinent answers may be provided by more detailed physiological comparison of neocortical areas with their apparent subcortical counterparts in nonmammalian forms. Obviously, comparative-anatomical studies, such as here discussed, cannot be expected to yield the answers. It can be said of neuroanatomical studies what was said of Coronado's children: It is perhaps not so important that they did not find the gold, as that they found a place in which to search for it.


## REFERENCES

ABPLANALP, P., 1968. An experimental neuroanatomical study of the visual system in tree shrews and squirrels. Massachusetts Institute of Technology, Cambridge, Massachusetts, doctoral thesis.

ALTMAN, J., and M. B. CARPENTER, 1961. Fiber projections of the superior colliculus in the cat. *J. Comp. Neurol.* 116: 157–177.

BAGLEY, C., JR., 1922. Cortical motor mechanism of the sheep brain. *Arch. Neurol. Psychiat.* 7: 417–453.

BOORD, R. L., 1969. The anatomy of the avian auditory system. *Ann. N.Y. Acad. Sci.* 167: 186–198.

COWAN, W. M., L. ADAMSON, and T. P. S. POWELL, 1961. An experimental study of the avian visual system. *J. Anat.* 95: 545–563.

HAARTSEN, A. B., and W. J. C. VERHAART, 1967. Cortical projections to brain stem and spinal cord in the goat by way of the pyramidal tract and the bundle of Bagley. *J. Comp. Neurol.* 129: 189–201.

HALL, W. C., and F. F. EBNER, 1969. Thalamotelencephalic projections in a turtle (*Pseudemys scripta*). *Anat. Rec.* 163: 193 (abstract).

HERRICK, C. J., 1922. Neurological Foundations of Animal Behavior. Henry Holt, New York.

HODOS, W., and H. J. KARTEN, 1966. Brightness and pattern discrimination deficits in the pigeon after lesions of nucleus rotundus. *Exp. Brain Res.* 2: 151–167.

HODOS, W., and H. J. KARTEN. Visual intensity and pattern discrimination deficits after lesions of ectostriatum in pigeons. *J. Comp. Neurol.* (in press).

HUBEL, D. H., and T. N. WIESEL, 1962. Receptive fields, binocular interaction and functional architecture in the cat's visual cortex. *J. Physiol.* (*London*) 160: 106–154.

JUORIO, A. V., and M. VOGT, 1967. Monoamines and their metabolites in the avian brain. *J. Physiol.* (*London*) 189: 489–518.

KÄLLÉN, B., 1962. Embryogenesis of brain nuclei in the chick telencephalon. *Ergebn. Anat. Entwickl.* 36: 62–82.

KARTEN, H. J., 1967. The organization of the ascending auditory pathway in the pigeon (*Columba livia*). I. Diencephalic projections of the inferior colliculus (nucleus mesencephali lateralis, pars dorsalis). *Brain Res.* 6: 409–427.

KARTEN, H. J., 1968. The ascending auditory pathway in the pigeon (*Columba livia*). II. Telencephalic projections of the nucleus ovoidalis thalami. *Brain Res.* 11: 134–153.

KARTEN, H. J., 1969. The organization of the avian telencephalon and some speculations on the phylogeny of the amniote telencephalon. *Ann. N.Y. Acad. Sci.* 167: 164–179.

KARTEN, H. J., and W. HODOS. Telencephalic projections of the nucleus rotundus in the pigeon (*Columba livia*). *J. Comp. Neurol.* (in press).

KARTEN, H. J., and W. J. H. NAUTA, 1968. Organization of the retinothalamic projections in the pigeon and owl. *Anat. Rec.* 160: 373 (abstract).

KARTEN, H. J., and A. M. REVZIN, 1966. The afferent connections of the nucleus rotundus in the pigeon. *Brain Res.* 2: 368–377.

KNAPP, H., and D. S. KANG, 1968. The retinal projections of the side-necked turtle (*Podocnemeis unifilis*) with some notes on the possible origin of the pars dorsalis of the lateral geniculate body. *Brain, Behav., Evol.* 1: 369–404.

KUHLENBECK, H., 1938. The ontogenetic development and phylogenetic significance of the cortex telencephali in the chick. *J. Comp. Neurol.* 69: 273–301.

MOREST, D. K., 1965. Identification of homologous neurons in the posterolateral thalamus of cat and Virginia opposum. *Anat. Rec.* 151: 390 (abstract).

NORTHCUTT, R. G., 1969. Discussion of the preceding paper (by H. J. Karten). *Ann. N. Y. Acad. Sci.* 167: 180–185.

RAMÓN-MOLINER, E., and W. J. H. NAUTA, 1966. The iso-dendritic core of the brain-stem. *J. Comp. Neurol.* 126: 311–335.

REVZIN, A. M., 1967. Unit responses to visual stimuli in the nucleus rotundus of the pigeon. *Fed. Proc.* 26: 656 (abstract).

REVZIN, A. M., 1969. A specific visual projection area in the hyperstriatum of the pigeon (*Columba livia*). *Brain Res.* 15: 246–249.

REVZIN, A. M., and H. J. KARTEN, 1966/67. Rostral projections of the optic tectum and the nucleus rotundus in the pigeon. *Brain Res.* 3: 264–276.

ROSE, M., 1914. Über die cytoarchitektonische Gliederung des Vorderhirns der Vögel. *J. Psychol. Neurol.* 21: 278–352.

SCHNEIDER, G. E., 1969. Two visual systems: Brain mechanisms for localization and discrimination are dissociated by tectal and cortical lesions. *Science* (*Washington*) 163: 895–902.

SCHNEIDER, G. E., and W. J. H. NAUTA, 1969. Formation of anomalous retinal projections after removal of the optic tectum in the neonate hamster. *Anat. Rec.* 163: 258 (abstract).

SNYDER, M., and I. T. DIAMOND, 1968. The organization and function of the visual cortex in the tree shrew. *Brain, Behav., Evol.* 1: 244–288.

STRONG, O. S., and A. ELWYN, 1964. Human Anatomy (R. C. Truex and M. B. Carpenter, editors), 5th edition, Williams and Wilkins, Baltimore, p. 8.

*J. psychiat. Res.*, 1971, Vol. 8, pp. 167–187. Pergamon Press. Printed in Great Britain.

# THE PROBLEM OF THE FRONTAL LOBE:
# A REINTERPRETATION

WALLE J. H. NAUTA

Department of Psychology, Massachusetts Institute of Technology, Cambridge, Mass.

## INTRODUCTION

THE FRONTAL lobe, despite decades of intensive research by physiologists, anatomists and clinicians, has remained the most mystifying of the major subdivisions of the cerebral cortex. Unlike any other of the great cerebral promontories, the frontal lobe appears not to contain a single sub-field that could be identified with any particular sensory modality, and its entire expanse must accordingly be considered association cortex. It should, perhaps, not be surprising in view of this circumstance alone that loss of frontal cortex, in primate forms in particular, leads to a complex functional deficit, the fundamental nature of which continues to elude laboratory investigators and clinicians alike. The purpose of this paper is, to review some aspects of this deficit in animals and man, and to inquire to what extent the consequences of frontal-lobe lesion can be evaluated in neurological terms.

It is unfortunate that the search for neural substrata of the frontal lobe syndrome is hampered by a nearly complete absence of data obtained by modern neurophysiological methods. As a consequence, no entirely satisfactory hypothesis concerning the neural mechanisms of the frontal lobe would seem attainable at present. On the other hand, largely as a result of several excellent anatomical studies reported within the past 5 years, the neural associations of the frontal lobe, in particular those with other regions of the cerebral cortex, have come into sharper focus, and it is on the basis of these anatomical findings that an attempt could be made to formulate a general and preliminary notion of the major neural mechanisms in which the frontal lobe is involved. Needless to say, no concept so exclusively based upon neuroanatomical data can be stated in any but very general terms, for the value of anatomical studies lies largely in the identification of specific questions that can be explored by physiological methods.

## FUNCTIONAL CONSIDERATIONS

### 1. *Observations in animals*

Monkeys subjected to extensive bilateral ablation of granular frontal cortex have long been noted to exhibit hypermotility and hyperreactivity to external stimuli. Little further information concerning their functional impairment was obtained until in 1935 JACOBSEN[1] reported his now classical observation that primates deprived of their prefrontal cortex

A

perform little better than at the level of chance in the so-called *delayed-response test*. In this simple task, that is quickly mastered by intact chimpanzees, monkeys, cats and dogs, the animal is shown under which one of two or more identical covers food is placed, but it is forced to postpone retrieval of the food for a short period ( usually $\frac{1}{2}$–2 min) during which the scene of the choice is hidden from its view by an opaque screen. The initial impression that the 'frontal animal' suffers from a memory loss, and while waiting for the screen to go up 'forgets' where he saw the food being hidden, has been effectively refuted, and it now seems certain that frontal-lobe ablation affects a response-guidance other than memory in the customary sense. In this test, as also in the somewhat related delayed-alternation (alternative baiting of the left and right one of an identical pair of containers) and multiple-choice tests (all covers marked by a differently-shaped object, one of which serves as the clue for bait in one block of trials, another one in a contiguous trial-block, and so on), 'frontal' animals exhibit a characteristic tendency to perseverate a particular choice, despite the obviously poor reward average. It has been suggested that this 'rigidity of central sets' actually constitutes the fundamental deficit[2] but KONORSKI and LAWICKA,[3] on the basis of a remarkable series of experiments in dogs and cats, arrived at a somewhat different interpretation according to which the *conditioning signal* (e.g. the being shown which box contains bait) undergoes an abnormally rapid decay during the delay, and is thus overcome by the essentially unguided *release response* triggered by the actual presentation of the choice. Perseveration, according to Konorski and Lawicka, is a relatively unspecific phenomenon that is encountered in a variety of functional impairments (often, for example, in aphasia). As an alternative hypothesis, Konorski and Lawicka suggest an abnormally strong release-response overriding a normal signal trace, a suggestion which would appear compatible with another observation made in 'frontal' monkeys[4,5] in the so-called *go–no go* tests. In this testing procedure the animal must learn not only to respond to a particular signal (either auditory or visual) but also, to withhold response to another signal that randomly alternates with the first. Under such conditions, monkeys deprived of their frontal cortex score more errors of commission than do normal monkeys. It must be remarked, however, that it still appears uncertain to what extent this deficiency is due to an actual perceptual impairment rather than to the nature of the go–no go test proper.

Last to be mentioned here is the observation that, in object discrimination tests, 'frontal' monkeys show an increased tendency to prefer any novel object over a tried, familiar one, even if the experiment is so arranged that the choice of a novel object is never rewarded. Although this novelty-preference would at first glance seem incompatible with the choice-reiteration noted above, both could conceivably result from one and the same guidance failure in a response that in normal subjects is more stably determined by the previously experienced consequence of the choice.

### 2. *The frontal-lobe syndrome in man*

Complex and difficult of analysis as are the consequences of frontal lobe ablation in animals, the frontal lobe syndrome in man is even more bewildering. As a result of its complexity, current concepts concerning the function of man's frontal lobes range between two widely separate extremes: on the one hand the notion that this part of the brain is crucially involved in man's highest mental faculties;[6,7] on the other extreme the conclusion that the only defects objectively demonstrable in frontal-lobe patients are of a perceptual nature, all other impairments being too variable and vague to serve as reliable characteristics of frontal-lobe dysfunction.[8]

The wide divergence between these two views may be attributable in part at least to the common difficulty of determining clinically the precise extent and location of the cerebral tissue damage in human patients. As emphasized by Teuber,[9] tumors of the frontal lobe are likely to have reached considerable size by the time the patient is seen by a neurologist, and in such cases it is difficult to determine whether the symptomatology may not have become complicated by a more general impairment of brain function. Penetrating traumatic lesions of the frontal lobes are only rarely explored surgically, and their result in tissue loss thus as a rule remains undetermined. In fact, the only category of patients in whom the extent of the tissue defect is usually known with reasonable accuracy are those who have been subjected to surgical ablations performed in order to alleviate epileptic disorders.

Most controversial among the symptoms of frontal-lobe disorder in man are the psychiatric signs, in particular mood changes (euphoria, irritability or its opposite: emotional indifference), and character changes (boastfulness, impetuousness, lack of initiative and other behavioral disorders). Significant and practically important though such general personality changes may be, they show much individual variation and for the present must therefore remain somewhat at the periphery of the search for common denominators in the frontal-lobe syndrome. This search is still directed largely toward an evaluation of the patient's perceptual and motor capacities, and his strategies in performing certain relatively simple tasks. Some results of such analyses will be reviewed in the following account.

In the course of detailed studies of patients with gunshot wounds of the frontal lobe, Teuber and his associates[8,9] have identified certain perceptual impairments, notably such affecting the maintenance of the perceived vertical during passive tilt of the body, the interpretation of line drawings with ambiguous perspective, and the capacity of reversing standpoint in dealing with mirror images of the body. Their tests, in addition, disclosed in frontal-lobe patients a significant impairment in the speed and efficiency with which a visual array is scanned for detail (see also Luria[10]). In discussing these functional deficits, Teuber[9] arrives at the important suggestion that all are attributable to the loss of a mechanism of 'corollary discharge', i.e. a flow of impulses from the central effector organization to central sensory structures, ". . . presetting the latter for those predictable changes of input that will be the consequences of the particular motor output."*

Among the results obtained with more complex testing procedures, the observations reported by Milner[11] and Luria[10,12] are particularly noteworthy. Milner's findings have demonstrated a profound and characteristic inability of frontal-lobe patients to achieve normal scores in the Wisconsin modification of the Weigl card-sorting test, a test in which the patient is asked to sort 128 cards on the basis of any one of three arbitrary criteria

---

* Some mechanism of this nature must be postulated to explain the well-known phenomenon that actively executed eye movements do *not* cause the illusion that the visual scene is moving, whereas by contrast, passive displacements of the eyeball, e.g. by finger push, do cause such an impression. In the 'active' case there must have been some form of forewarning to the effect that the perceived shift of images over the retina should be attributed to the impending eye movement.

(color, shape or number of the figures shown on the cards). The patient is given no verbal instruction as to which criterion is valid, and is required to determine the criterion pragmatically, on the basis of being told at each placement whether his choice is right or wrong. It is remarkable that frontal-lobe patients in this testing situation show no noteworthy difficulty in deducing the required strategy from the signals given by the observer. Their characteristic deficit appears when, at some point in the procedure, the observer arbitrarily changes the criterion (e.g. from color to number), thus requiring the patient to identify and follow a new ordering system. Under such conditions, frontal-lobe patients tend to perseverate the original strategy, even in the face of an ever-mounting score of errors. In the words of MILNER[11]: " . . . it does not seem accurate to attribute this poor test performance to a loss of abstract thought, since such patients often state spontaneously that 'it has to be the color, the form or the number,' although they seem unable to recognize the possibility of a change once a particular pattern of response has become established. They thus show a curious dissociation between the ability to verbalize the requirements of the test and the ability to use this verbalization (and other verbal cues) as a guide to action." Superfluous to say that the perseveration shown by frontal-lobe patients in the card-sorting test strongly recalls the high score of reiterative errors recorded for 'frontal' monkeys in various testing procedures.

LURIA and his colleagues[10,12] have identified a similar perseverative tendency in frontal-lobe patients. When asked to draw a sequence of simple figures, for example, "a cross, two circles and a triangle," the frontal-lobe patient may simply draw four crosses. This 'inertia' may declare itself dramatically when the patient is asked to perform a task requiring an orderly sequence of separate steps. In such cases, he may perseverate an early phase of the action, but it is important to note that he may, instead, be side-tracked in a direction that could have been appropriate in the context of another task. LURIA and HOMSKAYA[12] note that such patients evince no dismay at their failure to achieve the required goal, and suggest " . . . a deficit in matching of action carried out with the original intention . . ." as a central characteristic of the frontal-lobe syndrome. Particularly interesting among Luria and Homskaya's findings is their observation that this derailment of programmed behavior is remarkably refractory to the normally quite compelling effect of repeated verbalization of the required steps by the observer, or even by the patient himself, throughout the process. Such dissociation of the verbal signal from the subject's action appears closely related to MILNER's[11] observations in frontal-lobe patients in the card-sorting test. As an electrophysiological corollary of this behavioral disconnection, LURIA[10] cites the following important observation by his co-workers Homskaya and Simernitskaya: frontal-lobe patients—but not patients with more posterior cerebral lesions—fail to show an effect that in normal subjects characteristically follows a verbal instruction to await a visual or tactile signal, namely, an enhancement of the electrical potentials evoked in the corresponding sensory area by the signal. LURIA[10] concludes from this observation " . . . that the frontal lobes play a significant part in the regulation of the active states started by a verbal instruction."

Of considerable significance among the test results reported for frontal-lobe patients appear to be MILNER's[11] findings in the so-called stylus-maze test. According to Milner, the poor performance of frontal-lobe patients in this tracing task is attributable not to any form of spatial disorientation but, instead, to an apparent inability to comply reliably with the 'rules of the game'. Such patients often seem unable to curb a propensity for taking illegal shortcuts (such as moving the stylus diagonally across the board); as in the card-sorting test, they appear aware of their mistakes (in the sense that they can verbalize these) but unable to modify their strategy accordingly.

### 3. *Comment*

It is clear even from the foregoing highly fragmentary review that the frontal-lobe disorder is characterized foremost by a derangement of behavioral programming. One of the essential functional deficits of the frontal-lobe patient appears to lie in an inability to maintain in his behavior a normal stability-in-time: his action programs, once started, are likely to fade out, to stagnate in reiteration or to become deflected away from the intended goal. The fact that even his self-admitted awareness of a mis-match between the purpose and the result of his actions fails to affect his strategy suggests an inadequate 'internalization' of all those error—or error-approach signals, including even self-directed verbal commands, that normally modulate the evolvement of behavioral programs. The same failure to register (i.e. to furnish a stable constraint) could be thought to be involved in his tendency to break game-rules. If this indeed be the case, it would seem probable that 'feed-back' of opposite sign likewise would fail to become incorporated in the mechanisms of behavioral guidance. It could be asked, for example, if the reiterative phenomena illustrated so forcefully by Luria and his co-workers could not be due in part at least to a 'getting lost' of a message reporting successful accomplishment of initial or intermediate stages of the program.

A state of affairs as here suggested would amount to one in which action programs can be aimed at their intended goal, but must evolve under conditions of severely impaired functioning of some guidance mechanism associated (but not necessarily identical) with that determining the subject's affective responses to the apparent success or failure of his approach. Depending on circumstances, the effect of such inadequate guidance could be either an abnormal inflexibility of the program or an excessive vulnerability to interfering events.

### NEURAL ASSOCIATIONS OF THE FRONTAL LOBE

In view of the complex and evasive character of the frontal-lobe syndrome it is appropriate to ask what the nature might be of the neural information normally received and processed by the frontal cortex, and what efferent channels might account for its unique role in behavioral programming.

Obviously, nothing short of a comprehensive analysis of the response determinants of single cortical units can be expected to answer the question as to the informational content of neural impulses converging upon the frontal cortex. Studies of this sort (exemplified by Hubel's contribution to this volume) recently have provided extremely valuable insights into the feature-extracting mechanisms of the visual, somatosensory and auditory fields of the cerebral cortex. However, the problems encountered in exploring these modality-specific cortical fields are likely to be multiplied many times over in regions such as the frontal cortex which are certain to be associated with more than one sensory modality.

Especially the problem of identifying the natural stimuli consistently capable of eliciting responses by a sufficiently large number of cortical units in such regions might prove quite elusive. However that may be, explorations of this sort have not yet provided the data which alone could identify the nature of the information received, and the manner in which it is processed by the frontal cortex. Consequently, little more than anatomical data are currently available from which to draw inferences related to this question. This anatomical information, as the following account is intended to show, is by no means negligible.

## 1. *Anatomical definition*

The term, frontal lobe or 'prefrontal cortex', has come to denote the cortical field that extends forward of the 'premotor' area 6 and is projected upon by the mediodorsal nucleus of the thalamus. In primates and carnivores, this large cortical expanse covers the frontal pole of the hemisphere, but in the rat, and probably in other rodent species as well, the corresponding and relatively smaller region is composed of two widely separate sub-fields: one confined to the medial surface of the hemisphere, the other occupying the dorsal bank of the rhinal sulcus, and neither extending forward far enough to cover the frontal pole[13]. Apart from such differences of relative size and topography, there is another reason to believe that the prefrontal cortex reaches vastly different stages of development in different mammalian lineages: Only in primates does the larger rostral part of the field exhibit the cytoarchitectural features of a granular cortex. In the cat, dog, sheep, and rat, apparently the only non-primate species in which the prefrontal cortex has been delimited by experimental methods, the field is throughout of an agranular structure; in the rat its laminar pattern is so poorly differentiated that it is nearly indistinguishable from the anterior cingulate cortex with which its medial sub-field is caudally continuous. In fact, on cytoarchitectural grounds alone the existence of a non-primate homologue of the primate prefrontal cortex could be questioned, and it is largely the constancy of its afferent relationship with the mediodorsal nucleus of the thalamus that has allowed a prefrontal cortex to be identified in non-primate forms.

Since the anatomical account to follow is based very largely upon observations in the rhesus monkey, a brief description of the main anatomical features of the prefrontal cortex of this species would seem relevant. As shown in Fig. 1, the caudal border of the prefrontal cortex is marked by the deep *arcuate sulcus* which delimits it from the agranular 'motor cortex' of the precentral gyrus (areas 4 and 6 of Brodmann). The only other macroscopic features of importance are: (1) the general shape of the frontal lobe which allows a fairly flat medial surface to be distinguished from a concave ventral (or orbital) aspect and a lateral convexity, and (2) the deep *principal sulcus* which divides the lateral convexity into a dorsal and a ventral field.

The cortex forming the most caudal part of the prefrontal region, i.e. the rostral bank of the arcuate sulcus, is structurally speaking something of a transition zone between the agranular 'motor cortex' behind it and the distinctly granular cortex that occupies the large rostral remainder of the frontal lobe. This crescent-shaped 'dysgranular' zone of the prefrontal cortex, labelled area 8 by Brodmann, is customarily referred to as the 'frontal eye field'. The justification for this label lies in the fact that electrical stimulation of the region elicits conjugate contraversive eye movements, whereas its ablation is followed by a *transitory* inability of the monkey to turn its eyes to the contralateral side, and a concomitant failure to pay attention to stimuli delivered in the contralateral half of the visual field ('contralateral visual neglect'). Tempting as it would seem to assume that area 8 is the oculomotor area of the 'motor cortex', recent findings concerning the temporal relationship between eye movements and the activity patterns of single neurons of area 8 flatly contradict such a notion and appear to suggest that the area is involved in the *monitoring* rather than the effectuation of eye movements.[14,15] It is here included in the 'prefrontal' rather than the 'motor' cortex for the reason that its thalamo-cortical afferents come from the mediodorsal thalamic nucleus.[16]

The large remainder of the prefrontal region is distinctly granular in type. A varying number of cytoarchitectural subdivisions have been recognized in this large granular territory, but the structural differences have been too subtle to permit anything resembling agreement among individual observers (see AKERT[16]). Whether or not manifested by cytoarchitectural contrasts, however, a functional parcellation of the region appears virtually certain from the observation that different sub-fields of the granular frontal cortex have markedly different afferent and efferent relationships (see below).

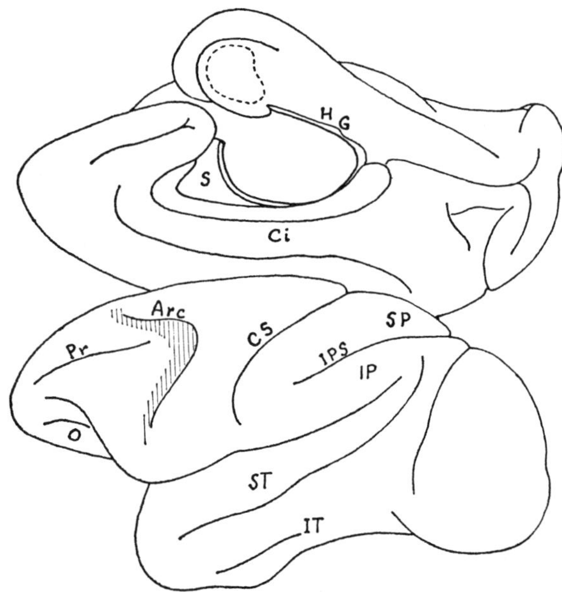


FIG. 1. Line drawing representing the lateral (lower half of figure) and medial aspect of the cerebral hemisphere of a rhesus monkey. The lateral view shows the arcuate sulcus (Arc) which marks the caudal boundary of the region here called 'the frontal lobe', and the principal sulcus (Pr) which divides the frontal convexity in a dorsal and ventral half. The shaded region forming the rostral bank of the arcuate sulcus represents the dysgranular area often referred to as 'frontal eye field.' A small rostral part of the orbital surface of the frontal lobe (O) is visible in the lateral view. The broken line in the upper half of the figure indicates the approximate position of the amygdala, a structure largely hidden from view by the overlying olfactory cortex. Other structures labelled are: Ci: cingulate gyrus; CS: central sulcus; HG: hippocampal gyrus; IP: inferior parietal lobule; IPS: intraparietal sulcus; IT: inferior temporal region; S: septum; SP: superior parietal lobule.

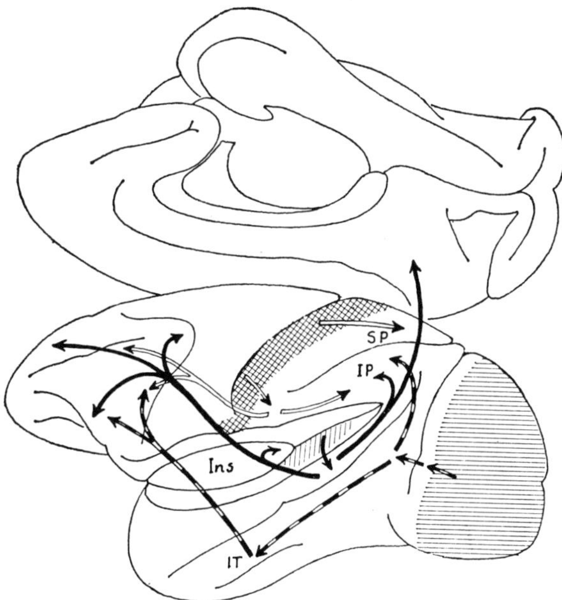


FIG. 2.   Schematic representation of some of the major association pathways extending away
from the great primary sensory fields of the cortex. The visual area 17 is indicated by horizontal
shading, the auditory area by vertical shading, and the somatic sensory cortex by cross-hatching.
Afferents to the frontal lobe presumably conveying auditory information are indicated by solid-
black arrows, those conveying visual information, by black and white arrows, and presumably
somatosensory afferents by open arrows. Note that pathways representing all of these three
modalities also converge upon the inferior parietal lobule (IP). The substantial projections
      from this parietal region to the frontal cortex are indicated separately in Fig. 3.

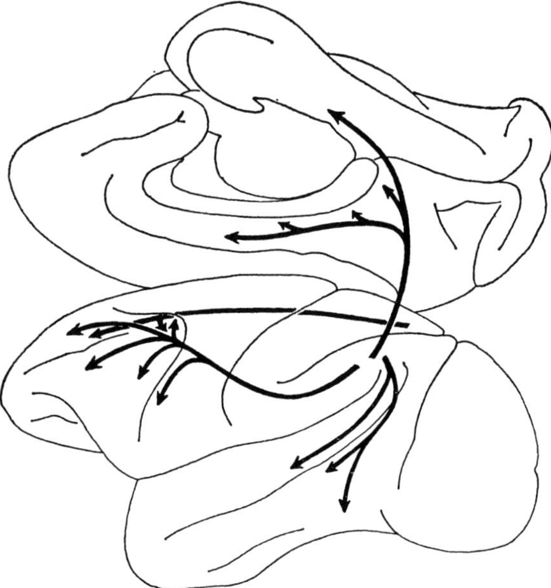

FIG. 3.   Schematic representation of parieto-frontal connections originating from the inferior and superior parietal lobules. Those from the inferior lobule can be thought to convey visual, auditory and somatosensory information. Note that the inferior parietal lobule also has substantial efferent connections with the cingulate and hippocampal gyri, and with all of the temporal gyri (see also Petras' contribution to this volume).

## 2. *Afferent connections of the frontal lobe*

Recent experimental findings with respect to the afferent connections of the frontal lobe have combined to suggest that the frontal cortex is a common end-point for generally long neuronal chains that extend away from the primary sensory regions of the cerebral cortex. Some such conduction routes are composed entirely of cortico-cortical links, and could therefore be classified as associative connections throughout. Others, by contrast, involve the mediodorsal nucleus of the thalamus and thus have their last link in the thalamo-frontal projection system.

2 (a) *Afferents from the visual, auditory and somatic sensory areas.* Associative chains linking the primary cortical fields of the visual, auditory and somatic sensory systems with the frontal cortex have been outlined in a series of recent experimental studies by KUYPERS *et al.*,[17] PANDYA and KUYPERS[18] and PANDYA *et al.*[19] in the rhesus monkey. As shown in extremely schematic form by Figs. 2–3, in each of these three sensory systems the connection in question initially involves one or more cortical zones adjoining the primary sensory field ('belt zones', for example areas 18 and 19 of the visual cortex), and spreads from there to either or both the inferior parietal lobule and the rostral half of the temporal neocortex. The latter two fields are major sources of direct cortico-cortical afferents to the frontal cortex (for a more detailed accounting of parieto- and temporo-frontal connections the reader is referred to Petras' contribution to this volume).

An interesting detail reported by the aforementioned authors is that, whereas virtually no fibers to the frontal cortex appear to originate in the primary sensory fields, some can be traced from the immediately adjoining field (such as area 18 of the visual cortex) and a considerably larger number from the field 'next in line' (e.g. area 19 of the visual cortex). It thus appears that the association of each of these three sensory systems, although largely organized so as to involve a lineal sequence of intermediate cortical processing stations, includes some additional conduction lines originating in parallel from such intercalated way-stations.

It is remarkable that the direct associative connection of the anterior temporal cortex to the frontal lobe by way of the uncinate bundle is paralleled by a pathway from lower temporal regions (middle and inferior temporal gyri) via the so-called inferior thalamic peduncle to the particular (medial, magnocellular) subdivision of the mediodorsal thalamic nucleus[20] that is known[16,21] to project to the cortex covering the orbital surface of the frontal lobe (Fig. 4). The field of origin of this transthalamic temporo-frontal conduction route overlaps that of the uncinate bundle to some extent at least, but whereas the latter, direct connection appears to involve a very large part of the frontal cortex,[18] the temporo-thalamo-frontal pathway must be limited in its distribution to the orbital surface. The functional significance of the mediodorsal nucleus as an intermediary in the path to the frontal lobe must for the present remain a matter of conjecture, but it is interesting that the lower temporal gyri share this efferent way-station with the olfactory cortex (see below).

It must be emphasized that the various cortical and thalamic intermediaries in these sensory-frontal conduction routes cannot be viewed as mere 'relay stations' along the path to the frontal lobe. There can be little doubt that fundamental input-transformations take place at each step along the transcortical way, and there is thus reason to suspect that the information content of the impulse flow arriving at the frontal cortex can be little more than a remote derivative of the neural events taking place in the primary sensory areas. Nonetheless, the systematic progression of associative conduction routes toward the frontal lobe suggests that all of the three major sensoria represented by modality-specific areas in the neocortex find some form of re-representation in the frontal cortex, however abstracted or compounded that form may be.

2 (b) *Afferents from the olfactory system.* It is of interest to note here that the frontal lobe is not the only neocortical region in which association systems related to the visual, somesthetic and auditory systems converge. For example, at earlier stages of cortico-cortical processing a similar confluence, or at least a partial spatial overlapping of these three modalities* appears to take place in the inferior parietal lobule. However, nowhere but in the frontal lobe does this convergence appear to be augmented by an afferent connection from the *olfactory system.* The unexpected evidence of a close relationship of the frontal lobe with the olfactory sensorium came from studies in the rat by SANDERS–WOUDSTRA[22] and POWELL *et al.*[23] that convincingly demonstrated a substantial projection from the prepirifrom (olfactory) cortex via the inferior thalamic peduncle to the medial, magnocellular subdivision of the mediodorsal nucleus of the thalamus. This olfacto-thalamic connection, schematically illustrated in Fig. 4, implies the theoretical possibility that the posterior orbitofrontal cortex is not more than a few synapses removed from the olfactory receptor neurons. No less interesting is the fact that the projection to the mediodorsal nucleus appears to be only one subdivision of a wider projection from the olfactory cortex that includes substantial connections to the lateral hypothalamic region.

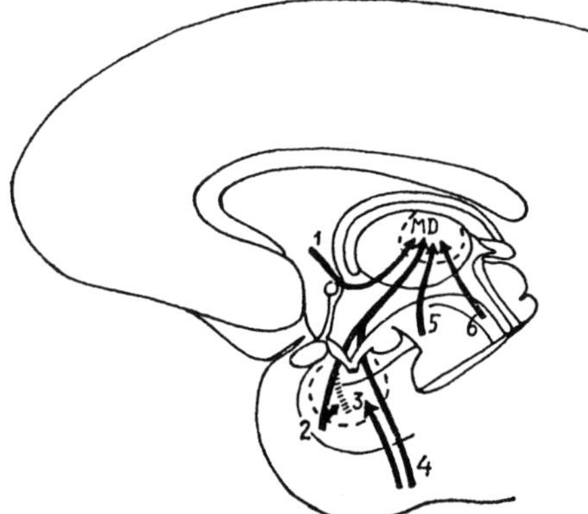


FIG. 4. Schematic drawing indicating some of the subcortical afferents of the mediodorsal nucleus of the thalamus (MD). Arrow 1 represents afferents from the septum, arrow 2: such from the olfactory cortex; arrow 3 (shaded because of uncertainty mentioned in the text): fibers from the amygdala; arrow 4: fibers from the inferior temporal region; arrow 5: fibers from the ventromedial area of the midbrain tegmentum, and arrow 6: fibers ascending through the dorsal region of the tegmentum. The connections 2, 3 and 4 are components of the inferior thalamic peduncle.

---

* The term, *overlap*, is here used without prejudice as to the question whether two or more afferent systems converging upon a given cortical region establish synaptic contacts with the same, or different, neuronal constituents of the region.

One could wonder whether perhaps, by these short routes to the frontal lobe and hypothalamus, the olfactory system—no less than other sensoria involved in learning and memory—may be giving us a basic anatomical 'flow-diagram' of telencephalic signal-processing, a model which other sensory modalities complicate by the interposition of a series of neocortical processing stations beyond the primary cortical receiving area.

2 (c) *Other afferent connections of the frontal lobe.* All of the evidence summarily reviewed above suggests the frontal lobe as a neocortical region representing the external environment as reported by all exteroceptive sensoria. But even this encompassing statement may not adequately describe its afferent relationships, for it does not take into account a remarkable diversity of further fiber systems connected to the frontal cortex by way of the mediodorsal nucleus.

Such additional afferents include connections to the medial subdivision of the mediodorsal thalamic nucleus from the septal region and from a ventral-paramedian zone of the mesencephalic tegmentum.[24] It is likely, but not certain, that the same medial component of the mediodorsal nucleus also receives afferents from the amygdala.*

Only sparse data are available with respect to the afferent connections of the considerably larger *lateral* subdivision of the mediodorsal thalamic nucleus, a cell territory known to project to the *convexity* of the frontal lobe.[16,21] The observation that it receives fibers originating in the intralaminar thalamic nuclei[26] holds little clue-value, for such fibers appear to be distributed widely among specific thalamic nuclei. Neither is much information to be derived from the fact that the nucleus receives numerous thalamic fibers from the frontal convexity: reciprocity appears to be a common, if not indeed universal, property of thalamo-cortical relationships. In a recent experimental study in the rat, however, CHI[27] has demonstrated a discrete fiber system that ascends beneath the ventrolateral border of the central grey substance of the midbrain, and terminates with dense arborizations in a circumscript region of the more lateral zone of the mediodorsal nucleus. The cells of origin of this well-defined, lemniscus-like fiber system have not been identified, but the position of the bundle in the midbrain suggests that it may be a transsynaptic continuation of a projection system ascending from the nucleus of the solitary tract, a projection which follows a comparable mesencephalic trajectory[28] but appears not to extend into the thalamus directly.

It is difficult to evaluate the functional nature of these various afferents of the mediodorsal nucleus on the basis of anatomical evidence alone. At present, only the projection from the olfactory cortex can be identified in terms of sensory modality. There is nonetheless ample reason to suspect that several at least of the remaining afferent fiber systems mentioned above convey information concerning the organism's internal milieu. Such a relationship would seem likely in particular for the fiber systems originating in the septum

---

* The lingering uncertainty concerning this point is due to the efferents from the overlying olfactory cortex that traverse the amygdaloid complex in passage to the hypothalamus and mediodorsal nucleus, and thus are unavoidably involved in lesions of the amygdala. Although, as a consequence, direct proof of an amygdalo-thalamic projection is lacking, the existence of this connection appears nonetheless likely in the context of current knowledge concerning temporal-lobe efferents contained in the ansa peduncularis and its thalamic extension, the inferior thalamic peduncle. As summarized in Fig. 4, such connections include efferents from both the inferior temporal region[20] and the olfactory cortex[23,25] to both the amygdaloid complex and mediodorsal thalamic nucleus.

and in paramedian zones of the midbrain tegmentum, brainstem regions prominently involved in the circuitry of the telencephalic limbic structures and hypothalamus. It would not seem far-fetched to interpret such conduction systems as conveyors of neural codes related to motivational states and their visceral concomitants.

### 3. *Efferent connections of the frontal lobe*

Much like most other subdivisions of the cerebral mantle, the frontal cortex has been found to be connected by a great variety of efferent pathways to other cortical regions as well as to subcortical structures. In both categories of connections, indications are found that the frontal lobe is closely associated with the limbic system, but there are other efferent relationships that cannot be so classified.

3 (a) *Efferent cortical associations of the frontal lobe.* Major cortico-cortical efferents connect the frontal lobe with the anterior temporal cortex, with the inferior parietal lobule, and with the cingulate and parahippocampal gyri (Fig. 5). The two former of these efferent connections to some extent at least appear to reciprocate the prominent afferent association

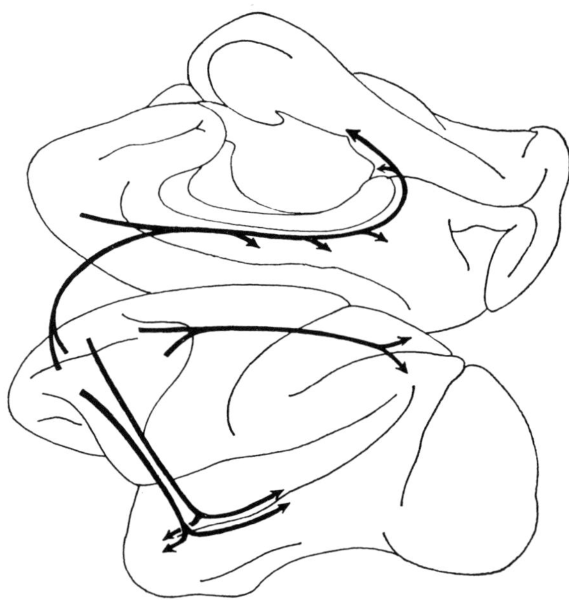

FIG. 5. Schematic representation of afferent connections of the frontal lobe with both parietal lobules, upper and middle temporal gyri, and cingulate and hippocampal gyri. This figure is based in parts on reports by PANDYA and KUYPERS[18], in part also on personal communications from Dr. D. N. Pandya.

of the frontal lobe with the anterior temporal and inferior parietal regions. The findings of PANDYA and KUYPERS[18] suggest that the connection to the inferior parietal lobule originates most massively in the caudal frontal region approximately corresponding to the 'frontal eye field', whereas fronto-temporal fibers appear to arise more evenly throughout the frontal convexity and are organized topographically in such fashion that the dorsal half of the convexity (i.e. the field dorsal to the principal sulcus) projects to the superior temporal gyrus, the ventral half to the dorsal bank of the middle temporal gyrus. It is important to note that a substantial projection to the amygdaloid complex arises from the inferior temporal region[20] (Fig. 4). The fronto-temporal connection could therefore be thought to permit the frontal cortex to modulate not only the activity patterns of the temporal cortex but also those of the amygdala.

The frontal efferents to the limbic cortex (cingulate and parahippocampal gyri) originate largely from dorsal convexity cortex, and there is evidence (Pandya, personal communication) that the field of origin has its center in the dorsal bank lining the middle one-third of the principal sulcus. The fronto-limbic association is established by a fairly compact fiber bundle that follows the white matter of the cingulate gyrus caudalward, issuing fibers to the overlying cortex throughout its extent. The longest fibers curve around the splenium of the corpus callosum into the hippocampal gyrus, and are distributed in this region as far as the presubiculum hippocampi. By virtue of this fronto-limbic connection the frontal cortex stands in very close relationship with the hippocampal formation, but it must be emphasized that the frontal lobe is not unique in this regard, for PANDYA and KUYPERS'[18] study has demonstrated that substantial direct projections to the cingulate and para-hippocampal gyri originate also from rostral temporal regions and from the inferior parietal lobule. The remarkable conclusion can be drawn from these observations that neocortical projections to the limbic (cingulate and juxtahippocampal) cortex originate largely at least from three interconnected cortical regions among which the frontal cortex appears to occupy a central position.

3 (b) *Subcortical projections of the frontal cortex.* Nowhere more remarkably than in its subcortical connections does the frontal lobe declare its close association with the limbic system. This relationship is most explicitly expressed by the substantial projections that have been traced from the frontal cortex[29,31] to the preoptic region and hypothalamus and, beyond these diencephalic structures, to a paramedian zone of the mesencephalic tegmentum that includes the ventral half of the central grey substance, the ventral tegmental area, and the nucleus centralis tegmenti superior of Bechterew. Together with the septal region, these subcortical structures and their interconnections compose a highly differentiated meso-diencephalic continuum that also is a major distribution zone for projections arising in various components of the limbic telencephalon, the hippocampus and amygdala in particular.[32,33]

By entertaining subcortical projections in such extensive overlap with those of the limbic forebrain, the frontal lobe distinguishes itself sharply from all other regions of the neocortical mantle. It must be noted, however, that this unique neocortical projection apparently does not arise from all subdivisions of the frontal lobe equally. A frontal projection to the lateral preopticohypothalamic region, for example, has been traced from caudal regions of the orbital surface,[30] whereas a second projection, apparently originating dorsal to the principal sulcus, distributes itself not only to lateral regions of the hypothalamus, but also to the so-called dorsal hypothalamic area and to the posterior hypothalamic nucleus, central grey substance, ventral tegmental area, and Bechterew's nucleus (personal, unpublished observations). In a recent article JOHNSON *et al.*[34] have described an additional projection passing from the dorsal bank of the principal sulcus to the septum. Few if any comparable projections appear to arise from convexity cortex ventral to the principal sulcus. It must be noted, however, that the fronto-hypothalamic and fronto-mesencephalic projections have not yet been analyzed in adequate detail with respect to either origin or distribution.

Besides these projections to the hypothalamus and associated brainstem structures, the frontal lobe has been found to emit fibers to the striatum, to the subthalamic region, and to the mesencephalic region lateral and dorsal to the red nucleus. The fronto-striatal projection involves a rostroventral part of the caput nuclei caudati as well as certain regions of the fundus striati (i.e. the zone of confluence of the caudate nucleus and putamen).

Any functional significance of the fronto-striatal connections must lie hidden in its topographic characteristics, for all or nearly all cortical regions project to particular zones of either the caudate nucleus or putamen, or both. No functional mosaic of the striatum appears to be known at present, but it may eventually prove significant that, in the cat, extensive lesions of the general region of the caudate nucleus receiving the frontal lobe projection have been found to cause a peculiar hyperkinetic dyskinesia[35] that has been compared tentatively to athetosis. The functional implications of the fronto-subthalamic and fronto-tegmental projections are obscure, except for the obvious likelihood that the frontocortical mechanism, by virtue of these connections, can affect general activity levels.


### DISCUSSION

In hindsight it is possible to argue that, even in the absence of any but elementary physiological knowledge, the general functional associations of many cortical regions could have been inferred from anatomical relationships. Such a statement would be not at all presumptuous in reference to the sensory areas or the sensorimotor cortex, and perhaps not even much so if it were made with respect to certain associative areas such as the parietal cortex. Whether it can be applied to cortical regions functionally as complex as the frontal cortex remains to be considered. In any event, in attempting to elucidate a unique functional attribute of a given cortical region, it would seem appropriate, in a first approach, to look for unusual features in the anatomical relationships of that region.

*The unique feature of the neural circuitry outlined in the foregoing account is, that it places the frontal cortex in a reciprocal relationship with two great functional realms, namely: (1) parietal and temporal regions of the cerebral cortex involved in the processing of visual, auditory and somatic sensory information, and (2) the telencephalic limbic system and its subcortical correspondents, in particular the hypothalamus and meso- and diencephalic structures associated with the hypothalamus.* The reciprocal nature of both of these two relationships deserves particular emphasis, for it entails a need to view the frontal lobe at once as a 'sensory' and as an 'effector' mechanism. In the following part of this discussion an attempt will be made to interpret part at least of the frontal-lobe syndrome as the consequence of loss of a senso-effector organization involved in mechanisms of both perceptual processing and behavioral programming.

*The frontal lobe is characterized so distinctly by its multiple associations with the limbic system, and in particular by its direct connections with the hypothalamus, that it would seem justified to view the frontal cortex as the major—although not the only—neocortical representative of the limbic system. The reciprocity in the anatomical relationship suggests that the frontal cortex both monitors and modulates limbic mechanisms.*

All of the experimental and clinical evidence leads to the conclusion that the telencephalic limbic structures and their subcortical correspondents are centrally involved in functions related to the organism's internal milieu. Not only do they receive neural and chemical signals emanating from that environment, they also modulate it by the medium of complex endocrine and neural effector systems. Moreover, it now appears certain that their functional states manifest themselves subjectively in the form of affects and motivations. In view of the fact that the frontal lobe is closely associated with the limbic system and hypothalamus, extensive lesions of the frontal cortex could be expected to have a marked effect upon the range and differentiation of the subject's viscero-endocrine, affective and motivational responses to his environment. This suggestion from anatomical data is compatible with Homskaya's observation (reported by LURIA[10]) of diagnostically significant anomalies in the conditioning of the galvanic skin reaction of frontal-lobe patients.

The physiological nature of the fronto-limbic relationship cannot be postulated *a priori* and would have to be identified by animal experimentation. It would seem of interest, for example, to determine whether perhaps the limbico-hypothalamic axis, no longer afferented by the frontal cortex, functions at a permanently changed level, a change that might reveal itself, among others, in anomalous viscero-endocrine responses even to vital-stress factors such as cold. Such a condition would be comparable to the enduring changes observed in certain kinesthetic and cutaneous reflexes following lesions of the sensorimotor cortex, a comparison that comes to mind readily because of the existence of a direct fronto-hypothalamic connection. However, whether or not loss of the frontal lobes so gravely affects the intrinsic mechanisms of the limbic system and hypothalamus, it is certain to eliminate important pathways by which the neocortex could be thought to modulate the organism's affective and motivational states. It may be of particular importance that such pathways include the fronto-hypothalamic connection, the only known direct route from the neocortex to the hypothalamus. The failure of the affective and motivational responses of the frontal-lobe patient to match environmental situations that he nonetheless can describe accurately could thus be tentatively interpreted as the consequence of a loss of a modulatory influence normally exerted by the neocortex upon the limbic mechanisms via the frontal lobe.

However, the reciprocity in the fronto-limbic association urges consideration of an opposite, perception-oriented (i.e., corticipetal) interpretation of the frontal-lobe syndrome. Viewed along that line, part at least of the behavioral effects of frontal-lobe destruction could be seen as the consequence of an 'interoceptive agnosia', i.e., an impairment of the subject's ability to integrate certain informations from his internal milieu with the environmental reports provided by his neocortical processing mechanisms. It may be recalled that the available anatomical evidence indicates the frontal cortex as one, and perhaps the only realm of the neocortex where neural pathways representing the internal milieu converge with conduction systems re-representing the external environment as reported by *all* exteroceptive modalities. The two opposite points of view are by no means mutually exclusive; in fact, both conditions are likely to exist together in the frontal-lobe patient.

It should be stressed at this point that even complete ablations of the frontal lobes are unlikely to block all impulse traffic between the neocortex and the limbico-hypothalamic complex. Under such conditions, interoceptive information would still be likely, by way of the anterior thalamic nucleus, to reach the cingulate and parahippocampal gyri, where it could conceivably be integrated with information reaching the same gyri (Fig. 5) from the temporal and inferior parietal regions. Although, therefore, complete ablations of the frontal lobes alone do not dissociate the neocortex from the limbic system and hypothalamus, such lesions may nevertheless severely impair the normal interplay between these two forebrain levels.

It is tempting to speculate that the reciprocal fronto-limbic relationship could be centrally involved in the phenomenon of *behavioral anticipation*, and elucidate the 'loss of foresight' that has so long been recognized as one of the most disabling consequences of massive frontal-lobe lesions. The normal individual decides upon a particular course of action by a thought-process in which a larger or smaller number of strategic alternatives are compared. It could be suggested—admittedly on purely introspective grounds—that the comparison in the final analysis is one between the affective responses evoked by each of the various alternatives.* The strategy ultimately elected would thus be one that has already passed censure by an interoceptive sensorium. It is entirely conceivable that this anticipatory selection process is severely impaired in the absence of the frontal cortex.

A somewhat similar, or at least related, frontal mechanism could be postulated to explain the phenomenon of *temporal stability* in behavioral programs. In an earlier section of this paper it was mentioned that TEUBER[9] has found reasons to attribute the visual and proprioceptive defects associated with frontal-lobe lesions to the loss of a mechanism of *corollary discharge* that pre-sets sensory processing mechanisms for such input changes as would predictably result from the impending motor output. Teuber's notion essentially postulates an effector function of the frontal lobe, even though that function would have its primary impact on perception rather than movement. The substantial efferent connections of the frontal lobe with the temporal cortex and the inferior parietal lobule could well form part of the anatomical substratum of such a function. It could be asked, however, if not perhaps a wider aspect of the frontal-lobe syndrome could be interpreted in terms of a corollary discharge. More specifically, it would seem possible to envisage a pre-setting not only of exteroceptive processing mechanisms, but also of those mechanisms dealing with interoceptive information. Such a pre-setting could be thought to establish a temporal sequence of affective reference points serving as 'navigational markers' and providing, by their sequential order, at once the general course and the temporal stability of complex goal-directed forms of behavior.

---

* The incorporation of an interoceptive intuitive element in decision-making, earlier suggested by Henri Bergson (for example, Chapter 2 of *L'Evolution Créatrice*), is emphasized by Charles DeGaulle in *Le Fil de l'Epée*. Collective awareness of this ultimate sensorium undoubtedly far antedates these explicit formulations, and is expressed in a variety of idioms (" . . . the mere thought of doing such a thing makes me ill").

B

It could be suspected—so far on no more than introspective grounds—that the itineraries of anticipated behavior require being registered and 'kept on hand' not only in somatic sensorimotor mechanisms but also in structures subserving the organism's affective responsiveness, and thus, that a plan for action cannot be kept in abeyance intact for any length of time unless it is represented in matching somatic and affective registries. If this were indeed the case, it would be readily understandable that loss of the frontal cortex as a major mediator of information exchange between the cerebral cortex and the limbic system is followed not only by an impairment of strategic choice-making, but also by a tendency of projected or current action programs to 'fade out' or become over-ridden by interfering influences. This notion would seem to be somewhat in line with KONORSKI and LAWICKA'S[3] suggestion that the poor performance of 'frontal' animals in the delayed-response task may be caused by an abnormally rapid decay of signal traces. In this context it could even be suggested that the 'frontal' animal has suffered a memory impairment after all, even though this loss affects the storage of its action plans rather than that of its external-perceptual images.

It cannot be correct to interpret the frontal lobe exclusively as a structure modulating sensory, viscero-endocrine, and affective mechanisms, for the frontal cortex also projects to structures that may be involved primarily in somatic effector functions. For example, as mentioned earlier in this paper, it has efferent connections with the caudoputamen, and PANDYA and KUYPERS'[18] findings suggest that it is also connected—even though perhaps not massively—to the premotor cortex. It would therefore seem likely that, as already suggested above, the role of the frontal lobe also manifests itself in the realm of somatic motor function. LURIA and HOMSKAYA[12] have observed patients with tumors causing massive frontal-lobe destruction who were often unable to initiate a movement required by the examiner, even though audibly repeating to themselves the examiner's instruction. It remains to be determined whether such akinetic and hypokinetic consequences of massive frontal-lobe tumors reflect a motivational loss or an effector deficit in the more restricted sense, and if the latter should be the case, whether they may not be due to a direct involvement of the caudoputamen or premotor cortex in the pathological process.

Up to this point in the discussion, the frontal lobe has been dealt with as if it were a homogeneous structure. There is, however, ample evidence that the input–output relationships of the monkey's frontal cortex vary considerably from one subregion of the field to the next, and some of these variations have been noted in the anatomical account. To recapitulate, it seems certain that those afferent associations likely to convey visual, auditory and somesthetic information primarily affect the caudal half of the monkey's frontal convexity, and particularly the region of the frontal eye-field. Olfactory information, by contrast, would seem to be distributed via the mediodorsal nucleus entirely to the caudal half of the orbital aspect of the lobe, and the same is probably true of the transthalamic afflux of impulses here interpreted as representing the internal milieu, except for that part conveyed through the mesencephalon by way of the lateral division of the mediodorsal nucleus to a yet undisclosed region of the frontal convexity. Further 'interoceptive' afferents to the frontal cortex are likely to come from the cingulate cortex, but their distribution remains to be determined. As to the efferent connections of the frontal lobe, the great conduction route to the hippocampal gyrus appears to originate largely from the dorsal bank of the sulcus principalis, whereas associations with the amygdala by way of the temporal cortex are more likely to arise from ventral convexity areas. The fronto-hypothalamic connection apparently originates from two widely separated fields: the caudal orbitofrontal region and some region dorsal to the sulcus principalis, whereas the 'feedback' association with the multimodal processing areas of the parietal lobe arises largely in and near the frontal eye-field. This anatomical mosaic suggests a great functional differentiation of the frontal region, a suggestion that has begun to be borne out by the results of some recent behavioral studies in the monkey. In one of these, the monkey's ability to perform at normal levels in the spatial delayed-alternation test was found to depend on the integrity of the cortex lining the principal sulcus, whereas his capacity to integrate auditory, visual and kinesthetic information was found impaired by lesions in the 'peri-arcuate' region, i.e. the general area of the frontal eye field.[37] In another study, the region crucially involved in the mechanisms required for the delayed-alternation task could be further localized to the cortex lining the middle one-third of the principal sulcus.[38] Further studies of this nature may disclose yet other instances of differential functional localization in the frontal lobe. One must hope that such and other laboratory experiments may eventually be extended so as to provide records not only of the overt behavior of animals with variously located frontal-lobe lesions but also of its visceral and endocrine concomitants, and of contemporaneous activity states in such structures as the hippo-campus, the amygdala, the caudate nucleus, and parietal and temporal regions of the cortex. Several tantalizing, but as yet anecdotal, findings pertinent to this general question have already been reported from both animal[36] and clinical[10] studies. From such inquiries a clearer picture of the physiological nature of frontal-lobe function may be expected eventually to emerge.

*Acknowledgements*—The author expresses his sincere appreciation to Drs. H.-L. Teuber and Stephen M. Shea for valuable guiding comments, to Dr. D. N. Pandya for providing several important data prior to publication, and to Miss Elizabeth B. Jones for her efficient technical assistance in the preparation of this article. His greatest debt, however, accumulated over two decades, is to Dr. David McKenzie Rioch for providing not onlymaterial support but, above all, an unexcelledi ntellectual milieu for inquiries into problems of brain and mind.

## REFERENCES

1. Jacobsen, C. F. Functions of the frontal association area in primates. *Archs Neurol. Psychiat.* **33,** 558, 1935.
2. Mishkin, M. Perseveration of central sets after frontal lesions in monkeys. In: *The Frontal Granular Cortex and Behavior,* Warren, J. M. and Akert, K. (Eds.), p. 219. McGraw-Hill, New York, 1964.
3. Konorski, J. and Lawicka, W. Analysis of errors by prefrontal animals on the delayed-response test. In: *The Frontal Granular Cortex and Behavior,* Warren, J. M. and Akert, K. (Eds.), p. 271. McGraw-Hill, New York, 1964.
4. Weiskrantz, L. and Mishkin, M. Effects of temporal and frontal cortical lesions on auditory discrimination in monkeys. *Brain* **81,** 406, 1958.
5. Bättig, K., Rosvold, H. E. and Mishkin, M. Comparison of the effects of frontal and caudate lesions on delayed response and alternation in monkeys. *J. comp. physiol. Psychol.* **53,** 400, 1960.
6. Rylander, G. *Personality Changes after Operations on the Frontal Lobes.* Oxford University Press, London, 1939.
7. Halstead, W. C. *Brain and Intelligence: a Quantitative Study of the Frontal Lobes.* The University of Chicago Press, Chicago, 1947.

8.  TEUBER, H.-L. Some alterations in behavior after cerebral lesions in man. In: *Evolution of Nervous Control from Primitive Organisms to Man*, BASS, A. D. (Ed.), p. 157. Am. Ass. Adv. Sci., Washington, D.C., 1959.

9.  TEUBER, H.-L. The riddle of frontal lobe function in man. In: *The Frontal Granular Cortex and Behavior*, WARREN, J. M. and AKERT, K. (Eds.), p. 410. McGraw-Hill, New York, 1964.

10. LURIA, A. R. *The Origin and Cerebral Organization of Man's Conscious Action*, Evening lecture to the XIX Int. Congress Psychol., London 1969. Moscow Univ. Press, Moscow, 1969.

11. MILNER, B. Some effects of frontal lobectomy in man. In: *The Frontal Granular Cortex and Behavior*, WARREN, J. M. and AKERT, K. (Eds.), p. 311. McGraw-Hill, New York, 1964.

12. LURIA, A. R. and HOMSKAYA, E. D. Disturbance in the regulative role of speech with frontal lobe lesions. In: *The Frontal Granular Cortex and Behavior*, WARREN, J. M. and AKERT, K. (Eds.), p. 352. McGraw-Hill, New York, 1964.

13. LEONARD, C. M. The prefrontal cortex of the rat. I. Cortical projection of the mediodorsal nucleus. II. Efferent connections. *Brain Res.* 12, 321, 1969.

14. BIZZI, E. Discharge of frontal eye field neurons during saccadic and following eye movements in unanesthetized monkeys. *Expl Brain Res.* 6, 69, 1968.

15. BIZZI, E. and SCHILLER, P. H. Single unit activity in the frontal eye fields of unanesthetized monkeys during eye and head movements. *Expl Brain Res.* 10, 151, 1970.

16. AKERT, K. Comparative anatomy of frontal cortex and thalamofrontal connections. In: *The Frontal Granular Cortex and Behavior*, WARREN, J. M. and AKERT, K. (Eds.), p. 372. McGraw-Hill, New York, 1964.

17. KUYPERS, H. G. J. M., SZWARCBART, M. K. and MISHKIN, M. Occipitotemporal cortico-cortical connections in the rhesus monkey. *Expl Neurol.* 11, 245, 1965.

18. PANDYA, D. N. and KUYPERS, H. G. J. M. Cortico-cortical connections in the rhesus monkey. *Brain Res.* 13, 13, 1969.

19. PANDYA, D. N., HALLETT, M. and MUKHERJEE, S. K. Intra- and interhemispheric connections of the neocortical auditory system in the rhesus monkey. *Brain Res.* 13, 49, 1969.

20. WHITLOCK, D. G. and NAUTA, W. J. H. Subcortical projections from the temporal neocortex in Macaca mulatta. *J. comp. Neurol.* 106, 183, 1956.

21. FREEMAN, W. and WATTS, J. W. Retrograde degeneration of the thalamus following prefrontal lobotomy. *J. comp. Neurol.* 86, 65, 1947.

22. SANDERS-WOUDSTRA, J. A. R. Experimenteel anatomisch onderzoek over de verbindingen van enkele telencefale hersengebieden bij de albino rat. Doctoral Dissertation, Groningen, 1961.

23. POWELL, T. P. S., COWAN, W. M. and RAISMAN, G. The central olfactory connexions. *J. Anat. (Lond.)* 99, 791, 1965.

24. GUILLERY, R. W. Afferent fibers to the dorsomedial thalamic nucleus in the cat. *J. Anat.* 93, 403, 1959.

25. VALVERDE, F. *Studies on the Piriform Lobe*. Harvard University Press, Cambridge, Mass., 1965.

26. NAUTA, W. J. H. and WHITLOCK, D. G. An anatomical analysis of the non-specific thalamic projection system. In: *Brain Mechanisms and Consciousness*, DELAFRESNAVE, J. F. (Ed.), p. 81. Blackwell, Oxford, 1954.

27. CHI, C. C. An experimental silver study of the ascending projections of the central grey substance and adjacent tegmentum in the rat, with observations in the cat. *J. comp. Neurol.* 139, 259, 1970.

28. MOREST, D. K. Experimental study of the projections of the nucleus of the tractus solitarius and the area postrema in the cat. *J. comp. Neurol.* 130, 277, 1967.

29. DE VITO, J. L. and SMITH, O. E. Subcortical projections of the prefrontal lobe of the monkey. *J. comp. Neurol.* 123, 413, 1964.

30. NAUTA, W. J. H. Neural associations of the amygdaloid complex in the monkey. *Brain* 85, 505, 1962.

31. NAUTA, W. J. H. Some efferent connections of the prefrontal cortex in the monkey. In: *The Frontal Granular Cortex and Behavior*, WARREN, J. M. and AKERT, K. (Eds.), p. 397. McGraw-Hill, New York, 1964.

32. NAUTA, W. J. H. Hippocampal projections and related neural pathways to the midbrain in the cat. *Brain* 81, 319, 1958.

33. NAUTA, W. J. H. Fibre degeneration following lesions of the amygdaloid complex in the monkey, *J. Anat.* 95, 515, 1961.

34. JOHNSON, T. N., ROSVOLD, H. E. and MISHKIN, M. Projections from behaviorally-defined sectors of the prefrontal cortex to the basal ganglia, septum, and diencephalon of the monkey. *Expl Neurol.* 21, 20, 1968.

35. Liles, S. L. and Davis, G. D. Permanent athetoid and choreiform movements after small caudate lesions in the cat. In: *Psychotropic Drugs and Dysfunctions of the Basal Ganglia*, Crane, G. E. and Gardner, R. (Eds.), p. 98. U.S. Public Health Service Publication No. 1938, 1969.
36. Hockman, C., Talesnik, J. and Livingston, K. E. Central nervous system modulation of cardiovascular reflexes. *Proc. Int. Un. physiol. Sci.* **7,** 196, 1968.
37. Goldman, P. and Rosvold, H. E. Localization of function within the dorsolateral prefrontal cortex of the rhesus monkey. *Expl Neurol.* **21,** 20, 1968.
38. Butters, N. and Pandya, D. N. Retention of delayed-alternation: effect of selective lesions of sulcus principalis. *Science* **165,** 1271, 1969.

*Brain Research*, 122 (1977) 393–413
© Elsevier/North-Holland Biomedical Press, Amsterdam – Printed in The Netherlands



# Research Reports

COLUMNAR DISTRIBUTION OF CORTICO-CORTICAL FIBERS IN THE FRONTAL ASSOCIATION, LIMBIC, AND MOTOR CORTEX OF THE DEVELOPING RHESUS MONKEY

PATRICIA S. GOLDMAN and WALLE J. H. NAUTA

*Laboratory of Neuropsychology, National Institute of Mental Health, Bethesda, Md. 20014 and Department of Psychology, Massachusetts Institute of Technology, Cambridge, Mass. 02139 (U.S.A.)*

(Accepted June 28th, 1976)

SUMMARY

The terminal distribution of cortico-cortical connections was examined by autoradiography 7–8 days following injections of tritium labeled amino acids into the dorsal bank of the principal sulcus, the posterior part of the medial orbital gyrus, or the hand and arm area of the primary motor cortex in monkeys ranging in age from 4 days to 5.5 months. Labeled axons originating in these various regions of the frontal lobe have topographically diverse ipsilateral and contralateral destinations but virtually all of these projections share a common mode of distribution: they terminate in distinct vertically oriented columns, 200–500 μm wide, that extend across all layers of cortex and alternate in regular sequence with columns of comparable width in which grains do not exceed background. Spatial periodicity in the pattern of transported label in such regions as the prefrontal association cortex, the retrosplenial limbic cortex and the motor cortex indicates that columniation in the intracortical distribution of afferent fibers is not unique to sensory specific cortex but is instead a general feature of neocortical organization.

A columnar mode of distribution of cortico-cortical projections is present in monkeys at all ages investigated but is especially well delineated in the youngest of them. Thus, grain concentrations within columns are very high in monkeys injected at 4 days of age, somewhat lower in monkeys injected at 39–45 days of age, and least dense in those injected at 5.5 months. The distinctness of the spatially segregated pattern of innervation in the cortex of neonates indicates that the columnar organization of association-fiber systems in the frontal and limbic cortex is achieved before or shortly after birth.

394

INTRODUCTION

It is well established by both physiological and anatomical observations that sensory areas of the mammalian cerebral cortex have a distinct vertical or columnar organization[1,12,13,24,25,29]. In the primate visual cortex, layer IV is innervated by geniculocortical fibers in a remarkable pattern of columns, each representing in regular alternation one or the other eye[12,13,29]. In the face area of the somatosensory cortex of the rat and mouse, such thalamocortical columns correspond to histologically distinct, barrel-shaped subdivisions of layer IV[17,30], each of which has been shown to represent the sensory innervation of a single contralateral whisker-follicle[28].

A columnar mode of distribution is not uniquely characteristic of thalamocortical fibers. In the somatosensory cortex of the monkey, a similar pattern of columns has been observed in both the distribution and origin of ipsilateral and contralateral cortico-cortical connections[16,27]. Together with the earlier physiological and anatomical evidence of thalamocortical columns, this finding would appear to suggest that columnar arrangements of neural elements may be a general organizational characteristic of the neocortex. Thus far, however, all of the reported anatomical evidence of cortical columns appears to have been obtained in studies of modally specific sensory areas of the cortex, in which areas, it could be argued, such stereometric parcellation could represent a particular adaptation to the functional "point-for-point" resolution of the peripheral receptor surface. It must be noted, however, that several physiological observations[2,15] have already suggested the existence of a columnar organization of neural elements also in the motor cortex and parietal association cortex, respectively. The present findings supply an anatomical corollary to such observations by demonstrating that a columnar mode of intracortical distribution of afferent fibers, instead of being unique to specific sensory regions, appears also in the cortico-cortical connections of the monkey's prefrontal association cortex, retrosplenial limbic cortex and precentral motor cortex. They show, moreover, that the association-fiber columns in these locations are particularly sharply defined during the first few weeks of the monkey's postnatal life.

METHODS

Autoradiographic studies were carried out in 7 rhesus monkeys ranging in age from 4 days to 5.5 months. Equal quantities of the tritiated amino acids, leucine and proline (spec. act., 30–50 Ci/mmole, New England Nuclear) were combined, concentrated to dry substance by vacuum evaporation, and redissolved in 0.9 % sterile saline to final concentrations of 10–30 $\mu$Ci/$\mu$l. In each experiment, about 40 $\mu$Ci of the isotopes were injected, by the aid of a Hamilton syringe, into any one of 3 different regions of frontal cortex: the midregion of the dorsal bank of the principal sulcus on the dorsolateral convexity (see Fig. 1); the medial orbital gyrus posterior to the juncture of the medial and lateral orbital sulci on the ventral surface of the lobe (see Fig. 6); and the hand and arm area of the primary motor cortex (see Fig. 11). Many of the monkeys received injections in 2 of the 3 areas, on opposite sides. The total quantity of tritiated amino acids injected into each area was usually divided

TABLE I

*Summary of experimental protocols*

| Monkey | Age | Injection sites (hemisphere) | Dose($\mu$Ci) |
|---|---|---|---|
| 6 | 4 days | Principal sulcus* (R) | 20 |
| | | Motor cortex* (L) | 40 |
| 9 | 4 days | Principal sulcus** (R) | 30 |
| 3 | 39 days | Principal sulcus* (R | 30 |
| | | Motor cortex* (L) | 10 |
| 10 | 5.5 months | Principal sulcus* (R) | 40 |
| 7 | 4 days | Medial orbital gyrus* (R) | 40 |
| | | Motor cortex* (L) | 40 |
| 8 | 45 days | Medial orbital gyrus* (R) | 40 |
| | | Motor cortex* (L) | 40 |
| 11 | 5.5 months | Medial orbital gyrus* (R) | 40 |
| | | Motor cortex* (L) | 40 |

* Multiple injections.
** Single injection.

over 4 or 5 single injections spaced approximately 1 mm apart. To be sure that the results reported below were not an artifact of this multiple-injection procedure, a 4-day-old monkey was given a single injection of 30 $\mu$Ci in the dorsal bank of the principal sulcus. A summary of the injection protocols is given in Table I.

In order to label both axons and their terminals, the animals were allowed to survive the operation for 7–8 days[3,20]. After vascular perfusion of the deeply anesthetized animals with 10% formalin, the brains were post-fixed for 1–2 weeks, embedded in an albumin–gelatin medium and cut frozen at 25–30 $\mu$m in the coronal plane. The sections were mounted on albumin-coated slides, air-dried, and then coated with Kodak NTB-2 nuclear-track emulsion that had been diluted with an equal volume of a 0.1% aqueous detergent solution (Dreft) and warmed to 43 °C. After drying, the slides were packed in light tight boxes containing Drierite desiccant, wrapped in aluminum foil and stored at −20 °C for 3–16 weeks. At the appropriate time, the slides were developed in Kodak D-19 at 16 °C, and counterstained with cresyl violet or thionin. Serial sections were examined microscopically under both bright-field and dark-field illumination, and photomicrographs were made to document the findings.

## RESULTS

*Principal sulcus cortex*

*Injection sites.* Four monkeys received isotope injections in the cortex of the dorsal bank of the principal sulcus, two at 4 days of age (P 6, P 9), one at 39 days (P 3), and one at 5.5 months (P 10). One 4-day-old monkey (P 9) was given the isotopes in a single penetration while all other animals received multiple injections (see the preceding description of methods).

A considerable difficulty with the autoradiographic tracing technique is en-

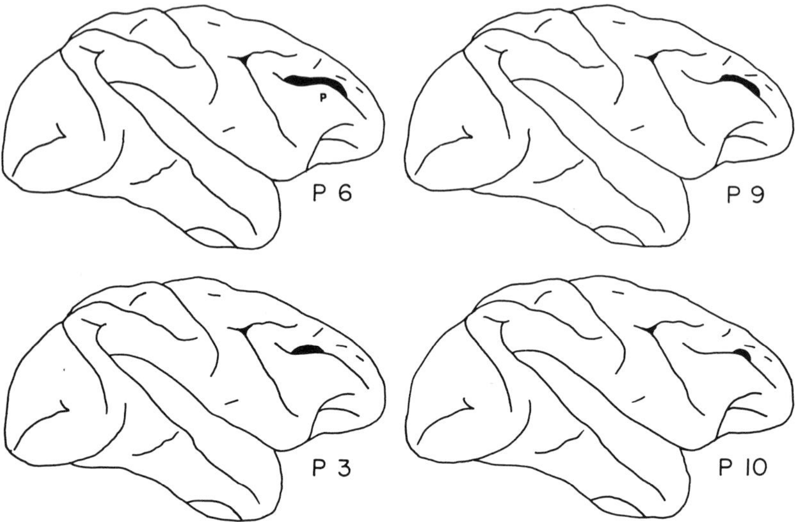

Fig. 1. Diagrammatic illustration of injection sites in the dorsal bank of the principal sulcus (P) on a standard lateral view of the monkey brain. Differences in absolute size of the brain at different ages have been taken into account in reconstructing the various areas of directly labeled cortex. Letters in this figure and in Figs. 6 and 11 refer to injection locus; numbers refer to the monkeys listed in Table I.

countered in attempting a precise delineation of the injection site, particularly at long survival times. The reconstruction of the region in which the isotope was actively taken up by neurons depends upon a number of factors including the survival time, the rate of protein synthesis within the cell and the rate of intracellular protein transport[5,8]. Especially at longer survival times it may be difficult to distinguish grains within the cell body from those marking the neuropil in which the cell bodies lie embedded. As the 7-day survival period allowed in the present study appears to have been sufficiently long to permit considerable depletion of intrasomatic label, the injection site, for the purpose of the present description, has been identified with that region in which the grain density was well above that associated with intra-axonal label transport. The site demarcated by this criterion is probably somewhat larger than the territory in which neurons actually assimilated the injected isotopes. With this qualification, the injection site in the dorsal bank of the principal sulcus can be described as a region of extremely dense labeling extending along the middle third of the sulcus and involving the latter's dorsal rim together with about half the height of its dorsal wall. No discontinuity of labeling appeared in the injected region even though individual injection loci could be recognized as saturated cannula tracks. The absolute size of the injection site was independent of the animal's age; however, because the depth and length of the sulcus increase with age, a proportionately larger region of sulcal cortex was labeled in the younger animals (Fig. 1).

*Ipsilateral associations.* A considerable variety of ipsilateral cortico-cortical connections could be traced from the injection site. Labeled axons tended to aggregate into fascicles as they descended into the white matter before dispersing in various directions. Many such fibers curved ventrally around the fundus of the principal sulcus and passed from here to the cortex lining the depth and ventral bank of the principal sulcus, to the convexity cortex ventral to this sulcus, to the medial orbital gyrus and, more caudally, as far as the insula and the cortex of the superior temporal sulcus and gyrus. Another group of labeled fibers passed dorsally from the injected area to the anterior bank of the arcuate sulcus and to the dorsal half of the frontal convexity. At levels rostral to the injection site, labeled axons spread over the entire extent of the cortex forming the medial surface of the frontal lobe. Farther caudally along the medial wall, labeled axons followed the fasciculus cinguli to the cingulate and retrosplenial regions of the gyrus fornicatus. The topographic details of these connections will be described elsewhere.

For the purpose of the present report it is essential that in virtually all of these cortical regions receiving fibers from the ipsilateral cortex of the principal sulcus, the labeled axons were distributed in distinct vertically oriented columns rather than in tangentially continuous sheets with a uniform distribution of grains (Fig. 2). In the pregenual medial frontal cortex, such columns appeared in individual sections in the form of labeled bands placed at wide and irregular intervals along the entire medial surface, an arrangement that could easily be interpreted as discrete widely spaced projections. By contrast, in most other regions columns of high grain concentration alternated in regular sequence with spaces in which the labeling did not or only slightly exceeded background levels. Such more regularly spaced columns were prominent in the cortex lining the depth and ventral bank of the principal sulcus, in the cortex of the arcuate sulcus, and in the retrosplenial area. Although in most areas the labeling marking the columns extended throughout the thickness of the cortex, it consistently was densest in the molecular layer in which, moreover, it tended to spread out horizontally well beyond the width of the column. The most striking example of regular periodicity in the intracortical distribution of ipsilateral cortico-cortical fibers was encountered in the retrosplenial limbic cortex of the 4-day-old monkey that had received a single injection of radioactive isotopes (Fig. 2). In this area, bands of grains 200–400 $\mu$m wide in the particular plane of section were seen to extend perpendicularly across all layers of the cortex and to alternate with 400–500 $\mu$m wide spaces in which the grain density hardly exceeded background. These linear measurements were made in selected coronal sections; a more precise determination of the stereometry of the labeled tissue-volumes would require reconstruction of more closely spaced serial sections.

It is noteworthy that the columnar pattern here described was in general considerably more pronounced in younger than in older animals, and that it was no less distinct in the 4-day-old monkey that had received only a single isotope injection than it was in the cases of clustered multiple injections.

*Contralateral associations.* Labeled axons could be traced from the injection site in the dorsal bank of the principal sulcus, through the underlying white matter and across the genu of the corpus callosum, into the opposite hemisphere (Fig. 3). These commissural fibers are situated mainly at intermediate depths of the genu. The labeling over this region of the corpus callosum is uniformly distributed, indicating that the commissural fibers are not segregated into fascicles while crossing the mid-

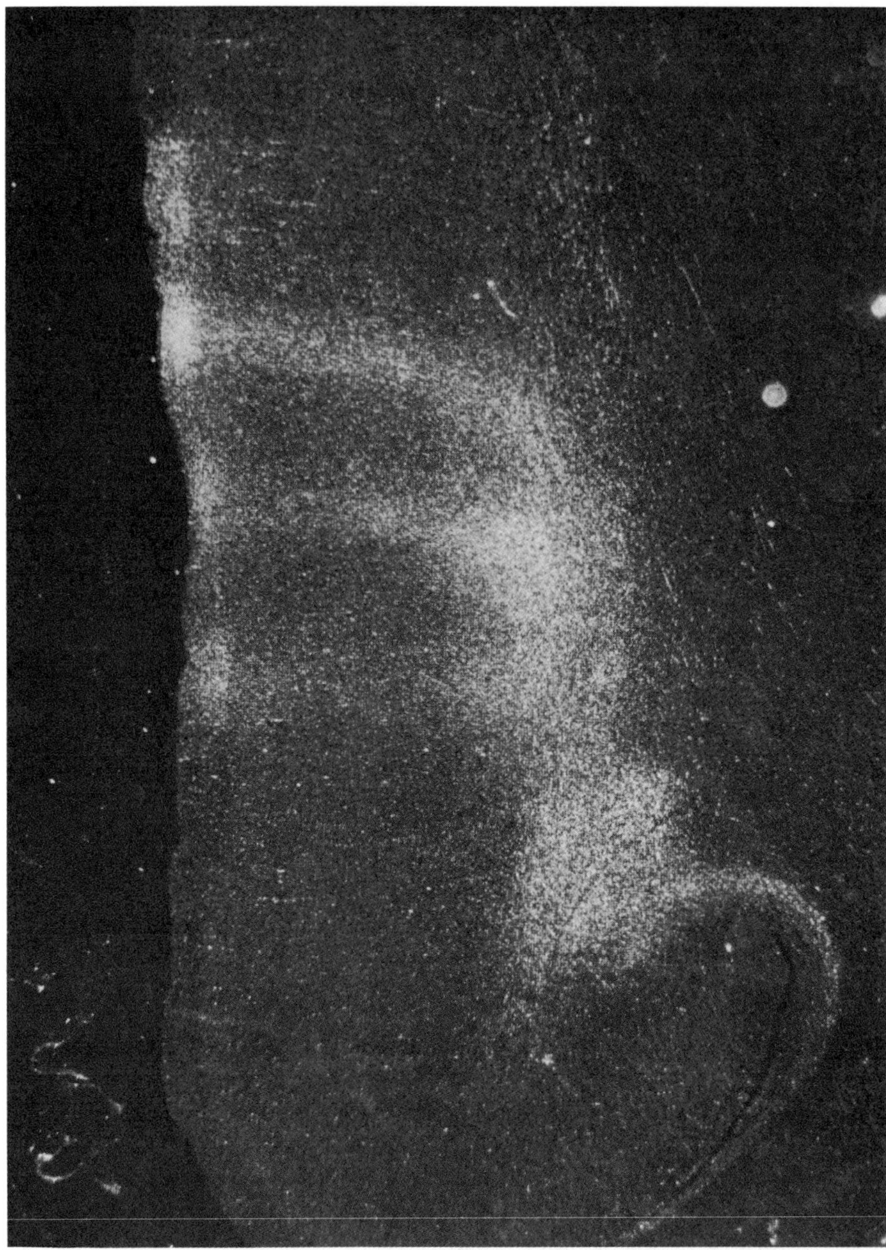

Fig. 2. Dark-field autoradiogram illustrating columns in the ipsilateral retrosplenial cortex following a single injection of [³H]leucine and [³H]proline into the dorsal bank cortex of the principal sulcus in a 4-day-old monkey (P 9). Cingulum bundle, containing labeled axons, is to the lower right of center. × 25.

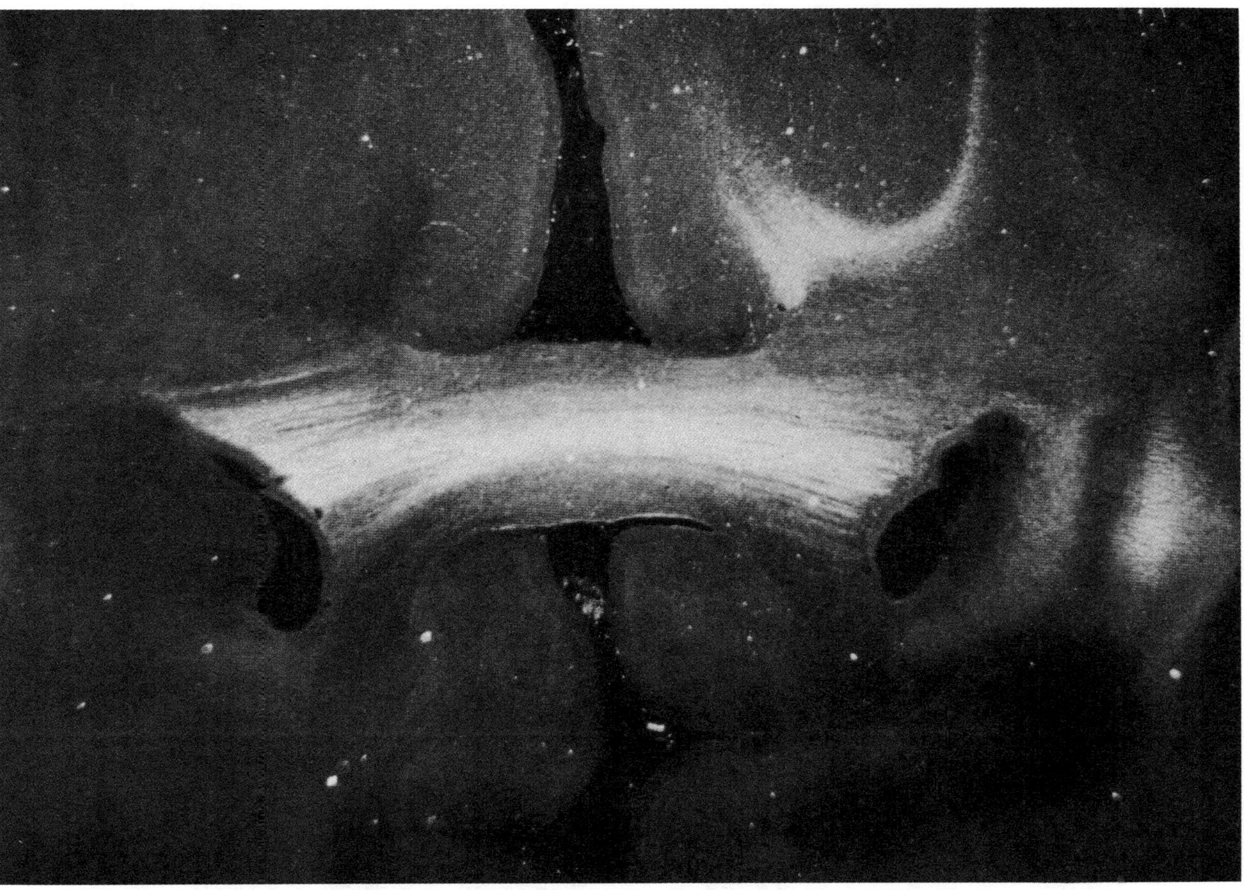

Fig. 3. Callosal pathway of projections from the middle third region of the dorsal bank of the principal sulcus in a 4-day-old monkey (P 9). Injected hemisphere is on the right, where cingulum bundle is heavily labeled. × 10.

line. Upon entering the contralateral hemisphere, the fibers ascend as a continuous band in the white matter beneath the cortex of the principal sulcus, following a course that is the mirror-image of their trajectory in the injected hemisphere.

After invading the homotopical cortex of the contralateral hemisphere, the callosal fibers are distributed over the entire expanse of the region in distinct, radially oriented columns that are separated from each other by variably wide spaces in which labeling is not above background (Fig. 4). In some sections an individual column may exhibit a somewhat greater grain concentration over particular layers, but no consistent pattern of laminar segregation similar to that observed by Jones et al.[16] in the cortico-cortical afferents of the monkey's somatosensory cortex appears in the cortex of the principal sulcus. Commissural columns in the 4-day-old monkey that received a single injection are 200–500 $\mu$m wide in the particular plane of section, and alternate with bands in which the grain concentration is at or near background level (Fig. 4). The space between columns of grains varies with the curvature of the sulcus. At its crown, where the cortex curves to an angle almost perpendicular to the dorsal convexity, the bands are about 500 $\mu$m apart, whereas in the wall of the sulcus where the curvature is less marked, the distance between bands is of the order of 1 mm. As many as 5 such bands of homotopical callosal fiber distribution were observed in two of the youngest animals, respectively 4 days and 2 months old at the time of operation. However, it was not possible to correlate the number of columns with age because of unavoidable variation in injection sites among animals. On the other hand, it is worth noting that the grain density in the columns was particularly high in the 4-day-old monkeys, somewhat less marked in the animal injected at 39 days, and least in the 5.5-month-old (Fig. 5).

In addition to the fibers distributed to the dorsal bank and rim of the principal sulcus, labeled callosal fibers were found to be distributed heterotopically to the cortex forming the ventral bank of this sulcus (Fig. 5) and, at more rostral levels, to the cortex of the medial wall of the hemisphere. As described above, these same cortical areas receive ipsilateral cortico-cortical fibers. Over the contralateral heterotopical areas, e.g., the ventral bank of the principal sulcus, the adjoining ventral-convexity cortex, and the medial frontal cortex, the label likewise was distributed in the form of columns. This columniation was, however, less prominent than that appearing in the homotopical cortex, and typically required longer exposure times to become noticeable in autoradiograms. It was most sharply defined in the 4-day-old infants, somewhat less pronounced in the 39-day-old monkey, and not even vaguely indicated in the 5.5-month-old, even at the longest exposure time of 16 weeks (Fig. 5).

*Posterior orbitofrontal cortex*

*Injection sites.* Three monkeys received multiple injections in the posterior part of the medial orbital gyrus at 4 days (OR 7), 45 days (OR 8) and 5.5-months (OR 11) of age, respectively. The injections yielded dense and apparently continuous labeling in an area bounded by the medial orbital sulcus on one side and the lateral orbital sulcus on the other. Slight variations in the size and placements of the injections were not correlated with age (Fig. 6).

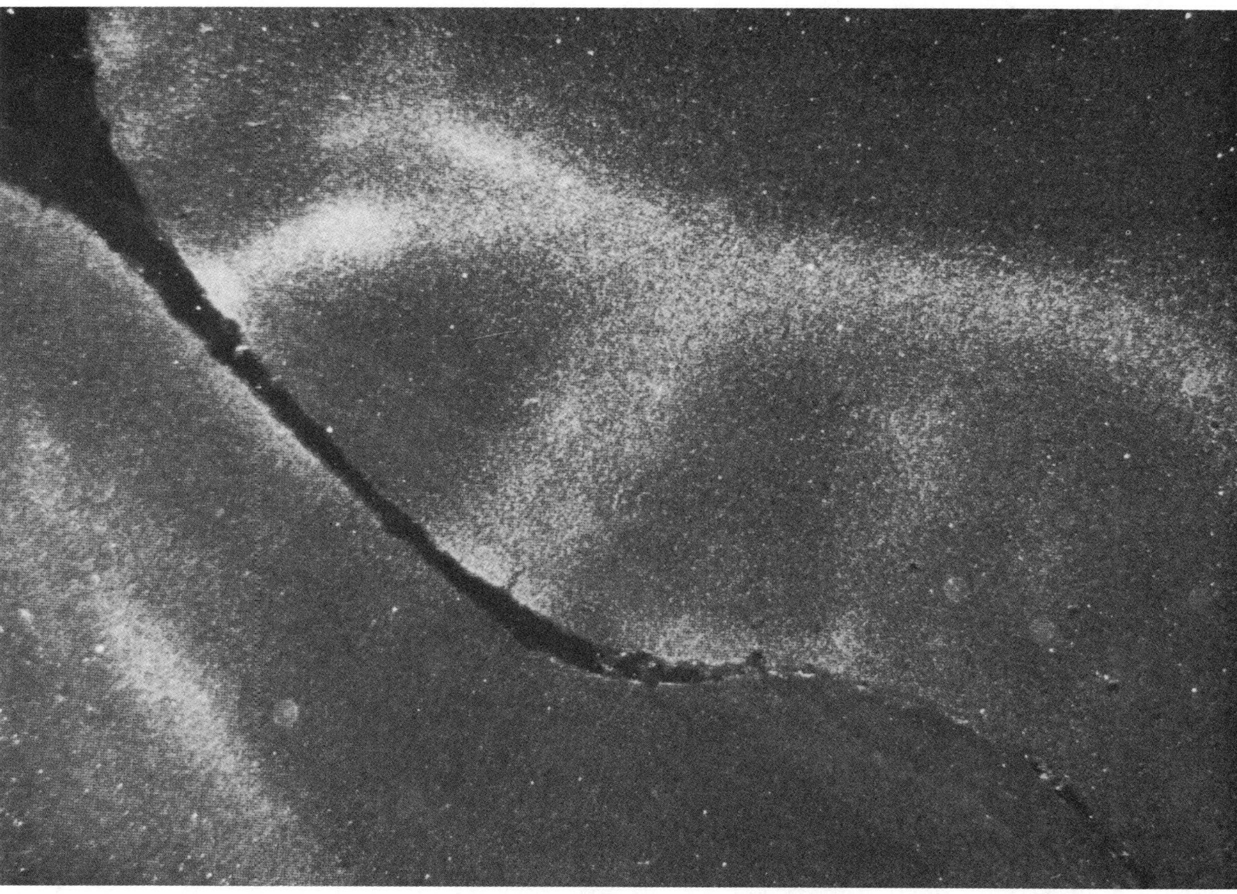

Fig. 4. Dark-field autoradiogram of columns in the homotopical region of the principal sulcus in the contralateral hemisphere of the 4-day-old monkey (P 9) given a single injection in the dorsal bank of the principal sulcus of the opposite hemisphere. × 25.

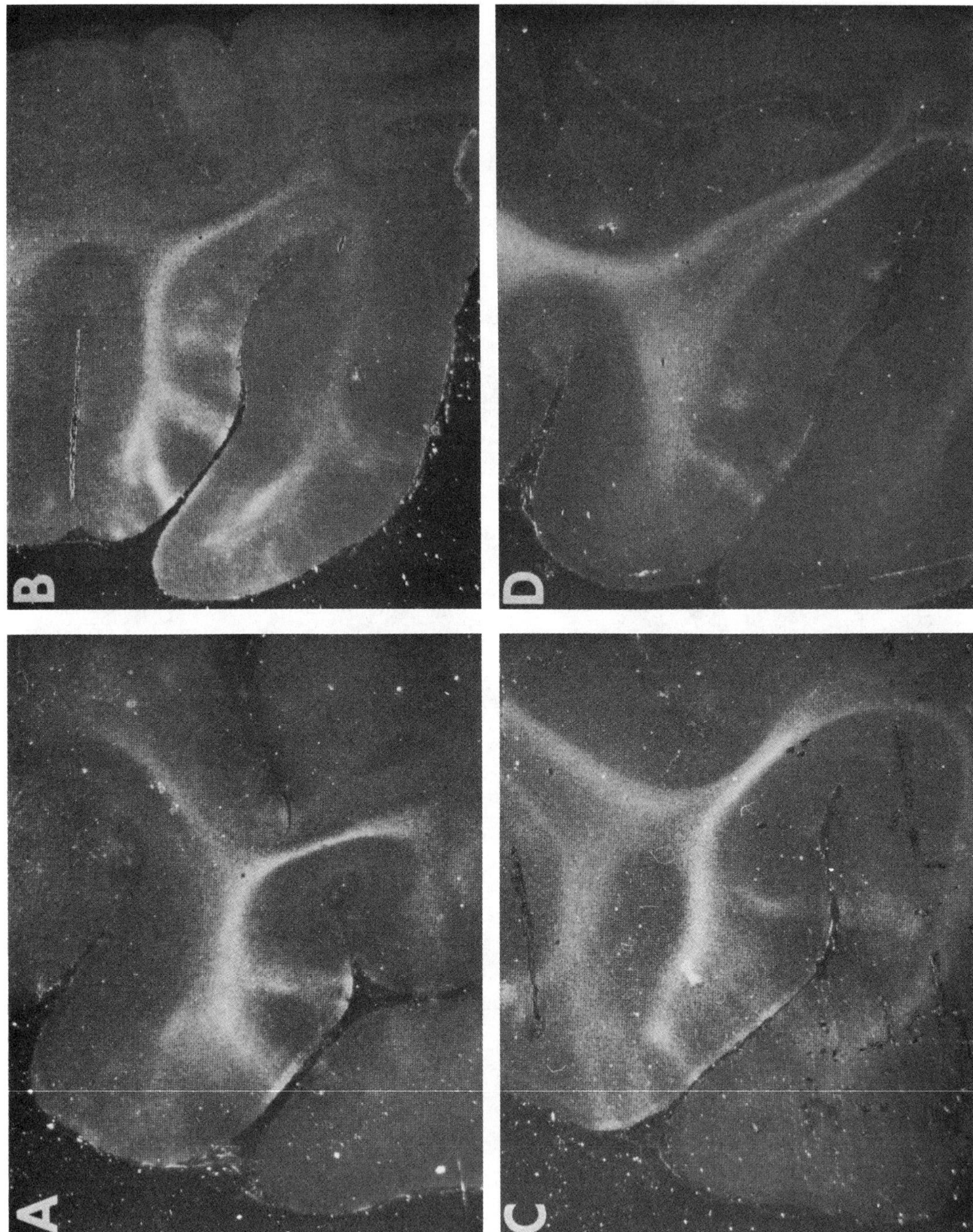

Fig. 5. Comparison of callosal terminations in the principal sulcus of monkeys at different ages: A, 4-day-old (P 6), multiple injections; B, 4-day-old (P 9), single injection; C, 39-day-old (P 3); D, 5.5-month-old (P 10). Note columnar distribution of callosal terminations in the ventral bank of the principal sulcus and in the inferior convexity cortex in A, B and C. × 20.

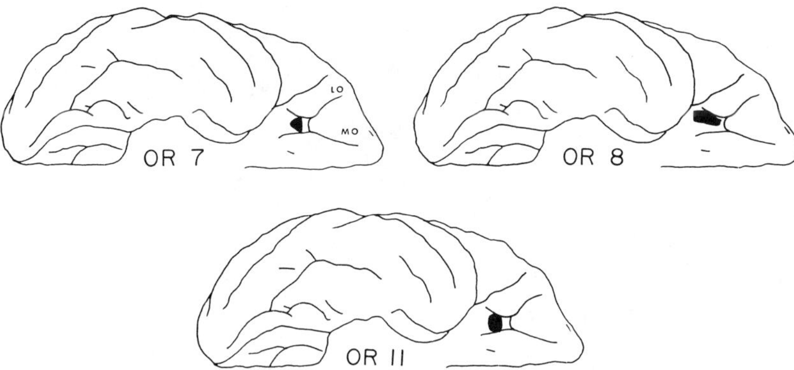

Fig. 6. Diagrammatic illustration of injection sites in the posterior part of the medial orbital gyrus on a standard ventral view of the monkey brain. MO, medial orbital sulcus; LO, lateral orbital sulcus.

*Ipsilateral associations.* In all cases, labeled axons passed from the injection field both lateralward to the lateral orbital gyrus and medially to the gyrus rectus. In both regions these fibers were distributed over all layers of the cortex in the form of columns which were most clearly defined in the two youngest animals. In the lateral side of the gyrus rectus only one such column exhibited a high grain density, whereas all of the several others were more faintly labeled (Fig. 7). It is possible that the latter would have been more densely labeled from a larger injection site.

*Contralateral associations.* Labeled commissural axons passed from the injection site through the rostrum of the corpus callosum (Fig. 8) to the posterior medial orbital gyrus of the contralateral hemisphere. In the monkey injected at 45 days of age their distribution in the homotopical region took the form of one well-defined narrow column approximately 400 $\mu$m in width, separated from a more laterally placed wider column by a relatively grain-free area 1.3 mm wide (Fig. 9A and B). Both columns extended across the full thickness of the cortex. The appearance of the wider column suggests that it may have been composed of several incompletely separated narrower columns.

Labeled fibers from the orbital cortex were distributed also to heterotopical regions of the contralateral cortex. A distinct narrow fascicle of such fibers entered the white matter of the contralateral gyrus rectus and terminated in an alternating pattern of columns over the lateral convexity of the gyrus (Fig. 10A), presenting almost a mirror image of the labeled cortico-cortical columns in the same gyrus of the injected hemisphere (Fig. 10B). The columnar distribution of the callosal projections of the orbital cortex, like that of the cortico-cortical efferents of the principal sulcus, was most clearly evident in the younger monkeys. Grain density over the various cortical target areas was least in the 5.5-month-old monkey.

### Motor cortex

*Injection sites.* Five monkeys that had received injections in either the principal sulcus or the medial orbital gyrus also received multiple injections in the hand and arm area of the contralateral motor cortex. These animals (M3, M6, M7, M8, M11)

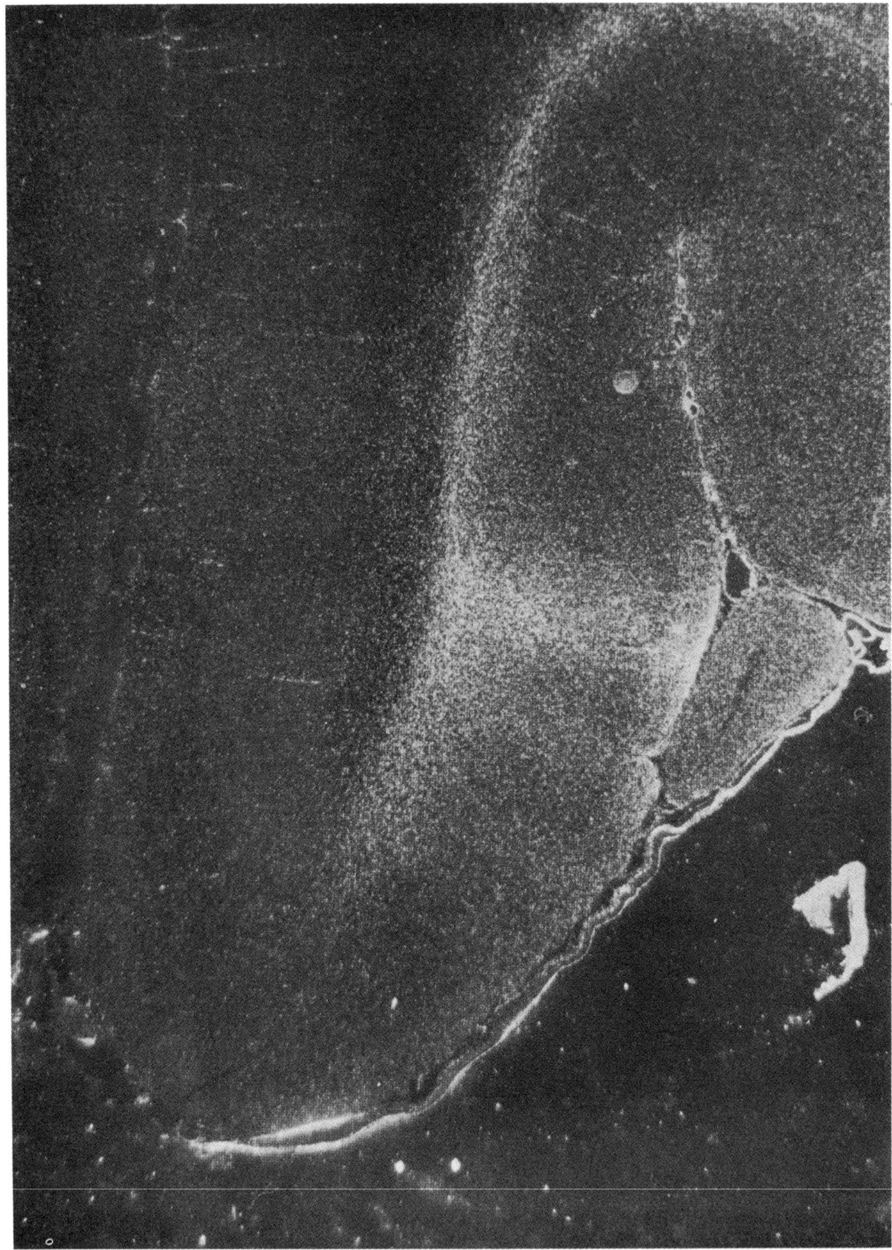

Fig. 7. Dark-field autoradiogram illustrating columnar projections to the ipsilateral gyrus rectus following injections of radioactive isotopes in the posterior part of the medial orbital gyrus in a 45-day-old monkey (OR 8). × 25.

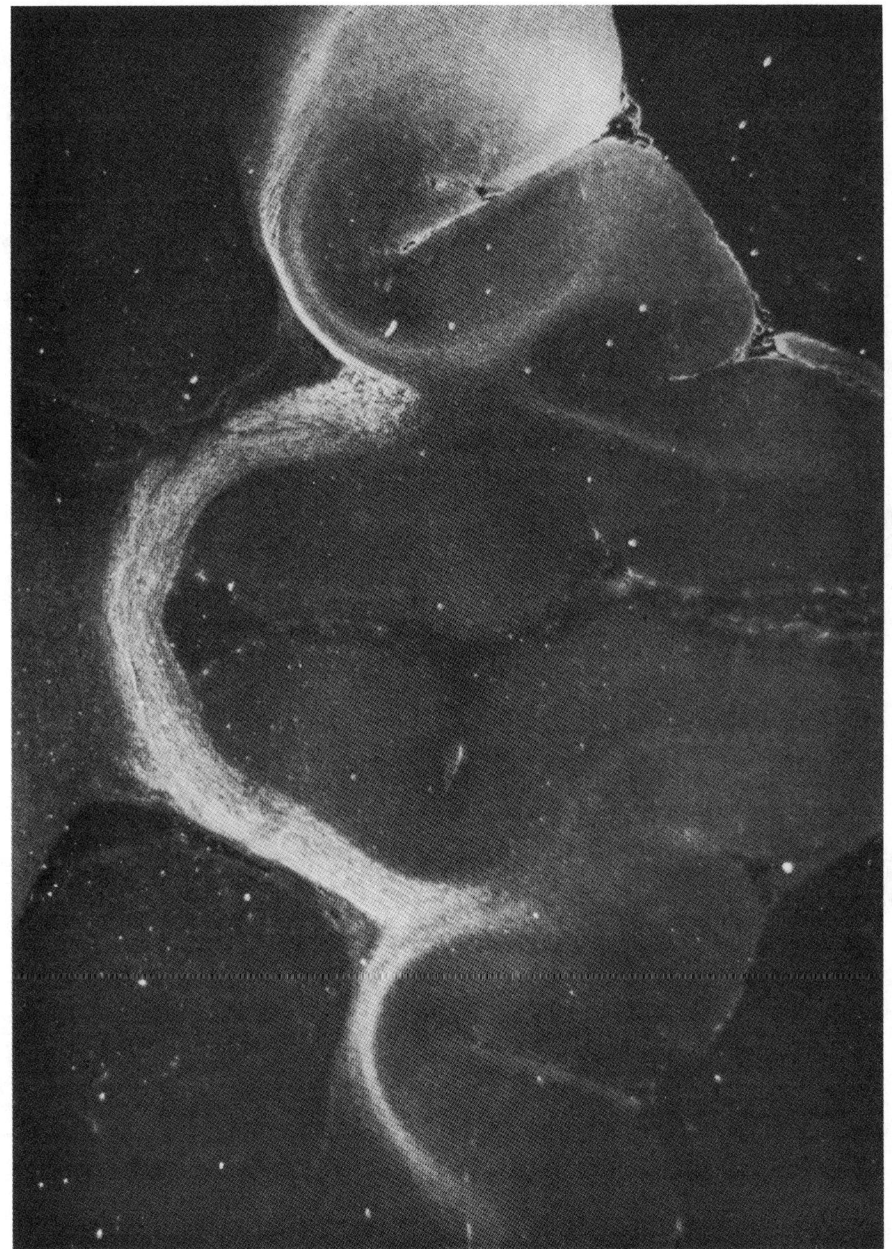

Fig. 8. Projections from the medial orbital gyrus crossing the hemispheres in the rostrum of the corpus callosum in a 45-day-old monkey (OR 8), injected hemisphere on the right. × 10.

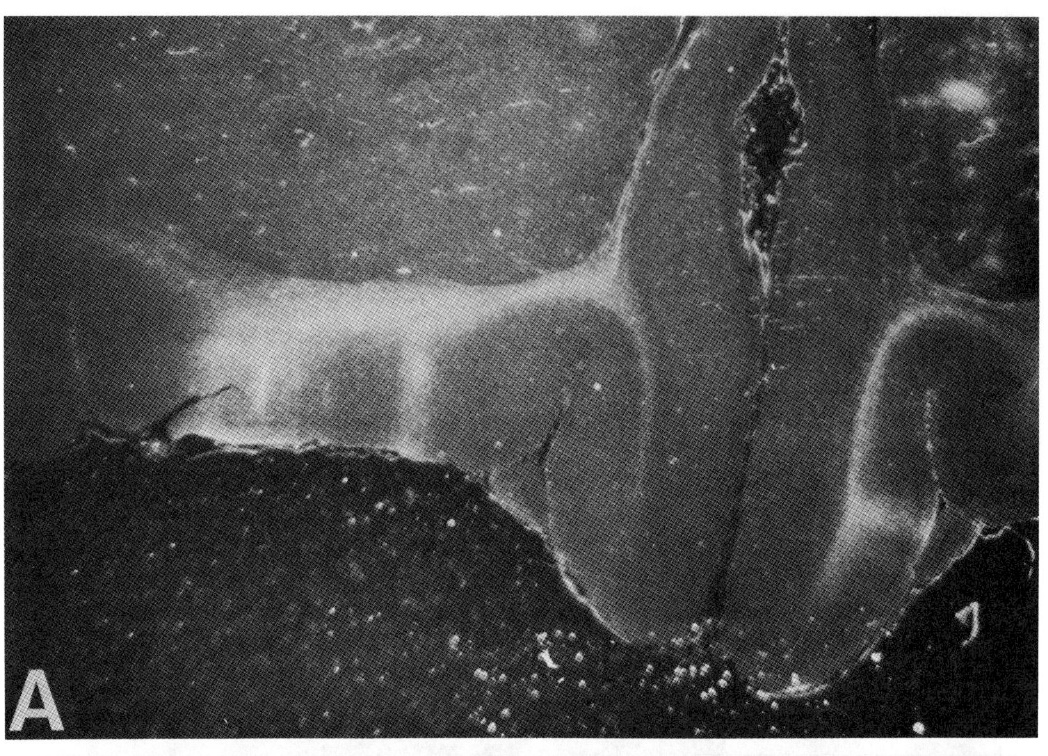

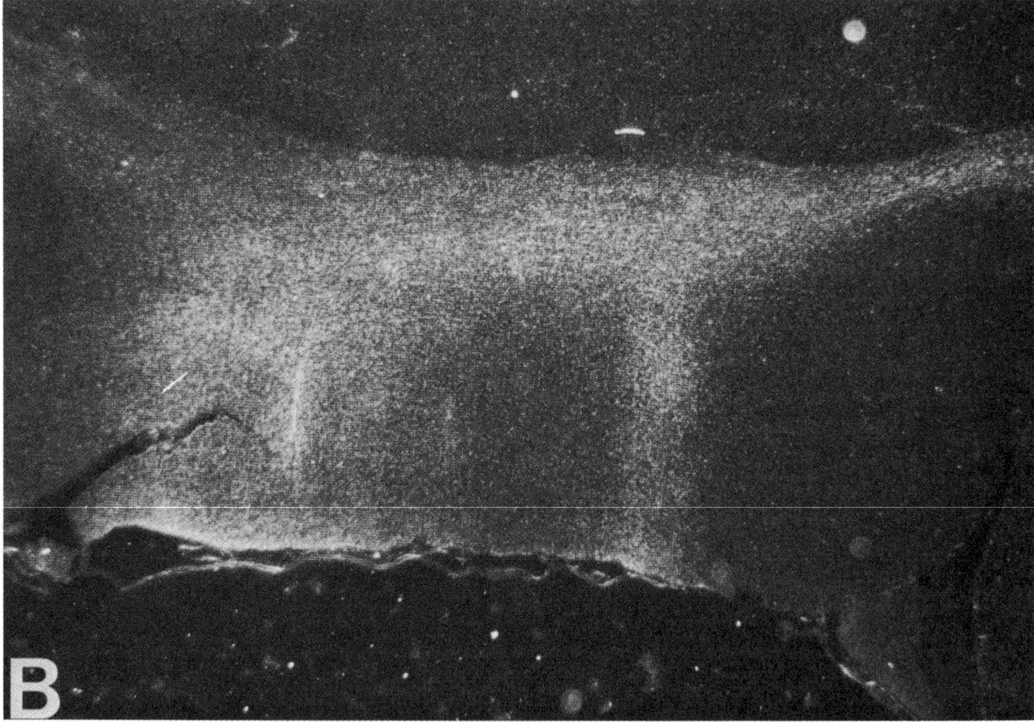

Fig. 9. A: low power autoradiogram of columns in the contralateral homotopical cortex of a 45-day-old monkey (OR 8) injected in the medial orbital gyrus of the opposite hemisphere. × 10. B: same area under higher magnification. × 25.

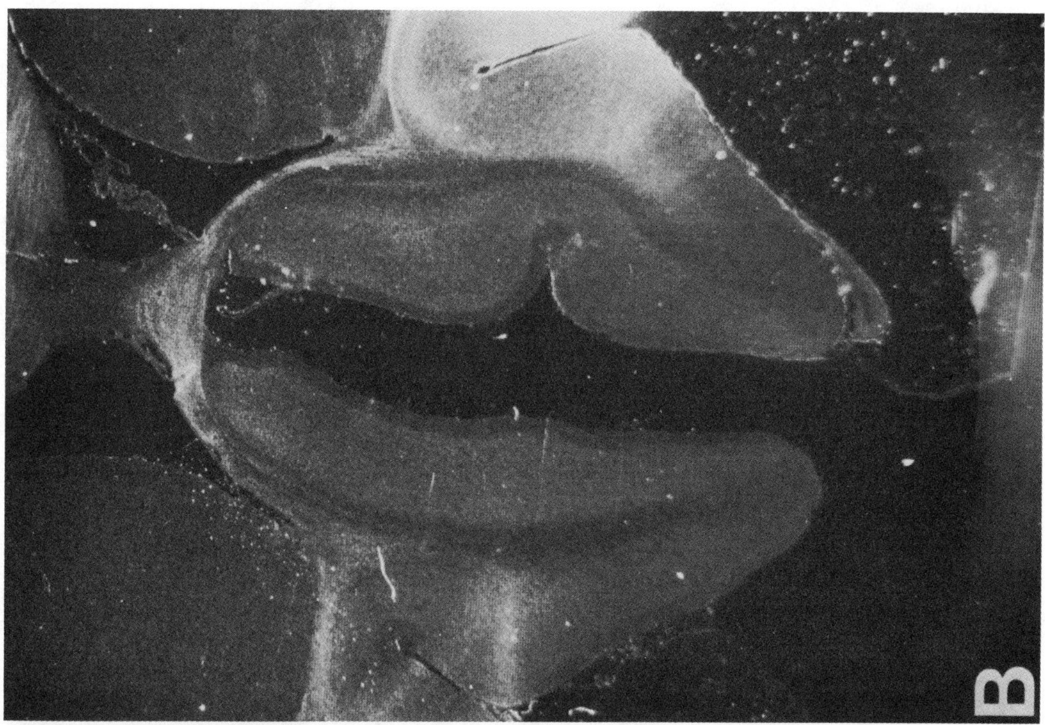

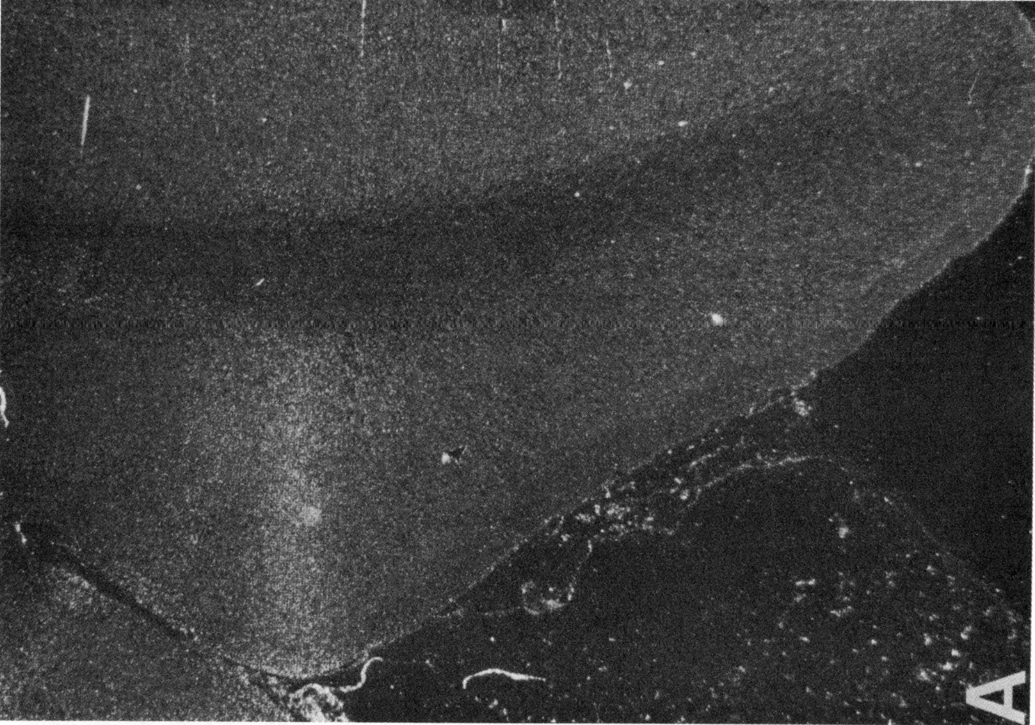

Fig. 10. A: dark-field autoradiogram of columns in a heterotopic region (the gyrus rectus) of the hemisphere contralateral to the injection site in the posterior part of the medial orbital gyrus, 4-day-old monkey (OR 7). × 25. B: lower power photograph illustrating the complementary projections from the medial orbital gyrus to the ipsilateral and to the contralateral gyrus rectus in the same 4-day-old monkey. × 10.

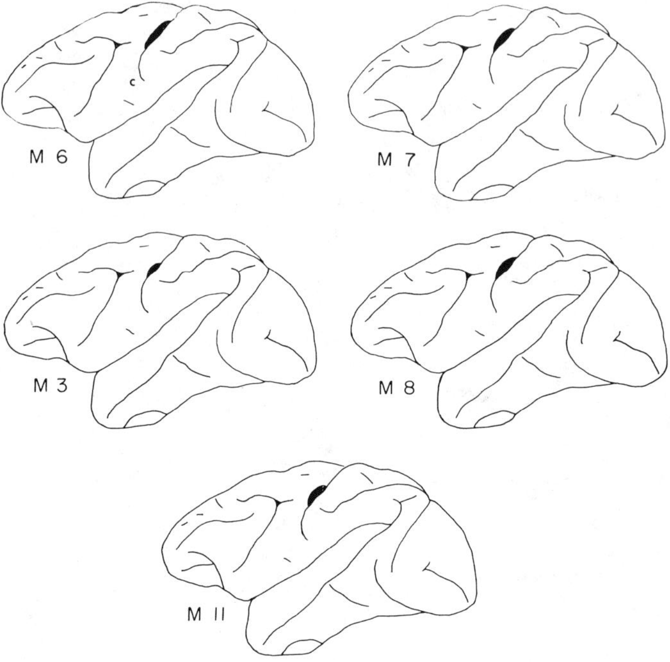

Fig. 11. Diagrammatic illustration of injection sites in the motor cortex on a standard lateral view of the monkey brain. c, central sulcus.

were distributed evenly among the various age groups. The injected area of motor cortex varied among cases, but the variations were not related to age (Fig. 11). The cortico-cortical connections of the motor cortex appeared not to overlap those of the prefrontal cortex.

*Ipsilateral associations.* The most prominent ipsilateral cortico-cortical connections revealed by the present method were those to the postcentral gyrus, the supplementary motor cortex on the medial aspect of the hemisphere, and the depths of the central sulcus, particularly the latter's anterior bank. A spatially periodical terminal distribution appeared to be somewhat less pronounced in these cortico-cortical connections than in those labeled by isotope injections in the prefrontal cortex, but labeled bands extending throughout the thickness of the cortex could be recognized nonetheless (Fig. 12A), especially clearly in the youngest monkeys.

*Commissural projections.* Labeled fibers crossed to the opposite hemisphere through the corpus callosum at levels caudal to the path of the prefrontal callosal fibers and in lesser concentration. A columnar distribution of these commissural fibers in the homotopical region of the motor cortex was plainly evident, and more so in the younger than in the older animals. Owing to the plane of sectioning, most columns were cut in oblique or transverse sections, and appeared in the form of round or oval grain aggregates which, in a 4-day-old animal, had a diameter of about 500 $\mu$m and were separated by label-free spaces of comparable width (Fig. 12B). The columns extended throughout the thickness of the cortex but, in contrast to

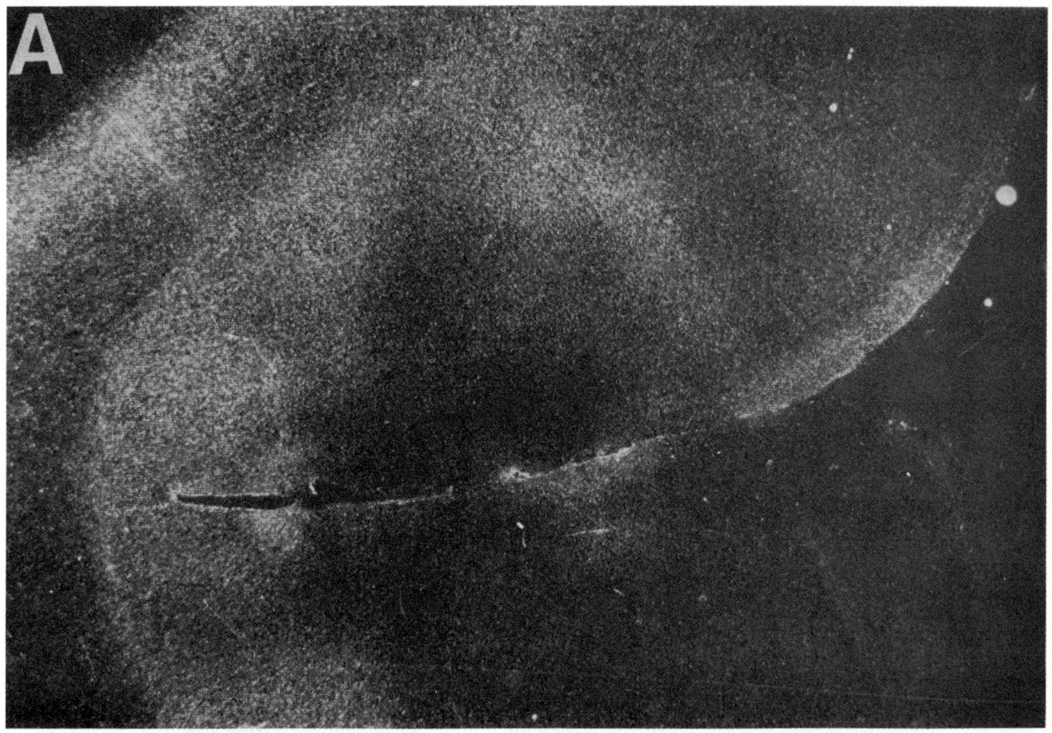

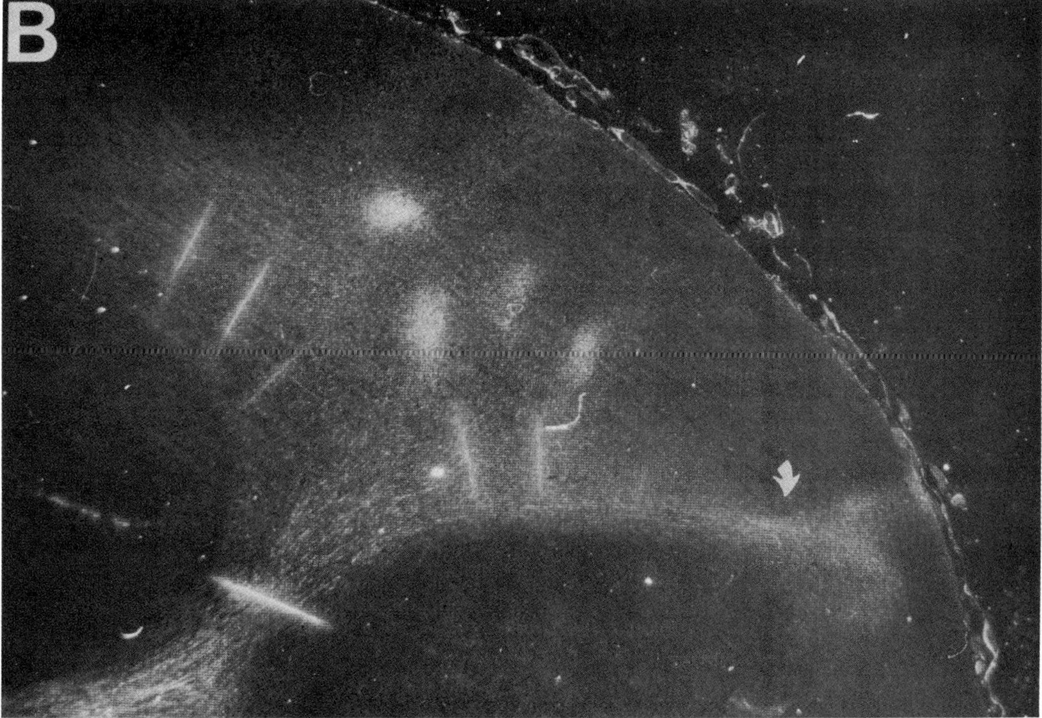

Fig. 12. A: dark-field autoradiogram illustrating ipsilateral cortico-cortical projections to the callosal marginal sulcus from the injection site in the hand and arm area of a 4-day-old monkey (M 7). × 25. B: dark-field autoradiogram illustrating columns in the contralateral homotopical cortex of the same monkey (M7). Arrow indicates single band of grains distributed over all layers of cortex at ventral boundary of the projection field. × 10.

the findings in the prefrontal and limbic cortex, the labeling appeared to be densest over layers III–VI and not particularly dense over layer I. It remains to be determined whether these differences in laminar distribution are genuine or only a reflection of the plane of section.

In the few instances in which a column extended over the full thickness of cortex in a single section, it was usually located at the ventral limit of the labeled region (Fig. 12B). As the boundaries between cytoarchitectonic subdivisions of the motor cortex are not distinct, it could not be determined whether these bands represented heterotopical or homotopical commissural associations.

DISCUSSION

*Columnar organization*

Lorente de Nó[22] was among the first to emphasize the vertical aspects of cortical architecture, and his observations on the arborization of thalamocortical fibers supplied an important morphological correlate of the later discovered columnar organization of sensory thalamocortical connections. It is not at all surprising that the first physiological evidence of cortical columniation emerged from studies of the sensory cortices, since columns in such locations appear to reflect the well-defined topographical relationships between peripheral receptors and their central representation that have made the sensory systems so uniquely accessible to detailed structure–function analysis. Even the more recent demonstration that in the somatosensory cortex cortico-cortical fibers likewise are distributed in a columnar pattern[16,27] could be interpreted as a consequence of such faithful re-representation of the peripheral receptor surface, if, namely, it were assumed that the cortico-cortical afferents of a modally specific sensory region must spatially 'fit in' with the columnar thalamo-cortical pattern. However, several observations, including those here reported, argue against such a categoric interpretation of cortical columns and suggest instead that columns represent a mode of intracortical fiber distribution that is encountered in all of the conventionally distinguished categories of cortex (specific sensory, motor and association) and, moreover, is common to both of the major fiber categories (thalamocortical and cortico-cortical) of which the cortical white matter is composed, including, it would seem, the 'U-fibers' interconnecting neighboring gyri. For the present, however, any statement on this subject should leave room for exception. Thalamocortical fibers, for instance, form a heterogeneous category that includes widely spread 'non-specific' connections, an extreme example of which is the recently demonstrated projection of the ventromedial thalamic nucleus to layer I of almost the entire neocortex[11]; no observations reported thus far have suggested a columnar distribution of such more diffuse thalamocortical projections. Moreover, evidence of a columnar thalamocortical fiber distribution in regions outside the sensory cortices is still lacking.

Although the cortico-cortical fibers originating in the frontal cortex are comparable to those terminating in the somatosensory cortex[16], in the sense that both are distributed in a columnar pattern, they appear to differ from the latter in the details of laminar distribution. Whereas cortico-cortical afferents in the somato-sensory cortex were reported to terminate primarily in layers I–IV and most massively in layers II–IV, prefrontal cortico-cortical efferents appear to be distributed rather uniformly over all layers within each column. However, a definitive analysis of the terminal distribution of such efferents must await more detailed morphological investigation.

*Developmental considerations*

In the present study a columnar distribution of cortico-cortical fibers originating in the prefrontal and motor cortex was plainly evident even in the youngest monkeys, sacrificed 11 days after birth. In fact, the columns consistently had higher grain concentrations and were more sharply delineated in the youngest than in the older monkeys, irrespective of injection locus or distribution area.

It is not possible on the basis of the experiments here reported to determine the actual number, volume and isotope content of the labeled association-fiber columns, or to state whether the decrease of grain density with age would follow a similar time course in all cortical association systems. In the absence of such data all explanations of the phenomenon must remain conjectural. First of all, the denser labeling of association-fiber columns in the younger monkeys could be due to age-dependent differences in the rate of axonal transport. In several neuronal systems of a variety of animal species the slow component of intra-axonal protein transport is more than twice as fast in neonatal as in adult animals[5,10,19,21]. By contrast, the rapid transport phase has been found to accelerate with age[4,9,10,23]. Detailed information on developmental changes in the slow and rapid phases of axonal transport in primates is not available. However, it seems possible that during the 7-day survival period, as a consequence of such rate differences, a greater amount of radioactively labeled proteins had reached the terminal arborizations of the fronto-fugal association fibers in the younger than in the older monkeys. Alternatively, it is conceivable that the observed differences reflect a gradual postnatal increase in the volume of the cortical neuropil. Such an increase, especially if due primarily to a proliferation of the terminal arbors of other fiber systems (intracortical or thalamocortical) could widen the spacing of cortico-cortical fibers and thereby reduce the density of their autoradiographic appearance. It is evident that further studies are required to determine the relative importance of these two — and probable additional — factors.

The remarkable distinctness of association-fiber columns in very young monkeys stands in contrast to the incomplete segregation of monocular dominance columns in the visual cortex of the 7-day-old monkey[14]. Contrary to what might have been expected on the basis of relative rates of functional development[6,7], association fibers originating in the frontal cortex would seem to achieve a distinct columnar distribution earlier than do the fibers of the optic radiation.

This unexpected time sequence raises questions with regard to the mechanisms underlying the development of cortical columns. It has been shown that the normal formation of discrete thalamocortical fiber columns in the rodent somatosensory cortex is prevented by destruction of vibrissa follicles at birth[18], but it appears that no reports published thus far have described the effect of such early interventions upon the distribution of cortico-cortical fibers which under normal conditions is limited to the intervals between the thalamocortical columns[26]. In the absence of such information it is not possible to state whether the distribution pattern of cortico-cortical afferents in the somatosensory cortex is entirely determined by the columniation of the thalamocortical input. Even should this indeed prove to be the case in modally specific sensory areas, it would remain to be seen whether the corticopetal projection of the mediodorsal thalamic nucleus exerts an equally dominant influence upon the development of cortico-cortical fiber columns in the frontal association cortex. Such a comparison of sensory with association cortex can be expected to contribute to an understanding of the conditions determining the ontogeny and the spatial interrelationships of the columns in which various categories of afferent fibers are distributed in the cerebral cortex.

ACKNOWLEDGEMENTS

This study was made possible by Public Health Service Grants NB 06542 and P01 NS 12336 and by intramural research funds of the National Institute of Mental Health.

REFERENCES

1 Abeles, M. and Goldstein, M. N., Jr., Functional architecture in cat primary auditory cortex: columnar organization according to depth, *J. Neurophysiol.*, 33 (1970) 172–187.
2 Asanuma, H. and Sakata, H., Functional organization of a cortical efferent system examined with focal depth stimulation in cats, *J. Neurophysiol.*, 30 (1967) 35–54.
3 Cowan, W. M., Gottlieb, D. I., Hendrickson, A. E., Price, J. E. and Woolsey, T. A., The autoradiographic demonstration of axonal connections in the central nervous system, *Brain Research*, 37 (1972) 21–51.
4 Crossland, W. J., Currie, J. R., Rogers, L. A. and Cowan, W. M., Evidence for a rapid phase of axoplasmic transport at early stages in the development of the visual system of the chick and frog, *Brain Research*, 78 (1974) 483–489.
5 Droz, B. and Leblond, C. P., Axonal migration of proteins in the central nervous system and peripheral nerves as shown by radioautography, *J. comp. Neurol.*, 121 (1963) 325–345.
6 Goldman, P. S., Age, sex and experience as related to the neural basis of cognitive development. In N. Buchwald and M. Brazier (Eds.), *Brain Mechanisms in Mental Retardation*, Academic Press, New York, 1975, pp. 379–392.
7 Goldman, P. S., Maturation of the mammalian nervous system and the ontogeny of behavior. In J. S. Rosenblatt, R. Hinde, E. Shaw and C. Beer (Eds.), *Advances in the Study of Behavior*, Academic Press, New York, 1976, pp. 1–90.
8 Graybiel, A. M., Wallerian degeneration and anterograde tracer methods. In W. M. Cowan and M. Cuénod (Eds.), *The Use of Axonal Transport for Studies of Neuronal Connectivity*, Elsevier, Amsterdam, 1975, pp. 173–216.
9 Gremo, F. and Marchisio, P. C., Dynamic properties of axonal transport of proteins and glycoproteins: a study based on the effects of metaphase blocking drugs in the developing optic pathway of chick embryos, *Cell Tiss. Res.*, 161 (1975) 303–316.
10 Hendrickson, A. E. and Cowan, W. M., Changes in the rate of axoplasmic transport during postnatal development of the rabbit's optic nerve and tract, *Exp. Neurol.*, 30 (1971) 403–422.
11 Herkenham, M., The nigro-thalamo-cortical connection mediated by the nucleus ventralis medialis thalami: evidence for a wide cortical distribution in the rat, *Anat. Rec.*, 184 (1976) 426.

12 Hubel, D. H. and Wiesel, T. N., Receptive fields, binocular interaction, and functional architecture in the cat's visual cortex, *J. Physiol. (Lond.)*, 160 (1962) 106–152.

13 Hubel, D. H. and Wiesel, T. N., Receptive fields and functional architecture of monkey striate cortex, *J. Physiol. (Lond.)*, 195 (1968) 215–243.

14 Hubel, D. H., Wiesel, T. N. and LeVay, S., Plasticity of ocular dominance columns in monkey striate cortex, *Phil. Trans. roy. Soc. B*, (1976) in press.

15 Hyvärinen, J. and Poranen, A., Function of the parietal associative area 7 as revealed from cellular discharges in alert monkeys, *Brain*, 97 (1974) 673–692.

16 Jones, E. G., Burton, H. and Porter, R., Commissural and cortico-cortical 'columns' in the somatic sensory cortex of primates, *Science*, 190 (1975) 572–574.

17 Killackey, H. P., Anatomical evidence for cortical subdivisions based on vertically discrete thalamic projections from the ventral posterior nucleus to cortical barrels in the rat, *Brain Research*, 51 (1973) 326–331.

18 Killackey, H. P., Belford, G., Ryugo, R. and Ryugo, D. K., Anomalous organization of thalamocortical projections consequent to vibrissae removal in the newborn rat and mouse, *Brain Research*, 104 (1976) 309–315.

19 Lasek, R. J., Axoplasmic transport in cat dorsal root ganglion cells as studied with L-[$^3$H]leucine, *Brain Research*, 7 (1968) 360–377.

20 Lasek, R. J., Protein transport in neurons, *Int. Rev. Neurobiol.*, 13 (1970) 289–324.

21 Lasek, R. J., Axonal transport of proteins in dorsal root ganglion cells of the growing cat: a comparison of growing and mature neurons, *Brain Research*, 20 (1970) 121–126.

22 Lorente de Nó, R., The cerebral cortex: architecture, intracortical connections and motor projections. In J. F. Fulton (Ed.), *Physiology of the Nervous System*, Oxford Univ. Press, New York, 1938, pp. 291–339.

23 Marchisio, P. C. and Sjöstrand, J., Axonal transport in the avian optic pathway during development, *Brain Research*, 26 (1971) 204–211.

24 Mountcastle, V. B., Modality and topographic properties of single neurons of cat's somatic sensory cortex, *J. Neurophysiol.*, 20 (1957) 408–434.

25 Mountcastle, V. B. and Powell, T. P. S., Central nervous mechanisms subserving position sense and kinesthesis, *Bull. Johns Hopk. Hosp.*, 105 (1959) 173–200.

26 Ryugo, R. and Killackey, H. P., Cortico-cortical connections of the barrel field of the rat somatosensory cortex, *Neurosci. Abstr.*, 1 (1975) 126.

27 Shanks, M. F., Rockel, A. J. and Powell, T. P. S., The commissural fiber connections of the primary somatic sensory cortex, *Brain Research*, 98 (1975) 166–171.

28 Van der Loos, H. and Woolsey, T. A., Somatosensory cortex: structural alterations following early injury to sense organs, *Science*, 179 (1973) 395–397.

29 Wiesel, T. N., Hubel, D. H. and Lam, D. M. K., Autoradiographic demonstration of ocular-dominance columns in the monkey's striate cortex by means of transneuronal transport, *Brain Research*, 79 (1974) 273–279.

30 Woolsey, T. A. and Van der Loos, H., The structural organization of layer IV in the somatosensory region (SI) of mouse cerebral cortex, *Brain Research*, 17 (1970) 205–242.

Reprinted from THE JOURNAL OF COMPARATIVE NEUROLOGY
Vol. 171, No. 3, February 1, 1977  © The Wistar Institute Press 1977

# An Intricately Patterned Prefronto-Caudate Projection in the Rhesus Monkey [1]

PATRICIA S. GOLDMAN AND WALLE J. H. NAUTA
*Laboratory of Neuropsychology, National Institute of Mental Health, Bethesda, Maryland 20014 and Department of Psychology, Massachusetts Institute of Technology, Cambridge, Massachusetts 02139*

*ABSTRACT*     The distribution of prefronto-caudate fibers in the caudate nucleus was studied autoradiographically in monkeys of various ages in which tritiated amino acids had been injected into the middle one-third of the length of the dorsal bank of the principal sulcus. The results indicate that, contrary to previous reports which had suggested a projection to only the head of the caudate nucleus, area 9 of Brodmann projects to the entire length of the nucleus.

In the head of the caudate nucleus the cortico-caudate fibers are distributed in a pattern which is remarkable in two respects. First, the grains are not uniformly distributed but rather are segregated into clusters separated from one another by territories in which grain density does not exceed background. Second, individual clusters of grains, circular or elliptical in shape, surround grain free cores. These patterns of fiber distribution within the head of the nucleus are more sharply defined in newborn than in older monkeys. Our findings suggest that the caudate nucleus is organized more as an anatomic and functional mosaic than as the homogeneously organized structure that it is commonly considered to be.

The results of fiber-degeneration studies in a variety of mammals have suggested that the caudate nucleus receives projections from virtually every region of the cerebral cortex in a topographic pattern such that the frontal cortex projects to the head of the caudate nucleus while the more posterior parietal, occipital and temporal areas project to progressively more caudal parts of the body and tail of the nucleus (Whitlock and Nauta, '56; Carman et al., '63; DeVito and Smith, '64; Nauta, '64; Webster, '61, '65; Johnson et al., '68, '76; Petras, '69; Kemp and Powell, '70, '71a).

In the course of an autoradiographic study of the postnatal development of the efferent connections of the prefrontal cortex in the rhesus monkey we found evidence of a fronto-caudate projection so different in its distribution from earlier descriptions that it seemed worth being reported separately.

## MATERIALS AND METHODS

The present account is based primarily on findings made in two 4-day-old Macaca mulatta monkeys that also served as subjects in a previous study of efferent cortico-cortical connections of the frontal cortex (Goldman and Nauta, '77). These animals had received injections of a mixture of tritiated leucine and proline in the middle one-third of the length of the dorsal bank of the principal sulcus. In one of the animals 30 µCi of the isotope had been injected in a single volume, whereas the other had received 20 µCi distributed over four injections spaced approximately 1 mm apart. Both animals were sacrificed seven days after the injection. Details of the injection procedure and the method of processing the material for autoradiography have been described elsewhere (Goldman and Nauta, '77).

The experimental material from which these two cases were selected also included animals at later stages of postnatal

[1] The research here reported was supported by USPHS Grants NS 06542 and PO1 NS 12336, and by intramural research funds of the National Institute of Mental Health.

development (up to 5.5 months), as well as one case in which the subject was a young adult monkey estimated at three years of age. The extent of the label-distribution was about the same in all of these cases, and the primary reason for documenting the findings in only the two youngest animals is, that the topographic pattern of the fronto-caudate connection was found to be far more sharply defined in these cases than in the older monkeys.

### RESULTS

In both cases radioactive label was found to have been transported from the mid-region of the dorsal bank of the principal sulcus (fig. 1) to the entire length of the ipsilateral caudate nucleus. Labelling was most massive in the head of the nucleus but also quite substantial in the body and tail.

The distribution of transported label in the head region showed a remarkable pattern. Especially at rostral levels the grains, instead of being evenly distributed, were concentrated in circumscript, irregularly shaped and locally confluent territories separated by areas in which grain density hardly exceeded background levels. Many of these territories exhibited an unlabelled center and thus appeared in many sections as circular or elliptical grain aggregates surrounding a grain-free core, in others as semicircular shapes incompletely

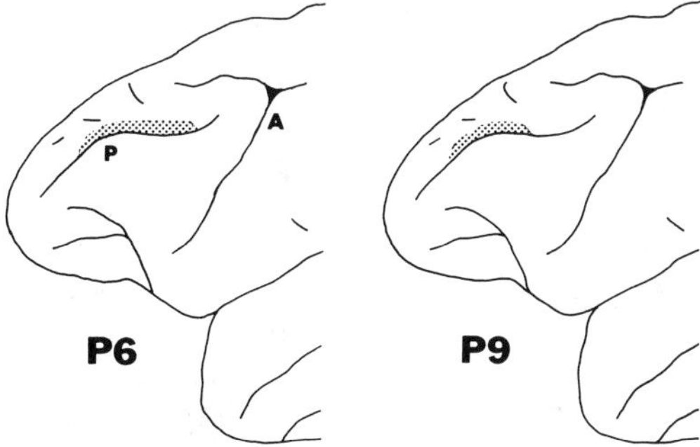


Fig. 1 Diagrammatic illustration of injection sites in P6, the monkey given multiple injections in the dorsal bank of the principal sulcus, and P9, the monkey given a single injection in the same area. P, principal sulculs; A, arcuate sulcus.

surrounding a label-free center (figs. 2,3).

At rostral levels, such grain clusters were disseminated throughout the cross-section of the caudate nucleus (fig. 2). Slightly more posteriorly in the head of the caudate nucleus, however, labelling was largely restricted to a dorsomedial sector of the nucleus near its ventricular surface (fig. 3). Although at these more caudal levels heavy labelling marked the dorsal stratum of the internal capsule, the concentration of grain over the ventrolateral region of the caudate nucleus did not exceed background. At these more posterior levels of the caput, likewise, the labelling pattern suggested a sharp segregation of the labelled fibers from other afferents: at least two irregularly shaped grain aggregates each surrounding an island of relatively label-free tissue appeared in the corresponding sections (figs. 3A,B). It must be noted that the stereometric complexity of the labelling in the head of the caudate nucleus was equally pronounced in both cases studied, and hence cannot be attributed to the multiple injections by which, in one case, the isotopes had been deposited in the cortex.

Grains were concentrated in a dorsomedial position also in the body of the caudate nucleus. However, in this part of the nucleus the label was rather uniformly distributed (fig. 4). As the caudate nucleus extends beneath the parietal lobe, it tapers in size and ultimately curves around the internal capsule to continue forward as a slender structure in the roof of the temporal horn of the lateral ventricle. Because of its semilunar curve, coronal sections involving the temporal horn display the tail of the nucleus twice in a single section (fig. 5). Both the dorsal and the ventral limb of the tail showed considerable labelling (figs. 6A,B). In the former, the grains were located ventrolaterally, whereas in the ventral, temporal limb they marked a medial region of the nucleus. The gradual shift in the relative position of the label could be readily traced in serial sections. Thus, in the head, the body, and the rostral extent of the tail, the grains were located over the dorsomedial region of the nucleus; as the tail extends caudalward they moved progressively more ventralward and finally came to occupy a ventrolateral position. Finally, in the region where the tail of the caudate nucleus begins to extend foward in the roof of the temporal horn of the lateral ventricle, the label shifted again and once more came to mark primarily the medial region of the nucleus. In no part of the tail did the label exhibit the complex configuration that characterized its distribution in the head of the nucleus.

### DISCUSSION

The present study revealed two novel facts concerning the projections of the prefrontal cortex to the caudate nucleus in the rhesus monkey: (1) Area 9 of Brodmann projects to the entire length of the nucleus, i.e., to the head, body and tail; (2) the distribution of these projections within the head of the caudate nucleus exhibits a complex stereometric organization characterized by irregularly shaped and locally confluent clusters most of which, especially in very young animals, have a sharply defined, somewhat spherical unlabelled core.

### Topographic distribution

The prefronto-caudate connection here outlined by autoradiography differs in major respects from earlier descriptions based on fiber-degeneration studies. In a case of large prefrontal decortication including the region of the principal sulcus Kemp and Powell ('70) found fiber degeneration confined to the lateral parts of the caudate nucleus, whereas Johnson et al. ('68) reported that ablation of the cortex lining the principal sulcus produced fiber degeneration in a mediodorsal sector of the caudate nucleus; in both studies, the fiber degeneration seemed to be confined to the head of the nucleus. It appears that a projection from the prefrontal cortex to the entire length of the caudate nucleus as here demonstrated by the autoradiographic method has not been suggested by earlier findings.

Neither does the remarkably patchy distribution of prefronto-caudate fibers in the head of the monkey's caudate nucleus appear to have been described before, but it may be recalled that Kemp and Powell ('71a) noted a cluster-pattern in the Nauta-Gygax degeneration elicited in the caudate nucleus by prefrontal decortication in the cat.

It is difficult to account for a discrepancy of such magnitude. In hindsight it seems likely that the Nauta-Gygax method could demonstrate only the largest of the generally thin corticostriatal fibers, and failed to bring out the terminal ramifications of these fibers. It is, moreover, possible that the prefrontal projection to the tail of the caudate nucleus is composed entirely of very thin fibers and thus escaped detection by the Nauta-Gygax method altogether. This speculation was suggested to us by our difficulties in tracing the full course of the prefronto-caudate connection in autoradiograms. The subcallosal fasciculus, the most likely conveyor bundle of corticocaudate fibers, exhibited considerable labelling at rostral levels (fig. 2), but farther caudally was relatively free of label (fig. 4). Since our findings suggested no other access route to the tail of the caudate nucleus, it seems possible that prefronto-caudate fibers enter the nucleus exclusively at the level of the caput, and can extend to the tail only by following an intranuclear trajectory in the course of which the fibers could become progressively thinner by repeatedly issuing collaterals.

A comparable discrepancy in the respective results of the autoradiographic and the ablation method has appeared in reports concerning the corticostriatal projections originating in the motor cortex. According to Kemp and Powell's ('70) findings in a study by the Nauta-Gygax method, the motor cortex of the monkey projects massively to the caudate nucleus, but autoradiographic evidence (Künzle, '75, confirmed by our own unpublished observations) suggests that it projects primarily to the putamen and only sparsely to the caudate nucleus. The autoradiographic findings suggest that the caudate nucleus may be more closely related anatomically and functionally to the prefrontal cortex than to the motor cortex.

### Developmental considerations

A somewhat similar discrepancy between results obtained with the silver degeneration method and those obtained by autoradiography appears in the study of the development of corticostriate connections. Johnson, et al. ('76) found pronounced quantitative changes in the density of Nauta-Gygax degeneration in the head of the caudate nucleus following comparable prefrontal lesions in rhesus monkeys at different stages of postnatal development. The virtual absence of such degeneration in monkeys operated at two months of age appears to be in conflict with the clear demonstration in the present study of plentiful corticostriatal afferents in newborn monkeys. The apparent conflict can be explained if it is assumed that silver degeneration methods are sensitive to such parameters of development as increase in fiber size while the autoradiographic method may be less dependent on such variables. Neither method, of course, is satisfactory for determining the maturity of synaptic relationships.

The present autoradiographic findings in newborn monkeys may also be compared with results obtained with the same method in mature monkeys. A "patchy" distribution of corticostriatal fibers somewhat comparable to that reported here was recently noted by Künzle ('75) in the putamen of the adult monkey. Künzle's description contains no mention of hollow-sphere or bell-shaped grain aggregates, and it is interesting that in our own case material such complex and well-defined stereometric shapes, common in the youngest animals, seemed to become progressively simplified with increasing age until, in the fully adult monkey, nearly all appeared to have been converted into solid clusters of various size and shape. Along with this simplification of configuration, the clusters became considerably less dense and sharply defined.

These postnatal changes naturally raise more questions than can be answered at present. Does the gradual loss of grain density in the clusters reflect an actual loss of fibers that had been laid down at or before birth, or is it perhaps caused by a "dilution" of the clusters by later-arriving fibers of other (e.g., thalamic, intrastriatal or dopamine-cell) origin? Is the unlabelled core of the clusters progressively filled up by postnatal sprouting of fibers already present in the shell of the cluster, or by other corticostriatal fibers arriving during postnatal development? In connection with the last question it is of interest to recall an intriguing observation by Olson et al. ('72) according to which in the newborn rat the nigral dopamine-fiber innervation of the striatum is largely confined to small, sharply defined islands that in later development become masked by what appears to be an "overlay" of diffusely distributed dopamine afferents, but that can be re-visualized even in the adult rat by the administration of a tyrosine-hydroxylase inhibitor.

### Discontinuity and stereometry of the prefronto-caudate projection

Discontinuities in afferent fiber distribution were first discovered in the afferent connections of the cerebral cortex where they take the form of patterns of columns, barrels, or ribbons, but the autoradiographic fiber-tracing method has disclosed a somewhat similar configuration also in the subcortical distribution of retinal fibers in the superior colliculus of both cat (Graybiel, '75) and monkey (Hubel et al., '75). In specific-sensory cortical areas and in superior colliculus such columnar patterns seem readily attributable to the anatomical and functional organization of the receptor surface at which the respective specific-afferent pathways originate. In other cortical regions such as the prefrontal and limbic cortex the recently discovered columnar distribution of afferent fiber systems (Goldman and Nauta, '77) is more difficult to interpret in functional terms, but even here it could be thought to reflect at least the pronounced vertical organization of cortical tissue in general. However, at present there seems to be no anatomical or functional basis whatever from which to evaluate the meaning of the intricate pattern of corticostriatal connections demonstrated in the putamen by Künzle ('75) and in the caudate nucleus by the present study. It can be expected that similar uncertainties will arise in evaluating the discontinuous patterns of innervation which have been demonstrated in other subcortical connections, e.g., those from the substantia nigra (Graybiel and Sciascia, '75) and from the prefrontal cortex (Goldman and Nauta, '77) to the superior colliculus. On the basis of such findings and those of the present study, however, it now appears that segregation in the distribution of afferent projections may be as common a feature of subcortical as of cortical structures, and hence a general property of central nervous system organization.

Reasoning on the basis of afferent connections only, the widely disseminated clusters in which the corticocaudate projection from only a small region (the dorsal bank of the principal sulcus) of the prefrontal cortex terminates in the head of the caudate nucleus would seem to suggest that the projection alternates spatially with other caudate afferents. It seems unlikely that the cluster pattern segregates corticocaudate from thalamocaudate fibers, since Kemp and Powell's ('71b) observations in a Golgi study indicate that these two fiber categories terminate on the dendrites of the same rather than of different cells. An interdigitation of corticostriatal fibers with other corticostriatal fibers of different origin, or with nigrostriatal or intrinsic striatal fibers would nonetheless remain to be considered possible.

On the other hand, however, the cluster pattern could have its primary basis in efferent rather than afferent relationships of the striatum. Until disproved, the possibility must be considered that the efferent connections of the primate striatum with the globus pallidus and substantia nigra originate from clusters of neurons spatially segregated from territories composed of intrinsic striatal neurons, and hence, that corticostriatal-fiber clusters reflect a particular mode of channeling of cortical impulse-efflux into striatopallidal and striatonigral pathways. The clusters of neurons within the caudate could, in turn, be a reflection of the columnar arrangement of neurons in the prefrontal cortex from which the fronto-caudate projections originate (Goldman and Nauta, '77). Such an arrangement could be conceived of as an expanded, more multicellular analogue of the thalamic glomerulus in which the "through-line" synaptic relationship between afferent (lemniscal) and efferent (thalamocortical) neurons is complicated by intrinsic thalamic neurons straddling the main synapse and leading off from, and back into, the "through-line" (for a review, see Széntágothai, '70).

Needless to say, there may be other models of the corticostriate system than those outlined here and further, the different possibilities need not be mutually exclusive. For example, the discontinuous distribution of prefronto-caudate connections might best be explained by one mode of organization while the stereometry within individual grain clusters is accounted for by the other. Although a better understanding of the anatomic and functional significance of configuration in corticostriate connections must await new discoveries, it is clear that the striatum is an anatomic and functional mosaic and not, as it has seemed until now, homogeneously organized.

## LITERATURE CITED

Carman, J. B., W. M. Cowan and T. P. S. Powell 1963 The organization of corticostriate connections in the rabbit. Brain, 86: 525-560.

DeVito, J. L., and O. A. Smith, Jr. 1964 Subcortical projections of the prefrontal lobe of the monkey. J. Comp. Neur., 123: 413-424.

Goldman, P. S., and W. J. H. Nauta 1977 Columnar distribution of cortico-cortical fibers in the motor, frontal association, and limbic cortex of the immature and adult monkey. Brain Res., in press.

—— 1976 Autoradiographic demonstration of a projection from prefrontal association cortex to the superior colliculus. Brain Res., 116: 145-149.

Graybiel, A. M. 1975 Anatomical organization of retinotectal afferents in the cat: an autoradiographic study. Brain Res., 96: 1-24.

Graybiel, A. M., and T. R. Sciascia 1975 Origin and distribution of nigrotectal fibers in the cat. Neurosci., 1: 174, (Abstract).

Hubel, D. H., S. LeVay and T. N. Wiesel 1975 Mode of termination of retino-tectal fibers in macaque monkey: an autoradiographic study. Brain Res., 96: 25-40.

Johnson, T. N., H. E. Rosvold, T. W. Galkin and P. S. Goldman 1976 Postnatal maturation of subcortical projections from the prefrontal cortex in the rhesus monkey. J. Comp. Neur., 166: 427-444.

Johnson, T. N., H. E. Rosvold and M. Mishkin 1968 Projections from behaviorally-defined sectors of the prefrontal cortex to the basal ganglia, septum and diencephalon of the monkey. Exp. Neurol., 21: 20-34.

Kemp., J. M., and T. P. S. Powell 1970 The corticostriate projection in the monkey. Brain, 93: 525-546.

—— 1971a The site of termination of afferent fibers in the caudate nucleus. Phil. Trans. R. Soc. Lond. (B), 262: 413-427.

—— 1971b The termination of fibers from the cerebral cortex and thalamus upon dendritic spines in the caudate nucleus: a study with the Golgi method. Phil. Trans. R. Soc., Lond. (B), 262: 429-439.

Künzle, H. 1975 Bilateral projections from precentral motor cortex to the putamen and other parts of the basal ganglia. Brain Res., 88: 195-210.

Nauta, W. J. H. 1964 Some efferent connections of the prefrontal cortex in the monkey. In: The Frontal Granular Cortex and Behavior. J. M. Warren and K. Akert, eds. McGraw-Hill, New York, pp. 397-409.

Olson, L., Å. Seiger and K. Fuxe 1972 Heterogeneity of striatal and limbic dopamine innervation: Highly fluorescent islands in developing and adult rats. Brain Res., 44: 283-288.

Petras, J. M. 1969 Some efferent connections of the superior and inferior parietal lobules with the basal ganglia, diencephalon and midbrain in the rhesus monkey. Anat. Rec., 163: 243-244.

Széntágothai, J. 1970 Glomerular synapses, complex synaptic arrangements, and their operational significance. In: The Neurosciences, Second Study Program. F. O. Schmitt, ed. The Rockefeller University Press, New York, pp. 427-443.

Webster, K. E. 1961 Cortico-striate interrelations in the albino rat. J. Anat. (London), 95: 532-545.

—— 1965 The cortico-striatal projection in the cat. J. Anat. (London), 99: 329-337.

Whitlock, D. G., and W. J. H. Nauta 1956 Subcortical projections from the temporal neocortex in Macaca mulatta. J. Comp. Neur., 106: 183-212.

## PLATE 1

### EXPLANATION OF FIGURE

2 Dark-field autoradiogram illustrating segregated clusters of grains in the rostral portion of the head of the caudate nucleus in P6. NC, nucleus caudatus; FSC, fasciculus subcallosus; CI, capsula interna; CC, corpus callosum; LV, lateral ventricle. × 25.

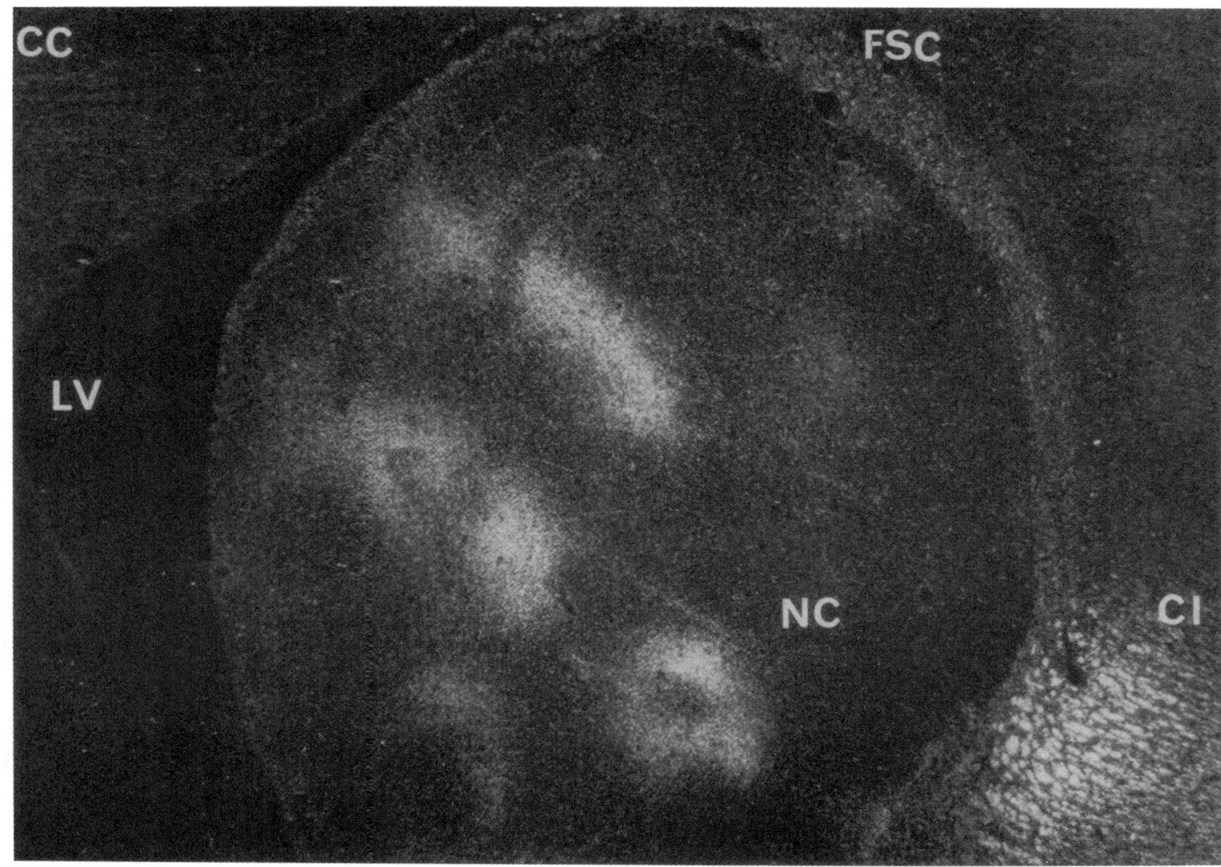

CC
FSC
LV
NC
CI

## PLATE 2

EXPLANATION OF FIGURES

3A  Dark-field autoradiogram illustrating clusters of grains surrounding annular islands free of grains in P9. CC, corpus callosum; CI, capsula interna. × 21.3.

3B  Stereometric pattern shown in A at higher magnification. × 51.

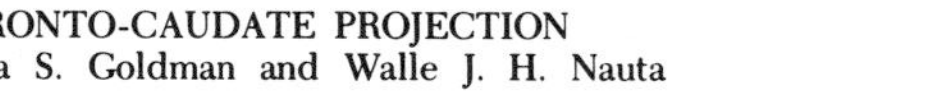

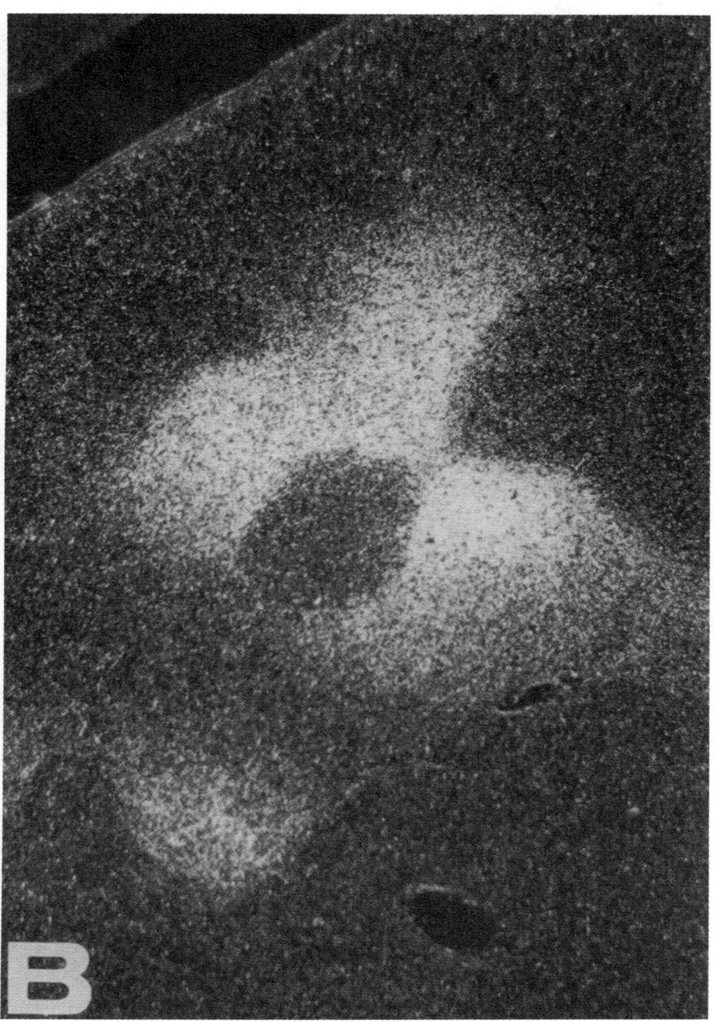
CC
CI
A

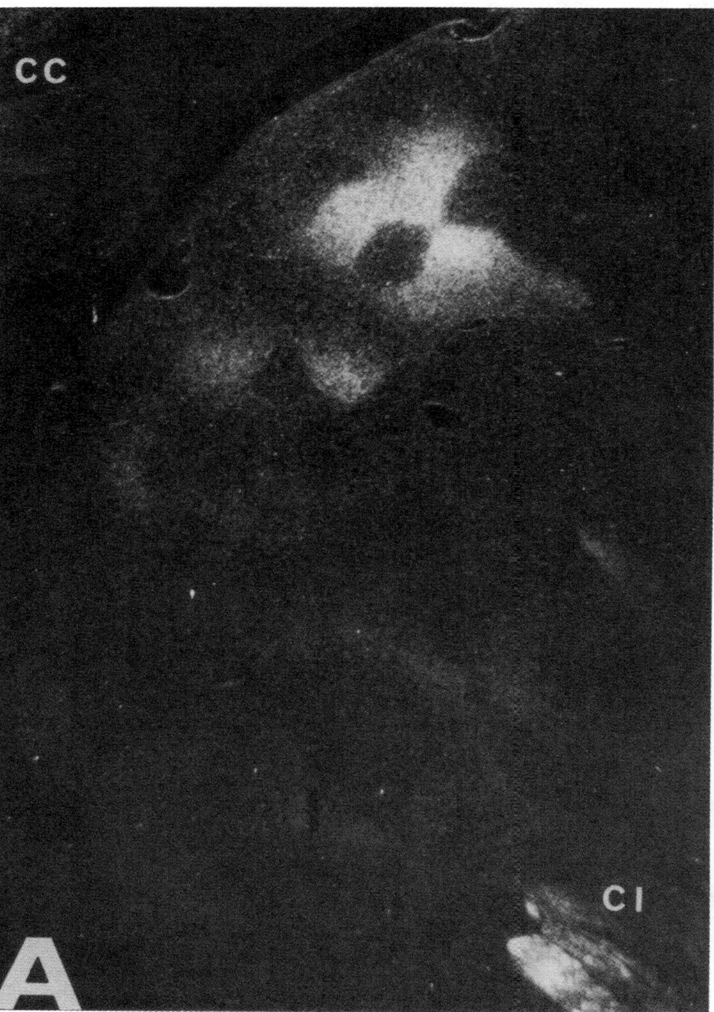
B

## PLATE 3

### EXPLANATION OF FIGURE

4   Dark-field autoradiogram illustrating grains in the dorsomedial sector of the body of the caudate nucleus, P9. CC, corpus callosum; CI, capsula interna. × 25.

PREFRONTO-CAUDATE PROJECTION
Patricia S. Goldman and Walle J. H. Nauta



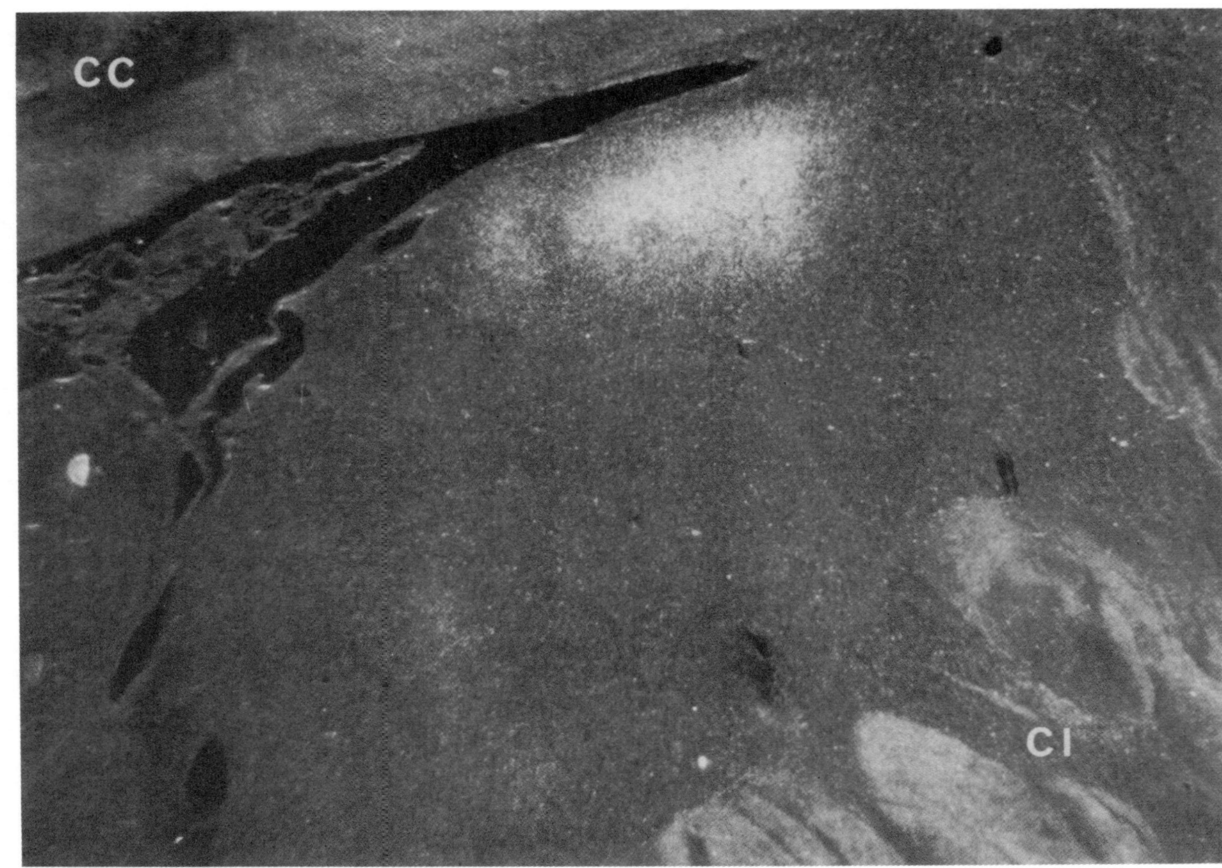

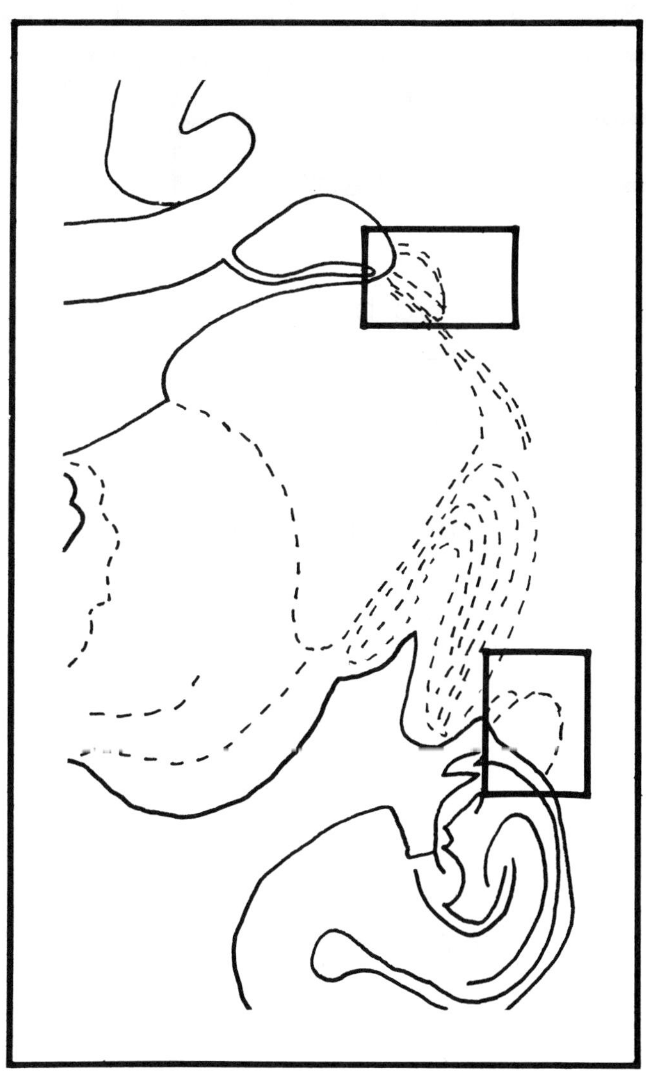

6  Dark-field autoradiograms from P9 illustrating grains over the tail of caudate. NC, nucleus caudatus; St, stria terminalis.

A  In the parietal lobe as shown in the upper inset of figure 5. × 54.

B  In the temporal lobe as illustrated in the lower inset in figure 5. × 54.

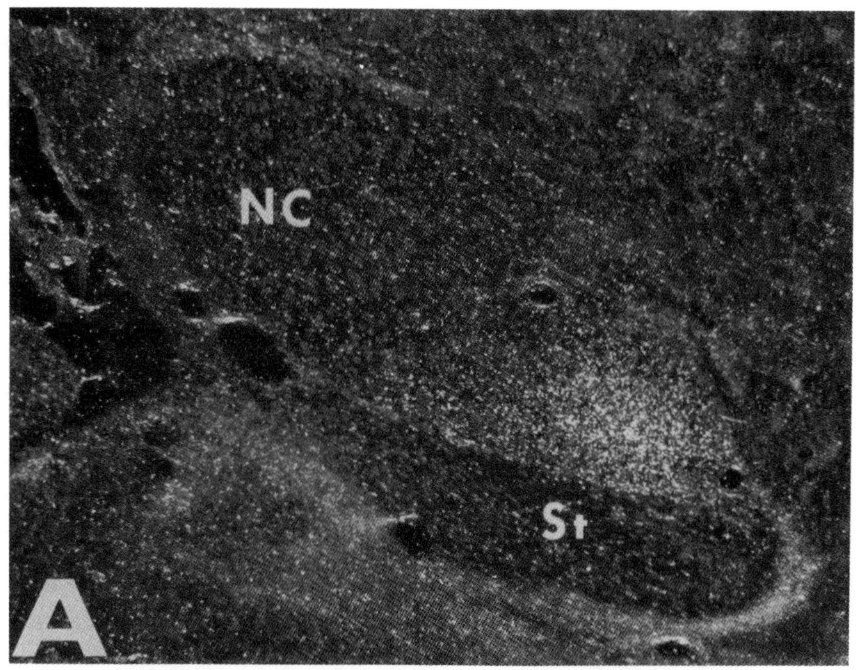
NC
St
A

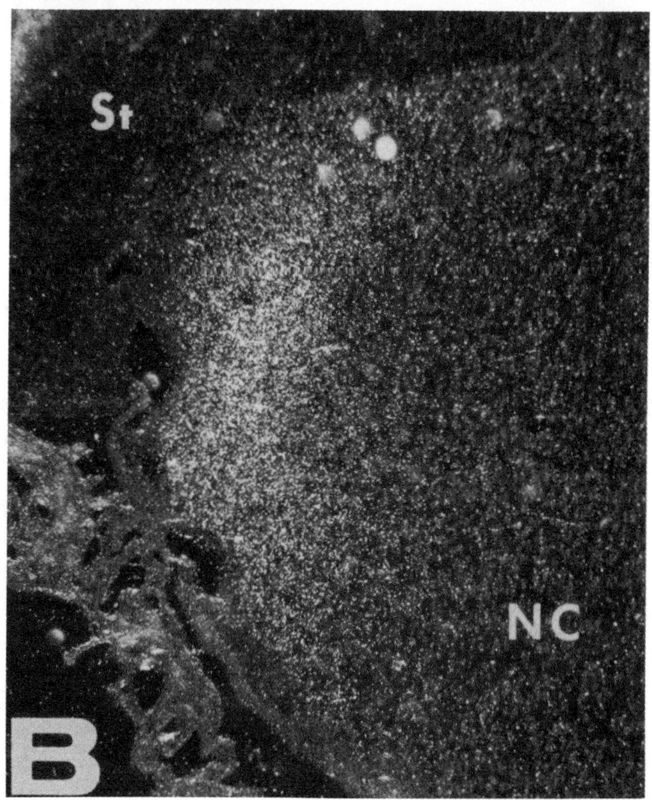
St
NC
B

Basel · München · Paris · London · New York · New Delhi · Singapore · Tokyo · Sydney

Mueller (ed.): Neurology and Psychiatry: A Meeting of Minds,
pp. 43–63 (Karger, Basel 1989)

# Reciprocal Links of the Corpus striatum with the Cerebral Cortex and Limbic System: A Common Substrate for Movement and Thought?

*Walle J. H. Nauta*

Department of Psychology and Brain Science, Massachusetts Institute of Technology, Cambridge, Mass., USA

Broadly speaking, the cerebral hemisphere of mammals can be said to be composed of three major anatomical and functional realms: the neocortex, the limbic system, and the extrapyramidal system (also referred to as basal ganglia or corpus striatum). Traditionally, and not without good physiological and clinicopathological reason, the neocortex is considered to embody the highest level of sensory analysis and perceptual integration, as well as certain of the more differentiated mechanisms subserving somatic motor function. Last but not least, it is regarded as the staging ground for ideational processes. The limbic system, by contrast, is generally viewed as the main cerebral representative of the internal milieu, expressing its functions, in part at least, in the form of affects and motivation. And finally, the basal ganglia are generally considered to be essential components of the brain's somatic motor system.

Useful as the notion of a tripartite composition has been (and still is) as a broad-scale first approach to the functional organization of the cerebral hemisphere, recent findings suggest that it may require some qualification. Among the more compelling reasons for revision is an anatomical one: observations accumulated over the past decade or so indicate that the three subdivisions are more closely interconnected than they once seemed to be. More in particular, several places have been identified in which their respective circuitries converge and thus could lead into communal or parallel effector channels. One, and apparently the most extensive of such confluences, occurs in the striatum. In recent years its remarkable pattern has begun to raise the question whether the extrapyramidal system, beside its well-documented role in skeletomotor function, could play a part in perceptual and ideational processes as well [Teuber, 1976].

In the following account the anatomical foundation of this notion will be dealt with in some detail. No comprehensive survey of extrapyramidal circuitry is intended here. For more encompassing reviews of that circuitry reference can be made to a recent Ciba Foundation symposium [Evered and O'Connor, 1984]. The striatum, however, being so much of a nodal intersection of cortical, limbic and extrapyramidal circuits, will be given somewhat more extensive coverage.


*General Terminological Remarks*

The term, extrapyramidal system, introduced by S. A. K. Wilson in 1912, has never been satisfactorily defined anatomically. Although literally the term can be interpreted so as to denote all of the brain's suprasegmental motor mechanisms not transmitting their discharge by way of the pyramidal tract, convention over the years has made it nearly synonymous with the basal ganglia and their efferent connections. The much older term, basal ganglia, originally subsumed all of the gray masses at the base of the cerebral hemisphere, including even the thalamus, but gradually became restricted to the corpus striatum (striatum and globus pallidus), and only later came to include also the subthalamic nucleus and substantia nigra, both linked to the corpus striatum by reciprocal fiber connections.


*The Striatum*

A survey of extrapyramidal circuitry most conveniently begins with the striatum, for it is this structure rather than the globus pallidus (or pallidum) that receives the principal extrinsic afferents converging upon the basal ganglia.

In all primates and many nonprimate mammals the striatum is subdivided by a plate-like internal capsule into two main districts, the dorsomedially situated caudate nucleus and the ventrolateral putamen. In other mammalian forms, including the rat, the internal capsule passes through the striatum in the form of a brush rather than a plate; in such species the striatum lacks an overt bisection and is therefore often referred to as caudatoputamen. In either case, however, the anterior part of the striatum includes two further subdivisions that require separate mention.

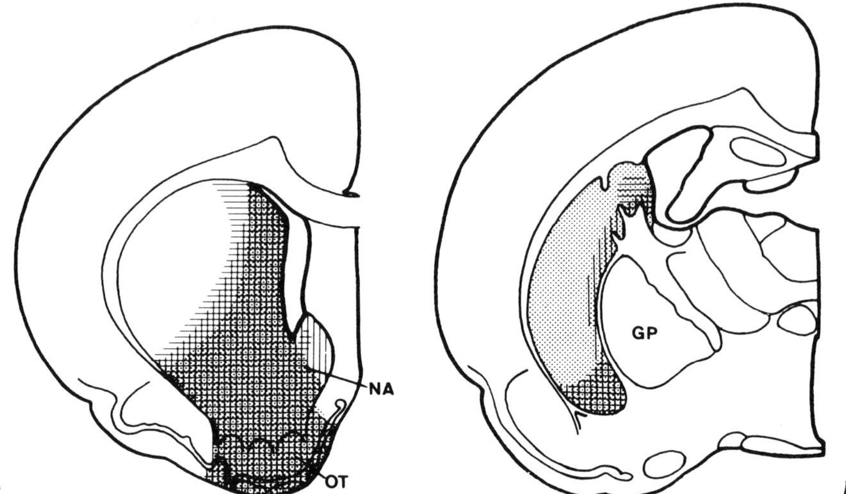


*Fig. 1.* Semidiagrammatic representation of three limbic system-associated projections to the striatum in rat brain: the dopamine projection from the ventral tegmental area (vertical lines), the projection from the frontal and cingulate cortices (horizontal lines), and the projection from the amygdala (stipple pattern). Not included in the figure (to avoid crowding) is the projection from the hippocampal formation to the nucleus accumbens. Left blank in the more rostral section *(a)* is the anterior, dorsolateral striatal region receiving only sparse limbic afferents but heavily projected upon by the sensorimotor cortex (cf. fig. 2). NA = Nucleus accumbens; OT = olfactory tubercle; GP = globus pallidus. Reproduced from Kelley et al. [1982] with permission of Pergamon Press, Oxford.

One of these, the ventromedial region nearest the base of the septum, and known as the nucleus accumbens septi (fig. 1a), was the first striatal region found to receive afferents from the limbic system, namely, fibers from the precommissural fornix [Fox, 1943; Sprague and Meyer, 1950; Swanson and Cowan, 1977]. On its ventral side the accumbens region extends into the olfactory tubercle, providing the latter with its core population of striatal cells [Heimer, 1978].

All districts of the striatum exhibit basically the same cytoarchitecture. Throughout its extent, more than 98% of its neurons are small to medium-sized, while the remaining 1–2% are large, cholinergic, multipolar cells that were long regarded as the source of the striatum's extrinsic efferents but are now believed to be intrinsic, or 'local-circuit' neurons.

On account of its lack of any broad-scale regional differentiation obvious on cursory inspection of preparations stained by routine survey techniques, the striatum has traditionally been described as a homogeneous structure. More recent studies, however, have disclosed a more covert, smaller-scale stereometric pattern of compartition that subdivides the striatum into an intricate labyrinth of small compartments ('striosomes') embedded in a more homogeneous cell matrix, and distinguished from the latter by peculiarities of histochemistry and cell-packing density [Graybiel and Ragsdale, 1983; Graybiel, 1984]. The relationship of this mosaic to the intrastriatal distribution of extrinsic afferents, to circuits intrinsic to the striatum itself, and to the sites of origin of striatal efferents is currently under intensive investigation. The findings reported thus far strongly suggest that the mosaic reflects a spatial ordering of striatal afferents and efferents of various neurotransmitter and/or neuropeptide content [Graybiel, 1984]. It appears likely, moreover, that intrinsic striatal neurons may provide an orderly system of cross-links between the striosomal and the matrix compartment, and thus between various classes of afferent and efferent conduction lines.


*Afferent Connections of the Striatum*

The striatum, the main entrance portal for neural inputs to the basal ganglia, receives its most voluminous afferent connections from the entire expanse of the cerebral cortex (including the hippocampal formation and the olfactory cortex), from the intralaminar thalamic nuclei, and from the substantia nigra (including the dopamine neurons of the ventral tegmental area). Additional substantial inputs originate from the amygdala, as well as from the serotonin neurons of the dorsal raphe nucleus of the midbrain. It is important to note that several of these afferent channels come from major limbic structures (hippocampal formation, amygdala, and cingulate cortex), whereas others arise from structures implicated in the subcortical circuitry of the limbic system: ventral tegmental area and dorsal raphe nucleus. The afferents from the cerebral cortex are distributed in the striatum in an intricate stereometric pattern that by and large preserves the topology of the cytoarchitectural subdivision of the cortical mantle but is considerably less sharply defined than the latter. Moreover, as will become evident below, the intrastriatal distribution of the projection from the neocortex as a whole must widely overlap that of projections from the limbic system, especially in the caudal half of the striatum.[1] A somewhat more detailed accounting of the afferentation of the striatum by the limbic system is necessary at this point. It is unfortunate that such an accounting can at present be given only for the rat brain. Even in that species, moreover, no complete description of the pattern of the corticostriatal projection can be given.

*The Limbic Innervation of the Striatum*

Recent studies in the rat by the aid of the autoradiographic fiber-tracing method [Kelley et al., 1982] have led to the conclusion that the anterior half of the striatum (the large part rostral to the level at which the anterior commissure crosses the midplane) can be subdivided into (a) a ventral and medial region including, but far exceeding, the nucleus accumbens and olfactory tubercle, and (b) a dorsolateral region. The reason for the distinction is that the former, anteroventromedial striatal subdivision is characterized by being densely innervated by several substantial and partially overlapping afferent fiber systems, each originating either from a major component of the limbic system such as the hippocampal formation, amygdala, or cingulate cortex, or from a structure involved in the circuitry of the limbic system, such as the ventral tegmental area and the prefrontal cortex. The remaining, anterodorsolateral striatal sector, by contrast, receives only sparse inputs of limbic origin and is, instead, massively innervated by the sensorimotor cortex. The subdivision of the anterior striatum here described is illustrated by figure 1a in which the limbic-afferented striatal region (for the sake of convenience to be called 'limbic striatum') is indicated by a variety of line and stipple patterns. The dorsolateral, 'nonlimbic' striatal sector heavily innervated by the sensorimotor cortex has been left open in the same drawing. At more caudal levels, as shown by figure 1b, the striatal region in which multiple limbic afferents overlap retains its medial position, but at these levels no 'nonlimbic' sector can be delineated, for the reason that the massive amygdalostriatal projection (indicated by stippling) here expands over the entire

---

[1] As here used, the term, overlap, solely refers to a convergence of afferent lines upon a large volume of striatal tissue; it carries no implications regarding communal or differential termination within either the matrix or striosomal compartment of the striatum.

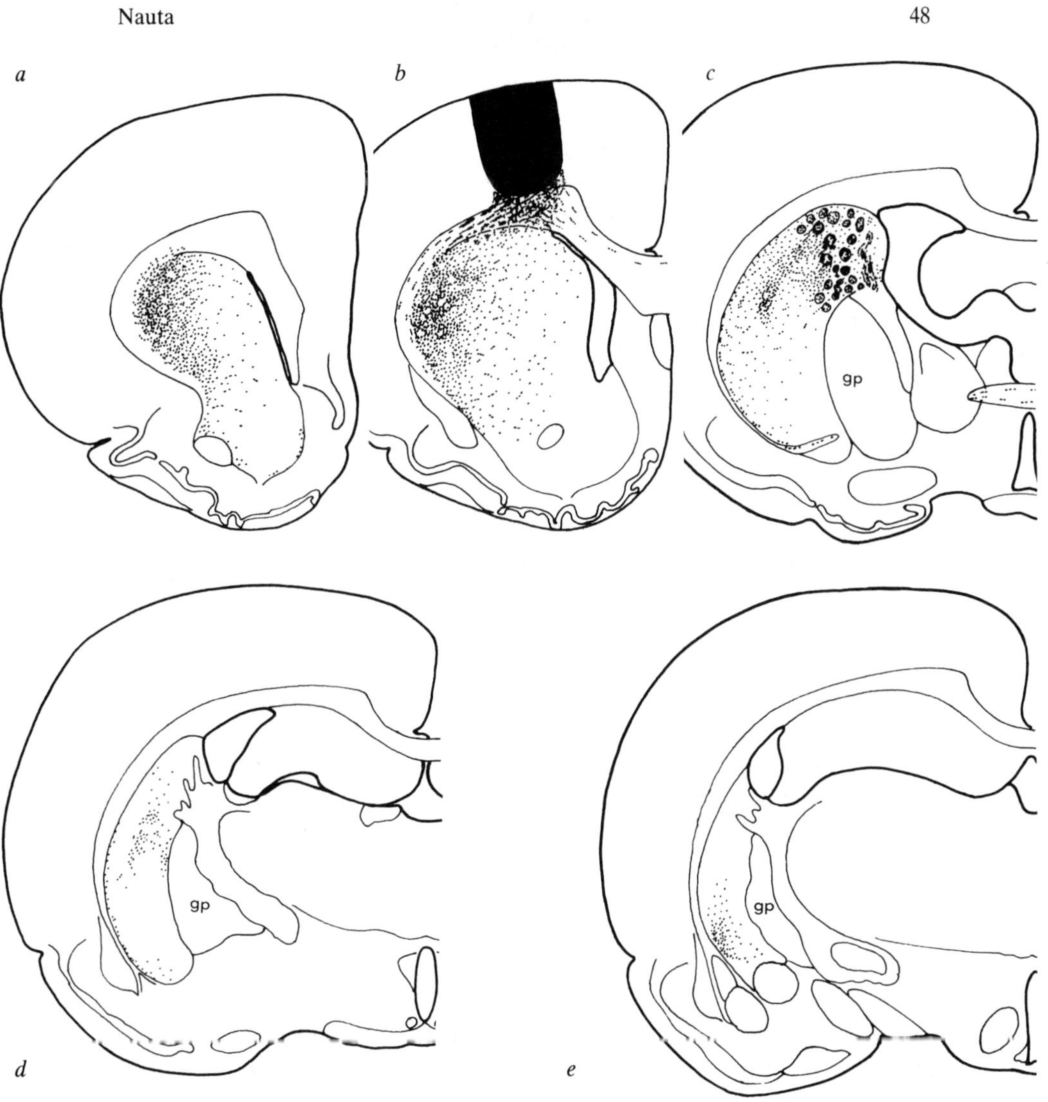

*Fig. 2.* Radioactive fiber labeling in the rat striatum following an injection of tritiated leucine-and-proline in the sensorimotor cortex. The injection site is shown in jet black in *b*. At the rostral levels shown in *a* and *b* note that the bulk of the projection involves the dorsolateral, relatively 'nonlimbic' striatal sector (left blank in figure 1a). Farther caudally *(d, e)* the motocortical projection overlaps the projection from the amygdala (cf. fig. 1b). gp = Globus pallidus. Reproduced from Kelley et al. [1982] with permission of Pergamon Press, Oxford.

cross-section of the striatum, and thereby overlaps the caudal extent of the corticostriatal projection arising from the motor cortex. (As shown by figure 2, the projection from the motor cortex continues from its major distribution area, a–b caudally and ventrally far enough to involve the entire antero-posterior dimension of the striatum.)

The evidence that the striatum – or at least its rostral half – comprises a limbic-innervated and a 'nonlimbic' district raised the question whether this dualism might not be reflected somehow in the distribution of the massive striatofugal projection to the globus pallidus. In more general terms, the question would be: Could the two striatal districts be the respective origins of two separate and distinct striatal exit lines to the pallidum?

Further study produced an affirmative answer. Comparison of a large number of cases, in each of which a small injection of tritiated leucine had been placed in one or the other striatal locus and the resulting anterograde fiber labeling had been recorded autoradiographically, clearly showed that the limbic-afferented anteroventromedial striatal region indicated in figure 1a projects exclusively to an only recently identified ventral component of the globus pallidus, the so-called *ventral pallidum* (vp). The 'nonlimbic' striatal sector, by contrast, projects to a larger pallidal subdivision, here to be called *dorsal pallidum* (dp) that forms most of the globus pallidus as conventionally outlined in atlases of the rat brain. This subdivision of the globus pallidus into two distinct parts needs some further comment.


*The Ventral Pallidum*


The vp was first described by Heimer and Wilson [1975] in the rat as an extension of the globus pallidus, stretching ventrally and rostrally beneath the anterior commissure. Its identification as a pallidal subdivision of the infracommissural forebrain region was originally based on cytological characteristics, as well as on the finding that it receives a massive striatal projection originating in particular from the nucleus accumbens [Wilson, 1972; Williams et al., 1977; Nauta et al., 1978]. Its pallidal nature was further documented by evidence that it also shares with the main mass of the globus pallidus a substantial projection from the subthalamic nucleus [Ricardo, 1980] as well as the histochemical characteristics of a high iron content [Switzer and Hill, 1979; Hill and Switzer, 1984] and a strong enkephalin-like [Haber and Nauta, 1981, 1983; Switzer et al., 1982] and glutamic acid decarboxylase-like immunoreactivity [Fallon and Ribak, 1980].

Despite these similarities, the vp must be viewed as distinct from the main mass of the globus pallidus. A principal reason for this distinction is histochemical: Whereas both the vp and the main mass of the globus pallidus are infiltrated by a dense plexus of enkephalin-positive fibers, the vp is pervaded by an additional and almost equally dense plexus of substance-P-positive fibers that extends over only a short distance into the main pallidal mass. For a proper understanding of the criteria followed in defining the vp for the purposes of the present account, a brief elaboration of this last point is necessary.

It is important to note that intrapallidal enkephalin and substance P are not intrinsic to the globus pallidus proper, but are associated with the striatopallidal projection. As first observed by Fox et al. [1974] in Golgi material, striatopallidal fibers enmesh the dendrites of pallidal neurons individually in sleeve-like plexuses. The characteristic histological picture reflecting this junctional arrangement in material immunoreacted for enkephalin ['woolly fibers', Haber and Nauta, 1983] provides a light-microscopic criterion for determining the extent of the globus pallidus. In such material the dense plexus of enkephalin-positive 'woolly fibers' filling the main mass of the globus pallidus is seen to continue without interruption ventrally into the infracommissural region where it extends rostrally over a wide area bordering the ventral side of the striatum and bed nucleus of the stria terminalis [Switzer et al., 1982; Haber and Nauta, 1983]. Interestingly, such material also shows that this infracommissural plexus extends around the lateral border of the nucleus of the diagonal band into the deep, fibrocellular layer of the olfactory tubercle [Switzer et al., 1982; Haber and Nauta, 1983]: this is in agreement with an earlier suggestion, made by Heimer [1978] on cytoarchitectural grounds, that the tubercle is a striatopallidal complex, as well as with the description by Ribak and Fallon [1983] of large cells associated with the islands of Calleja and electron microscopically indistinguishable from pallidal neurons.

Since the density of the enkephalin-positive intrapallidal fiber plexus varies little throughout its extent, it cannot serve to delineate the vp from the main, supra- and retrocommissural mass of the globus pallidus. Such a demarcation is, however, possible in sections processed for substance-P-like immunoreactivity. Such material reveals a dense plexus of substance-P-positive 'woolly fibers' that, with minor exceptions, is coextensive with the infracommissural enkephalin-positive plexus. But, unlike the latter, it extends over only a short distance into the main pallidal mass, then abruptly declines in density and only sparsely extends through the large remainder of the globus pallidus. The curved line along which this decline in density of the substance-P-positive 'woolly-fiber' plexus occurs suggests itself as the border between the vp and the remaining part of the globus pallidus, here for purposes of contrast to be referred to as dp [Haber and Nauta, 1981; cf. also the present fig. 3 in which the extent of the substance-P-positive plexus is indicated by a stippled pattern].

In summary, the vp is here defined as the largely infracommissural region massively pervaded by enkephalin-positive and substance-P-positive (and probably additional) striatofugal fibers terminating in the peridendritic-plexus fashion characteristic of the striatopallidal projection.

One further peculiarity of the vp must be mentioned here. The region corresponds to that part of the globus pallidus which contains the densest population of large, cholinergic neurons (fig. 3). Such cells are more sparsely scattered throughout the remaining parts of the globus pallidus; collectively, they form the intrapallidal component – or subgroup Ch 4 [Mesulam et al., 1983] – of the magnocellular nucleus of the basal forebrain.[2]

## Differential Projections from Dorsal and Ventral Pallidum

The evidence that the subdivision of the striatum into a limbic-afferented and a 'nonlimbic' district is mirrored in a ventral versus dorsal dualism in the intrapallidal distribution of the striatopallidal projection raises the question whether the separateness of the two striatal exit lines might not extend even beyond the globus pallidus. To investigate this question, a small deposit of tritiated leucine was placed in either the dorsal or the vp in a large number of rats; the distributions of fibers radioactively

---

[2] In the human and other primates the magnocellular nucleus of the basal forebrain forms a large, compact group of large cholinergic cells, the basal nucleus of Meynert, located in the substantia innominata just ventral to the globus pallidus. Even in these more highly differentiated species, however, trails of cholinergic cells extend from the basal nucleus dorsally into the medullary laminae of the overlying globus pallidus. In rodents and carnivores the cholinergic cells of the magnocellular nucleus are more widely and diffusely distributed, and lack the circumscript local aggregation that enabled Meynert to recognize a basal nucleus in the human brain.

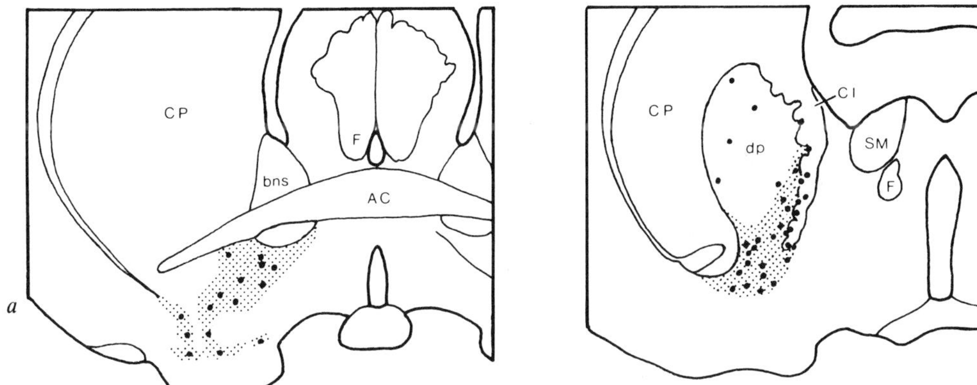

*Fig. 3.* Drawings of two frontal sections at the level of, respectively, the anterior commissure *(a)* and rostral part of the retrocommissural globus pallidus *(b)*. The extent of the dense substance-P-positive fiber plexus, coextensive with the vp of Heimer and Wilson [1975], is indicated by stippling (cf. Haber and Nauta, 1983]. Coarser black dots each represent a strongly acetylcholinesterase (AChE)-positive intrapallidal neuron; together, these neurons compose the intrapallidal subdivision, celled subgroup Ch4 by Mesulam et al. [1983], of the magnocellular nucleus of the basal forebrain. The numerous further cells of the magnocellular nucleus, distributed in the nucleus of the diagonal band and other sublenticular structures [cf. Butcher, 1983; Mesulam et al., 1983] have been omitted in these drawings. Note that such cells are much sparser in the dorsal (dp) than in the vp. AC = Anterior commissure; bns = bed nucleus of stria terminalis; CI = capsula interna; CP = caudatoputamen; dp = dorsal pallidum; F = fornix; SM = stria medullaris. Reproduced from Haber et al. [1985] with permission from Alan R. Liss, New York.

---

*Fig. 4.* Semischematic illustration of the two partly segregated striatofugal conduction lines described in the text. *a* 'Limbic' (shaded) and 'nonlimbic' striatal subdivisions and their respective main telencephalic afferents. *b* The striatofugal line leading from the motor cortex-innervated 'nonlimbic' striatal subdivision to the dorsal pallidum (heavy arrow) which in turn projects to the subthalamic nucleus and substantia nigra. *c* The line leading from the limbic-afferented striatal subdivision to the vp (heavy arrow) and from there not only to the subthalamic nucleus and substantia nigra but also to the frontocingulate and medial sensorimotor cortex, lateral habenular and mediodorsal thalamic nucleus, amygdala, hypothalamus, ventral tegmental area and more caudal regions of the midbrain tegmentum. abl = Amygdala; AC = anterior commissure; CC = cerebral peduncle; cgs = central gray substance; CI = capsula interna; dp = dorsal pallidum; ep = entopeduncular nucleus (internal segment of globus pallidus); FCC = frontocingulate cortex; H = hippocampus; hl = lateral habenular nucleus; ht = hypothalamus; md = mediodorsal nucleus; na = nucleus accumbens; rt = reticular nucleus of thalamus; SMC = sensorimotor cortex; sn = substantia nigra; st = subthalamic nucleus; vp = ventral pallidum; vta = ventral tegmental area. Reproduced from Haber et al. [1985] with permission from Alan R. Liss, New York.

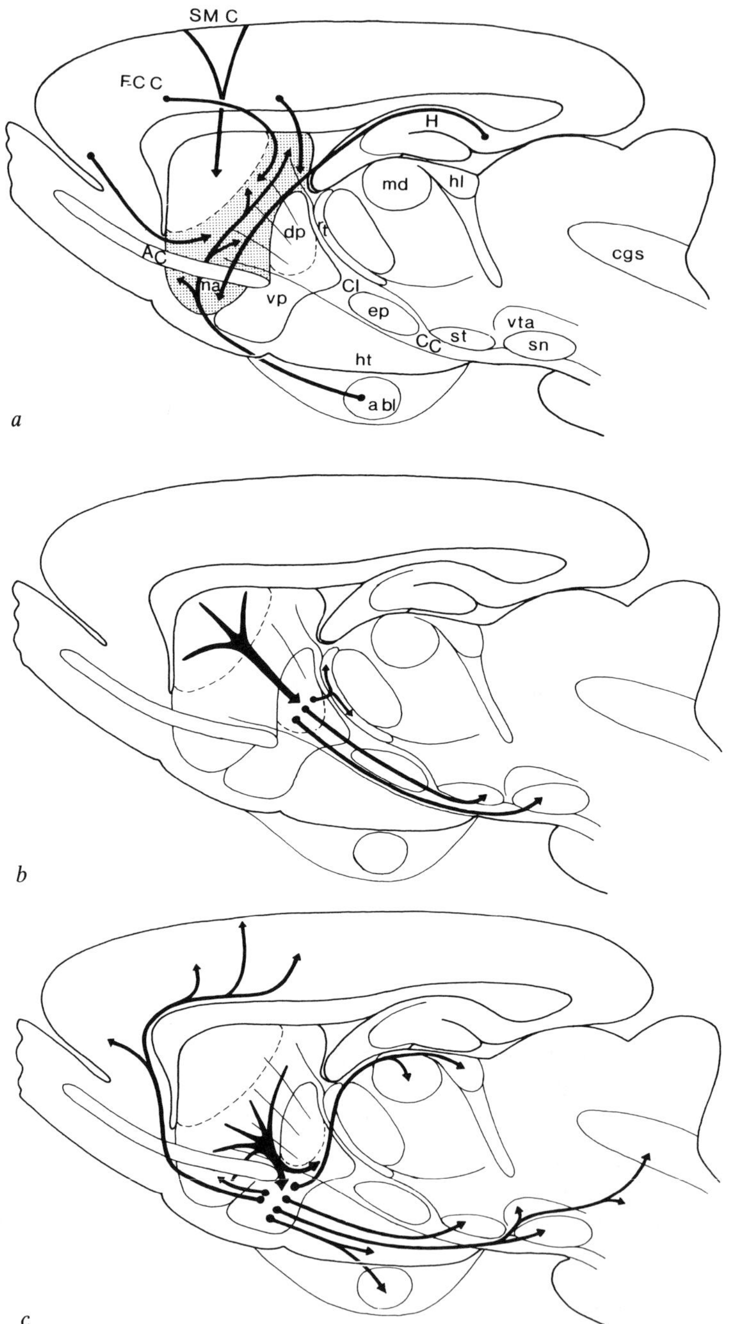
SM C
FC C
H
md
hl
dp
rf
AC
cgs
na
vp
Cl
ep
st
vta
sn
CC
ht
a bl
a
b
c

labeled by each of these isotope placements were then recorded auto-radiographically, carefully charted, and compared. Since a documented account of the findings has been published previously elsewhere [Haber et al., 1985], only the main conclusions will be reported here.

The results of the study showed pronounced differences between the exit lines from the dorsal and vp, respectively. The findings are illustrated by figure 4 and can be summarized as follows:

(1) The efferent connections of the dorsal pallidum (fig. 4b) compose the classical projection from the external pallidum segment and involve primarily the subthalamic nucleus and substantia nigra. This pallidal exit line appears to lie in the extension of the striatopallidal projection arising from the 'nonlimbic' dorsolateral striatal sector (heavy arrow in fig. 4b), the striatal region that in turn receives its main telencephalic input from the sensorimotor cortex (fig. 4a).

(2) The vp projects, like the dp, to the subthalamic nucleus and substantia nigra, but it has further projections to several structures implicated in the circuitry of the limbic system, particularly the amygdala, the lateral habenular nucleus, and the mediodorsal nucleus of the thalamus (fig. 4c). It projects more sparsely to the hypothalamus, to the ventral tegmental area, and to more caudal regions of the midbrain tegmentum, including the region of the pedunculopontine nucleus. Of some special interest are those projections of the vp that ascend to the prefrontal cortex and an adjoining region of the sensorimotor cortex: There is evidence suggesting that this projection, as well as the one to the amygdala, originates largely, at least, from the cholinergic cells of the so-called magnocellular basal nucleus that lie embedded in the vp [Grove et al., 1983]. All these exit lines from the vp, however, appear to lie in the trans-synaptic extension of that component of the striatopallidal projection (heavy arrow in fig. 4c) which originates from the limbic-afferented striatum, that is to say, the large striatal region densely innervated by a variety of afferents from within the circuitry of the limbic system (the shaded region in fig. 4a).


*Discussion*

One of the principal realizations to emerge from the anatomical findings of the past few years is that the striatum, a common target for voluminous projections from the cerebral cortex and the limbic system, has multiple exit channels available that could serve as 'return loops' to the cerebral cortex. Up until about ten years ago only one such loop was generally recognized: the extrapyramidal motor circuit leading from the striatum over, successively, the pallidum and the thalamic VA-VL complex back to areas within the motor domain of the neocortex. Naturally, the discovery of this loop (a step-wise disclosure that was not completed until 1945; cf. Nauta and Mehler, 1966] added further support to the prevalent view of the basal ganglia as a central somatic motor mechanism.

Evidence of further trans-striatal loops to the cortex, not specifically targeted on the motor cortex, is of more recent date. The first such loop to be recognized turned out to lead from the striatum over the pars reticulata of the substantia nigra – rather than over the globus pallidus proper[3] – to the ventromedial nucleus of the thalamus and from there to most of the extent of the neocortex [Cole et al., 1964; Carpenter et al., 1976; Herkenham, 1976]. A second potential return loop to the cortex was found to lead from the striatum over the globus pallidus to the reticular nucleus of the thalamus [Nauta, 1979a; Haber et al., 1985], a thalamic cell group that is unique in the sense that it lacks a projection to the cortex and instead projects to many – if not indeed all – other thalamic nuclei [Scheibel and Scheibel, 1966]; it thus could be suspected of being able to affect the functional state of virtually the entire expanse of the cerebral cortex by modulating the latter's inputs from the thalamus. (The pallido-reticular connection is schematically indicated in fig. 4b.)

It must be emphasized that neither of these two striato-pallido-thalamo-cortical return loops is at present known to be directed at any functionally specified cortical area in particular. On the contrary, both seem likely to involve most if not all of the neocortex. Since it appears to be unknown whether they originate throughout the striatum or from any particular striatal subdivision, nothing can be said at present concerning the functional nature of the corticostriatal input the two loops could 'feed back' to the cortex. This is not as categorically true for the two cortico-striato-pallido-thalamo-cortical circuits to be discussed next.

A distinction between the latter two circuits became possible only after Heimer and Wilson [1975] had recognized the vp, and Kelley et al. [1982] had noted that the anterior half of the striatum comprises a predominantly

---

[3] On both anatomical and physiological grounds the pars reticulata of the substantia nigra can be interpreted as a caudalmost component of the globus pallidus [cf. Nauta, 1979b; DeLong and Georgopoulos, 1981].

motor cortex-innervated (or 'nonlimbic') and a convergently limbic-innervated subdivision. The evidence presented in a foregoing section of this paper indicates that the limbic versus nonlimbic dichotomy in the innervation of the striatum reflects itself in the existence of two largely segregated striatofugal conduction lines. The present evidence of this dualism is schematically illustrated by figure 4, and can be summarized as follows.

A dorsolateral sector of the anterior striatum receiving only sparse afferents of limbic origin but instead heavily innervated by the sensorimotor cortex [Kelley et al., 1982] projects to an antero-dorsolateral region of the main, supra- and retrocommissural (i.e. conventionally recognized) mass of the globus pallidus; this pallidal region ('dorsal pallidum') in turn gives rise to the classical pallidofugal pathway descending to the subthalamic nucleus and substantia nigra (fig. 4b). By contrast, the heavily limbic-innervated antero-ventromedial striatal region (shaded in fig. 4a) projects to the largely infracommissural, but partly also supra- and retrocommissural vp (fig. 3), from which in turn efferents pass not only to the subthalamic nucleus and substantia nigra but also to various limbic-system-associated structures, in particular the amygdala, frontocingulate (and adjoining sensorimotor) cortex, mediodorsal thalamic nucleus, lateral habenular nucleus, hypothalamus, ventral tegmental area, and more caudal and dorsal regions of the midbrain tegmentum (fig. 4c).

As thus defined, the two striato-pallidofugal lines share the subthalamic nucleus and substantia nigra as common distribution targets. If it is assumed that both these target structures are first and foremost components of the somatic motor system, the present findings suggest both as sites where central skeletomuscular mechanisms are accessible to striatopallidal output of limbic antecedents, conveyed by the line leading from the limbic afferented striatum over the vp. It is likely that further sites of such access lie farther caudally in the midbrain. A small part of the great exit pathway from the globus pallidus (the ansa lenticularis), instead of recurving dorsally into the thalamic VA-VL complex, continues on a straight course into the midbrain tegmentum where it terminates mainly in the region of the so-called pedunculopontine nucleus. On the basis of physiological findings Grillner and Shik [1973] have referred to this tegmental area as the 'mesencephalic locomotor region'. Interestingly, fibers from the region of the vp have been found distributed here as well; Swanson et al. [1984] have interpreted this finding as evidence of a limbic innervation of Grillner and Shik's somatic motor mechanism.

A third route by which the limbic system could be thought to have access to central somatic motor mechanisms could lead from the limbic-innervated striatum over the strongly AChE-positive cells embedded in the vp. The present anterograde evidence suggesting that at least some of these cells project to the sensorimotor cortex only confirms earlier findings in retrograde-labeling experiments [McKinney et al., 1983; Saper, 1984].

Even more difficult to interpret from a functional point of view are the limbico-striato-pallidofugal lines leading to various limbic and limbic-system-associated structures, summarized in figure 4c. From one vantage point, these connections could be viewed collectively as a differentiated 'return loop' enabling the limbic system to monitor the effect of its outputs to the striatum (and thus possibly acting as an aid in adjusting the organism's motivational set to the output efficacy of the central somatic-motor system). But other possibilities must be considered. The question arises, for example, if one significance of the much-interrupted limbico-striato-pallido-limbic circuit here proposed may not lie in the fact that the striatum is the site of an unusually wide variety of intricately interspersed neurotransmitters and neuromodulators [Graybiel and Ragsdale, 1983], particular combinations of which might be required no less stringently for optimal functioning of limbic and cognitive-cortical mechanisms than the same or other combinations are required for optimal somatic-motor function. The possibility of such a more general functional role of the corpus striatum suggests itself from a remarkable parallel between the 'limbic' and 'nonlimbic' trans-striatal conduction lines: both prominently include channels over which part of their impulse flow can be led back to the domain of its origin in the cerebral hemisphere. To be more explicit: the familiar 'extrapyramidal motor circuit', motor cortex $\rightarrow$ striatum $\rightarrow$ pallidum $\rightarrow$ thalamic VA-VL complex $\rightarrow$ premotor cortex $\rightarrow$ motor cortex (fig. 5a) appears to have limbic counterparts in the sequences: limbic system $\rightarrow$ striatum $\rightarrow$ vp $\rightarrow$ amygdala, and: limbic system $\rightarrow$ striatum $\rightarrow$ vp $\rightarrow$ thalamic mediodorsal nucleus $\rightarrow$ frontal and anterior limbic cortex (fig. 5b). If it is considered likely on the basis of long-known anatomical connections that the motor function of the corpus striatum expresses itself in part at least through a major transthalamic circuit affecting the mechanisms of the motor cortex, the more recent anatomical data here reviewed make it appear no less likely that the corpus striatum, through analogous though less massive circuits involving the amygdala and prefrontal cortex, can affect a class of functions that is cognitive rather than primarily skeletomuscular.

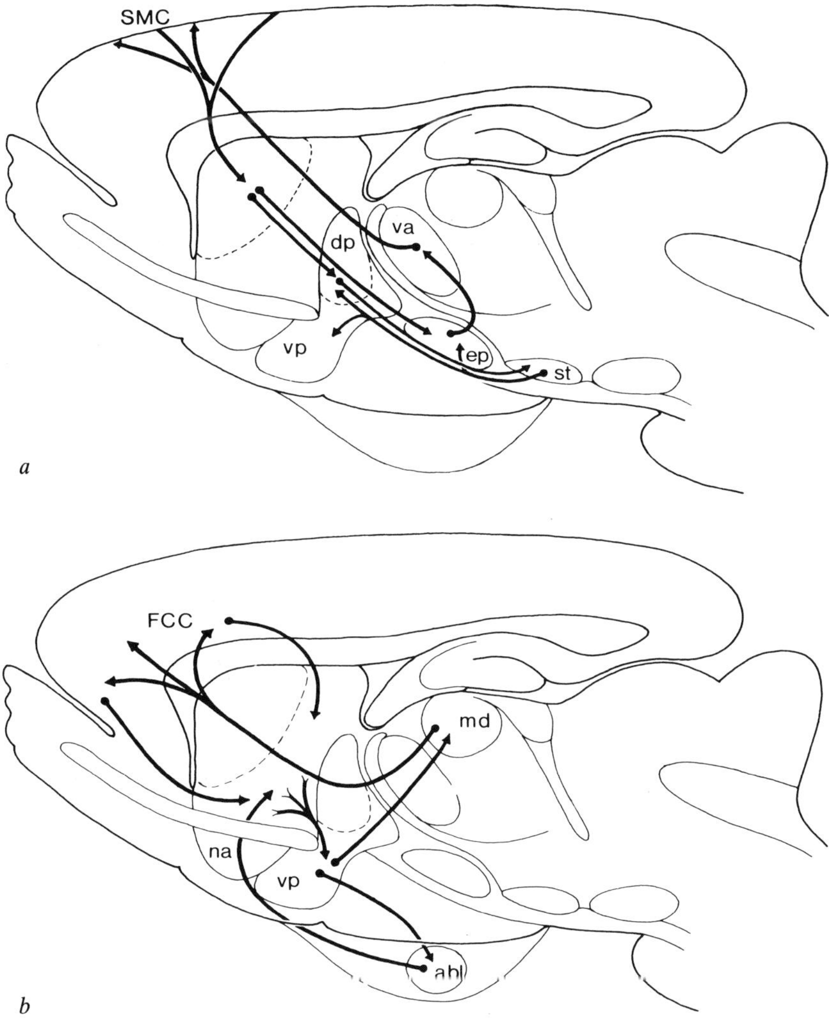



*Fig. 5.* Semidiagrammatic illustration of the two cortico-striato-pallido-thalamo-cortical circuits associated with, respectively, the motor cortex *(a)* and the frontocingulate cortex *(b).* Note that the extrapyramidal motor circuit shown in *a* sequentially involves the 'nonlimbic' striatum, dp as well as internal pallidal segment (ep), and thalamic VA-VL complex (va). By contrast, the analogous circuit associated with the frontocingulate cortex, shown in *b,* involves the 'limbic' striatum, vp and thalamic mediodorsal nucleus (md); note also the shorter loop

*General Comment*

The recent evidence that the corpus striatum has various transthalamic exit lines available over which it could be thought to modulate the function not only of the motor cortex, but of most or all of the remaining cortical areas as well, does not come entirely unexpected. That the basal ganglia are implicated not only in motor but also in cognitive functions had been suspected for some time from findings in neurobehavioral experiments [Teuber, 1976], as well as from the frequent concurrence of extrapyramidal-motor and mental disorder. Most unambiguously perhaps, from detailed behavioral studies evidence had been reported of impaired frontal-lobe function in Parkinson's disease [Bowen, 1976] and, more recently, certain lesions of the striatum have been found to cause aphasia with marked impairment of language comprehension [Damasio, 1985].

Exactly which aspects of cognitive function could be influenced by the basal ganglia is at present unknown. One would be inclined to search for a functional attribute that is shared in common by cognition and movement, and affected in common by basal ganglia disorder. It is of interest in this connection to recall John Hughlings Jackson's vigorous argumentation against the notion, prevalent before the discovery of the motor cortex by Fritsch and Hitzig in 1870, that the 'convolutions' (i. e. the cerebral cortex) were the 'organ of mind', and thus exclusively for 'ideas', while 'movements' were represented solely in lower parts of the brain. Jackson's views are illustrated by the following quotations from his selected writings [Taylor, 1958]:

'But of what "substance" can the organ of mind be composed, unless of processes representing movements and impressions; and how can the convolutions differ from the inferior centres, except as parts representing more intricate coordinations of impressions and movements in time and space than they do?' [Hughlings Jackson, 1870].

'There is . . . proof, although indirect, of the opinion that faint and central excitation of sensori-*motor* processes suffices in thought' [italics as in original; Hughlings Jackson, 1872].

---

associated with the amygdala (abl). To avoid excessive complexity of the diagrams, only one of several possible cross-links between the *a* and *b* circuits has been indicated: as shown in *a,* the subthalamic nucleus – innervated by both dp and vp – projects back to both dp and vp, as well as to the internal pallidal segment (ep). For further labels see legend to figure 4. Reproduced from Haber et al. [1985] with permission from Liss, New York.

It is, of course, difficult to say how literally Jackson's view should be taken. Is ideation really the internal expression of the same sensori-motor mechanisms which, when activated more forcibly, would manifest themselves as movements? On the other hand, it is easy to see at least parallels between thought and movement. To begin with, both seem to require a motivational vis a tergo to be initiated and to be kept in progress. It takes an effort to change from a state of relaxed daydreaming – a sort of resting tonus perhaps – to one of goal-directed thinking, and it is something like an act of endurance to keep such a thought process going. Movement and thought share with each other a need for 'feedback' allowing for corrective adjustments. Thought often spills over into the motor sphere, in the form of unintentional changes in body posture and facial expression, pacing the floor, or thinking aloud. More similarities could be mentioned, including the evidence that the physiological substrates of both movement and thought embody a dopamine-dependent mechanism. The 'primary thought disorder' of schizophreniform psychoses may respond well to dopamine-receptor blocking agents, and the protracted use of such drugs can result in dyskinesia.

From all this it could be suspected, as Hughlings Jackson did, that the central motor mechanism and the neural apparatus of ideation have elements in common and may even be in partial overlap. At present, however, any suggestion as to where the two mechanisms might intersect must remain very largely hypothetical. A main reason for this is that it is still impossible to state where in the brain the impetus to a deliberate movement arises, and by what sequence of steps it is ultimately transduced into the requisite pattern of motoneuronal activation and inhibition. The neural substrate for thought processes, needless to say, is even more obscure. It seems reasonable to assume that ideation is primarily a function of the neocortex, but even if it were exclusively a cortical phenomenon it would be continuously affected by subcortical inputs varying from the selectively targeted specific-sensory thalamocortical radiations on the one extreme to the pervasively distributed serotonergic and cholinergic innervations of the cortex by, respectively, the mesencephalic raphe nuclei and the basal nucleus of Meynert on the other extreme. The striato-pallido-thalamo-cortical circuits are among this diversity of afferents to the cortex. Two of the four such circuits described in the foregoing account – the ones leading over, respectively, the ventromedial and the reticular nucleus of the thalamus – are at least potentially diffuse in the sense that they appear to be distributed to much if not all of the neocortex. The two remaining circuits, however, are more selectively targeted. One of these (fig. 5a) is the classical 'extrapyramidal motor circuit' leading over the thalamic VA-VL complex to the motor domain of the cortex, the other (fig. 5b) the more recently identified parallel loop leading over the mediodorsal thalamic nucleus to the frontal and anterior limbic cortex. Even though hypothetical, it seems reasonable to postulate that the loop involving the frontal cortex could play a role in cognitive functions that is comparable to the role of the 'extrapyramidal motor circuit' in bodily movement. In line with this hypothesis, it is tempting to suggest that movement and ideation share with each other a need for propulsion and timing, and that the basal ganglia might be for both functions a source of such dynamic properties.

## References

Bowen, F.P.: Behavioral alterations in patients with basal ganglia lesions; in Yahr, The basal ganglia (Raven Press, New York 1976).

Butcher, L.L.: Acetylcholinesterase histochemistry; in Björklund, Hökfelt, Handbook of chemical neuroanatomy, vol. 1 (Elsevier, Amsterdam 1983).

Carpenter, M.B.; Nakano, K.; Kim, R.: Nigrothalamic projections in the monkey demonstrated by autoradiographic technics. J. comp. Neurol. *165:* 401–416 (1976).

Cole, M.; Nauta, W.J.H.; Mehler, W.R.: The ascending efferent projections of the substantia nigra. Trans. Am. Neurol. Ass. *1964:* 74–78.

Damasio, A.R.: Language and the basal ganglia; in Evarts, Wise, Bousfield, The motor system in neurobiology (Elsevier Biomedical Press, Amsterdam 1985).

DeLong, M.A.; Georgopoulos, A.P.: Motor functions of the basal ganglia; in Brooks, Motor control, part 2, pp. 1017–1061 (Am. Physiological Society, Bethesda 1981).

Evered, D.; O'Connor, M. (eds): Functions of the basal ganglia. Ciba Fdn Symp., No. 107 (Pitman, London 1984).

Fallon, J.H.; Ribak, C.E.: Multiple neurotransmitter studies in the islands of Calleja complex of the basal forebrain. III. Connections, correlations and reservations. Soc. Neurosci. Abstr. *6:* 114 (1980).

Fox, C.A.: The stria terminalis, longitudinal association bundle and precommissural fornix fibers in the cat. J. comp. Neurol. *79:* 277–295 (1943).

Fox, C.A.; Andrade, A.N.; Lu Qui, I.J.; Rafols, J.A.: The primate globus pallidus: a Golgi and electron microscopic study. J. Hirnforsch. *15:* 75–93 (1974).

Graybiel, A.M.: Neurochemically specified subsystems in the basal ganglia; in Evered, O'Connor, Functions of the basal ganglia. Ciba Fdn Symp., No. 107, pp. 114–144 (Pitman, London 1984).

Graybiel, A.M.; Ragsdale, C.W.: Biochemical anatomy of the striatum; in Emson, Chemical neuroanatomy (Raven Press, New York 1983).

Grillner, S.; Shik, M.T.: On descending control of lumbosacral spinal cord from the 'mesencephalic locomotor region'. Acta physiol. scand. *87:* 320–333 (1973).

Grove, E.A.; Haber, S.N.; Domesick, V.B.; Nauta, W.J.H.: Differential projections from AChE-positive and AChE-negative ventral-pallidum cells in the rat. Soc. Neurosci. Abstr. *9:* 16 (1983).

Haber, S.N.; Groenewegen, H.J.; Grove, E.A.; Nauta, W.J.H.: Efferent connections of the ventral pallidum: evidence of a dual striato-pallidofugal pathway. J. comp. Neurol. *235:* 322–335 (1985).

Haber, S.N.; Nauta, W.J.H.: Substance P, but not enkephalin, immunoreactivity distinguishes ventral from dorsal pallidum. Soc. Neurosci. Abstr. *7:* 916 (1981).

Haber, S.N.; Nauta, W.J.H.: Ramifications of the globus pallidus in the rat as indicated by patterns of immunohistochemistry. Neuroscience *9:* 245–260 (1983).

Heimer, L.: The olfactory cortex and the ventral striatum; in Livingston, Hornykiewicz, Limbic mechanisms, the continuing evolution of the limbic system concept (Plenum Press, New York 1978).

Heimer, L.; Wilson, R.D.: The subcortical projections of the allocortex: Similarities in the neural associations of the hippocampus, the piriform cortex, and the neocortex; in Santini, Golgi Centennial Symp., pp. 177–193 (Raven Press, New York 1975).

Herkenham, M.: The nigro-thalamo-cortical connection mediated by the nucleus ventralis medialis thalami: evidence for a wide cortical distribution in the rat (Abstract). Anat. Rec. *184:* 426 (1976).

Hill, J.M.; Switzer, R.C.: The regional distribution and cellular localization of iron in the rat brain. Neuroscience *11:* 595–603 (1984).

Hughlings Jackson, J.: see Taylor (1958).

Kelley, A.E.; Domesick, V.B.; Nauta, W.J.H.: The amygdalostriatal projection in the rat – an anatomical study by anterograde and retrograde tracing methods. Neuroscience *7:* 615–630 (1982).

McKinney, M.; Coyle, J.T.; Hedreen, J.C.: Topographic analysis of the innervation of the rat neocortex and hippocampus by the basal forebrain cholinergic system. J. comp. Neurol. *217:* 103–121 (1983).

Mesulam, M.-M.; Mufson, E.J.; Wainer, B.H.; Levey, A.I.: Central cholinergic pathways in the rat: An overview based on an alternative nomenclature (Ch 1-Ch 6). Neuroscience *10:* 1185–1201 (1983).

Nauta, H.J.W.: Projections of the pallidal complex: an autoradiographic study in the cat. Neuroscience *4:* 1853–1873 (1979a).

Nauta, H.J.W.: A proposed conceptual reorganization of the basal ganglia and telencephalon. Neuroscience *4:* 1875–1881 (1979b).

Nauta, W.J.H.; Mehler, W.R.: Projections of the lentiform nucleus in the monkey. Brain Res. *1:* 3–42 (1966).

Nauta, W.J.H.; Smith, G.P.; Faull, R.L.M.; Domesick, V.B.: Efferent connections and nigral afferents of the nucleus accumbens septi in the rat. Neuroscience *3:* 385–401 (1978).

Ribak, C.E.; Fallon, J.H.: The islands of Calleja complex of rat basal forebrain. I. Light and electron microscopic observations. J. comp. Neurol. *205:* 207–218 (1983).

Ricardo, J.: Efferent connections of the subthalamic region in the rat. I. The subthalamic nucleus of Luys. Brain Res. *202:* 257–271 (1980).

Saper, C.B.: Organization of cerebral cortical afferent systems in the rat. II. Magnocellular basal nucleus. J. comp. Neurol. *222:* 313–342 (1984).

Scheibel, M.E.; Scheibel, A.B.: The organization of the nucleus reticularis thalami: a Golgi study. Brain Res. *1:* 43–62 (1966).

Sprague, J.M.; Meyer, M.: An experimental study of the fornix in the rabbit. J. Anat. *84:* 354–368 (1950).

Swanson, L.W.; Cowan, W.M.: An autoradiographic study of the organization of the efferent connections of the hippocampal formation in the rat. J. comp. Neurol. *172:* 49–84 (1977).

Swanson, L.W.; Mogenson, G.J.; Gerfen, C.R.; Robinson, P.: Evidence for a projection from the lateral preoptic area and substantia innominata to the 'mesencephalic locomotor region' in the rat. Brain Res. *295:* 161–178 (1984).

Switzer, R.C.; Hill, J.: Globus pallidus component in the olfactory tubercle: evidence based on iron distribution. Soc. Neurosci. Abstr. *5:* 79 (1979).

Switzer, R.C.; Hill, J.; Heimer, L.: The globus pallidus and its rostroventral extension into the olfactory tubercle of the rat: a cyto- and chemoarchitectural study. Neuroscience *7:* 1891–1904 (1982).

Taylor, J.: Selected writings of John Hughlings Jackson, vol. I (Basic Books, New York 1958).

Teuber, H.-L.: Complex functions of the basal ganglia; in Yahr, The basal ganglia (Raven Press, New York 1976).

Williams, D.J.; Crossman, A.R.; Slater, P.: The efferent projections of the nucleus accumbens in the rat. Brain Res. *130:* 217–227 (1977).

Wilson, R.D.: Efferent connections of the nucleus accumbens in the rat; Master thesis, MIT, Cambridge, Mass. (1972).

W.J.H. Nauta, MD, PhD, 196 Kent Road, Newton, MA 02168 (USA)

# Permissions

Birkhäuser Boston would like to thank the original publishers of the papers of Walle J.H. Nauta for granting permission to reprint specific papers in this collection.

[23] Reprinted from *Neuroscience* **9**, © 1983 by Pergamon Press plc.

[24] Reprinted from *Neuroscience* **7**, © 1982 by Pergamon Press plc.

[25] Reprinted from *Brain Research* **5**, © 1967 by Elsevier Science Publishers.

[26] Reprinted from *The Neurosciences: Second Study Program*, edited by F. O. Schmitt, © 1970 by the Rockefeller University Press.

[27] Reprinted from *The Journal for Psychiatric Research* **8**, © 1971 by Pergamon Press plc.

[28] Reprinted from *Brain Research* **122**, © 1977 by Elsevier Science Publishers.

[29] Reprinted from *The Journal of Comparative Neurology* **171**, © 1977 by Wiley-Liss, a division of John Wiley & Sons, Inc.

[30] Reprinted from *Neurology and Psychiatry: A Meeting of the Minds*, edited by J. Mueller, © 1989 by S. Karger.